PHYSICS

Fourth Edition

PHYSICS

Fourth Edition

Paul E. Tippens

Professor of Physics
Department of Physics and Chemistry
Southern College of Technology
Marietta, Georgia

GLENCOE
Macmillan/McGraw-Hill

New York, New York   Columbus, Ohio   Mission Hills, California   Peoria, Illinois

Sponsoring Editor: **Stephen M. Zollo**
Editing Supervisor: **Allen Appel**
Design and Art Supervisor: **Joseph Piliero**
Production Supervisor: **Kathleen Donnelly**

Text Designer: **NSG Design**
Cover Designer: **NSG Design**

Library of Congress Cataloging-in-Publication Data

Tippens, Paul E.
    Physics/Paul E. Tippens.—4th ed.
         p.   cm.
    Rev. ed. of: Applied physics. 3rd ed. c1985.
    ISBN 0-07-065028-4
    1. Physics.   I. Tippens, Paul E. Applied physics.   II. Title.
QC21.2.T55   1991
530—dc20

Physics, Fourth Edition

Imprint 1992

Send all inquiries to:

GLENCOE DIVISION
Macmillan/McGraw-Hill
936 Eastwind Drive
Westerville, Ohio 43081

ISBN 0-07-065028-4

2 3 4 5 6 7 8 9 10 11 12 13 14 15   RRC   00 99 98 97 96 95 94 93 92

CONTENTS

CONTENTS   **vii**

PART FOUR ➣→

MODERN PHYSICS   737

PREFACE

Physics, Fourth Edition, is designed for the noncalculus physics course taken by those who are pursuing careers in science or engineering technology. It has also been used as an introductory text for many other disciplines in which a clear emphasis on the applications of physical principles is desired. An understanding of very simple algebra and right-triangle trigonometry is assumed, but adequate review is given both in the text and in the appendixes. A major goal of this text, as with the previous editions, has been to produce a readable, friendly text that gives a solid foundation in physics without discouraging those who have traditionally been frightened by the subject. Ideally, the student will consider the text as a study companion, and the instructor will, at the same time, know that important concepts are not compromised.

Regardless of the abilities and charisma of various physics instructors, no one can doubt that a student is ultimately responsible for his or her own learning. The primary goal of any text should therefore be to provide the organization, examples, and exercises that lead to understanding and applications. The following features of this edition help to meet this goal:

1. Clearly stated objectives at the beginning of each chapter.
2. Introductory paragraphs that set forth the rationale and applications.
3. Extensive use of examples with detailed solutions designed to develop problem-solving skills.
4. Careful selection and use of over 800 line drawings to illustrate concepts in a way that complements the board work of the instructor.
5. An informative writing style that does not compromise learning objectives.
6. Detailed summaries of important concepts and formulas at the end of every chapter.
7. Over 500 end-of-chapter questions designed to stimulate thought and to test verbal understanding of basic concepts.
8. Over 1100 carefully selected problems organized by subject and level of difficulty.

The organization of the text is convenient for those who are on semester or quarter schedules. The treatment of optics is such that it can either precede or follow electromagnetic theory. Typically, those on a semester schedule would cover mechanics, heat, and thermodynamics during the first semester, and electricity, magnetism, sound, light, and modern physics during the second semester. On a quarterly basis, the division might be mechanics for the first quarter; heat, light, and sound for the second quarter; and electricity, magnetism, and modern physics for the last quarter. Where possible, each chapter has been organized to stand as a complete unit so that maximum flexibility is given to the selection and organization of topics.

The text is developed in such a way as to maximize the learning process. For example, the treatment of statics precedes the treatment of kinematics. Newton's first and third laws are covered early, but the second law is delayed until the concepts of free-body diagrams and static equilibrium are understood. Care is taken in later chapters to avoid misunderstandings and to clear up mistaken concepts. This treatment allows the students to develop their understanding in a logical and continuous fashion. Often, when statics is presented in later chapters in the text, a review of the treatment of forces and vectors is needed. Moreover, a background with free-body diagrams, gained from an early treatment of statics, permits the presentation of more detailed examples with Newton's second law. Users who prefer an alternative approach will have no difficulty in rearranging topics so that kinematics is presented before statics. Flexibility in the teaching approach is also achievable through the choice of examples and problems.

A companion study guide provides additional examples, questions, and problems to supplement the text. An independent series of computer tutorials is available to those who have access to IBM and compatible personal computers. *Tutorials in Physics* supplements the learning process through the interaction of sound and graphics, providing learning aids that are not often available in the classroom environment.

Paul E. Tippens

PART ONE
Mechanics

Introduction

A knowledge of physics is essential to an understanding of our world. No other science has been as active in revealing the causes and effects of natural events. A casual glance at our past demonstrates a continuum of experiment and discovery ranging from early measurements of gravity to later conquests of space. By studying objects at rest and in motion, scientists have been able to derive fundamental laws which apply for many applications in mechanical engineering. The investigation of electricity and magnetism produced new sources of energy and new methods of distributing power for the use of mankind. An understanding of the physical principles which govern the production of heat, light, and sound has added countless applications which have served to make us more comfortable and more able to cope with our environment.

It is difficult to imagine a single product available today that does not represent an application of some physical principle. This means that regardless of your career choice, you will in some way need to understand physics. Granted there are some occupations and professions which do not require the depth of understanding necessary for engineering applications, but all fields of work utilize and apply these concepts. With a thorough understanding of mechanics, heat, sound, and electricity, you carry with you the building blocks for almost any career. Should you find it necessary or desirable to change careers either before or after graduation, you will be able to draw from a general base of science and mathematics. By taking this course seriously and by devoting an unusual amount of time and energy, you will have less trouble in the future. In your later coursework, and on the job, you will be riding the crest of the wave instead of merely staying afloat in an angry sea.

WHAT IS PHYSICS?

Even if you have previously taken courses in high school physics, you probably still have a rather "fuzzy" idea of what *physics* really means and how it might differ, for example, from *science*. For our purpose, the sciences may be divided between *biological* and *physical*. The biological sciences deal with living things. The physical sciences deal primarily with the nonliving sides of nature.

Physics can be defined as the science which investigates the fundamental concepts of matter, energy, and space, and the relationships between them.

In terms of this broad definition, there are no clear boundaries between the many physical sciences. This is evident from the overlapping fields of biophysics, chemical physics, astrophysics, geophysics, electrochemistry, and so forth.

The goal of this textbook is to provide a very basic introduction to the world of physics. The emphasis is on applications, and the broad field of physics will be narrowed to the essential concepts which underlie all technical knowledge. You will study mechanics, heat, light, sound, electricity, and atomic structure. The most basic of these topics, and probably the most important for beginning students, is mechanics.

Mechanics is concerned with the position *(statics)* and motion *(dynamics)* of matter in space. Statics represents the study of physics associated with bodies at rest. Dynamics is concerned with a description of motion and its causes. In each case, the engineer or technician is concerned with measuring and describing physical quantities in terms of their cause and effect.

An engineer, for example, uses physical principles to determine which type of bridge structure will be the most efficient for a given situation. The concern is for the *effect* of forces. If the completed bridge should fail, the *cause* of failure must be analyzed in order to apply this knowledge to future construction. It is important to note that by *cause* the scientist means the sequence of physical events leading to an *effect*.

WHAT PART IS PLAYED BY MATHEMATICS?

Mathematics serves many purposes. It is philosophy, art, metaphysics, and logic. However, these values are subordinate to its main value as a tool for the scientist, engineer, or technician. One of the rewards of a first course in physics is the growing awareness of the relevance of mathematics. A study of physics reveals very specific applications of basic mathematics.

Suppose you wish to predict how long it takes to stop a car traveling at a given speed. You might record the initial speed, the change in speed, and the distance and time required to stop the car. When all these facts are recorded, the data might be used to establish a tentative relationship. We cannot do this without the tools of mathematics.

If you completed such an experiment, your measurements plus a knowledge of mathematics would lead you to the following relationship:

$$s = vt + \tfrac{1}{2}at^2$$

where s = stopping distance
v = initial speed
t = stopping time
a = rate of change in speed

This statement is a *workable hypothesis*. From this equation we can predict the stopping distance for other vehicles. When it has been used long enough for us to be reasonably sure that it is true, we call it a *scientific theory*. In other words, any scientific theory is just a workable hypothesis which has stood the test of time.

Thus, we can see that mathematics is useful in deriving formulas which describe physical events accurately. Mathematics plays an even larger role in solving such formulas for specific quantities.

For example, in the above formula, it would be a relatively simple matter to find values for s, v, and a when the other quantities are given, but the solution for t involves a more specialized knowledge of mathematics. How easily you derive or solve a theoretical relationship depends on your background in mathematics.

A review of the mathematics required for this text is presented as App. A. If you are unfamiliar with any of the topics discussed, you should study this appendix carefully. Particular attention should be given to the sections on powers of 10, literal equations, and trigonometry. Skill in applying the tools of mathematics will largely determine your success in any physics course.

HOW SHOULD I STUDY PHYSICS?

Reading technical material is different from other reading. Attention to specific meanings of words must be given if the material is to be understood. Graphs, drawings, charts, and photographs are often included in technical literature. They are always helpful and may even be essential to the description of physical events. You should study them thoroughly so that you understand the principles clearly.

Much of what you learn will be from classroom lectures and experiments. The beginning student often asks: How can I concentrate fully on the lecture and at the same time take good notes? Of course, it may not be possible to understand fully all the concepts presented and still take complete notes. You must learn to note only the significant portions of each lesson. Make sure you listen attentively to the explanation of various topics. Learn to recognize key words like *work, force, energy,* and *momentum.*

Adequate preparation before class should give you a good idea of which portions of the lecture are covered in the text and which are not. If a problem or definition is in the text, it is usually better to jot down a key word and concentrate fully on what the instructor is saying. These notes can be expanded later.

Organization skills and study habits are essential to success in a beginning physics course. Therefore, you should compile a neat and comprehensive notebook using both lectures and the textbook as source material. At the risk of being too specific, the author recommends a loose-leaf filler notebook with dividers. Possible sections might be labeled as *handouts, exams, problems, labs,* and *notes.* Study and review for tests and for the final exam may still be difficult, but if your study materials are not well organized, it is often impossible.

2 Technical Measurement and Vectors

OBJECTIVES

After completing this chapter, you should be able to:

1. Write the base units for mass, length, and time in SI and U.S. Customary System (USCS) units.
2. Define and apply the SI prefixes which indicate multiples of base units.
3. Convert from one unit to another unit for the same quantity when given the necessary definitions.
4. Define a vector quantity and a scalar quantity and give examples of each.
5. Determine the components of a given vector.
6. Find the resultant of two or more vectors.

The application of physics, whether in the shop or in a technical laboratory, always requires measurements of some kind. An automobile mechanic might be measuring the diameter or *bore* of an engine cylinder. Refrigeration technicians might be concerned with volume, pressure, and temperature measurements. Electricians use instruments which measure electric resistance and current, and mechanical engineers are concerned about the effects of forces whose magnitudes must be accurately determined. In fact, it is difficult to imagine any occupation that is not involved with the measurement of some physical quantity.

In the process of physical measurement, we are often concerned with the direction as well as the magnitude of a particular quantity. The length of a wooden rafter is determined by the angle it makes with the horizontal. The direction of an applied force determines its effectiveness in producing a displacement. The direction in which a conveyor belt moves is often just as important as the speed with which it moves. Such physical quantities as *displacement, force,* and *velocity* are often encountered in industry. In this chapter, the concept of *vectors* is introduced to permit the study of both the magnitude and the direction of physical quantities.

The language of physics and technology is universal. Facts and laws must be expressed in an accurate and consistent manner if everyone is to mean exactly the same thing by the same term. For example, suppose an engine is said to have a piston displacement of 3.28 liters (200 cubic inches). Two questions must be answered if this statement is to be understood: (1) How is the *piston displacement* measured, and (2) what is the *liter?*

Piston displacement is the volume that the piston displaces, or "sweeps out," as it moves from the bottom of the cylinder to the top. It is really not a displacement in the usual sense of the word; it is a volume. A standard measure for volume that is easily recognized throughout the word is the liter. Therefore, when an engine has a label on it that reads "piston displacement = 3.28 liters," all mechanics will give the same meaning to the label.

In the above example, the piston displacement (volume) is an example of a *physical quantity.* Notice that this quantity was defined by describing the procedure for its measurement. In physics, all quantities are defined in this manner. Other examples of physical quantities are length, weight, time, speed, force, and mass.

A physical quantity is measured by comparison with some known standard. For example, we might need to know the length of a metal bar. With appropriate instruments we might determine the length of the bar to be 4 meters. The bar does not contain 4 things called "meters"; it is merely compared with the length of some standard known as a "meter." The length could also be represented as 13.1 ft or 4.37 yards if we used other known measures.

The *magnitude* of a physical quantity is given by a *number* and a *unit* of measure. Both are necessary because either the number or the unit by itself is meaningless. Except for pure numbers and fractions, it is necessary to include the unit with the number when listing the magnitude of any quantity.

> *The **magnitude** of a physical quantity is completely specified by a number and a unit, e.g., 20 meters or 40 liters.*

Since there are many different measures for the same quantity, we need a way of keeping track of the exact size of particular units. To do this, it is necessary to establish standard measures for specific quantities. A standard is a permanent or easily determined physical record of the size of a unit of measurement. For example, the standard for measuring electrical resistance, the *ohm,* might be defined by comparison with a standard resistor whose resistance is accurately known. Thus, a resistance of 20 ohms would be 20 times as great as that of a standard 1-ohm resistor.

Remember that every physical quantity is defined by telling how it is measured. Depending on the measuring device, each quantity can be expressed in a number of different units. For example, some distance units are *meters, kilometers, miles,* and *feet,* and some speed units are *meters per second, kilometers per hour, miles per hour,* and *feet per second.* Regardless of the units chosen, however, distance must be a *length* and speed must be *length* divided by *time.* Thus, *length* and *length/time* are the *dimensions* of the physical quantities *distance* and *speed.*

Note that speed is defined in terms of two more fundamental quantities (length and time). It is convenient to establish a small number of base quantities such as length and time from which all other physical quantities can be derived. Thus, we might say that speed is a *derived* quantity and that length or time is a *base* quantity. If we reduce all physical measurements to a small number of quantities with standard base units, there will be less confusion in their application.

2-2
THE INTER-NATIONAL SYSTEM

The international system of units is called *Système International d'Unités* (SI) and is essentially the same as what we have come to know as the *metric system*. The International Committee on Weights and Measures has established seven base quantities and has assigned official base units to each quantity. A summary of these quantities, their base units, and the symbols for the base units is given in Table 2-1.

Table 2-1   The SI Base Units for Seven Fundamental Quantities and Two Supplemental Quantities

Quantity	Unit	Symbol
Base units		
Length	meter	m
Mass	kilogram	kg
Time	second	s
Electric current	ampere	A
Temperature	kelvin	K
Luminous intensity	candela	cd
Amount of substance	mole	mol
Supplemental units		
Plane angle	radian	rad
Solid angle	steradian	sr

Each of the units listed in Table 2-1 has a specific measurable definition which can be duplicated anywhere in the world. Of these base units only one, the *kilogram*, is currently defined in terms of a single physical sample. This standard specimen is kept at the International Bureau of Weights and Measures in France. Copies of the original specimen have been made for use in other nations. The United States prototype is shown in Fig. 2-1. All other units are defined in terms of reproducible physical events and can be accurately determined at a number of locations throughout the world.

We can measure many quantities, such as volume, pressure, speed, and force, which are combinations of two or more fundamental quantities. However, no one has ever encountered a measurement that cannot be expressed in terms of length, mass, time, current, temperature, luminous intensity, or amount of substance. Combinations of these quantities are referred to as *derived* quantities, and they are measured in derived units. Several common derived units are listed in Table 2-2.

Fig. 2-1 The U.S. standard kilogram, a platinum-iridium cylinder housed in the National Bureau of Standards.

Table 2-2 Derived Units for Common Physical Quantities

Quantity	Derived units	Symbol	
Area	square meter	m^2	
Volume	cubic meter	m^3	
Frequency	hertz	Hz	s^{-1}
Mass density (density)	kilogram per cubic meter	kg/m^3	
Speed, velocity	meter per second	m/s	
Angular velocity	radian per second	rad/s	
Acceleration	meter per second squared	m/s^2	
Angular acceleration	radian per second squared	rad/s^2	
Force	newton	N	$kg \cdot m/s^2$
Pressure (mechanical stress)	pascal	Pa	N/m^2
Kinematic viscosity	square meter per second	m^2/s	
Dynamic viscosity	newton-second per square meter	$N \cdot s/m^2$	
Work, energy, quantity of heat	joule	J	$N \cdot m$
Power	watt	W	J/s
Quantity of electricity	coulomb	C	
Potential difference, electromotive force	volt	V	J/C
Electric field strength	volt per meter	V/m	
Electric resistance	ohm	Ω	V/A
Capacitance	farad	F	C/V

Table 2-2 (Continued)

Quantity	Derived units	Symbol	
Magnetic flux	weber	Wb	V · s
Inductance	henry	H	V · s/A
Magnetic flux density	tesla	T	Wb/m^2
Magnetic field strength	ampere per meter	A/m	
Magnetomotive force	ampere	A	
Luminous flux	lumen	lm	cd · sr
Luminance	candela per square meter	cd/m^2	
Illuminance	lux	lx	lm/m^2
Wave number	1 per meter	m^{-1}	
Entropy	joule per kelvin	J/K	
Specific heat capacity	joule per kilogram kelvin	J/(kg · K)	
Thermal conductivity	watt per meter kelvin	W/(m · K)	
Radiant intensity	watt per steradian	W/sr	
Activity (of a radioactive source)	1 per second	s^{-1}	

Unfortunately, the SI units are not fully implemented in many industrial applications. The United States is making progress toward the adoption of SI units. However, wholesale conversions are costly, particularly in many mechanical and thermal applications, and total conversion to the international system will require some time. For this reason it is necessary to be familiar with older units for physical quantities. The USCS units for several important quantities are listed in Table 2-3.

Table 2-3 U.S. Customary System Units

Quantity	SI unit	USCS unit
Length	meter (m)	foot (ft)
Mass	kilogram (kg)	slug (slug)
Time	second (s)	second (s)
Force (weight)	newton (N)	pound (lb)
Temperature	kelvin (K)	degree Rankine (R)

It should be noted that even though the foot, pound, and other units are frequently used in the United States, they have been redefined in terms of the SI standard units. Thus, all measurements are presently based on the same standards.

2-3
MEASUREMENT OF LENGTH AND TIME

The standard SI unit of length, the *meter* (m), was originally defined as one ten-millionth of the distance from the North Pole to the equator. For practical reasons, this distance was marked off on a standard platinum-iridium bar. In 1960, the standard was changed to allow greater access to a more accurate measure of the meter based on an atomic standard. One meter was said to be equal to exactly 1,650,763.73 wavelengths of orange-red light from krypton 86. The number was chosen so that the new standard would be very close to the older standard. However, even this standard was not without its problems. The wavelength of light emitted from krypton suffered uncertainties due to the processes occurring within the atom during emission. Also, the development of stabilized lasers allowed one to measure a wavelength much more accurately in terms of time and the velocity of light. In 1983, the latest (and probably the final) standard for the meter was adopted:

> One **meter** is the length of path traveled by a light wave in a vacuum in a time interval of 1/299,792,458 second.

The new standard for the meter is more precise, but it also has other advantages. Its definition depends on the standard for time (s) as given below, and it hinges on a standard value for the velocity of light. The velocity of light is now exactly:

$$c = 2.99792458 \times 10^8 \text{ m/s (exact by definition)}$$

A standard value for the speed of light makes sense because, according to Einstein's theory, the speed of light is a fundamental constant. Moreover, any future refinements to the standard for measuring time will automatically improve the standard for length.

Of course, we do not usually need to know the precise definition of length in order to make accurate practical measurement. Many tools, such as simple meter sticks and calipers, are calibrated to agree with the standard measure.

The original definition of time depended on the idea of a solar day as the time interval between two successive passages of the sun over a given meridian on the earth. One second was then 1/86,400 of a mean solar day. It's not hard to image the difficulties and the inconsistencies associated with such a standard. In 1976 the SI standard for time was defined as follows:

> One **second** is the time needed for 9,192,631,770 vibrations of a cesium atom.

Thus the atomic standard of one second is the period of vibration of an atom of cesium. The better cesium clocks are so accurate that they gain or lose no more than 1 second in 300,000 years.

Because this measure of time tends to drift ahead of mean solar time, the National Bureau of Standards periodically inserts a *leap second,* usually once a year on December 31. Therefore, the last minute of each year often contains 61 seconds, rather than 60 seconds.

A distinct advantage of the metric system over other systems of units is the use of prefixes to indicate multiples of the base unit. Table 2-4 defines the accepted prefixes and demonstrates their use to indicate multiples and subdivisions of the meter.

From the table, you can determine that

$$1 \text{ meter (m)} = 1000 \text{ millimeters (mm)}$$
$$1 \text{ meter (m)} = 100 \text{ centimeters (cm)}$$
$$1 \text{ kilometer (km)} = 1000 \text{ meters (m)}$$

The relationship between the centimeter and the *inch* can be seen in Fig. 2-2. By definition, 1 inch is equal to exactly 25.4 millimeters. This definition and other

Table 2-4   Multiples and Submultiples for SI Units

Prefix	Symbol	Multiplier	Use
tera	T	$1{,}000{,}000{,}000{,}000 = 10^{12}$	1 terameter (Tm)
giga	G	$1{,}000{,}000{,}000 = 10^{9}$	1 gigameter (Gm)
mega	M	$1{,}000{,}000 = 10^{6}$	1 megameter (Mm)
kilo	k	$1{,}000 = 10^{3}$	1 kilometer (km)
centi	c	$0.01 = 10^{-2}$	1 centimeter (cm)*
milli	m	$0.001 = 10^{-3}$	1 millimeter (mm)
micro	μ	$0.000001 = 10^{-6}$	1 micrometer (μm)
nano	n	$0.000000001 = 10^{-9}$	1 nanometer (nm)
—	Å	$0.0000000001 = 10^{-10}$	1 angstrom (Å)*
pico	p	$0.000000000001 = 10^{-12}$	1 picometer (pm)

*The use of the centimeter and the angstrom is discouraged, but they are still widely used.

useful definitions are as follows (symbols for the units are in parentheses):

$$1 \text{ inch (in.)} = 25.4 \text{ millimeters (mm)}$$
$$1 \text{ foot (ft)} = 0.3048 \text{ meter (m)}$$
$$1 \text{ yard (yd)} = 0.914 \text{ meter (m)}$$
$$1 \text{ mile (mi)} = 1.61 \text{ kilometers (km)}$$

In reporting data, it is preferable to use the prefix that will allow the number to be expressed in the range from 0.1 to 1000. For example, 7,430,000 meters should be expressed as 7.43×10^{6} m, and then it should be reported as 7.43 megameters, abbreviated 7.43 Mm. It would not usually be desirable to write this measurement as 7430 kilometers (7430 km) unless the distance is being compared with other distances measured in kilometers. In the case of the quantity 0.00064 ampere, it is proper to write either 0.64 milliampere (0.64 mA) or 640 microamperes (640 μA). Normally, prefixes are chosen for multiples of a thousand.

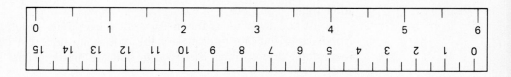

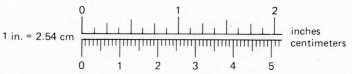

1 in. = 2.54 cm

Fig. 2-2 A comparison of the inch to the centimeter as a measure of length.

Some numbers are exact while others are approximate. If we select 20 bolts from a bin and use only one-fourth of them, the numbers 20 and ¼ are considered to be exact quantities. However, if we measured the length or the width of a rectangular plate, the accuracy of the measurement would be determined by the precision of the instrument and by the skill of the observer. Suppose we measure the width of such a plate with a vernier caliper and find it to be 3.42 cm. The last digit is estimated and is therefore susceptible to error. The actual width is between 3.40 cm and 3.50 cm. To write this width as 3.420 cm would imply greater accuracy than was justified. We say that the number 3.42 has three *significant* figures, and we are careful not to write more numbers or zeros than are meaningful.

All physical measurements are assumed to be approximate, with the last significant digit having been determined by an estimation of some kind. When writing such numbers, we often include some zeros to properly indicate the location of the decimal point. However, except for these zeros, all other digits are considered significant digits. For example, the distance 76,000 m has only two significant digits. We assume that the three zeros following the 6 were needed to locate the decimal unless there is information to the contrary. Other examples are as follows:

4.003 cm	4 significant figures
0.34 cm	2 significant figures
60,400 cm	3 significant figures
0.0450 cm	3 significant figures

Note that zeros which are not specifically needed to locate the decimal are significant (as in the last two examples).

With the widespread use of calculators, students often report their results with a greater accuracy than is warranted. For example, suppose we measure the length of a rectangular plate to be 9.54 cm and its width to be 3.4 cm. The area of such a plate might be calculated as 32.436 cm^2 (five significant figures). But a chain is only as strong as its weakest link. Since the width was accurate to only two significant figures, we cannot report a greater accuracy in our result. The area should be given as 32 cm^2. The number in our calculator gives a false indication of the accuracy and is misleading to others who did not participate in the measurement.

> *Rule 1. When approximate numbers are multiplied or divided, the number of significant digits in the final answer is the same as the number of significant digits in the least accurate of the factors.*

Another problem arises when approximate numbers are added or subtracted. In such cases consideration should be given to the precision indicated by each measurement. The sum of a group of measurements may have more significant digits than some of the individual measurements, but it cannot be more *precise*. For example, suppose we determine the perimeter of the rectangular plate. We might write:

$$9.54 \text{ cm} + 3.4 \text{ cm} + 9.54 \text{ cm} + 3.4 \text{ cm} = 25.9 \text{ cm}$$

The least precise measurement was to the nearest tenth of a centimeter; therefore, the perimeter should be given to the nearest tenth of a centimeter (even though it has three significant figures).

Rule 2. When approximate numbers are added or subtracted, the number of decimal places in the result should equal the smallest number of decimal places of any term in the sum.

Throughout this text, we will generally assume that the data given are precise enough to give answers rounded to three significant figures. Thus, an 8-in. rod would be treated as though it were 8.00 in. long. If converted to metric units, the length would be reported as 203 mm. This practice is common since it avoids the necessity of cluttering the text with zeros.

2-5
MEASURING INSTRUMENTS

The choice of a measuring instrument is determined by the accuracy required and by the physical conditions surrounding the measurement. A basic choice for the mechanic or machinist is most often the steel rule, such as the one shown in Fig. 2-3. This rule is often accurate enough when you are measuring openly accessible lengths. Steel rules may be graduated as fine as thirty-seconds or even sixty-fourths of an inch. Metric rules are usually graduated in millimeters.

For the measurement of inside and outside diameters, calipers such as those shown in Fig. 2-4 may be used. The caliper itself cannot be read directly and therefore must be set to a steel rule or a standard size gauge.

Fig. 2-3 Some 6-in. (15-cm) steel scales. *(a)* Scales $\frac{1}{32}$ in. and 0.5 mm. (The L.S. Starrett Company.) *(b)* Scales 1/1000 and 1/50 in. (The L.S. Starrett Company.)

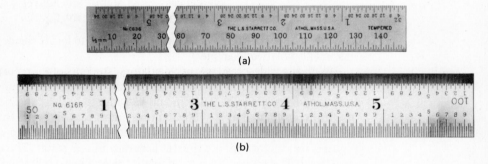

(a)

(b)

Fig. 2-4 Using calipers to measure an inside diameter.

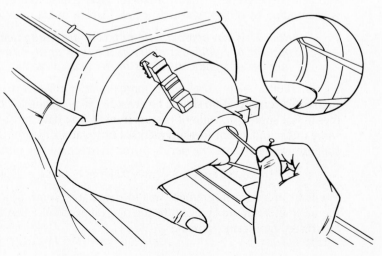

The best accuracy possible with a steel rule is determined by the size of the smallest graduation and is of the order of 0.01 in. or 0.1 mm. For greater accuracy the machinist often uses a standard micrometer caliper, such as the one shown in Fig. 2-5 or a standard vernier caliper, as in Fig. 2-6. These instruments make use of sliding scales to record very accurate measurements. Micrometer calipers can measure to the nearest ten-thousandth of an inch (0.002 mm), and vernier calipers are used to measure within 0.001 in. or 0.02 mm.

The depth of blind holes, slots, and recesses is often measured with a micrometer depth gauge. Figure 2-7 shows such a gauge being used to measure the depth of a shoulder.

Fig. 2-5 A micrometer caliper, showing a reading of 0.250 in. (The L.S. Starrett Company.)

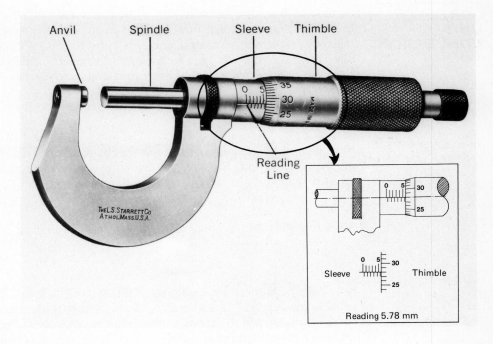

Fig. 2-6 The vernier caliper. (The L.S. Starrett Company.)

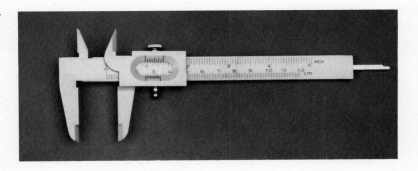

Fig. 2-7 Measuring the depth of a shoulder with a micrometer depth gauge.

2-6
UNIT CONVERSIONS

Because so many different units are required for a variety of jobs, it is often necessary to convert a measurement from one unit to another. For example, suppose that a machinist records the outside diameter of a pipe as 1³⁄₁₆ in. To order a fitting for the pipe, the machinist may need to know this diameter in millimeters. Such conversions can easily be accomplished by treating units algebraically and using the principle of cancellation.

In the above case, the machinist should first convert the fraction to a decimal.

$$1\tfrac{3}{16} \text{ in.} = 1.19 \text{ in.}$$

Next, the machinist should write down the quantity to be converted, giving both the number and the unit (1.19 in.). The definition which relates inches to millimeters is recalled:

$$1 \text{ in.} = 25.4 \text{ mm}$$

Since this statement is an equality, we can form two ratios, each equal to 1.

$$\frac{1 \text{ in.}}{25.4 \text{ mm}} = 1 \qquad \frac{25.4 \text{ mm}}{1 \text{ in.}} = 1$$

Note that the number 1 does not equal the number 25.4, but the *length* of 1 in. is equal to the *length* of 25.4 mm. Thus, if we multiply some other length by either of these ratios, we will get a new number, but we will not change the length. Such ratios are called *conversion factors*. Either of the above conversion factors may be multiplied by 1.19 in. without changing the length represented. Multiplication by the first ratio does not give a meaningful result. Note that units are treated as algebraic quantities.

$$(1.19 \text{ in.})\left(\frac{1 \text{ in.}}{25.4 \text{ mm}}\right) = \left(\frac{1.19}{25.4}\right)\left(\frac{\text{in.}^2}{\text{mm}}\right) \qquad \textit{Wrong!}$$

Multiplication by the second ratio, however, gives the following result:

$$(1.19 \text{ in.})\left(\frac{25.4 \text{ mm}}{1 \text{ in.}}\right) = \frac{(1.19)(25.4)}{(1)} \text{ mm} = 30.2 \text{ mm}$$

Therefore, the outside diameter of the pipe is 30.2 mm.

The following procedure is used in unit conversions:

1. Write down the quantity to be converted.
2. Define each unit appearing in the quantity to be converted in terms of the desired unit(s).
3. For each definition, form two conversion factors, one being the reciprocal of the other.
4. Multiply the quantity to be converted by those factors which will cancel all but the desired units.

Sometimes it is necessary to work with quantities which have multiple units. For example, *speed* is defined as *length* per unit *time* and may have units of *meters per second* (m/s), *feet per second* (ft/s), or other units. The same algebraic procedure can help with conversion of multiple units.

EXAMPLE 2-1   Convert a speed of 60 km/h to units of meters per second.

Solution   We recall two definitions which might result in four possible conversion factors.

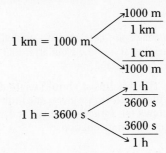

$$
1 \text{ km} = 1000 \text{ m}
\begin{cases}
\dfrac{1000 \text{ m}}{1 \text{ km}} \\[2mm]
\dfrac{1 \text{ cm}}{1000 \text{ m}}
\end{cases}
$$

$$
1 \text{ h} = 3600 \text{ s}
\begin{cases}
\dfrac{1 \text{ h}}{3600 \text{ s}} \\[2mm]
\dfrac{3600 \text{ s}}{1 \text{ h}}
\end{cases}
$$

We write down the quantity to be converted, then choose conversion factors which will cancel nondesired units.

$$
60 \, \frac{\cancel{\text{km}}}{\cancel{\text{h}}} \left(\frac{1000 \text{ m}}{1 \, \cancel{\text{km}}} \right) \left(\frac{1 \, \cancel{\text{h}}}{3600 \text{ s}} \right) = 16.7 \, \frac{\text{m}}{\text{s}}
$$

Additional examples of the procedure are as follows:

$$
30 \, \frac{\cancel{\text{mi}}}{\cancel{\text{h}}} \left(\frac{5280 \text{ ft}}{1 \, \cancel{\text{mi}}} \right) \left(\frac{1 \, \cancel{\text{h}}}{60 \, \cancel{\text{min}}} \right) \left(\frac{1 \, \cancel{\text{min}}}{60 \text{ s}} \right) = 44 \text{ ft/s}
$$

$$
20 \, \frac{\cancel{\text{lb}}}{\cancel{\text{in.}^2}} \left(\frac{1550 \, \cancel{\text{in}^2}}{1 \text{ m}^2} \right) \left(\frac{0.454 \text{ kg}}{1 \, \cancel{\text{lb}}} \right) = 1.41 \times 10^4 \text{ kg/m}^2
$$

The required definitions can be found in App. B, if they are not available in this chapter.

When working with technical formulas, it is always helpful to substitute units as well as numbers. For example, the formula for speed v is

$$
v = \frac{s}{t}
$$

where s is the distance traveled in a time t. Thus, if a car travels 400 m in 10 s, its speed will be

$$v = \frac{400 \text{ m}}{10 \text{ s}} = 40 \frac{\text{m}}{\text{s}}$$

Notice that the units of velocity are meters per second, written m/s.

Whenever velocity appears in a formula, it must always have units of *length* divided by *time*. These are said to be the *dimensions* of velocity. There may be many different units for a given physical quantity, but the dimensions result from a definition, and they do not change.

In working with formulas, it will be useful to remember two rules concerning dimensions:

> **Rule 1** *If two quantities are to be added or subtracted, they must be of the same dimensions.*

> **Rule 2** *The quantities on both sides of an equals sign must be of the same dimensions.*

EXAMPLE 2-2   Show that the formula

$$s = v_0 t + \tfrac{1}{2}at^2$$

is dimensionally correct. The symbol s represents the distance traveled in a time t while accelerating at a rate a from an initial speed v_0. Assume that acceleration has units of meters per second squared (m/s^2).

Solution   Since the units of a are given, the unit of length must be meters (m) and the unit of time must be seconds (s). Ignoring the factor $\tfrac{1}{2}$, which has no dimensions, we substitute units for the quantities in the equation.

$$\text{m} = \frac{\text{m}}{\cancel{s}}(\cancel{s}) + \frac{\text{m}}{\cancel{s^2}}(\cancel{s})^2$$

Notice that both Rule 1 and Rule 2 are satisfied. Therefore, the equation is dimensionally correct.

The fact that an equation is dimensionally correct is a valuable check. Such an equation still may not be a *true* equation, but at least it is consistent dimensionally.

2-7 VECTOR AND SCALAR QUANTITIES

Some quantities can be totally described by a number and a unit. Only the *magnitudes* are of interest in an area of 12 m^2, a volume of 40 ft^3, or a distance of 50 km. Such quantities are called *scalar* quantities.

> *A scalar quantity is completely specified by its magnitude—a number and a unit. Examples are speed (15 mi/h), distance (12 km), and volume (200 cm^3).*

Scalar quantities which are measured in the same units may be added or subtracted in the usual way. For example,

$$14 \text{ mm} + 13 \text{ mm} = 27 \text{ mm}$$
$$20 \text{ ft}^2 - 14 \text{ ft}^2 = 6 \text{ ft}^2$$

Some physical quantities, such as force and velocity, have direction as well as magnitude. In such cases, they are called *vector* quantities. The direction must be a part of any calculations involving such quantities.

> A **vector quantity** is completely specified by a magnitude and a direction. It consists of a number, a unit, and a direction. Examples are displacement (20 m N) and velocity (40 mi/h, 30°N of W).

The direction of a vector may be given by reference to conventional north, east, west, and south directions. Consider, for example, the vectors 20 m, W and 40 m at 30°N of E, as shown in Fig. 2-8. The expression "north of east" indicates that the angle is formed by rotating a line northward from the easterly direction.

Another method of specifying direction which will be particularly useful later on is to make reference to perpendicular lines called *axes*. These imaginary lines are usually chosen to be horizontal and vertical, but they may be oriented along other directions as long as the two lines remain perpendicular. An imaginary horizontal line is usually called the x axis, and an imaginary vertical line is called the y axis. (See Fig. 2-9.) Directions are given by angles measured counterclockwise from the positive x axis. The vectors 40 m at 60° and 50 m at 210° are shown in the figure.

Assume a person travels by car from Atlanta to St. Louis. The *displacement* from Atlanta can be represented by a line segment drawn to scale from Atlanta to St. Louis

Fig. 2-8 Indicating the direction of a vector by reference to north (N), south (S), east (E), and west (W).

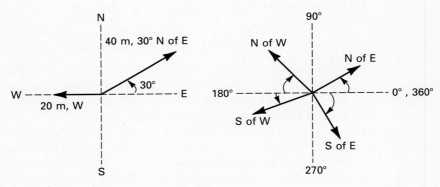

Fig. 2-9 Indicating the direction of a vector as an angle measured from the positive x axis.

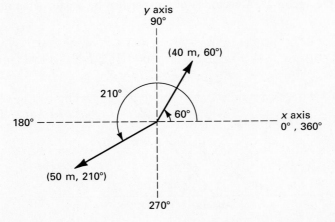

(see Fig. 2-10). An arrowhead is drawn on the St. Louis end to denote the direction. It is important to note that the displacement, represented by the vector D_1, is completely independent of the actual path or the mode of transportation. The odometer would show that the car had actually traveled a scalar distance s_1 of 541 mi. The magnitude of the displacement is only 472 mi.

Another important difference is that the vector displacement has a constant direction of 140° (or 40°N of W). However, the direction of the car at any instant on the trip is not important when considering the scalar distance.

Now, let us suppose that our traveler continues the drive to Washington. This time the vector displacement D_2 is 716 mi at a constant direction of 10°N of E. The corresponding ground distance s_2 is 793 mi. The total distance traveled for the entire trip from Atlanta is the arithmetic sum of the scalar quantities s_1 and s_2.

$$s_1 + s_2 = 541 \text{ mi} + 793 \text{ mi} = 1334 \text{ mi}$$

However, the *vector sum* of the two displacements D_1 and D_2 must take note of direction as well as magnitudes. The question now is not the distance traveled but the resulting displacement from Atlanta. This vector sum is represented in Fig. 2-10 by the symbol **R**, where

$$R = D_1 + D_2$$

Methods we will discuss in the next section will allow us to determine the magnitude and direction of **R**. Using a ruler and a device for measuring angles, we would see that

$$R = 545 \text{ mi}, 51°$$

Remember that in performing vector additions, both the magnitude and direction of the displacements must be considered. The additions are geometric instead of algebraic. It is possible for the magnitude of a vector sum to be less than the magnitude of either of the component displacements.

A vector is usually denoted in print by boldface type. For example, the symbol D_1 denotes a displacement vector in Fig. 2-10. A vector can be denoted conveniently in handwriting by underscoring the letter or by putting an arrow over it. In print, the magnitude of a vector is usually indicated by italics; thus, D denotes the magnitude of the vector **D**. A vector is often specified by a pair of numbers (R, θ). The first

Fig. 2-10 Displacement is a vector quantity. Its direction is indicated by a solid arrow. Distance is a scalar quantity, indicated above by a dotted line.

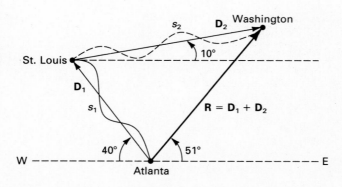

number and unit give the magnitude, and the second number gives the angle measured counterclockwise from the positive x axis. For example,

$$\mathbf{R} = (R, \theta) = (200 \text{ km}, 114°)$$

Note that the magnitude R of a vector is always positive. A negative sign before the symbol of a vector merely reverses its direction; i.e., it interchanges the arrow tip without affecting the length. If $\mathbf{A} = (10 \text{ m E})$, then $-\mathbf{A}$ would be (10 m W).

2-8
ADDITION OF VECTORS BY GRAPHICAL METHODS

In this section we discuss two common graphical methods for finding the geometric sum of vectors. The *polygon method* is the more useful, since it can be readily applied to more than two vectors. The *parallelogram method* is useful for the addition of two vectors at a time. In each case the magnitude of a vector is indicated to scale by the length of a line segment. The direction is denoted by an arrow tip at the end of the line segment.

EXAMPLE 2-3   A ship travels 100 mi due north on the first day of a voyage, 60 mi northeast on the second day, and 120 mi due east on the third day. Find the resultant displacement by the polygon method.

Solution   A suitable scale might be 20 mi = 1 cm, as in Fig. 2-11. Using this scale, we note that

$$100 \text{ mi} = 100 \text{ mi} \times \frac{1 \text{ cm}}{20 \text{ mi}} = 5 \text{ cm}$$

$$60 \text{ mi} = 60 \text{ mi} \times \frac{1 \text{ cm}}{20 \text{ mi}} = 3 \text{ cm}$$

$$120 \text{ mi} = 120 \text{ mi} \times \frac{1 \text{ cm}}{20 \text{ mi}} = 6 \text{ cm}$$

Fig. 2-11 The polygon method of vector addition.

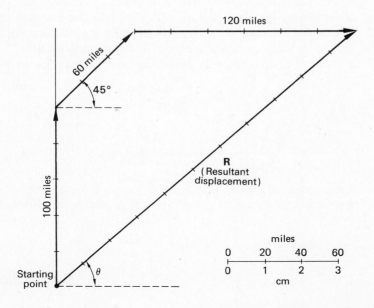

By measuring with a ruler, we find from the scale diagram that the arrow for the resultant is 10.8 cm long. Therefore, the magnitude is

$$10.8 \ \text{cm} = 10.8 \ \text{cm} \times \frac{20 \ \text{mi}}{1 \ \text{cm}} = 216 \ \text{mi}$$

Measuring the angle θ with a protractor shows the direction to be 41°. The resultant displacement is therefore

$$R = (216 \ \text{mi}, 41°)$$

Note that the order in which the vectors are added does not change the resultant in any way. We could have begun with any of the three distances traveled by the ship.

The polygon method can be summarized as follows:

1. Choose a scale and determine the length of the arrows which correspond to each vector.
2. Draw to scale an arrow representing the magnitude and direction of the first vector.
3. Draw the arrow of the second vector so that its tail is joined to the tip of the first vector.
4. Continue the process of joining tail to tip until the magnitude and direction of all vectors have been represented.
5. Draw the resultant vector with its tail at the origin (starting point) and its tip joined to the tip of the last vector.
6. Measure with ruler and protractor to determine the magnitude and direction of the resultant.

Graphical methods can be used to find the resultant of all kinds of vectors. They are not restricted to measuring displacement. In the next example we determine the resultant of two *forces* by the parallelogram method.

In the parallelogram method, which is useful only for two vectors at a time, these vectors are drawn to scale with their tails at a common origin (see Fig. 2-12). The two arrows then form two adjoining sides of a parallelogram. The other two sides are constructed by drawing parallel lines of equal length. The resultant is represented by the diagonal of the parallelogram included between the two vector arrows.

Fig. 2-12 The parallelogram method of vector addition.

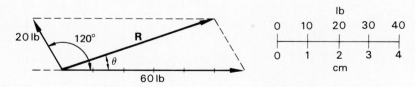

EXAMPLE 2-4　A rope is wrapped around a telephone pole, making an angle of 120°. If one end is pulled with a force of 60 lb and the other with a force of 20 lb, what is the resultant force on the telephone pole?

Solution　Using a scale of 1 cm = 10 lb gives

$$60 \ \text{lb} \times \frac{1 \ \text{cm}}{10 \ \text{lb}} = 6 \ \text{cm}$$

$$20 \; \cancel{lb} \times \frac{1 \; cm}{10 \; \cancel{lb}} = 2 \; cm$$

A parallelogram is constructed in Fig. 2-12 by drawing the two forces to scale from a common origin with 120° between them. Completing the parallelogram allows the resultant to be drawn as a diagonal from the origin. Measurement of R and θ with a ruler and protractor gives 53 lb for the magnitude and 19° for the direction. Hence,

$$\mathbf{R} = (53 \; lb, \; 19°)$$

2-9

FORCE AND VECTORS

A push or pull that tends to cause motion is called a *force*. A stretch spring exerts forces on the objects to which its ends are attached; compressed air exerts forces on the walls of its container; and a tractor exerts a force on the trailer it is pulling. Probably the most familiar force is the force of gravitational attraction exerted on every body by the earth. This force is called the *weight* of the body. A definite force exists even though there is no contact between the earth and the bodies it attracts. Weight is a vector quantity that is directed toward the center of the earth.

The SI unit of force is the *newton* (N). Its relationship to the U.S. customary unit, the *pound* (lb), is

$$1 \; N = 0.225 \; lb \qquad 1 \; lb = 4.45 \; N$$

A 120-lb woman has a weight of 534 N. If the weight of a pipe wrench is 20 N, it would weigh about 4.5 lb in USCS units. Until all industries have converted completely to SI units, the pound will still be with us, and frequent conversions are necessary.

Two of the measurable effects of forces are (1) changing the dimensions or shape of a body and (2) changing a body's motion. If in the first case there is no resultant displacement of the body, the push or pull causing the change in shape is called a *static force*. If a force changes the motion of a body, it is called a *dynamic force*. Both types of forces are conveniently represented by vectors, as in Example 2-4.

The effectiveness of any force depends on the direction in which it acts. For example, it is easier to pull a sled along the ground with an inclined rope, as shown in Fig. 2-13, than to push it. In each case, the applied force is producing more than a single effect. That is, the pull on the cord is both lifting the sled and moving it forward. Similarly, pushing the sled would have the effect of adding to the weight of the sled. We are thus led to the idea of *components* of a force, i.e., the effective values of a force in directions other than that of the force itself. In Fig. 2-13, the force F can be replaced by its horizontal and vertical components, $\mathbf{F}_x$ and $\mathbf{F}_y$.

Fig. 2-13 The force F acting at an angle θ can be replaced by its horizontal and vertical components.

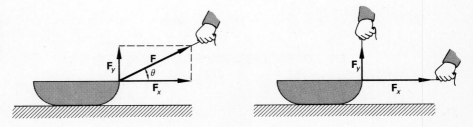

If a force is represented graphically by its magnitude and an angle (R, θ), its components along the x and y directions can be determined. A force F acting at an angle θ above the horizontal is drawn in Fig. 2-14. The meaning of the x and y components, F_x and F_y, can be seen in this diagram. The segment from O to the perpendicular dropped from A to the x axis is called the x component of F and is labeled F_x. The segment from O to the perpendicular line from A to the y axis is called the y component of F and is labeled F_y. By drawing the vectors to scale, we can determine the magnitudes of the components graphically. These two components acting together would have the same effect as the original force F.

Fig. 2-14 Graphical representation of the x and y components of F.

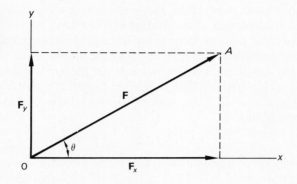

EXAMPLE 2-5   A lawn mower is pushed downward with a force of 40 N at an angle of 50° with the horizontal. What is the magnitude of the horizontal effect of this force?

Solution   A sketch is drawn, as in Fig. 2-15a, to translate the word problem into a picture. This approach often aids in understanding the problem. The 40-N force is transmitted through the handle to the body of the lawn mower. A vector diagram is shown in Fig. 2-15b. A ruler and protractor are used to draw the diagram accurately. A scale of 1 cm = 10 N is convenient for this exam-

Fig. 2-15 Finding the components of a force by the graphical method.

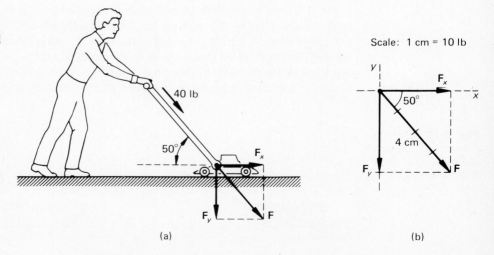

(a)

(b)

ple. The horizontal effect of the 40-N force is the x component, as labeled in the figure. Measurement of this line segment gives

$$F_x = 2.57 \text{ cm}$$

Since 1 cm = 10 N, we obtain

$$F_x = 2.57 \text{ cm} \frac{10 \text{ N}}{1 \text{ cm}} = 25.7 \text{ N}$$

Notice that the effective force is quite a bit less than the applied force. As an additional exercise, you should show that the magnitude of the *downward* component of the 40-N force is $F_y = 30.6$ N.

2-10
THE RESULTANT FORCE

When two or more forces act at the same point on an object, they are said to be *concurrent forces.* The combined effect of such forces is called the *resultant force.*

> The **resultant force** is that single force which will produce the same effect in both magnitude and direction as two or more concurrent forces.

Resultant forces may be calculated graphically by representing each concurrent force as a vector. The polygon or parallelogram method of vector addition will then give the resultant force.

Often forces act in the same line, either together or in opposition to each other. If two forces act on a single object in the same direction, the resultant force is equal to the sum of the magnitudes of the forces. The direction of the resultant would be the same as that of either force. Consider, for example, a 15-N force and a 20-N force acting in the same easterly direction. Their resultant is 35 N, E, as demonstrated in Fig. 2-16*a*.

If the same two forces act in opposite directions, the magnitude of the resultant force is equal to the *difference* of the magnitudes of the two forces, and it acts in the direction of the larger force. Suppose the 15-N force in our example were changed so that it pulled to the west. As seen in Fig. 2-16*b*, the resultant would be 5 N, E.

If forces act at an angle between 0 and 180° to each other, their resultant is the vector sum. The polygon method or the parallelogram method of vector addition may be used to find the resultant force. In Fig. 2-16*c*, our two forces of 15 and 20 N act at an angle of 60° with each other. The resultant force, calculated by the parallelogram method, is found to be 30.4 N at 34.7°.

2-11
TRIGO- NOMETRY AND VECTORS

Graphical treatments of vectors are good for visualizing forces, but they are usually not very accurate. A much more useful approach is to take advantage of simple right-triangle trigonometry, and today's calculators have simplified the process considerably. A brief review is provided in App. A-4 for those who feel such a review is necessary. Familiarity with the *pythagorean theorem* and some experience with the *sine, cosine,* and *tangent* functions is all you will need for this unit of study.

Trigonometric methods can improve your accuracy and speed in determining the resultant vector or in finding the components of a vector. In most cases, it is helpful to use imaginary x and y axes when working with vectors in an analytical

Fig. 2-16 The effect of direction on the resultant of two forces.

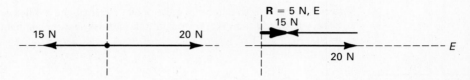

(a) Forces in same direction.

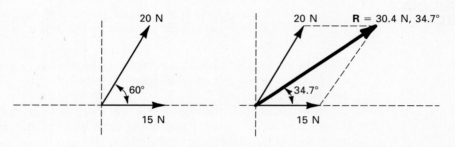

(b) Forces acting in opposite directions.

(c) Forces acting at an angle of 60° with respect to each other.

way. Any vector can then be drawn with its tail at the center of these imaginary lines. Components of the vector might be seen as effects along the x and y axis.

EXAMPLE 2-6 What are the x and y components of a force of 200 N at an angle of 60°?

Solution A diagram is drawn placing the tail of the 200-N vector at the center of the x and y axes (see Fig. 2-17).

Fig. 2-17 Using trigonometry to find the x and y components of a vector.

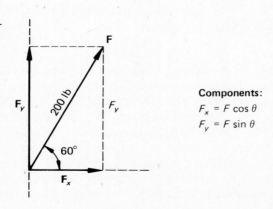

Components:

$F_x = F \cos \theta$

$F_y = F \sin \theta$

MECHANICS

We first compute the x component, F_x, by noting that it is the side adjacent. The 200-N vector is the hypotenuse. Using the cosine function, we obtain

$$\cos 60° = \frac{F_x}{200 \text{ N}}$$

from which

$$F_x = (200 \text{ N}) \cos 60° = 100 \text{ N}$$

For purposes of calculation, we recognize that the side opposite to 60° is equal in length to F_y. Thus, we may write

$$\sin 60° = \frac{F_y}{200 \text{ N}}$$

or

$$F_y = (200 \text{ N}) \sin 60° = 173.2 \text{ N}$$

In general we may write the x and y components of a vector in terms of its magnitude F and direction θ:

$$\boxed{\begin{array}{l} \mathbf{F}_x = F \cos \theta \\ \mathbf{F}_y = F \sin \theta \end{array}} \qquad \begin{array}{l} \textit{Components} \\ \textit{of a Vector} \end{array} \quad (2\text{-}1)$$

where θ is the angle between the vector and the positive x axis, measured in a counterclockwise direction.

The sign of a given component can be determined from a vector diagram. The four possibilities are shown in Fig. 2-18. The magnitude of the component can be found by using the small angle ϕ when angle θ is greater than 90°.

Fig. 2-18 *(a)* In the first quadrant, angle θ is between 0 and 90°; both F_x and F_y are positive. *(b)* In the second quadrant, angle θ is between 90 and 180°; F_x is negative and F_y is positive. *(c)* In the third quadrant, angle θ is between 180 and 270°; F_x and F_y are negative. *(d)* In the fourth quadrant, angle θ is between 270 and 360°; F_x is positive and F_y is negative.

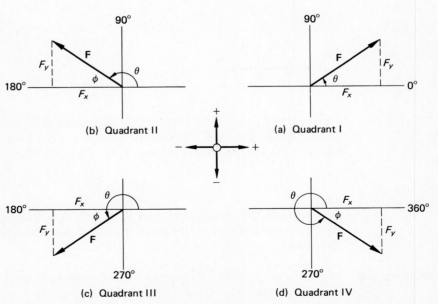

EXAMPLE 2-7   Find the x and y components of a 400-N force at an angle of 220° from the positive x axis.

Solution   Refer to Fig. 2-18c, which describes this problem when $\theta = 220°$. The small angle ϕ is found by reference to 180°.

$$\phi = 220° - 180° = 40°$$

From the figure, both F_x and F_y will be negative.

$$F_x = -|F \cos \theta| = -(400 \text{ N}) \cos 40°$$
$$= -(400 \text{ N})(0.766) = -306 \text{ N}$$
$$F_y = -|F \sin \theta| = -(400 \text{ N}) \sin 40°$$
$$= -(400 \text{ N})(0.643) = -257 \text{ N}$$

Note that the signs were determined from the figure. With many electronic calculators, both the magnitude and the sign of $\mathbf{F}_x$ and $\mathbf{F}_y$ can be found directly from Eq. (2-1) using $\theta = 220°$. You should verify this fact.

Trigonometry is also useful in calculating the resultant force. In the special case for two forces $\mathbf{F}_x$ and $\mathbf{F}_y$ at right angles to each other, as in Fig. 2-19, the resultant (R, θ) may be found from

$$R = \sqrt{F_x^2 + F_y^2} \qquad \tan \theta = \frac{F_y}{F_x} \qquad\qquad (2\text{-}2)$$

Fig. 2-19 The resultant of perpendicular vectors.

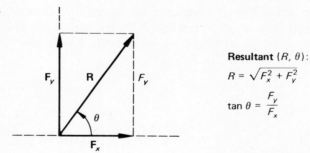

Resultant (R, θ):

$$R = \sqrt{F_x^2 + F_y^2}$$

$$\tan \theta = \frac{F_y}{F_x}$$

If either F_x or F_y is negative, it is usually easier to determine the small angle ϕ, as described in Fig. 2-18. The sign (or direction) of the forces F_x and F_y determines which of the four quadrants is used. Then, Eq. (2-2) becomes

$$\tan \phi = \left| \frac{F_y}{F_x} \right|$$

Only the absolute values of F_x and F_y are needed. If desired, the angle θ from the positive x axis may be determined. In any case, the direction must be clearly identified.

EXAMPLE 2-8   What is the resultant of a 5-N force directed horizontally to the right and a 12-N force directed vertically downward?

Solution   Label the two forces $F_x = 5$ N and $F_y = -12$ N (downward). Draw a diagram for the situation described by Fig. 2-18d. The magnitude of the resultant R is found from Eq. (2-2)

$$R = \sqrt{F_x^2 + F_y^2} = \sqrt{(5\ \text{N})^2 + (-12\ \text{N})^2}$$
$$= \sqrt{25\ \text{N}^2 + 144\ \text{N}^2} = \sqrt{169\ \text{N}^2} = 13\ \text{N}$$

To find the direction of R, we first find the small angle ϕ:

$$\tan \phi = \left| \frac{-12\ \text{N}}{5\ \text{N}} \right| = 2.4$$

$$\phi = 67.4° \text{ below positive } x \text{ axis}$$

The angle θ measured counterclockwise from the positive x axis is

$$\theta = 360° - 67.4° = 292.6°$$

The resultant force is 13 N at 292.6°.

2-12

THE COMPONENT METHOD OF VECTOR ADDITION

Often a body is acted on by a number of forces having different magnitudes, directions, and points of application. Forces which intersect at a common point or have the same point of application are said to be *concurrent* forces. When such forces are not at right angles with each other, calculating the resultant can be more difficult. The vectors would not fall along either the x or the y axis in all cases. The component method of addition of vectors is needed for these more general cases. Consider the vectors A, B, and C in Fig. 2-20. The resultant R is the vector sum A + B + C.

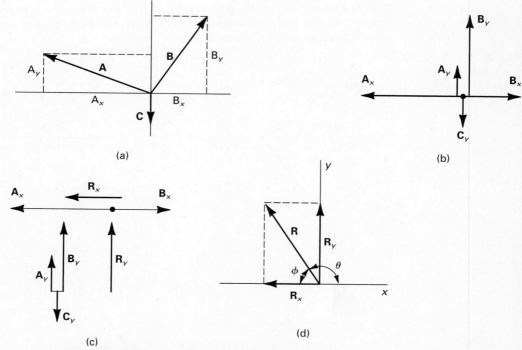

Fig. 2-20 The component method of vector addition.

However, A and B are not along an axis and cannot be added in the usual manner. The following procedure may be used:

1. Draw each vector from the center of imaginary x and y axes.
2. Find the x and y components of each vector.
3. Find the x component of the resultant by adding the x components of all vectors. (Components to the right are positive and those to the left are negative.)

$$R_x = A_x + B_x + C_x$$

4. Find the y component of the resultant by adding the y components of all vectors. (Upward components are positive and downward components are negative.)

$$R_y = A_y + B_y + C_y$$

5. Find the magnitude and direction of the resultant from its perpendicular components R_x and R_y.

$$R = \sqrt{R_x^2 + R_y^2} \qquad \tan \phi = \frac{R_y}{R_x}$$

The above steps are shown graphically in Fig. 2-20.

EXAMPLE 2-9 Three ropes are tied to a stake, and the following forces are exerted: A = 20 lb, E; B = 30 lb, 30°N of W; and C = 40 lb, 52°S of W. Determine the resultant force.

Solution Follow the steps described above.

1. Draw a figure representing each force (Fig. 2-21a). Two things should be noticed from the figure: (a) all angles are determined from the x axis, and (b) the components of each vector are labeled opposite and adjacent to known angles.

Fig. 2-21 Calculating the x and y components of all vectors.

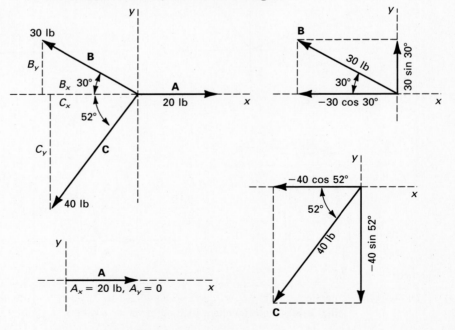

MECHANICS

2. Find the x and y components of each vector. (Refer to Fig. 2-21b, c, and d.) Note that the force A has no y component. Care must be taken to obtain the correct sign for each component. For example, B_x, C_x, and C_y are each negative. The results are listed in Table 2-5.

Table 2-5   A Table of Components

Force	ϕ_x	x component	y component
$A = 20$ lb	$0°$	$A_x = 20$ lb	$A_y = 0$
$B = 30$ lb	$30°$	$B_x = -(30 \text{ lb})(\cos 30°)$	$B_y = (30 \text{ lb})(\sin 30°)$
		$= -26.0$ lb	$= 15$ lb
$C = 40$ lb	$52°$	$C_x = (40 \text{ lb})(\cos 52°)$	$C_y = (-40 \text{ lb})(\sin 52°)$
		$= -24.6$ lb	$= -31.5$ lb
		$R_x = \sum F_x = -30.6$ lb	$R_y = \sum F_y = -16.5$ lb

3. Add the x components to obtain R_x.

$$R_x = A_x + B_x + C_x$$
$$= 20 \text{ lb} - 26 \text{ lb} - 24.6 \text{ lb} = -30.6 \text{ lb}$$

4. Add the y components to obtain R_y.

$$R_y = A_y + B_y + C_y$$
$$= 0 + 15 \text{ lb} - 31.5 \text{ lb} = -16.5$$

5. Now we find R and θ from R_x and R_y (see Fig. 2-22)

$$R = \sqrt{R_x^2 + R_y^2} = \sqrt{(-30.6)^2 + (-16.5)^2}$$
$$= \sqrt{936.4 + 272.2} = 34.8 \text{ lb}$$

$$\tan \phi = \left| \frac{R_y}{R_x} \right| = \left| \frac{-16.5}{-30.6} \right| = 0.539$$

$$\phi = 28.3°S \text{ of E (or } 208.3°)$$

Thus, the resultant force is 34.8 lb at 28.3°S of E.

Fig. 2-22

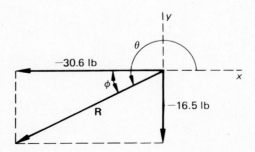

2-13

VECTOR DIFFERENCE

When we study relative velocity, acceleration, and certain other quantities, we must be able to find the difference of two vector quantities. The difference between two vectors is obtained by adding one vector to the negative of the other. The negative of a vector is found by constructing a vector equal in magnitude but opposite in direction. For example, if A is a vector whose magnitude is 40 m and whose direction is east, then the vector $-A$ is a displacement of 40 m directed to the west. Just as we have in algebra that

$$a - b = a + (-b)$$

we have in vector subtraction that

$$A - B = A + (-B)$$

The process of subtracting vectors is illustrated in Fig. 2-23. The given vectors are shown in Fig. 2-23a; Fig. 2-23b shows the vectors A and $-B$. The vector sum by the polygon method is pictured in Fig. 2-23c.

Fig. 2-23 Finding the difference of two vectors.

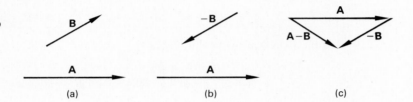

(a) (b) (c)

SUMMARY

Technical measurement is essential for the application of physics. You have learned that there are seven fundamental quantities and that each has a single approved SI unit. In mechanics, the three fundamental quantities of length, mass, and time are the only ones required for most applications. Some of these applications involve *vectors* and some involve *scalars*. Since vector quantities have direction, they must be added and subtracted by special methods. The following points summarize this unit of study:

- The SI prefixes used to express multiples and subdivisions of the base units are given below:

  giga (G) = 10^9 centi (c) = 10^{-2} nano (n) = 10^{-9}
  mega M) = 10^6 milli (m) = 10^{-3} pico (p) = 10^{-12}
  kilo (k) = 10^3 micro (μ) = 10^{-6}

- To convert one unit to another.
  a. Write down the quantity to be converted (number and unit).
  b. Recall the necessary definitions.
  c. Form two conversion factors for each definition.
  d. Multiply the quantity to be converted by those conversion factors which cancel all but the desired units.
- The **polygon method** of vector addition: The **resultant** vector is found by drawing each vector to scale, placing the tail of one vector to the tip of another until all vectors are drawn. The resultant is the straight line drawn from the starting point to the tip of the last vector (Fig. 2-24).
- The **parallelogram method** of vector addition: The resultant of two vectors is the diagonal of a parallelogram formed by the two vectors as adjacent sides. The direction is away from the common origin of the two vectors (Fig. 2-25).
- The x and y **components** of a vector (R, θ):

Fig. 2-24

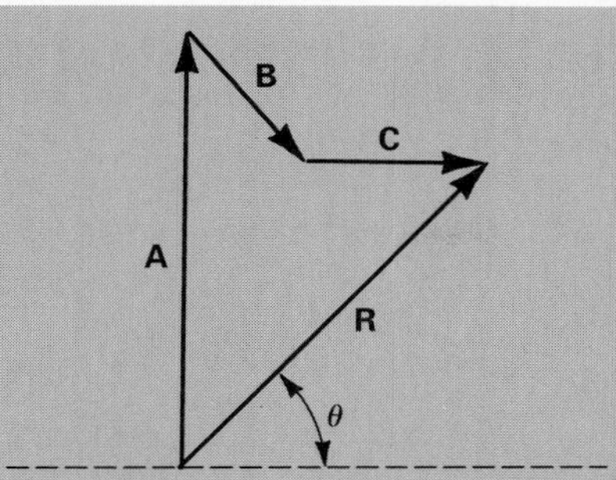

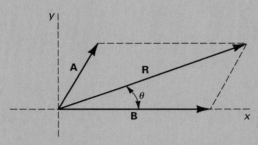

Fig. 2-25

$$R_x = R \cos \theta \qquad R_y = R \sin \theta$$

- The resultant of two perpendicular vectors (R_x, R_y):

$$R = \sqrt{R_x^2 + R_y^2} \qquad \tan \phi = \frac{R_y}{R_x}$$

- The component method of vector addition:

$$R_x = A_x + B_x + C_x + \cdots \qquad R_y = A_y + B_y + C_y + \cdots$$

$$R = \sqrt{R_x^2 + R_y^2} \qquad \tan \phi = \frac{R_y}{R_x}$$

QUESTIONS

2-1. Define the following terms:
   a. fundamental quantity
   b. SI units
   c. conversion factor
   d. dimensions
   e. vector quantity
   f. scalar quantity
   g. components
   h. resultant vector
   i. concurrent forces
   j. polygon method
   k. parallelogram method
   l. component method

2-2. Express the following measurements in proper SI form using the appropriate prefixes. The symbol for the base unit is given in parentheses:
   a. 298,000 meters (m)
   b. 7600 volts (V)
   c. 0.000067 amperes (A)
   d. 0.0645 newtons (N)
   e. 43,000,000 grams (g)
   f. 0.00000065 farads (F)

2-3. What three fundamental quantities appear in the definition of most laws of mechanics? Name the three fundamental units associated with each quantity in the SI and USCS units.

2-4. A unit of specific heat capacity is cal/g · C°. How many definitions are needed to convert these units to their corresponding units in the USCS, where the units are Btu/lb · F°? Show by a series of products how you would perform such a conversion.

2-5. Given that the units of s, v, a, and t are meters (m), meters per second (m/s), meters per second squared (m/s^2), and seconds (s), respectively, what are the dimensions of each quantity? Accept or reject the following equations on the basis of dimensional analysis:

a. $s = vt + \frac{1}{2}at^2$ c. $v_f = v_0 + at^2$

b. $2as = v_f^2 - v_0^2$ d. $s = vt + 4at^2$

2-6. Distinguish between vector and scalar quantities, and give examples of each. Explain the difference between adding vectors and adding scalars. Is it possible for the sum of two vectors to have a magnitude less than either of the original vectors?

2-7. What are the minimum and maximum resultants of two forces of 10 N and 7 N if they act on the same object?

2-8. Look up a section on rectangular and polar coordinates in a math book. What are the similarities between components of a vector and the rectangular and polar coordinates of a point?

2-9. If a vector has a direction of 230° from the positive x axis, what are the signs of its x and y components? If the ratio R_y/R_x is negative, what are the possible angles for R as measured from the positive x axis?

PROBLEMS

Unit Conversions

2-1. A soccer field is 100 m long and 60 m across. What are the length and width of the field in feet?

Ans. 328 ft, 197 ft

2-2. A wrench has a handle 8 in. long. What is the length of the handle in centimeters?

2-3. A cube has 5 in. on a side. What is the volume of the cube in SI units?

Ans. 0.00205 m^3

2-4. The speed limit on an interstate highway is posted as 65 mi/h. (a) What is this speed in kilometers per hour? (b) in feet per second?

2-5. A Nissan engine has a position displacement (volume) of 1600 cm^3 and a bore diameter of 84 mm. Express these measurements in cubic inches and in inches.

Ans. 97.6 in.^{3}, 3.31 in.

2-6. An electrician must install an underground cable from the highway to a home. If the home is located 1.2 mi into the woods, how many feet of cable will be needed? How many meters?

2-7. One U.S. gallon is a volume equivalent to 231 in.^{3}. Suppose a gasoline tank in an automobile is roughly equivalent to a rectangular solid 18 in. long, 16 in. wide, and 12 in. high. How many gallons will this tank hold?

Ans. 15.0 gal

2-8. A piece of metal is 40 cm long and 20 cm wide. Express the area of this piece in SI units.

Addition of Vectors by Graphical Methods

2-9. A woman walks 4 km east and then 8 km north. **(a)** Use the polygon method to find her resultant displacement. **(b)** Verify the resultant by the parallelogram method.

Ans. 8.94 km, 63.4° N of E

2-10. A downward force of 200 N acts simultaneously with a 500-N force directed to the left. Find the resultant force using the polygon method. Verify your answer with the parallelogram method.

2-11. The following three forces act simultaneously on the same object: A = 300 N, 30° N of E; B = 600 N, 270°; and C = 100 N due east. Represent each force as a vector and determine the resultant from the polygon method of vector addition.

Ans. 576 N, 51.4° S of E

2-12. A boat travels west a distance of 200 m, then turns north for a distance of 400 m. If it then moves 100 m in a direction 30°S of E, what is the resultant displacement by the polygon method?

2-13. Two ropes are attached to the same hook on the ceiling. The force in the right rope is 80 lb and the force in the left rope is 120 lb. If the ropes make an angle of 60° with each other, use the parallelogram method to find the magnitude of the resultant force on the hook.

Ans. 174 lb

2-14. A 40-lb force acts due west simultaneously with an unknown force F. If the resultant force is 50 lb at 36.9° N of W, what are the magnitude and direction of the unknown force? (Draw the resultant and the 40-lb force graphically. Then draw and measure the force F.)

Trigonometry and Vectors

2-15. Find the x and y components of: **(a)** a displacement of 200 km, at 34°; **(b)** a velocity of 40 km/h, at 120°; and **(c)** a force of 50 N at 330°.

Ans. 166 km, 112 km; −20 km/h, 34.6 km/h; 43.3 N, −25 N

2-16. A sled is pulled with a rope making an angle of 40° with the horizontal. The tension in the rope is 540 N. What are the vertical and horizontal components of this pull?

2-17. In removing a nail, a force of 260 N is applied by a hammer in the direction shown in Fig. 2-26. What are the horizontal and vertical components of this force?

Ans. −67.3 N, 251 N

Fig. 2-26

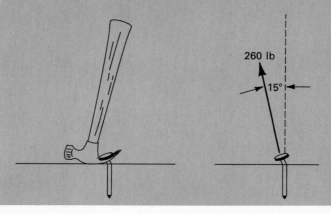

260 lb

15°

2-18. A jogger runs 2.0 mi west and then 6.0 mi north. Find the magnitude and direction of the resultant displacement.

2-19. A river flows south with a speed of 20 km/h. A certain boat has a maximum speed of 50 km/h in still water. What is the maximum speed when the boat heads directly west in this river? In what direction will the boat travel? (Represent each velocity as a vector and find the resultant.)

Ans. 53.9 km/h, 21.8° S of W

* 2-20. It is found that a horizontal force of 40 lb is needed to drag a crate along a concrete floor. If an attached rope makes an angle of 30° with the horizontal, what rope pull is required along the rope?

* 2-21. An upward lift of 80 N is required to lift a window. What force exerted along a pole, making an angle of 34° with the wall, is needed to raise the window?

Ans. 96.5 N

* 2-22. The resultant of forces A and B is 400 N at 210°. If force A is 200 N at 270°, what are the magnitude and direction of force B?

The Component Method of Vector Addition

2-23. Find the resultant of the following perpendicular forces: (a) 400 N, 0°; (b) 820 N, 270°; (c) 650 N, 180°; and (d) 500 N, 90°.

Ans. 406 N, 232°

2-24. Four ropes are attached to a ring at right angles to each other. The tensions, in order, are 40 lb, 80 lb, 70 lb, and 20 lb. What is the magnitude of the resultant force on the ring?

* 2-25. Two forces act on the car as shown in Fig. 2-27. Force A is equal to 120 N, west and force B is equal to 200 N at 60° N of W. What is the resultant of these two forces?

Ans. 280 N, 38.2° N of W

Fig. 2-27

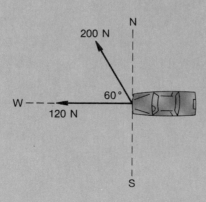

* 2-26. Suppose that the direction of force B in Prob. 2-25 is reversed (+180°), if other parameters are unchanged, what is the new resultant force? This result is the vector difference (A − B).

* 2-27. Determine the resultant force on the bolt in Fig. 2-28.

Ans. 69.6 lb, 154.1°

Fig. 2-28

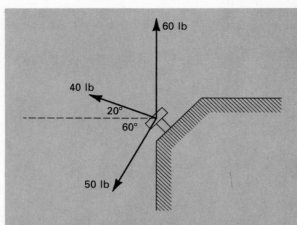

60 lb

40 lb

20°

60°

50 lb

* 2-28. Determine the resultant of the following forces by the component method of vector addition: A = (200 lb, 30°); B = (300, 330°); and C = (400 lb, 250°).

* 2-29. Three boats exert forces on a mooring hook. What is the resultant force on the hook if boat *A* exerts a force of 420 N, boat *B* exerts a force of 150 N, and boat *C* exerts a force of 500 N? (Refer to Fig. 2-29.)

   Ans. 853 N, 101.7°

Fig. 2-29

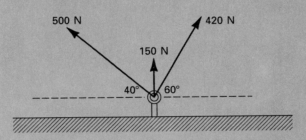

500 N   420 N

150 N

40°   60°

Additional Problems

2-30. Find the horizontal and vertical components of the following vectors: A = (400 lb, 37°); B = (90 m, 320°); C = (70 km/h, 150°).

2-31. Find the resultant *R* = A + B for the following pairs of forces: **(a)** A = 520 N, south, B = 269 N, west; **(b)** A = 18 m/s, north and B = 15 m/s, west.

   Ans. (a) 585 N, 242.6°; (b) 23.4 m/s, 129.8°

* 2-32. Determine the vector difference (A − B) for the pairs of forces in Prob. 2-31.

** 2-33. What third force must be added to the following two forces in order that the resultant force is zero: 40 N, 110° and 80 N, 185°?

   Ans. 98.3 N, 341.8°

** 2-34. What are the magnitude **F** and direction θ of the force needed to pull the car directly east with a resultant force of 400 lb? (Refer to Fig. 2-30.)

2-35. A cable is attached to the end of a beam. What pull at an angle of 40° with the beam is required to produce an effective force of 200 N along the beam?

   Ans. 261 N

Fig. 2-30

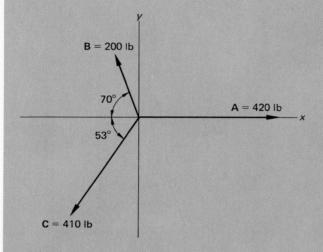

2-36. Determine the resultant of the forces shown in Fig. 2-31.

Fig. 2-31

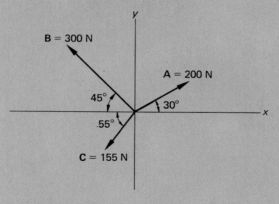

2-37. Determine the resultant of the forces shown in Fig. 2-32.

Ans. 225 N, 124.6°

Fig. 2-32

2-38. A 200-N block rests on a 30° inclined plane. If the weight of the block can be represented as acting vertically downward, what are the components of the weight down the plane and perpendicular to the plane?

2-39. A traffic light is attached to the midpoint of a rope so that each segment makes an angle of 10° with the horizontal. The resultant force is zero. What must be the weight of the traffic light if the tension in each rope segment is 200 N?

Ans. 69.5 N

* 2-40. Compute the resultant of the following forces: 220 lb, 60°; 125 lb, 210°; and 175 lb, 340°.

* 2-41. A force table is a laboratory apparatus with a circular top graduated from 0 to 360°. Weights A, B, and C are suspended from pulleys attached to the edge of the table at various angles. Determine the resultant (R, θ) for each case:

  (a) $A = 10$ N at 0° $B = 30$ N at 120° $C = 20$ N at 323°
  (b) $A = 20$ N at 0° $B = 30$ N at 225° $C = 10$ N at 300°
  (c) $A = 20$ N at 0° $B = 10$ N at 127° $C = 30$ N at 225°
  (d) $A = 20$ N at 0° $B = 30$ N at 150° $C = 15$ N at 233°

  Ans. (a) 17.8 N, 51.8°; (b) 30.1 N, 277.2°; (c) 15.1 N, 241.3°; (d) 15.3 N, 168.7°

** 2-42. An airplane is trying to keep on a due west course toward an airport. The airspeed of this plane is 600 km/h. If the wind has a speed of 40 km/h and is blowing in a direction of 30° S of W, what direction should the aircraft be pointed and what will be its speed relative to the ground?

3
Translational Equilibrium and Friction

OBJECTIVES

After completing this chapter, you should be able to:

1. Demonstrate by example or experiment your understanding of Newton's first and third laws of motion.
2. State the first condition for equilibrium, give a physical example, and demonstrate graphically that the first condition is satisfied.
3. Construct a free-body diagram representing all forces acting on an object which is in translational equilibrium.
4. Solve for unknown forces by applying the first condition for equilibrium.
5. Apply your understanding of kinetic and static friction to the solution of equilibrium problems.

Forces may act in such a manner as to cause motion or to prevent motion. Large bridges must be designed so that the overall effect of forces is to prevent motion. Every truss, girder, beam, and cable must be in *equilibrium*. In other words, the resultant forces acting at any point on the entire structure must be balanced. Shelves, chain hoists, hooks, lifting cables, and even large buildings must be constructed so that the effects of forces are controlled and understood. In this chapter we will continue our study of forces by studying objects at rest. The friction force which is so essential for equilibrium in many applications will also be introduced in this chapter as a natural extension of our work with all forces.

3-1
NEWTON'S FIRST LAW

We know from experience that a stationary object remains at rest unless acted on by some outside force. A can of oil will stay on a workbench until someone tips it over. A suspended weight will hang until it is released. We know that forces are necessary to cause anything to move if it is originally at rest.

Less obvious is the fact that an object in motion will continue in motion until an outside force changes the motion. For example, a steel bar that slides on the shop floor soon comes to rest because of its interaction with the floor. The same bar would slide much farther on ice before stopping. This is because the horizontal interaction, called *friction,* between the floor and the bar is much greater than the friction between the ice and the bar. This leads to the idea that a sliding bar on a perfectly frictionless horizontal plane would stay in motion forever. These ideas are a part of Newton's first law of motion.

> *Newton's First Law: A body at rest remains at rest and a body in motion remains in uniform motion in a straight line unless acted on by an external unbalanced force.*

Due to the existence of friction, no actual body is ever completely free from external forces. But there are situations in which it is possible to make the resultant force zero or approximately zero. In such cases, the body will behave in accordance with the first law of motion. Since we recognize that friction can never be eliminated completely, we also recognize that Newton's first law is an expression of an *ideal* situation. A flywheel rotating on lubricated ball bearings tends to keep on spinning, but even the slightest friction will eventually bring it to rest.

Newton called the property of a particle that allows it to maintain a constant state of motion or rest *inertia.* His first law is sometimes called the *law of inertia.* When an automobile is accelerated, the passengers obey this law by tending to remain at rest until the external force of the seat compels them to move. Similarly, when the automobile stops, the passengers continue in motion with constant speed until they are restrained by their seat belts or through their own efforts. All matter has inertia. The concept of *mass* is introduced later as a measure of a body's inertia.

3-2 NEWTON'S THIRD LAW

There can be no force unless two bodies are involved. When a hammer strikes a nail, it exerts an "action" force on the nail. But the nail must also "react" by pushing back against the hammer. In all cases there must be an *acting* force and a *reacting* force. Whenever two bodies interact, the force exerted by the second body on the first (the reaction force) is equal in magnitude and opposite in direction to the force exerted by the first body on the second (the action force). This principle is stated in *Newton's third law:*

> *Newton's Third Law: To every action there must be an equal and opposite reaction.*

Therefore, there can never be a single isolated force. Consider the examples of action and reaction forces in Fig. 3-1.

Note that the acting and reacting forces do not cancel each other. They are equal in magnitude and opposite in direction, but they act on *different* objects. For two forces to cancel, they must act on the same object. It might be said that the action forces create the reaction forces.

For example, someone starting to climb a ladder begins by putting one foot on the rung and pushing on it. The rung must exert an equal and opposite force on the foot if it is not to collapse. The greater the force exerted by the foot on the rung, the greater the reaction against the foot must be. Of course the rung cannot create a reaction force until the force of the foot is applied. The action force acts on the object, and the reacting force acts on the agent which applies the force.

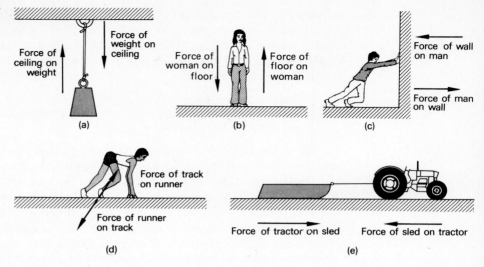

Fig. 3-1 Examples of action and reaction forces.

(a) Force of ceiling on weight / Force of weight on ceiling

(b) Force of woman on floor / Force of floor on woman

(c) Force of wall on man / Force of man on wall

(d) Force of track on runner / Force of runner on track

(e) Force of tractor on sled / Force of sled on tractor

The resultant force was defined as a single force whose effect is the same as a given system of forces. If the tendency of a number of forces is to cause motion, the resultant will also produce this tendency. A condition of equilibrium exists where the resultant of all external forces is zero. This is the same as saying that each external force is balanced by the sum of all the other external forces when equilibrium exists. Therefore, according to Newton's first law, a body in equilibrium must be either at rest or in motion with constant velocity since there is no unbalanced force.

Consider the system of forces in Fig. 3-2*a*. The vector polygon solution shows that regardless of the sequence in which the vectors are added, their resultant is always zero. The tip of the last vector lands on the tail of the first vector (see Sec. 2-7).

Fig. 3-2 Forces in equilibrium.

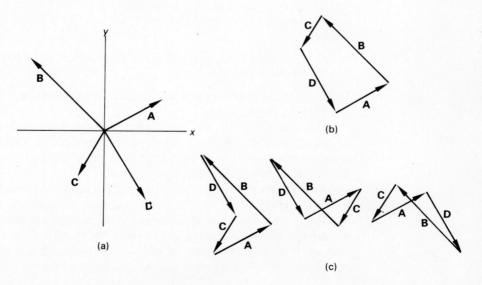

(a)

(b)

(c)

A system of forces not in equilibrium can be put in equilibrium by replacing their resultant force with an equal but opposite force called the *equilibrant*. For instance, the two forces A and B in Fig. 3-3a have a resultant R in a direction 30° above the horizontal. If we add **E**, which is equal in magnitude to **R** but which has an angle 180° greater, the system will be in equilibrium, as shown in Fig. 3-3b.

Fig. 3-3 The equilibrant.

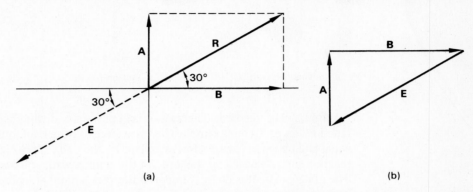

(a)   (b)

We have shown in the previous chapter that the magnitudes of the x and y components of any resultant **R** are given by

$$R_x = \sum F_x = A_x + B_x + C_x + \cdots$$
$$R_y = \sum F_y = A_y + B_y + C_y + \cdots$$

When a body is in equilibrium, the resultant of all forces acting on it is zero. In this case, both R_x and R_y must be zero; hence, for a body in equilibrium,

$$\boxed{\sum F_x = 0 \qquad \sum F_y = 0}$$ (3-1)

These two equations represent a mathematical statement of the *first condition for equilibrium,* which can be stated as follows:

> *A body is in translational equilibrium if and only if the vector sum of the forces acting upon it is zero.*

The term *translational equilibrium* is used to distinguish the first condition from the second condition for equilibrium, which involves rotational motion, discussed in Chap. 4.

3-4
FREE-BODY DIAGRAMS

Before applying the first condition for equilibrium to the solution of physical problems, you must be able to construct vector diagrams. Consider, for example, the 40-lb weight suspended by ropes shown in Fig. 3-4a. There are three forces acting on the knot—those exerted by the ceiling, by the wall, and by the earth (weight). If each of these forces is labeled and represented as a vector, we can draw a vector diagram such as the one in Fig. 3-4b. Such a diagram is called a *free-body diagram*.

A free-body diagram is a vector diagram which describes all forces acting on a particular body or object. Note that in the case of concurrent forces, all vectors point away from the center of the x and y axes, which cross at a common origin.

Fig. 3-4 Free-body diagrams showing action and reaction forces.

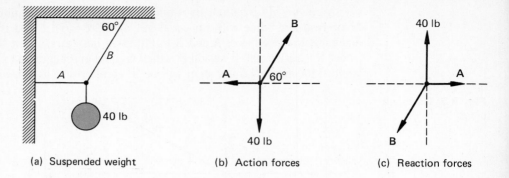

(a) Suspended weight   (b) Action forces   (c) Reaction forces

In drawing free-body diagrams, it is important to distinguish between action and reaction forces. In our example, there are forces *on* the knot, but there are also three equal and opposite reaction forces exerted *by* the knot. From Newton's third law, the reaction forces exerted *by* the knot *on* the ceiling, wall, and earth are shown in Fig. 3-4c. To avoid confusion, it is important to pick a point at which all forces are acting and draw those forces which act *on* the body at that point.

When solving equilibrium problems, we will find it useful to also indicate the components of forces on the vector diagram. A complete procedure for drawing a free-body diagram is as follows:

1. From the given conditions of a problem, draw a neat sketch representing the situation. Label it sufficiently to indicate all known and unknown forces.
2. Isolate each body of the system to be studied. Do this either mentally or by drawing a light circle around the point of application of the forces.
3. Construct a force diagram for each body to be studied. The forces are represented as vectors with their tails placed at the center of a rectangular coordinate system. (See examples in Figs. 3-5 and 3-7.)
4. Represent the x and y axes with dotted lines. These axes need not necessarily be drawn horizontally and vertically, as we shall see.

Fig. 3-5 (a) A sketch is drawn to clarify the problem. (b) A free-body diagram is constructed.

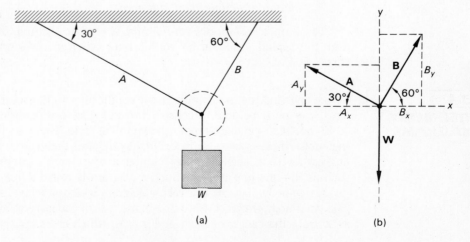

(a)   (b)

5. Dot in rectangles corresponding to the x and y components of each vector, and determine known angles from given conditions.
6. Label all known and unknown components opposite and adjacent to known angles.

Although this process may appear laborious, it is helpful and sometimes necessary for a clear understanding of a problem. As you gain practice drawing free-body diagrams, their use will become routine.

The two types of forces which act on a body are *contact forces* and *field forces*. Both must be considered in the construction of a force diagram. For example, the gravitational attraction on a body by the earth, called its *weight*, does not have a point of contact with the body. Nevertheless, it exerts a very real force and must be considered an important factor in any force problem. The direction of the weight vector should always be assumed to be downward.

EXAMPLE 3-1   A block of weight W hangs from a cord which is knotted to two other cords, A and B, fastened to the ceiling. If cord B makes an angle of 60° with the ceiling and cord A forms a 30° angle, draw the free-body diagram of the knot.

Solution   Following the procedure above, we construct the diagram shown in Fig. 3-5. This free-body diagram is valid and workable, but choosing the x and y axes along the vectors B and A, instead of horizontally and vertically, simplifies the diagram. Thus, in Fig. 3-6 we need only resolve one force W into components since A and B would now lie entirely along a particular axis.

Fig. 3-6 Rotation of x and y axes to coincide with the perpendicular vectors A and B.

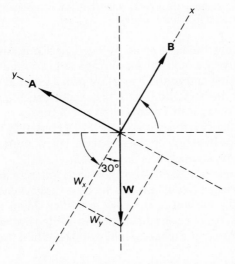

Hint: Whenever possible choose x and y axes so that as many forces as possible are completely specified along them.

Probably the most difficult part of constructing vector diagrams is the visualization of forces. In drawing free-body diagrams, it is helpful to imagine that the forces are acting on *you*. Become the knot in a rope or the block on a table and try to see the forces you would experience. Two additional examples are shown in Fig. 3-7. Note

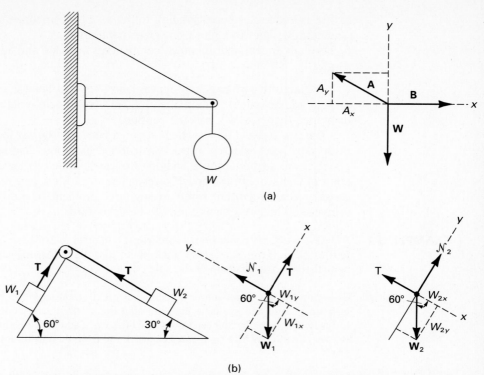

Fig. 3-7 Examples of free-body diagrams. Note that the components of vectors are labeled opposite and adjacent to known angles.

that the force exerted by the light boom in Fig. 3-7*a* is outward and not toward the wall. This is because we are interested in forces exerted *on* the end of the boom and not those exerted *by* the end of the boom. We pick a point at the end of the boom where the two ropes are attached. The 60-N weight and the tension **T** are action forces exerted by the ropes at this point. If the end of the boom is not to move, these forces must be balanced by a third force—the force exerted by the wall (through the boom). This third force **B**, acting on the end of the boom, must not be confused with the inward reaction force which acts *on* the wall.

The second example (Fig. 3-7*b*) also shows action forces acting on two weights connected by a light cord. Friction forces, which are discussed later, are not included in these diagrams. The tension in the cord on either side is shown as *T*, and the normal forces $\mathcal{N}_1$ and $\mathcal{N}_2$ are perpendicular forces exerted by the plane on the blocks. If these forces were absent, the blocks would swing together. (Note the choice of axes in each diagram.)

3-5
SOLUTION OF EQUILIBRIUM PROBLEMS

In Chap. 2 we discussed a procedure for finding the resultant of a number of forces by rectangular resolution. A similar procedure can be used to add forces which are in equilibrium. In this case the first condition for equilibrium tells us that the resultant is zero, or

$$R_x = \sum F_x = 0 \qquad R_y = \sum F_y = 0 \qquad\qquad (3\text{-}2)$$

Thus we have two equations which can be used to find unknown forces.

The following steps should be followed in solving for unknown forces in equilibrium:

1. Sketch and label the conditions of the problem.
2. Draw a free-body diagram (Sec. 3-4).
3. Resolve all forces into their x and y components even though they may contain unknown factors, such as $A \cos 30°$ or $B \sin 45°$. (You may wish to construct a table of forces like Table 3-1.)
4. Use the first condition for equilibrium [Eq. (3-1)] to set up two equations in terms of the unknown forces.
5. Solve algebraically for the unknown factors.

Table 3-1

Force	θ_x	x component	y component
A	60°	$A_x = -A \cos 60°$	$A_y = A \sin 60°$
B	0°	$B_x = B$	$B_y = 0$
W	−90°	$W_x = 0$	$W_y = -100$ N
		$\sum F_x = B - A \cos 60°$	$\sum F_y = A \sin 60° - 100$ N

EXAMPLE 3-2 A 100-N ball suspended by a rope A is pulled aside by a horizontal rope B and held so that rope A forms an angle of 30° with the vertical wall. (See Fig. 3-8.) Find the tensions in ropes A and B.

Solution We solve by following the above steps illustrated in Fig. 3-8.

1. Draw a sketch. (Fig. 3-8a.)
2. Draw a free-body diagram. (Fig. 3-8b.)
3. Resolve all forces into their components (Table 3-1). Note from the figure that A_x and W_y are negative.

Fig. 3-8 Forces acting on the knot are represented in a free-body diagram.

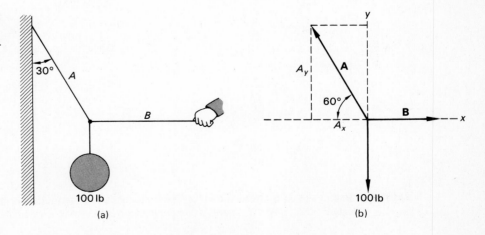

(a) (b)

4. We now apply the first condition for equilibrium. Summing the forces along the x axis yields

$$\sum F_x = B - A \cos 60° = 0$$

from which we obtain

$$B = A \cos 60° = 0.5A \qquad (3\text{-}3)$$

since $\cos 60° = 0.5$. A second equation results from summing the y components.

$$\sum F_y = A \sin 60° - 100 \text{ N} = 0$$

from which

$$A \sin 60° = 100 \text{ N} \qquad (3\text{-}4)$$

5. Finally, we solve for the unknown forces. Since $\sin 60° = 0.866$, we have from Eq. (3-4)

$$0.866A = 100 \text{ N}$$

or

$$A = \frac{100 \text{ N}}{0.866} = 115 \text{ N}$$

Now that the value of A is known, Eq. (3-3) can be solved for B as follows:

$$B = 0.5A = (0.5)(115 \text{ N})$$
$$B = 57.5 \text{ N}$$

EXAMPLE 3-3   A 200-N ball hangs from a cord knotted to two other cords, as shown in Fig. 3-9. Find the tensions in ropes A, B, and C.

Fig. 3-9

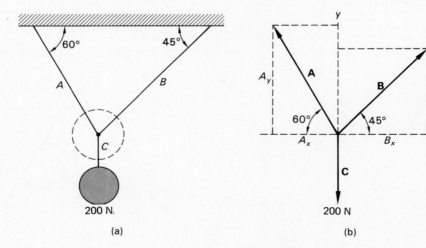

(a)   (b)

Solution   Since a sketch is already provided, the first step is to construct a free-body diagram, as illustrated in Fig. 3-9b. The x and y components for each vector as calculated from the figure are:

x component	y component
$A_x = -A \cos 60°$	$A_y = A \sin 60°$
$B_x = B \cos 45°$	$B_y = B \sin 45°$
$C_x = 0$	$C_y = -200 \text{ N}$

Summing the forces along the x axis, we obtain

$$\sum F_x = -A \cos 60° + B \cos 45° = 0$$

which can be simplified by substituting known trigonometric functions. Hence

$$-0.5A + 0.707B = 0 \qquad (3\text{-}5)$$

More information is needed to solve this equation. We obtain a second equation by summing the forces along the y axis giving

$$0.866A + 0.707B = 200 \text{ N} \qquad (3\text{-}6)$$

Equations (3-4) and (3-5) are now solved simultaneously for A and B by the process of substitution. Solving for A in Eq. (3-4) gives

$$A = \frac{0.707B}{0.5} \qquad \text{or} \qquad A = 1.414B \qquad (3\text{-}7)$$

Now we substitute this equality into Eq. (3-6) obtaining

$$0.866(1.414B) + 0.707B = 200 \text{ N}$$

which can be solved for B as follows:

$$1.225B + 0.707B = 200 \text{ N}$$
$$1.93B = 200 \text{ N}$$
$$B = \frac{200 \text{ N}}{1.93} = 104 \text{ N}$$

The tension A can now be found by substituting $B = 104$ N into Eq. (3-7):

$$A = 1.414(104 \text{ N}) = 146 \text{ N}$$

The tension in cord C is, of course, 200 N since it must be equal to the weight.

EXAMPLE 3-4 A 200-lb block rests on a frictionless inclined plane of slope angle 30°. A cord attached to the block passes over a frictionless pulley at the top of the plane and is attached to a second block. What must the weight of the second block be if the system is to be in equilibrium? (Neglect the weight of the cord.)

Solution After sketching the system, a free-body diagram is constructed for each body, as shown in Fig. 3-10. Applying the first condition for equilibrium to the second block (Fig. 3-10c), we find that

Fig. 3-10 A free-body diagram is drawn for each body in the problem.

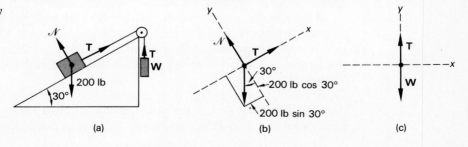

(a)　　　　　　(b)　　　　　　(c)

$$T - W = 0$$

or

$$T = W$$

Since the rope is continuous and the system is frictionless, the tension in Fig. 3-10*b* for the 200-lb block must also be equal to the weight W.

Considering the diagram for the 200-lb block, we determine the components of each force as follows:

x component	y component
$T_x = T = W$	$T_y = 0$
$\mathcal{N}_x = 0$	$\mathcal{N}_y = \mathcal{N}$
$(200 \text{ lb})_x = (-200 \text{ lb})(\sin 30°)$	$(200 \text{ lb})_y = (-200 \text{ lb})(\cos 30°)$

Applying the first condition yields

$$\sum F_x = 0 \qquad T - (200 \text{ lb})(\sin 30°) = 0 \tag{3-8}$$
$$\sum F_y = 0 \qquad \mathcal{N} - (200 \text{ lb})(\cos 30°) = 0 \tag{3-9}$$

From Eq. (3-8) we obtain

$$T = (200 \text{ lb})(\sin 30°) = 100 \text{ lb}$$

and, therefore, $W = 100$ lb since $T = W$. Thus a 100-lb weight is required to maintain equilibrium.

The normal force exerted by the plane on the 200-lb block can be found from Eq. (3-9), although this calculation was not necessary to determine the weight W. Hence

$$\mathcal{N} = (200 \text{ lb})(\cos 30°) = (200 \text{ lb})(0.866)$$
$$= 173 \text{ lb}$$

3-6
FRICTION

Whenever a body moves while it is in contact with another object, frictional forces oppose the relative motion. These forces are caused by the adhesion of one surface to the other and by the interlocking of irregularities in the rubbing surfaces. It is friction that holds a nail in a board, allows us to walk, and makes automobile brakes work. In all these cases friction has a desirable effect.

In many other instances, however, friction must be minimized. For example, it increases the work necessary to operate machinery, it causes wear, and it generates heat, which often causes additional damage. Automobiles and airplanes are streamlined in order to decrease air friction, which is large at high speeds.

Whenever one surface moves past another, the frictional force exerted by each body on the other is parallel or tangent to the two surfaces and acts in such a manner as to oppose relative motion. It is important to note that these forces exist not only when there is relative motion but even when one object only *tends* to slide past another.

Suppose a force is exerted on a block which rests on a horizontal surface, as shown in Fig. 3-11. At first the block will not be moved because of the action of a force called the *force of static friction* $\mathscr{F}_s$. But as the applied force is increased,

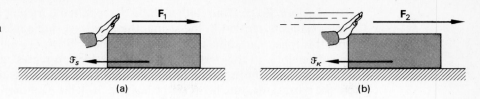

Fig. 3-11 (a) In static friction, motion is impending; (b) in kinetic friction, the two surfaces are in relative motion.

motion eventually occurs, and the frictional force exerted by the horizontal surface while the block is moving is called the *force of kinetic friction* $\mathcal{F}_k$.

The laws governing frictional forces can be determined experimentally in the laboratory by using an apparatus similar to the one shown in Fig. 3-12*a*. A box of weight *W* is placed on a horizontal table, and a string attached to the box is passed over a light frictionless pulley and attached to a weight hanger. All forces acting on the box and hanger are shown in their corresponding free-body diagrams (Fig. 3-12*b* and *c*).

Let us consider that the system is in equilibrium, which requires the box to be stationary or moving with a constant velocity. In either case, we may apply the first condition for equilibrium. Consider the force diagram of the box as shown in Fig. 3-12*c*.

$$\sum \mathcal{F}_x = 0 \qquad \mathcal{F} - T = 0$$
$$\sum \mathcal{F}_y = 0 \qquad \mathcal{N} - W = 0$$

or

$$\mathcal{F} = T \qquad \text{and} \qquad \mathcal{N} = W$$

Thus the force of friction is equal in magnitude to the tension in the string, and the normal force exerted by the table on the box is equal to the weight of the box. Note that the tension in the string is determined by the weight of the hanger plus the weights placed on the hanger.

We begin the experiment by slowly adding weights to the hanger, thus gradually increasing the tension in the string. As the tension is increased, the equal but oppositely directed force of static friction is also increased. If T is increased sufficiently, the box will start to move, indicating that T has overcome the *maximum* force of

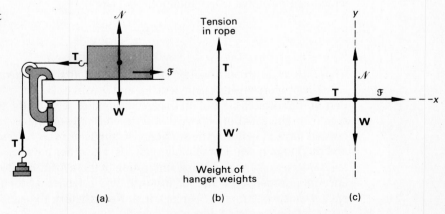

Fig. 3-12 Experiment to determine the force of friction.

(a)

(b)

(c)

static friction $\mathcal{F}_{s,max}$. Thus, although the force of static friction $\mathcal{F}_s$ will vary according to values of tension in the string, there exists a single maximum value $\mathcal{F}_{s,max}$. Only this maximum value will be useful in the solution of problems involving friction. Therefore, in this text $\mathcal{F}_s$ will be understood to represent $\mathcal{F}_{s,max}$.

To continue the experiment, suppose we add weights to the box, thereby increasing the normal pressure between the box and the table. Our normal force will now be

$$\mathcal{N} = W + \text{added weights}$$

Repeating the above procedure will show that a *proportionately* larger value of T will be necessary to overcome $\mathcal{F}_s$. In other words, if we double the normal force between two surfaces, the maximum force of static friction which must be overcome is also doubled. If $\mathcal{N}$ is tripled, $\mathcal{F}_s$ is tripled, and so it will be for other factors. Therefore it can be said that the maximum force of static friction is directly proportional to the normal force between the two surfaces. We can write this proportionality as

$$\mathcal{F}_s \propto \mathcal{N}$$

which can be stated as an equation:

$$\boxed{\mathcal{F}_s = \mu_s \mathcal{N}} \tag{3-10}$$

where μ_s is a proportionality constant called the *coefficient of static friction*. Since μ_s is the constant ratio of two forces, it is a dimensionless quantity.

In the above experiment it will be noticed that after $\mathcal{F}_s$ has been overcome, the box will increase its speed, or accelerate, until it is stopped by the pulley. This indicates that a lesser value of T would be necessary to keep the box moving with a constant speed. Thus the force of kinetic friction $\mathcal{F}_k$ must be smaller than the value of $\mathcal{F}_s$ for the same surfaces. In other words, it requires more force to start a block moving than it does to keep it moving with constant speed. In the latter case, the first condition for equilibrium is still satisfied; hence the same reasoning which led to Eq. (3-10) for static friction will yield the following proportionality for kinetic friction:

$$\mathcal{F}_k \propto \mathcal{N}$$

which can be stated as an equation. As before,

$$\boxed{\mathcal{F}_k = \mu_k \mathcal{N}} \tag{3-11}$$

where μ_k is a proportionality constant called the *coefficient of kinetic friction*.

The proportionality coefficients μ_s and μ_k can be shown to depend on the roughness of the surfaces but not on the area of contact between the two surfaces. It can be seen from the equations above that μ depends only on the frictional force $\mathcal{F}$ and the normal force $\mathcal{N}$ between the surfaces. Of course, it must be realized that Eqs. (3-10) and (3-11) are not fundamentally rigorous, like other physical equations. Many variables interfere with the general application of these formulas. No one who has experience in automobile racing, for instance, will believe that the friction force is *completely* independent of the contact area. Nevertheless, the equations are useful tools for estimating resistive forces in specific cases.

Table 3-2 shows some representative values for the coefficients of static and kinetic friction for different types of surfaces. These values are approximate and depend upon the condition of the surfaces.

Table 3-2 Approximate Coefficients of Friction

Material	μ_s	μ_k
Wood on wood	0.7	0.4
Steel on steel	0.15	0.09
Metal on leather	0.6	0.5
Wood on leather	0.5	0.4
Rubber on concrete, dry	0.9	0.7
wet	0.7	0.57

Problems involving friction are solved like other force problems, except that the following points should be considered:

1. Frictional forces are parallel to the surfaces and directly oppose motion of the surfaces across each other.
2. The force of static friction is larger than the force of kinetic friction for the same materials.
3. In drawing free-body diagrams, it is usually more expedient to choose the x axis along the plane of motion and the y axis normal to the plane of motion.
4. The first condition for equilibrium can be applied to set up two equations representing forces along the plane of motion and normal to it. (The more complicated problem, which involves a resultant force, will be treated in a later chapter.)
5. The relations $\mathscr{F}_s = \mu_s \mathscr{N}$ and $\mathscr{F}_k = \mu_k \mathscr{N}$ can be applied to solve for the desired quantity.

EXAMPLE 3-5 A 50-lb block rests on a horizontal surface. A horizontal pull of 10 lb is required to just start the block moving. After motion is started, only a 5-lb force is required to move the block with a constant speed. Find the coefficients of static and kinetic friction.

Solution The key words which should be recognized are *just start moving* and *move with constant speed*. The former implies static friction whereas the latter implies kinetic friction. In each case a condition of equilibrium exists. The correct free-body diagrams are shown in Fig. 3-13a

Fig. 3-13 *(a)* A force of 10 lb is required to overcome the maximum force of static friction. *(b)* A force of only 5 lb is required to move the block with constant speed.

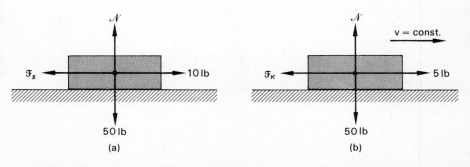

and *b*. Let us first consider the force which overcomes static friction. Applying the first condition for equilibrium to Fig. 3-13*a* gives

$$\sum F_x = 0 \qquad\qquad 10 \text{ lb} - \mathscr{F}_s = 0$$
$$\sum F_y = 0 \qquad\qquad \mathscr{N} - 50 \text{ lb} = 0$$

from which we note that

$$\mathscr{F}_s = 10 \text{ lb} \qquad \mathscr{N} = 50 \text{ lb}$$

Thus we can find the coefficient of static friction from Eq. (3-8).

$$\mu_s = \frac{\mathscr{F}_s}{\mathscr{N}} = \frac{10 \text{ lb}}{50 \text{ lb}} = 0.2$$

The force which overcomes kinetic friction is 5 lb. Hence, summing the forces along the *x* axis yields

$$5 \text{ lb} - \mathscr{F}_k = 0$$

or

$$\mathscr{F}_k = 5 \text{ lb}$$

The normal force is still 50 lb, and so

$$\mu_k = \frac{\mathscr{F}_k}{\mathscr{N}} = \frac{5 \text{ lb}}{50 \text{ lb}} = 0.1$$

EXAMPLE 3-6   What force T at an angle of 30° above the horizontal is required to drag a 40-lb block to the right at constant speed if $\mu_k = 0.2$?

Solution   We shall first sketch the problem and then construct a free-body diagram, as shown in Fig. 3-14. Applying the first condition for equilibrium yields

$$\sum F_x = 0 \qquad T_x - \mathscr{F}_k = 0$$
$$\sum F_y = 0 \qquad \mathscr{N} + T_y - 40 \text{ lb} = 0 \qquad\qquad (3\text{-}12)$$

The latter equation shows the normal force to be

$$\mathscr{N} = 40 \text{ lb} - T_y \qquad\qquad (3\text{-}13)$$

It should be noted that the normal force is decreased by the *y* component of T. Substituting $\mathscr{F}_k = \mu_k \mathscr{N}$ into Eq. (3-12) gives

$$T_x - \mu_k \mathscr{N} = 0$$

Fig. 3-14 The force *T* at an angle above the horizontal reduces the normal force, resulting in a smaller force of friction.

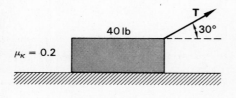

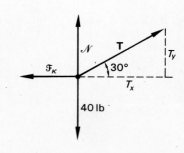

But $\mathcal{N} = 40 \text{ lb} - T_y$ from Eq. (3-13), and so

$$T_x - \mu_k(40 \text{ lb} - T_y) = 0 \qquad (3\text{-}14)$$

From the free-body diagram it is noted that

$$T_x = T \cos 30° = 0.866T$$

and

$$T_y = T \sin 30° = 0.5T$$

Thus, recalling that $\mu_k = 0.2$, we can write Eq. (3-14) as

$$0.866T - (0.2)(40 \text{ lb} - 0.5T) = 0$$

which can be solved for T as follows:

$$0.866T - 8 \text{ lb} + 0.1T = 0$$
$$0.966T - 8 \text{ lb} = 0$$
$$0.966T = 8 \text{ lb}$$
$$T = \frac{8 \text{ lb}}{0.966} = 8.3 \text{ lb}$$

Therefore, a force of 8.3 lb is required to pull the block with constant speed if the rope makes an angle of 30° above the horizontal.

EXAMPLE 3-7 A 100-lb block rests on a 30° inclined plane. If $\mu_k = 0.1$, what push **P** parallel to the plane and directed up the plane will cause the block to move *(a)* up the plane with constant speed and *(b)* down the plane with constant speed?

Solution (a) The general problem is sketched in Fig. 3-15a. For motion up the plane, the force of friction is directed down the plane, as shown in Fig. 3-15b. Applying the first condition for equilibrium, we obtain

$$\sum F_x = 0 \qquad P - \mathcal{F}_k - W_x = 0 \qquad (3\text{-}15)$$
$$\sum F_y = 0 \qquad \mathcal{N} - W_y = 0 \qquad (3\text{-}16)$$

From the figure, the x and y components of the weight are

$$W_x = (100 \text{ lb})(\sin 30°) = 50 \text{ lb}$$
$$W_y = (100 \text{ lb})(\cos 30°) = 86.6 \text{ lb}$$

Fig. 3-15 Friction on the inclined plane.

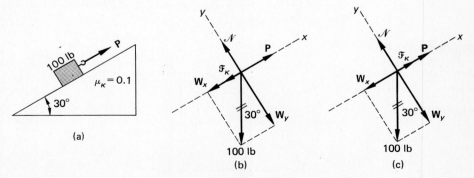

(a)

(b)

(c)

Substitution of the latter into Eq. (3-16) allows us to solve for the normal force. Hence

$$\mathcal{N} - 86.6 \text{ lb} = 0$$
$$\mathcal{N} = 86.6 \text{ lb}$$

The push required to move up the plane is, from Eq. (3-15),

$$P = \mathcal{F}_k + W_x$$

But $\mathcal{F}_k = \mu_k \mathcal{N}$, so that

$$P = \mu_k \mathcal{N} + W_x$$

Substituting known values for μ_k, $\mathcal{N}$, and W_x, we obtain

$$P = (0.1)(86.6 \text{ lb}) + 50 \text{ lb}$$
$$= 58.7 \text{ lb}$$

Note that the push up the plane in this case must overcome both the frictional force of 8.66 lb and the 50-lb component of the weight down the plane.

Solution (b)   Now we must consider the push P required to retard the downward motion. The only difference between this problem and the problem in part *a* is that the friction force is now directed up the plane. The normal force does not change, and the components of the weight do not change. Therefore, if we sum the forces along the *x* axis in Fig. 3-15c, we have

$$\sum F_x = 0 \qquad P + \mathcal{F}_k - W_x = 0$$

from which

$$P = W_x - \mathcal{F}_k$$

or

$$P = 50 \text{ lb} - 8.66 \text{ lb}$$
$$= 41.3 \text{ lb}$$

The force of 41.3 lb directed up the plane retards the downward motion of the block so that its speed is constant. If this force P were not exerted, the block would accelerate down the plane of its own accord.

EXAMPLE 3-8   What is the maximum slope angle θ for an inclined plane such that a block of weight W will not slide down the plane?

Solution   A sketch and free-body diagram are constructed as shown in Fig. 3-16. The maximum value of θ would be that value which overcomes static friction $\mathcal{F}_s = \mu_s \mathcal{N}$. Applying the first condition for equilibrium gives

Fig. 3-16 The limiting angle of repose.

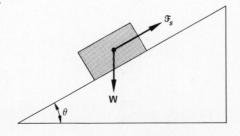
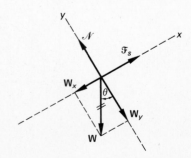

$$\sum F_x = 0 \qquad \mathscr{F}_s - W_x = 0$$
$$\sum F_y = 0 \qquad \mathscr{N} - W_y = 0$$

from which we transpose to obtain

$$\mathscr{F}_s = W_x \qquad \mathscr{N} = W_y \qquad\qquad (3\text{-}17)$$

From Fig. 3-16 we note that θ is an angle whose tangent is W_x/W_y; hence, from Eq. (3-17), we have

$$\tan\,\theta = \frac{W_x}{W_y} = \frac{\mathscr{F}_x}{\mathscr{N}}$$

But $\mathscr{F}_s/\mathscr{N}$ is equal to the coefficient of static friction μ_s. Hence

$$\boxed{\tan\,\theta = \mu_s}$$

Thus a block, regardless of its weight, will remain at rest on an inclined plane unless $\tan\,\theta$ equals or exceeds μ_s. The angle θ in this case is called the *limiting angle* or the *angle of repose*.

It is left as an exercise for you to show that a block will slide down the plane with the constant speed if $\tan\,\theta = \mu_k$.

SUMMARY

In this chapter, we have defined objects which are at rest or in motion with constant speed to be in equilibrium. Through the use of vector diagrams and Newton's laws, we have found it possible to determine unknown forces for systems which are known to be in equilibrium. The following items will summarize the more important concepts to be remembered:

- *Newton's first law of motion* states that an object at rest and an object in motion with constant speed will maintain the state of rest or constant motion unless acted on by a resultant force.
- *Newton's third law of motion* states that every action must produce an equal and opposite reaction. The action and reaction forces do not act on the same body.
- Free-body diagrams: From the conditions of the problem a neat sketch is drawn and all known quantities are labeled. Then a force diagram indicating all forces and their components is constructed. All information such as that given in Fig. 3-17 should be a part of the diagram.
- Translational equilibrium: A body in translational equilibrium has no resultant force acting on it. In such cases, the sum of all the x components is zero, and the sum of all the y components is zero. This is known as the first condition for equilibrium and is written

$$\boxed{R_x = \sum F_x = 0} \qquad \boxed{R_y = \sum F_y = 0}$$

- Applying these conditions to Fig. 3-17, for example, we obtain two equations in two unknowns:

$$B\cos 45° - A\cos 60° = 0$$
$$B\sin 45° + A\sin 60° - 200\text{ N} = 0$$

These equations can be solved to find A and B.

Fig. 3-17

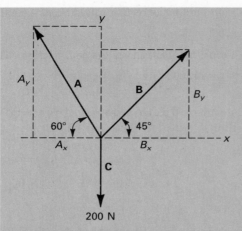

200 N

- Static friction exists between two surfaces when motion is impending; kinetic friction occurs when the two surfaces are in relative motion. In either case, the friction forces are proportional to the normal force. They are given by

$$\mathcal{F}_s = \mu_s \mathcal{N} \qquad \mathcal{F}_k = \mu_k \mathcal{N}$$

These forces must often be considered in equilibrium problems.

QUESTIONS

3-1. Define the following terms:
   a. Inertia
   b. Reaction force
   c. Equilibrium
   d. Equilibrant
   e. Free-body diagram
   f. Friction force
   g. Coefficient of friction
   h. Normal force
   i. Angle of repose

3-2. A popular magic trick consists of placing a coin on a card and the card on the top of a glass. The edge of the card is flipped briskly with the forefinger, causing the card to fly off the top of the glass as the coin drops into the glass. Explain. What law does this illustrate?

3-3. When the head of a hammer becomes loose, you can reseat it by holding the hammer vertically and tapping the base of the handle against the floor. Explain. What law does this illustrate?

3-4. Explain the part played by Newton's third law of motion in the following activities: (a) walking, (b) rowing, (c) rockets, and (d) parachuting.

3-5. Can a moving body be in equilibrium? Give several examples.

3-6. According to Newton's third law of motion, every force has an equal and oppositely directed reaction force. Therefore, the concept of a resultant unbalanced force must be an illusion that really does not hold under close examination. Do you agree with this statement? Give the reasons for your answer.

3-7. A brick is suspended from the ceiling by a light string. A second identical string is attached to the bottom of the brick and hangs within the reach of a student. When the student pulls the lower string slowly, the upper string breaks, but when the lower string is jerked, it breaks. Explain?

MECHANICS

3-8. A long steel cable is stretched between two buildings. Show by diagrams and discussion why it is not possible to pull the cable so taut that it will be perfectly horizontal with no sag in the middle.

3-9. We have seen that it is always advantageous to choose the x and y axes so that as many forces as possible are completely specified along an axis. Suppose that no two forces are perpendicular to each other. Will there still be an advantage to rotating axes to align an unknown force with an axis as opposed to aligning a known force? Test this approach by applying it to one of the text examples.

3-10. Discuss a few beneficial uses of the force of friction.

3-11. Why do we speak of a *maximum* force of static friction? Why do we not discuss a maximum force of kinetic friction?

3-12. Why is it easier to pull a sled at an angle than it is to push a sled at the same angle? Draw free-body diagrams to show what the normal force would be in each case.

3-13. Is the normal force acting on a body always equal to its weight?

3-14. When walking across a frozen pond, should you take short steps or long ones? Why? If the ice were completely frictionless, would it be possible for you to get off the pond? Explain.

PROBLEMS

Free-body Diagrams

3-1. Draw a free-body diagram for the arrangements shown in Fig. 3-18. Isolate a point where the important forces are acting, and represent each force acting on that point as a vector. (Neglect the weight of the strut.) Determine a reference angle and label the components opposite and adjacent to a reference angle. Rotate your coordinate axes if it results in a simpler diagram.

Fig. 3-18

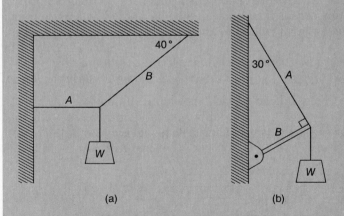

(a)   (b)

3-2. Study each force acting *on* the block or *on* the end of the very light struts in Fig. 3-19. Construct free-body diagrams with angles and components indicated. Take care not to pick up the reaction forces. Equilibrium should demand a balanced vector diagram. Is there an advantage to choosing the axes other than horizontal and vertical?

Fig. 3-19

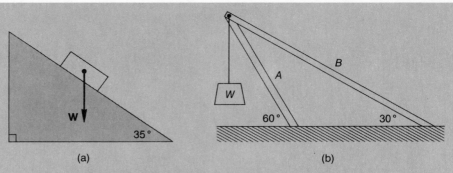

(a)   (b)

Solution of Equilibrium Problems

3-3. If the weight of the block in Fig. 3-18a is 80 N, what are the tensions in ropes A and B?

Ans. $A = 95.3$ N, $B = 124$ N

3-4. If rope B in Fig. 3-18a will break for tensions greater than 200 lb, what is the maximum weight W that can be supported?

*

3-5. Determine the compression in the boom and the tension in the cord for Fig. 3-18b when the weight is equal to 600 N. Neglect the weight of the boom.

Ans. $A = 520$ N, $B = 300$ N

3-6. The block in Fig. 3-19a weighs 70 N. What are the magnitudes of the friction force directed up the plane and of the normal force directed perpendicular to the plane?

**

3-7. The two-by-four studs A and B in Fig. 3-19b are used to support a 400-lb weight. Neglecting the weights of the studs, find the unknown forces and state whether each stud is under compression or tension.

Ans. $A = 693$ lb, compression; $B = 400$ lb, tension

3-8. An 80-N traffic light is supported at the midpoint of a line between two poles that are 30 m apart. If the light sags a vertical distance of 1 m at the midpoint, what is the tension in the supporting line? (Hint: First find the angle the line makes with the horizontal; then draw a free-body diagram assuming the same tension in each line segment.)

3-9. A truck is removed from the mud by attaching a line between the truck and a tree. When the angles are as shown in Fig. 3-20, a force of 40 lb is exerted at the midpoint of the line. What force is exerted on the truck?

Ans. 58.5 lb

Fig. 3-20

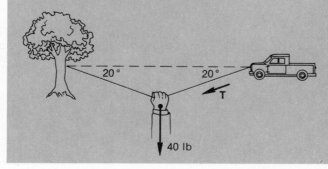

Fig. 3-21

* **3-10.** A 20-N picture is hung from a nail, as in Fig. 3-21, so that the supporting cords make an angle of 60°. What is the tension of each cord segment?

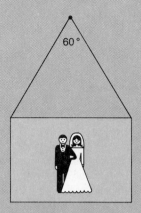

Friction

3-11. A horizontal force of 40 N will just start an empty 600-N sled moving across packed snow. After motion is begun, only 10 N is needed to keep it moving at constant speed. (a) What are the coefficients of static and kinetic friction? (b) If 200 N of supplies are added to the sled, what new force is required to drag the sled at constant speed?

Ans. (a) 0.0667, 0.0167; (b) 13.3 N

3-12. A dockworker finds that a force of 60 lb is needed to drag a 150-lb crate across the deck at constant speed. What is the coefficient of kinetic friction? If a smaller crate of similar construction can be moved with a force of only 40 lb, what is the weight of this crate?

3-13. The coefficient of static friction between wood and wood is 0.7 and the coefficient of kinetic friction is 0.4. What horizontal force is required to just start a 50-N wooden block moving along a wooden floor? What force will keep it moving after static friction has been overcome?

Ans. 35 N, 20 N

3-14. A steel block weighing 240 N rests on a level steel beam. What horizontal force will move the block at constant speed if the coefficient of kinetic friction is 0.09?

* **3-15.** A 60-N block is dragged along the horizontal floor at constant speed. The attached rope makes an angle of 35° with the floor and has a tension of 40 N. Draw a free-body diagram of all forces acting on the block. Assuming equilibrium (constant speed), determine the friction force and the normal force. What is the coefficient of kinetic friction?

Ans. 32.8 N, 37.1 N, 0.885

* **3-16.** A 200-N sled is pushed at constant speed with a force that is 28° below the horizontal. If the magnitude of the push is 50 N, what is the coefficient of kinetic friction?

* **3-17.** Assume that $W = 60$ N, $\theta = 43°$, and $\mu_k = 0.3$ in Fig. 3-22. What is the normal force on the block? What is the component of the weight acting down the plane?

Fig. 3-22

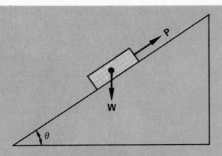

What push P directed up the plane will cause the block to move up the plane with constant speed?

Ans. 43.9 N, 40.9 N, 54.1 N

3-18. If the block in Prob. 3-17 is released, it will move rapidly down the incline. What force up the plane is needed to retard the downward motion of the sliding block until it travels at constant speed?

3-19. The coefficient of static friction for wood on wood is 0.7. What is the maximum angle for an inclined wooden plane if a wooden block is to remain at rest on the plane?

Ans. 35.0°

3-20. A wooden roof is sloped at a 40° angle. What is the minimum coefficient of static friction between the sole of a roofer's shoe and the roof to avoid slipping?

Additional Problems

3-21. Determine the tension in rope A and the compression B in the strut for Fig. 3-23. The compression in the strut is equal in magnitude but opposite in direction to the force exerted by the strut on its end.

Ans. $A = 231$ N, $B = 462$ N

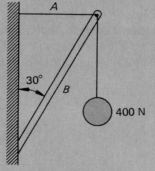

Fig. 3-23

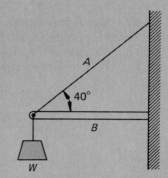

Fig. 3-24

3-22. If the breaking strength of cable A in Fig. 3-24 is 200 N, what is the maximum weight W which can be supported by this apparatus?

3-23. A 90-N wagon with frictionless wheels is rolled up a 37° inclined plane by a push P directed parallel to the plane. Draw a free-body diagram of the forces acting on the wagon. What is the minimum push P required to roll the wagon up the plane?

Ans. 54.2 N

3-24. What push P directed horizontally is needed to hold the wagon of Prob. 3-23 at rest on the incline?

3-25. A 70-lb block of steel is at rest on a 41° incline. What is the magnitude of the static friction force directed up the plane? Is this necessarily the maximum force of static friction? What is the normal force at this angle?
Ans. 45.9 N, No, 52.8 N

3-26. A cake of ice slides with constant speed across a wooden floor ($\mu_k = 0.1$) when a horizontal force of 8 lb is applied. What is the weight of the cake of ice?

3-27. Find the tension in ropes A and B for the arrangements shown in Fig. 3-25.
Ans. (a) $A = 170$ N, $B = 294$ N; (b) $A = 134$ N, $B = 209$ N

Fig. 3-25

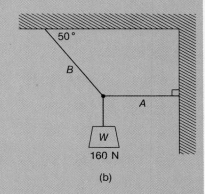

(a)

(b)

3-28. Find the tension in the cable and the compression in the boom for the arrangements of Fig. 3-26.

Fig. 3-26

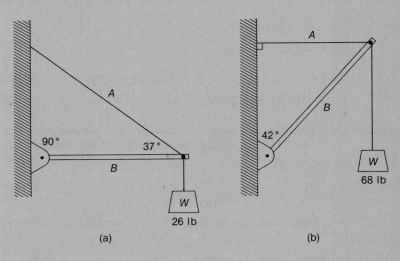

(a)

(b)

3-29. Determine the tension in the ropes for Fig. 3-27a and the forces in the boards for Fig. 3-27b.
Ans. (a) $A = 1410$ N, $B = 1150$ N; (b) $A = 33.7$ lb, $B = 23.8$ lb

Fig. 3-27

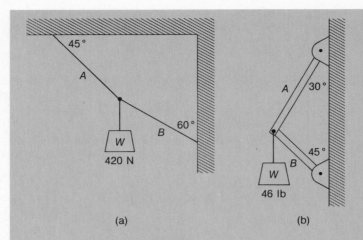

(a)   (b)

3-30. A 2-N eraser is pressed against a vertical blackboard with a horizontal push of 12 N into the board. The coefficient of static friction is 0.25. (a) Find the force required to move the eraser parallel to the base of the blackboard. (b) Find the upward force required to start vertical motion.

** **3-31.** It is determined experimentally that a 20-lb horizontal force will move a 60-lb lawn mower at constant speed. The handle of the mower makes an angle of 40° with the ground. (a) What is the coefficient of kinetic friction? (b) What force along the handle will move the mower forward at constant speed? (c) What is the normal force during the forward motion?

Ans. 0.333, 36.2 lb, 83.3 lb

** **3-32.** Suppose the lawn mower of Prob. 3-31 is to be moved backward at constant speed by pulling along the handle at 40° with the ground. What force along the handle is required in this case? What is the normal force?

3-33. A cable is stretched horizontally across the top of two vertical poles which are separated a distance of 20 m. A sign weighing 250 N is attached to the midpoint of the cable, causing the center to sag a distance of 1.2 m. What is the tension in the cable?

Ans. 1049 N

* **3-34.** Assume that the cable in Prob. 3-33 has a breaking strength of 1200 N. What is the maximum weight of the sign if the system is not to fail?

* **3-35.** Determine the compression in the center strut B and the tension in rope A for the situation described by Fig. 3-28.

Ans. $A = 643$ N, $B = 940$ N

Fig. 3-28

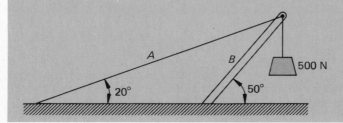

3-36. Find the tension in each cord of Fig. 3-29 if the weight W is 476 N.

Fig. 3-29

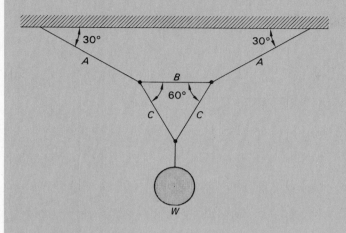

3-37. What horizontal push P is required to just hold a 200-N block on a 60° inclined plane if $\mu_s = 0.4$?

Ans. 157 N

3-38. What horizontal push P will just start the 200-N block of Prob. 3-37 moving up the plane?

3-39. A 46-N sled has a pole attached to it at an angle of 30° above the horizontal. The coefficient of kinetic friction is 0.1. **(a)** Find the force required to pull the sled at constant speed. **(b)** Find the force required to push the sled with constant speed at the same angle.

Ans. 5.02 N, 5.64 N

3-40. Two weights are hung over two frictionless pulleys as shown in Fig. 3-30. What weight W will cause the 300-lb block to just start moving to the right?

Fig. 3-30

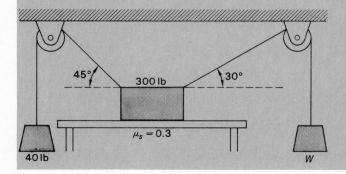

Fig. 3-31

3-41. In the arrangement of Fig. 3-31, assume that the coefficient of static friction between the 200-lb block and the surface is 0.3. Find the maximum weight which can be hung at point O without upsetting the equilibrium.

Ans. 21.8 lb

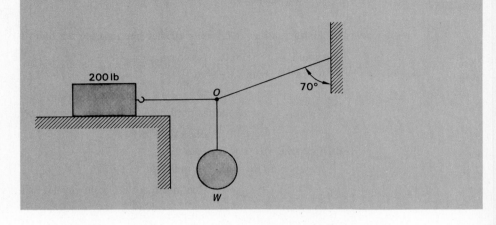

4 Torque and Rotational Equilibrium

OBJECTIVES

After completing this chapter, you should be able to:

1. Illustrate by example and definition your understanding of the terms *moment arm* and *torque.*
2. Calculate the *resultant torque* about any axis when given the magnitude and position of forces on an extended object.
3. Solve for unknown forces or distances by applying the *first and second conditions for equilibrium.*
4. Define and illustrate by example what is meant by the *center of gravity.*

In previous chapters, we have discussed forces which act at a single point. Translational equilibrium exists when the vector sum of forces is zero. However, there are many cases in which the forces acting on an object do not have a common point of application. Such forces are said to be *nonconcurrent.* For example, a mechanic exerts a force on the handle of a wrench to tighten a bolt. A carpenter uses a long lever to pry the lid from a wooden box. The engineer considers twisting forces which tend to snap a beam attached to a wall. The steering wheel of an automobile is turned by forces which do not have a common point of application. In such cases, there may be a *tendency to rotate* which we will define as *torque.* If we learn to measure or predict the torques produced by certain forces, we can obtain desired rotational effects. If no rotation is desired, there must be no resultant torque. This leads naturally to a condition for *rotational equilibrium* which is very important for industrial and engineering applications.

4-1
CONDITIONS FOR EQUILIBRIUM

When a body is in equilibrium, it is either at rest or in uniform motion. According to Newton's first law, only the application of a resultant force can change this condition. We have seen that if all forces acting on such a body intersect at a single point and their vector sum is zero, the system must be in equilibrium. When a body is acted on by forces which do not have a common *line of action,* it may be in translational equilibrium but not in rotational equilibrium. In other words, it may not move to the right or left or up or down, but it may still rotate. In studying equilibrium, we must consider the point of application of each force as well as its magnitude.

Consider the forces exerted on the lug wrench in Fig. 4-1a. Two equal opposing forces F are applied to the right and to the left. The first condition for equilibrium tells us that the vertical and horizontal forces are balanced. Hence, the system is said to be in equilibrium. However, if the same two forces are applied as shown in Fig. 4-1b, the wrench has a definite tendency to rotate. This is true even though the vector sum of the forces is still zero. Clearly, we need a second condition for equilibrium to cover rotational motion. A formal statement of this condition will be given later. First, we need to define some terms.

Fig. 4-1 *(a)* Equilibrium exists; the forces have the same line of action. *(b)* Equilibrium does not exist because opposing forces do not have the same line of action.

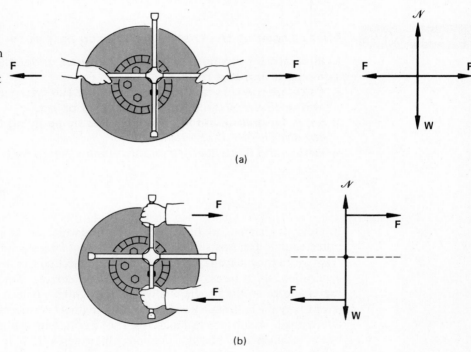

(a)

(b)

In Fig. 4-1b, the forces F do not have the same *line of action*.

> *The **line of action** of a force is an imaginary line extended indefinitely along the vector in both directions.*

When the lines of action of forces do not intersect at a common point, rotation may occur about a point called the *axis of rotation*. In our example, the axis of rotation is an imaginary line passing through the stud perpendicular to the page.

4-2

THE MOMENT ARM

The perpendicular distance from the axis of rotation to the line of action of a force is called the *moment arm* of that force. It is the moment arm that determines the effectiveness of a given force in causing rotational motion. For example, if we exert a force **F** at increasing distances from the center of a large wheel, it becomes easier and easier to rotate the wheel about its center. (See Fig. 4-2.)

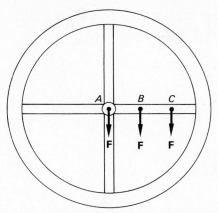

Fig. 4-2 The unbalanced force F has no rotational effect at point A but becomes increasingly effective as the moment arm gets longer.

> *The **moment arm** of a force is the perpendicular distance from the line of action of the force to the axis of rotation.*

If the line of action of a force passes through the axis of rotation (point *A* of Fig. 4-2), the moment arm is zero. No rotational effect is observed regardless of the magnitude of the force. In this simple example, the moment arms at points *B* and *C* are simply the distance from the axis of rotation to the point of application of the force. Note, however, that the line of action of a force is a mere geometrical construction. The moment arm is drawn perpendicular to this line. It may be equal to the distance from the axis to the point of application of a force, but this is true only when the applied force is directed perpendicular to this distance. In the examples of Fig. 4-3, *r* represents the moment arm and *O* represents the axis of rotation. Study each example, observing how the moment arms are drawn and reasoning whether the rotation is clockwise or counterclockwise about *O*.

4-3
TORQUE

Force has been defined as a push or pull that tends to cause motion. *Torque* τ can be defined as the tendency to produce a change in rotational motion. It is also called the *moment of force* in some textbooks. As we have seen, rotational motion is affected by both the magnitude of a force *F* and by its moment arm *r*. Thus, we will define torque as the product of a force and its moment arm.

Torque = force × moment arm

$$\tau = Fr \qquad (4\text{-}1)$$

It must be understood that *r* in Eq. (4-1) is measured perpendicular to the line of action of the force **F**. The units of torque are the units of force times distance, for example, *newton-meters* (N · m) and *pound-feet* (lb · ft).

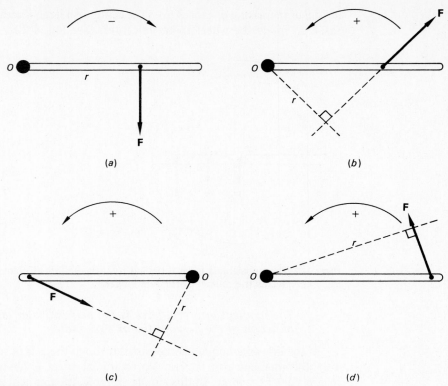

Fig. 4-3 Some examples of moment arms.

Earlier, we established a sign convention to indicate the direction of forces. The direction of torque depends on whether it tends to produce clockwise (cw) or counterclockwise (ccw) rotation. We shall follow the same convention we used for measuring angles. If the force **F** tends to produce counterclockwise rotation about an axis, the torque will be considered positive. Clockwise torques will be considered negative. In Fig. 4-3, all the torques are positive (ccw) except for that in Fig. 4-3*a*.

Fig. 4-4 The tangential force exerted by a cable wrapped around a drum.

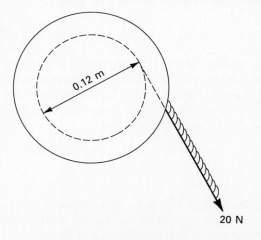

EXAMPLE 4-1   A force of 20 N is exerted on a cable wrapped around a drum which has a diameter of 120 mm. What is the torque produced about the center of the drum? (See Fig. 4-4.)

Solution   Notice that the line of action of the 20-N force is perpendicular to the diameter of the drum. The moment arm, therefore, is equal to the radius of the drum. Converting the diameter to meters (0.12 m), the radius is 0.06 m. The torque is found from Eq. (4-1):

$$\tau = Fr = -(20 \text{ N})(0.06 \text{ m}) = -1.20 \text{ N} \cdot \text{m}$$

The torque is negative because it tends to cause clockwise rotation.

EXAMPLE 4-2   A mechanic exerts a 20-lb force at the end of a 10-in. wrench, as shown in Fig. 4-5. If this pull makes an angle of 60° with the handle, what is the torque produced on the nut?

Fig. 4-5 Calculating torque.

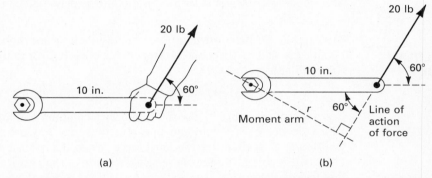

(a)   (b)

Solution   First, we draw a neat sketch, extend the line of action of the 20-lb force, and draw in the moment arm as shown. Note that the moment arm r is perpendicular both to the line of action of the force and to the axis of rotation. You must remember that the moment arm is a geometrical construction and may or may not be along some physical structure such as the handle of the wrench. From the figure we obtain

$$r = (10 \text{ in.}) \sin 60° = 8.66 \text{ in.}$$
$$\tau = Fr = (20 \text{ lb})(8.66 \text{ in.}) = 173 \text{ lb} \cdot \text{in.}$$

If desired, this torque can be converted to 14.4 lb · ft.

In some applications, it is more useful to work with the *components* of a force to obtain the resultant torque. In the preceding example, for instance, we could have resolved the 20-lb vector into its horizontal and vertical components. Instead of finding the torque of a single force, we would then need to find the torque of two component forces. As shown in Fig. 4-6, the 20-lb vector has components F_x and F_y which are found from trigonometry:

Fig. 4-6 The component method of calculating torque.

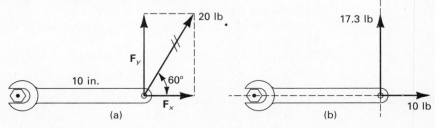

(a)   (b)

$$F_x = (20 \text{ lb})(\cos 60°) = 10 \text{ lb}$$
$$F_y = (20 \text{ lb})(\sin 60°) = 17.3 \text{ lb}$$

Notice from Fig. 4-6b that the line of action of the 10-lb force passes through the axis of rotation. It does not produce any torque because its moment arm is zero. The entire torque is, therefore, due to the 17.3-lb component perpendicular to the handle. The moment arm of this force is the length of the wrench, and the torque is

$$\tau = Fr = (17.3 \text{ lb})(10 \text{ in}) = 173 \text{ lb} \cdot \text{in.}$$

Note that the same result is obtained using this method. No more calculations are required because the horizontal component has a zero moment arm. If we choose the components of a force along and perpendicular to a known distance, we need only concern ourselves with the torque of this perpendicular component.

4-4
RESULTANT TORQUE

In Chap. 2, we demonstrated that the resultant of a number of forces could be obtained by adding the x and y components of each force to find the components of the resultant.

$$R_x = A_x + B_x + C_x + \cdots \qquad R_y = A_y + B_y + C_y + \cdots$$

This procedure applies to forces which have a common point of intersection. Forces which do not have a common line of action may produce a resultant torque in addition to a resultant translational force. When the applied forces act in the same plane, the resultant torque is the algebraic sum of the positive and negative torques due to each force.

$$\boxed{\tau_R = \sum \tau = \tau_1 + \tau_2 + \tau_3 + \cdots} \qquad (4\text{-}2)$$

Remember that counterclockwise torques are positive and clockwise torques are negative.

One essential element of good problem-solving techniques is organization. The following procedure is useful for calculating resultant torque.

CALCULATING RESULTANT TORQUE

1. Read the problem; then draw and label a figure.
2. Construct a free-body diagram showing all forces, distances, and the axis of rotation.
3. Extend lines of action of each force with dotted lines.
4. Draw and label the moment arms for each force.
5. Calculate the moment arms if necessary.
6. Calculate the torques due to each force independent of others and affix appropriate sign.
7. The resultant torque is the algebraic sum of the torques due to each force. See Eq. (4-2).

EXAMPLE 4-3 A piece of angle iron is hinged at point A as shown in Fig. 4-7. Determine the resultant torque at A due to the 60- and 80-N forces.

Fig. 4-7

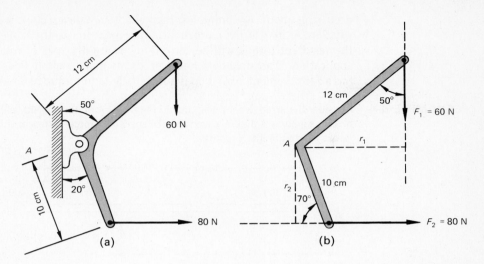

(a)   (b)

Solution   A free-body diagram is drawn and the moment arms r_1 and r_2 are constructed as in Fig. 4-7b. The lengths of the moment arms are:

$$r_1 = (12 \text{ cm}) \sin 50° = 9.19 \text{ cm}$$
$$r_2 = (10 \text{ cm}) \sin 70° = 9.40 \text{ cm}$$

Considering A as the axis of rotation, the torque due to F_1 is negative (cw) and the torque due to F_2 is positive (ccw). The resultant torque is found as follows:

$$\begin{aligned}
\tau_R = \tau_1 + \tau_2 &= F_1 r_1 + F_2 r_2 \\
&= -(60 \text{ N})(9.19 \text{ cm}) + (80 \text{ N})(9.40 \text{ cm}) \\
&= -552 \text{ N} \cdot \text{cm} + 752 \text{ N} \cdot \text{cm} \\
&= 200 \text{ N} \cdot \text{cm}
\end{aligned}$$

The resultant torque is $200 \text{ N} \cdot \text{cm}$, counterclockwise. This answer is best expressed as $2.00 \text{ N} \cdot \text{m}$ in SI units.

4-5
EQUILIBRIUM

We are now ready to discuss the necessary condition for rotational equilibrium. The condition for translational equilibrium was stated in equation form as

$$\boxed{\sum F_x = 0 \qquad \sum F_y = 0} \tag{4-3}$$

If we are to ensure that the rotational effects are also balanced, we must stipulate that there is no resultant torque. Hence the second condition for equilibrium is:

The algebraic sum of all the torques about any axis must be zero.

$$\boxed{\sum \tau = \tau_1 + \tau_2 + \tau_3 + \cdots = 0} \tag{4-4}$$

The second condition for equilibrium simply tells us that the clockwise torques are exactly balanced by the counterclockwise torques. Moreover, since rotation is not occurring about any point, we may choose whatever point we wish as an axis of rotation. As long as the moment arms are measured to the same point for each force, the resultant torque will be zero. By choosing the axis of rotation at the point of application of an unknown force, problems may be simplified. If a particular force has a zero moment arm, it does not contribute to torque regardless of its magnitude.

EXAMPLE 4-4   Consider the arrangement diagramed in Fig. 4-8. A uniform beam weighing 200 N is held up by two supports A and B. Given the distances and forces listed in this figure, what are the forces exerted by the supports?

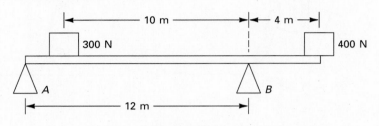

(a) Sketch of system

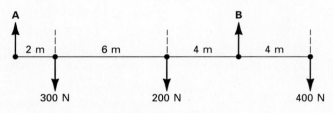

(b) Free-body diagram

Fig. 4-8 The solution of torque problems is aided by first drawing a rough sketch of the system. A free-body diagram can then be drawn to indicate forces and distances.

Solution   A free-body diagram is drawn to show clearly all forces and distances between forces. Note that all of the weight of the uniform beam is considered as acting at the center of the board. Next we will apply the first condition for equilibrium, Eq. (4-3):

$$\sum F_y = A + B - 300 \text{ N} - 200 \text{ N} - 400 \text{ N} = 0$$

or

$$A + B = 900 \text{ N}$$

Since this equation has two unknowns, we must have more information. Therefore, we will apply the second condition for equilibrium.

First, we must select an axis about which we can measure moment arms. The logical choice is at the point of application of some unknown force. Choosing the axis at B gives this force a zero moment arm. The sum of the torques about B results in the following equation.

$$\sum \tau_B = -A(12 \text{ m}) + (300 \text{ N})(10 \text{ m}) + (200 \text{ N})(4 \text{ m}) - (400 \text{ N})(4 \text{ m}) = 0$$

Note that the 400-N force and the force A tend to produce clockwise rotation about B. (Their torques were negative.) Simplifying gives

$$-(12 \text{ m})A + 3000 \text{ N} \cdot \text{m} - 1600 \text{ N} \cdot \text{m} + 800 \text{ N} \cdot \text{m} = 0$$

Adding (12 m)A to both sides and simplifying, we obtain

$$2200 \text{ N} \cdot \text{m} = (12 \text{ m})A$$

Dividing both sides by 12 m, we find that

$$A = 183 \text{ N}$$

Now to find the force exerted by support B, we can return to the equation obtained from the first condition for equilibrium.

$$A + B = 900 \text{ N}$$

Solving for B, we obtain

$$B = 900 \text{ N} - A = 900 \text{ N} - 183 \text{ N}$$
$$= 717 \text{ N}$$

As a check on this solution, we could choose the axis of rotation at A, then apply the second condition for equilibrium to find B.

EXAMPLE 4-5 A uniform 200-lb boom, 24 ft long, is supported by a cable as shown in Fig. 4-9. The boom is hinged at the wall, and the cable makes a 30° angle with the boom, which is horizontal. If a load of 500 lb is hung from the right end, what is the tension T in the cable? What are the horizontal and vertical components of the force exerted by the hinge?

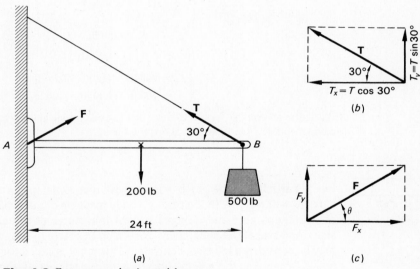

Fig. 4-9 Forces on a horizontal boom.

Solution Let us consider the boom as the object in equilibrium. Of the two unknown forces F and T, we know least about the force F. Therefore, it is logical to choose the hinge as our axis of rotation when summing torques. In this manner, the unknown force F will have zero moment arm, rendering its torque about point A zero also. (Do not make the mistake of assuming that the force exerted by the hinge is entirely along the boom.) We can determine the cable tension from the second condition of equilibrium.

$$\sum \tau_A = F(0) - (200 \text{ lb})(12 \text{ ft}) - (500 \text{ lb})(24 \text{ ft}) + T_x(0) + T_y(24 \text{ ft}) = 0$$
$$= 0 - 2400 \text{ lb} \cdot \text{ft} - 12{,}000 \text{ lb} \cdot \text{ft} + T_y(24 \text{ ft}) = 0$$
$$T_y(24 \text{ ft}) = 14{,}400 \text{ lb} \cdot \text{ft}$$

From Fig. 4-9*b*,

$$T_y = T \sin 30° = 0.5 \, T$$

so that

$$(0.5T)(24 \text{ ft}) = 14{,}400 \text{ lb} \cdot \text{ft}$$
$$12T = 14{,}400 \text{ lb}$$
$$T = 1200 \text{ lb}$$

In order to find the horizontal and vertical components of **F**, we can apply the first condition for equilibrium. The horizontal component is found by summing forces along the *x* axis.

$$\sum F_x = 0 \qquad F_x - T_x = 0$$

from which

$$F_x = T_x = T \cos 30°$$
$$= (1200 \text{ lb})(\cos 30°) = 1040 \text{ lb}$$

The vertical component is found by summing the forces along the *y* axis.

$$\sum F_y = 0 \qquad F_y + T_y - 200 \text{ lb} - 500 \text{ lb} = 0$$

Solving for F_y, we obtain

$$F_y = 700 \text{ lb} - T_y$$

or

$$F_y = 700 \text{ lb} - (1200 \text{ lb})(\sin 30°)$$
$$= 700 \text{ lb} - 600 \text{ lb} = 100 \text{ lb}$$

It is left as an exercise for you to show that the magnitude and direction of the force **F** from its components is 1045 lb at 5.5° above the boom.

4-6
CENTER OF GRAVITY

Every particle on the earth has at least one force in common with every other particle—its *weight*. In the case of a body made up of many particles, these forces are essentially parallel and directed toward the center of the earth. Regardless of the shape and size of the body, there exists a point at which the entire weight of the body may be considered to be concentrated. This point is called the *center of gravity* of the body. Of course, the weight does not in fact all act at this point. But we would calculate the same torque about a given axis if we considered the entire weight to act at that point.

The center of gravity of a regular body, such as a uniform sphere, cube, rod, or beam, is located at its geometric center. This fact was used in the examples of the previous section, where we considered the weight of an entire beam as acting at its center. Although the center of gravity is a fixed point, it does not necessarily lie within the body. For example, a hollow sphere, a circular hoop, and a rubber tire all have centers of gravity outside the material of the body.

From the definition of the center of gravity, it is recognized that any body which is suspended at this point will be in equilibrium. This is true because the weight vector, which represents the sum of forces acting on each portion of the body, will have a zero moment arm. Thus, we can compute the center of gravity of a body by determining the point at which an upward force will produce rotational equilibrium.

EXAMPLE 4-6   Compute the center of gravity of the two spheres in Fig. 4-10 if they are connected by a 30-in. rod of negligible weight.

Fig. 4-10 Computing the center of gravity.

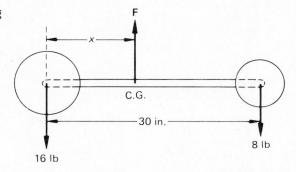

Solution   We first draw an upward vector indicating the force at the center of gravity which would balance the system. Suppose we choose to locate this vector at a distance from the center of the 16-lb sphere. The distance x would be drawn and labeled on the figure. Since the upward force must equal the sum of the downward forces, the first condition for equilibrium tells us that

$$F = 16 \text{ lb} + 8 \text{ lb} = 24 \text{ lb}$$

Now we will choose our axis of rotation at the center of the 16-lb sphere. This is the best choice because the distance x is measured from this point. The second condition for equilibrium is applied as follows:

$$\sum \tau = (24 \text{ lb})x - (8 \text{ lb})(30 \text{ in.}) = 0$$
$$(24 \text{ lb})x = 240 \text{ lb} \cdot \text{in.}$$
$$x = 10 \text{ in.}$$

If the rod joining the two spheres is suspended from the ceiling at a point 10 in. from the center of the 16-lb sphere, the system will be in equilibrium. This point is the center of gravity. You should show that the same conclusion follows if you sum torques about the right end or about any other point.

SUMMARY

When forces acting on a body do not have the same line of action or do not intersect at a common point, rotation may occur. In this chapter, we introduced the concept of torque as a measure of the tendency to rotate. The major concepts are summarized below:

- The **moment arm** of a force is the perpendicular distance from the line of action of the force to the axis of rotation.
- The **torque** about a given axis is defined as the product of the magnitude of a force and its moment arm:

Torque = force × moment arm

$$\tau = Fr$$

It is positive if it tends to produce counterclockwise motion and negative if the motion produced is clockwise.

+ ccw

− cw

- The resultant torque τ_R about a particular axis A is the algebraic sum of the torques produced by each force. The signs are determined by the above convention.

$$\tau_R = \sum \tau_A = F_1r_1 + F_2r_2 + F_3r_3 + \cdots$$

- **Rotational equilibrium:** A body in rotational equilibrium has no resultant torque acting on it. In such cases, the sum of all the torques about *any* axis must equal zero. The axis may be chosen anywhere because the system is not tending to rotate about any point. This is called the second condition for equilibrium and may be written

$$\sum \tau = 0 \quad \textit{the sum of all torques about any point is zero}$$

- **Total equilibrium** exists when the first and second conditions are satisfied. In such cases, three independent equations can be written:

$$\text{(a) } \sum F_x = 0 \qquad \text{(b) } \sum F_y = 0 \qquad \text{(c) } \sum \tau = 0$$

By writing these equations for a given situation, unknown forces, distances, or torques can be found.

- The center of gravity of a body is the point through which the resultant weight acts regardless of how the body is oriented. For applications involving torque, the entire weight of the object may be considered as acting at this point.

QUESTIONS

4-1. Define the following terms:
   a. line of action
   b. axis of rotation
   c. moment arm
   d. torque
   e. rotational equilibrium
   f. center of gravity

4-2. You lift a heavy suitcase with your right hand. Describe and explain the position of your body.

4-3. A parlor trick consists of asking you to stand against a wall with your feet together so that the side of your right foot rests against the wall. You are then asked to raise your left foot off the floor. Why can't you do this without falling?

4-4. Why is a Volkswagen bus more likely to turn over than a Corvette or some other sports car?

4-5. If you know the weight of a brick to be 6 lb, describe how you could use a meterstick and a pivot to measure the weight of a baseball.

4-6. Describe and explain the arm and leg motions used by a tightrope walker to maintain balance.

4-7. Discuss the following items and their use of the principle of torque: (a) screwdriver, (b) wrench, (c) pliers, (d) wheelbarrow, (e) nutcracker, and (f) crowbar.

PROBLEMS

The Moment Arm

4-1. Draw and label the moment arms for the situations described in Fig. 4-11. Assume the axis to be at point A in each case. What is the magnitude of the moment arm in each case?

Ans. 0.845 ft, 1.73 m

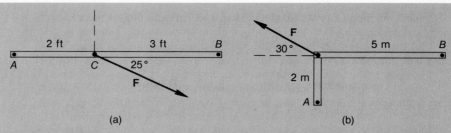

Fig. 4-11

(a)   (b)

4-2. Draw and label the new moment arms if each axis in Fig. 4-11 is changed to point *B*. What are the magnitudes of the moment arms?

Torque

4-3. A leather belt is wrapped around a pulley 20 cm in diameter. A force of 60 N is applied to the belt. What is the torque at the center of the shaft?

Ans. 12 N · m

4-4. The light rod in Fig. 4-12 is 20 in. long and the axis of rotation is *A*. Determine the magnitude and the sign of the torque due to the 200-lb force if the angle θ is **(a)** 90°, **(b)** 60°, **(c)** 30°, **(d)** 0°.

Fig. 4-12

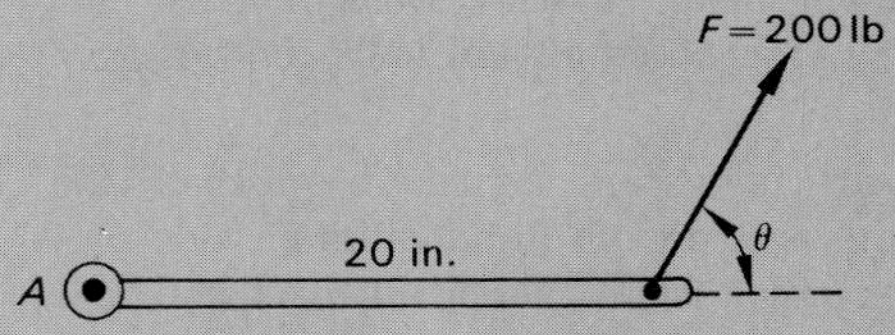

4-5. Assume that the force *F* in Fig. 4-11*a* is equal to 80 lb. What are the magnitude and sign of the torque **(a)** about point *A* and **(b)** about point *B*? Neglect the weight of the rod in each case.

Ans. −67.6 lb · ft, +101 lb · ft

4-6. The force *F* in Fig. 4-11*b* is 400 N and the angle iron is of negligible weight. What are the magnitude and the sign of the torque about axis *A* and about axis *B*?

Resultant Torque

4-7. In Fig. 4-13, what is the resultant torque **(a)** about point *A* and **(b)** about the left end of the beam? Neglect the weight of the beam.

Ans. **(a)** 90 N · m., **(b)** −120 N · m

Fig. 4-13

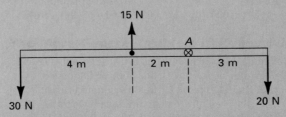

4-8. Assume the axis to be at point *A* in Fig. 4-13. What perpendicular force acting at the right end of the light beam will give a resultant torque equal to zero?

Fig. 4-14

* 4-9. Determine the resultant torque about point A in Fig. 4-14.

Ans. 6.19 lb · ft

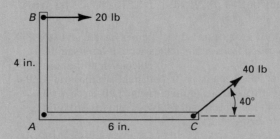

* 4-10. What is the resultant torque about point C in Fig. 4-14?

Equilibrium

4-11. Determine the unknown forces for the arrangements in Fig. 4-15. Assume that equilibrium exists and that the weight of the bar is negligible in each case.

Ans. $A = 26.7$ lb, $F = 107$ lb; $F_1 = 198$ lb, $F_2 = 87.5$ lb; $A = 50.9$ N, $B = 49.1$ N

Fig. 4-15

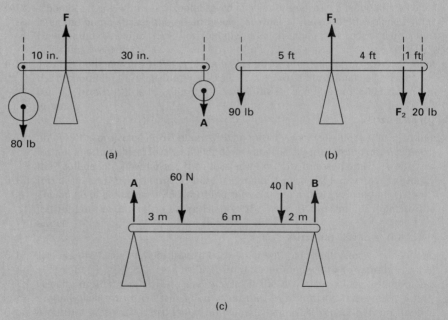

4-12. A V-belt is wrapped around a pulley 16 in. in diameter. If a resultant torque of 4 lb · ft is required, what force must be applied along the belt?

4-13. A bridge 20 m long is supported at each end. Consider the weight of the bridge to be 4500 N. Find the forces exerted at each end when a 1600-N mower is located 8 m from the left end of the bridge.

Ans. 2890 N, 3210 N

4-14. A 10-ft platform is placed across two stepladders, one at each end. The platform weighs 40 lb, and a painter located 4 ft from the right end weighs 180 lb. What are the forces exerted by each of the ladders?

* 4-15. Consider the boom arrangement of Fig. 4-16. The weight of the boom is 400 N and it is 6 m in length. The cable is attached at a point located a distance of 4.5 m from the wall. If the weight W is 1200 N, what is the tension in the cable? **Ans.** 2340 N

Fig. 4-16

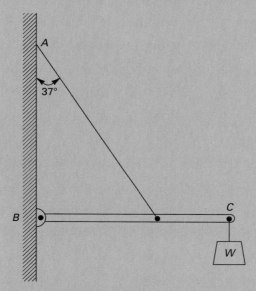

* 4-16. For Prob. 4-15, determine the horizontal and vertical components of the force exerted by the hinge on the boom. What are the magnitude and direction of this force?

Center of Gravity

4-17. A uniform metal bar has a length of 6 m and a weight of 30 N. A 50-N weight is suspended from the left end and a 20-N weight is hung at the right end. At what distance from the left end will a single upward force produce equilibrium? **Ans.** 2.10 m

4-18. Find the center of gravity for two spheres connected by a very light rod 200 mm in length. The left sphere weighs 40 N and the right sphere weighs 12 N.

* 4-19. Weights of 2, 5, 8, and 10 N are hung from a 10-m rod at distances of 2, 4, 6, and 8 m from the left end. Neglecting the weight of the rod, at what point from the left end will a single upward force balance the system? **Ans.** 6.08 m

4-20. Compute the center of gravity of a sledgehammer if the metal head weighs 12 lb and the 32-in. supporting handle weighs 2 lb. Assume that the handle is of uniform construction and weight.

Additional Problems

* 4-21. What is the resultant torque about the hinge in Fig. 4-17? Neglect the weight of the curved bar. **Ans.** -3.42 N · m

Fig. 4-17

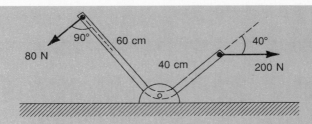

Fig. 4-18

*

4-22. What horizontal force applied to the left end of the curved bar in Fig. 4-17 will produce rotational equilibrium?

4-23. Weights of 100, 200, and 500 lb are placed on a board resting on two supports as shown in Fig. 4-18. Neglecting the weight of the board, what are the forces exerted by the supports?

Ans. 425 lb, 375 lb

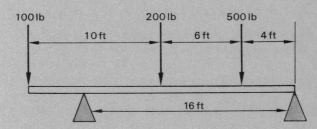

4-24. A uniform bar 8 m long and weighing 2400 N is supported by a fulcrum 3 m from the right end. If a 9000-N weight is placed on the right end, what downward force must be exerted at the left end in order to balance the system?

*

4-25. A 30-lb box and a 50-lb box are positioned at opposite ends of a 16-ft board. The board is supported only at its midpoint. Where should a third box weighing 40 lb be placed to balance the system?

Ans. 4 ft left of center

4-26. A uniform horizontal bar is 800 mm long and is of negligible weight. A weight of 40 N is hung from the left end of the rod, and an 84-N weight is hung from the right end. Where should a single upward support be positioned to balance the system?

*

4-27. Find the forces F_1, F_2, and F_3 such that the system in Fig. 4-19 is in equilibrium.
Ans. $F_1 = 212$ lb, $F_2 = 254$ lb, $F_3 = 83.9$ lb

Fig. 4-19

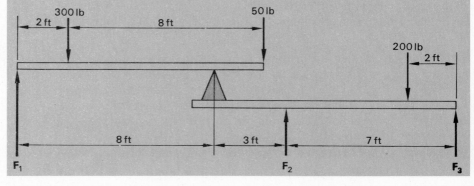

4-28. Find the resultant torque about point B in Fig. 4-14.

4-29. Find the resultant torque about point A in Fig. 4-20.

Ans. $-3.87\,\text{N}\cdot\text{m}$

Fig. 4-20

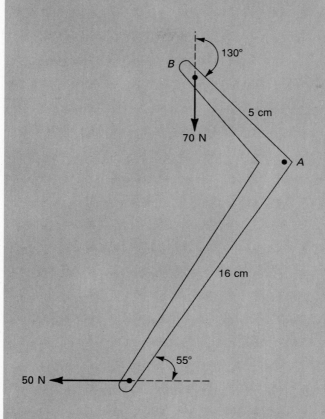

4-30. What is the resultant torque about point B in Fig. 4-20?

4-31. If the weight of the boom in Fig. 4-21 is neglected, what weight W will produce a tension of 400 N in the horizontal cable?

Ans. 154 N

Fig. 4-21

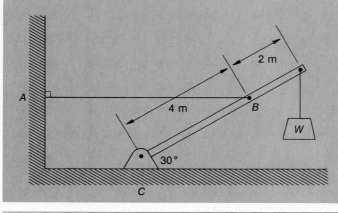

★ ★ 4-32. For the situation described in Prob. 4-31, what are the magnitude and direction of the force acting on the boom at its lower end?

★ 4-33. The weight of the boom in Fig. 4-21 is 200 N and the suspended weight W is equal to 800 N. What is the tension in the attached cable?

Ans. 2340 N

★ ★ 4-34. What are the magnitude and direction of the force exerted by the wall on the boom for the arrangement described in Prob. 4-33?

★ ★ 4-35. Determine the tension in the cable for the arrangement described by Fig. 4-22. The boom has a weight of 300 N and is of unknown length.

Ans. 360 N

Fig. 4-22

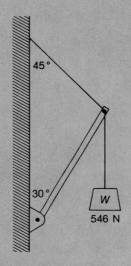

★ 4-36. A particular automobile has a distance of 3.4 m between front and rear axles. If 60 percent of the weight rests on the front wheels, how far is the center of gravity located from the front axle?

MECHANICS

5 Uniformly Accelerated Motion

After completing this chapter, you should be able to:

1. Define and give formulas for *average speed* and *average acceleration*.
2. Solve problems involving *time, distance, average speed,* and *average acceleration*.
3. Apply one of the four general equations for uniformly accelerated motion to solve for one of the five parameters: *initial speed, final speed, acceleration, time,* and *distance*.
4. Solve acceleration problems involving free-falling bodies in a gravitational field.

Everything in the physical world is in motion, from the largest galaxies in the universe to the elementary particles within atoms. We must study the motions of objects if we are to understand their behavior and learn to control them. Uncontrolled or erratic motion, as in falling buildings, destructive vibrations, or a runaway car, can create dangerous situations, but controlled motion often serves our convenience. It is important to be able to analyze motion and to represent it in terms of fundamental formulas.

5-1
SPEED AND VELOCITY

The simplest kind of motion an object can experience is uniform motion in a straight line. If the object covers the same distances in each successive unit of time, it is said to move with *constant speed*. For example, if a train covers 26 ft of track every second that it moves, we say that it has a constant speed of 26 ft/s. Whether the speed is constant or not, the average speed of a moving object is defined by

$$Average\ speed = \frac{distance\ traveled}{time\ elapsed}$$

$$\bar{v} = \frac{s}{t} \qquad (5\text{-}1)$$

The bar over the symbol v means that the speed represents an average value for the time interval t.

Remember that the dimension of speed is the ratio of a length to a time interval. Hence, the units of miles per hour, feet per second, meters per second, and centimeters per second are all typical units of speed.

EXAMPLE 5-1 A golfer sinks a putt 3 s after the ball leaves the club face. If the ball traveled with an average speed of 0.8 m/s, how long was the putt?

Solution Solving Eq. (5-1) for s, we have

$$s = \bar{v}t = (0.8 \text{ m/s})(3 \text{ s})$$

Therefore, the distance of the putt is

$$s = 2.4 \text{ m}$$

It is important to recognize that speed is a scalar quantity which is completely independent of direction. In Example 5-1, it was not necessary for us to know either the speed of the golf ball at any instant or the nature of its path. Similarly, the average speed of a car traveling from Atlanta to Chicago is a function only of the distance registered on its odometer and the time required to make the trip. It makes no difference as far as computation is concerned whether the driver of the car took the direct or scenic route or even if he stopped for meals.

We must make a clear distinction between the scalar quantity *speed* and its directional counterpart *velocity*. This is best done by recalling the difference between *distance* and *displacement,* as discussed in Chap. 2. Suppose, in Fig. 5-1, an object

Fig. 5-1 Displacement and velocity are vector quantities, whereas distance and speed are independent of direction: s, distance; D, displacement; v, velocity; t, time.

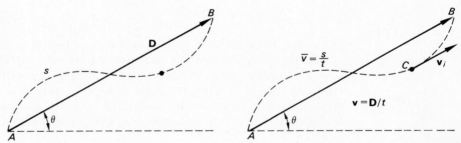

moves along the broken path from A to B. The actual distance traveled is denoted by s, whereas the displacement is represented by the polar coordinates

$$\mathbf{D} = (D, \theta)$$

As an example, suppose the distance s in Fig. 5-1 is 500 mi, and the displacement is 350 mi at 45°. If the actual traveling time is 8 h, the average speed is

$$\bar{v} = \frac{s}{t} = \frac{500 \text{ mi}}{8 \text{ h}} = 62.5 \text{ mi/h}$$

However, the average *velocity* must consider the displacement magnitude and direction. The average velocity is given by

$$\bar{v} = \frac{D}{t} = \frac{350 \text{ mi, } 45°}{8 \text{ h}}$$

$$\bar{v} = 43.8 \text{ mi/h, } 45°$$

Therefore, if the path of the moving object is curved, the difference between speed and velocity is one of magnitude as well as direction.

Automobiles cannot always travel at constant speeds for long periods of time. In traveling from point A to B, we may be required to slow down or speed up because of road conditions. For this reason, it is sometimes useful to talk of *instantaneous speed* or *instantaneous velocity*.

> The **instantaneous speed** is a scalar quantity representing the speed at the instant the car is at an arbitrary point C. It is, therefore, the time rate of change in distance.

> The **instantaneous velocity** is a vector quantity representing the velocity at any point C. It is the time rate of change in displacement.

In this chapter we shall be concerned with motion along a straight line so that the magnitudes of speed and velocity are the same at any instant. If the direction does not change, the instantaneous speed is the scalar part of the instantaneous velocity. However, it is a good practice to reserve the term *velocity* for the more complete description of motion. As we shall see in the coming sections, a change in velocity may result in a change of direction as well. In such cases, the terms velocity and displacement are more appropriate than the terms speed and distance.

5-2 ACCELERATED MOTION

In most cases, the velocity of a moving object changes as motion continues. This type of motion is called *accelerated motion*. The time rate at which velocity changes is called the *acceleration*. For example, suppose we observe the motion of a body for a period of time t. The initial velocity v_0 of the body will be defined as its velocity at the beginning of the time interval, i.e., when $t = 0$. The final velocity is defined as the velocity v_f the body has at the end of the time interval, when $t = t$. Thus, if we are able to measure the initial and final velocities of a moving object, we can say that its acceleration is given by

$$Acceleration = \frac{change \ in \ velocity}{time \ interval}$$

$$a = \frac{v_f - v_0}{t} \qquad (5\text{-}2)$$

Acceleration written in the above manner is a vector quantity and thus depends upon changes in direction as well as changes in magnitude. If the direction of motion is in a straight line, only the speed of the object is changing. If it follows a curved path, both magnitude and directional changes occur and the acceleration is no longer in the same direction as the motion. In fact, if the curved path follows a perfect circle, the acceleration will always be at right angles to the motion. In this case, only the direction of motion is changing while the speed at any point on the circle is constant. The latter type of motion will be treated in a later chapter.

5-3
UNIFORMLY ACCELERATED MOTION

The simplest kind of acceleration is motion in a straight line in which the speed changes at a constant rate. This special kind of motion is generally referred to as *uniformly accelerated motion or constant acceleration.* Since there is no change in direction, the vector difference in Eq. (5-2) becomes simply the algebraic difference between the magnitude of the final velocity v_f and the magnitude of the initial velocity v_0. Hence, for uniform acceleration

$$a = \frac{v_f - v_0}{t} \qquad (5\text{-}3)$$

For example, consider a car which moves with constant acceleration from point A to point B, as in Fig. 5-2. The car's speed at A is 40 ft/s, and its speed at B is 60 ft/s.

Fig. 5-2 Uniformly accelerated motion.

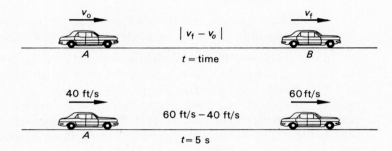

If the increase in speed requires 5 s, the acceleration can be determined from Eq. (5-3). Hence

$$a = \frac{v_f - v_0}{t} = \frac{60 \text{ ft/s} - 40 \text{ ft/s}}{5 \text{ s}}$$

$$= \frac{20 \text{ ft/s}}{5 \text{ s}} = 4 \text{ ft/s}^2$$

The answer is read as *four feet per second per second* or *four feet per second squared.* It means that every second the car increases its speed by 4 ft/s. Since the car had already reached a speed of 40 ft/s when we started our time ($t = 0$), 1, 2, and 3 s later it would have speeds of 44, 48, and 52 ft/s, respectively.

EXAMPLE 5-2   A train reduces its speed from 60 to 20 mi/h in 8 s. Find the acceleration.

Solution   We first note that there is an inconsistency of units between velocity in miles per hour and time in seconds. The velocity is converted to feet per second as follows:

$$60 \, \frac{\text{mi}}{\text{h}} \times \frac{5280 \text{ ft}}{1 \text{ mi}} \times \frac{1 \text{ h}}{3600 \text{ s}} = 88 \text{ ft/s}$$

Similarly, it is determined that 20 mi/h is equal to 29.3 ft/s. Substitution into Eq. (5-3) gives

$$a = \frac{v_f - v_0}{t} = \frac{29.3 \text{ ft/s} - 88 \text{ ft/s}}{8 \text{ s}}$$

$$a = -7.33 \text{ ft/s}^2$$

The minus sign tells us that the speed is *reduced* by 7.33 ft/s every second. This type of acceleration is sometimes referred to as *deceleration*.

Many times the same equation is used to solve for different quantities. You should, therefore, solve each equation literally for each symbol in the equation. A very convenient form arises for Eq. (5-3) when it is solved explicitly for the final velocity. Thus

$$v_f = v_0 + at \qquad (5\text{-}4)$$
Final velocity = initial velocity + change in velocity

EXAMPLE 5-3 An automobile maintains a constant acceleration of 8 m/s². If its initial velocity was 20 m/s, what will its velocity be after 6 s?

Solution The final velocity is obtained from Eq. (5-4).

$$v_f = v_0 + at = 20 \text{ m/s} + (8 \text{ m/s}^2)(6 \text{ s})$$

or

$$v_f = 20 \text{ m/s} + 48 \text{ m/s}$$

Hence the final speed is

$$v_f = 68 \text{ m/s}$$

Now that the concept of initial and final velocities is understood, let us return to the equation for average velocity and express it in terms of initial and final states. The average velocity of an object moving with uniform acceleration is found just like the arithmetic mean of two numbers. Given an initial velocity and a final velocity, the average speed is simply

$$\bar{v} = \frac{v_f + v_0}{2} \qquad (5\text{-}5)$$

Utilizing this relation in Eq. (5-1) gives us a more useful expression for computing distance:

$$\boxed{s = \bar{v}t = \frac{v_f + v_0}{2}t} \qquad (5\text{-}6)$$

EXAMPLE 5-4 A moving object increases its speed uniformly from 200 to 400 cm/s in 2 min. (*a*) What is its average velocity, and (*b*) how far did it travel in the 2 min?

Solution The average velocity is found by direct substitution into Eq. (5-5).

$$\bar{v} = \frac{v_f + v_0}{2} = \frac{400 \text{ cm/s} + 200 \text{ cm/s}}{2}$$

or

$$\bar{v} = \frac{600 \text{ cm/s}}{2} = 300 \text{ cm/s}$$

The distance traveled is then found from Eq. (5-1).

$$s = \bar{v}t = (300 \text{ cm/s})(2 \text{ min})$$

The units of time are inconsistent, but since 2 min = 120 s, we have

$$s = (300 \text{ cm/s})(120 \text{ s}) = 36,000 \text{ cm}$$

5-4
OTHER USEFUL RELATIONS

Thus far we have presented two fundamental relations. One arises from the definition of velocity and the other from the definition of acceleration. They are

$$s = \bar{v}t = \frac{v_f + v_0}{2} t \tag{5-6}$$

and

$$v_f = v_0 + at \tag{5-4}$$

Although these are the only formulas necessary to attack the many problems presented in this chapter, two other very useful relationships can be obtained from them. The first is derived by eliminating the final velocity from Eqs. (5-6) and (5-4). Substituting the latter into the former yields

$$s = \frac{(v_0 + at) + v_0}{2} t$$

Simplifying gives

$$s = \frac{(2v_0 + at)t}{2} = \frac{2v_0t + at^2}{2}$$

or

$$\boxed{s = v_0t + \tfrac{1}{2}at^2} \tag{5-7}$$

The second relation is derived by eliminating t from the basic equations. Solving Eq. (5-4) for t yields

$$t = \frac{v_f - v_0}{a}$$

which on substitution into Eq. (5-6) gives

$$s = \left(\frac{v_f + v_0}{2}\right)\left(\frac{v_f - v_0}{a}\right)$$

from which

$$\boxed{2as = v_f^2 - v_0^2} \tag{5-8}$$

Although these two equations add no new information, they are very useful in solving problems in which either the final velocity or time is not given and you need to find one of the other parameters.

5-5
SOLUTION OF ACCELERATION PROBLEMS

Although the solution to problems involving constant acceleration depends primarily upon choosing the correct formula and substituting known values, there are several suggestions to help the beginning student. Physical problems of this kind frequently involve motion which either started from rest or is brought to a stop from some initial velocity. In either case, the formulas discussed can be simplified by the substitution of either $v_0 = 0$ or $v_f = 0$, as the case may be. Table 5-1 summarizes the general formulas.

Table 5-1 Summary of Accelerated Formulas

(1)	$s = \dfrac{v_f + v_0}{2} t$
(2)	$v_f = v_0 + at$
(3)	$s = v_0 t + \frac{1}{2}at^2$
(4)	$2as = v_f^2 - v_0^2$

A close look at the four general equations will reveal a total of five parameters: s, v_0, v_f, a, and t. Given any three of these quantities, the remaining two can be found from the general equations. Therefore, a good starting point in solving any problem is to read it thoroughly with a view to establishing the three quantities required for solution. It is also important to choose a direction to call positive and apply it consistently to velocity, displacement, and acceleration when inserting the values into equations.

If you have difficulty in deciding which equation should be used, it may help to recall the conditions such an equation must satisfy. First, it must contain the unknown parameter. Second, all other parameters that appear in the equation must be known. For example, if a problem gives you values for v_f, v_0, and t, you may solve for a in Eq. (2) of Table 5-1.

The following procedure illustrates a technique which might be followed for solving acceleration problems:

1. Read the problem; then draw and label a rough sketch.
2. Indicate the consistent positive direction.
3. Establish three given parameters and two that are unknown. Make sure the signs and units are consistent.

Given: _____   Find: _____
_____
_____

4. Select an equation which contains one of the unknown parameters but not the other one.
5. Substitute known quantities and solve the equation.

The examples which follow are abbreviated and do not include figures, but they do illustrate the approach.

EXAMPLE 5-5   A motorboat starting from rest attains a velocity of 30 mi/h in 15 s. What was its acceleration, and how far did it travel?

$$\text{Given: } v_0 = 0 \qquad\qquad \text{Find: } a = ?$$
$$v_f = 30 \text{ mi/h} = 44 \text{ ft/s} \qquad s = ?$$
$$t = 15 \text{ s}$$

Solution   In solving for acceleration, we must choose an equation which contains a but not s. Equation (2) can be used where $v_0 = 0$. Hence

$$v_f = at$$

from which

$$a = \frac{v_f}{t} = \frac{44 \text{ ft/s}}{15 \text{ s}}$$
$$= 2.93 \text{ ft/s}^2$$

The displacement can be found from Eq. (1), as follows:

$$s = \frac{v_f}{2} t = \frac{44 \text{ ft/s}}{2} (15 \text{ s})$$
$$= 330 \text{ ft}$$

EXAMPLE 5-6   An airplane lands on a carrier deck at 200 mi/h and is brought to a stop in 600 ft. Find the acceleration and the time required to stop.

$$\text{Given: } v_0 = 200 \text{ mi/h} = 294 \text{ ft/s} \qquad \text{Find: } a = ?$$
$$v_f = 0 \qquad\qquad\qquad\qquad t = ?$$
$$s = 600 \text{ ft}$$

Solution   Choosing Eq. (4), we solve for a as follows:

$$2as = v_f^2 - v_0^2$$
$$(2a)(600 \text{ ft}) = 0 - (294 \text{ ft/s})^2$$
$$a = \frac{-(294 \text{ ft/s})^2}{(2)(600 \text{ ft})} = \frac{-86{,}400 \text{ ft}^2/\text{s}^2}{1200 \text{ ft}}$$
$$= -72 \text{ ft/s}^2$$

Then, solving for the time in Eq. (2) yields

$$t = \frac{v_f - v_0}{a} = \frac{-v_0}{a} = \frac{-294 \text{ ft/s}}{-72 \text{ ft/s}^2}$$

or

$$t = 4.08 \text{ s}$$

EXAMPLE 5-7   A train traveling initially at 16 m/s is under constant acceleration of 2 m/s². How far will it travel in 20 s? What will its final velocity be?

$$\text{Given: } v_0 = 16 \text{ m/s} \qquad \text{Find: } s = ?$$
$$a = 2 \text{ m/s}^2 \qquad\qquad v_f = ?$$
$$t = 20 \text{ s}$$

Solution   From Eq. (3) we have

$$s = v_0 t + \tfrac{1}{2}at^2$$
$$= (16 \text{ m/s})(20 \text{ s}) + \tfrac{1}{2}(2 \text{ m/s}^2)(20 \text{ s})^2$$
$$= 320 \text{ m} + 400 \text{ m} = 720 \text{ m}$$

The final velocity is found from Eq. (2).

$$v_f = v_0 + at = 16 \text{ m/s} + (2 \text{ m/s}^2)(20 \text{ s})$$
$$= 56 \text{ m/s}$$

5-6 SIGN CONVENTION IN ACCELERATION PROBLEMS

The signs of acceleration, displacement, and velocity are mutually independent and are each determined by different criteria. Probably no other point is more confusing for beginning students. Whenever the direction of motion changes such as with an object tossed into the air or with an object attached to an oscillating spring, the signs of displacement and acceleration are particularly difficult to visualize. Only the sign of velocity is determined by the direction of motion. The sign of displacement depends on the location or position of the object, and the sign of acceleration is determined by the force which produces the acceleration.

Once a choice is made for the positive direction, the following convention will determine the signs of velocity, displacement, and acceleration:

Velocity is positive or negative depending on whether the direction of motion is with or against the chosen positive direction.

Acceleration is positive or negative depending on whether the resultant force is with or against the chosen positive direction.

Displacement is positive or negative depending on the location or position of the object relative to its zero position.

You will recall from Newton's first law that an object will not experience a change in velocity unless acted on by a resultant force. It should be apparent, then, that the resultant force is what produces the acceleration. The direction and, therefore, the sign of acceleration are determined by this force. An object can be moving to the right and have an acceleration to the left. An object may be increasing its speed and still have a negative acceleration. These events are confusing only if we confuse the motion of the object with the resultant force on the object.

As an object moves along a straight line, the *direction* of its velocity may change. For example, a stone thrown upward will eventually reach its highest point and then return in a downward direction. (The acceleration is constant and downward as the velocity is first positive and then negative.) The parameter s in such cases refers to the *displacement* of the object rather than to the total distance traveled. The value of s may be +8 m on the way up, and also, later, on the way down. The position s only becomes negative when it drops below the point of release (s = 0).

5-7 GRAVITY AND FREELY FALLING BODIES

Much of our knowledge about the physics of falling bodies originated with the Italian scientist Galileo Galilei (1564–1642). He was the first to demonstrate that in the absence of friction all bodies, large or small, heavy or light, fall to the earth with the same acceleration. This was a revolutionary idea, for it contradicted what a person might expect. Until the time of Galileo, people followed the teachings of Aristotle that heavy objects fall proportionally faster than lighter objects. The classic explanation for the paradox rests with the fact that heavier bodies are proportionately more difficult to accelerate. This resistance to motion is a property of a body called its *inertia*. Thus, in a vacuum, a feather will fall at the same rate as a steel ball because the larger inertial effect of the steel ball compensates exactly for its larger weight. (See Fig. 5-3.)

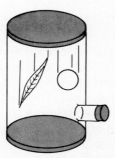

Fig. 5-3 All bodies fall with the same acceleration in a vacuum.

In the treatment of falling bodies given in this chapter, the effects of air friction are neglected entirely. Under these circumstances, gravitational acceleration is uniformly accelerated motion. At sea level and 45° latitude, this acceleration has been measured to be 32.17 ft/s^2, or 9.806 m/s^2, and is denoted by g. For our purposes, the following values will be sufficiently accurate:

$$g = 32 \text{ ft/s}^2$$
$$g = 9.8 \text{ m/s}^2$$

(5-9)

Since gravitational acceleration is constant acceleration, the same general equations of motion apply. However, one of the parameters is known in advance and need not be stated in the problem. If the constant g is inserted into the general equations (Table 5-1) the following modified forms will result:

(1a) $s = \dfrac{v_f + v_0}{2} t$ $s = \bar{v} t$

(2a) $v_f = v_0 + gt$

(3a) $s = v_0 t + \frac{1}{2} g t^2$

(4a) $2gs = v_f^2 - v_0^2$

Before utilizing these equations, a few general comments are in order. In problems dealing with free-falling bodies, it is extremely important to choose a direction to call positive and to follow through consistently in the substitution of known values. The sign of the answer is necessary to determine the location of a point or the direction of the velocity at specific times. For example, the distance s in the above equations represents the distance above or below the origin. If the upward direction is chosen as positive, a positive value for s indicates a distance above the starting point; if s is negative, it represents a distance below the starting point. Similarly, the signs of v_0, v_f, and g indicate their directions.

EXAMPLE 5-8 A rubber ball is dropped from rest, as shown in Fig. 5-4. Find its velocity and position after 1, 2, 3, and 4 s.

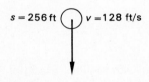

$s = 0$ $v = 0$

$s = 16\,\text{ft}$ $v = 32$ ft/s

$s = 64\,\text{ft}$ $v = 64$ ft/s

$s = 144\,\text{ft}$ $v = 96$ ft/s

$s = 256\,\text{ft}$ $v = 128$ ft/s

Fig. 5-4 A freely falling body has a constant downward acceleration of 32 ft/s².

Solution Since all parameters will be measured downward, it will be more convenient to choose the downward direction as positive. Organizing the data, we have

Given: $v_0 = 0$ Find: $v_f = ?$

$g = +32$ ft/s² $s = ?$

$t = 1, 2, 3,$ and 4 s

The velocity as a function of time is given by Eq. (2a), where $v_0 = 0$.

$$v_f = v_0 + gt = gt$$
$$= (32 \text{ ft/s}^2)t$$

After 1 s we have

$$v_f = (32 \text{ ft/s}^2)(1 \text{ s}) = 32 \text{ ft/s} \quad \text{downward}$$

Similar substitution of $t = 2$, 3, and 4 s will yield final velocities of 64, 96, and 128 ft/s. All these velocities are directed downward since that direction was chosen as positive.

The position as a function of time is calculated from Eq. (3a). Since the initial velocity is zero, we write

$$s = v_0 t + \tfrac{1}{2}gt^2 = \tfrac{1}{2}gt^2$$

from which

$$s = \tfrac{1}{2}(32 \text{ ft/s}^2)t^2 = (16 \text{ ft/s}^2)t^2$$

After 1 s the body will fall a distance

$$s = (16 \text{ ft/s}^2)(1 \text{ s})^2 = (16 \text{ ft/s}^2)(1 \text{ s}^2)$$
$$= 16 \text{ ft}$$

After 2 s

$$s = (16 \text{ ft/s}^2)(2 \text{ s})^2 = (16 \text{ ft/s}^2)(4 \text{ s}^2)$$
$$= 64 \text{ ft}$$

Similarly, calculations give 144 and 256 ft for the positions after 3 and 4 s. The above results are summarized in Table 5-2.

Table 5-2

Time t, s	Speed at the end of time t, ft/s	Position at the end of time t, ft
0	0	0
1	32	16
2	64	64
3	96	144
4	128	256

EXAMPLE 5-9   Assuming a ball is projected upward with an initial velocity of 96 ft/s, explain without using equations how its upward motion is just the reverse of its downward motion.

Solution   We shall assume the upward direction to be positive, making the acceleration due to gravity equal to -32 ft/s^2. The negative sign indicates that the speed of an object projected vertically will have its velocity reduced by 32 ft/s every second it rises. (Refer to Fig. 5-5.) If its initial velocity is 96 ft/s, its velocity after 1 s will be reduced to 64 ft/s. After 2 s its velocity will be 32 ft/s, and after 3 s its velocity will be reduced to zero. When the velocity becomes zero, the ball has reached its maximum height and begins to fall freely from rest. However, now the velocity of the ball will be *increasing* by 32 ft/s every second since both the direction of motion and the acceleration of gravity are in the negative direction. Its velocity after 4, 5, and 6 s will be -32, -64, and -96 ft/s, respectively. Except for the sign, which indicates the direction of motion, the velocities are the same at equal heights above the ground.

EXAMPLE 5-10   A baseball thrown vertically upward from the roof of a tall building has an initial velocity of 20 m/s. (a) Calculate the time required to reach its maximum height. (b) Find the maximum height. (c) Determine its position and velocity after 1.5 s. (d) What are its position and velocity after 5 s? (See Fig. 5-6.)

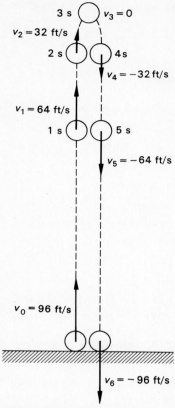

Fig. 5-5 A ball that is thrown vertically upward returns to the ground with the same speed.

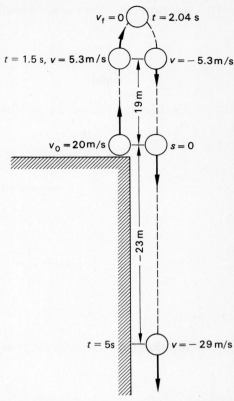

Fig. 5-6 A ball projected vertically upward rises until its velocity is zero; then it falls with increasing downward velocity.

Solution (a)   Let us choose the upward direction as positive since the initial velocity is directed upward. At the highest point, the final velocity of the ball will be zero. Organizing the known data, we have

Given: $v_0 = 20$ m/s   Find: $t = ?$

$v_f = 0$ $s = ?$

$g = -9.8$ m/s^2

The time required to reach the maximum height can be determined from Eq. (2a):

$$t = \frac{v_f - v_0}{g} = -\frac{v_0}{g}$$

$$= \frac{-20 \text{ m/s}}{-9.8 \text{ m/s}^2} = 2.04 \text{ s}$$

Solution (b)   The maximum height is found by setting $v_f = 0$ in Eq. (1a).

$$s = \frac{v_f + v_0}{2}\, t = \frac{v_0}{2}\, t$$

$$= \frac{20 \text{ m/s}}{2}(2.04 \text{ s}) = 20.4 \text{ m}$$

Solution (c)   To find the position and velocity after 1.5 s, we must establish new conditions.

Given: $v_0 = 20$ m/s   Find: $s = ?$

$g = -9.8$ m/s^2 $v_f = ?$

$t = 1.5$ s

We can now calculate the position as follows:

$$s = v_0 t + \tfrac{1}{2}gt^2$$
$$= (20 \text{ m/s})(1.5 \text{ s}) + \tfrac{1}{2}(-9.8 \text{ m/s}^2)(1.5 \text{ s})^2$$
$$= 30 \text{ m} - 11 \text{ m} = 19 \text{ m}$$

The velocity after 1.5 s is given by

$$v_f = v_0 + gt$$
$$= 20 \text{ m/s} + (-9.8 \text{ m/s}^2)(1.5 \text{ s})$$
$$= 20 \text{ m/s} - 14.7 \text{ m/s} = 5.3 \text{ m/s}$$

Solution (d)   The same equations apply to find the position and velocity after 5 s. Thus

$$s = v_0 t + \tfrac{1}{2}gt^2$$
$$= (20 \text{ m/s})(5 \text{ s}) + \tfrac{1}{2}(-9.8 \text{ m/s}^2)(5 \text{ s})^2$$
$$= 100 \text{ m} - 123 \text{ m} = -23 \text{ m}$$

The negative sign indicates that the ball is located 23 m below the point of release. The speed after 5 s is given by

$$v_f = v_0 + gt$$
$$= 20 \text{ m/s} + (-9.8 \text{ m/s}^2)(5 \text{ s})$$
$$= 20 \text{ m/s} - 49 \text{ m/s} = -29 \text{ m/s}$$

In this case, the negative sign indicates that the ball is traveling downward.

SUMMARY

A convenient way of describing objects in motion is to discuss their *velocity* or their acceleration. In this chapter, a number of applications have been presented which involve these physical quantities.

- **Average velocity** is the distance traveled per unit of time, and **average acceleration** is the change in velocity per unit of time.

$$\bar{v} = \frac{s}{t} \qquad a = \frac{v_f - v_0}{t}$$

- The definitions of velocity and acceleration result in the establishment of four basic equations involving uniformly accelerated motion:

$$s = \left(\frac{v_0 + v_f}{2}\right)t$$

$$v_f = v_0 + at$$

$$s = v_0t + \tfrac{1}{2}at^2$$

$$2as = v_f^2 - v_0^2$$

Given any three of the five parameters (v_0, v_f, a, s, and t), the other two can be determined from one of these equations.

- To solve acceleration problems, read the problem with a view to establishing the three given parameters and the two which are unknown. You might set up columns like

Given: $a = 4$ m/s^2 Find: $v_f = $?

$s = 500$ m $v_0 = $?

$t = 20$ s

This procedure helps you choose the appropriate equation. Remember to choose a direction as positive and to apply it consistently throughout the problem.

- **Gravitational acceleration:** Problems involving gravitational acceleration can be solved like other acceleration problems. In this case, one of the parameters is known in advance to be

$$a = g = 9.8 \text{ m/s}^2 \text{ or } 32 \text{ ft/s}^2$$

The sign of the gravitational acceleration is $+$ or $-$ depending on whether you choose up or down as the positive direction.

QUESTIONS

5-1. Define the following terms:
   a. Constant speed
   b. Average speed
   c. Velocity
   d. Acceleration
   e. Instantaneous velocity
   f. Instantaneous speed
   g. Uniformly accelerated motion
   h. Acceleration due to gravity

5-2. Distinguish clearly between the terms *speed* and *velocity*. A stock-car racer drives 500 laps around a 1-mi track in a time of 5 h. What was the average speed? What was the average velocity?

5-3. A bus driver travels a distance of 200 mi in 4 h. At the same time, a tourist travels the 200 mi by car but stops twice for a 30-min rest along the route. Nevertheless, the tourist arrives at the destination at the same instant as the bus driver. Compare the average speed of the bus driver with the average speed of the tourist.

5-4. Give some examples of motion in which the speed is constant but the velocity is not.

5-5. Two evenly spaced bowling balls are rolling along a level return trough at constant speed. Describe their later speed and separation in view of the fact that the first ball starts uphill to the rack before the second ball.

5-6. A sports announcer states, "The newly designed race car negotiated the 500-mi obstacle course in a record-breaking endurance run, reaching average speeds of 150 mi/h along the way." What is wrong with this statement?

5-7. A long strip of pavement is marked off in 100-ft intervals. Students stationed on a nearby ridge use their stopwatches to measure the time required for a car to cover the distance. The following data are obtained:

Distance, ft	0	100	200	300	400	500
Time, s	0	2.1	4.3	6.4	8.4	10.5

Plot a graph with distance as the ordinate and time as the abscissa. What is the significance of the slope of the curve? What is the speed of the car?

5-8. The acceleration of gravity on a distant planet is one-fourth of the acceleration experienced on the earth. Does this mean that a rock dropped from a height of 4 ft will hit the ground in one-fourth the time required on the earth? Explain.

5-9. A spring-loaded gun fires a Ping-Pong ball vertically upward. On the moon the ball is observed to rise to a height of six times that observed on the earth. What can we say about the acceleration of gravity on the surface of the moon?

5-10. The symbol g is sometimes referred to as *gravity* or the *acceleration of gravity*. Comment on the appropriateness of these expressions.

5-11. A rock is thrown vertically upward from the edge of a roof. It moves to its highest point, returns to its starting point, and then drops to the street below. Imagine that you take a snapshot at various times during the flight of the rock. Discuss the signs of velocity, acceleration, and distance at each instant.

5-12. Assume that the upward direction is positive in Question 5-11. What is the acceleration at the instant the rock leaves the hand? When it reaches its highest point? Just before it strikes the street below?

5-13. In the absence of air resistance, the same time is required for a projectile to reach its maximum height as is required for it to return to its starting position. Will this fact still be true if air resistance is not neglected? Draw free-body force diagrams for each situation.

PROBLEMS

Speed and Velocity

5-1. A car travels a distance of 86 km. If the average speed was 8 m/s, how many hours were required for the trip?

Ans. 2.99 h

5-2. Sound travels through the air at an average speed of 340 m/s. A girl drops a rock from a bridge to the water 20 m below. After the rock strikes the water, how much time will be required for the sound of the splash to reach the girl's ear?

5-3. A truck traveled 640 mi on a run from Atlanta to Philadelphia. The entire trip took 14 h, but the driver made two 30-min stops for meals. What was the average velocity?

Ans. 45.7 mi/h

5-4. A car travels along a curved mountain road for a distance of 80 km in 4 h. The straight-line distance from beginning to end is only 60 km. What is the average speed? What is the average velocity?

** 5-5. A jogger runs for 15 min at an average speed of 12 mi/h. What must be his average speed for the next 45 min if the average speed for the entire hour is to be 9 mi/h?

Ans. 8 mi/h

5-6. A woman walks a distance directly north at an average velocity of 6 km/h for 4 min. She then walks directly east at an average velocity of 4 km/h for 10 min. **(a)** What is the total distance traveled? **(b)** What is the net displacement? **(c)** What is the average speed for the entire trip? **(d)** What is the average velocity for the trip?

Uniform Acceleration

5-7. An arrow leaves the bow 0.5 s after being released from a cocked position. If it has reached a speed of 40 m/s in this time, what was the average acceleration?

Ans. 80 m/s^2

5-8. An automobile traveling at a constant speed of 50 km/h accelerates at a rate of 4 m/s^2 for 3 s. What is its speed at the end of the 3-s interval?

5-9. A truck traveling at a speed of 60 mi/h suddenly brakes to a stop. The skid marks are observed to be 180 ft long. What was the average acceleration? What was the stopping time?

Ans. -21.5 ft/s^2, 4.09 s

5-10. An arresting device is used to land airplanes on a carrier deck. The average acceleration produced by this device is 150 ft/s^2. If a plane is brought to a stop in 1.5 s, what was the stopping distance and the speed just before contact with the arresting device?

5-11. In a braking test, a car is observed to come to rest in 3 s. If its initial speed was 60 km/h, what was the acceleration and what was the stopping distance?

Ans. -5.56 m/s^2, 25.0 m

5-12. A bullet leaves a 28-in. rifle barrel with a muzzle velocity of 2700 ft/s. What was its average acceleration in the barrel, assuming that it started from rest? How long did the bullet remain in the barrel?

5-13. A ball rolls up an incline as shown in Fig. 5-7. It is given an initial velocity of 16 m/s at the bottom. Two seconds later it is still traveling up the plane with a velocity of only 4 m/s. **(a)** What is the acceleration? **(b)** What is the maximum displacement from the bottom for these conditions? **(c)** What is its velocity 4 s after it starts up the plane?

Ans. **(a)** -6 m/s^2, **(b)** 21.3 m, **(c)** -8 m/s

Fig. 5-7

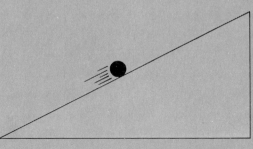

5-14. The ball in Fig. 5-7 rolls freely with an acceleration of 4 m/s^2 directed down the plane. At time $t = 0$, its velocity is 8 m/s up the plane. How much later will its velocity be 6 m/s directed down the plane?

5-15. A monorail train traveling at 80 km/h must be brought to a stop in a distance of 40 m. What average acceleration is required? What is the stopping time?

Ans. -6.17 m/s^2, 3.6 s

Gravity and Free-falling Bodies

5-16. A ball is dropped from rest and falls for 5 s. What are its position and velocity at that instant?

5-17. A rock is dropped from rest near the earth's surface. When will its position be 18 m below the point of release? What will be its velocity at that instant?

Ans. 1.92 s after release, -18.8 m/s

5-18. A bolt is dropped accidentally from the top of a building. Five seconds later it strikes the street below. How high is the building? What was the velocity of the bolt when it struck the street?

5-19. A brick is thrown downward from the top of a building 80 ft above the ground. Just before it strikes the earth below, it has a downward velocity of 90 ft/s. What was the initial velocity of the brick?

Ans. 54.6 ft/s

5-20. A projectile is thrown vertically upward with an initial speed of 15 m/s. What are its position and velocity after 1 s and after 4 s?

5-21. An arrow is shot vertically upward with an initial velocity of 80 ft/s. **(a)** How long will it rise? **(b)** How high will it rise? **(c)** What are its position and velocity after 2 s? **(d)** What are its position and velocity after 6 s?

Ans. **(a)** 2.5 s; **(b)** 100 ft; **(c)** 96 ft, 16 ft/s; **(d)** -96 ft, -112 ft/s

*   **5-22.** An object projected upward returns to its starting position in 5 s. What was its initial speed and how high did it rise?

5-23. A hammer is thrown vertically upward to the top of a roof 50 ft high. **(a)** What was the minimum initial velocity required? **(b)** How much time was required?

Ans. **(a)** 56.6 ft/s, **(b)** 1.77 s

Additional Problems

5-24. A rocket traveling in space at 60 m/s is given a sudden acceleration. If it reaches a speed of 140 m/s in 8 s, what was its average acceleration? How far did it travel in this time?

5-25. A railroad car loaded with coal starts from rest and coasts freely down a gentle slope. If the average acceleration was 4 ft/s^2, what will be the speed of the car in 5 s? What distance will be covered in this period of time?

Ans. 20 ft/s, 50 ft

*   **5-26.** A truck travels north at a constant speed of 40 ft/s, and at $t = 0$ it is located a distance of 500 ft ahead of a certain car. If the car starts at rest and accelerates at 10 ft/s^2, when will it overtake the truck? How far is the point of contact from the initial position of the car?

**   **5-27.** Consider the two balls A and B shown in Fig. 5-8. Ball A has a constant acceleration of 4 m/s^2 directed to the right, and ball B has a constant acceleration of 2 m/s^2 directed to the left. Ball A is initially traveling to the left at 2 m/s, while

Fig. 5-8

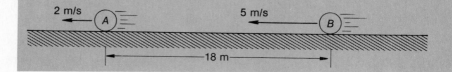

ball B is traveling to the left initially at 5 m/s. At what time t after $t = 0$ will the balls collide?

Ans. 2.00 s

5-28. A ball is thrown vertically upward with an initial velocity of 23 m/s. What are its position and velocity after 2 s, after 4 s, and after 8 s?

5-29. An object is tossed vertically upward and rises to a maximum height of 16 m. What was the initial speed and how much time was required?

Ans. 17.7 m/s, 1.81 s

★ 5-30. A ball is dropped from rest at the top of a 100-m-tall building. At the same instant a second ball is thrown upward from the base of the building with an initial speed of 50 m/s. When will the two balls collide and at what distance above the street?

★ 5-31. A ball is rolled up an inclined plane so that it just reaches the top of the incline in 3 s. The speed of the ball changes by 4 m/s every second that it travels. What is the length of the incline, and what was the initial speed?

Ans. 12 m/s, 18 m

★★ 5-32. A ball is dropped from the window of a skyscraper, and 2 s later a second ball is thrown vertically downward. What must be the initial velocity of the second ball if it is to overtake the first ball just as it strikes the ground 400 m below?

★★ 5-33. A balloonist rising vertically with a velocity of 4 m/s releases a sandbag at the instant when the balloon is 16 m above the ground. **(a)** Compute the position and velocity of the sandbag (relative to the ground) after 0.3 s and after 2 s. **(b)** How many seconds after its release will it strike the ground? **(c)** With what speed does it strike the ground?

Ans. **(a)** 16.8 m, 1.06 m/s, 4.40 m, -15.6 m/s; **(b)** 2.26 s; **(c)** -18.2 m/s

5-34. A stone is thrown vertically downward from the top of a bridge. Four seconds later it strikes the water below with a final velocity of 60 m/s. What was the initial velocity of the stone? How high is the bridge?

★★ 5-35. An arrow is shot vertically upward with a speed of 40 m/s. Three seconds later another arrow is shot upward with a velocity of 60 m/s. At what time and position will they meet?

Ans. 4.54 s, 80.6 m

★★ 5-36. A baggage lift is accelerating upward at 5 ft/s². At the instant its upward speed is 8.0 ft/s, a bolt drops from the top of the lift 10 ft from its floor. **(a)** Find the time until the bolt strikes the floor. **(b)** What distance has the bolt fallen relative to the ground?

Projectile Motion

OBJECTIVES

After completing this chapter, you should be able to:

1. **Explain with equations and diagrams the horizontal and vertical motion of a projectile launched at various angles.**
2. **Determine the position and velocity of a projectile when its initial velocity and position are given.**
3. **Determine the range, the maximum height, and the time of flight for a projectile when the initial speed and angle of projection are given.**

We have seen that objects projected vertically upward or downward or dropped from rest are accelerated uniformly in the earth's gravitational field. In this chapter, we consider the more general problem of an object projected freely into a gravitational field in a nonvertical direction, as shown in Fig. 6-1.

An object launched into space without motive power of its own is called a *projectile*. If we neglect air resistance, the only force acting on a projectile is its weight **W**, which causes its path to deviate from a straight line. It receives constant downward acceleration due to gravity, but it differs from the motion studied previously. The direction of the acceleration usually differs from that of the initial velocity. The projectile has a constant horizontal velocity and a vertical velocity that changes uniformly under the influence of gravity.

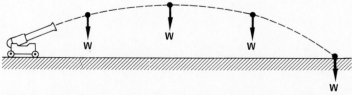

Fig. 6-1 A projectile is an object launched freely into space under the influence of gravity alone. The only force acting on such an object is its weight.

HORIZONTAL PROJECTION

If an object is projected horizontally, its motion can best be described by considering its horizontal and vertical motion separately. For example, in Fig. 6-2, a ball is dropped vertically at the same instant that another ball is projected horizontally. The horizontal velocity for the latter ball is unchanged, as indicated by arrows of the same length throughout its trajectory. The vertical velocity, on the other hand, is initially zero and increases uniformly, in accordance with equations derived in Chap. 5 for freely falling bodies. The balls will strike the water at the same instant, even though one is also moving horizontally. Thus problems are greatly simplified by finding separate solutions for horizontal and vertical components.

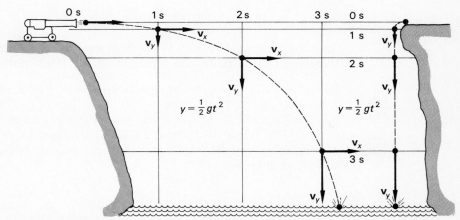

Fig. 6-2 Motion of a projectile fired horizontally. The vertical velocity and position increase with time like those of a free-falling body. Note that the horizontal distance increases linearly with time, indicating a constant horizontal velocity.

A comparison of the general equations for uniformly accelerated motion with those used for projectile motion is given in Table 6-1. For example, the equation relating distance to initial velocity and time

$$s = v_0 t + \tfrac{1}{2}at^2$$

can be written

$$y = v_{0y}t + \tfrac{1}{2}gt^2$$

where y = vertical position
v_{0y} = initial vertical velocity
g = acceleration of gravity

For problems in which the initial velocity is horizontal, the final position will be below the origin, and the final velocity will be directed downward.

Table 6-1 Uniformly Accelerated Motion and Projectiles

Uniformly accelerated motion	Projectile motion	
(1) $s = \bar{v}t = \dfrac{v_0 + v_f}{2}t$	$x = \bar{v}_{0x}t$	$y = \dfrac{v_{0y} + v_y}{2}t$
(2) $v_f = v_0 + at$	$v_y = v_{0y} + gt$	
(3) $s = v_0t + \frac{1}{2}at^2$	$y = v_{0y}t + \frac{1}{2}gt^2$	
(4) $2as = v_f^2 - v_0^2$	$2gy = v_y^2 - v_{0y}^2$	

Since gravity is also directed vertically downward, it is more convenient to choose the downward direction as positive. It should also be noted that for horizontal projection

$$v_{0x} = \bar{v}_x \qquad v_{0y} = 0$$

since the horizontal velocity is constant and the initial vertical velocity is zero. Therefore, the horizontal and vertical positions at any instant can be found from

$$\boxed{\begin{aligned} x &= v_{0x}t \\ y &= \tfrac{1}{2}gt^2 \end{aligned}} \qquad \begin{aligned} &\textit{Horizontal Position} \\ &\textit{Vertical Position} \end{aligned} \quad (6\text{-}1)$$

Similarly, the horizontal and vertical components of the final velocity at any instant are given by

$$\boxed{\begin{aligned} \bar{v}_x &= v_{0x} \\ v_y &= gt \end{aligned}} \qquad \begin{aligned} &\textit{Horizontal Velocity} \\ &\textit{Vertical Velocity} \end{aligned} \quad (6\text{-}2)$$

The final position and velocity can be found from their components. In all the above equations, the positive value must be substituted for g if we choose the downward direction as positive.

EXAMPLE 6-1   A cannonball is projected horizontally with an initial velocity of 120 m/s from the top of a cliff 250 m above a lake, as illustrated in Fig. 6-3. (*a*) In what time will it strike the water at the foot of the cliff? (*b*) What is the horizontal distance from the foot of the cliff to the point of impact in the lake? (*c*) What are the horizontal and vertical components of its final velocity?

Solution (a)   The time required to strike the water is a function only of vertical parameters. The initial velocity in the y direction is zero, and the ball must fall through a distance of 250 m. Therefore,

$$y = \tfrac{1}{2}gt^2$$

which yields, on substitution,

$$250 \text{ m} = \tfrac{1}{2}(9.8 \text{ m/s}^2)t^2$$

Simplifying, we obtain

$$(4.9 \text{ m/s}^2)t^2 = 250 \text{ m}$$

from which

$$t = \sqrt{\frac{250 \text{ s}^2}{4.9}} = 7.14 \text{ s}$$

Fig. 6-3

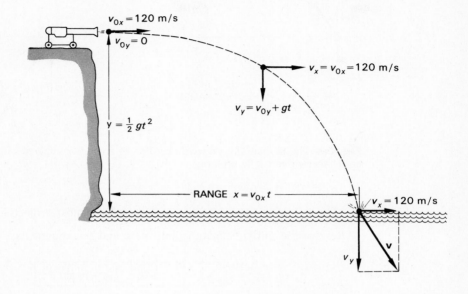

Solution (b)   The horizontal distance is a function only of the initial horizontal velocity and the time required to strike the water. Hence

$$x = v_{0x}t = (120 \text{ m/s})(7.14 \text{ s}) = 857 \text{ m}$$

Solution (c)   The horizontal component of the velocity is unchanged and equal to 120 m/s. The vertical component, on the other hand, is given by

$$\begin{aligned} v_y &= gt = (9.8 \text{ m/s})(7.14 \text{ s}) \\ &= 70 \text{ m/s} \end{aligned}$$

It is left as an exercise to show that the final velocity at the point of impact is 139 m/s in a direction 30.3° below the horizontal.

6-2

THE MORE GENERAL PROBLEM OF TRAJECTORIES

The more general case of projectile motion occurs when the projectile is fired at an angle. This problem is illustrated in Fig. 6-4, where the motion of a projectile fired at an angle θ with an initial velocity v_0 is compared with the motion of an object projected vertically upward. Once again it is easy to see the advantage of treating the horizontal and vertical motions separately. In this case the equations listed in Table 6-1 may be used, and we should consider the upward direction as positive. Thus, if the vertical position y is above the origin, it will be positive; if it is below the origin, it will be negative. Similarly, upward velocities will be positive. Since the acceleration is always directed downward, we must substitute a negative value for g.

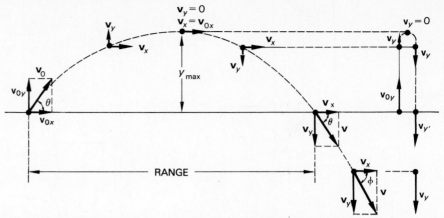

Fig. 6-4 The motion of a projectile fired at an angle is compared with the motion of an object thrown vertically upward.

The following procedure is useful in solving most projectile problems:

1. Resolve the initial velocity v_0 into its x and y components:

$$v_{0x} = v_0 \cos \theta \qquad v_{0y} = v_0 \sin \theta \qquad\qquad (6\text{-}3)$$

2. The horizontal and vertical components of its position at any instant are given by

$$x = v_{0x}t$$
$$y = v_{0y}t + \tfrac{1}{2}gt^2 \qquad\qquad (6\text{-}4)$$

3. The horizontal and vertical components of its velocity at any instant are given by

$$v_x = v_{0x}$$
$$v_y = v_{0y} + gt \qquad\qquad (6\text{-}5)$$

4. The final position and velocity can then be obtained from their components.

The important point to remember in applying these equations is to be consistent throughout with units and sign convention.

EXAMPLE 6-2   An artillery shell is fired with an initial velocity of 400 ft/s at an angle of 30° above the horizontal. Find (*a*) its position and velocity after 8 s, (*b*) the time required to reach its maximum height, and (*c*) the horizontal range R as indicated in Fig. 6-4.

Solution (a)   The horizontal and vertical components of the initial velocity are

$$v_{0x} = v_0 \cos \theta = (400 \text{ ft/s})(\cos 30°) = 346 \text{ ft/s}$$
$$v_{0y} = v_0 \sin \theta = (400 \text{ ft/s})(\sin 30°) = 200 \text{ ft/s}$$

The x component of its position after 8 s is

$$x = v_{0x}t = (346 \text{ ft/s})(8 \text{ s}) = 2770 \text{ ft}$$

The y component of its position at this time is given by

$$y = v_{0y}t + \tfrac{1}{2}gt^2$$

from which

$$y = (200 \text{ ft/s})(8 \text{ s}) + \tfrac{1}{2}(-32 \text{ ft/s}^2)(8 \text{ s})^2$$
$$= 1600 \text{ ft} - 1024 \text{ ft} = 576 \text{ ft}$$

Hence, the position of the shell after 8 s is 2770 ft down range and 576 ft above its initial position.

In computing its velocity at this point, we first recognize that the x component of the velocity does not change. Thus

$$v_x = v_{0x} = 346 \text{ ft/s}$$

The y component must be computed from

$$v_y = v_{0y} + gt$$

so that

$$v_y = 200 \text{ ft/s} + (-32 \text{ ft/s})(8 \text{ s})$$
$$= 200 \text{ ft/s} - 256 \text{ ft/s}$$
$$= -56 \text{ ft/s}$$

The minus sign indicates that the projectile is on its way down. Finally, we can compute the resultant velocity from its components, as shown in Fig. 6-5. The angle ϕ is calculated from

$$\tan \phi = \frac{v_y}{v_x} = \frac{56 \text{ ft/s}}{346 \text{ ft/s}} = 0.162$$

from which

$$\phi = 9.2°$$

The magnitude of the velocity is

$$v = \frac{v_y}{\sin \phi} = \frac{56 \text{ ft/s}}{\sin 9.2°} = 350 \text{ ft/s}$$

Fig. 6-5

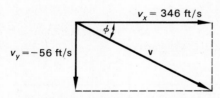

$v_x = 346$ ft/s

$v_y = -56$ ft/s

ϕ

v

Solution (b)   At the maximum point in the shell's trajectory, the y component of its velocity is zero. Hence, the time to reach this height can be computed from

$$v_y = v_{0y} + gt$$

Substituting given values for v_{0y} and g, we obtain

$$0 = 200 \text{ ft/s} + (-32 \text{ ft/s})t$$

from which

$$t = \frac{200 \text{ ft/s}}{32 \text{ ft/s}^2} = 6.25 \text{ s}$$

As an exercise, it is suggested that the student use this result to show that the maximum height in the trajectory is 625 ft.

Solution (c) The range of the projectile can be computed by recognizing that the total time t' of the entire flight is equal to twice the time to reach the highest point. Hence

$$t' = 2t = (2)(6.25 \text{ s}) = 12.5 \text{ s}$$

and the range is

$$R = v_{0x}t' = (346 \text{ ft/s})(12.5 \text{ s})$$
$$= 4325 \text{ ft}$$

Therefore, the projectile rises for 6.25 s to a height of 625 ft and then falls to the ground at a horizontal distance of 4325 ft from the point of release.

SUMMARY

The key to problems involving projectile motion is to treat the horizontal motion and the vertical motion separately. Most projectile problems are solved using the following approach:

- Resolve the initial velocity v_0 into its x and y components:

$$v_{0x} = v_0 \cos \theta \qquad v_{0y} = v_0 \sin \theta$$

- The horizontal and vertical components of its position at any instant are given by

$$x = v_{0x}t \qquad y = v_{0y}t + \tfrac{1}{2}gt^2$$

- The horizontal and vertical components of its velocity at any instant are given by

$$v_x = v_{0x} \qquad v_y = v_{0y} + gt$$

- The final position and velocity can then be obtained from their components.

The important point to remember in applying these equations is to be consistent throughout with units and sign conversion.

QUESTIONS

6-1. Define the following terms:
a. Projectile
b. Trajectory
c. Range

6-2. Is the motion of a projectile fired at an angle an example of uniformly accelerated motion? What if the projectile is fired vertically upward? Explain.

6-3. At what angle should a baseball be thrown if it is to reach its maximum range? At what angle must it be thrown to attain maximum height? Plot the trajectories you would expect for a ball thrown at (a) 90°, (b) 60°, (c) 45°, (d) 30°, and (e) 0°.

6-4. Suppose a hunter shoots an arrow directly at a squirrel on a tree limb, and the squirrel begins to fall at the instant the arrow leaves the bow. Will the arrow hit the squirrel? Why? Draw a sketch showing their relative motions.

6-5. A child drops a rubber ball from the window of a car traveling at a constant speed of 60 mi/h. What is the initial velocity of the ball relative to the ground? Describe its motion.

6-6. A toy car is pulled across the floor with uniform speed. If a marble is projected vertically upward from the car, describe its motion in relation to the car. Describe its motion in relation to the floor.

6-7. Explain the adjustment of sights on a rifle for increasing target distances.

6-8. Explain the reasoning behind the use of high or low trajectories for punting in a football game.

PROBLEMS

Horizontal Projection

6-1. A box of supplies is dropped from an airplane which is located a vertical distance of 340 m above a lake. The plane has a horizontal velocity of 70 m/s relative to the ground. What horizontal distance will the box travel before striking the water?

Ans. 583 m

6-2. Logs are discharged horizontally from a greased chute which is located 20 m above a millpond. If the logs leave the chute with a horizontal speed of 15 m/s, how far will the logs travel before striking the water?

6-3. An overhead crane has an electromagnet which is 80 ft above the ground. The crane operator wants to drop a piece of scrap metal at a spot 60 ft beyond the end of the crane's track. What horizontal speed must the crane have when it reaches the end of its track?

Ans. 26.8 ft/s

6-4. A steel ball rolls off the edge of a table top 6 ft high. If the ball strikes the floor at a distance of 5 ft from the base of the table, what was its velocity at the instant it left the table?

* 6-5. A stone is thrown horizontally from the top of a building with an initial velocity of 200 m/s. At the same instant another stone is dropped from rest. (a) Compute the position and velocity of the second stone after 3 s. (b) How far has the first stone traveled horizontally during this 3 s? (c) How far has it traveled vertically? (d) What are the horizontal and vertical components of the velocity of the first stone after 3 s?

Ans. (a) −44.1 m, −29.4 m/s; (b) 600 m; (c) −44.1 m; (d) 200 m/s, −29.4 m/s

The More General Problem of Trajectories

* 6-6. A stone is thrown at an angle of 58° with an initial velocity of 20 m/s. What are its horizontal position and its vertical position after 3 s? What are the horizontal and vertical components of its velocity after 3 s?

* 6-7. An arrow is shot into the air with a velocity of 120 ft/s at an angle of 37° with the horizontal. (a) What are the horizontal and vertical components of its initial velocity? (b) What is its position (horizontally and vertically) after 2 s? (c) Find the components of its velocity after 2 s. **(d)** What are the magnitude and direction of its resultant velocity after 2 s?

Ans. (a) 95.8 ft/s, 72.2 ft/s; (b) 192 ft, 80.4 ft; (c) 95.8 ft/s, 8.23 ft/s; (d) 96.2 ft/s, 4.91°

6-8. Assume that an arrow is shot at an angle of 40° with the horizontal at a velocity of 30 m/s. Find the horizontal and vertical positions of the arrow after 4 s. What is the resultant velocity?

6-9. A projectile is launched at an angle of 30° with an initial velocity of 20 m/s. **(a)** What is the highest point in the trajectory? **(b)** What is its horizontal range? **(c)** If the ground is level for the entire flight, how long is the projectile in the air?

Ans. (a) 5.10 m, (b) 35.3 m, (c) 2.04 s

6-10. A baseball leaves a bat with a velocity of 35 m/s at an angle of 32°. What is the highest point in its path?

6-11. The baseball of Prob. 6-10 rises and falls, striking a billboard at a point 8 m above the playing field. What was the time of flight? How far was the ball hit horizontally?

Ans. 3.29 s, 97.6 m

Additional Problems

6-12. A bomber flying with a horizontal speed of 500 mi/h releases a bomb. Six seconds later the bomb strikes the ocean below. At what altitude was the plane flying? How far did the bomb travel horizontally? What are the magnitude and direction of its final velocity?

6-13. Two tall buildings are 400 ft apart. A ball is thrown horizontally from the roof of the first building 1700 ft from the ground. With what horizontal speed must the ball be thrown if it is to enter a window of the second building 800 ft from the ground?

Ans. 53.3 ft/s

6-14. An object is projected horizontally with an initial velocity of 800 ft/s. At the same instant another object located 300 ft down range is dropped from rest. When will they collide and how far are they located below the release point?

6-15. It is desired to strike a target, whose horizontal range is 12 km, with a projectile. The angle of elevation is 35°. **(a)** Find the required muzzle velocity for the projecting gun assuming no air resistance. **(b)** What is the time of flight to the target?

Ans. (a) 354 m/s, (b) 41.4 s

6-16. Two children ride bicycles toward each other with constant velocities of 20 and 10 ft/s. Each child releases a ball 6 ft from the ground with a forward horizontal velocity of 40 ft/s relative to the bicycle. If the two balls collide 1 ft from the ground, what was their separation when they were initially released?

6-17. A putting green is located 240 ft horizontally and 64 ft vertically from the tee. What must be the magnitude and direction of the initial velocity if a ball is to strike the green at this location after a time 4 s?

Ans. 100 ft/s, 53.1°

6-18. A wild boar charges directly toward a hunter with a constant speed of 60 ft/s. At the instant the boar is 100 yd away, the hunter fires an arrow at 30° with the ground. With what velocity must the arrow leave the bow at 30° if it is to strike its target? Assume the bow and the boar to be at the same vertical position.

6-19. A shot leaves a putter's hand 2 m from the ground with an initial velocity of 14 m/s at an angle of 42° with the horizontal. How long will it rise? What is its velocity at the highest point in the trajectory? How high is it from the ground at this instant? What is its horizontal range from the foot of the putter?

Ans. 0.956 s, 10.4 m/s, 6.48 m, 21.9 m

7 Newton's Second Law

OBJECTIVES

After completing this chapter, you should be able to:

1. Describe the relationship between force, mass, and acceleration and give the consistent units for each in the metric and U.S. customary systems of units.
2. Define the units *newton* and *slug* and explain why they are derived units rather than fundamental units.
3. Demonstrate by definition and example your understanding of the distinction between mass and weight.
4. Determine mass from weight and weight from mass at a point where the acceleration due to gravity is known.
5. Draw a free-body diagram for objects in motion with constant acceleration, set the resultant force equal to the total mass times the acceleration, and solve for unknown parameters.

According to Newton's first law of motion, an object will undergo a change in its state of rest or motion *only* when acted on by a resultant, unbalanced force. We now know that a change in motion, i.e., a change in speed, results in *acceleration*. In many industrial applications, we need to be able to predict the acceleration that will be produced by a given force. For example, the forward force required to accelerate a car from rest to a speed of 60 km/h in 8 s is of interest to the automotive industry. In this chapter, you will study the relationships between force, mass, and acceleration.

7-1

NEWTON'S SECOND LAW OF MOTION

Before studying the relationship between a resultant force and acceleration in a formal way, let us first consider a simple experiment.

A linear air track is an apparatus for studying the motion of objects under conditions that approximate zero friction. Hundreds of tiny air jets create an upward force which balances the weight of the glider in Fig. 7-1. A string is attached to the front of the glider, and a spring scale of negligible weight is used to measure the applied horizontal force as shown. The acceleration the glider receives can be measured by determining the change in speed for a known time interval. The first applied force F_1 in Fig. 7-1a causes an acceleration a_1. Twice the force $2F_1$ will produce twice the acceleration $2a_1$, and three times the force $3F_1$ gives three times the acceleration $3a_1$.

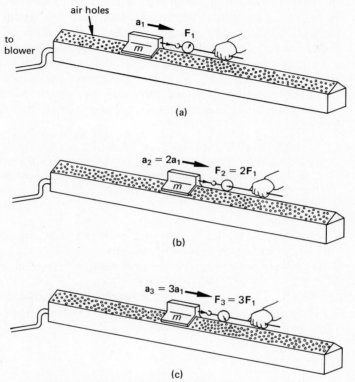

Fig. 7-1 Variation of acceleration with force.

Such observations show that the acceleration of a given body is directly proportional to the applied force. This means that the ratio of force to acceleration is always constant:

$$\frac{F_1}{a_1} = \frac{F_2}{a_2} = \frac{F_3}{a_3} = \text{constant}$$

The constant is a measure of how effective a given force is in producing acceleration. We shall see that this ratio is a property of the body called its mass m, where

$$m = \frac{F}{a}$$

A mass of one kilogram (1 kg) was defined in Chap. 2 by comparison with a standard artifact. Keeping this definition, we can now define a new force unit such that it will give a unit mass a unit acceleration.

A force of one **newton** (1 N) is that resultant force which will give a 1-kg mass an acceleration of 1 m/s^2.

The *newton* is declared the SI unit of force. A resultant force of 2 N will produce an acceleration of 2 m/s^2, and a force of 3 N will give an acceleration of 3 m/s^2 to a 1-kg mass.

Now let's return to our air track experiment to see how acceleration is affected by increasing the mass. This time we shall keep the applied force F constant. The mass will be changed by adding in succession more gliders of equal size and weight. Note from Fig. 7-2 that if the force is unchanged, an increase in mass will result in a proportionate *decrease* in the acceleration. Applying a constant force of 12 N in succession to 1-, 2-, and 3-kg masses will produce accelerations of 12 m/s^2, 6 m/s^2, and 4 m/s^2, respectively. These three cases are shown in Fig. 7-2a, b, and c.

From the above observations, we are now prepared to state Newton's second law of motion.

> *NEWTON'S SECOND LAW OF MOTION: Whenever an unbalanced force acts on a body, it produces in the direction of the force an acceleration that is directly proportional to the force and inversely proportional to the mass of the body.*

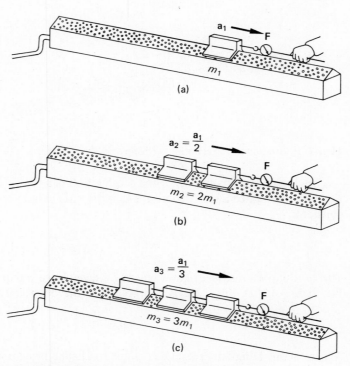

Fig. 7-2 Variation of acceleration with mass.

By using our newly defined unit, the newton, we can write this law as the following equation:

Resultant force = mass × acceleration

$$\boxed{F = ma}$$

Newton's Second Law (7-1)

Since this relationship depends on the definition of a new unit, we may substitute only units consistent with that definition. For example, if the mass is in kilograms (kg), the unit of force must be newtons (N) and the unit of acceleration must be meters per second squared (m/s^2).

Force (N) = *mass* (kg) × *acceleration* (m/s^2)

In USCS, a new mass unit is defined from the chosen units of *pound* (lb) for force and *feet per second squared* (ft/s^2) for acceleration. The new unit of mass is called the *slug* (from the sluggish or inertial property of mass).

A mass of one **slug** is that mass to which a resultant force of 1 lb will give an acceleration of 1 ft/s^2.

Force (lb) = *mass* (slugs) × *acceleration* (ft/s^2)

The SI unit of force is less than the USCS unit, and a mass of one slug is much greater than a mass of one kilogram. The following conversion factors might be helpful:

1 lb = 4.448 N   1 slug = 14.59 kg

A 1-lb bag of apples might contain four or five apples—each weighing about a newton. A person weighing 160 lb on earth would have a mass of 5 slugs or 73 kg.

It is important to recognize that the F in Newton's second law represents a *resultant* or unbalanced force. If more than one force acts on an object, it will be necessary to determine the resultant force *along the direction of motion*. The resultant force will always be along the direction of motion, since it is the *cause* of the acceleration. All components of forces which are perpendicular to the acceleration will be balanced. If the x axis is chosen along the direction of motion, we can determine the x component of each force and write

$$\boxed{\sum F_x = ma_x}$$

(7-2)

A similar equation could be written for the y components if the y axis were chosen along the direction of motion.

7-2
THE RELATIONSHIP BETWEEN WEIGHT AND MASS

Before we consider examples of Newton's second law, we must have a clear understanding of the difference between the *weight* of a body and its *mass*. Perhaps no other point is more confusing to the beginning student. The *pound* (lb), which is a unit of force, is often used as a mass unit, the pound-mass (lb_m). The kilogram, which is a unit of mass, is often used in industry as a unit of force, the kilogram-force (kgf). These seemingly inconsistent units result from the fact that there are many different systems of units in use. In this text, there should be less cause for confusion because we use only the SI and U.S. customary (British gravitational) units. Therefore, in this book the *pound* (lb) will always refer to *weight,* which is a force, and the unit *kilogram* (kg) will always refer to the *mass* of a body.

The weight of any body is the force with which the body is pulled vertically downward by gravity. When a body falls freely to the earth, the only force acting on it is its weight **W**. This net force produces an acceleration g, which is the same for all falling bodies. Thus, from Newton's second law we can write the relationship between a body's weight and its mass:

$$W = mg \qquad \text{or} \qquad m = \frac{W}{g} \qquad\qquad (7\text{-}3)$$

In either system of units (1) the mass of a particle is equal to its weight divided by the acceleration of gravity; (2) weight has the same units as the unit of force; and (3) the acceleration of gravity has the same units as acceleration.

Therefore, we can summarize as follows:

$$\text{SI: } W \text{ (N)} = m \text{ (kg)} \times g \text{ (9.8 m/s}^2)$$
$$\text{USCS: } W \text{ (lb)} = m \text{ (slug)} \times g \text{ (32 ft/s}^2)$$

The values for *g*, and hence the weights, in the above relations apply only at points on the earth near sea level, where *g* has these values.

Two things must be remembered in order to understand fully the difference between mass and weight:

Mass is a universal constant equal to the ratio of a body's weight to the gravitational acceleration due to its weight.

Weight is the *force* of gravitational attraction and varies depending upon the acceleration of gravity.

Therefore, the mass of a body is only a measure of its inertia and is not in any way dependent upon gravity. In outer space, a hammer has negligible weight, but it serves to drive nails just the same since its mass is unchanged.

In USCS units a body is usually described by stating its weight **W** in *pounds.* The mass, if desired, is computed from this weight and has the unit of *slugs.* In SI units a body is usually described in terms of its mass in *kilograms.* The weight, if desired, is computed from the given mass and has the units of *newtons.* In the following examples, all parameters are measured at points where $g = 32$ ft/s^2 or 9.8 m/s^2.

EXAMPLE 7-1   Find the mass of a person who weighs 150 lb.

Solution

$$m = \frac{W}{g} = \frac{150 \text{ lb}}{32 \text{ ft/s}^2} = 4.69 \text{ slugs}$$

EXAMPLE 7-2   Find the weight of an 18-kg block.

Solution

$$W = mg = 18 \text{ kg}(9.8 \text{ m/s}^2) = 176 \text{ N}$$

EXAMPLE 7-3   Find the mass of a body whose weight is 100 N.

Solution

$$m = \frac{W}{g} = \frac{100 \text{ N}}{9.8 \text{ m/s}^2} = 10.2 \text{ kg}$$

7-3

APPLICATION OF NEWTON'S SECOND LAW TO SINGLE-BODY PROBLEMS

The primary difference between the problems discussed in this chapter and earlier problems is that a net, unbalanced force is acting to produce an acceleration. Thus, after constructing the free-body diagrams which describe the situation, the first step is to determine the unbalanced force and set it equal to the product of mass and acceleration. The desired quantity can then be determined from the relation

Resultant force = mass × acceleration

$$\boxed{F \text{ (resultant)} = ma} \qquad (7\text{-}1)$$

The following examples will serve to demonstrate the relationship between force, mass, and acceleration.

EXAMPLE 7-4   What acceleration will a force of 20 N impart to a 10-kg body?

Solution   There is only one force acting, and so

$$F = ma \qquad \text{or} \qquad a = \frac{F}{m}$$

and

$$a = \frac{20 \text{ N}}{10 \text{ kg}} = 2 \text{ m/s}^2$$

EXAMPLE 7-5   What resultant force will give a 32-lb body an acceleration of 5 ft/s²?

Solution   In order to find the resultant force we must first determine the mass of the body from its given weight.

$$m = \frac{W}{g} = \frac{32 \text{ lb}}{32 \text{ ft/s}^2} = 1 \text{ slug}$$

then

$$F = ma = (1 \text{ slug})(5 \text{ ft/s}^2) = 5 \text{ lb}$$

EXAMPLE 7-6 What mass has a body if a force of 60 N will give it an acceleration of 4 m/s²?

Solution Solving for m in Newton's law, we have

$$m = \frac{F}{a} = \frac{60 \text{ N}}{4 \text{ m/s}^2} = 15 \text{ kg}$$

In the preceding examples, the unbalanced forces were easily determined. However, as the number of forces acting on a body increases, the problem of determining the resultant force is less obvious. In these cases, perhaps it is wise to point out a few considerations.

According to Newton's second law, a resultant force always produces an acceleration *in the direction of the resultant force*. This means that the net force and the acceleration it causes are of the same algebraic sign, and they each have the same line of action. Therefore, if the direction of motion (acceleration) is considered positive, less negative factors will be introduced into the equation $F = ma$. For example, in Fig. 7-3b, it is preferable to choose the direction of motion (left) as positive since the equation

$$P - \mathscr{F}_K = ma$$

is preferable to the equation

$$\mathscr{F}_K - P = -ma$$

which would result if we chose the right direction as positive.

Fig. 7-3 The direction of acceleration should be chosen as positive.

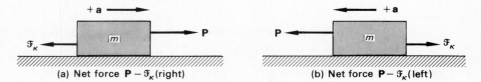

(a) Net force $P - \mathscr{F}_K$ (right)   (b) Net force $P - \mathscr{F}_K$ (left)

Another consideration which results from the above discussion is that the forces acting normal to the line of motion will be in equilibrium if the resultant force is constant. Thus, in problems involving friction, normal forces can be determined from the first condition for equilibrium.

In summary, we apply the following equations to acceleration problems:

$$\sum F_x = ma_x \qquad \sum F_y = ma_y = 0 \qquad\qquad (7\text{-}4)$$

where $\sum F_x$ and a_x are taken as positive and along the line of motion and $\sum F_y$ and a_y are taken normal to the line of motion.

EXAMPLE 7-7 A horizontal force of 20 N drags a 4-kg block across a level floor. If $\mu_k = 0.2$, find the acceleration of the block.

Solution A free-body diagram is superimposed on the sketch in Fig. 7-4. We will choose right as positive. To avoid confusing mass with weight, it is often desirable to calculate each in advance. The mass (4 kg) is given, and the weight is found from $W = mg$.

$$m = 4 \text{ kg} \qquad W = (4 \text{ kg})(9.8 \text{ m/s}^2) = 39.2 \text{ N}$$

Fig. 7-4

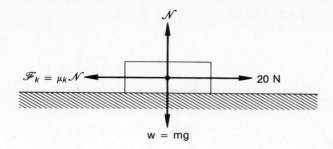

The resultant force is along the direction of motion. The vertical forces are balanced. Applying Newton's second law,

$$\text{Resultant force} = \text{mass} \times \text{acceleration}$$
$$20\ \text{N} - \mathscr{F}_K = ma$$

Recalling that $\mathscr{F}_K = \mu_k \mathscr{N}$, we may write

$$20\ \text{N} - \mu_k \mathscr{N} = ma$$

Since the vertical forces are balanced, we see from the figure that

$$\mathscr{N} = W = 39.2\ \text{N}$$

Substituting $\mathscr{N} = 39.2$ N, $\mu_k = 0.2$, and $m = 4$ kg, we have

$$20\ \text{N} - (0.2)(39.2\ \text{N}) = (4\ \text{kg})\,a$$
$$a = \frac{12.2\ \text{N}}{4\ \text{kg}} = 3.04\ \text{m/s}^2$$

You should verify that *newtons per kilogram* is equivalent to *meters per second squared*.

7-4
PROBLEM SOLVING TECHNIQUES

Solving all physics problems requires an ability to organize the given data and to apply formulas in a consistent manner. Often a procedure is helpful to the beginning student, and this is particularly true for the problems in this chapter. A logical sequence of operations for problems involving Newton's second law is outlined below.

1. Read the problem carefully for general understanding.
2. Draw a sketch, labeling given information.
3. Draw a free-body diagram with an axis along the direction of motion.
4. Indicate the positive direction of acceleration.
5. Determine the mass and weight of each object:

$$W = mg \qquad m = \frac{W}{g}$$

6. From the free-body diagram, determine the resultant force along the direction of motion. (ΣF).
7. Determine total mass ($m_t = m_1 + m_2 + \cdots$)

8. Set the resultant force ΣF equal to total mass (m_t) times acceleration a:

$$\boxed{\Sigma F = m_t a}$$

9. Solve for the unknown quantity.

EXAMPLE 7-8 A 2000-lb elevator is lifted upward with an acceleration of 4 ft/s². What is the tension in the supporting cable?

Solution Read the problem, then draw a sketch from which a free-body diagram may be drawn (see Fig. 7-5). Notice that the positive direction of acceleration (upward) is indicated on the free-body diagram.

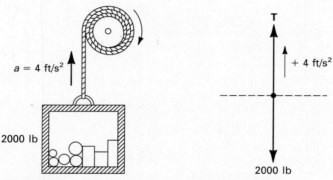

Fig. 7-5 Upward acceleration in a gravitational field.

Now we determine the mass and the weight of the 2000-lb elevator. The weight, of course, is 2000 lb. The mass must be calculated from $m = W/g$.

$$W = 2000 \text{ lb} \qquad m = \frac{2000 \text{ lb}}{32 \text{ ft/s}^2} = 62.5 \text{ slugs}$$

Since the elevator is the only object moving, the 62.5 slugs represents the *total* mass m_t. The resultant force from the free-body diagram is

$$\Sigma F = T - 2000 \text{ lb}$$

From Newton's second law, we now write

$$\text{Resultant force} = \text{total mass} \times \text{acceleration}$$
$$T - 2000 \text{ lb} = (62.5 \text{ slugs})(4 \text{ ft/s}^2)$$
$$T - 2000 \text{ lb} = 250 \text{ lb}$$

Finally, we solve for the unknown T by adding 2000 lb to both sides of the equation.

$$T = 250 \text{ lb} + 2000 \text{ lb}$$
$$T = 2250 \text{ lb}$$

MECHANICS

EXAMPLE 7-9   A 100-kg ball is lowered by means of a cable with a downward acceleration of 5 m/s². What is the tension in the cable?

Solution   As before, we construct a sketch and a free-body diagram (Fig. 7-6). Note that the downward direction is chosen as positive since that is the direction of motion.

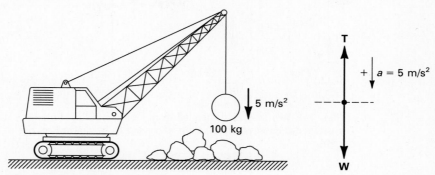

Fig. 7-6 Downward acceleration.

This time the *mass* is given and the *weight* must be calculated from $W = mg$.

$$m = 100 \text{ kg} \qquad W = (100 \text{ kg})(9.8 \text{ m/s}^2) = 980 \text{ N}$$

The resultant force is the *net* downward force, or

$$\sum = W - T \qquad \text{(remember, down is positive)}$$

Now, from Newton's second law, we write

$$\textit{Net downward force} = \textit{total mass} \times \textit{downward acceleration}$$
$$W - T = ma$$

Substituting known quantities, we obtain

$$980 \text{ N} - T = (100 \text{ kg})(5 \text{ m/s}^2)$$
$$980 \text{ N} - T = 500 \text{ N}$$

from which we solve for T by adding T to both sides and subtracting 500 N from both sides:

$$980 \text{ N} - 500 \text{ N} = T$$
$$T = 480 \text{ N}$$

EXAMPLE 7-10   An Atwood machine consists of a single pulley with masses suspended on each side. It is a simplified version of many industrial systems in which counterweights are used for balance. Assume that the mass on the right side is 10 kg and that the mass on the left side is 2 kg. (*a*) What is the acceleration of the system? (*b*) What is the tension in the cord?

Solution (a)   We first draw a sketch and a free-body diagram for each mass (Fig. 7-7). The weight and mass of each object are determined.

$$m_1 = 2 \text{ kg} \qquad W_1 = m_1g = (2 \text{ kg})(9.8 \text{ m/s}^2) \qquad \text{or} \qquad W_1 = 19.6 \text{ N}$$
$$m_2 = 10 \text{ kg} \qquad W_2 = m_2g = (10 \text{ kg})(9.8 \text{ m/s}^2) \qquad \text{or} \qquad W_2 = 98 \text{ N}$$

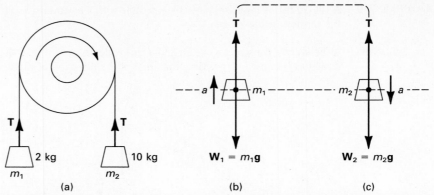

Fig. 7-7 Two masses suspended from a single pulley. Free-body diagrams are drawn; the positive direction of acceleration is chosen to be upward on the left and downward on the right.

Now the problem is to determine the net unbalanced force on the entire system. Note that the pulley merely changes the direction of the forces. The unbalanced force is, therefore, just the difference in the weights. This is just what we would expect from experience. Notice the tension T is the same on either side, since there is only one rope. Thus, the tension cancels out and does not figure in the resultant force, which may be written as follows:

$$\sum F = W_2 - T + T - W_1$$

$$\sum F = W_2 - W_1$$

The total mass of the system is simply the sum of all the masses in motion.

$$m_t = m_1 + m_2 = 2 \text{ kg} + 10 \text{ kg}$$
$$m_t = 12 \text{ kg} \qquad \text{(total mass)}$$

From Newton's second law of motion, we have

Resultant force = total mass × acceleration
$$W_2 - W_1 = (m_1 + m_2)a$$

Substituting for W_2, W_1, m_1, and m_2, we have

$$98 \text{ N} - 19.6 \text{ N} = (2 \text{ kg} + 10 \text{ kg})a$$

From which we may solve for a as follows:

$$78.4 \text{ N} = (12 \text{ kg})a$$
$$a = \frac{78.4 \text{ N}}{12 \text{ kg}} = 6.53 \text{ m/s}^2$$

Solution (b) In order to solve for the tension in the cord, we must consider either of the masses by itself, since considering the system as a whole would not involve cord tension. Suppose we consider the forces acting on m_1:

$$\textit{Resultant force = mass} \times \textit{acceleration}$$
$$T - W_1 = m_1 a$$

But $a = 6.53$ m/s^2 and the mass and weight are known, so that we have

$$T - 19.6 \text{ N} = (2 \text{ kg})(6.53 \text{ m/s}^2)$$
$$T - 19.6 \text{ N} = 13.06 \text{ N}$$
$$T = 32.7 \text{ N}$$

We would obtain the same value for the tension if we applied Newton's law to the second mass. You should demonstrate this fact as an additional exercise.

EXAMPLE 7-11   A 64-lb block rests on a frictionless table top. A rope attached to it passes over a light frictionless pulley and is attached to a weight W, as shown in Fig. 7-8a. (a) What must the value of W be to give the system an acceleration of 16 ft/s^2? (b) What is the tension in the rope?

Fig. 7-8

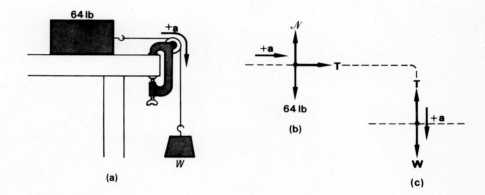

(a)

(b)

(c)

Solution (a)   Draw free-body diagrams for each body in the system, as shown in Fig. 7-8b and c. Since the vertical forces on the 64-lb block are balanced, the net force on the entire system is simply the weight W. Hence, applying Newton's law yields

$$\textit{Resultant force on system = total mass} \times \textit{acceleration}$$

$$W = \left(\frac{64 \text{ lb}}{g} + \frac{W}{g} \right) a$$

$$W = \frac{64 \text{ lb} + W}{g} a = (64 \text{ lb} + W)\frac{a}{g}$$

$$W = (64 \text{ lb} + W)\frac{16 \text{ ft/s}^2}{32 \text{ ft/s}^2}$$

$$W = \frac{64 \text{ lb} + W}{2}$$

$$2W = 64 \text{ lb} + W$$

$$2W - W = 64 \text{ lb}$$

$$W = 64 \text{ lb}$$

Solution (b)   To solve for the tension in the rope, we may choose either Fig. 7-8b or c since each diagram involves the unknown T. The better choice is the former because the net force on the 64-lb body is the tension T. Thus

$$Resultant\ force = mass \times acceleration$$

$$T = \frac{64\ lb}{32\ ft/s^2}(16\ ft/s^2)$$

$$= 32\ lb$$

One more example will be given in this section to allow you to become more familiar with the reasoning process involved with more complex systems. Since the foundation has been laid in previous examples, we include only the significant steps in the solution.

EXAMPLE 7-12   Consider the masses $m_1 = 20$ kg and $m_2 = 18$ kg in the system represented by Fig. 7-9. If the coefficient of kinetic friction is 0.1 and the inclination angle θ is 30°, find (a) the acceleration of the system and (b) the tension in the cord joining the two masses.

Fig. 7-9

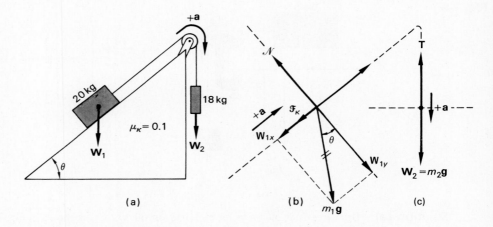

(a)   (b)   (c)

Solution (a)   Using the symbols as defined in Fig. 7-9, we apply Newton's law to the system:

$$Resultant\ force\ on\ system = total\ mass \times acceleration$$

$$W_2 - W_{1x} - \mathscr{F}_K = (m_1 + m_2)a$$

The symbols on the left side are found as follows:

$$W_2 = m_2g = (18\ kg)(9.8\ m/s^2) = 176\ N$$
$$W_{1x} = m_1g \sin\theta = (20\ kg)(9.8\ m/s^2)(\sin 30°) = 98\ N$$
$$W_{1y} = m_1g \cos\theta = (20\ kg)(9.8\ m/s^2)(\cos 30°) = 170\ N$$
$$\mathscr{F}_K = \mu_K N = \mu_K W_{1y} = (0.1)(170\ N) = 17\ N$$

Substitution into the equation of motion yields

$$176\ N - 98\ N - 17\ N = (20\ kg + 18\ kg)a$$

from which we obtain

$$a = 1.61\ m/s^2$$

Solution (b)   To find the tension in the cord, we apply Newton's law to the 18-kg mass, as shown in

MECHANICS

Fig. 7-9c:

$$\text{Resultant force} = \text{mass} \times \text{acceleration}$$
$$m_2 g - T = m_2 a$$
$$T = m_2 g - m_2 a = m_2(g - a)$$
$$= (18 \text{ kg})(9.8 \text{ m/s}^2 - 1.61 \text{ m/s}^2)$$
$$= 147 \text{ N}$$

SUMMARY

In this chapter, we have considered the fact that a resultant force will always produce an acceleration in the direction of the force. The magnitude of the acceleration is directly proportional to the force and inversely proportional to the mass, according to Newton's second law of motion. The following concepts are essential to applications of this fundamental law:

- The mathematical formula which expresses Newton's second law of motion may be written as follows:

$$\text{Force} = \text{mass} \times \text{acceleration}$$
$$F = ma \qquad m = \frac{F}{a} \qquad a = \frac{F}{m}$$

In SI units: $1 \text{ N} = (1 \text{ kg})(1 \text{ m/s}^2)$
In USCS units: $1 \text{ lb} = (1 \text{ slug})(1 \text{ ft/s}^2)$

- Weight is the force due to a particular acceleration g. Thus, weight W is related to mass m by Newton's second law:

$$W = mg \qquad m = \frac{F}{g} \qquad g = 9.8 \text{ m/s}^2 \text{ or } 32 \text{ ft/s}^2$$

For example, a mass of 1 kg has a weight of 9.8 N. A weight of 1 lb has a mass of $\frac{1}{32}$ slug. In a given problem you must determine whether weight or mass is given. Then you must determine what is needed in an equation. Conversions of mass to weight and weight to mass are common.

- Application of Newton's second law:
  1. Construct a free-body diagram for each body undergoing an acceleration. Indicate on this diagram the direction of positive acceleration.
  2. Determine an expression for the net resultant force on a body or a system of bodies.
  3. Set the resultant force equal to the total mass of the system multiplied by the acceleration of the system.
  4. Solve the resulting equation for the unknown quantity.

QUESTIONS

7-1. Define the following terms:
   a. Newton's second law
   b. Mass
   c. Weight
   d. Slug
   e. Newton

7-2. In a laboratory experiment the acceleration of a small car is measured by the separation of spots burned at regular intervals in a paraffin-coated tape. Larger and larger weights are transferred from the car to a hanger at the end of a tape which passes over a light, frictionless pulley. In this manner, the mass of the entire system is kept constant. Since the car moves on a horizontal air track with negligible friction, the resultant force is equal to the weights at the end of the tape. The following data are recorded:

Weight, W	2	4	6	8	10	12
Acceleration, m/s^2	1.4	2.9	4.1	5.6	7.1	8.4

Plot a graph of weight vs. acceleration. What is the significance of the slope of this curve? What is the mass of the system?

7-3. In the experiment described in Question 7-2, the student places a constant weight of 4 N at the free end of the tape. Several runs are made, increasing the mass of the car each time by adding weights. What happens to the acceleration as the mass of the system is increased? What should the value of the product of the mass of the system and the acceleration be for each run? Is it necessary to include the mass of the constant 4-N weight in these experiments?

7-4. Distinguish clearly between the mass of an object and its weight and give the appropriate units for each in the SI and USCS systems of units.

7-5. What exactly do we mean when we describe an athlete as a 160-lb person. What would be the mass of this person on the moon?

7-6. A round piece of brass found in the laboratory is labeled 500 g. Is this its weight or its mass? How can you be sure?

7-7. A state of equilibrium is maintained on a force table by hanging masses from pulleys mounted at various locations on the circular edge. In calculating the equilibrant, we sometimes use grams instead of newtons. Are we justified in doing this?

7-8. When drawing free-body diagrams, why is it usually to our advantage to choose either the x or y axis along the direction of motion even if it means rotated axes? Use the example of motion along an inclined plane as an illustration.

7-9. In the example of an Atwood machine (Example 7-10), we neglected the mass of the cord which connects the two masses. Discuss how this problem is altered if the mass of the cord is large enough to affect the motion.

7-10. In industry we often hear of a kilogram-force (kg$_f$) unit which is defined as a force equivalent to the weight of a 1-kg mass near the earth's surface. In the United States, we also speak often of the pound-mass (lb$_m$) unit, which is the mass of an object which has a weight of 1 lb near the surface of the earth. Find the value of these quantities in appropriate SI units, and discuss the problems caused by their use.

PROBLEMS

Newton's Second Law

7-1. A 4-kg mass is acted on by a resultant force of (a) 4 N, (b) 8 N, and (c) 12 N. What are the resulting accelerations?

Ans. (a) 1 m/s^2, (b) 2 m/s^2, (c) 3 m/s^2

7-2. A constant force of 20 N acts on a mass of (a) 2 kg, (b) 4 kg, and (c) 6 kg. What are the resulting accelerations?

7-3. A constant force of 60 lb acts on each of three objects, producing accelerations of 4, 8, and 12 ft/s^2. What are the masses?

Ans. 15, 7.5, and 5 slugs

7-4. What resultant force is necessary to give a 4-kg hammer an acceleration of 6 m/s^2?

The Relationship between Weight and Mass

7-5. Find the mass and the weight of a body if a resultant force of 16 N will give it an acceleration of 5 m/s^2.

Ans. 3.20 kg, 31.4 N

7-6. Find the mass and the weight of a body if a resultant force of 200 lb causes its speed to increase from 20 ft/s to 60 ft/s in 5 s.

7-7. The acceleration due to gravity on the surface of the moon is only 1.6 m/s^2. In an experiment, it is found that a resultant force of 40 N causes a ball to accelerate at 4 m/s^2. What are the mass and the weight of this ball **(a)** on moon's surface, and **(b)** on the earth's surface?

Ans. **(a)** 10 kg, 16 N; **(b)** 10 kg, 98 N

7-8. A resultant force of 200 lb produces an acceleration of 5 ft/s^2. What is the mass of the object being accelerated? What is its weight?

Applications for Single-Body Problems

7-9. A 2500-lb automobile is speeding at 55 mi/h. What retarding force is required to stop the car in 200 ft on a level road?

Ans. −1270 lb

7-10. What horizontal push is required to drag a 6-kg sled with an acceleration of 4 m/s^2? Assume that a horizontal friction force of 20 N opposes the motion.

7-11. A horizontal force of 100 N pulls an 8-kg block across a level floor. If the coefficient of kinetic friction between the block and the floor is 0.2, find the acceleration of the block.

Ans. 10.5 m/s^2

7-12. A 64-lb load hangs at the end of a rope. Find the acceleration of the load if the tension in the cable is **(a)** 64 lb, **(b)** 40 lb, and **(c)** 96 lb.

7-13. A 10-kg mass is lifted upward by a light cable. What is the tension in the cable if the acceleration is **(a)** zero, **(b)** 6 m/s^2 upward, and **(c)** 6 m/s^2 downward?

Ans. **(a)** 98 N, **(b)** 158 N, **(c)** 38 N

7-14. An 800-kg elevator is lifted vertically by a strong rope. Find the acceleration of the elevator if the tension in the rope is **(a)** 9000 N, **(b)** 7840 N, and **(c)** 2000 N.

7-15. In Fig. 7-10, an unknown mass slides down the 30° inclined plane against a constant friction force. If the coefficient of sliding friction is 0.2, what is the acceleration?

Ans. 3.20 m/s^2

Fig. 7-10

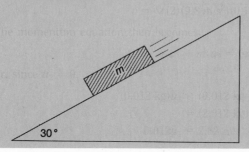

7-16. A 25-kg sled is dragged by a rope making an angle of 30° with the horizontal. When the tension in the rope is 100 N, the horizontal acceleration is 2 m/s². What is the coefficient of kinetic friction?

Applications for Multi-body Problems

7-17. A light cord passes over a light frictionless pulley, as in Fig. 7-7. Masses m_1 and m_2 are attached to each end of the cord. What will be the acceleration of the system and the tension in the cord if (a) $m_1 = 12$ kg and $m_2 = 10$ kg; (b) $m_1 = 20$ g and $m_2 = 50$ g?

Ans. (a) 0.891 m/s², 107 N; (b) 4.2 m/s², 0.28 N

7-18. Suppose the masses in Prob. 7-17 are replaced with weights $W_1 = 24$ lb and $W_2 = 16$ lb. What are the resulting acceleration and tension in the cord?

7-19. A 10-kg mass and a 5-kg mass are tied together with a horizontal rope A. The system is dragged horizontally by another rope B attached to the 10-kg mass. The coefficient of kinetic friction for all surfaces is 0.3. If the tension in rope B is 100 N, what is the acceleration of the system and what is the tension in cord A?

Ans. 3.73 m/s², 33.3 N

7-20. Consider the system shown in Fig. 7-11. Consider that block A has a mass of 16 kg and block B has a mass of 10 kg. Neglecting friction, what is the resultant force on the system? What is the total mass of the system? What are the acceleration of the system and the tension in the cord?

Fig. 7-11

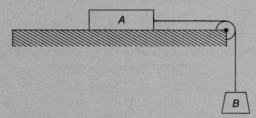

7-21. Assume that $\mu_k = 0.3$ and that block A weighs 20 lb and block B weighs 16 lb in Fig. 7-11. What are the acceleration of the system and the tension in the cord?

Ans. 8.89 ft/s², 11.6 lb

7-22. Consider that the masses A and B in Fig. 7-11 are equal. In the absence of friction, what would be the acceleration?

7-23. The three-mass system is connected as shown in Fig. 7-12. If $m_1 = 10$ kg, $m_2 = 8$ kg, and $m_3 = 6$ kg, what is the acceleration of the system neglecting friction?

Ans. 1.63 m/s²

7-24. Assume that the coefficient of friction between mass m_2 and the table in Fig. 7-12 is 0.3. In order to obtain the same acceleration as in Prob. 7-23, what new mass must be attached as m_1?

Fig. 7-12

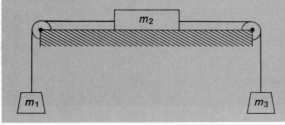

Additional Problems

7-25. A 2000-lb elevator is lifted vertically with an acceleration of 8 ft/s². Find the minimum breaking strength of a single cable which lifts the elevator. A 200-lb man stands on a scale which records his weight. What will the scales read as the elevator is lifted at 8 ft/s²?

Ans. 2500 lb, 250 lb

7-26. A 9-kg load is accelerated upward with a cord whose breaking strength is 200 N. What is the maximum upward acceleration such that the cord does not break?

7-27. The coefficient of friction between the tire and the road is 0.7. What is the minimum horizontal stopping distance for a 1600-kg car traveling at 60 km/h?

Ans. 20.2 m

* **7-28.** A 400-lb sled slides down a hill inclined at an angle of 60°. The coefficient of kinetic friction is 0.2. **(a)** What is the normal force on the sled? **(b)** What is the force of kinetic friction? **(c)** What is the resultant force down the hill? **(d)** What is the acceleration? **(e)** Was it necessary to know the weight of the sled in order to determine its acceleration?

** **7-29.** A block of unknown mass is given a push up a 40° inclined plane and then released. It continues to move *up* the plane at an acceleration of −9 m/s². What is the coefficient of kinetic friction?

Ans. 0.360

* **7-30.** Block *A* has a weight of 64 lb. What must be the weight of block *B* in Fig. 7-13 in order to cause block *A* to move up the plane with an acceleration of 6 ft/s²? Neglect friction.

* **7-31.** The mass of block *B* in Fig. 7-13 is 4 kg. What must be the mass of block *A* if it is to move down the plane at an acceleration of 2 m/s²? Neglect friction.

Ans. 7.28 kg

Fig. 7-13

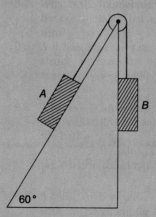

* **7-32.** Assume that the mass of block *A* is 6 kg and that the mass of block *B* is 10 kg. What are the acceleration and tension in the cord for Fig. 7-13? Neglect friction.

* **7-33.** A block of unknown mass is pulled horizontally across a surface where $\mu_k = 0.3$. If a horizontal pull of 40 N causes an acceleration of 6 m/s², what must be the mass of the block?

Ans. 4.47 kg

** **7-34.** Assume that the weight of block *A* in Fig. 7-13 is 64 lb and that $\mu_k = 0.4$. What must be the weight of block *B* in order that block *A* will move up the plane with an acceleration of 6 ft/s²?

8 Work, Energy, and Power

OBJECTIVES

After completing this chapter, you should be able to:

1. Define and write mathematical formulas for work, potential energy, kinetic energy, and power.
2. Apply the concepts of work, energy, and power to the solution of problems similar to those given as examples in the text.
3. Define and demonstrate by example your understanding of the following units: *joule, foot-pound, watt, horsepower,* and *foot-pound per second.*
4. Discuss and apply your knowledge of the relationship between the performance of work and the corresponding change in kinetic energy.
5. Discuss and apply your knowledge of the principle of conservation of mechanical energy.
6. Determine the power of a system and understand its relationship to time, force, distance, and velocity.

The principal reason for the application of a resultant force is to cause a displacement. For example, a large crane lifts a steel beam to the top of a building; the compressor in an air conditioner forces a fluid through its cooling cycle; and electromagnetic forces move electrons across a television screen. Whenever a force acts through a distance, we will learn, *work* is done in a way that can be measured or predicted. The capacity for doing work will be defined as *energy,* and the rate at which it is accomplished will be defined as *power.* The control and use of energy is probably the major concern of industry today, and a thorough understanding of the three concepts of work, energy, and power is essential.

8-1
WORK

When we attempt to drag a block with a rope, as in Fig. 8-1a, nothing happens. We are exerting a force, but the block has not moved. On the other hand, if we continually increase our pull, eventually the block will be displaced. In this case we have

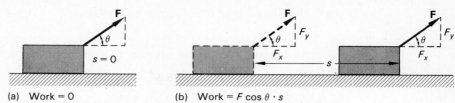

(a) Work = 0   (b) Work = $F \cos \theta \cdot s$

Fig. 8-1 The work done by a force F undergoing a displacement s.

actually accomplished something in return for our efforts. This accomplishment is defined in physics as *work*. The term *work* has an explicit, quantitative, operational definition. In order for work to be done, three things are necessary:

1. There must be an applied force.
2. The force must act through a certain distance, called the *displacement*.
3. The force must have a component along the displacement.

Assuming that we are given these conditions, a formal definition of work may be stated:

> **Work** is a scalar quantity equal to the product of the magnitudes of the displacement and the component of the force in the direction of the displacement.

$$Work = Force\ component \times displacement$$

$$\boxed{Work = F_x s} \tag{8-1}$$

In this equation F_x is the component of **F** along the displacement *s*. In Fig. 8-1, only F_x contributes to work. Its magnitude can be found from trigonometry, and work can be expressed in terms of the angle θ between **F** and *s*:

$$Work = (F \cos \theta)s \tag{8-2}$$

Quite often the force causing the work is directed entirely along the displacement. This happens when a weight is lifted vertically or when a horizontal force drags an object along the floor. In these simple cases, $F_x = F$, and the work is the simple product of force and displacement:

$$Work = Fs \tag{8-3}$$

Another special case occurs when the applied force is perpendicular to the displacement. In this instance, the work will be zero, since $F_x = 0$. An example is motion parallel to the earth's surface in which gravity acts vertically downward and is perpendicular to all horizontal displacements. Then the force of gravity does no work.

EXAMPLE 8-1   What work is done by a 60-N force in dragging the block of Fig. 8-1 a distance of 50 m when the force is transmitted by a rope making an angle of 30° with the horizontal?

Solution   We must first determine the component F_x of the 60-N force **F**. Only this component contributes to work. Graphically, this is accomplished by drawing the 60-N vector to scale at a 30°

angle. Measuring F_x and converting to newtons gives

$$F_x = 52.0 \text{ N}$$

With trigonometry, we could accomplish the same calculation using the cosine function

$$F_x = (60 \text{ N})(\cos 30°) = 52.0 \text{ N}$$

Now, applying Eq. (8-1), we obtain the work

$$\text{Work} = F_x \cdot s = (52.0 \text{ N})(50 \text{ m})$$
$$= 2600 \text{ N} \cdot \text{m}$$

Note that the units of work are the units of force times distance. Thus, in SI units, work is measured in *newton-meters* (N · m). By agreement, this combination unit is renamed the *joule,* which is denoted by the symbol J.

> One **joule** (1 J) is equal to the work done by a force of one newton in moving an object through a parallel distance of one meter.

In Example 8-1, the work done in dragging the block would be written as 2600 J.

In the United States, work is sometimes also given in USCS units. When the force is given in *pounds* (lb) and the displacement is given in *feet* (ft), the corresponding work unit is called the *foot-pound* (ft·lb).

> One **foot-pound** (1 ft·lb) is equal to the work done by a force of one pound in moving an object through a parallel distance of one foot.

No special name is given to this unit.

The following conversion factors will be useful when comparing work units in the two systems:

$$1 \text{ J} = 0.7376 \text{ ft·lb} \qquad 1 \text{ ft·lb} = 1.356 \text{ J}$$

8-2
RESULTANT WORK

When we consider the work of several forces acting on the same object, it is often useful to distinguish between positive and negative work. In this text, we will follow the convention that the work of a particular force is positive if the force component is in the same direction as the displacement. Negative work is done by a force component which opposes the actual displacement. Hence, work done by a crane in lifting a load is positive, but the gravitational force exerted by the earth on the load is doing negative work. Similarly, when we stretch a spring, the work on the spring is positive; the work on the spring is negative when the spring contracts, pulling us back. Another important example of negative work is that performed by a frictional force which is opposite to the direction of displacement.

If several forces act on a body in motion, the *resultant work* (total work) is the algebraic sum of the works of the individual forces. This will also be equal to the work of the resultant force. It is seen that the accomplishment of net work requires the existence of a resultant force. These ideas are clarified in the following example:

EXAMPLE 8-2 A push of 80 N moves a 5-kg block up a 30° inclined plane, as shown in Fig. 8-2. The coefficient of kinetic friction is 0.25, and the length of the plane is 20 m. *(a)* Compute the work done by each force acting on the block. *(b)* Show that the net work done by these forces is the same as the work of the resultant force.

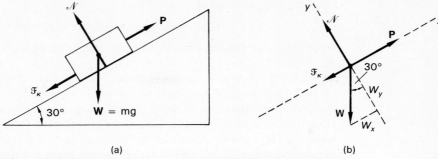

(a)   (b)

Fig. 8-2 The work required to push a block up a 30° inclined plane.

Solution (a)   There are four forces acting on the block: $\mathcal{N}$, **P**, $\mathcal{F}_k$, and **W**. (See Fig. 8-2*b*). The normal force $\mathcal{N}$ does no work because it has no component along the displacement.

$$(\text{Work})_{\mathcal{N}} = 0$$

The push **P** is entirely along the displacement and in the direction of the displacement. Hence

$$(\text{Work})_P = Ps = (80\text{ N})(20\text{ m}) = 1600\text{ J}$$

To find the work of the friction force $\mathcal{F}_k$ and the work of the weight **W** we must first determine the components of the weight along and perpendicular to the plane.

$$W = mg = (5\text{ kg})(9.8\text{ m/s}^2) = 49.0\text{ N}$$
$$W_x = (49.0\text{ N})\sin 30° = 24.5\text{ N}$$
$$W_y = (49.0\text{ N})\cos 30° = 42.4\text{ N}$$

Now the friction force $\mathcal{F}_k = \mu_k \mathcal{N}$ and $\mathcal{N} = W_y$, so that

$$\mathcal{F}_k = \mu_k \mathcal{N} = \mu_k W_y$$
$$= -(0.25)(42.4\text{ N}) = -10.6\text{ N}$$

The minus sign indicates that the friction force is down the plane. The work will therefore be negative since the displacement is up the plane.

$$(\text{Work})_{\mathcal{F}} = \mathcal{F}_k s = (-10.6\text{ N})(20\text{ m}) = -212\text{ J}$$

The weight W of the block also does negative work since its component W_x is directed opposite to the displacement.

$$(\text{Work})_W = -(24.5\text{ N})(20\text{ m}) = -490\text{ J}$$

Solution (b)   The net work is obtained by summing the works of the individual forces.

$$\text{Net work} = (\text{work})_N + (\text{work})_P + (\text{work})_{\mathcal{F}} + (\text{work})_W$$
$$= 0 + 1600\text{ J} - 212\text{ J} - 490\text{ J}$$
$$= 898\text{ J}$$

To show that this is also the work of the resultant force, we first compute the resultant force. According to methods introduced in earlier chapters,

$$F_R = P - \mathcal{F}_k - W_x$$
$$= 80\text{ N} - 10.6\text{ N} - 24.5\text{ N} = 44.9\text{ N}$$

The work of F_R is therefore

$$\text{Net work} = F_R s = (44.9 \text{ N})(20 \text{ m}) = 898 \text{ J}$$

which compares with the value obtained by computing the work of each force separately.

It is important to distinguish between the *resultant* or *net* work and the work of an individual force. If we speak of the work required to move a block through a distance, the work done by the pulling force is not necessarily the resultant work. Work may be done by a friction force or by other forces. Resultant work is simply the work done by a resultant force. If the resultant force is zero, then the resultant work is zero even though individual forces may be doing positive or negative work.

<table>
<tr><td>

8-3

ENERGY

</td></tr>
</table>

Energy may be thought of as *anything which can be converted into work*. When we say that an object has energy, we mean that it is capable of exerting a force on another object in order to do work on it. Conversely, if we do work on some object, we have added to it an amount of energy equal to the work done. The units of energy are the same as those for work; the *joule* and the *foot-pound*.

In mechanics we shall be concerned with two kinds of energy:

Kinetic energy E_k: Energy possessed by a body by virtue of its motion.

Potential energy E_p: Energy possessed by a system by virtue of position or condition.

One can readily think of many examples of each kind of energy. For instance, a moving car, a moving bullet, and a rotating flywheel all have the ability to do work because of their motion. Similarly, a lifted object, a compressed spring, and a cocked rifle have the potential for doing work because of position. Several examples are provided in Fig. 8-3.

<table>
<tr><td>

8-4

WORK AND KINETIC ENERGY

</td></tr>
</table>

We have defined kinetic energy as the capability for performing work as a result of the motion of a body. To see the relationship between motion and work, let's consider a constant force **F** acting on the block in Fig. 8-4. Consider that the block has an initial speed v_0 and that the force **F** acts through a distance s causing the speed to increase to a final value v_f. If the body has a mass m, Newton's second law tells us that it will gain speed, or accelerate, at a rate given by

$$a = \frac{F}{m} \qquad (8\text{-}4)$$

until it reaches the final speed v_f. From Chap. 5, we recall

$$2as = v_f^2 - v_0^2$$

from which

$$a = \frac{v_f^2 - v_0^2}{2s}$$

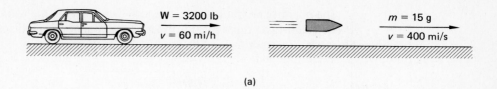

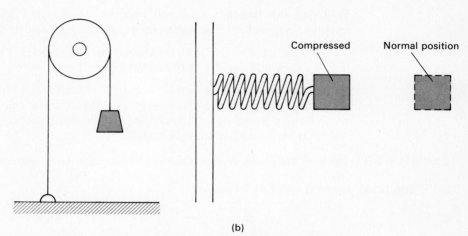

Fig. 8-3 Examples of *(a)* kinetic energy and *(b)* potential energy.

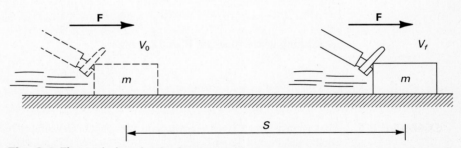

Fig. 8-4 The work done by the force F results in a change in the kinetic energy of the mass *m*.

Substitution of this into Eq. (8-4) yields

$$\frac{F}{m} = \frac{v_f^2 - v_0^2}{2s}$$

which can be solved for the product *Fs* to obtain

$$Fs = \tfrac{1}{2}mv_f^2 - \tfrac{1}{2}mv_0^2 \qquad (8\text{-}5)$$

The quantity on the left side of Eq. (8-5) represents the work done on the mass m. The quantity on the right side must be the change in kinetic energy as a result of this work. Thus, we can define kinetic energy E_K as

$$\boxed{E_K = \tfrac{1}{2}mv^2}\tag{8-6}$$

Following this notation, $\tfrac{1}{2}mv_f^2$ and $\tfrac{1}{2}mv_0^2$ would represent final and initial kinetic energies, respectively. The important result can be stated as follows:

The work of a resultant external force on a body is equal to the change in kinetic energy of the body.

A close look at Eq. (8-5) will show that an *increase* in kinetic energy ($v_f > v_0$) will result from *positive work, whereas a decrease in kinetic energy* ($v_f < v_0$) will result from *negative* work. In the special case in which zero work is done, the kinetic energy is a constant, given by Eq. (8-6).

EXAMPLE 8-3   Compute the kinetic energy of a 4-kg sledgehammer at the instant its velocity is 24 m/s.

Solution   Applying Eq. (8-6), we obtain

$$E_k = \tfrac{1}{2}mv^2 = \tfrac{1}{2}(4\text{ kg})(24\text{ m/s})^2$$
$$= 1152\text{ N}\cdot\text{m} = 1152\text{ J}$$

EXAMPLE 8-4   Compute the kinetic energy of a 3200-lb automobile traveling at 60 mi/h (88 ft/s).

Solution   We compute as before except that we must determine the mass from the weight.

$$E_K = \frac{1}{2}mv^2 = \frac{1}{2}\left(\frac{W}{g}\right)v^2$$

Substituting given values for W and v, we have

$$E_K = \frac{1}{2}\left(\frac{3200\text{ lb}}{32\text{ ft/s}^2}\right)(88\text{ ft/s})^2$$
$$= 3.87\times10^5\text{ ft·lb}$$

EXAMPLE 8-5   What average force F is necessary to stop a 16-g bullet traveling at 260 m/s as it penetrates into wood a distance of 12 cm?

Solution   The total work required to stop the bullet will be equal to the change in kinetic energy. (See Fig. 8-5.) Since the bullet is stopped, $v_f = 0$, and so Eq. (8-5) yields

$$Fs = -\tfrac{1}{2}mv_0^2$$

Substituting gives

$$F(0.12\text{ m}) = -\tfrac{1}{2}(0.016\text{ kg})(260\text{ m/s})^2$$

Dividing by 0.12 m, we have

$$F = \frac{-(0.016\text{ kg})(260\text{ m/s})^2}{(2)(0.12\text{ m})}$$
$$= -4510\text{ N}$$

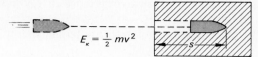

Fig. 8-5 The work done in stopping the bullet is equal to the change in kinetic energy of the bullet.

The minus sign indicates that the force was opposite to the displacement. It should be noted that this force is about 30,000 times the weight of the bullet.

8-5
POTENTIAL ENERGY

The energy that systems possess by virtue of their positions or conditions is called *potential energy.* Since energy expresses itself in the form of work, potential energy implies that there must be a potential for doing work. For example, suppose the pile driver in Fig. 8-6 is used to lift a body of weight W to a height h above the ground stake. We say that the body–earth system has gravitational potential energy. When such a body is released, it will do work when it strikes the stake. If it is heavy enough and if it has fallen from a great enough height, the work done will result in driving the stake through a distance s.

The external force **F** required to lift the body must at least be equal to the weight **W**. Thus, the work done on the system is given by

$$\text{Work} = Wh = mg \cdot h$$

This amount of work can also be done *by* the body after it has dropped a distance h. Thus, the body has potential energy equal in magnitude to the external work required to lift it. This energy does not come from the earth–body system, but results from work done on the system by an external agent. Only *external* forces, such as **F**

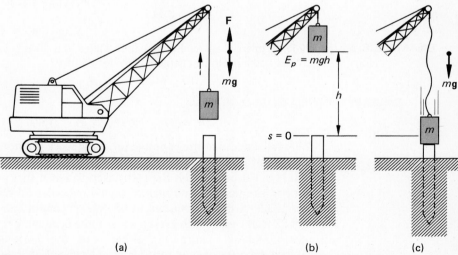

(a)   (b)   (c)

Fig. 8-6 *(a)* Lifting a mass m to a height h requires the work mgh. *(b)* The body-earth system, therefore, has a potential energy $E_p = mgh$. *(c)* When the mass is released, it has the capacity for doing the work mgh on the stake.

in Fig. 8-6 or friction, can add energy to or remove energy from the system made up of the body and the earth.

Note from the preceding discussion that potential energy E_p can be found from

$$\boxed{E_p = Wh = mgh}$$ *Potential Energy* (8-7)

where W and *m* are the weight and the mass of an object located a distance *h* above some reference point.

The potential energy depends on the choice of a particular reference level. The gravitational potential energy for an airplane is quite different when measured with respect to a mountain peak, a skyscraper, or sea level. The capacity for doing work is much greater if the aircraft falls to sea level. Potential energy has physical significance only in the event that a reference level is established.

EXAMPLE 8-6 A 250-g carburetor is held 200 mm above a workbench which is 1 m above the floor. Compute the potential energy relative to *(a)* the bench top and *(b)* the floor.

Solution (a) The height *h* of the carburetor above the bench is 200 mm or (0.2 m), and the mass is 250 g or (0.25 kg). Thus, the potential energy relative to the bench is

$$E_p = mgh = (0.25 \text{ kg})(9.8 \text{ m/s}^2)(0.2 \text{ m})$$
$$= 0.49 \text{ J}$$

Notice that kilograms, meters, and seconds are the only units of mass, length, and time which are consistent with the definition of a joule.

Solution (b) The potential energy with reference to the floor is based on a different value of *h*.

$$E_p = mgh = (0.25 \text{ kg})(9.8 \text{ m/s}^2)(1.2 \text{ m})$$
$$= 2.94 \text{ J}$$

EXAMPLE 8-7 An 800-lb commercial air-conditioning unit is lifted by a chain hoist until it is 22 ft above the floor. What is the potential energy relative to the floor?

Solution Applying Eq. (8-7), we obtain

$$E_p = Wh = (800 \text{ lb})(22 \text{ ft}) = 17,600 \text{ ft·lb}$$

We have stated that the potential for doing work is a function only of the weight and the height *h* above some reference point. The potential energy at a particular position is not dependent on the path taken to reach that position. This is because the same work must be done against gravity regardless of the path. In Example 8-7, work of 17,600 ft·lb was required to lift the air conditioner through a vertical distance of 22 ft. If we chose to exert a lesser force by moving it up an incline, a greater distance would be required. In either case, the work done against gravity is 17,600 ft·lb, because the end result is the placement of an 800-lb weight at a height of 22 ft.

CONSERVATION OF ENERGY

Quite often, at relatively low speeds, an interchange takes place between kinetic and potential energies. For example, suppose a mass m is lifted to a height h and dropped, as shown in Fig. 8-7. An external force has increased the energy of the system, giving it a potential energy $E_p = mgh$ at the highest point. This is the total energy available to the system, and it cannot change unless an external resistive force

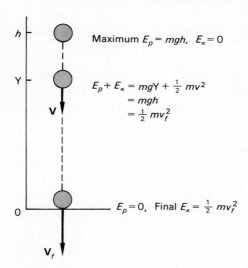

Fig. 8-7 In the absence of friction, the total mechanical energy is constant. At any point, it is equal to either the potential energy at the top or the kinetic energy at the bottom.

Maximum $E_p = mgh$, $E_\kappa = 0$

$E_p + E_\kappa = mg\text{Y} + \frac{1}{2}mv^2$
$= mgh$
$= \frac{1}{2}mv_f^2$

$E_p = 0$, Final $E_\kappa = \frac{1}{2}mv_f^2$

is encountered. As the mass falls, its potential energy decreases because its height above the ground is reduced. The lost potential energy reappears in the form of kinetic energy of motion. In the absence of air resistance, the total energy $(E_p + E_k)$ remains the same. Potential energy continues to be converted into kinetic energy until the mass reaches the ground ($h = 0$). At this final position, the kinetic energy is equal to the total energy, and the potential energy is zero. The important point to be made is that the sum of E_p and E_k is the same at any point during the fall (see Fig. 8-7).

$$\text{Total energy} = E_p + E_k = \text{constant}$$

We say that mechanical energy is *conserved.* In our example the total energy at the top is mgh and the total energy at the bottom is $\frac{1}{2}mv^2$ if we neglect air resistance. We are now prepared to state the principle of *conservation of mechanical energy:*

> *Conservation of Mechanical Energy: In the absence of air resistance or other dissipative forces, the sum of the potential and kinetic energies is a constant provided that no energy is added to the system.*

In applying this principle it is useful to think of beginning and ending points for any process. At either point if the velocity is not zero, there is kinetic energy, and if the height is not zero, there is potential energy. We can write

$$(E_p + E_k)_{\text{BEG}} = (E_p + E_k)_{\text{END}}$$
$$mgh_o + \tfrac{1}{2}mv_o^2 = mgh_f + \tfrac{1}{2}mv_f^2 \qquad (8\text{-}8)$$

The subscripts o and f refer to initial and final values. Equation (8-8), of course, applies when no friction forces are involved.

In the example of an object falling from rest at an initial position h_o, the total initial energy is $mgh_o (v_o = 0)$ and the total final energy is $\frac{1}{2}mv_f^2$ ($h = 0$).

$$mgh_o = \tfrac{1}{2}mv_f^2$$

Solving this relationship for v_f gives a useful equation for determining the final velocity from energy considerations for an object falling from rest with no friction.

$$v_f = \sqrt{2gh_o}$$

A great advantage of this method is that the final velocity is determined from the initial and final energy states. The actual path taken does not matter in the absence of friction. For example, the same final velocity would result if the object followed a curved path from the same initial height h_o.

EXAMPLE 8-8   In Fig. 8-8, a 40-kg ball is pulled to one side until it is 1.6 m above its lowest point. Neglecting friction, what will be its velocity as it passes through its lowest point?

Fig. 8-8 The velocity of a suspended mass as it passes through its lowest point can be found from energy considerations.

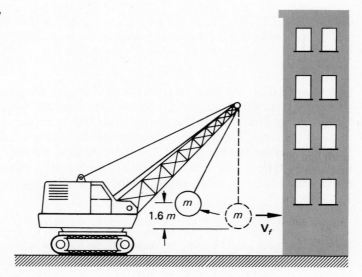

Solution   Conservation of total energy requires that $(E_p + E_k)$ be the same at the beginning and at the end. Thus,

$$mgh_o + 0 = 0 + \tfrac{1}{2}mv_f^2$$

from which we can divide out the mass m and obtain

$$v_f = \sqrt{2gh_o} = \sqrt{2(9.8 \text{ m/s}^2)(1.6 \text{ m})}$$
$$= 5.60 \text{ m/s}$$

As an additional example, you should show that the total energy at the beginning and at the end is 627 J.

ENERGY AND FRICTION FORCES

It is helpful to consider the conservation of mechanical energy as an accounting process in which one keeps track of what happens to the energy of a system from beginning to end. For example, suppose you withdraw $1000 from a bank, then pay $400 for an airlines ticket to New York. You would have $600 left to spend on entertainment. The $400 was lost, but it still must be accounted for. Now consider a sled at the top of a hill. The initial total energy is 1000 J. If 400 J of energy was lost due to friction, the sled would arrive at the bottom with a total energy of only 600 J. By accounting for friction or other dissipative forces, we can give a more general statement of the conservation of energy principle:

> *Conservation of energy: The total energy of a system is always constant although energy changes from one form to another may occur within the system.*

In real world applications, it is not possible to remove external forces from consideration. Therefore, a more general statement of the principle of conservation of energy accounts for losses due to friction:

$$(E_p + E_k)_{\text{BEG}} = (E_p + E_k)_{\text{END}} + |\text{energy losses}| \qquad (8\text{-}9)$$

The absolute value signs attached to the energy-losses term are a reminder that we are not concerned with the sign of the work done against friction forces. We are simply accounting for the disposition of all of the initial energy. If we represent the work of a friction force by the product $\mathscr{F}_k s$, we can write

$$mgh_o + \tfrac{1}{2}mv_o^2 = mgh_f + \tfrac{1}{2}mv_f^2 + |\mathscr{F}_k s| \qquad (8\text{-}10)$$

Of course, if an object begins at rest a height h_o above its final position, this equation simplifies to

$$mgh_o = \tfrac{1}{2}mv_f^2 + |\mathscr{F}_k s| \qquad (8\text{-}11)$$

In working problems, it is useful to establish the sum of potential and kinetic energies at some beginning point. The absolute value of any energy losses must then be added to the total energy of the system at the ending point in such a manner that energy is conserved.

EXAMPLE 8-9

A 20-kg sled rests at the top of a 30° slope 80 m in length as shown in Fig. 8-9. If $\mu_k = 0.2$, what is the velocity at the bottom of the incline?

Fig. 8-9 Some of the initial potential energy at the top of the incline is lost in doing work against friction as the block slides down.

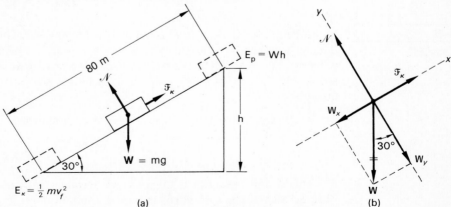

(a)　　　(b)

Solution The total energy at the beginning is potential energy since the initial velocity was zero. The height h_o is given by

$$h_o = (80 \text{ m}) \sin 30° = 40 \text{ m}$$

which allows us to compute the initial total energy

$$
\begin{aligned}
(E_p + E_k)_{\text{BEG}} &= mgh_o + 0 \\
&= (20 \text{ kg})(9.8 \text{ m/s}^2)(40 \text{ m}) \\
&= 7840 \text{ J}
\end{aligned}
$$

Thus we have 7840 J that must be accounted for as the sled moves to the bottom. In order to determine how much of this is lost in work against friction, we must first find the normal force $\mathscr{N}$ exerted by the plane on the block. From Fig. 8-9b,

$$\mathscr{N} = W_y = (20 \text{ kg})(9.8 \text{ m/s}^2) \cos 30° = 170 \text{ N}$$

from which the friction force is

$$\mathscr{F}_k = \mu_k \mathscr{N} = (0.2)(170 \text{ N}) = 34.0 \text{ N}$$

The absolute value of the work done by the friction force is

$$\mathscr{F}_k s = (34.0 \text{ N})(80 \text{ m}) = 2720 \text{ J}$$

We can now determine how much energy is left for velocity from Eq. (8-11)

$$
\begin{aligned}
mgh_o &= \tfrac{1}{2}mv_f^2 + |\mathscr{F}_k s| \\
7840 \text{ J} &= \tfrac{1}{2}mv_f^2 + 2720 \text{ J}
\end{aligned}
$$

and

$$\tfrac{1}{2}mv_f^2 = 7840 \text{ J} - 2720 \text{ J} = 5120 \text{ J}$$

Substituting $m = 20$ kg, we have

$$\tfrac{1}{2}(20 \text{ kg})v_f^2 = 5120 \text{ J}$$

Finally solving for v_f, we obtain

$$v_f = 22.6 \text{ m/s}$$

You should show that the final velocity would have been 28 m/s if there had been no friction forces present.

8-8
POWER

In our definition of work, *time* is not involved in any way. The same amount of work is done whether the task takes an hour or a year. Given enough time, even the weakest motor can lift an enormous load. However, if we wish to perform a task efficiently, the *rate* at which work is done becomes a very important engineering quantity.

Power is the rate at which work is accomplished.

$$P = \frac{\text{work}}{t} \tag{8-12}$$

The SI unit for power is the *joule per second,* which is renamed the *watt* (W). Thus an 80-W light bulb burns energy at the rate of 80 J/s.

$$1 \text{ W} = 1 \text{ J/s}$$

In USCS units, we use the *foot-pound per second* (ft·lb/s). No special name is given to this unit of power.

The watt and the foot-pound per second are inconveniently small units for most industrial purposes. Therefore, the *kilowatt* (kW) and the *horsepower* (hp) are defined:

$$1 \text{ kW} = 1000 \text{ W}$$
$$1 \text{ hp} = 550 \text{ ft·lb/s}$$

In the United States, the watt and kilowatt are used almost exclusively in connection with electric power; horsepower is reserved for mechanical power. This practice is purely a convention and by no means necessary. It is perfectly proper to speak of an 0.08-hp light bulb or to brag about a 238-kW engine. The conversion factors are

$$1 \text{ hp} = 746 \text{ W} = 0.746 \text{ kW}$$
$$1 \text{ kW} = 1.34 \text{ hp}$$

Since work is frequently done in a continuous fashion, an expression for power which involves velocity is useful. Thus

$$P = \frac{\text{work}}{t} = \frac{Fs}{t} \qquad (8\text{-}13)$$

from which

$$P = F\frac{s}{t} = Fv \qquad (8\text{-}14)$$

where v is the velocity of the body on which the parallel force F is applied.

EXAMPLE 8-10   A 40-kg load is raised to a height of 25 m. If the operation requires 1 min, find the power required. What is the power in units of horsepower?

Solution   The work done in lifting the load is

$$\text{Work} = Fs = mgh = (40 \text{ kg})(9.8 \text{ m/s}^2)(25 \text{ m})$$
$$= 9800 \text{ J}$$

The power is then

$$P = \frac{\text{work}}{t} = \frac{9800 \text{ J}}{60 \text{ s}} = 163 \text{ W}$$

Since 1 hp = 746 W, the horsepower developed is

$$P = (163 \text{ W})\frac{1 \text{ hp}}{746 \text{ W}} = 0.219 \text{ hp}$$

EXAMPLE 8-11   A 60-hp motor provides power for the elevator of a hotel. If the weight of the elevator is 2000 lb, how much time is required to lift the elevator 120 ft?

Solution The work required is given by

$$\text{Work} = Fs = (2000 \text{ lb})(120 \text{ ft})$$
$$= 2.4 \times 10^5 \text{ ft·lb}$$

Since 1 hp = 550 ft·lb/s, the power developed is

$$P = (60 \text{ hp})\frac{550 \text{ ft·lb/s}}{1 \text{ hp}} = 3.3 \times 10^4 \text{ ft·lb/s}$$

From Eq. (8-13)

$$P = \frac{Fs}{t}$$

so that

$$t = \frac{Fs}{P} = \frac{2.4 \times 10^5 \text{ ft·lb}}{3.3 \times 10^4 \text{ ft·lb/s}}$$
$$= 7.27 \text{ s}$$

Equation (8-12) can be solved for work: Work = *Pt*. Therefore, the *kilowatthour* (kW · h) unit used by electric companies in billing is a unit of *power* (kilowatt) times *time* (hour), or a unit of work. Quite reasonably, the bill is for the amount of work that has been done. However, the price per kilowatthour may also be determined by the peak power demand of the consumer.

SUMMARY

The concepts of work, energy, and power have been discussed in this chapter. The major points to remember are summarized below:

- The *work* done by a force *F* acting through a distance *s* is found from the following equations (refer to Fig. 8-1):

$$\text{Work} = F_x s \qquad \text{Work} = (F \cos \theta)s$$

SI units: *joule* (J)   USCS unit: *foot-pound* (ft·lb)

- *Kinetic energy* E_k is the capacity for doing work as a result of motion. It has the same units as work and is found from

$$E_k = \frac{1}{2}mv^2 \qquad E_k = \frac{1}{2}\left(\frac{W}{g}\right)v^2$$

- Gravitational *potential energy* is the energy which results from the position of an object relative to the earth. Potential energy E_p has the same units as work and is found from

$$E_p = Wh \qquad E_p = mgh$$

where *W* or *mg* is the weight of the object and *h* is the height above some reference position.

- Net work is equal to the change in kinetic energy.

$$Fs = \tfrac{1}{2}mv_f^2 - \tfrac{1}{2}mv_0^2$$

- *Conservation of mechanical energy with no friction:*

$$(E_p + E_k)_{\text{BEG}} = (E_p + E_k)_{\text{END}}$$
$$mgh_o + \tfrac{1}{2}mv_o^2 = mgh_f + \tfrac{1}{2}mv_f^2$$

- Conservation of energy including friction:

$$(E_p + E_k)_{\text{BEG}} = (E_p + E_k)_{\text{END}} + |\text{energy losses}|$$
$$mgh_o - \tfrac{1}{2}mv_o^2 = mgh_f + \tfrac{1}{2}mv_f^2 + |\mathscr{F}_k s|$$

- *Power is the rate at which work is done:*

$$P = \frac{\text{work}}{t} \qquad P = \frac{Fs}{t} \qquad P = Fv$$

SI unit: watt (W)   USCS unit: ft·lb/s

Other units 1 kW = 10^3 W   1 hp = 550 ft·lb/s

QUESTIONS

8-1. Define the following terms:
   a. Work
   b. Joule
   c. Potential energy
   d. Kinetic energy
   e. Conservation of energy
   f. Power
   g. Horsepower
   h. Watt
   i. Kilowatthour

8-2. Distinguish clearly between the physicist's concept of work and the general concept of work.

8-3. Two teams are engaged in a tug of war. Is work done? When?

8-4. Whenever net work is done on a body, will the body necessarily undergo acceleration? Discuss.

8-5. A diver stands on a board 10 ft above the water. What kind of energy results from this position? What happens to this energy as she dives into the water? Is work done? If so, what is doing the work, and on what is the work done?

8-6. Compare the potential energies for two bodies A and B if (a) A is twice as high as B but of the same mass; (b) B is twice as heavy as A but at the same height; and (c) A is twice as heavy as B, but B is twice as high as A.

8-7. Compare the kinetic energies of two bodies A and B if (a) A has twice the speed of B; (b) A has half the mass of B; and (c) A has twice the mass and half the speed of B.

8-8. In stacking 8-ft boards, you lift an entire board at its center and lay it on the pile. Your helper lifts one end, rests it on the pile, and then lifts the other end. Compare the work done.

8-9. In the light of what you have learned of work and energy, describe the most efficient procedure for ringing the bell with a sledgehammer at the fair. What precautions should you take?

8-10. A roller coaster at the fair boasts "a maximum height of 100 ft with a maximum speed of 60 mi/h." Do you believe the advertisement? Explain.

WORK, ENERGY, AND POWER

8-11. A man mows his yard for years using a 4-hp mower. He then buys a 6-hp mower. After using the new mower for a while he proclaims, "I've got more than twice the power that I had with the old mower." Is his statement correct? Why do you think he is convinced of the large increase in power?

8-12. Write an algebraic expression for the conservation of energy under the following two circumstances. (a) A block of mass m slides from rest at the top of an incline whose slope distance is s. It encounters a friction force $\mathcal{F}_k$ on the way down and arrives at the bottom with a velocity v_f. (b) The same block is now given an initial speed v_o at the bottom of the slope such that it just comes to rest at the top. (c) Discuss the differences in the two equations.

PROBLEMS

Work

8-1. A tugboat exerts a constant force of 4000 N on a ship which it moves for a distance of 15 m through a harbor. What work is done by the tugboat?

Ans. 60,000 J

8-2. An external work of 400 ft·lb is applied in lifting a 30-lb motor at constant speed. If all this work contributes to the displacement, how high will it be lifted?

8-3. A 12-lb hammer has a mass of about 5.44 kg. If the hammer is lifted to a height of 3 m, what minimum work was required in joules and in foot-pounds?

Ans. 160 J, 118 ft·lb

8-4. A trunk is pulled 24 m across the floor by a rope making an angle θ with the horizontal, as shown in Fig. 8-10. The tension in the rope is 8 N. Compare the work done for angles of 0, 30, and 60°.

Fig. 8-10

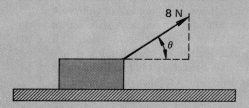

8-5. A force of 30 lb is applied along the handle of a lawn mower causing the mower to move a distance of 40 ft along the ground. If the handle makes an angle of 30° with the ground, what work was done by the 30-lb force?

Ans. 1040 ft·lb

Resultant Work

8-6. A horizontal force of 20 N drags a small sled across the ground at constant speed. The speed is constant because the force of friction exactly balances the 20-N pull. If a distance of 42 m is covered, what is the work done by the pulling force? What is the work of the friction force? What is the total or net work accomplished?

8-7. An average force of 40 N shortens a coiled spring by 6 cm. (a) What is the work done by the 40-N force? (b) What is the work done by the reaction force of the spring?

Ans. (a) 2.4 J, (b) −2.4 J

* 8-8. A 10-kg block is pushed 8 m along a horizontal surface by a constant force of 26 N. If $\mu_k = 0.2$, what is the resultant work? What acceleration will the block receive?

* 8-9. A rope drags a 10-kg block for a distance of 20 m across a floor against a constant friction force of 30 N. The rope makes an angle of 35° with the floor and has a tension of 60 N. (a) What work is done by the 60-N force? (b) What work is done by the friction force? (c) What resultant work is done? (d) What is the coefficient of friction?

          Ans. (a) 983 J, (b) −600 J, (c) 383 J, (d) 0.472

* 8-10. A 40-kg sled is dragged over a horizontal surface a distance of 500 m. The coefficient of friction between the sled and the snow is 0.2 and the dragging force is parallel to the ground. (a) What is the work done by the dragging force? (b) What work is done by the friction force?

* 8-11. A 12-kg crate is pushed along a 32° inclined plane until it reaches the top a distance of 16 m above its lowest point. At the same time an identical 12-kg crate is lifted vertically to the same height. In the absence of friction, show that the work of the external force is the same for each case. Would the same work be required if friction forces were considered?

          Ans. 1880 J, no

Work and Kinetic Energy

8-12. What is the kinetic energy of (a) a 5-g bullet moving with a velocity of 200 m/s, (b) a 64-lb projectile when its speed is 40 ft/s, (c) a 6-kg hammer moving at 4 m/s?

8-13. What is the change in kinetic energy when a 2400-lb car increases its speed from 30 mi/h to 60 mi/h? What resultant work was required? For purposes of comparison, what is the equivalent work in joules?

          Ans. 218,000 ft·lb, 218,000 ft·lb, 295,000 J

8-14. A 0.6-kg hammer head is moving with a speed of 30 m/s at the instant it strikes the head of a spike. What was the kinetic energy of the hammer head just before it struck the spike? What work can the hammer head do on the spike?

* 8-15. What average force is required to cause the speed of a 2-kg object to increase from 5 m/s to 12 m/s over a distance of 8 m? Verify your answer by first finding the acceleration and then applying Newton's second law.

          Ans. 14.9 N

* 8-16. A 12-lb hammer is moving at 80 ft/s as it strikes a nail. If the nail penetrates into the wall a distance of 1/4 in., what was the average stopping force?

* 8-17. A 1500-kg car is moving along a level road at a speed of 60 km/h. What work must be done by the brakes to bring the car to a stop? If $\mu_k = 0.7$, what was the stopping distance?

          Ans. 208,000 J, 20.2 m

Gravitational Potential Energy

8-18. A 2-kg book rests on a table top 80 cm from the floor. Find the potential energy of the book relative to (a) the floor, (b) the seat of a chair 40 cm from the floor, and (c) the ceiling 3 m from the floor.

* 8-19. A 96-lb safe is pushed up a 30° incline for a parallel distance of 12 ft. What is the increase in potential energy? Would the change in potential energy be the same if a 10-lb friction force acted for the entire distance?

          Ans. 576 ft·lb, Yes

8-20. At a particular instant a mortar shell has a velocity of 60 m/s. If its potential energy is one-half of its kinetic energy at that instant, what is its height above the earth?

8-21. A 20-kg sled is pushed up a 34° slope to a vertical height of 140 m. A constant friction force of 50 N acts for the entire trip. What is the increase in potential energy? What external work was required?

Ans. 27,400 J; 40,000 J

Conservation of Energy (No Friction)

8-22. In Fig. 8-6, a 64-lb weight is lifted to a height of 10 ft and then released to fall freely. What is the sum of potential and kinetic energies at the highest point? When the weight reaches a point 3 ft from the ground, what is its kinetic energy? What will be its velocity at the bottom of the path?

8-23. What initial velocity must be given to a 5-kg mass if it is to rise to a height of 10 m? What is the total energy at any point during its motion?

Ans. 14.0 m/s, 490 J

8-24. A simple pendulum 1 m long has an 8-kg bob. (a) How much work is required to move the pendulum from its vertical position to a horizontal position? (b) What is the total energy as the bob returns to its lowest point? (c) What is the velocity of the bob at the bottom of its path?

8-25. A ballistic pendulum is a laboratory device (Fig. 8-11) which might be used to calculate the velocity of a projectile. A 40-g ball is caught by a 500-g suspended mass. After impact, the two masses move together until they stop at a point 45 mm above the point of impact. From energy considerations, what was the velocity of the combined masses just after impact? (Neglect friction.)

Ans. 0.939 m/s

Fig. 8-11

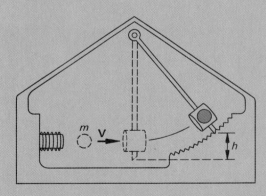

8-26. A 100-lb sled slides from rest at the top of a 37° incline whose height is 80 ft. In the absence of friction what is the velocity when the sled reaches the bottom?

8-27. The block in Fig. 8-12 has a mass of 8 kg and an initial velocity of 7 m/s at point A. What will be its velocity when it reaches point B? What will be its velocity at point C? Neglect friction forces.

Ans. 21.0 m/s, 16.9 m/s

Fig. 8-12

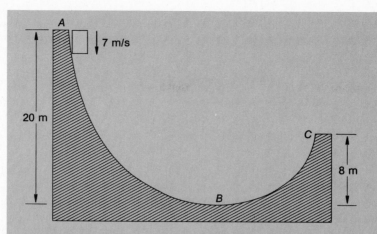

8-28. An 80-lb girl sits in a swing of negligible weight. If she is given an initial velocity of 20 ft/s, to what height will she rise?

Energy and Friction Forces

* 8-29. A 500-g block is released from the top of a 30° incline and slides 160 cm down the incline to the bottom. A constant friction force of 0.9 N acts for the entire distance. (a) What is the total energy at the top? (b) How much work is done against friction? (c) What is the final velocity of the block?

Ans. (a) 3.92 J, (b) 1.44 J, (c) 3.15 m/s

* 8-30. What initial velocity must be given to the 500-g block in Prob. 8-29 if it is to just reach the top of the same slope?

* 8-31. A 64-lb cart starts up a 37° incline with a velocity of 60 ft/s. If it comes to a stop after a distance of 70 ft, how much energy was lost due to friction forces?

Ans. 904 ft-lb

* 8-32. A 0.4-kg ball is dropped a vertical distance of 40 m. If it rebounds to a height of 16 m, how much energy was lost in collision with the floor?

** 8-33. A 200-lb sled is given an initial speed of 10 ft/s down a 34° slope. The coefficient of kinetic friction is 0.2. How far must the sled travel down the slope until its speed reaches 30 ft/s?

Ans. 31.8 m

Power

8-34. A power-station conveyor belt lifts 500 tons of ore per hour to a height of 90 ft. What average horsepower is required?

8-35. A 40-kg mass is lifted through a distance of 20 m in a time of 3 s. What average power is employed?

Ans. 2.61 kW

8-36. A 300-kg elevator is raised with a constant velocity through a vertical distance of 100 m in 2 min. What is the useful output power of the hoist?

8-37. At what constant speed can a 40-hp hoist lift a 2-ton load if all of the output power is utilized?

Ans. 5.50 ft/s

8-38. A 0.5-kW motor is 90 percent efficient and it drives a pulley 8 cm in diameter. If the shaft rotates at 1800 revolutions per minute (rpm), how large a force is the belt running on the pulley able to deliver?

Additional Problems

8-39. An unknown mass is tied to the end of a cord that is 4 m long. The cord and mass are then moved until they are in a horizontal line and released from rest. What will be the velocity of the mass when it passes through its lowest point?

> **Ans.** 8.85 m/s

8-40. A 10-kg mass is lifted to a height of 20 m. What are the potential energy, the kinetic energy, and the total energy at that height. The mass is released and falls freely. What are the total energy, potential energy, and kinetic energy when the mass is 5 m above the ground? What is its velocity at that point? (Neglect air resistance.)

8-41. The hammer of a pile driver weighs 800 lb and falls a distance of 16 ft before striking the pile. The impact drives the pile 6 in. deeper into the ground. Based on work-energy considerations, what was the average force driving the pile?

> **Ans.** 25,600 lb

8-42. A 20 N force drags an 8-kg block by a rope which makes an angle of 37° with a horizontal surface ($\mu_k = 0.2$). The block starts from rest and is dragged a distance of 40 m in 1 min. **(a)** What resultant work is done? **(b)** What resultant power is expended? **(c)** What is the final velocity?

8-43. A 2-kg ball is suspended from a 3-m cable attached to a spike in the wall. As the ball hangs vertically it just makes contact with the wall. The ball is pulled out so that the cable makes an angle of 70° with the wall and then released. If 10 J of energy are lost during the collision with the wall, what is the maximum angle between the cable and the wall after the first rebound?

> **Ans.** 59.2°

8-44. A crate is lifted at a constant speed of 5 m/s by an engine whose output power is 40 kW. What is the mass of the crate?

8-45. A 5-lb hammer is moving horizontally at 25 ft/s when it strikes a nail. If the nail meets an average resistive force of 1200 lb, compute the penetration depth.

> **Ans.** 0.488 in

8-46. A 3-kg ball dropped from a height of 12 m has a velocity of 10 m/s just before it hits the ground. What is the average retarding force due to the air resistance? If the ball rebounds from the surface with a speed of 8 m/s, what energy was lost on impact? How high will it rebound if the average air resistance is the same as before?

Impulse and Momentum

OBJECTIVES

After completing this chapter, you should be able to:

1. Define and give examples of *impulse* and *momentum* as vector quantities.
2. Write and apply a relationship between *impulse* and the resulting *change in momentum*.
3. State the law of *conservation of momentum* and apply it to the solution of physical problems.
4. Define and be able to calculate the *coefficient of restitution* for two surfaces.
5. Distinguish by example and definition between elastic and inelastic collisions.
6. Predict the velocities of two colliding bodies after impact when the coefficient of restitution, masses, and initial speeds are given.

Energy and work are scalar quantities which say absolutely nothing about direction. The law of conservation of energy describes only the relationship between initial and final states; it says nothing about how the energies are distributed.

For example, when two objects collide, we can say that the total energy before collision must equal the energy after collision if we neglect friction and other heat losses. But we need a new concept if we are to determine how the total energy is divided between the objects or even their relative directions after impact.

The concepts of impulse and momentum presented in this chapter add a vector description to our discussion of energy and motion.

9-1
IMPULSE AND MOMENTUM

When a golf ball is driven from the ground, as in Fig. 9-1, a large average force **F** acts on the ball during a very short interval of time Δt, causing the ball to accelerate from rest to a final velocity v_f. It is extremely difficult to measure either the force or its

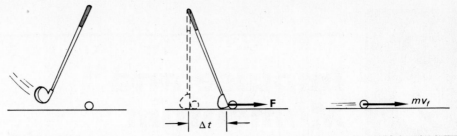

Fig. 9-1 When a golf club strikes the ball, a force F acting through the time interval Δt results in a change in its momentum.

duration, but their product F Δt can be determined from the resulting change in velocity of the golf ball. From Newton's second law we have

$$F = m a = m\frac{v_f - v_0}{\Delta t}$$

Multiplying by Δt gives

$$F \Delta t = m(v_f - v_0)$$

or

$$\boxed{F \Delta t = mv_f - mv_0} \tag{9-1}$$

This equation is so useful in solving problems involving impact that the terms are given special names.

> The **impulse** F Δt is a vector quantity equal in magnitude to the product of the force and the time interval in which it acts. Its direction is the same as that of the force.

> The **momentum p** of a particle is a vector quantity equal in magnitude to the product of its mass m and its velocity v.

$$\boxed{p = mv}$$

Thus Eq. (9-1) can be stated verbally:

Impulse (FΔt = *change in momentum* ($mv_f - mv_0$)

The SI unit of impulse is the *newton-second* (N · s). The unit of momentum is the *kilogram-meter per second* (kg · m/s). It is convenient to distinguish these units even though they are actually the same:

$$N \cdot s = \frac{kg \cdot m}{s^2} \times s = kg \cdot m/s$$

The corresponding units in the USCS are the *pound-second* (lb · s) and *slug-foot per second* (slug · ft/s).

EXAMPLE 9-1   A 3-kg sledgehammer is moving at a speed of 14 m/s as it strikes a steel spike. It is brought to a stop in 0.02 s. Determine the average force on the spike.

Solution   Since $v_f = 0$, we have from Eq. (9-1),

$$F\Delta t = -mv_0$$

If we consider that the hammer is moving downward, we substitute $v_0 = -14$ m/s, giving

$$F = \frac{-mv_0}{\Delta t} = \frac{-(3 \text{ kg})(-14 \text{ m/s})}{0.02 \text{ s}}$$

$$= 2100 \text{ N}$$

This force, exerted on the hammer, is equal in magnitude but opposite in direction to the force exerted on the spike. It must be emphasized that the force found in this manner is an *average* force. At some instants the force will be much greater than 2100 N.

EXAMPLE 9-2   A 0.2-kg baseball moving toward the batter with a velocity of 30 m/s is struck with a bat which causes it to move in a reverse direction with a velocity of 50 m/s. (Refer to Fig. 9-2.) Find the impulse and the average force exerted on the ball if the bat is in contact with the ball for 0.008 s.

Fig. 9-2 Impact of a bat with a baseball.

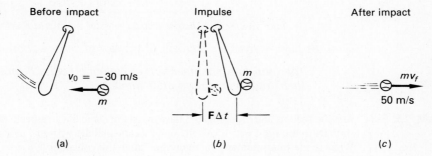

Before impact   Impulse   After impact

$v_0 = -30$ m/s

m

m

$F\Delta t$

mv_f

50 m/s

(a)   (b)   (c)

Solution   Consider the direction of the final velocity as positive. Applying Eq. (9-1), we solve for the impulse as follows:

$$F \Delta t = mv_f - mv_0 = m(v_f - v_0)$$
$$F \Delta t = (0.2 \text{ kg})[50 \text{ m/s} - (-30 \text{ m/s})]$$
$$= (0.2 \text{ kg})(80 \text{ m/s}) = 16 \text{ kg} \cdot \text{m/s}$$
$$\text{Impulse} = F \Delta t = 16 \text{ N} \cdot \text{s}$$

The average force is then found by substituting $t = 0.008$ s.

$$F(0.008 \text{ s}) = 16 \text{ N} \cdot \text{s}$$
$$F = 2000 \text{ N}$$

9-2

THE LAW OF CONSERVATION OF MOMENTUM

Let us consider the *head-on* collision of the masses m_1 and m_2, as shown in Fig. 9-3. We denote their velocities before impact as u_1 and u_2 and after impact as v_1 and v_2. The impulse of the force F_1 acting on the right mass is

$$F_1 \Delta t = m_1 v_1 - m_1 u_1$$

Fig. 9-3 Head-on collision of two masses.

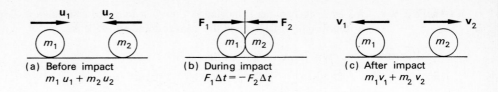

(a) Before impact
$m_1 u_1 + m_2 u_2$

(b) During impact
$F_1 \Delta t = -F_2 \Delta t$

(c) After impact
$m_1 v_1 + m_2 v_2$

Similarly, the impulse of the force F_2 on the left mass is

$$F_2 \Delta t = m_2 v_2 - m_2 u_2$$

During the time Δt, $F_1 = -F_2$, so that

$$F_1 \Delta t = -F_2 \Delta t$$

or

$$m_1 v_1 - m_1 u_1 = -(m_2 v_2 - m_2 u_2)$$

and finally, after rearranging,

$$\boxed{m_1 u_1 + m_2 u_2 = m_1 v_1 + m_2 v_2} \tag{9-2}$$

Total momentum before impact = total momentum after impact

Thus we have derived a statement of the law of *conservation of momentum:*

> *The total linear momentum of colliding bodies before impact is equal to their total momentum after impact.*

EXAMPLE 9-3   Assume that m_1 and m_2 of Fig. 9-3 have masses of 8 and 6 kg, respectively. The mass m_1 moves initially to the right with a velocity of 4 m/s and collides with m_2, moving to the left at 5 m/s. What is the total momentum before and after the collision?

Solution   We choose the direction to the right as positive, taking care to affix the proper sign to the velocities.

$$p_0(\text{before impact}) = m_1 u_1 + m_2 u_2$$
$$p_0 = (8 \text{ kg})(4 \text{ m/s}) + (6 \text{ kg})(5 \text{ m/s})$$
$$= 32 \text{ kg} \cdot \text{m/s} - 30 \text{ kg} \cdot \text{m/s} = 2 \text{ kg} \cdot \text{m/s}$$

The same total momentum must exist after impact, and so we can write

$$p_f = m_1 v_1 + m_2 v_2 = 2 \text{ kg} \cdot \text{m/s}$$

If either v_1 or v_2 can be measured after the collision, the other can be determined from this relation.

EXAMPLE 9-4   A rifle weighs 8 lb and fires a bullet weighing 0.02 lb at a muzzle velocity of 2800 ft/s. Compute the recoil velocity if the rifle is freely suspended.

Solution   Since both the rifle m_1 and the bullet m_2 are initially at rest, the total momentum before firing must equal zero. The momentum is unaltered, and so it must also be zero after firing. Hence Eq. (9-2) gives

$$0 = m_1v_1 + m_2v_2$$
$$m_1v_1 = -m_2v_2$$
$$v_1 = -\frac{m_2v_2}{m_1}$$
$$= -\frac{(0.02 \text{ lb}/32 \text{ ft/s}^2)(2800 \text{ ft/s})}{8 \text{ lb}/32 \text{ ft/s}^2}$$
$$= -7 \text{ ft/s}$$

An interesting experiment which demonstrates the conservation of momentum can be performed with eight small marbles and a grooved track, as shown in Fig. 9-4.

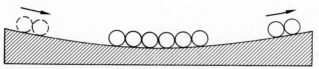

Fig. 9-4 Conservation of momentum.

If one marble is released from the left, it will be stopped upon colliding with the others, and one on the right end will roll out with the same velocity. Similarly, when two, three, four, or five marbles are released from the left, the same number will roll out to the right with the same velocity, the others remaining at rest in the center.

You might reasonably ask why two marbles roll off in Fig. 9-4 instead of one with twice the velocity, since this condition would also conserve momentum. For example, if each marble has a mass of 50 g, and if two marbles approach from the left at a velocity of 20 cm/s, the total momentum before impact is 2000 g · cm/s. This same momentum can be achieved after impact if only one marble left, assuming it had a velocity of 40 cm/s. The answer lies in the fact that energy must be conserved. If one marble came off with twice the velocity, its kinetic energy would be much greater than that available from the other two. The kinetic energy put into the system would be

$$E_0 = \tfrac{1}{2}mv^2 = \tfrac{1}{2}(0.1 \text{ kg})(0.2 \text{ m/s})^2$$
$$= 2 \times 10^{-3} \text{ J}$$

The kinetic energy of one marble traveling at 40 cm/s is exactly twice this value.

$$E_f = \tfrac{1}{2}mv^2 = \tfrac{1}{2}(0.05 \text{ kg})(0.4 \text{ m/s})^2$$
$$= 4 \times 10^{-3} \text{ J}$$

Therefore, it is seen that energy as well as momentum is important in describing impact phenomena.

9-3
ELASTIC AND INELASTIC IMPACTS

From the experiment in the previous section, the student might assume that the kinetic energy as well as the momentum is unchanged by a collision. Although this assumption is approximately true for hard bodies like marbles and billiard balls, it is not true for soft bodies, which rebound much more slowly upon impact. During impact, all bodies become slightly deformed, and small amounts of heat are liberated. The vigor with which a body restores itself to its original shape after deformation is a measure of its *elasticity,* or restitution.

If the kinetic energy remains constant in a collision (an ideal case), the collision is said to be *completely elastic.* In this instance no energy is lost through heat or deformation in collision. A hardened steel ball dropped on a marble slab would approximate a completely elastic collision.

If the colliding bodies stick together and move off as a unit afterward, the collision is said to be *completely inelastic.* A bullet which becomes embedded in a wooden block is an example of this type of impact. The majority of collisions fall between these two extremes.

In a completely elastic collision between two masses m_1 and m_2, we can say that both energy and momentum will be unaltered. Hence, we can apply two equations:

Energy:
$$\tfrac{1}{2}m_1u_1^2 + \tfrac{1}{2}m_2u_2^2 = \tfrac{1}{2}m_1v_1^2 + \tfrac{1}{2}m_2v_2^2$$

Momentum:
$$m_1u_1 + m_2u_2 = m_1v_1 + m_2v_2$$

which can be simplified to give

$$m_1(u_1^2 - v_1^2) = m_2(u_2^2 - v_2^2)$$
$$m_1(u_1 - v_1 = m_2(u_2 - v_2)$$

Dividing the first equation by the second gives

$$\frac{u_1^2 - v_1^2}{u_1 - v_1} = \frac{u_2^2 - v_2^2}{u_2 - v_2}$$

Factoring the numerators and dividing out, we obtain

$$u_1 + v_1 = u_2 + v_2$$

or

$$v_1 - v_2 = u_2 - u_1 = -(u_1 - u_2) \tag{9-3}$$

Thus, in the ideal case of a completely elastic collision, the relative velocity after collision, $v_1 - v_2$, is equal to the negative of the relative velocity before collision. The closer these quantities are to being equal, the more elastic the collision. The negative ratio of the relative velocity after collision to the relative velocity before collision provides a means of measuring the elasticity of a collision.

> The **coefficient of restitution** e is the negative ratio of the relative velocity after impact to the relative velocity before impact.

$$e = -\frac{v_1 - v_2}{u_1 - u_2}$$

Incorporating the minus sign into the numerator of this equation yields

$$\boxed{e = \frac{v_2 - v_1}{u_1 - u_2}} \tag{9-4}$$

If the collision is completely elastic, $e = 1$. If the collision is completely inelastic, $e = 0$. In the inelastic case the two bodies move off with the same velocity, so that $v_2 = v_1$. In general, the coefficient of restitution has some value between 0 and 1.

A simple method of determining the coefficient of restitution is shown in Fig. 9-5. A sphere of the material being measured is dropped onto a fixed plate from a

Fig. 9-5

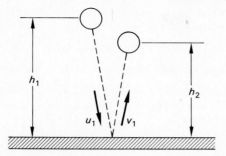

height h_1. Its rebound is measured to a height h_2. In this case, the mass of the plate is so large that v_2 is approximately 0. Therefore,

$$e = \frac{v_2 - v_1}{u_1 - u_2} = -\frac{v_1}{u_1}$$

The velocity u_1 is simply the velocity acquired in falling from a height h_1, which is found from

$$u_1^2 - u_0^2 = 2gh_1$$

But the initial velocity $u_0 = 0$, so that

$$u_1^2 = 2gh_1$$

or

$$u_1 = \sqrt{2gh_1}$$

We have considered the downward direction as positive. If the ball rebounds to a height h_2, its rebound velocity v_1 must be $-\sqrt{2gh_2}$. (The minus sign indicates the change in direction.) Hence, the coefficient of restitution is given by

$$e = -\frac{v_1}{u_1} = -\frac{-\sqrt{2gh_2}}{\sqrt{2gh_1}}$$

or

$$\boxed{e = \sqrt{\frac{h_2}{h_1}}} \tag{9-5}$$

The resulting coefficient is a joint property of the ball and the rebounding surface.

For a very elastic surface, e has a value of 0.95 or greater (steel or glass), whereas for less resilient substances e may be extremely small. It is interesting to note that the rebound height is a function of the vigor with which the impact deformation is restored. Contrary to popular conception, a steel ball or glass marble will bounce to a greater height than most rubber balls.

EXAMPLE 9-5   A 2-kg ball traveling to the left with a speed of 24 m/s collides head-on with a 4-kg ball traveling to the right at 16 m/s. (*a*) Find the resulting velocity if the two balls stick together on impact. (*b*) Find their final velocities if the coefficient of restitution is 0.80.

Solution (a)   In this instance $v_2 = v_1$ and $e = 0$. Let us denote their final velocity by v. The law of conservation of momentum tells us that

$$m_1 u_1 + m_2 u_2 = m_1 v_1 + m_2 v_2 = (m_1 + m_2)v$$

since $v_1 = v_2 = v$. Choosing the right direction as positive, we have on substitution

$$(2 \text{ kg})(-24 \text{ m/s}) + (4 \text{ kg})(16 \text{ m/s}) = (2 \text{ kg} + 4 \text{ kg})v$$
$$-48 \text{ kg} \cdot \text{m/s} + 64 \text{ kg} \cdot \text{m/s} = (6 \text{ kg})v$$
$$16 \text{ kg} \cdot \text{m/s} = (6 \text{ kg})v$$

from which

$$v = \tfrac{16}{6} \text{ m/s} = 2.67 \text{ m/s}$$

The fact that this velocity is positive indicates that both bodies move together to the right after collision.

Solution (b)   In this case e is not zero, and the balls rebound after collision with different velocities. Therefore, more information is needed than that derived from the momentum equation alone. We resort to the given value of $e = 0.80$ and Eq. (9-4) to give us more information.

$$e = 0.80 = \frac{v_2 - v_1}{u_1 - u_2}$$

or

$$v_2 - v_1 = (0.80)(u_1 - u_2)$$

Substituting the known values for u_1 and u_2, we obtain

$$v_2 - v_1 = (0.80)(-24 \text{ m/s} - 16 \text{ m/s})$$
$$= (0.80)(-40 \text{ m/s})$$

or finally

$$v_2 - v_1 = -32 \text{ m/s}$$

We can now use the momentum equation to arrive at another relation between v_2 and v_1, allowing us to solve the two equations simultaneously.

$$m_1 u_1 + m_2 u_2 = m_1 v_1 + m_2 v_2$$

The left side of this equation has already been found in part a to equal 16 kg · m/s. Therefore, we have on substitution for m_1 and m_2 on the right side

$$16 \text{ kg} \cdot \text{m/s} = (2 \text{ kg})v_1 + (4 \text{ kg})v_2$$

from which

$$2v_1 + 4v_2 = 16 \text{ m/s}$$

or

$$v_1 + 2v_2 = 8 \text{ m/s}$$

Hence we have two equations:

$$v_2 - v_1 = -32 \text{ m/s} \qquad v_1 + 2v_2 = 8 \text{ m/s}$$

Solving simultaneously, we obtain

$$v_1 = 24 \text{ m/s} \qquad v_2 = -8 \text{ m/s}$$

Thus, after colliding, the balls reverse their directions, m_1 moving to the right with a velocity of 24 m/s and m_2 moving to the left with a velocity of 8 m/s.

EXAMPLE 9-6   A 12-g bullet is fired into a 2-kg block of wood suspended from a cord, as shown in Fig. 9-6. The impact of the bullet causes the block to swing 10 cm above its original level. Compute the velocity of the bullet as it strikes the block.

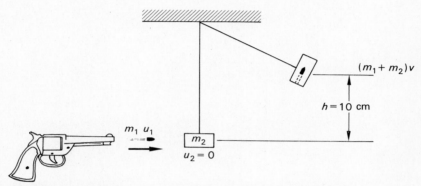

Fig. 9-6 Computing the muzzle velocity u_1 from energy and momentum considerations.

Solution   We can compute the combined velocity after impact from energy considerations. The kinetic energy of the block and bullet immediately after the collision is converted into potential energy as they rise to a height h. Hence, if v is the initial velocity of the block and bullet, we have

$$\tfrac{1}{2}(m_1 + m_2)v^2 = (m_1 + m_2)gh$$

Dividing by $m_1 + m_2$, we obtain

$$v^2 = 2gh$$

from which

$$v = \sqrt{2gh}$$

Therefore, the combined velocity after collision is

$$v = \sqrt{(2)(9.8 \text{ m/s}^2)(0.1 \text{ m})} = 1.4 \text{ m/s}$$

The momentum equation then becomes

$$m_1 u_1 + m_2 u_2 = (m_1 + m_2)v$$

or, since $u_2 = 0$,

$$(0.012 \text{ kg})u_1 = (0.012 \text{ kg} + 2 \text{ kg})(1.4 \text{ m/s})$$
$$= (2.012 \text{ kg})(1.4 \text{ m/s})$$
$$0.012u_1 = 2.82 \text{ m/s}$$

which gives an entrance velocity of

$$u_1 = 235 \text{ m/s}$$

SUMMARY

In this chapter you learned the relationship between impulse and momentum. Physical problems were then introduced to deal with elastic and inelastic collisions. The major concepts are summarized below:

- The *impulse* is the product of the average force F and the time interval Δt through which it acts.

  Impulse $= F \, \Delta t$ SI units: N · s   USCS units: lb · s

- The *momentum* of a particle is its mass times its velocity.

  Momentum $p = mv$ SI units: kg · m/s   USCS units: slug · ft/s

- The impulse is equal to the change in momentum:

$$F \, \Delta t = mv_f - mv_0$$

  Note: N · s = kg · m/s (equivalent units)

- *Conservation of momentum:* The total momentum before impact is equal to the total momentum after impact (see Fig. 9-3).

$$m_1u_1 + m_2u_2 = m_1v_1 + m_2v_2$$

- The *coefficient of restitution* is found from relative velocities before and after collision or from the rebound height:

$$e = \frac{v_2 - v_1}{u_1 - u_2} \qquad e = \sqrt{\frac{h_2}{h_1}}$$

- For a completely elastic collision, $e = 1$.
  For a completely inelastic collision, $e = 0$.

QUESTIONS

9-1. Define the following terms:
  a. Impulse   d. Elastic impact
  b. Momentum   e. Inelastic impact
  c. Conservation of momentum   f. Coefficient of restitution

9-2. Show the equivalence of the units of impulse with the units of momentum in USCS units.

9-3. Discuss the vector nature of impulse and momentum.

9-4. How does the magnitude of the impulse 1 lb · s compare with the magnitude of the impulse 1 N · s?

9-5. Discuss the conservation of energy and momentum for (a) an elastic collision and (b) an inelastic collision.

9-6. If you hold a weapon loosely when firing, it appears to give a greater kick than when you hold it tight against your shoulder. Explain. What effect does the weight of the weapon have?

9-7. A mortar shell explodes in midair. How is momentum conserved? How is energy conserved?

9-8. Suppose you hit a tennis ball into the air with a racket. The ball first strikes the concrete court and then bounds over the fence, landing in the grass. How many impulses were involved, and which impulse was the greatest?

9-9. A father and his daughter stand facing each other on a frozen pond. If the girl pushes her father backward, describe their relative motion and velocities. Would they differ if the father pushed the daughter?

9-10. Two small cars of masses m_1 and m_2 are tied together with a compressed spring between them. The cord is burned with a match, releasing the spring and imparting an equal impulse to each car. Compare the ratio of their displacements to the ratio of their masses at some later instant.

PROBLEMS

Impulse and Momentum

9-1. A 0.5-kg wrench is dropped from a height of 10 m. What is its momentum just before it strikes the floor?

Ans. 7 kg m/s, down

9-2. Compute the momentum and kinetic energy of a 2400-lb car moving north at 55 mi/h.

9-3. A 2500-kg truck traveling at 40 km/h strikes a brick wall and comes to a stop in 0.2 s. (a) What is the change in momentum? (b) What is the impulse? (c) What is the average force on the wall during the crash?

Ans. (a) -2.78×10^4 kg · m/s, (b) -2.78×10^4 N · s, (c) 1.39×10^5 N

* 9-4. What is the momentum of a 3-g bullet moving at 600 m/s in a direction 30° above the horizontal? What are the horizontal and vertical components of this momentum?

* 9-5. A 0.2-kg baseball reaches the batter with a speed of 20 m/s. After it has been struck, it leaves the bat at 35 m/s in a reverse direction. If the ball exerts an average force of 8400 N, how long was it in contact with the bat?

Ans. 1.31 ms

* 9-6. A bat exerts an average force of 248 lb on a 0.6-lb ball for a time of 0.01 s. If the ball reaches the bat with a velocity of 44 ft/s, what will be its velocity on leaving the bat?

* 9-7. An 8×10^6-lb train traveling at 60 mi/h brakes to a stop in a distance of 600 ft. (a) What is the stopping impulse? (b) What is the required braking force?

Ans. (a) 2.20×10^7 lb · s, (b) 1.61×10^6 lb

* 9-8. A 400-g rubber ball is dropped a vertical distance of 12 m. It remains in contact with the pavement for 0.01 s and rebounds to a height of 10 m. What is the total change in momentum? What average force is exerted by the floor on the ball?

Conservation of Momentum

9-9. A spring is tightly compressed between a 6-kg block and a 2-kg block which are then tied together with a string. The blocks are resting on frictionless surface. When the string breaks, the 2-kg block moves off with a speed of 9 m/s. What is the speed of the 6-kg block?

Ans. 3 m/s

9-10. Two children, weighing 80 and 50 lb, are at rest on roller skates. If the larger child pushes the other so that the smaller one moves away at a speed of 6 mi/h, what will be the velocity of the larger child?

9-11. A 60-g firecracker explodes sending a 45-g piece flying to the left and another piece flying to the right with a velocity of 40 m/s. What was the velocity of the first piece?

Ans. -13.3 m/s

9-12. A 24-g bullet is fired with a muzzle velocity of 900 m/s from a 5-kg rifle. Find the recoil velocity of the rifle. Find the ratio of the kinetic energy of the bullet to that of the rifle.

9-13. A 4-kg ball moving with a velocity of 8 m/s collides head-on with another ball of 2-kg mass which is initially at rest. After impact, the first mass is still moving in the same direction but with a velocity of only 4 m/s. **(a)** What is the velocity of the 2-kg mass after collision? **(b)** How much energy was lost during the impact?

Ans. (a) 8 m/s, (b) -32 J

Elastic and Inelastic Collisions

9-14. A empty truck weighing 3 tons rolls freely at 5 ft/s along a level road and collides with a loaded truck weighing 5 tons, standing at rest and free to move. If the two trucks couple together, find their velocity after collision. Compare the kinetic energy before collision with that after impact. How do you account for the loss in energy?

9-15. A 30-kg child stands on a frictionless ice surface. His father throws a 0.8-kg football with a velocity of 15 m/s. What velocity will the child have after catching the football?

Ans. 0.390 m/s

9-16. A 2-kg wad of clay is tied to the end of a string as shown in Fig. 9-7. A 0.5-kg steel ball moving horizontally at an unknown speed embeds itself into the clay causing both ball and clay to rise to a height of 20 cm. What was the entrance velocity of the steel ball?

Fig. 9-7 A ballistic pendulum.

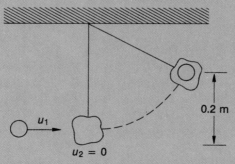

0.2 m

u_1

$u_2 = 0$

9-17. In Prob. 9-16, suppose the 0.5-kg ball passes entirely through the clay and emerges with a velocity of 10 m/s. What must be the new entrance velocity if the wad of clay is to reach the same height of 20 cm?

Ans. 17.9 m/s

9-18. A billiard ball moving to the left with a velocity of 30 cm/s collides head-on with another ball moving to the right at 20 cm/s. The masses of the balls are identical. If the collision is perfectly elastic, what is the velocity of each ball after impact?

9-19. The coefficient of restitution of steel is 0.90. If a steel ball is dropped from a height of 7 m, how high will it rebound?

Ans. 5.67 m

* **9-20.** What is the time between the first contact with the surface and the second contact for Prob. 9-19?

* **9-21.** A ball dropped from rest onto a fixed horizontal plate rebounds to a height which is 81 percent of its original height. **(a)** Find the coefficient of restitution. **(b)** Determine the vertical velocity of impact required to cause the ball to rebound to a height of 8 m.

Ans. **(a)** 0.9, **(b)** 13.9 m/s

* **9-22.** A 300-g block moving north with a velocity of 50 cm/s collides with a 200-g block moving south at 100 cm/s. **(a)** What are their final velocities if they stick together on impact? **(b)** What is the loss of kinetic energy in the impact? **(c)** What are the final velocities if the collision is completely elastic?

** **9-23.** A 5-lb ball and a 12-lb ball approach each other with equal speeds of 25 ft/s. **(a)** What will their combined speed be after impact if the collision is completely inelastic? **(b)** What will be their respective velocities after impact if the collision is perfectly elastic?

Ans. **(a)** 10.3 ft/s; **(b)** −45.6 ft/s, 4.41 ft/s

Additional Problems

** **9-24.** A 600-g ball and a 200-g ball are suspended by 2-m cords so that their edges are in contact when each cord is vertical. The 600-kg ball is then moved out so that its cord makes an angle of 30° with the vertical. When this ball is released, how high will the 200-g ball rise above its lowest position? Assume a perfectly elastic collision.

* **9-25.** A 10-kg block rests on a frictionless surface. A 20-g bullet moving at 200 m/s strikes the block and passes entirely through it, exiting with a velocity of 10 m/s. What is the velocity of the block after impact? How much kinetic energy was lost in the process?

Ans. 0.380 m/s, 399 J

* **9-26.** A 60-g body has an initial velocity of 100 cm/s to the right, and a 150-g body has an initial velocity of 30 cm/s to the left. If the coefficient of restitution is 0.80, find their respective speeds and directions after impact. What percentage of the initial kinetic energy is lost in the collision?

** **9-27.** Two wooden 2-kg balls rest on a frictionless track 5 m apart. If a third ball of the same mass strikes the first with a velocity of 30 m/s, how long will it take for the first ball to strike the second? Assume $e = 1$.

Ans. 0.167 s

* **9-28.** If the block in Fig. 9-6 weighs 36 lb, how high will it be raised above its initial level if a 1-lb projectile enters it with a velocity of 200 ft/s and remains in the block?

* **9-29.** An atomic particle of mass 20×10^{-28} kg moving with a velocity of 4×10^6 m/s collides head-on with a particle of mass 12×10^{-28} kg initially at rest. Assuming that the collision is completely elastic, find the velocity of the incident particle after the impact.

Ans. 1×10^6 m/s

* **9-30.** An astronaut in orbit outside his capsule uses a revolver to control his motion. The astronaut with all his gear weighs 200 lb on earth. If the revolver fires 0.05-lb bullets with a muzzle velocity of 2700 ft/s, and if the astronaut fires 10 shots, what will be his final velocity? Compute the kinetic energy of the astronaut and the kinetic energy of the ten bullets. Account for the difference between the two energies.

9-31. A 0.30-kg baseball moving horizontally at 40 m/s is struck by a bat. If the ball leaves the bat with a speed of 60 m/s at an angle of 30°, what are the horizontal and vertical components of the average force exerted by the bat? Assume that the bat was in contact with the ball for 0.005 s.

Ans. 5520 N, 1800 N

9-32. Two small toy cars of masses m_1 and m_2 are tied together with a compressed spring between them. The cord is burned with a match, releasing the spring and imparting an equal impulse to each car. Compare the ratio of their displacements to the ratio of their masses at some later instant.

10 Uniform Circular Motion

OBJECTIVES

After completing this chapter, you should be able to:

1. Demonstrate by definition and examples your understanding of the concepts of *centripetal acceleration* and *centripetal force.*
2. Apply your knowledge of centripetal force and centripetal acceleration to the solution of problems similar to those in this text.
3. Define and apply the concepts of *frequency* and *period* of rotation, and relate them to the linear speed of an object in uniform circular motion.
4. Apply your knowledge of centripetal force to problems involving *banking angles,* the *conical pendulum,* and motion in a *vertical circle.*
5. State and apply the universal law of gravitation.

In previous chapters we considered primarily motion in a straight line. This approach is sufficient to describe and apply most mechanical concepts. Unfortunately, bodies in the natural world generally move along curved paths. Artillery shells travel along parabolic paths under the influence of the earth's gravitational field. Planets revolve about the sun in paths which are nearly circular. On the atomic level electrons circle about the nucleus of an atom. In fact, it is difficult to imagine any phenomenon in physics which does not involve motion in at least two dimensions.

10-1 MOTION IN A CIRCULAR PATH

Newton's first law tells us that all bodies moving in a straight line with constant speed will maintain their velocity unaltered unless acted on by an external force. The velocity of a body is a vector quantity consisting of both its speed and its direction. Just as a resultant force is required to change its speed, a resultant force must be applied to change its direction. Whenever this force acts in a direction other than the original direction of motion, the path of a moving particle is changed.

The simplest kind of two-dimensional motion occurs when a constant external force always acts at right angles to the path of a moving particle. In this case the resultant force will produce an acceleration which alters only the direction of motion, leaving the speed constant. This simple type of motion is referred to as *uniform circular motion.*

> **Uniform circular motion** is motion in which there is no change in speed, only a change in direction.

An example of uniform circular motion is afforded by swinging a rock in a circular path with a string, as shown in Fig. 10-1. As the rock revolves with constant speed, the inward force of the tension in the string constantly changes the direction of the rock, causing it to move in a circular path. If the string should break, the rock would fly off at a tangent perpendicular to the radius of its circular path.

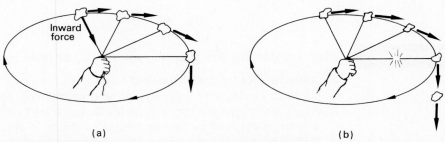

(a)   (b)

Fig. 10-1 *(a)* The inward pull of the string on the rock causes it to move in a circular path. *(b)* If the string breaks, the rock will fly off at a tangent to the circle.

10-2
CENTRIPETAL ACCELERATION

Newton's second law of motion states that a resultant force must produce an acceleration in the direction of the force. In uniform circular motion, the acceleration changes the velocity of a moving particle by altering its direction.

The position and velocity of a particle moving in a circular path of radius R are shown at two instants in Fig. 10-2. When the particle is at point A, its velocity is represented by the vector v_1. After a time interval Δt, its velocity is represented by

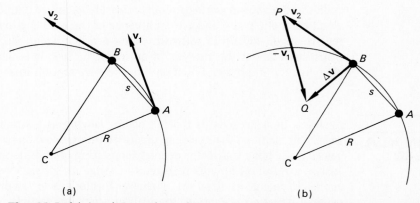

(a)   (b)

Fig. 10-2 *(a)* A and B are the positions at two instants separated by a time interval Δt. *(b)* The change in velocity Δv is represented graphically. The vector will point directly toward the center if Δt is made small enough for the chord s to equal the arc joining points A and B.

the vector v_2. The acceleration by definition is the change in velocity per unit time. Thus

$$\mathbf{a} = \frac{\Delta v}{\Delta t} = \frac{v_2 - v_1}{\Delta t} \tag{10-1}$$

The change in velocity Δv is represented graphically in Fig. 10-2b. The difference between the two vectors v_2 and v_1 is constructed according to the methods introduced in Chap. 2. Since the velocities v_2 and v_1 have the same magnitude, they form the legs of an isosceles triangle BPQ whose base is Δv. If we construct a similar triangle ABC, it can be noted that the magnitude of Δv has the same relationship to the magnitude of either velocity as the chord s has to the radius R. This proportionality is stated symbolically as

$$\frac{\Delta v}{v} = \frac{s}{R} \tag{10-2}$$

where v represents the absolute magnitude of either v_1 or v_2.

The distance the particle actually covers in traveling from point A to point B is not the distance s but the length of the arc from A to B. The shorter the time interval Δt, the closer these points are until in the limit the chord length becomes equal to the arc length. In this case, the length s is given by

$$s = v\,\Delta t$$

which when substituted into Eq. (10-2) yields

$$\frac{\Delta v}{v} = \frac{v\,\Delta t}{R}$$

Since the acceleration from Eq. (10-1) is $\Delta v / \Delta t$, we can rearrange terms and obtain

$$\frac{\Delta v}{\Delta t} = \frac{v^2}{R}$$

Therefore, the rate of change in velocity, or the *centripetal acceleration*, is given by

$$\boxed{a_c = \frac{v^2}{R}} \tag{10-3}$$

where v is the linear speed of a particle moving in a circular path of radius R.

The term *centripetal* means that the acceleration is always directed toward the center. Notice in Fig. 10-2b the vector Δv does not point toward the center. This is because we have considered a long time interval between the measurements at A and B. If we restrict the separation of these points to an infinitesimal distance, the vector Δv would point toward the center.

The units of centripetal acceleration are the same as those for linear acceleration. For example, in the SI, v^2/R would have the units

$$\frac{(m/s)^2}{m} = \frac{m^2/s^2}{m} = m/s^2$$

EXAMPLE 10-1 A 2-kg body is tied to the end of a cord and whirled in a horizontal circle of radius 1.5 m. If the body makes three complete revolutions every second, determine its linear speed and its centripetal acceleration.

Solution If the body makes 3 rev/s, the time to travel one complete circle (a distance $2\pi R$) is $\frac{1}{3}$ s. Hence, the linear speed is

$$v = \frac{2\pi R}{0.333 \text{ s}} = \frac{2\pi(1.5 \text{ m})}{0.333 \text{ s}}$$
$$= 28.3 \text{ m/s}$$

Therefore, the centripetal acceleration, from Eq. (10-3), is

$$a_c = \frac{v^2}{R} = \frac{(28.3 \text{ m/s})^2}{1.5 \text{ m}} = 534 \text{ m/s}^2$$

The procedure used to compute linear speed in the above example is so useful that it is worth remembering. If we define the *period* as the time for one complete revolution and designate it by the letter T, the linear speed can be computed by dividing the period into the circumference. Thus

$$v = \frac{2\pi R}{T} \qquad (10\text{-}4)$$

Another useful parameter in engineering problems is the rotational speed, expressed in *revolutions per minute* (rpm) or *revolutions per second* (rev/s). This quantity is called the *frequency* of rotation and is given by the reciprocal of the period.

$$f = \frac{1}{T} \qquad (10\text{-}5)$$

The validity of this relation is demonstrated by noting that the reciprocal of seconds per revolution (s/rev) is revolutions per second (rev/s). Substitution of this definition into Eq. (10-4) yields an alternative equation for determining the linear speed.

$$v = 2\pi fR \qquad (10\text{-}6)$$

For example, if the frequency is 1 rev/s and the radius 1 ft, the linear speed would be 2π ft/s.

10-3
CENTRIPETAL FORCE

The inward force necessary to maintain uniform circular motion is defined as the *centripetal force*. From Newton's second law of motion, the magnitude of this force must equal the product of mass and centripetal acceleration. Thus

$$F_c = ma_c = \frac{mv^2}{R} \qquad (10\text{-}7)$$

where m is the mass of an object moving with a speed v in a circular path of radius R. The units chosen for the quantities F_c, m, v, and R must be consistent for the system chosen. For example, the SI units for mv^2/R are

$$\frac{\text{kg} \cdot \text{m}^2/\text{s}^2}{\text{m}} = \text{kg} \cdot \text{m/s}^2 = \text{N}$$

An inspection of Eq. (10-7) reveals that the inward force F_c is directly proportional to the square of the velocity of the moving object. This means that increasing the linear speed to twice its original value will require four times the original force. Similar reasoning will show that doubling the mass or halving the radius will require twice the original centripetal force.

For problems in which the rotational speed is expressed in terms of the frequency, the centripetal force can be determined from

$$F_c = \frac{mv^2}{R} = 4\pi^2 f^2 mR \qquad (10\text{-}8)$$

This relation results from substitution of Eq. (10-6) which expresses the linear speed in terms of the frequency of revolution.

EXAMPLE 10-2   A 4-kg ball is swung in a horizontal circle by a cord 2 m long. What is the tension in the cord if the period is 0.5 s?

Solution   The tension in the cord will be equal to the centripetal force necessary to hold the 4-kg body in a circular path. The linear speed is obtained by dividing the period into the circumference.

$$v = \frac{2\pi R}{T} = \frac{2\pi(2 \text{ m})}{0.5 \text{ s}} = 25.1 \text{ m/s}$$

from which the centripetal force is

$$F_c = \frac{mv^2}{R} = \frac{(4 \text{ kg})(25.1 \text{ m/s})^2}{2 \text{ m}}$$
$$= 1260 \text{ N}$$

EXAMPLE 10-3   Two 4-lb weights rotate about the center axis at 12 rev/s, as shown in Fig. 10-3. (a) What is the resultant force acting on each weight? (b) What is the tension in the rod?

Solution (a)   The total downward force of the weights and rod is balanced by the upward force of the center support. Therefore, the resultant force acting on each revolving weight is a pull toward the center equal to the centripetal force. The mass of each weight is

$$m = \frac{W}{g} = \frac{4 \text{ lb}}{32 \text{ ft/s}^2} = 0.125 \text{ slug}$$

Substituting the given values of frequency, mass, and radius into Eq. (10-8), we obtain

$$F_c = 4\pi^2 f^2 mR = 4\pi^2 (12 \text{ rev/s})^2 (0.125) \text{ slug})(1.5 \text{ ft})$$
$$= 1066 \text{ lb}$$

The same calculations hold for the other weight.

Solution (b)   The resultant force just computed represents the centripetal force exerted by the rod on the weights. According to Newton's third law, there must be an equal and opposite reaction force

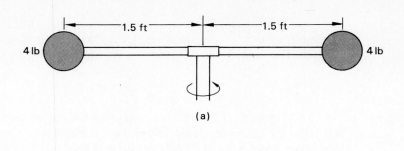

(a)

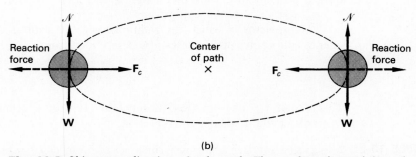

(b)

Fig. 10-3 Objects traveling in a circular path. The resultant force of the rod on the objects provides the necessary centripetal force. According to Newton's third law, the objects exert an equal and opposite reaction force called the *centrifugal force*. These forces do not cancel each other because they act on different objects.

exerted by the weight on the rod. Remember that although these forces are equal in magnitude and opposite in direction, they do not act on the same body. Because the outward force exerted on the rod is *fleeing the center,* it is sometimes referred to as the *centrifugal force*. It is this centrifugal force that causes the tension in the rod. Since it is equal in magnitude to the centripetal force, the tension in the rod must also be 1066 lb.

10-4
BANKING CURVES

When an automobile is driven around a sharp turn on a perfectly level road, friction between the tires and the road provides centripetal force (see Fig. 10-4). If this centripetal force is not adequate, the car may slide off the road. The maximum value

Fig. 10-4 The centripetal force of friction. Note that there is no outward force on the car.

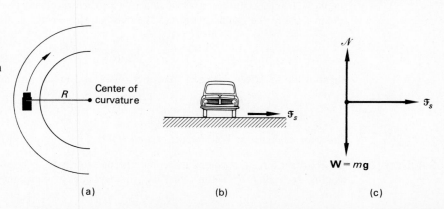

(a)   (b)   (c)

of the force of friction determines the maximum speed with which a car can negotiate a turn of a given radius.

EXAMPLE 10-4   What is the maximum speed at which an automobile can negotiate a curve of radius 100 m without sliding if the coefficient of static friction is 0.7?

Solution   As the car increases its speed, the force of static friction required to hold it gets larger. Finally, the car attains a speed so great that the centripetal force equals the maximum force of static friction. At that instant

$$\mathscr{F}_s = F_c = \frac{mv^2}{R}$$

and since $\mathscr{F}_s = \mu_s \mathscr{N}$, we can write

$$\frac{mv^2}{R} = \mu_s \mathscr{N}$$

Applying the first condition for equilibrium to the vertical forces in Fig. 10-4 reveals that the normal force is equal to the weight of the car

$$\mathscr{N} = W = mg$$

Hence

$$\frac{mv^2}{R} = \mu_s mg \qquad or \qquad v^2 = \mu_s gR$$

from which

$$\boxed{v = \sqrt{\mu_s gR}} \qquad\qquad (10\text{-}9)$$

Substituting known values for μ_s, g, and R, we can now compute the maximum speed.

$$v = \sqrt{(0.7)(9.8 \text{ m/s}^2)(100 \text{ m})} = 26.2 \text{ m/s}$$

or approximately 94.3 km/h (58.6 mi/h).

It is quite surprising that the weight of the automobile was not considered in determining the maximum speed. Our own experience tends to contradict the independence of weight. The answer to this paradox does not lie in the validity of the above equations but in the fact that the friction force acts on the wheels below the center of gravity of the car. There is also an uneven distribution of weight. Since the center of gravity of heavy cars is usually lower than for light cars, the difference in stability and distribution of weight will affect the turning speed. However, if we stipulate that the weight is uniformly distributed and neglect problems of vertical stability, the above equations are valid.

Now let us consider the effects of banking a turn to eliminate the friction force. As seen from Fig. 10-5, a road can be banked in such a manner that the normal force $\mathscr{N}$ has vertical and horizontal components:

$$\mathscr{N}_x = \mathscr{N} \sin \theta \qquad \mathscr{N}_y = \mathscr{N} \cos \theta$$

The horizontal component provides the necessary centripetal force. Therefore, if we represent the linear velocity by v and the radius of the turn by R, the banking angle θ

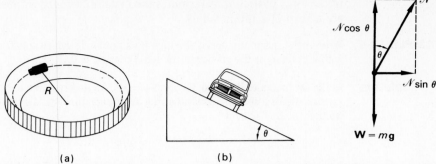

(a) (b)

Fig. 10-5 Effects of banking a turn. The horizontal component of the normal force, N $\sin\theta$, provides the necessary centripetal acceleration.

required to eliminate the need for friction is obtained from

$$\mathcal{N}\sin\theta = \frac{mv^2}{R}$$

Since the vertical forces are balanced,

$$\mathcal{N}\cos\theta = mg$$

Dividing the first equation by the second yields

$$\tan\theta = \frac{v^2}{Rg} \tag{10-10}$$

EXAMPLE 10-5 Find the required banking angle for a curve of radius 300 m if the curve is to be negotiated at a speed of 80 km/h without the need of a friction force.

Solution The linear speed is first converted to meters per second.

$$v = 80 \text{ km/h} = 22.2 \text{ m/s}$$

Substituting into Eq. (9-8) yields

$$\tan\theta = \frac{v^2}{Rg} = \frac{(22.2 \text{ m/s})^2}{(300 \text{ m})(9.8 \text{ m/s}^2)} = 0.168$$

from which the required banking angle is

$$\theta = 9.5°$$

10-5
THE CONICAL PENDULUM

A conical pendulum consists of a mass m revolving in a horizontal circle with constant speed v at the end of a cord of length L. A comparison of Fig. 10-6 with Fig. 10-5 shows that the formula derived for the banking angle also applies for the angle the cord makes with the vertical in a conical pendulum. In the latter problem, the necessary centripetal force is provided by the horizontal component of the tension in

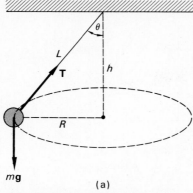
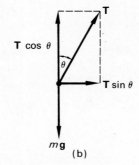

Fig. 10-6 The conical pendulum.

the cord. The vertical component is equal to the weight of the revolving mass. Hence

$$T \sin \theta = \frac{mv^2}{R} \qquad T \cos \theta = mg$$

from which

$$\tan \theta = \frac{v^2}{Rg} \tag{10-10}$$

is obtained as in the previous section.

A careful study of Eq. (10-10) will show that as the linear speed increases, the angle that the cord makes with the vertical also increases. Therefore, the vertical position of the mass (indicated in Fig. 10-6) is raised, causing a reduction in the distance h below the point of support. If we wish to express Eq. (10-10) in terms of the vertical position h, we should note that

$$\tan \theta = \frac{R}{h}$$

from which we obtain

$$\frac{R}{h} = \frac{v^2}{gR}$$

Thus, the distance of the weight below the support is a function of the linear speed and is given by

$$h = \frac{gR^2}{v^2} \tag{10-11}$$

This principle operates governors used on some engines (Fig. 10-7). The position of the weight can be used to open or close fuel valves. Note that all the forces do not act at a single point in this case.

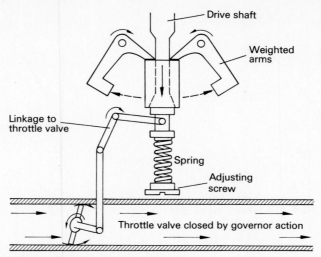

Fig. 10-7 The centrifugal governor.

A more useful form for Eq. (10-11) can be obtained by expressing the linear speed in terms of the rotational frequency. Since $v = 2\pi f R$, we can write

$$h = \frac{gR^2}{4\pi^2 f^2 R^2} = \frac{g}{4\pi^2 f^2}$$

Solving for f, we obtain

$$f = \frac{1}{2\pi}\sqrt{\frac{g}{h}} \tag{10-12}$$

This form has the advantage of eliminating the radius of revolution R from consideration.

10-6
MOTION IN A VERTICAL CIRCLE

Motion in a vertical circle is very different from the circular motion discussed in earlier sections. Since gravity always acts downward, the direction of the weight is the same at the top of the path as it is at the bottom. However, the forces which maintain the circular motion must always be directed toward the center of the path. When more than one force acts on an object, it is the *resultant* force that produces the centripetal force.

Consider a mass m tied to the end of a rope and swung in a circle of radius R as shown by Fig. 10-8. We assign v_1 as the velocity at the top of the circular path and v_2 as the velocity at the bottom. Let us first consider the resultant force on the object as it passes through its highest point. The weight mg and the tension T_1 in the rope are each directed downward. The resultant of these forces is the centripetal force. Hence

$$T_1 + mg = \frac{mv_1^2}{R} \tag{10-13}$$

Fig. 10-8

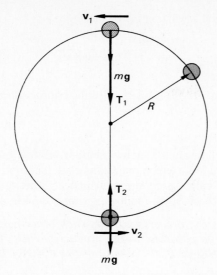

On the other hand, when the mass passes through its lowest point, the weight mg is still directed downward, but the tension T_2 is directed upward. The resultant is still the necessary centripetal force, so that we have

$$T_2 - mg = \frac{mv_2{}^2}{R} \tag{10-14}$$

It should be apparent from these equations that the tension in the cord at the bottom is greater than it is at the top. In one case, the weight adds to the tension, whereas in the other, the weight subtracts from the tension. The centripetal force (resultant) is a function of velocity, mass, and radius at any location.

EXAMPLE 10-6   In Fig. 10-6, assume that a 2-kg ball has a velocity of 5 m/s when it rounds the top of a circle whose radius is 80 cm. (*a*) What is the tension in the cord at that instant? (*b*) What is the minimum speed at the top necessary to maintain circular motion?

Solution (a)   Solving Eq. (10-13) for T_1, we obtain

$$T_1 = \frac{mv_1{}^2}{R} - mg$$

$$= \frac{(2 \text{ kg})(5 \text{ m/s})^2}{0.80 \text{ m}} - (2 \text{ kg})(9.8 \text{ m/s}^2)$$

$$= 42.9 \text{ N}$$

You should show that the centripetal force at this point is 62.5 N.

Solution (b)   The critical situation at the top of the path occurs when the tension drops to zero. At this instant, the centripetal force is provided only by the weight mg. Thus, Eq. (10-13) reduces to

$$mg = \frac{mv_1{}^2}{R}$$

Dividing out the mass m and solving for v_1, we obtain

$$v_1 = \sqrt{gR} \qquad (10\text{-}15)$$
$$= \sqrt{(9.8 \text{ m/s}^2)(0.8 \text{ m})}$$
$$= 2.8 \text{ m/s}$$

The velocity found from Eq. (10-15) is often called the *critical velocity*.

10-7
GRAVITATION

The earth and planets follow approximately circular orbits around the sun. Newton suggested that the inward force which maintains planetary motion is only one example of a universal force called *gravitation*, which acts between all masses in the universe. He stated his thesis in the following *universal law of gravitation*:

> *Each particle in the universe attracts each other particle with a force that is directly proportional to the product of their masses and inversely proportional to the square of the distance between them.*

The proportionality is usually stated in the form of an equation:

$$\boxed{F = G\,\frac{m_1 m_1}{r^2}} \qquad (10\text{-}15)$$

where m_1 and m_2 are the masses of any two particles separated by a distance r, as shown in Fig. 10-9.

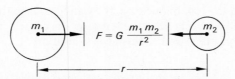

Fig. 10-9 The universal law of gravitation.

The proportionality constant G is a universal constant equal to

$$G = 6.67 \times 10^{-11} \text{ N} \cdot \text{m}^2/\text{kg}^2$$
$$= 3.44 \times 10^{-8} \text{ lb} \cdot \text{ft}^2/\text{slug}^2$$

EXAMPLE 10-7   A 4-kg ball and a 2-kg ball are positioned so that their centers are 40 cm apart. With what force do they attract each other?

Solution   The force of attraction is found from Eq. (10-15):

$$F = G\,\frac{m_1 m_2}{R^2} = \frac{(6.67 \times 10^{-11} \text{ N m}^2/\text{kg}^2)(4 \text{ kg})(2 \text{ kg})}{(0.40 \text{ m})^2}$$
$$F = 3.34 \times 10^{-9} \text{ N}$$

It is seen that the gravitational force is actually a very small force. Because of the relatively large mass of the earth in comparison with objects on its surface, we often assume that gravitational forces are strong. However, if we consider two marbles closely separated on a horizontal surface, our experience certainly substantiates that the gravitational attraction is very weak.

EXAMPLE 10-8   At the surface of the earth, the acceleration of gravity is 9.8 m/s². If the radius of the earth is 6.38×10^6 m, compute the mass of the earth.

Solution   Suppose we consider a mass m near the surface of the earth. The gravitational force which attracts this mass is equal to its weight given by

$$W = mg = G\,\frac{mm_e}{r^2}$$

Solving for m_e, we obtain

$$m_e = \frac{gr^2}{G}$$

from which the mass of the earth is found to be

$$m_e = \frac{(9.8 \text{ m/s}^2)(6.38 \times 10^6 \text{ m})^2}{6.67 \times 10^{-11} \text{ N} \cdot \text{m}^2/\text{kg}^2} = 5.98 \times 10^{24} \text{ kg}$$

or approximately 6.59×10^{21} tons.

SUMMARY

We have defined uniform circular motion as motion in a circle where the speed is constant and only the direction changes. The change in direction produced by a central force is referred to as *centripetal acceleration*. The major concepts presented in this chapter are as follows:

- The linear speed v of an object in uniform circular motion can be calculated from the period T or frequency f:

$$v = \frac{2\pi R}{T} \qquad\qquad v = 2\pi fR$$

- The centripetal acceleration a_c is found from the linear speed, the period, or the frequency as follows:

$$a_c = \frac{v^2}{R} \qquad a_c = \frac{4\pi^2 R}{T^2} \qquad a_c = 4\pi^2 f^2 R$$

- The centripetal force F_c is equal to the product of the mass m and the centripetal acceleration a_c. It is given by

$$F_c = \frac{mv^2}{R} \qquad\qquad F_c = 4\pi^2 f^2 mR$$

- Other useful formulas are as follows:

$$v = \sqrt{\mu_s g R}$$

Maximum Speed without Slipping

$$\tan \theta = \frac{v^2}{gR}$$

Banking Angle or Conical Pendulum

$$f = \frac{1}{2\pi} \sqrt{\frac{g}{h}}$$

Frequency of a Conical Pendulum

QUESTIONS

10-1. Define the following terms:
   a. Uniform circular motion
   b. Centripetal acceleration
   c. Centripetal force
   d. Gravitational constant
   e. Universal law of gravitation
   f. Linear speed
   g. Period
   h. Frequency
   i. Critical speed
   j. Conical pendulum

10-2. Explain with diagrams why the acceleration of a body moving in a circle at constant speed is directed toward the center.

10-3. A bicyclist leans to the side when negotiating a turn. Why? Describe with a free-body diagram the forces acting on the rider.

10-4. In negotiating a circular turn, a car hits a patch of ice and skids off the road. According to Newton's first law, the car will move forward in a direction tangent to the curve, not outward at right angles to it. Why?

10-5. If the force causing circular motion is directed toward the center of rotation, why is water thrown off clothes during the spin cycle of a washing machine?

10-6. When a ball tied at the end of a string is revolved in a circle at constant speed, the inward centripetal force is equal in magnitude to the outward centrifugal force. Does this represent a condition of equilibrium? Explain.

10-7. What factors contribute to the most desirable banking angles on roadways?

10-8. Does the centripetal force do work in uniform circular motion?

10-9. A motorcyclist negotiates a circular track at constant speed. What exerts the centripetal force, and on what does the force act? What exerts the centrifugal reaction force, and on what does it act?

10-10. A rock at the end of a string moves in a vertical circle. Under what conditions can its linear speed be constant? On what does the centripetal force act? On what does the centrifugal force act?

10-11. What is the value of the gravitational constant G on the moon?

10-12. Given the mass of the earth, its distance from the sun, and its orbital speed, explain how you would determine the mass of the sun.

PROBLEMS

Centripetal Acceleration

10-1. A ball is attached to the end of a cord 1.5 m long. The ball swings in a horizontal circle with a speed of 8 m/s. What is the centripetal acceleration? What are the period and the frequency of rotation?

> **Ans.** 42.7 m/s², 1.18 s, 0.849 rev/s

10-2. An object revolves in a circle of diameter 3 m at a frequency of 6 rev/s. What is the period of revolution? What is the linear speed? What is the centripetal acceleration?

10-3. A drive pulley 6 cm in diameter is set to rotate at 9 rev/s. What is the centripetal acceleration at the edge of the pulley? What is the linear speed of a belt attached to the pulley?

> **Ans.** 95.9 m/s², 1.70 m/s

10-4. A merry-go-round revolves with a period of 6 s. How far from the center should one sit in order to experience a centripetal acceleration of 12 ft/s²?

Centripetal Force

10-5. A 3-kg rock is swung in a horizontal circle by a 2-m cord so that it makes one complete revolution in 0.3 s. What is the centripetal force on the rock? Is there also an outward force on the rock?

> **Ans.** 2630 N, no

10-6. An 8-lb weight swings in a horizontal circle with a linear speed of 95 ft/s. What is the radius of the path if the centripetal force is 2000 lb?

10-7. A 4-lb object is tied to a cord and swings in a circle of radius 3 ft. Neglect the effects of gravity and assume a frequency of 80 rpm. (a) What is the centripetal acceleration? (b) What is the tension in the cord? (c) What happens if the cord breaks?

> **Ans.** (a) 211 ft/s², (b) 26.3 lb

10-8. Two 8-kg masses are attached to the end of a thin rod 400 mm long. The rod is supported in the middle and whirled in a circle. Assume that the rod cannot support a tension greater than 80 N. What is the maximum frequency of revolution?

10-9. A coin rests on a rotating platform at a distance of 12 cm from the center of rotation. If the coefficient of static friction is 0.6, what is the maximum frequency of rotation such that the coin does not slip?

> **Ans.** 1.11 rev/s

10-10. A 4-lb damp shirt is being rotated on the spin cycle of a washer at 300 rpm. The diameter of the rotating drum is 27 in. What is the centripetal force? Is the force on the shirt directed outward or inward?

Flat Curves and Banked Curves

10-11. On a rainy day the coefficient of friction between the tires and roadway is 0.4. What is the maximum speed at which a car can negotiate a curve of 80 m radius?

> **Ans.** 63.7 km/h

10-12. A bus negotiates a turn of radius 400 ft while traveling at a speed of 60 mi/h. If slipping just begins at this speed, what is the coefficient of static friction between the tires and the road?

10-13. Find the required banking angle if a car is to make a 180° U-turn in a circular distance of 600 m at a speed of 50 km/h.

Ans. 5.88°

10-14. What are the required banking angles to negotiate the curves of Probs. 10-11 and 10-12 without the need for friction?

$*$ **10-15.** A curve in a road 9 m wide has a radius of 96 m. How much higher than the inside edge should the outside edge be for an automobile to travel safely at 40 km/h?

Ans. 1.17 m

The Conical Pendulum

10-16. A conical pendulum swings in a horizontal circle of radius 30 cm. What angle does the supporting cord make with the vertical when the linear speed of the mass is 12 m/s?

10-17. What is the linear speed of the flyweights in Fig. 10-10 if $L = 20$ cm and $\theta = 60°$? What is the frequency of revolution?

Ans. 1.71 m/s, 1.58 rev/s

Fig. 10-10

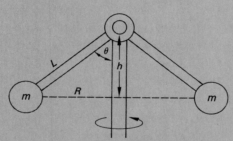

$*$ **10-18.** Each of the flyweights in Fig. 10-10 has a mass of 2 kg. The length of each arm is 40 cm, and the shaft rotates at 80 rpm. What is the tension in each arm? What is the angle θ? What is the distance h?

$*$ **10-19.** In Fig. 10-10, assume that $L = 6$ in., that each flyweight is 1.5 lb, and that the shaft is rotating at 100 rpm. What is the tension in each arm? What is the angle θ? What is the distance h?

Ans. 2.57 lb, 54.3°, 0.292 ft

Motion in a Vertical Circle

10-20. A rock rests on the bottom of a bucket. If the bucket is moved in a vertical circle of radius 70 cm, what is the least speed the stone must have as it rounds the top of the circle in order that the rock will stay in the bucket?

10-21. A 36-kg girl rides on the seat of a swing attached to chains that are 20 m long. If she is released from a position 8 m lower than the top of the swing, what is her linear speed when she passes the lowest point? What force does she exert on the swing at this point? Should we call this the centripetal force or the centrifugal force?

Ans. 15.3 m/s, 776 N, centrifugal

$*$ **10-22.** A test pilot goes into a dive at 620 ft/s and pulls out in a curve of radius 2800 ft. (See Fig. 10-11.) If the pilot weighs 160 lb, what force is exerted on her by the seat? What acceleration will be experienced at the lowest point in the dive? How many times greater is this value than the acceleration due to gravity g?

10-23. Since the actual force acting on a pilot depends on the pilot's own weight, the centripetal effects are usually measured in terms of acceleration. If the desired maximum acceleration for a human is 7 times that of gravity ($7g$), what is the maximum velocity for pulling out of a dive of radius 1 km?

Ans. 943 km/h

10-24. A 3-kg ball swings in a vertical circle at the end of an 8-m cord. When it reaches the top of its circular path, its velocity is 16 m/s. What is the tension in the cord? What is the minimum speed at the top to maintain circular motion?

Gravitation

10-25. A 4-kg mass is separated from a 2-kg mass by a distance of 8 cm. Compute the gravitational force of attraction between the two masses.

Ans. 8.34×10^{-8} N

10-26. How far apart should a 2-ton weight be from a 3-ton weight if their mutual force of attraction is to be 4×10^{-4} lb?

10-27. Assume that the radius of the earth is 6.38×10^6 m, and recall that the weight of an object is the gravitational force on the object due to the earth. Use the law of gravitation to estimate the mass of the earth.

Ans. 5.98×10^{24} kg

10-28. A 3-kg mass is located 10 cm away from a 6-kg mass. What is the resultant gravitational force on a 2-kg mass placed midway between the other masses?

10-29. The mass of the earth is about 81 times the mass of the moon. If the radius of the earth is 4 times that of the moon, what is the acceleration of gravity on the moon's surface?

Ans. 1.94 m/s^2

Additional Problems

10-30. At what frequency should a 6-lb ball be revolved in a radius of 3 ft in order to produce a centripetal acceleration of 12 ft/s^2? What will be the tension in the cord?

10-31. What centripetal acceleration is required to move a 2.6-kg mass in a horizontal circle of radius 300 mm if its linear speed is 15 m/s? What is the centripetal force?

Ans. 750 m/s^2, 1950 N

10-32. What must be the speed of a satellite located 1000 mi above the surface of the earth if it is to travel in a circular path? The radius of the earth is 4000 mi. What is the nature of the centripetal force in this case?

10-33. A 9-kg block rests on the bed of a truck as it turns a curve of radius 86 m. Assume that $\mu_k = 0.3$ and that $\mu_s = 0.4$. **(a)** Does the friction force act toward the center of the curve or away from it? **(b)** What is the maximum speed with which the truck can make the turn if the block is to remain at rest? **(c)** If the truck makes the turn at a much greater speed, what would be the resulting acceleration of the block?

Ans. **(a)** toward the center, **(b)** 18.4 m/s, **(c)** 2.94 m/s^2

10-34. A 10-in.-diameter record turns on a record player at 78 rpm. A bug rests on the record 1 in. from the outside edge. **(a)** If the bug weights 0.02 lb, what force acts on him? **(b)** What exerts this force? **(c)** Where should the bug crawl in order to reduce this force by one-half?

10-35. What frequency of revolution is required to raise the flyweights in Fig. 10-10 a vertical distance of 25 mm above their equilibrium position? Assume that $L = 150$ mm.

Ans. 84.6 rpm

* **10-36.** Consider the apparatus shown in Fig. 10-10. How many revolutions per second are required to make the angle $\theta = 30°$? What is the tension in the cord at that instant? Assume that $L = 50$ cm and $m = 2$ kg.

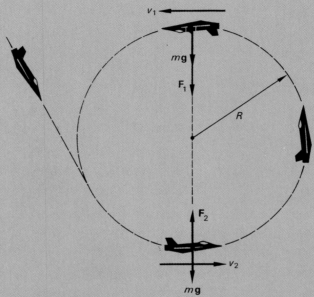

Fig. 10-11 Forces exerted on an airplane at the upper and lower limits of a vertical loop.

* **10-37.** A ball of mass m revolves in a vertical circle of radius 6 m. What velocity at the top of the circle is needed in order that the tension in the cord will be equal to the weight of the mass? What must be the velocity at the bottom to produce a cord tension equal to twice the weight?

Ans. 10.8 m/s, 7.67 m/s

* **10-38.** The combined mass of a motorcycle and driver is 210 kg. If the driver is to negotiate a loop-the-loop of 6 m radius, what must be the minimum speed at the top of the loop? If the speed at the top is 12 m/s, what force is exerted by the top of the loop?

Rotation of Rigid Bodies

OBJECTIVES

After completing this chapter, you should be able to:

1. Define *angular displacement, angular velocity,* and *angular acceleration,* and apply these concepts to the solution of physical problems.
2. Draw analogies relating angular-motion parameters (θ, ω, α) to linear-motion parameters, and solve angular acceleration problems in a manner similar to that learned in Chap. 5 for linear acceleration problems (refer to Table 11-1).
3. Write and apply the relationships between linear speed or acceleration and angular speed or acceleration.
4. Define the *moment of inertia* of a body and describe how this quantity and the angular speed can be used to calculate *rotational kinetic energy.*
5. Apply the concepts of *Newton's second law, rotational work, rotational power,* and *angular momentum* to the solution of physical problems.

We have been considering only translational motion, in which an object's position is changing along a straight line. But it is possible for an object to move in a curved path, or to undergo rotational motion. For example, wheels, shafts, pulleys, gyroscopes, and many other mechanical devices rotate about their axes without translational motion. The generation and transmission of power is nearly always dependent on rotational motion of some kind. It is essential for you to be able to predict and control such motion. The concepts and formulas presented in this chapter are designed to provide you with these essential skills.

11-1
ANGULAR DISPLACE- MENT

The angular displacement of a body describes the amount of rotation. If point A on the rotating disk in Fig. 11-1 rotates on its axis to point B, the angular displacement is denoted by the angle θ. There are several ways of measuring this angle. We are already familiar with the units of degrees and revolutions, which are related according to the definition

$$1 \text{ rev} = 360°$$

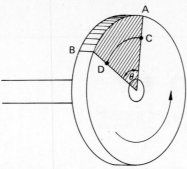

Fig. 11-1 Angular displacement θ is indicated by the shaded portion of the disk. The angular displacement is the same from C to D as it is from A to B for a rigid body.

Neither of these units is very useful in describing rotation of rigid bodies. A more applicable measure of angular displacement is the *radian* (rad). An angle of 1 rad is a central angle whose arc s is equal in length to the radius R. (See Fig. 11-2.) More generally, the radian is defined by the equation

$$\theta = \frac{s}{R} \tag{11-1}$$

where s is the arc of a circle described by the angle θ. Since the ratio of s to R is the ratio of two distances, the radian is a unitless quantity.

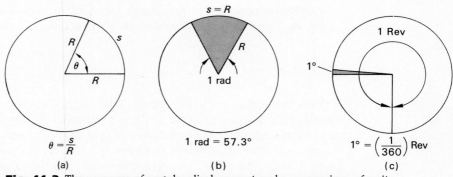

Fig. 11-2 The measure of angular displacement and a comparison of units.

The conversion factor which relates radians to degrees is found by considering an arc length s equal to the circumference of a circle $2\pi R$. Such an angle in radians is given from Eq. (11-1) by

$$\theta = \frac{2\pi R}{R} = 2\pi \text{ rad}$$

Hence

$$1 \text{ rev} = 360° = 2\pi \text{ rad}$$

from which we note that

$$1 \text{ rad} = \frac{360°}{2\pi} = 57.3°$$

EXAMPLE 11-1 If the arc length s is 6 ft and the radius is 10 ft, compute the angular displacement θ in radians, degrees, and revolutions.

Solution Substituting directly into Eq. (11-1), we obtain

$$\theta = \frac{s}{R} = \frac{6 \text{ ft}}{10 \text{ ft}} = 0.6 \text{ rad}$$

Converting to degrees yields

$$\theta = (0.6 \text{ rad}) \frac{57.3°}{1 \text{ rad}} = 34.4°$$

and since 1 rev = 360°

$$\theta = (34.4°) \frac{1 \text{ rev}}{360°} = 0.0956 \text{ rev}$$

EXAMPLE 11-2 A point on the edge of a rotating disk of radius 8 m moves through an angle of 37°. Compute the length of the arc described by the point.

Solution Since Eq. (11-1) was defined for an angle measured in radians, we must first convert 37° to radian units

$$\theta = (37°) \frac{1 \text{ rad}}{57.3°} = 0.646 \text{ rad}$$

The arc length is given by

$$s = R\theta = 8 \text{ m } (0.646 \text{ rad}) = 5.17 \text{ m}$$

The unit radian is dropped because it represents a ratio of length to length (m/m = 1).

11-2 ANGULAR VELOCITY

The time rate of change in angular displacement is called the *angular velocity.* Thus, if an object rotates through an angle θ in a time t, its angular velocity is given by

$$\boxed{\overline{\omega} = \frac{\theta}{t}} \qquad (11\text{-}2)$$

The symbol ω, the Greek letter *omega,* is used to denote rotational velocity. Although angular velocity may be expressed in *revolutions per minute* or *revolutions per second,* in most physical problems it is necessary to use *radians per second* to conform with convenient formulas. Since the rate of rotation in many technical problems is given in terms of the frequency of revolutions, the following relation will be useful:

$$\omega = 2\pi f \qquad (11\text{-}3)$$

where ω is measured in *radians per second* and f is measured in *revolutions per second*.

EXAMPLE 11-3   A bicycle wheel having a diameter of 66 cm makes 40 revolutions in 1 min. (*a*) What is its angular velocity? (*b*) What linear distance will the wheel travel?

Solution (a)   The angular velocity depends only on the frequency of rotation. Since 1 rev = 2π radians,

$$f = \left(\frac{40 \text{ rev}}{\text{min}}\right)\left(\frac{1 \text{ min}}{60 \text{ s}}\right) = 0.667 \text{ rev/s}$$

Substitution into Eq. (11-3) gives the angular velocity.

$$\omega = \left(\frac{2\pi \text{ rad}}{\text{min}}\right)\left(\frac{0.667 \text{ rev}}{\text{s}}\right) = 4.19 \text{ rad/s}$$

Solution (b)   The linear displacement s can be found from the angular displacement θ in radians.

$$\theta = \left(\frac{2\pi \text{ rad}}{1 \text{ rev}}\right)(40 \text{ rev}) = 251 \text{ rad}$$

Solving Eq. (11-1) for s, we obtain

$$s = \theta R = (251 \text{ rad})(0.66 \text{ m}) = 166 \text{ m}$$

It is important to realize that the angular velocity discussed in this section represents an *average* velocity. The same distinction must be made between the average and the instantaneous angular velocities as that discussed in Chap. 5 for average and instantaneous linear velocities.

11-3
ANGULAR ACCEL-ERATION

Like linear motion, angular motion may be uniform or accelerated. The rate of rotation may increase or decrease under the influence of a resultant torque. For example, if the angular velocity changes from an initial value ω_0 to a final value ω_f in a time t, the angular acceleration is given by

$$\alpha = \frac{\omega_f - \omega_0}{t}$$

The Greek letter α (*alpha*) denotes angular acceleration. A more useful form for this equation is

$$\boxed{\omega_f = \omega_0 + \alpha t} \qquad (11\text{-}4)$$

A comparison of Eq. (11-4) with Eq. (5-4) for linear acceleration will show that their forms are identical if we draw analogies between angular and linear parameters.

Now that the concept of initial and final angular velocities has been introduced, we can express the average angular velocity in terms of its initial and final values:

$$\overline{\omega} = \frac{\omega_f + \omega_0}{2}$$

Substituting this equality for $\bar{\omega}$ in Eq. (11-2) yields a more useful expression for the angular displacement:

$$\boxed{\theta = \bar{\omega}t = \frac{\omega_f + \omega_0}{2}t}$$

(11-5)

This equation is also similar to an equation derived for linear motion. In fact, the equations for angular acceleration have the same basic form as those derived in Chap. 5 for linear acceleration if we draw the following analogies:

$$s \text{ (m)} \leftrightarrow \theta \text{ (rad)}$$
$$v \text{ (m/s)} \leftrightarrow \omega \text{ (rad/s)}$$
$$a \text{ (m/s}^2) \leftrightarrow \alpha \text{ (rad/s}^2)$$

Time, of course, is the same for both types of motion and is measured in seconds. Table 11-1 illustrates the similarities between angular and linear motion.

Table 11·1 Comparison of Linear Acceleration and Angular Acceleration

Constant linear acceleration	Constant angular acceleration
$s = \bar{v}t = \dfrac{v_f + v_0}{2}t$	$\theta = \bar{\omega}t = \dfrac{\omega_f + \omega_0}{2}t$
$v_f = v_0 + at$	$\omega_f = \omega_0 + \alpha t$
$s = v_0 t + \frac{1}{2}at^2$	$\theta = \omega_0 t + \frac{1}{2}\alpha t^2$
$2as = v_f^2 - v_0^2$	$2\alpha\theta = \omega_f^2 - \omega_0^2$

In applying these formulas, we must be careful to choose the appropriate units for each quantity. It is also important to choose a direction (clockwise or counterclockwise) as positive and to follow through consistently in affixing the appropriate sign to each quantity.

EXAMPLE 11-4   A flywheel increases its rate of rotation from 6 to 12 rev/s in 8 s. What is its angular acceleration?

Solution   We will first compute the initial and final angular velocities:

$$\omega_0 = 2\pi f_0 = \left(\frac{2\pi \text{ rad}}{\text{rev}}\right)\left(\frac{6 \text{ rev}}{\text{s}}\right) = 12\pi \text{ rad/s}$$

$$\omega_f = 2\pi f_0 = \left(\frac{2\pi \text{ rad}}{\text{rev}}\right)\left(\frac{12 \text{ rev}}{\text{s}}\right) = 24\pi \text{ rad/s}$$

$$\alpha = \frac{\omega_f - \omega_0}{t} = \frac{(24\pi - 12\pi) \text{ rad/s}}{8 \text{ s}}$$

$$= 1.5\pi \text{ rad/s}^2 = 4.71 \text{ rad/s}^2$$

EXAMPLE 11-5   A grinding disk rotating initially with an angular velocity of 6 rad/s receives a constant acceleration of 2 rad/s^2. (a) What angular displacement will it describe in 3 s? (b) How many revolutions will it make? (c) What is its final angular velocity?

Solution (a)   The angular displacement is given by

$$\theta = \omega_0 t + \tfrac{1}{2}\alpha t^2$$
$$= (6 \text{ rad/s})(3 \text{ s}) + \tfrac{1}{2}(2 \text{ rad/s}^2)(3 \text{ s})^2$$
$$= 18 \text{ rad} + (1 \text{ rad/s}^2)(9 \text{ s}^2)$$
$$= 27 \text{ rad}$$

Solution (b)   Since 1 rev = 2π rad, we obtain

$$\theta = (27 \text{ rad})\frac{1 \text{ rev}}{2\pi \text{ rad}} = 4.30 \text{ rev}$$

Solution (c)   The final velocity is equal to the initial velocity plus the change in speed. Thus

$$\omega_f = \omega_0 + \alpha t = 6 \text{ rad/s} + (2 \text{ rad/s}^2)(3 \text{ s})$$
$$= 12 \text{ rad/s}$$

11-4

RELATION BETWEEN ANGULAR AND LINEAR MOTION

The *axis of rotation* of a rigid rotating body can be defined as that line of particles which remains stationary during rotation. This may be a line through the body, as with a spinning top, or it may be a line through space, as with a rolling hoop. In any case, our experience tells us that the farther a particle is from the axis of rotation, the greater its linear speed. This fact was expressed in Chap. 10 by the formula

$$v = 2\pi f R$$

where f is the frequency of rotation. We now derive a similar relation in terms of angular speed. The rotating particle in Fig. 11-3 turns through an arc s, which is

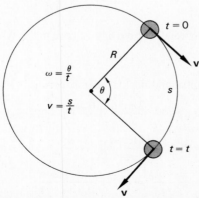

Fig. 11-3 The relation between angular speed and linear speed.

---

given by

$$s = \theta R$$

from Eq. (11-1). If this distance is traversed in a time t, the linear speed of the particle is given by

$$v = \frac{s}{t} = \frac{\theta R}{t}$$

Since $\theta/t = \omega$, the linear speed can be expressed as a function of the angular speed.

$$\boxed{v = \omega R} \tag{11-6}$$

This result also follows from Eq. (11-3), in which the angular velocity is expressed as a function of the frequency of revolution.

EXAMPLE 11-6   A drive shaft has an angular speed of 60 rad/s. At what distance should flyweights be positioned from the axis if they are to have a linear speed of 120 ft/s?

Solution   Solving for R in Eq. (11-6), we obtain

$$R = \frac{v}{\omega} = \frac{120 \text{ ft/s}}{60 \text{ rad/s}} = 2 \text{ ft}$$

Let us now return to a particle moving in a circle of radius R and assume that the linear speed changes from some initial value v_0 to a final value v_f in a time t. The *tangential acceleration* a_T of such a particle is given by

$$a_T = \frac{v_f - v_0}{t}$$

Because of the close relationship between linear speed and angular speed, as represented by Eq. (11-6), we can also express the tangential acceleration in terms of a change in angular velocity.

$$a_T = \frac{\omega_f R - \omega_0 R}{t} = \frac{\omega_f - \omega_0}{t} R$$

or

$$\boxed{a_T = \alpha R} \tag{11-7}$$

where α represents the angular acceleration.

We must be careful to distinguish between the tangential acceleration, as defined in Eq. (11-7), and the centripetal acceleration defined by

$$a_c = \frac{v^2}{R} \tag{11-8}$$

The tangential acceleration represents a change in linear speed whereas the centripetal acceleration represents only a change in the direction of motion. The distinction is shown graphically in Fig. 11-4. The resultant acceleration can be found by computing the vector sum of the tangential and centripetal accelerations.

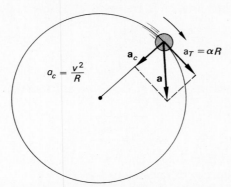

Fig. 11-4 The relationship between tangential and centripetal acceleration.

EXAMPLE 11-7   Find the resultant acceleration of a particle moving in a circle of radius 0.5 m at the instant when its angular speed is 3 rad/s and its angular acceleration is 4 rad/s².

Solution   The tangential acceleration from Eq. (11-7) is

$$a_T = \alpha R = (4 \text{ rad/s}^2)(0.5 \text{ m}) = 2 \text{ m/s}^2$$

Since $v = \omega R$, the centripetal acceleration is given by

$$a_c = \frac{v^2}{R} = \frac{\omega^2 R^2}{R} = \omega^2 R$$

from which we obtain

$$a_c = (3 \text{ rad/s})^2(0.5 \text{ m}) = 4.5 \text{ m/s}^2$$

Finally, the magnitude of the resultant acceleration is obtained from Pythagoras' theorem.

$$a = \sqrt{a_T^2 + a_c^2} = \sqrt{2^2 + 4.5^2}$$
$$= 4.92 \text{ m/s}^2$$

The direction of the resultant acceleration can be found from its components in the usual manner.

11-5
ROTATIONAL KINETIC ENERGY; MOMENT OF INERTIA

We have seen that a particle moving in a circle of radius R has a linear speed given by

$$v = \omega R$$

If the particle has a mass m, it will have a kinetic energy given by

$$E_K = \tfrac{1}{2}mv^2 = \tfrac{1}{2}m\omega^2 R^2$$

A rigid body like that in Fig. 11-5 can be considered as consisting of many particles of varying masses located at different distances from the axis of rotation O. The total

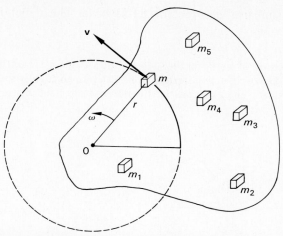

Fig. 11-5 Rotation of an extended body. The body can be thought of as many individual masses all rotating with the same angular velocity.

kinetic energy of such an extended body will be the sum of the kinetic energies of each particle making up the body. Hence

$$E_K = \sum \tfrac{1}{2}m\omega^2 r^2$$

Since the constant ½ and the angular velocity ω are the same for all particles, we can rearrange to obtain

$$E_K = \tfrac{1}{2}\left(\sum mr^2\right)\omega^2$$

The quantity in parentheses, $\sum mr^2$, has the same value for a given body regardless of its state of motion. We define this quantity as the *moment of inertia* and represent it by I

$$I = m_1 r_1^2 + m_2 r_2^2 + m_3 r_3^2 + \cdots$$

or

$$\boxed{I = \sum mr^2} \tag{11-9}$$

The SI unit for I is the *kilogram-square meter* and the USCS unit is the *slug-square foot*.

Using this definition, we can express the rotational kinetic energy of a body in terms of its moment of inertia and its angular velocity:

$$\boxed{E_K = \tfrac{1}{2}I\omega^2} \tag{11-10}$$

Note the similarity between the terms m for linear motion and I for rotational motion.

EXAMPLE 11-8 Find the moment of inertia for the system illustrated in Fig. 11-6. The rods joining the masses are of negligible weight, and the system rotates with an angular speed of 6 rad/s. What is the rotational kinetic energy? (Consider the masses to be point masses.)

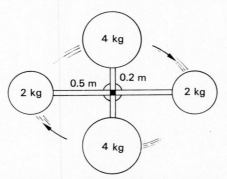

Fig. 11-6 Computing the moment of inertia.

Solution The moment of inertia from Eq. (11-9) is

$$I = m_1r_1^2 + m_2r_2^2 + m_3r_3^2 + m_4r_4^2$$
$$= (2 \text{ kg})(0.5 \text{ m})^2 + (4 \text{ kg})(0.2 \text{ m})^2 + (2 \text{ kg})(0.5 \text{ m})^2 + (4 \text{ kg})(0.2 \text{ m})^2$$
$$= (0.50 + 0.16 + 0.50 + 0.16) \text{kg} \cdot \text{m}^2$$
$$= 1.32 \text{ kg} \cdot \text{m}^2$$

The rotational kinetic energy is given by

$$E_K = \tfrac{1}{2}I\omega^2 = \tfrac{1}{2}(1.32 \text{ kg} \cdot \text{m}^2)(6 \text{ rad/s})^2$$
$$= 23.8 \text{ J}$$

For bodies which are not composed of separate masses but are actually continuous distributions of matter, the calculation of moments of inertia is more difficult and usually involves calculus. A few simple cases are given in Fig. 11-7, along with the formulas for computing their moments of inertia.

Sometimes it is desirable to express the rotational inertia of a body in terms of its *radius of gyration k*. This quantity is defined as the radial distance from the center of rotation to a circumference at which the total mass of the body might be concentrated without changing its moment of inertia. According to this definition, the moment of inertia can be found from the formula

$$\boxed{I = mk^2} \qquad (11\text{-}11)$$

where m represents the total mass of the rotating body and k is its radius of gyration.

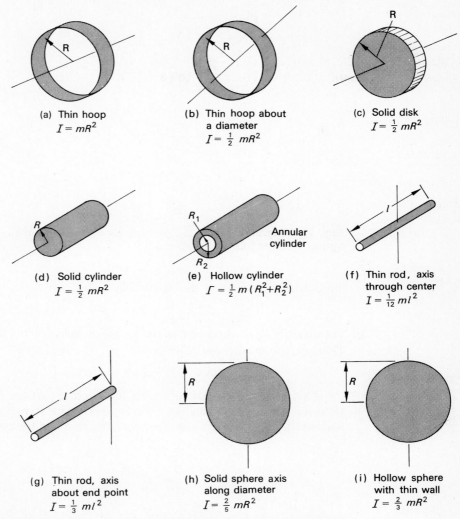

(a) Thin hoop
$$I = mR^2$$

(b) Thin hoop about
a diameter
$$I = \frac{1}{2}mR^2$$

(c) Solid disk
$$I = \frac{1}{2}mR^2$$

(d) Solid cylinder
$$I = \frac{1}{2}mR^2$$

(e) Hollow cylinder
$$I = \frac{1}{2}m(R_1^2 + R_2^2)$$

Annular cylinder

(f) Thin rod, axis
through center
$$I = \frac{1}{12}ml^2$$

(g) Thin rod, axis
about end point
$$I = \frac{1}{3}ml^2$$

(h) Solid sphere axis
along diameter
$$I = \frac{2}{5}mR^2$$

(i) Hollow sphere
with thin wall
$$I = \frac{2}{3}mR^2$$

Fig. 11-7 Moments of inertia for bodies about their indicated axes.

<table>
<tr><td>

11-6

**THE SECOND
LAW OF
MOTION IN
ROTATION**

</td><td>

Suppose we analyze the motion of the rotating rigid body in Fig. 11-8. Consider a force **F** acting on the small mass m, indicated by a shaded portion of the object, at a distance r from the axis of rotation.

The force **F** applied perpendicular to r causes the body to rotate with a tangential acceleration

$$a_T = \alpha r$$

where α is the angular acceleration. From Newton's second law of motion

$$F = ma_T = m\alpha r$$

</td></tr>
</table>

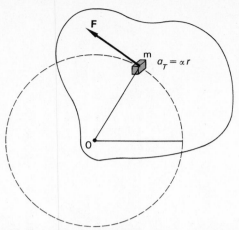

Fig. 11-8 Newton's second law for rotational motion states the relationship between torque Fr and angular acceleration α.

Multiplying both sides of this relation by r yields

$$Fr = (mr^2)\alpha$$

The quantity Fr is recognized as the torque produced by the force **F** about the axis. Thus, for the mass m we write

$$\tau = (mr^2)\alpha$$

A similar equation may be derived for all other portions of the rotating object. However, the angular acceleration will be constant for every portion, regardless of its mass or distance from the axis. Therefore, the resultant torque on the whole body is

$$\tau = \left(\sum mr^2\right)\alpha$$

or

$$\boxed{\tau = I\alpha} \tag{11-12}$$

Torque = moment of inertia × angular acceleration

Note the similarity of Eq. (11-12) with the second law for linear motion, $F = ma$. Newton's *law for rotational motion* is as follows:

> *A resultant torque applied to a rigid body will always result in an angular acceleration that is directly proportional to the applied torque and inversely proportional to the body's moment of inertia.*

In applying Eq. (11-12), it is important to recall that the torque produced by a force is equal to the product of its distance from the axis and the perpendicular component of the force. It must also be remembered that the angular acceleration is expressed in radians per second per second.

EXAMPLE 11-9   A circular grinding disk of radius 0.6 m and mass 90 kg is rotating at 460 rpm. What frictional force, applied tangent to the edge, will cause the disk to stop in 20 s?

Solution   We first calculate the moment of inertia of the disk from the formula given in Fig. 11-7.

$$I = \tfrac{1}{2}mR^2 = \tfrac{1}{2}(90\text{ kg})(0.6\text{ m})^2$$
$$= 16.2\text{ kg} \cdot \text{m}^2$$

Converting the rotational speed to radians per second, we obtain

$$\omega = \left(\frac{2\pi \text{ rad}}{\text{rev}}\right)\left(\frac{460 \text{ rev}}{\text{min}}\right)\left(\frac{1 \text{ min}}{60 \text{ s}}\right)$$
$$= 48.2\text{ rad/s}$$

Thus the angular acceleration is

$$\alpha = \frac{\omega_f - \omega_0}{t}$$
$$= \frac{0 - (48.2 \text{ rad/s})}{20 \text{ s}} = -2.41 \text{ rad/s}^2$$

Applying Newton's second law gives

$$\tau = Fr = I\alpha$$

from which

$$F = \frac{I\alpha}{r} = \frac{(16.2 \text{ kg} \cdot \text{m}^2)(-2.41 \text{ rad/s}^2)}{0.6 \text{ m}}$$
$$= -65.0 \text{ N}$$

The negative sign appears because the force must be directed opposite to the direction of rotation of the disk.

11-7
ROTATIONAL WORK AND POWER

Work was defined in Chap. 8 as the product of a displacement and the component of the force in the direction of the displacement. We now consider the work done in rotational displacement under the influence of a resultant torque.

Consider a force F acting at the edge of a pulley of radius r, as shown in Fig. 11-9. The effect of such a force is to rotate the pulley through an angle θ while the point

Fig. 11-9 Work and power in rotation.

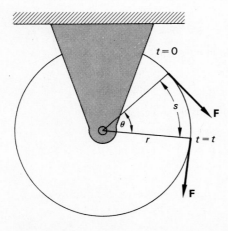

at which the force is applied moves through a distance s. The arc distance s is related to θ by

$$s = r\theta$$

Hence the work of the force $\mathbf{F}$ is by definition

$$\text{Work} = Fs = Fr\theta$$

but Fr is the torque due to the force, so that we obtain

$$\boxed{\text{Work} = \tau\theta} \tag{11-13}$$

The angle θ must be expressed in radians for either system of units in order that the work can be expressed in foot-pounds or joules.

Mechanical energy is usually transmitted in the form of rotational work. When we speak of the power output of engines, we are concerned with the rate at which rotational work is done. Thus rotational power can be determined by dividing both sides of Eq. (11-13) by the time t required for the torque τ to effect a displacement θ

$$\text{Power} = \frac{\text{work}}{t} = \frac{\tau\theta}{t} \tag{11-14}$$

Since θ/t represents the average angular velocity $\overline{\omega}$, we write

$$\boxed{\text{Power} = \tau\overline{\omega}} \tag{11-15}$$

Notice the similarity of this relation with its analog, $P = Fv$, derived earlier for linear motion. Both are measures of *average* power.

EXAMPLE 11-10　A wheel of radius 60 cm has a moment of inertia of 5 kg · m². A constant force of 60 N is applied at the rim. (*a*) Assuming it starts from rest, how much work is done in 4 s? (*b*) What power is developed?

Solution (a)　The work is a product of torque and angular displacement. We first calculate the applied torque:

$$\tau = Fr = (60 \text{ N})(0.6 \text{ m}) = 36 \text{ N} \cdot \text{m}$$

Next, we find α from Newton's second law.

$$\alpha = \frac{\tau}{I} = \frac{36 \text{ N} \cdot \text{m}}{5 \text{ kg} \cdot \text{m}^2} = 7.2 \text{ rad/s}^2$$

Now the angular displacement θ can be found.

$$\theta = \omega_0 t + \tfrac{1}{2}\alpha t^2$$
$$= (0) + \tfrac{1}{2}(7.2 \text{ m/s}^2)(4 \text{ s})^2 = 57.6 \text{ rad}$$

The work is, therefore,

$$\text{Work} = \tau\theta = (36 \text{ N} \cdot \text{m})(57.6 \text{ rad})$$
$$= 2070 \text{ J}$$

Solution (b) The average power is

$$P = \frac{\text{work}}{t} = \frac{2070 \text{ J}}{4 \text{ s}} = 518 \text{ W}$$

The same result could be found by calculating the average angular speed w and using Eq. (11-15). As an added example, you might show that the work done is equal to the change in rotational kinetic energy.

11-8 ANGULAR MOMENTUM

Consider a particle of mass m moving in a circle of radius r, as shown in Fig. 11-10a. If its linear velocity is v, it will have a linear momentum $p = mv$. With reference to the fixed axis of rotation, we define the *angular momentum L* of the particle as the

Fig. 11-10 Defining angular momentum.

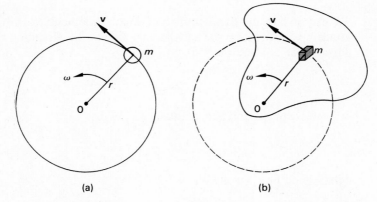

(a)   (b)

product of its linear momentum and the perpendicular distance from the axis to the revolving particle.

$$L \text{ (particle)} = mvr \qquad (11\text{-}16)$$

Now let us consider the definition of angular momentum as it applies to an extended rigid body. Figure 11-10b describes such a body, which is rotating about its axis O. Each particle in the body has an angular momentum given by Eq. (11-16). Substituting $v = \omega r$, each particle has an angular momentum given by

$$mvr = m\omega r^2 = (mr^2)\omega$$

Since the body is rigid, all particles within the body have the same angular velocity, and the total angular momentum of the body is

$$L = \left(\sum mr^2\right)\omega$$

Thus the total angular momentum is equal to the product of a body's angular velocity and its moment of inertia

$$\boxed{L = I\omega} \qquad (11\text{-}17)$$

EXAMPLE 11-11 A thin uniform rod is 1 m long and has a mass of 6 kg. If the rod is pivoted at its center and set into rotation with an angular speed of 16 rad/s, compute its angular momentum.

Solution The moment of inertia of a thin rod is, from Fig. 11-7,

$$I = \frac{ml^2}{12} = \frac{(6 \text{ kg})(1 \text{ m})^2}{12} = 0.5 \text{ kg} \cdot \text{m}^2$$

Hence its angular momentum is

$$L = I\omega = (0.5 \text{ kg} \cdot \text{m}^2)(16 \text{ rad/s})$$
$$= 8 \text{ kg} \cdot \text{m}^2/\text{s}$$

Notice that the SI unit of angular momentum is $\text{kg} \cdot \text{m}^2/\text{s}$. The USCS unit is $\text{slug} \cdot \text{ft}^2/\text{s}$.

11-9
CONSER-VATION OF ANGULAR MOMENTUM

We can understand the definition of angular momentum better if we return to the fundamental equation for angular motion, $\tau = I\alpha$. Recalling the defining equation for angular acceleration

$$\alpha = \frac{\omega_f - \omega_0}{t}$$

we can write Newton's second law as

$$\tau = I\frac{\omega_f - \omega_0}{t}$$

Multiplying by t, we obtain

$$\tau t = I\omega_f - I\omega_0 \qquad (11\text{-}18)$$

Angular impulse = change in angular momentum

The product τt is defined as the *angular impulse*. Note the similarity between this equation and the one derived in Chap. 9 for linear impulse.

If no external torque is applied to a rotating body, we can set $\tau = 0$ in Eq. (11-18), yielding

$$0 = I\omega_f - I\omega_0$$
$$I\omega_f = I\omega_0 \qquad (11\text{-}19)$$

Final angular momentum = initial angular momentum

Thus we are led to a statement of the *conservation of angular momentum*:

> *If the sum of the external torques acting on a body or system of bodies is zero, the angular momentum remains unchanged.*

This statement holds true even if the rotating body is not rigid but is altered so that its moment of inertia changes. In this case, the angular speed also changes so that the product $I\omega$ is always constant. Skaters, divers, and acrobats all control the rate at which their bodies rotate by extending or retracting their limbs to decrease or increase their angular speed.

An interesting experiment illustrating the conservation of angular momentum is shown in Fig. 11-11. A woman stands on a rotating platform with heavy weights in each hand. She is first set into rotation with her arms fully extended. By drawing her hands closer to her body, she decreases her moment of inertia. Since her angular momentum cannot change, she will notice a considerable increase in her angular speed. By extending her arms, she will be able to decrease her angular speed.

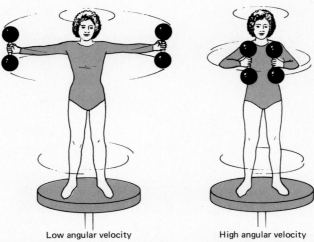

Low angular velocity   High angular velocity

Fig. 11-11 Experiment to demonstrate the conservation of angular momentum. The woman controls her rate of rotation by moving the weights inward to increase rotational speed or outward to decrease rotational speed.

EXAMPLE 11-12   Assume the woman in Fig. 11-11 has a constant moment of inertia equal to 3 slug · ft^2 and that she holds a 16-lb weight in each hand. While holding the weights at a distance $r_1 = 3$ ft from the axis of rotation, she is given an initial velocity $\omega_0 = 3$ rad/s. If she pulls the weights inward until they are at a distance $r_2 = 1$ ft from the center of rotation, what will her resulting angular speed be?

Solution   If we consider the 16-lb weights as point masses, the moment of inertia I_m of the woman will be increased in each case by Mr^2, where $M = 32$ lb/(32 ft/s^2) = 1 slug is the total mass of the two weights and r is the distance of the weights from the center of rotation. Since the final momentum must equal the initial momentum, we can write

$$(I_m + Mr^2)\omega_f = (I_m + Mr_1^2)\omega_0$$

We now solve for the final angular speed as follows:

$$[3 \text{ slug} \cdot \text{ft} + (1 \text{ slug})(1 \text{ ft})^2]\omega_f = [3 \text{ slug} \cdot \text{ft}^2 + (1 \text{ slug})(3 \text{ ft})^2](3 \text{ rad/s})$$
$$(4 \text{ slug} \cdot \text{ft}^2)\omega_f = (12 \text{ slug} \cdot \text{ft}^2)(3 \text{ rad/s})$$
$$\omega_f = 9 \text{ rad/s}$$

Her angular speed has been tripled by bringing the weights in to her chest.

In this chapter, we extended the concept of circular motion to include the rotation of a rigid body composed of many particles. We found that many problems can be solved by methods discussed earlier for linear motion. The essential concepts are summarized below:

- The similarities between angular and linear motion:

Rotational	θ	ω	α	I	$I\omega$	τ	$I\alpha$	$\tau\theta$	$\frac{1}{2}I\omega^2$	$\tau\omega$
Linear	s	v	a	m	mv	F	ma	Fs	$\frac{1}{2}mv^2$	Fv

- The angle in radians is the ratio of the arc distance s to the radius R of the arc. Symbolically we write:

$$\theta = \frac{s}{R} \qquad s = \theta R$$

The radian is a unitless ratio of two lengths.
- Angular velocity, which is the rate of angular displacement, can be calculated from θ or from the frequency of rotation:

$$\overline{\omega} = \frac{\theta}{t} \qquad \overline{\omega} = 2\pi f \qquad \text{\textit{Average Angular Velocity}}$$

- Angular acceleration is the time rate of change in angular speed:

$$\alpha = \frac{\omega_f - \omega_0}{t} \qquad \text{\textit{Angular Acceleration}}$$

- By comparing θ to s, ω to v, and α to a, the following equations can be utilized for angular acceleration problems:

$$\theta = \frac{\omega_f - \omega_0}{2}t$$
$$\omega_f = \omega_0 + \alpha t$$
$$\theta = \omega_0 t + \frac{1}{2}\alpha t^2$$
$$2\alpha\theta = \omega_f^2 - \omega_0^2$$

When any three of the five parameters θ, α, t, ω_f and ω_0 are given, the other two can be found from one of these equations. Choose a direction of rotation as being positive throughout your calculations.

- The following equations are useful when comparing linear motion with rotational motion:

$$v = \omega R \qquad a_T = \alpha R$$

- Other useful relationships:

$I = \sum mR^2$ *Moment of Inertia*		$E_K = \frac{1}{2}I\omega^2$ *Rotational Kinetic Energy*	
$I = m\kappa^2$ *Radius of Gyration*		$\tau = I\alpha$ *Newton's Law*	
Work $= \tau\theta$ *Work*		$P = \tau\omega$ *Power*	
$L = I\omega$ *Angular Momentum*		$I\omega_f = I\omega_0$ *Conservation of Angular Momentum*	

QUESTIONS

11-1. Define the following terms:
   a. Radius of gyration
   b. Radian
   c. Angular speed
   d. Angular momentum
   e. Moment of inertia
   f. Tangential acceleration
   g. Angular displacement
   h. Angular acceleration
   i. Rotational kinetic energy

11-2. Make a list of the SI and USCS units for angular velocity, angular acceleration, moment of inertia, torque, and rotational kinetic energy.

11-3. State the angular analogies for the following translational equations:
   a. $\omega_f = v_0 + at$
   b. $s = v_0t + \frac{1}{2}at^2$
   c. $F = ma$
   d. $E_\kappa = \frac{1}{2}mv^2$
   e. Work $= Fs$
   f. Power $= $ work$/t = Fv$

11-4. A sphere, a cylinder, a disk, and a hollow ring all have identical masses and are rotating with constant angular speed about a common axis. Compare their individual rotational kinetic energies, assuming their outside diameters are equal.

11-5. Explain how a diver controls his motion so that he can strike the water feet first or head first.

11-6. A cat held feet up and dropped from any elevation above a few feet will always land feet first. How does he accomplish this?

11-7. When energy is supplied to a body so that both translation and rotation result, its total kinetic energy is given by

$$E_\kappa = \frac{1}{2}mv^2 + \frac{1}{2}I\omega^2$$

How the energy is divided between rotational and translational effects is determined by the distribution of mass (moment of inertia). From these statements, which of the following objects will roll to the bottom of an inclined plane first?
   a. A solid disk of mass M b. A circular hoop of mass M

11-8. Refer to Question 11-7. If a solid sphere, a solid disk, a solid cylinder, and a hollow cylinder are released simultaneously from the top of an inclined plane, in what order will they reach the bottom?

11-9. A disk whose moment of inertia is I_1 and whose angular velocity is ω_1 is meshed with a disk whose moment of inertia is I_2 and whose angular velocity is ω_2. Write the conservation equation by denoting their combined angular velocity by ω.

11-10. Refer to Question 11-9. Suppose $\omega_1 = \omega_2$ and $I_1 = 2I_2$. How would their combined velocity compare with their initial velocity? Suppose $\omega_1 = 3\omega_2$ and $I_1 = I_2$?

Angular Displacement

11-1. A cable is wrapped around a drum 80 cm in diameter. How many revolutions of this drum will cause an object attached to the cable to move a linear distance of 2 m? What is the angular displacement in radians?

Ans. 0.796 rev, 5 rad

11-2. A bicycle wheel is 26 in. in diameter. If the wheel makes 60 revolutions, what linear distance will it travel?

11-3. A point at the edge of a large wheel of radius 3 m moves through an angle of 37°. Find the length of the arc described by the point.

Ans. 1.94 m

11-4. A point on the edge of a rotating platform 6 ft in diameter moves through a distance of 2 ft. Compute the angular displacement (a) in radians, (b) in degrees, and (c) in revolutions.

Angular Velocity

11-5. An electric motor turns at 600 rpm. What is its angular speed? What is the angular displacement after 6 s?

Ans. 62.8 rad/s, 377 rad

11-6. A rotating pulley completes 12 rev in 4 s. Determine the average angular speed (a) in revolutions per second, (b) in revolutions per minute, and (c) in radians per second.

Angular Acceleration

11-7. A rotating flywheel starts from rest and reaches a final rotational speed of 900 rpm in 4 s. Determine the angular acceleration and the angular displacement after 4 s.

Ans. 23.6 rad/s^2, 188 rad

11-8. The spin cycle on a washing machine slows from 900 rpm to 300 rpm in 50 rev. What are the angular acceleration and the time required?

* **11-9.** A wheel revolving initially at 6 rev/s has an angular acceleration of 4 rad/s^2. What is the angular speed after 5 s? How many revolutions will it make?

Ans. 57.7 rad/s, 38.0 rev

* **11-10.** A grinding disk is brought to a stop in 40 rev. If the braking acceleration was −6 rad/s^2, what was the initial frequency of revolution in rev/s?

Relation between Angular and Linear Motion

11-11. A cylindrical piece of stock 6 in. in diameter rotates in a lathe at 800 rpm. What is the linear velocity at the surface of the cylinder?

Ans. 20.9 ft/s

11-12. The proper tangential velocity for machining steel stock is about 70 cm/s. At what rpm should a steel cylinder 8 cm in diameter be turned in a lathe?

* **11-13.** A pulley 32 cm in diameter and rotating initially at 4 rev/s receives a constant angular acceleration of 2 rad/s^2. (a) What is the linear velocity of a belt wrapped around the pulley after 8 s? (b) What is the tangential acceleration of the belt?

Ans. 6.58 m/s, 0.320 m/s^2

11-14. A woman standing 4 m from the center of a rotating platform covers a distance of 100 m in 20 s. If she started from rest, what is the angular acceleration of the platform? What is the angular velocity after 20 s?

Rotational Kinetic Energy; Moment of Inertia

11-15. A 2-kg mass and a 6-kg mass are connected by a 30-cm light steel bar. The system is then rotated horizontally at 300 rpm about an axis 10 cm from the 6-kg mass. **(a)** What is the moment of inertia about this axis? **(b)** What is the rotational kinetic energy?

Ans. **(a)** 0.140 kg m^2, **(b)** 69.1 J

11-16. A 1.2-kg bicycle wheel has a radius of 70 cm with spokes of negligible weight. If it starts from rest and receives an angular acceleration of 3 rad/s^2, what will be its rotational kinetic energy after 4 s?

11-17. A 16-lb grinding disk is rotating at 400 rpm. What is the radius of the disk if its kinetic energy is 54.8 ft · lb? What is the moment of inertia?

Ans. 6 in., 0.0625 slug · ft^2

Newton's Second Law

11-18. The flywheel of an engine has a moment of inertia of 24 slug · ft^2. What torque is required to accelerate the wheel from rest to an angular speed of 400 rpm in 10 s?

11-19. An unbalanced torque of 150 N · m imparts an angular acceleration of 12 rad/s^2 to the rotor of a generator. What is the moment of inertia?

Ans. 12.5 kg · m^2

11-20. A large 120-kg turbine wheel has a radius of gyration of 1 m. A frictional torque of 80 N · m opposes the rotation of the shaft. What torque must be applied to accelerate the wheel from rest to 300 rpm in 10 s?

11-21. An 8-kg grinding disk has a diameter of 60 cm and is rotating at 600 rpm. What braking force must be applied tangentially to the edge of the disk to bring it to a stop in 5 s?

Ans. 15.1 N

11-22. A rope is wrapped several times around a cylinder which has a mass of 30 kg and a radius of 0.2 m. If the rope is pulled with a force of 40 N, what is the angular acceleration of the cylinder? What is the linear acceleration of the rope?

Rotational Work and Power

11-23. A 70-N force is applied tangentially to the rim of a circular hoop 1 m in diameter. If the hoop turns through 30 rev in 5 s, what are the work and power delivered?

Ans. 6600 J, 1.32 kW

11-24. A cord is wrapped around the rim of a 4-kg flywheel whose radius of gyration is 0.4 m. A steady pull of 50 N acts on the cord, unwinding it a distance of 5 m. **(a)** Show that the linear work is equal to the rotational work. **(b)** Show that the work done is equal to the change in kinetic energy.

11-25. The driving wheel attached to an electric motor has a diameter of 12 in. and makes 1400 rpm. When a belt drive is connected around the wheel, the tension in the belt on the slack side is 40 lb and the tension on the right side is 110 lb. Find the horsepower transmitted by the belt.

Ans. 9.33 hp

11-26. A 1.2-kW motor acts for 8 s on an wheel having a moment of inertia of $2 \ kg \cdot m^2$. Assuming the wheel was initially at rest, what angular speed will be developed?

Angular Momentum

11-27. A 500-g steel rod 30 cm in length is pivoted about its center and rotated at 300 rpm. What is the angular momentum?

Ans. $0.118 \ kg \cdot m^2/s$

11-28. In Prob. 11-27, what braking torque must be applied to stop the rotation in 2 s?

11-29. A block is attached to a cord passing over a pulley through a hole in a horizontal frictionless surface, as shown in Fig. 11-12. Initially the block is revolving at 4 rad/s at a distance r from the center of the hole. If the cord is pulled from below until its radius is $r/4$, what is the new angular velocity?

Ans. 64 rad/s

Fig. 11-12

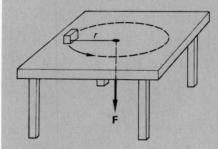

11-30. Suppose in Fig. 11-12 the block has a mass of 2 kg and is rotating at 3 rad/s at a distance of 1 m from the center of the hole. At what distance r from the hole will the tension in the cord be 45 N?

11-31. A 6-kg disk A rotating at 400 rpm engages with a 3-kg disk B initially at rest. The radius of disk A is 0.4 m, and the radius of disk B is 0.2 m. (See Fig. 11-13.) What is the combined angular speed after the disks are meshed?

Ans. 37.2 rad/s

Fig. 11-13

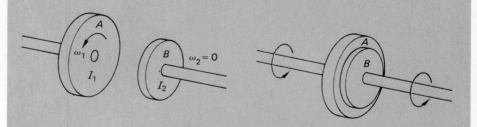

11-32. Assume the same conditions as in Prob. 11-31 except that disk B is rotating at 60 rad/s initially in the direction opposite to disk A. What will be the resultant angular velocity when the two disks are meshed?

Additional Problems

* **11-33.** A solid 12-kg cylinder whose radius of gyration is 10 cm has a constant force of 45 N applied to the rim by a wide leather belt. If the force acts for 4 s, show that the work done is equal to the change in kinetic energy.

    Ans. 2700 J

* **11-34.** A uniform ring weighing 16 lb revolves about its center at 6 rev/s. If its rotational kinetic energy is 400 ft·lb, what must be the radius of the ring?

* **11-35.** A wheel of radius 2 ft has a moment of inertia of 8.2 slug · ft². A constant force of 12 lb acts tangentially at the edge of the wheel. Assume that the wheel is initially at rest. What is the angular acceleration? How much work is done in 5 s? What horsepower is needed?

    Ans. 2.93 rad/s², 878 ft·lb, 0.319 hp

* **11-36.** The radius of gyration of an 8-kg wheel is 50 cm. Find its moment of inertia and kinetic energy if it is rotating at 400 rpm. How much work is required to slow the rotation to 100 rpm?

** **11-37.** A light rope is wrapped several times around a solid cylinder of mass 10 kg and radius 50 cm which rotates without friction. The free end of the rope is attached to a 2-kg mass, which is released from rest a distance 6 m above the floor. (a) What is the linear velocity of the mass just before it strikes the floor? (b) What is the angular velocity of the wheel at this instant?

    Ans. (a) 5.80 m/s, (b) 11.6 rad/s

** **11-38.** For Prob. 11-37, find the linear acceleration of the suspended mass and the angular acceleration of the solid cylinder.

Simple Machines

OBJECTIVES

After completing this chapter, you should be able to:

1. Describe a simple machine and its operation in general terms to the extent that *efficiency* and *conservation of energy* are explained.
2. Write and apply formulas for computing the efficiency of a simple machine in terms of work or power.
3. Distinguish by definition and example between *ideal* mechanical advantage and *actual* mechanical advantage.
4. Draw a diagram of each of the following simple machines and beside each diagram write a formula for computing the ideal mechanical advantage: (*a*) lever, (*b*) inclined plane, (*c*) wedge, (*d*) gears, (*e*) pulley systems, (*f*) wheel and axle, (*g*) screw jack, (*h*) belt drive.
5. Compute the mechanical advantage and the efficiency of each of the simple machines listed in the previous objective.

A simple machine is any device which transmits the application of a force into useful work. With a chain hoist we can transmit a small downward force into a very large upward force for lifting. In industry delicate samples of radioactive material are handled by machines that allow an applied force to be reduced significantly. Single pulleys may be used to change the direction of an applied force without affecting its magnitude. A study of machines and their efficiency is essential for the productive use of energy. In this chapter, you will become familiar with levers, gears, pulley systems, inclined planes, and other machines routinely used for many industrial applications.

12-1
SIMPLE
MACHINES
AND
EFFICIENCY

In a simple machine input work is done by the application of a single force, and the machine performs output work by means of a single force. During any such operation (Fig. 12-1), three processes occur:

1. Work is supplied to the machine.
2. Work is done against friction.
3. Output work is done by the machine.

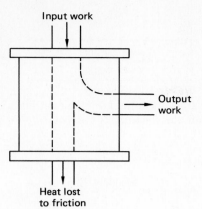

Fig. 12-1 Three processes occur in the operation of a machine: (1) the input of a certain amount of work, (2) the loss of energy in doing work against friction, and (3) the output of useful work.

Input work

Output work

Heat lost to friction

According to the principle of conservation of energy, these processes are related as follows:

Input work = work against friction + output work

The amount of useful work performed by a machine can never be greater than the work supplied to it. There will always be some loss due to friction or other dissipative forces. For example, in pumping up a bicycle tire with a small hand pump we exert a downward force on the plunger, causing air to be forced into the tire. That some of our input work is lost to friction can easily be verified by feeling how warm the wall of the hand pump becomes. The smaller we can make the friction loss in a machine, the greater the return for our effort. In other words, the effectiveness of a given machine can be measured by comparing its output work with the work supplied to it.

The **efficiency** E of a machine is defined as the ratio of the work output to the work input.

$$E = \frac{\text{work output}}{\text{work input}} \qquad (12\text{-}1)$$

The efficiency as defined in Eq. (12-1) will always be a number between 0 and 1. Common practice is to express this decimal as a percentage by multiplying by 100. For example, a machine which does 40 J of work when 80 J of work is supplied to it has an efficiency of 50 percent.

Another useful expression for efficiency can be noted from the definition of power as work per unit time. We can write

$$P = \frac{\text{work}}{t} \qquad \text{or} \qquad \text{Work} = Pt$$

The efficiency in terms of power input P_i and power output P_o is given by

$$E = \frac{\text{work output}}{\text{work input}} = \frac{P_o t}{P_i t}$$

or

$$E = \frac{\text{power output}}{\text{power input}} = \frac{P_o}{P_i}$$ (12-2)

EXAMPLE 12-1   A 60-hp motor winds a cable around a drum. (*a*) If the cable lifts a 3-ton load of bricks to a height of 12 ft in 3 s, calculate the efficiency of the motor. (*b*) At what rate is work done against friction?

Solution (a)   First compute the output power.

$$P_o = \frac{Fs}{t} = \frac{(6000 \text{ lb})(12 \text{ ft})}{3 \text{ s}}$$

$$= (24{,}000 \text{ ft} \cdot \text{lb/s})\frac{1 \text{ hp}}{550 \text{ ft} \cdot \text{lb/s}}$$

$$= 43.6 \text{ hp}$$

The efficiency is then found from Eq. (12-2):

$$E = \frac{P_o}{P_i} = \frac{43.6 \text{ hp}}{60 \text{ hp}} = 0.727$$

$$= 72.7\%$$

Solution (b)   The rate at which work is done against friction is the difference between the input power and the output power, or 16.4 hp.

12-2
MECHANICAL ADVANTAGE

Simple machines like the lever, block and tackle, chain hoist, gears, inclined plane, and the screw jack all play important roles in modern industry. We can illustrate the operation of any of these machines by the general diagram in Fig. 12-2. An input force F_i acts through a distance s_i, accomplishing the work $F_i s_i$. At the same time, an output force F_o acts through a distance s_o, performing the useful work $F_o s_o$.

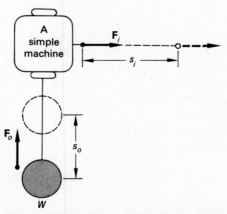

Fig. 12-2 During the operation of any simple machine, an input force F_i acts through a distance s_i while an output force F_o acts through a distance s_o.

MECHANICS

The **actual mechanical advantage** M_A of a machine is defined as the ratio of the output force F_o to the input force F_i.

$$M_A = \frac{\text{output force}}{\text{input force}} = \frac{F_o}{F_i} \qquad (12.3)$$

An actual mechanical advantage greater than 1 indicates that the output force is greater than the input force. Although most machines have values of M_A greater than 1, this is not always the case. In handling small, fragile objects, it is sometimes desirable to make the output force smaller than the input force.

In the previous section, we noted that the efficiency of a machine increases as frictional effects become small. Applying the conservation-of-energy principle to the simple machine in Fig. 12-2 yields

Work input = work against friction + work output

$$F_i s_i = (\text{work})_\mathscr{F} + F_o s_o$$

The most efficient engine possible would realize no losses due to friction. We can represent this *ideal* case by setting $(\text{work})_\mathscr{F} = 0$ in the above equation. Thus

$$F_o s_o = F_i s_i$$

Since this equation represents an ideal case, we define the *ideal mechanical advantage* M_I as

$$M_I = \frac{F_o}{F_i} = \frac{s_i}{s_o} \qquad (12\text{-}4)$$

The **ideal mechanical advantage** of a simple machine is equal to the ratio of the distance the input force moves to the distance the output force moves.

The efficiency of a simple machine is the ratio of output work to input work. Therefore, for the general machine of Fig. 12-2 we have

$$E = \frac{F_o s_o}{F_i s_i} = \frac{F_o/F_i}{s_i/s_o}$$

Finally, utilizing Eqs. (12-3) and (12-4), we obtain

$$E = \frac{M_A}{M_I} \qquad (12\text{-}5)$$

All the above concepts have been treated as they apply to a general machine. In the following sections we shall apply them to specific machines.

Possibly the oldest and most generally useful machine is the simple lever. A *lever* consists of any rigid bar pivoted at a certain point called the *fulcrum*. Figure 12-3 illustrates the use of a long rod to lift a weight W. We can calculate the ideal mechanical advantage of such a device in two ways. The first method involves the principle of equilibrium, and the second uses the principle of work, as discussed in the previous section. Since the equilibrium method is easier for the lever, we shall apply it first.

Fig. 12-3 The lever.

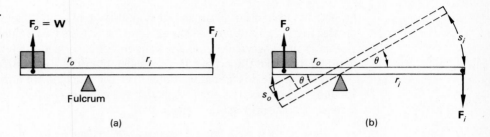

(a)   (b)

Because no translational motion is involved during the application of a lever, the condition for equilibrium is that the input torque equal the output torque:

$$F_i r_i = F_o r_o$$

The ideal mechanical advantage can be found from

$$\boxed{M_I = \frac{F_o}{F_i} = \frac{r_i}{r_o}} \qquad (12\text{-}6)$$

The ratio F_o/F_i is considered the *ideal* case because no friction forces are considered.

The same result is obtained from work considerations. Note from Fig. 12-3b that the force $\mathbf{F}_i$ moves through the arc distance s_i while the force $\mathbf{F}_o$ moves through the arc distance s_o. However, the two arcs are subtended by the same angle θ, and so we can write the proportion

$$\frac{s_i}{s_o} = \frac{r_i}{r_o}$$

Substitution into Eq. (12-4) will verify the result obtained from equilibrium considerations, that is, $M_I = r_i/r_o$.

EXAMPLE 12-2   An iron bar 3 m long is used to lift a 60-kg block. The bar is used as a lever, as shown in Fig. 12-3. The fulcrum is placed 80 cm from the block. What is the ideal mechanical advantage of the system, and what input force is required?

Solution   The distance $r_o = 0.8$ m; and the distance $r_i = 3$ m $- 0.8$ m $= 2.2$ m. Therefore, the ideal mechanical advantage is

$$M_I = \frac{r_i}{r_o} = \frac{2.2 \text{ m}}{0.8 \text{ m}} = 2.75$$

The output force in this case is equal to the weight of the 60-kg block ($W = mg$). Therefore, the required input force is given by

$$F_i = \frac{F_o}{M_I} = \frac{(60 \text{ kg})(9.8 \text{ m/s}^2)}{2.75}$$
$$= 214 \text{ N}$$

Before leaving the subject of the lever, it should be noted that very little of the input work is lost to friction forces. For all practical purposes, the actual mechanical advantage for a simple lever is equal to the ideal mechanical advantage. Other examples of the lever are illustrated in Fig. 12-4.

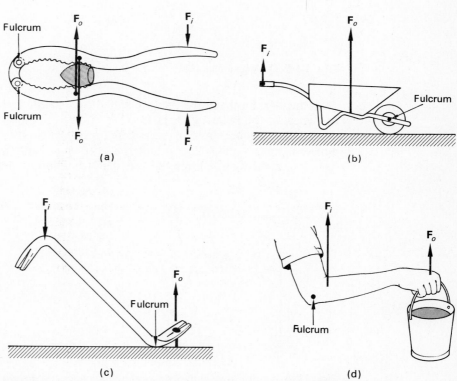

Fig. 12-4 The lever forms the operating principle of many simple machines.

12-4
APPLICATIONS OF THE LEVER PRINCIPLE

A serious limitation of the elementary lever is the fact that it operates through a small angle. There are many ways of overcoming this restriction by allowing for continuous rotation of the lever arm. For example, the *wheel and axle* (Fig. 12-5) allows for the continued action of the input force F_i. Applying the reasoning described in Sec. 12-2 for a general machine, it can be shown that

$$M_I = \frac{F_o}{F_i} = \frac{R}{r}$$

(12-7)

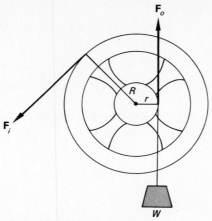

Fig. 12-5 The wheel and axle.

Thus the ideal mechanical advantage of a wheel and axle is the ratio of the radius of the wheel to the radius of the axle.

Another application of the lever concept is through the use of *pulleys*. A single pulley, as shown in Fig. 12-6, is simply a lever whose input moment arm is equal to its output moment arm. From the principle of equilibrium, the input force will equal

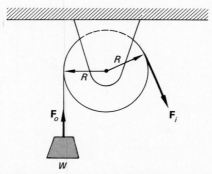

Fig. 12-6 A single fixed pulley serves only to change the direction of the input force.

the output force, and the ideal mechanical advantage will be

$$M_I = \frac{F_o}{F_i} = 1 \tag{12-8}$$

The only advantage of such a device lies in its ability to change the direction of an input force.

A single movable pulley (Fig. 12-7), on the other hand, has an ideal mechanical advantage of 2. Note that the two supporting ropes must each be shortened by 1 ft in order to lift the load through a distance of 1 ft. Therefore, the input force moves through a distance of 2 ft while the output force only moves a distance of 1 ft.

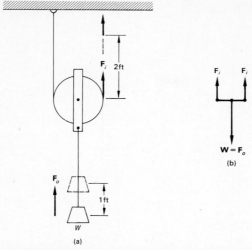

Fig. 12-7 A single movable pulley. *(a)* The input force moves through twice the distance that the output force travels. *(b)* The free-body diagram shows that $2F_i = F_o$.

Applying the principle of work, we have

$$F_i(2 \text{ ft}) = F_o(1 \text{ ft})$$

from which the ideal mechanical advantage is

$$M_I = \frac{F_o}{F_i} = 2 \qquad\qquad (12\text{-}9)$$

The same result can be shown by constructing a free-body diagram, as in Fig. 12-7*b*. From the figure it is evident that

$$2F_i = F_o$$

or

$$M_I = \frac{F_o}{F_i} = 2$$

The latter method is usually applied to problems involving movable pulleys since it allows one to associate M_I with the number of strands supporting the movable pulley.

EXAMPLE 12-3 Calculate the ideal mechanical advantage of the block-and-tackle arrangement shown in Fig. 12-8.

Solution We first construct a free-body diagram, as shown in Fig. 12-8*b*. From the figure we note that

$$4F_i = F_o$$

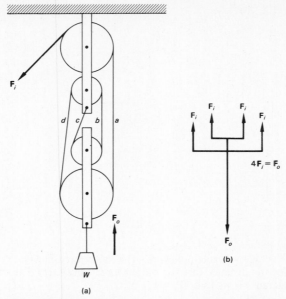

Fig. 12-8 The block and tackle. This arrangement has an ideal mechanical advantage of 4, since four strands support the movable block.

from which

$$M_I = \frac{F_o}{F_i} = \frac{4F_i}{F_i} = 4$$

Note that the uppermost pulley serves only to change the direction of the input force. The same M_I would result if F_i were applied upward at point a.

12-5
THE TRANS-MISSION OF TORQUE

The simple machines discussed so far are used to transmit and apply forces in order to move loads. In most mechanical applications, work is done by transmitting torque from one drive to another. For example, the belt drive (Fig. 12-9) transmits the torque from a driving pulley to an output pulley. The mechanical advantage of such a system is the ratio of the torques between the output pulley and the driving pulley:

$$M_I = \frac{\text{output torque}}{\text{input torque}} = \frac{\tau_o}{\tau_i}$$

From the definition of torque, we can write this expression in terms of the radii of the pulleys

$$M_I = \frac{\tau_o}{\tau_i} = \frac{F_o r_o}{F_i r_i}$$

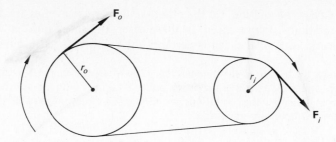

Fig. 12-9 The belt drive.

If there is no slippage between the belt and pulleys, it is safe to say that the tangential input force F_i is equal to the tangential output force F_o. Thus

$$M_I = \frac{F_o r_o}{F_i r_i} = \frac{r_o}{r_i}$$

Since the diameters of pulleys are usually specified instead of their radii, a more convenient expression is

$$M_I = \frac{D_o}{D_i} \tag{12-10}$$

where D_i is the diameter of the driving pulley and D_o is the diameter of the output pulley.

Suppose we now apply the principle of work to the belt drive. Remember that work is defined in rotary motion as the product of torque τ and angular displacement θ. For the belt drive, assuming ideal conditions, the input work equals the output work. Thus

$$\tau_i \theta_i = \tau_o \theta_o$$

The power input must also be equal to the power output. Dividing the above equation by the time t required to rotate through the angles θ_i and θ_o, we obtain

$$\tau_i \frac{\theta_i}{t} = \tau_o \frac{\theta_o}{t} \qquad \text{or} \qquad \tau_i \omega_i = \tau_o \omega_o$$

where ω_i and ω_o are the angular speeds of the input and output pulleys. Note that the ratio τ_o / τ_i represents the ideal mechanical advantage. Therefore, we can add another expression to Eq. (12-10), to obtain

$$M_I = \frac{D_o}{D_i} = \frac{\omega_i}{\omega_o} \tag{12-11}$$

This important result shows that the mechanical advantage is achieved at the expense of rotary motion. In other words, if the mechanical advantage is 2, the input shaft must rotate with twice the angular speed of the output shaft. The ratio ω_i / ω_o is sometimes referred to as the *speed ratio*.

If the speed ratio is greater than 1, the machine produces an output torque which is greater than the input torque. As we have seen, this feat is accomplished at the expense of rotation. On the other hand, many machines are designed to increase the rotational output speed. In these cases, the speed ratio is less than 1, and the increased rotational speed is accomplished with reduced torque output.

EXAMPLE 12-4   Consider the belt drive illustrated in Fig. 12-9, where the diameter of the small driving pulley is 6 in. and the diameter of the driven pulley is 18 in. A 6-hp motor drives the input pulley at 600 rpm. Calculate the revolutions per minute and torque delivered to the driven wheel if the system is 75 percent efficient.

Solution   We first calculate the ideal mechanical advantage (100 percent efficiency) of the system. From Eq. (12-11),

$$M_I = \frac{D_o}{D_i} = \frac{18 \text{ in.}}{6 \text{ in.}} = 3$$

Since the efficiency is 75 percent, the actual mechanical advantage is given from Eq. (12-5).

$$M_A = EM_I = (0.75)(3) = 2.25$$

Now the actual mechanical advantage is the simple ratio of output torque τ_o to input torque τ_i. Recalling that the power in rotational motion is equal to the product of torque and angular velocity, we can solve for τ_i as follows:

$$\tau_i = \frac{P_i}{\omega_i} = \frac{(6 \text{ hp})[(550 \text{ ft} \cdot \text{lb/s})/\text{hp}]}{(600 \text{ rev/min})(2\pi \text{ rad/rev})(1 \text{ min/60 s})}$$

$$= \frac{(6)(550 \text{ ft} \cdot \text{lb/s})}{20\pi \text{ rad/s}} = 52.5 \text{ ft} \cdot \text{lb}$$

Since $M_A = \tau_o/\tau_i$, the output torque is given by

$$\tau_o = M_A\tau_i = (2.25)(52.5 \text{ ft} \cdot \text{lb})$$

$$= 118 \text{ ft} \cdot \text{lb}$$

Assuming the belt does not slip, it will move with the same linear velocity v around each pulley. Since $v = \omega r$, we can write the equality

$$\omega_i r_i = \omega_o r_o \qquad \text{or} \qquad \omega_i D_i = \omega_o D_o$$

from which

$$\omega_o = \frac{\omega_i D_i}{D_o} = \frac{(600 \text{ rpm})(6 \text{ in.})}{18 \text{ in.}} = 200 \text{ rpm}$$

Note that the ratio of ω_i to ω_o yields the ideal mechanical advantage and not the actual mechanical advantage. The difference between M_I and M_A is due to friction, both in the belt and in the shaft bearings. Since greater tension on the belt will result in greater friction forces, maximum efficiency is obtained by reducing the belt tension until it just prevents the belt from slipping on the pulleys.

Before leaving our discussion of the transmission of torque, we must consider the application of *gears*. A gear is simply a notched wheel which can transmit torque by meshing with another notched wheel, as shown in Fig. 12-10. A pair of meshing gears differs from a belt drive only in the sense that the gears rotate in opposite directions. The same relationships derived for the belt drive hold for gears:

$$M_I = \frac{D_o}{D_i} = \frac{\omega_i}{\omega_o} \tag{12-12}$$

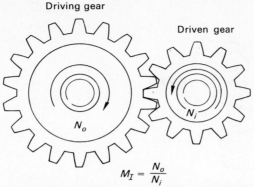

Driving gear

Driven gear

N_o

N_i

$$M_I = \frac{N_o}{N_i}$$

Fig. 12-10 Spur gears. The ideal mechanical advantage is the ratio of the number of teeth on the output gear to the number of teeth on the input gear.

A more useful expression makes use of the fact that the number of teeth N on the rim of a gear is proportional to its diameter D. Because of this dependence, the ratio of the number of teeth on the driven gear N_o to the number of teeth on the driving gear N_i is the same as the ratio of their diameters. Hence we can write

$$M_I = \frac{N_o}{N_i} = \frac{D_o}{D_i} \qquad (12\text{-}13)$$

The use of gears avoids the problem of slippage, which is common with belt drives. It also conserves space and allows for a greater torque to be transmitted.

In addition to the *spur* gears illustrated in Fig. 12-10, there are several other types of gears. Four common types are worm gears, helical gears, bevel gears, and planetary gears. Examples of each are shown in Fig. 12-11. The same general relationships apply for all these gears.

12-6
THE INCLINED PLANE

The only machines we have discussed so far involve application of the lever principle. A second fundamental machine is the *inclined plane*. Suppose you have to move a heavy load from the ground to a truck bed without hoisting equipment. You would probably select a few long boards and form a ramp from the ground to the bed of the truck. Experience has taught you that it takes less effort to push a load up a small elevation than it does to lift the load directly. Since a smaller input force results in the same output force, a mechanical advantage is realized. However, the smaller input force is accomplished at the cost of greater distance.

Consider the movement of a weight **W** up the inclined plane in Fig. 12-12. The slope angle θ is such that the weight must be moved through a distance s to reach a height h at the top of the incline. If we neglect friction, the work required to push the weight up the plane is the same as the work required to lift it up vertically. We can express this equality as

Work input = work output

$$F_i s = Wh$$

SIMPLE MACHINES **219**

Fig. 12-11 Four common types of gears: *(a)* helical, *(b)* planetary, *(c)* bevel, *(d)* worm. (The spur gear, which is the most common type, is shown in Fig. 10-10.)

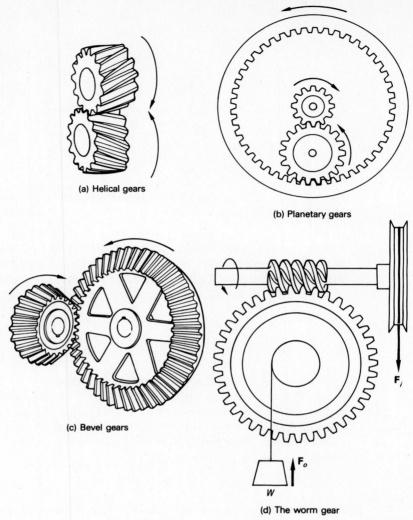

(a) Helical gears

(b) Planetary gears

(c) Bevel gears

(d) The worm gear

Fig. 12-12 The inclined plane. The input force represents the effort required to push the block up the plane; the output force is equal to the weight of the block.

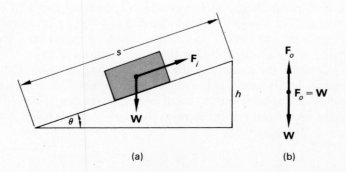

(a)

(b)

where F_i is the input force and W is the output force. The ideal mechanical advantage will be the ratio of the weight to the input force. Stating this symbolically, we have

$$M_I = \frac{W}{F_i} = \frac{s}{h} \tag{12-14}$$

EXAMPLE 12-5 The 200-lb wooden crate in Fig. 12-13 is to be raised to a loading platform 6 ft high. A ramp 12 ft long is used to slide the crate from the ground to the platform. Assume that the coefficient of friction is 0.3. (*a*) What is the ideal mechanical advantage of the ramp? (*b*) What is the actual mechanical advantage?

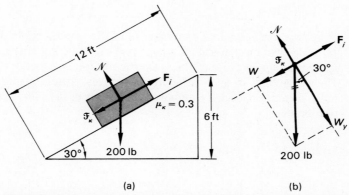

(a)   (b)

Fig. 12-13 Comparing the actual mechanical advantage of an inclined plane.

Solution (a) The ideal mechanical advantage, from Eq. (12-14), is

$$M_I = \frac{s}{h} = \frac{12 \text{ ft}}{6 \text{ ft}} = 2$$

This value represents the mechanical advantage of the ramp if it were frictionless.

Solution (b) The actual mechanical advantage is the ratio of the weight lifted to the required input force, considering friction. Applying the first condition for equilibrium to the free-body diagram (Fig. 12-13*b*), we find that the normal force $\mathcal{N}$ is given by

$$\mathcal{N} = W_y = (200 \text{ lb})(\cos 30°) = 173 \text{ lb}$$

from which the friction force must be

$$\mathcal{F} = \mu\mathcal{N} = (0.3)(173 \text{ lb}) = 51.9 \text{ lb}$$

Summing the forces along the plane, we obtain

$$F_i - \mathcal{F} - W_x = 0$$

But $W_x = (200 \text{ lb})(\sin 30°) = 100 \text{ lb}$, so that we have

$$F_i - 51.9 \text{ lb} - 100 \text{ lb} = 0$$
$$F_i = 51.9 \text{ lb} + 100 \text{ lb} = 152 \text{ lb}$$

We can now compute the actual mechanical advantage:

$$M_A = \frac{F_o}{F_i} = \frac{200 \text{ lb}}{152 \text{ lb}} = 1.32$$

It is left as an exercise for the student to show that the efficiency of the ramp is 66 percent.

12-7

APPLICATIONS OF THE INCLINED PLANE

Many machines apply the principle of the inclined plane. The simplest is the *wedge* (Fig. 12-14), which is actually a double inclined plane. In the ideal case, the mechanical advantage of a wedge of length L and thickness t is given by

$$M_I = \frac{L}{t} \tag{12-15}$$

This equation is a direct consequence of the general relation expressed by Eq. (12-14). The ideal mechanical advantage is always much greater than the actual mechanical advantage because of the large friction forces between the surfaces in contact. The wedge finds its application in axes, knives, chisels, planers, and all other cutting tools. A cam is a kind of rotary wedge which is used to lift valves in internal combustion engines.

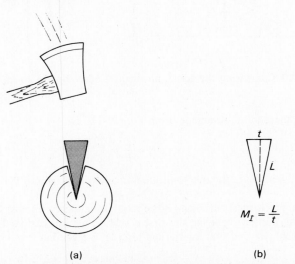

$$M_I = \frac{L}{t}$$

(a)   (b)

Fig. 12-14 The wedge is actually a double inclined plane.

One of the most useful applications of the inclined plane is the *screw*. This principle can be explained by examining a common tool known as the *screw jack* (Fig. 12-15). The threads are essentially an inclined plane wrapped continuously around a cylindrical shaft. When the input force F_i turns through a complete revolution ($2\pi R$), the output force F_o will advance through the distance p. This distance p is actually the distance between two adjacent threads, and it is called the *pitch* of the screw. The ideal mechanical advantage is the ratio of the input distance to the output distance.

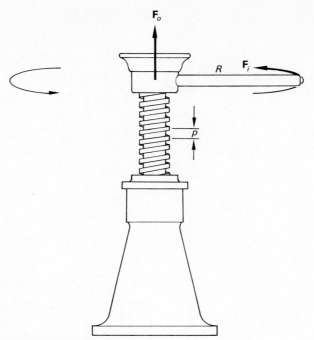

Fig. 12-15 The screw jack.

$$M_I = \frac{s_i}{s_o} = \frac{2\pi R}{p} \qquad (12\text{-}16)$$

The screw is an example of a very inefficient machine, but in this case it is usually an advantage since friction forces are needed to hold the load in place while the input force is not being applied.

SUMMARY

A simple machine has been defined as a device which converts a single input force F_i into a single output force F_o. In general, the input force moves through a distance s_i, and the output force moves through a distance s_o. The purpose is to accomplish useful work in a manner suited to a particular application. The major concepts are given below:

- A simple machine is a device which converts a single input force F_i into a single output force F_o. The input force moves through a distance s_i, and the output force moves a distance s_o. There are two mechanical advantages:

$$M_A = \frac{F_o}{F_i} \qquad \text{actual mechanical advantage (friction considered)}$$

$$M_I = \frac{s_o}{s_i} \qquad \text{ideal mechanical advantage (assumes no friction)}$$

- The efficiency of a machine is a ratio of output work to input work. It is normally expressed as a percentage and can be calculated from any of the following relations:

$$E = \frac{\text{work output}}{\text{work input}} \qquad E = \frac{\text{power output}}{\text{power input}} \qquad E = \frac{M_A}{M_I}$$

- The ideal mechanical advantages for a number of simple machines are given below.

$$M_I = \left(\frac{F_o}{F_i}\right)_{\text{ideal}} = \frac{r_i}{r_o} \qquad \textit{Lever}$$

$$M_I = \left(\frac{F_o}{F_i}\right)_{\text{ideal}} = \frac{R}{r} \qquad \textit{Wheel and Axle}$$

$$M_I = \frac{D_o}{D_i} = \frac{\omega_i}{\omega_o} \qquad \textit{Belt Drive}$$

$$M_I = \frac{W}{F_i} = \frac{s}{h} \qquad \textit{Inclined Plane}$$

$$M_I = \frac{L}{t} \qquad \textit{Wedge}$$

$$M_I = \frac{N_o}{N_i} = \frac{D_o}{D_i} \qquad \textit{Gears}$$

$$M_I = \frac{s_i}{s_o} = \frac{2\pi R}{P} \qquad \textit{Screw Jack}$$

QUESTIONS

12-1. Define the following terms:
   a. Machine
   b. Efficiency
   c. Lever
   d. Pulley
   e. Gears
   f. Wedge
   g. Screw
   h. Pitch
   i. Inclined plane
   j. Wheel and axle
   k. Actual mechanical advantage
   l. Ideal mechanical advantage
   m. Belt drive

12-2. What is meant by *useful work* or *output work*? What is meant by *input work*? Write the general relationship between input work and output work.

12-3. Two jacks are operated simultaneously to lift the front end of a car. Immediately afterward, it is noted that the left jack feels warmer than the right one. Which jack is more efficient? Explain.

12-4. A machine may alter the magnitude and/or the direction of an input force. (a) Name several examples in which both changes occur. (b) Give examples in which only the magnitude of the input force is altered. (c) Give some examples in which only the direction is altered.

12-5. A machine lifts a load through a vertical distance of 4 ft while the input force moves through a distance of 2 ft. Would this machine be helpful in lifting large weights? Explain.

12-6. A bicycle can be operated in three gear ranges. In *low range* the pedals describe two complete revolutions while the rear wheel turns through one revolution. In *medium range* the pedals and the wheels turn at the same rate. In *high range* the rear wheel of the bicycle completes two revolutions for every complete pedal revolution. Discuss the advantages and disadvantages of each range.

12-7. What happens to the ideal mechanical advantage if a simple machine is operated in reverse? What happens to its efficiency?

12-8. Give several examples of machines which have an actual mechanical advantage less than 1.

12-9. Why do buses and trucks often use larger steering wheels than those found on automobiles? What principle is used?

12-10. Draw diagrams of pulley systems which have ideal mechanical advantages of 2, 3, and 5.

12-11. Usually the road to the top of a mountain winds around the mountain instead of going straight up the side. Why? If we neglect friction, is more work required to reach the top along the spiral road? Is more power required? If we consider friction, would it require less work to drive straight up the side of the mountain? Explain.

PROBLEMS

Simple Machines, Efficiency, and Mechanical Advantage

12-1. A 25 percent efficient machine performs external work of 200 J. What input work is required?

Ans. 800 J

12-2. During the operation of a 300-hp engine, energy is lost at the rate of 200 hp because of friction. What are the useful output power and the efficiency of the engine?

12-3. A 60-W motor lifts a 2-kg mass to a height of 4 m in 3 s. (a) Compute the output power. (b) What is the efficiency of the motor? (c) What is the rate at which work is done against friction?

Ans. (a) 26.1 W, (b) 43.6%, (c) 33.9 W

12-4. A frictionless machine lifts a 200-lb load through a distance of 10 ft. If the input force moves through a distance of 300 ft, what is the ideal mechanical advantage of the machine? What is the magnitude of the input force?

12-5. A 60 percent efficient machine lifts a 10-kg mass to a height of 9 m in 3 s. What is the required input power?

Ans. 490 W

Applications of the Lever Principle

12-6. One edge of a 200-lb safe is lifted with a 4-ft steel rod. What input force is required at the end of the rod if a fulcrum is placed 6 in. from the safe? (To lift one edge, a force equal to one-half the weight of the safe is required.)

12-7. A 60-N weight is lifted in the three different ways shown in Fig. 12-16. Compute the ideal mechanical advantage and the required input force for each application.

Ans. (a) 2, 30 N; (b) 3, 20 N; (c) 0.33, 180 N

Fig. 12-16

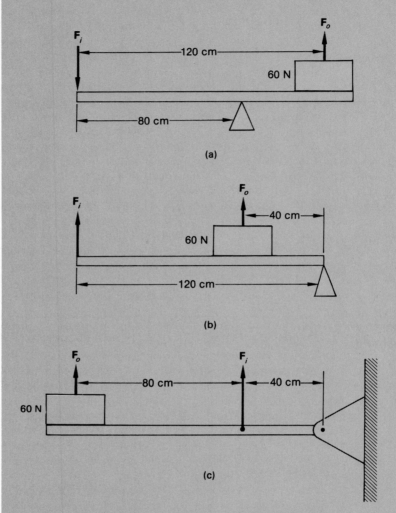

(a)

(b)

(c)

12-8. A 20-kg mass is to be lifted with a rod 2 m long. If you can exert a downward force of 40 N on one end of the rod, where should you place a block of wood to act as a fulcrum?

12-9. A wheel 20 cm in diameter is attached to an axle with a diameter of 6 cm. If a weight of 400 N is attached to the axle, what force must be applied to the rim of the wheel to lift the weight at constant speed? Neglect friction.

Ans. 120 N

12-10. What is the mechanical advantage of a screwdriver used as a wheel and axle if its blade is 0.3 in. wide and the handle diameter is 0.8 in.?

12-11. Determine the force F required to lift a 200-N load w with each of the four pulley systems shown in Fig. 12-17.

Ans. (a) 100 N, (b) 50 N, (c) 40 N, (d) 50 N

Fig. 12-17

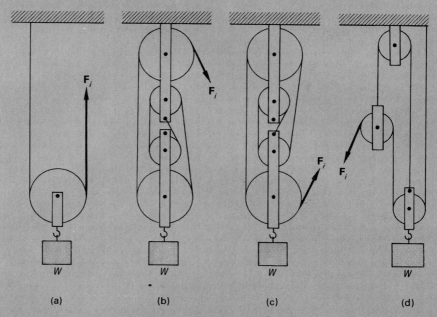

(a) (b) (c) (d)

** **12-12.** The *chain hoist* (Fig. 12-18) is a combination of the wheel and axle and the block and tackle. Show that the ideal mechanical advantage of such a device is given by

$$M_I = \frac{2R}{R - r}$$

Fig. 12-18 The chain hoist.

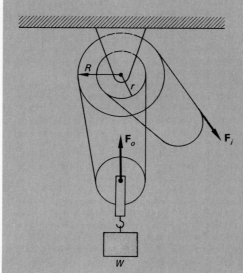

12-13. Assume that the larger radius in Fig. 12-18 is three times the smaller radius. What input force is required to lift a 10-kg load if there is no friction?

Ans. 32.7 N

The Transmission of Torque

12-14. An 8-hp motor drives the input pulley of a belt drive at 600 rpm. Compute the rpm and torque delivered to the driven pulley if the system is 60 percent efficient. The diameters of the input and output pulleys are 4 and 8 in., respectively.

12-15. A pair of step pulleys (Fig. 12-19) makes it possible to change output speeds merely by shifting the belt. If an electric motor turns the input pulley at 2000 rpm, find the possible angular speeds of the output shaft. The pulley diameters are 4, 6, and 8 in.

Ans. (a) Small input pulley: 2000, 1330, 1000 rpm; (b) middle input pulley: 3000, 2000, 1500 rpm; (c) large input pulley: 4000, 2670, 2000 rpm

Fig. 12-19 The step pulley.

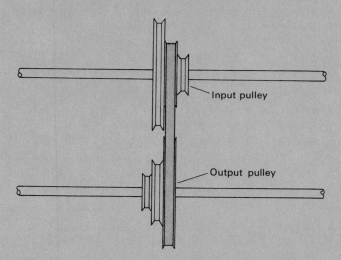

Input pulley

Output pulley

12-16. A worm drive similar to that shown in Fig. 12-11 has n teeth in the gear wheel. (If $n = 80$, one complete turn of the worm will advance the wheel one-eightieth of a revolution.) Derive an expression for the ideal mechanical advantage of the worm gear in terms of the radius of the input pulley R, the radius of the output shaft r, and the number of teeth n in the gear wheel.

12-17. The worm drive of Prob. 12-16 has 80 teeth in the gear wheel. If the radius of the input wheel is 30 cm and the radius of the output shaft is 5 cm, what input force is required to lift a 1200-kg load? Assume an efficiency of 80 percent.

Ans. 30.6 N

Applications of the Inclined Plane

12-18. It is desired that a wedge have an ideal mechanical advantage of 10. What must be the thickness of the base if the wedge is 20 cm long?

12-19. A 10-kg crate is moved from the ground to a loading platform by means of a ramp 6 m long and 2 m high. Assume that $\mu_k = 0.25$. (a) What is the ideal mechanical advantage of the ramp? **(b)** What is the actual mechanical advantage? **(c)** What is the efficiency of the ramp?

Ans. (a) 3, (b) 1.76, (c) 59%

* 12-20. The lever of a screw jack is 24 in. long. If the screw has six threads per inch, what is its ideal mechanical advantage? If the jack is 15 percent efficient, what force is needed to lift 2000 lb?

* 12-21. A wrench with a 6-in. handle acts to tighten a ¼-in.-diameter bolt having 10 threads per inch. What is the pitch of the bolt? Compute the ideal mechanical advantage. If an input force of 20 lb actually results in a 600-lb force on the nut, what is the efficiency?

Ans. 0.1 in., 377, 8%

Additional Problems

* 12-22. A shaft rotating at 800 rpm delivers a torque of 240 N · m to an output shaft which is rotating at 200 rpm. If the efficiency of the machine is 70 percent, compute the output torque. What is the output power?

12-23. A screw jack has a screw whose pitch is 0.25 in. Its handle is 16 in. long, and a load of 1.9 tons is being lifted. **(a)** Neglecting friction, what force is required at the end of the handle? **(b)** What is the ideal mechanical advantage of this jack?

Ans. (a) 9.45 lb, (b) 402

12-24. A certain refrigeration compressor comes equipped with a 250-mm-diameter pulley and is designed to operate at 600 rpm. What should be the diameter of the motor pulley if the motor speed is 2000 rpm?

* 12-25. In a fan belt, the driving wheel is 20 cm in diameter and the driven wheel is 50 cm in diameter. The power input comes from a 4-kW motor which causes the driving wheel to rotate at 300 rpm. If the efficiency is 80 percent, calculate the rpm and the torque delivered to the driven wheel.

Ans. 120 rpm, 255 N · m

Elasticity

OBJECTIVES

After completing this chapter, you should be able to:

1. Demonstrate by example and discussion your understanding of *elasticity, elastic limit, stress, strain,* and *ultimate strength.*
2. Write and apply formulas for calculating Young's modulus, shear modulus, and bulk modulus.
3. Define and discuss the meanings of *hardness, malleability,* and *ductility* as applied to metals.

Until now we have been discussing objects in motion or at rest. The objects have been assumed to be rigid and absolutely solid. But we know that wire can be stretched, that rubber tires will compress, and that bolts will sometimes break. A more complete understanding of nature requires a study of the mechanical properties of matter. The concepts of *elasticity, tension,* and *compression* are analyzed in this chapter. As the kinds of alloys increase and the demands on them become greater, our knowledge of such concepts becomes more important. For example, the stress placed on space ships or on cables in modern bridges is of a magnitude unheard of a few years ago.

13-1
ELASTIC PROPERTIES OF MATTER

We define an *elastic* body as one which returns to its original size and shape when a deforming force is removed. Rubber bands, golf balls, trampolines, diving boards, footballs, and springs are common examples of elastic bodies. Putty, dough, and clay are examples of inelastic bodies. For all elastic bodies, we shall find it convenient to establish a cause-and-effect relationship between a deformation and the deforming forces.

Consider the coiled spring of length *l* shown in Fig. 13-1. We can study its elasticity by adding successive weights and observing the increase in length. A 2-lb weight lengthens the spring by 1 in.; a 4-lb weight lengthens the spring by 2 in.; and a 6-lb weight lengthens the spring by 3 in. Evidently, there is a direct relationship between the elongation of a spring and the applied force.

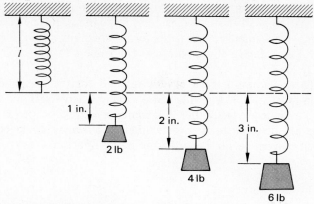

Fig. 13-1 The uniform elongation of a spring.

Robert Hooke first stated this relationship in connection with the invention of a balance spring for a clock. In general, he found that a force F acting on a spring (Fig. 13-2) produces an elongation s that is directly proportional to the magnitude of the

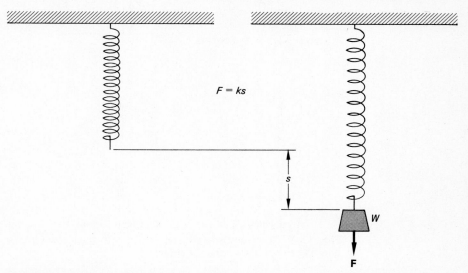

Fig. 13-2 The relation between a stretching force F and the elongation it produces.

force. *Hooke's law* can be written

$$F = ks \tag{13-1}$$

The proportionality constant k varies extremely with the type of material and is called the *spring constant*. For the example illustrated in Fig. 13-1, the spring constant is

$$k = \frac{F}{s} = 2 \text{ lb/in.}$$

Hooke's law is by no means restricted to coiled springs; it will apply to the deformation of all elastic bodies. In order to make the law more generally applicable, it will be convenient to define the terms *stress* and *strain*. Stress refers to the *cause* of an elastic deformation, whereas strain refers to the *effect*, i.e., the deformation itself.

Three common types of stresses and their corresponding deformations are shown in Fig. 13-3. A *tensile* stress occurs when equal and opposite forces are directed away from each other. A *compressive* stress occurs when equal and opposite forces are directed toward each other. A *shearing* stress occurs when equal and opposite forces do not have the same line of action.

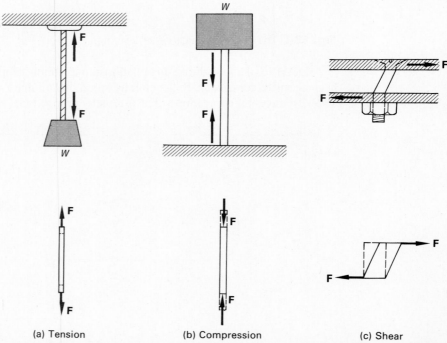

<table>
<tr><td>(a) Tension</td><td>(b) Compression</td><td>(c) Shear</td></tr>
</table>

Fig. 13-3 Three common stresses shown with their corresponding deformations: *(a)* tension, *(b)* compression, and *(c)* shear.

The effectiveness of any force producing a stress is highly dependent upon the area over which the force is distributed. For this reason, a more complete definition of stress can be stated as follows:

> **Stress** is the ratio of an applied force to the area over which it acts, e.g., newtons per square meter or pounds per square foot.

As mentioned earlier, the term *strain* must represent the effect of a given stress. The general definition of strain might be as follows:

> **Strain** is the relative change in the dimensions or shape of a body as the result of an applied stress.

In the case of a tensile or compressive stress, the strain may be considered a change in length per unit length. A shearing stress, on the other hand, may alter only the shape of a body without changing its dimensions. Shearing strain is usually measured in terms of an angular displacement.

The *elastic limit* is the maximum stress a body can experience without becoming permanently deformed. For example, an aluminum rod whose cross-sectional area is 1 in.^{2} will become permanently deformed by the application of a tensile force greater than 19,000 lb. This does not mean that the aluminum rod will break at this point; it means only that the rod will not return to its original size. In fact, the tension can be increased to about 21,000 lb before the rod breaks. It is this property of metals which allows them to be drawn out into wires of smaller cross sections. The greatest stress a wire can withstand without breaking is known as its *ultimate strength.*

If the elastic limit of a material is not exceeded, we can apply Hooke's law to any elastic deformation. Within the limits of a given material, it has been experimentally verified that the ratio of a given stress to the strain it produces is a constant. In other words, the stress is directly proportional to the strain. *Hooke's law* states:

> *Provided that the elastic limit is not exceeded, an elastic deformation (strain) is directly proportional to the magnitude of the applied force per unit area (stress).*

If we call the proportionality constant the *modulus of elasticity,* we can write Hooke's law in its most general form:

$$\text{Modulus of elasticity} = \frac{stress}{strain} \tag{13-2}$$

In the following sections we shall discuss the specific applications of this fundamental relation.

<table>
<tr><td>

13-2

YOUNG'S MODULUS

</td><td>

In this section we consider longitudinal stresses and strains as they apply to wires, rods, or bars. For example, in Fig. 13-4 a force **F** applied to the end of a wire of cross-sectional area A. The longitudinal stress is given by

</td></tr>
</table>

$$\text{Longitudinal stress} = \frac{F}{A}$$

The metric unit for stress is the *newton per square meter,* which is identical to the *pascal* (Pa).

$$1 \text{ Pa} = 1 \text{ N/m}^2$$

The USCS unit for stress is the *pound per square inch* (lb/in.^{2}). Since the pound per square inch remains in common use, it will be helpful to compare it with the SI unit:

$$1 \text{ lb/in.}^2 = 6895 \text{ Pa} = 6.895 \text{ kPa}$$

The effect of such a stress is to stretch the wire, i.e., to increase its length. Hence the longitudinal strain can be represented by the change in length per unit length. We can write

$$\text{Longitudinal strain} = \frac{\Delta l}{l}$$

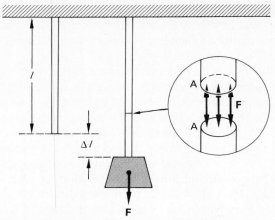

Fig. 13-4 Computing Young's modulus for a wire of cross section A. The elongation Δl is exaggerated for clarity.

where l is the original length and Δl is the elongation. Experimentation has shown that a comparable decrease in length occurs for a compressive stress. The same equations will apply whether we are discussing an object under tension or an object under compression.

If we define the longitudinal modulus of elasticity as *Young's modulus Y*, we can write Eq. (13-2) as

$$\text{Young's modulus} = \frac{longitudinal\ stress}{longitudinal\ strain}$$

$$Y = \frac{F/A}{\Delta l/l} = \frac{Fl}{A\ \Delta l} \qquad (13\text{-}3)$$

The units of Young's modulus are the same as the units of stress, i.e., pounds per square inch or pascals. This follows since the longitudinal strain is a unitless quantity. Representative values for some of the most common materials are listed in Tables 13-1 and 13-2.

Table 13-1 Elastic Constants for Various Materials in SI Units

Material	Young's modulus Y, MPa*	Shear modulus S, MPa	Bulk modulus B, MPa	Elastic limit, MPa	Ultimate strength, MPa
Aluminum	68,900	23,700	68,900	131	145
Brass	89,600	35,300	58,600	379	455
Copper	117,000	42,300	117,000	159	338
Iron	89,600	68,900	96,500	165	324
Steel	207,000	82,700	159,000	248	489

*(1 MPa = 10^6 Pa)

Table 13-2 Elastic Constants for Various Materials in USCS Units

Material	Young's modulus Y, lb/in.^{2}	Shear modulus S, lb/in.^{2}	Bulk modulus B, lb/in.^{2}	Elastic limit, lb/in.^{2}	Ultimate strength, lb/in.^{2}
Aluminum	10×10^6	3.44×10^6	10×10^6	19,000	21,000
Brass	13×10^6	5.12×10^6	8.5×10^6	55,000	66,000
Copper	17×10^6	6.14×10^6	17×10^6	23,000	49,000
Iron	13×10^6	10×10^6	14×10^6	24,000	47,000
Steel	30×10^6	12×10^6	23×10^6	36,000	71,000

EXAMPLE 13-1 A telephone wire 120 m long and 2.2 mm in diameter is stretched by a force of 380 N. What is the longitudinal stress? If the length after stretching is 120.10 m, what is the longitudinal strain? Determine Young's modulus for the wire.

Solution The cross-sectional area of the wire is

$$A = \frac{\pi D^2}{4} = \frac{\pi (2.2 \times 10^{-3} \text{ m})^2}{4} = 3.8 \times 10^{-6} \text{ m}^2$$

$$\text{Stress} = \frac{F}{A} = \frac{380 \text{ N}}{3.8 \times 10^{-6} \text{ m}^2}$$

$$= 100 \times 10^6 \text{ N/m}^2 = 100 \text{ MPa}$$

$$\text{Strain} = \frac{\Delta l}{l} = \frac{0.10 \text{ m}}{120 \text{ m}} = 8.3 \times 10^{-4}$$

$$Y = \frac{\text{stress}}{\text{strain}} = \frac{100 \text{ MPa}}{8.3 \times 10^{-4}} = 120,000 \text{ MPa}$$

EXAMPLE 13-2 What is the maximum load which can be hung from a steel wire $\frac{1}{4}$ in. in diameter if its elastic limit is not to be exceeded? Determine the increase in length under this load if the original length is 3 ft.

Solution From Table 13-2, the elastic limit for steel is 36,000 lb/in.^{2} Since this value represents the limiting stress, we write

$$\frac{F}{A} = 36,000 \text{ lb/in.}^2$$

where A is given by

$$A = \frac{\pi D^2}{4} = \frac{\pi (0.25 \text{ in.})^2}{4} = 0.0491 \text{ in.}^2$$

Thus the limiting load is

$$F = (36,000 \text{ lb/in.}^2)A$$
$$= (36,000 \text{ lb/in.}^2)(0.0491 \text{ in.}^2) = 1770 \text{ lb}$$

The increase in length under such a load is found from Eq. (13-3) as follows:

$$\Delta l = \frac{l}{Y} \frac{F}{A} = \frac{36 \text{ in.}}{30 \times 10^6 \text{ lb/in.}^2} (36,000 \text{ lb/in.}^2)$$

$$= 0.0432 \text{ in.}$$

13-3
SHEAR **MODULUS**

Compressive and tensile stresses produce a slight change in volume as a result of altered dimensions. As mentioned earlier, a shearing stress alters only the shape of a body, leaving its volume unchanged. For example, consider the parallel noncurrent forces acting on the cube in Fig. 13-5. The applied force causes each successive layer of atoms to slip sideways, much like the pages of a book under similar stress. The interatomic forces restore the block to its original shape when the stress is relieved.

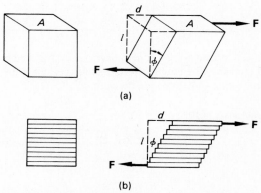

Fig. 13-5 Shearing stress and shearing strain.

The shearing stress is defined as the ratio of the tangential force F to the area A over which it is applied. The shearing strain is defined as the angle ϕ (in radians), which is called the *shearing angle* (refer to Fig. 13-5b). Applying Hooke's law, we can now define the *shear modulus S* as follows:

$$S = \frac{\text{shearing stress}}{\text{shearing strain}} = \frac{F/A}{\phi} \qquad (13\text{-}4)$$

The angle ϕ is usually so small that it is approximately equal to tan ϕ. Making use of this fact, we can rewrite Eq. (13-4) in the form

$$S = \frac{F/A}{\tan \phi} = \frac{F/A}{d/l} \qquad (13\text{-}5)$$

Since the value of S is an indication of the rigidity of a body, it is sometimes referred to as the *modulus of rigidity*.

EXAMPLE 13-3   A steel stud (Fig. 13-6) 1 in. in diameter projects 1.5 in. out from the wall. If the end of the bolt is subjected to a shearing force of 8000 lb, compute its downward deflection.

Solution   The cross-sectional area is

$$A = \frac{\pi D^2}{4} = \frac{\pi (1 \text{ in.})^2}{4} = 0.785 \text{ in.}^2$$

If we represent the downward deflection by d, we can solve as follows:

$$S = \frac{F/A}{d/l} = \frac{Fl}{Ad}$$

$$d = \frac{Fl}{AS} = \frac{(8000 \text{ lb})(1.5 \text{ in.})}{(0.785 \text{ in.}^2)(12 \times 10^6 \text{ lb/in.}^2)}$$

$$= 1.27 \times 10^{-3} \text{ in.}$$

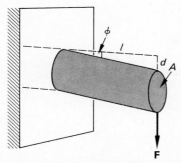

Fig. 13-6 The downward deflection of a stud is an example of shearing strain.

13-4
VOLUME ELASTICITY; BULK MODULUS

So far we have considered stresses which cause a change in the shape of an object or result in primarily one-dimensional strains. In this section we shall be concerned with changes in volume. For example, consider the cube in Fig. 13-7 on which forces are applied uniformly over the surface. The initial volume of the cube is denoted by V, and the area of each face is represented by A. The resultant force F applied normal to each face causes a change in volume $-\Delta V$. The minus sign indicates that the

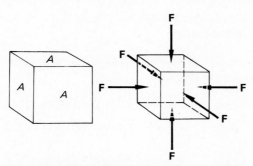

Fig. 13-7 The bulk modulus. The original volume is reduced by the application of a uniform compressive force on each face.

change represents a volume reduction. The *volume stress F/A* is the normal force per unit area, whereas the volume strain $-\Delta V/V$ is the change in volume per unit volume. Applying Hooke's law, we define the modulus of volume elasticity, or *bulk modulus,* as follows:

$$B = \frac{\text{volume stress}}{\text{volume strain}} = -\frac{F/A}{\Delta V/V} \tag{13-6}$$

This type of strain applies to liquids as well as solids. Table 13-3 lists the bulk moduli for a few of the more common liquids. When working with liquids, it is sometimes more convenient to represent the stress as pressure P, which is defined as the force per unit area F/A. With this definition, we can rewrite Eq. (13-6):

$$B = \frac{-P}{\Delta V/V} \qquad (13\text{-}7)$$

Table 13-3 Bulk Moduli for Liquids

Liquid	Bulk modulus B	
	lb/in.^{2}	MPa
Benzene	1.5×10^5	1,050
Ethyl alcohol	1.6×10^5	1,100
Mercury	40×10^5	27,000
Oil	2.5×10^5	1,700
Water	3.1×10^5	2,100

The reciprocal of the bulk modulus is called the *compressibility k*. Often it is more convenient to study the elasticity of materials by measuring their compressibilities. By definition,

$$k = \frac{1}{B} = -\frac{1}{P}\frac{\Delta V}{V_0} \qquad (13\text{-}8)$$

Equation (13-8) indicates that the compressibility is the fractional change in volume per unit increase in pressure.

EXAMPLE 13-4   A hydrostatic press contains 5 liters of water. Find the decrease in volume of the water when it is subjected to a pressure of 2000 kPa.

Solution   The decrease in volume is found by solving for ΔV from Eq. (13-7)

$$\Delta V = -\frac{PV}{B} = -\frac{(2 \times 10^6 \text{ Pa})(5 \text{ L})}{2.1 \times 10^9 \text{ Pa}}$$
$$= -0.00476 \text{ L} = -4.76 \text{ mL}$$

13-5
OTHER PHYSICAL PROPERTIES OF METALS

In addition to the elasticity, the tensile strength, and the shearing strength of materials, there are other important properties of metals. A solid consists of molecules arranged so closely together that they attract each other very strongly. This attraction, called *cohesion*, gives a solid a definite shape and size. It also affects its usefulness to industry as a working material. Such properties as hardness, ductility, malleability, and conductivity must be understood before metals are chosen for specific applications. Three of these properties are illustrated in Fig. 13-8.

Hardness is an industrial term used to describe the ability of metals to resist forces that tend to penetrate them. Hard materials resist being scratched, worn away,

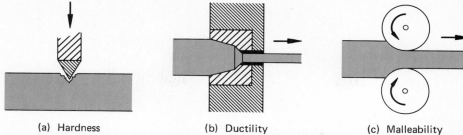

| (a) Hardness | (b) Ductility | (c) Malleability |

Fig. 13-8 Illustrating the working properties of metals: *(a)* A hard metal resists penetration; *(b)* a ductile metal can be drawn into a wire; and *(c)* a malleable metal can be rolled into sheets.

penetrated, or otherwise damaged physically. Some metals such as sodium and potassium are very soft, while iron and steel are two of the hardest materials. The hardness of metals is tested with machines that push a coneshaped diamond point into test materials. The penetration is measured, and a hardness reading is taken directly from a dial.

Two other special properties of materials are *ductility* and *malleability*. The meaning of each of these terms is seen from Fig. 13-8. Ductility is defined as the ability of a metal to be drawn out into a wire. Tungsten and copper are very ductile. Malleability is the property which enables us to hammer or bend metals into any desired shape or to roll them into sheets. Most metals are malleable, gold being the most malleable.

Conductivity refers to the ability of metals to permit the flow of electricity. The best conductors are silver, copper, gold, and aluminum, in that order. More will be said about this property in later chapters.

SUMMARY

In industry, we must utilize materials effectively and for appropriate situations. Otherwise, metal failures will result in costly damage or serious injury to employees. In this chapter, we have discussed the elastic properties of matter and some of the formulas used to predict the effects of stress on certain solids. The following points will summarize this chapter:

- According to *Hooke's law,* an elastic body will deform or elongate an amount *s* under the application of a force *F*. The constant of proportionality *k* is the *spring constant:*

$$F = ks \qquad k = \frac{F}{s} \qquad \text{\textit{Hooke's Law}}$$

- *Stress* is the ratio of an applied force to the area over which it acts. *Strain* is the relative change in dimensions which results from the stress. For example,

$$\text{Longitudinal stress} = \frac{F}{A} \qquad \text{Longitudinal strain} = \frac{\Delta l}{l}$$

- The *modulus of elasticity* is the constant ratio of stress to strain:

$$\text{Modulus of elasticity} = \frac{stress}{strain}$$

- *Young's modulus Y* is for longitudinal deformations:

$$Y = \frac{F/A}{\Delta l/l} \qquad \text{or} \qquad Y = \frac{Fl}{A \cdot \Delta l} \qquad \textit{Young's Modulus}$$

- A shearing strain occurs when an angular deformation ϕ is produced:

$$S = \frac{F/A}{\tan \phi} \qquad \text{or} \qquad S = \frac{F/A}{d/l} \qquad \textit{Shear Modulus}$$

- Whenever an applied stress results in a change in volume ΔV, you will need the *bulk modulus B,* given by

$$B = -\frac{-F/A}{\Delta V/V} \qquad \textit{Bulk Modulus}$$

The reciprocal of the bulk modulus is called the compressibility.

QUESTIONS

13-1. Define the following terms:
   a. Elasticity
   b. Hooke's law
   c. Spring constant
   d. Tensile stress
   e. Compressive stress
   f. Shear stress
   g. Strain
   h. Elastic limit
   i. Ultimate strength
   j. Young's modulus
   k. Shear modulus
   l. Bulk modulus

13-2. Explain clearly the relationship between stress and strain.

13-3. Two wires have the same length and cross-sectional area but are not of the same material. Each wire is hung from the ceiling with a 2000-lb weight attached. The wire on the left stretches twice as far as the one on the right. Which has the greater Young's modulus?

13-4. Does Young's modulus depend on the length and cross-sectional area? Explain.

13-5. Two wires, A and B, are made of the same material and are subjected to the same loads. Discuss their relative elongations when (a) wire A is twice as long as wire B and has twice the diameter of wire B and (b) wire A is twice as long as wire B and has one-half the diameter of wire B.

13-6. After studying the various elastic constants given in Tables 13-1 and 13-2, would you say it was usually easier to stretch a material or to shear a material? Explain.

13-7. A 400-lb weight is evenly supported by three wires of the same dimensions, one of copper, one of aluminum, and one of steel. Which wire experiences the greatest stress? Which experiences the least stress? Which wire experiences the greatest strain? Which experiences the least strain?

13-8. Discuss the various stresses resulting when a machine screw is tightened.

13-9. Give several practical examples of longitudinal, shearing, and volume strains.

13-10. For a given metal, would you expect there to be any relation between its modulus of elasticity and its coefficient of restitution? Discuss.

13-11. Which has the greater compressibility, steel or water?

PROBLEMS

Elastic Properties of Matter

13-1. When a mass of 500 g is hung from a spring, the spring stretches 3 cm. (a) What is the spring constant? (b) How much farther will it stretch if an additional 500-g mass is hung from it?

Ans. (a) 163 N/m, (b) an additional 3 cm

13-2. A 6-in. spring has a 4-lb weight hung from one end causing the new length to be 6.5 in. (a) What is the spring constant? (b) What is the strain?

13-3. A coil spring 12 cm long is used to support a 1.8-kg mass producing a strain of 0.10. How far did the spring stretch? What is the spring constant? What total mass should be hung from this spring if an elongation of 4 cm is desired?

Ans. 1.2 cm, 14.7 N/cm, 6 kg

Young's Modulus

13-4. A metal wire has a diameter of 1 mm and a length of 2 m. A 500-kg mass is hung from the end, and the wire stretches by 1.40 cm. (a) What is the stress? (b) What is the strain? (c) What is Young's modulus for this metal?

13-5. A wire 15 ft long and 0.1 in.^{2} in cross section is found to increase its length by 0.01 ft under a tension of 2000 lb. What is Young's modulus for this wire?

Ans. 30×10^6 lb/in.^{2}

* 13-6. A wire whose cross section is 4 mm^2 is stretched 0.1 mm by a certain weight. How far will a wire of the same material and length stretch if its cross-sectional area is 8 mm^2 and the same weight is attached?

* 13-7. A no. 18 copper wire has a diameter of 0.04 in. and is originally 10 ft long. (a) What is the greatest load that can be supported by this wire without exceeding its elastic limit? (b) Compute the change in length of the wire under this load. (c) What is the maximum load that can be supported without breaking the wire?

Ans. (a) 28.9 lb, (b) 0.0135 ft, (c) 61.6 lb

Shear Modulus

13-8. A steel rod projects 1.0 in. above a floor and is 0.5 in. in diameter. The shearing force F is 6000 lb and the shear modulus is 11.6×10^6 lb/in.^{2}. What is the shearing stress? What is the shearing angle? What is the horizontal deflection?

13-9. A 1500-kg load is supported at the end of a 5-m aluminum beam as shown in Fig. 13-9. The beam has a cross-sectional area of 26 cm^2 and the shear modulus is 23,700 MPa. (a) What is the shearing stress? (b) What is the downward deflection of the beam?

Ans. (a) 5.65×10^6 Pa, (b) 1.19 mm

Fig. 13-9

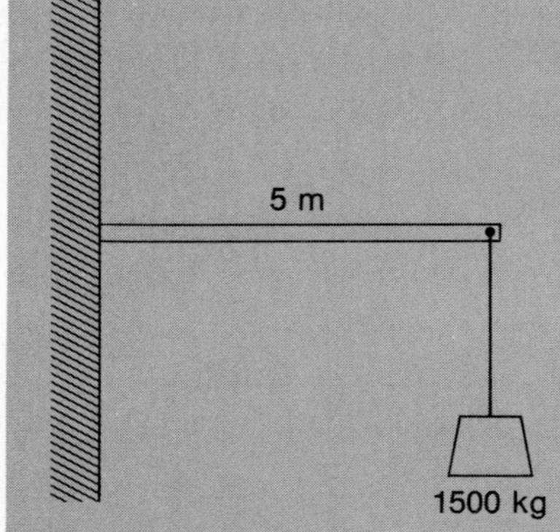

5 m

1500 kg

* 13-10. A steel plate 0.5 in. thick has an ultimate shearing strength of 50,000 lb/in.^{2}. What force must be applied to punch a $\frac{1}{4}$-in. hole through the plate?

Bulk Modulus

13-11. A solid brass sphere of volume 0.8 m^3 is dropped into the ocean to a depth where the water pressure is 20 MPa. The bulk modulus of brass is 35,000 MPa. What is the change in volume of the sphere?

Ans. -4.57×10^{-4} m^3

* 13-12. A cylinder 10 in. in diameter is filled to a height of 6 in. with glycerin. A piston of the same diameter pushes downward on the liquid with a force of 800 lb. The compressibility of glycerin is 1.50×10^{-6} in.^{2}/lb. (a) What is the stress on the liquid? (b) How far down does the piston move?

13-13. What is the percent decrease in volume of water when it is subjected to a pressure of 15 MPa?

Ans. 0.714%

Additional Problems

13-14. A steel piano wire has an ultimate strength of about 35,000 lb/in.^{2}. How large a load can a 0.5-in.-diameter steel wire hold without breaking?

* 13-15. The shear modulus for copper is about 4.2×10^{10} Pa. A 3000-N shearing force is applied to the upper surface of a copper cube which is 40 mm on a side. How large a shearing angle is caused by the force?

Ans. 4.46×10^{-5} rad

13-16. The twisting of a cylindrical shaft (Fig. 13-10) through an angle θ is an example of a shearing strain. An analysis of the situation shows that the angle of twist in radians is given by

$$\theta = \frac{2\tau l}{\pi S R^4}$$

where τ = applied torque
l = length of cylinder
R = radius of cylinder
S = shear modulus

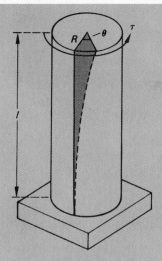

Fig. 13-10 A torque τ applied at one end of a solid cylinder causes it to twist through an angle θ.

If a torque of 100 lb · ft is applied to the end of a cylindrical steel shaft 10 ft long and 2 in. in diameter, what will be the angle of twist in radians?

13-17. An aluminum shaft 1 cm in diameter and 16 cm tall is subjected to a torsional shearing stress. What applied torque will cause a twist of 1° as defined in Fig. 13-10?

Ans. 2.54 N · m

13-18. A piston, 8 cm in diameter, exerts a force of 2000 N on 1 liter of benzene. What is the decrease in volume of the benzene?

13-19. How much will a 600-mm length of brass wire stretch when a 4-kg mass is hung from its end? The diameter of the wire is 1.2 mm.

Ans. 2.32×10^{-4} m

13-20. A solid cylindrical steel column is 12 ft high and 6 in. in diameter. What will be its decrease in length when supporting a load of 90 tons?

13-21. What is the minimum diameter of a brass rod if it is to support a 400-N load without exceeding the elastic limit?

Ans. 1.16 mm

13-22. Two sheets of aluminum on an aircraft wing are to be held together by aluminum rivets of cross-sectional area 0.25 in.^{2}. The shearing stress on each rivet must not exceed one-tenth of the elastic limit for aluminum. How many rivets are needed if each rivet supports the same fraction of a total shearing force of 25,000 lb?

14 Simple Harmonic Motion

OBJECTIVES

After completing this chapter, you should be able to:

1. Describe and apply the relationship between force and displacement in simple harmonic motion.
2. Use the reference circle to describe the variation in magnitude and direction of displacement, velocity, and acceleration for simple harmonic motion.
3. Write and apply formulas for the determination of displacement x, velocity v, or acceleration a in terms of time, frequency, and amplitude.
4. Write and apply a relationship between the frequency of motion and the mass of a vibrating object when the spring constant is known.
5. Compute the frequency or period in simple harmonic motion when the position and acceleration are given.
6. Describe the motion of a simple pendulum and calculate the length required to produce a given frequency.

Until now we have discussed the motion of objects under the influence of a constant, unchanging force. Such motion was described by calculating the position and velocity as functions of time. However, the real world often consists of varying forces. Common examples are swinging pendulums, balance wheels of watches, tuning forks, a mass vibrating at the end of a coiled spring, and vibrating air columns in musical instruments. In these cases and many others, we need a more complete description of motion caused by a resultant force which varies in a predictable manner.

14-1 PERIODIC MOTION

Whenever an object is deformed, an elastic restoring force appears which is proportional to the deformation. When released, such an object will vibrate back and forth about its equilibrium position. For example, after a swimmer springs from a diving board (Fig. 14-1), the board continues to vibrate about its normal position for a definite length of time.

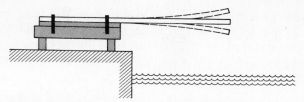

Fig. 14-1 The periodic vibration of a diving board.

This type of motion is said to be *periodic* because the position and velocity of the moving particles are repeated as a function of time. Since the restoring force is reduced after each vibration, the board eventually comes to rest.

> **Periodic motion** is that motion in which a body moves back and forth over a fixed path, returning to each position and velocity after a definite interval of time.

An ice puck coupled to a spring is shown in Fig. 14-2. The ice puck is a laboratory device which rides on a cushion of carbon dioxide, approximating frictionless motion. Suppose we pull the ice puck to the side and release it so that it oscillates

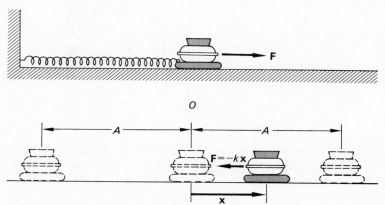

Fig. 14-2 The frictionless ice puck can be used to illustrate simple harmonic motion. The restoring force F is always directed toward the center of oscillation *O*.

about its initial position *O* without friction. According to Hooke's law, the restoring force F is directly proportional to the displacement x of the ice puck from its equilibrium position *O*. Since the restoring force is always opposed to the displacement, a negative sign must be introduced. Thus we write

$$F = -kx \qquad (14\text{-}1)$$

The maximum displacement of the ice puck from its equilibrium position is called the *amplitude A*. It is at this position that the ice puck experiences the maximum restoring force and consequently its maximum acceleration. The force decreases as the puck approaches its center of oscillation, becoming zero at that point. However, the momentum of the puck carries it on past the center, bringing the restoring force of the spring back into play again. This force increases until the puck comes to a stop at its maximum displacement, whereupon it begins its return trip. If there were no loss of energy due to friction, the back-and-forth motion would continue indefinitely. This type of oscillatory motion in the absence of friction is known as *simple harmonic motion* (SHM).

> Simple harmonic motion is periodic motion in the absence of friction produced by a restoring force which is directly proportional to the displacement and oppositely directed.

The *period T* is defined as the time for one complete trip, or oscillation. For example, if the ice puck is released from the right at a distance A from its equilibrium position, the time required for it to return to that position is its period of vibration. It must be pointed out, however, that every point in the vibratory path must be covered in measuring a complete oscillation. The time required to move from the center of oscillation to the distance A and back represents only one-half of a period.

The *frequency f* is the number of complete oscillations per unit time. Since the period is the time for one oscillation, the frequency must be the reciprocal of the period, or

$$f = \frac{1}{T}$$

The unit of frequency is often expressed as vibrations per second or as inverse seconds (s^{-1}); the SI unit for frequency is the *hertz* (Hz).

$$1 \text{ Hz} = \frac{1}{s}$$

Thus, a frequency of 500 vibrations per second becomes 500 Hz.

14-2
THE REFERENCE CIRCLE

It has been shown that an object vibrating with SHM is influenced by a restoring force that is directly proportional to its displacement. If we apply Newton's second law to Eq. (14-1), we shall see that the acceleration is also proportional to the displacement. Thus $F = ma = -kx$ from which

$$a = -\frac{k}{m}x \tag{14-2}$$

As long as the mass m remains constant, the ratio k/m will also be constant, indicating that the magnitude of the acceleration increases with the displacement. The minus sign appears because the acceleration is always directed toward the center of oscillation.

Since the acceleration in SHM is not constant, the equations derived in earlier chapters will not apply. To determine new relationships directly requires the use of calculus. Fortunately, these equations can be deduced by geometrical methods. When a body moves in a circular path with uniform speed, its linear projection moves with SHM. This fact is illustrated in Fig. 14-3, where the shadow of a ball attached to a rotating disk oscillates back and forth with periodic motion. This experiment suggests that our knowledge of uniform circular motion may be helpful in describing SHM.

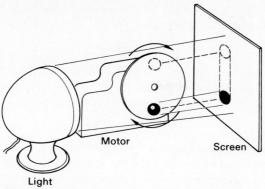

Light

Fig. 14-3 The projection or shadow of a ball attached to a rotating disk moves with simple harmonic motion.

The *reference circle* in Fig. 14-4 is used to compare the motion of an object moving in a circle with its horizontal projection. Since it is the motion of the projection that we wish to study, we shall refer to the position P of the object moving in a circle as the *reference point*. The radius of the reference circle is equal to the

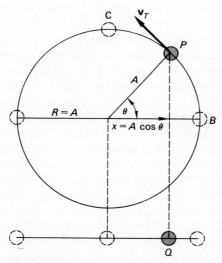

Fig. 14-4 Displacement in simple harmonic motion.

amplitude of the horizontal oscillation. If the linear speed v_T and angular speed ω of the reference point are constant, the projection Q will move back and forth with SHM. Time is assigned a zero value when the reference point is at B in Fig. 14-4. At any later time t, the reference point P will have moved through an angle θ. The displacement x of the projection Q is therefore

$$x = A \cos \theta$$

Recalling that the angle $\theta = \omega t$, we can now write the displacement as a function of the angular velocity of the reference point.

$$x = A \cos \theta = A \cos \omega t \qquad (14\text{-}3)$$

Although the angular velocity ω is useful for describing motion of the reference point P, it is not directly applied to the projection Q. However, we should recall that the angular velocity is related to the frequency of revolution by

$$\omega = 2\pi f$$

where ω is expressed in radians per second and f is the number of revolutions per second. It should also be recognized that the projection Q will describe one complete oscillation while the reference point describes one complete revolution. Thus the frequency f is the same for each point. Substituting $\omega = 2\pi f$ into Eq. (14-3), we obtain

$$\boxed{x = A \cos 2\pi ft} \qquad (14\text{-}4)$$

This equation can be applied to compute the displacement of a body moving with SHM of amplitude A and frequency f. Remember that the displacement x is always measured from the center of oscillation.

14-3
VELOCITY IN SIMPLE HARMONIC MOTION

Let us consider a body moving back and forth with SHM under the influence of a restoring force. Since the direction of the vibrating body is reversed at the end points of its motion, its velocity must be zero when its displacement is a maximum. It is then accelerated toward the center by the restoring force until it reaches its maximum speed at the center of oscillation, i.e., when its displacement is zero.

In Fig. 14-5 the velocity of a vibrating body is compared at three instants with corresponding points on a reference circle. It will be noticed that the velocity v of the vibrating body at any instant is the horizontal component of the tangential velocity v_T of the reference point. At point B the reference point is moving vertically upward and has no horizontal velocity. This point therefore corresponds to the zero velocity of the vibrating body when it reaches its amplitude A. At point C the horizontal component of v_T is equal to its entire magnitude. This point corresponds to a position of maximum velocity for the vibrating body, i.e., at its center of oscillation. In general, the velocity of the vibrating body at any point Q is determined from the reference circle to be

$$v = -v_T \sin \theta = -v_T \sin \omega t \qquad (14\text{-}5)$$

The minus sign appears since the direction of the velocity is toward the left. We can make this equation more useful by recalling the relationship between the tangential velocity v_T and the angular velocity:

$$v_T = \omega A = 2\pi fA$$

Substitution into Eq. (14-5) yields

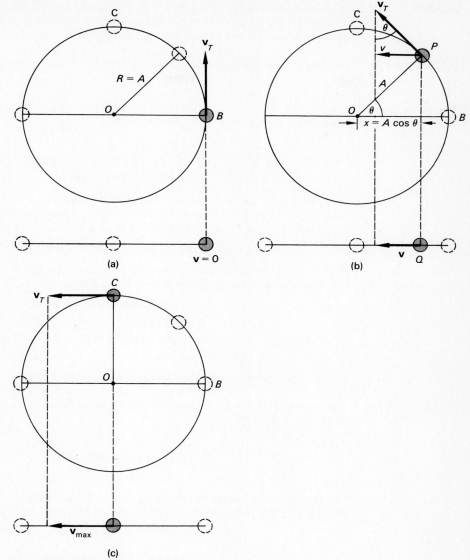

(a)

(b)

(c)

Fig. 14-5 Velocity and the reference circle.

$$v = -2\pi fA \sin 2\pi ft \qquad (14\text{-}6)$$

This equation will give the velocity of a vibrating body at any instant if it is remembered that $\sin \theta$ is negative when the reference point lies below the diameter of the reference circle.

EXAMPLE 14-1   An ice puck attached to a spring, as in Fig. 14-2, is pulled to the right a distance of 6 cm and released. (*a*) If it returns to the point of release in 2 s and continues to vibrate with SHM, compute its position and velocity after 5.2 s. (*b*) What is its maximum velocity?

Solution (a)   The time for one complete vibration is 2 s. Thus the frequency is given by

$$f = \frac{1}{T} = \frac{1}{2 \text{ s}} = 0.5 \text{ Hz}$$

The position after 5.2 s is

$$x = A \cos 2\pi f t$$
$$= (6 \text{ cm}) \cos [(2\pi)(0.5 \text{ Hz})(5.2 \text{ s})]$$
$$= (6 \text{ cm}) \cos (16.34 \text{ rad})$$

Before evaluating cos (16.34 rad) make sure that the calculator is set to read angles in radians (rad). It is also a good idea to avoid rounding of numbers until your final answer. A small error in the radian measure is significant. Taking these precautions, the displacement x becomes

$$x = (6 \text{ cm})(-0.809) = -4.85$$

The minus sign indicates that the ice puck is 4.85 cm to the left of its equilibrium position. The velocity of the puck after 5.2 s is found from

$$v = -2\pi f A \sin 2\pi f t$$
$$= -2\pi(0.5 \text{ Hz})(6 \text{ cm})(\sin 16.34 \text{ rad})$$
$$= (-18.8 \text{ cm/s})(-0.588)$$
$$= 11.1 \text{ cm/s}$$

The velocity is positive, indicating that it is moving to the right.

Solution (b)   The maximum velocity occurs when the displacement is zero or when the reference angle is 90 or 270°. Thus

$$v_{max} = -2\pi f A \sin 90° = -2\pi f A$$
$$= -2\pi(0.5 \text{ Hz})(6 \text{ cm})$$
$$= -18.8 \text{ cm/s}$$

This represents the maximum velocity directed to the left since we chose 90° as a reference angle. If 270° had been chosen, a positive value would have resulted.

14-4

ACCELERATION IN SIMPLE HARMONIC MOTION

The velocity of a vibrating body is never constant. Therefore, although it was not often mentioned, acceleration plays a very important role in the equations derived in the previous section. We now attempt to obtain an expression which will allow us to determine the acceleration of objects under the influence of a restoring force.

At the position of maximum displacement, the velocity of a vibrating object is zero. It is at this instant that the body is acted on by the maximum restoring force. Thus the acceleration of the body is a maximum when its velocity is zero. As the object approaches its equilibrium position, the restoring force (and therefore the acceleration) is reduced until it reaches zero at the center of oscillation. At the equilibrium position, the acceleration is zero and the velocity has its maximum value.

Figure 14-6 demonstrates that the acceleration a of a particle moving with SHM is equal to the horizontal component of the centripetal acceleration a_c of the reference point. From the figure,

$$a = -a_c \cos \theta = -a_c \cos \omega t \qquad (14\text{-}7)$$

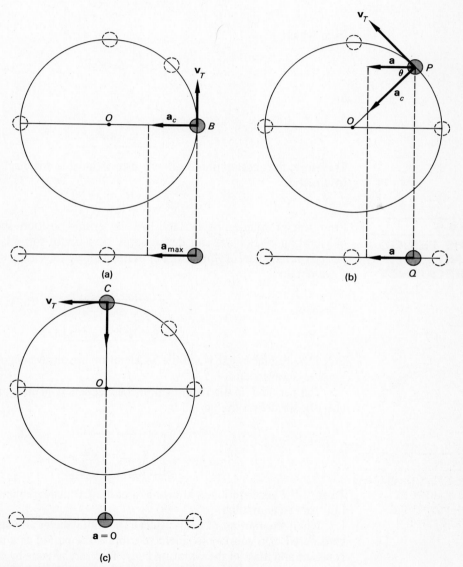

Fig. 14-6 Acceleration and the reference circle.

The minus sign indicates that the acceleration is directed toward the left. Recalling that $a_c = \omega^2 R$ and that $R = A$, we can rewrite Eq. (14-7) as

$$a = -\omega^2 A \cos \omega t$$

Substituting $\omega = 2\pi f$, as in the previous section, we obtain

$$a = -4\pi^2 f^2 A \cos 2\pi f t \qquad (14\text{-}8)$$

Equation (14-8) can be simplified by noting from Eq. (14-3) that

$$\cos \theta = \cos 2\pi f t = \frac{x}{A}$$

from which

$$a = -4\pi^2 f^2 A \frac{x}{A}$$

or

$$\boxed{a = -4\pi^2 f^2 x} \qquad (14\text{-}9)$$

Therefore, the acceleration is directly proportional to the displacement and opposite in direction.

14-5
THE PERIOD AND FREQUENCY

From the information now established concerning position, speed, and acceleration of vibrating bodies, we can derive some very useful formulas for computing the period or frequency of vibration. For example, if we solve Eq. (14-9) for the frequency f, we obtain

$$\boxed{f = \frac{1}{2\pi} \sqrt{-\frac{a}{x}}} \qquad (14\text{-}10)$$

Since the displacement x and the acceleration are always opposite in sign, the term $-a/x$ is always positive.

The period T is the reciprocal of the frequency. Making use of this fact in Eq. (14-10), we define the period by

$$\boxed{T = 2\pi \sqrt{-\frac{x}{a}}} \qquad (14\text{-}11)$$

Thus, if the acceleration is known at a particular displacement, the period of vibration can be computed.

When the motion of bodies under the influence of an elastic restoring force is considered, it is more convenient to express the period as a function of the spring constant and mass of the vibrating body. This can be done by comparing Eqs. (14-2) and (14-9):

$$a = -\frac{k}{m}x \qquad a = -4\pi^2 f^2 x$$

Combining these relations, we obtain

$$4\pi^2 f^2 = \frac{k}{m}$$

from which the frequency is

$$f = \frac{1}{2\pi}\sqrt{\frac{k}{m}} \tag{14-12}$$

Finally, the period T is given by the reciprocal of the frequency. Thus

$$T = 2\pi\sqrt{\frac{m}{k}} \tag{14-13}$$

Note that neither the period nor the frequency depends upon the amplitude (maximum displacement) of the vibrating body. They depend only on the spring constant and the mass of the vibrating body.

EXAMPLE 14-2 A 2-kg steel ball is attached to the end of a flat strip of metal that is clamped at its base, as shown in Fig. 14-7. (a) If a force of 5 N is required to displace the ball by 16 cm, what will its period of vibration be upon release? (b) What is its maximum acceleration?

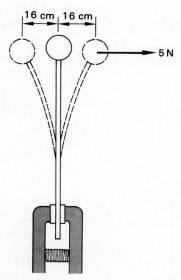

Fig. 14-7

Solution (a) A force of 5 N displaces the mass by 16 cm. Thus the spring constant is

$$k = \frac{F}{x} = \frac{5 \text{ N}}{0.16 \text{ m}} = 31.2 \text{ N/m}$$

The period is found from Eq. (14-13).

$$T = 2\pi\sqrt{\frac{m}{k}} = 2\pi\sqrt{\frac{2 \text{ kg}}{31.2 \text{ N/m}}}$$

$$= 2\pi(0.253) = 1.59 \text{ s}$$

Solution (b) The maximum acceleration occurs when the displacement is a maximum, i.e., when $x = 0.16$ m. Thus

$$a = -4\pi^2 f^2 x = -\frac{4\pi^2}{T^2}x$$

$$= -\frac{4\pi^2(0.16 \text{ m})}{(1.59 \text{ s})^2} = -2.5 \text{ m/s}^2$$

14-6
THE SIMPLE PENDULUM

When a heavy pendulum bob is swinging at the end of a light cord or rod, as in Fig. 14-8, it approximates SHM. If we assume that all the mass is concentrated at the center of gravity of the bob and that the restoring force acts at a single point, we refer to this apparatus as a *simple* pendulum. Although this assumption is never strictly true, a close approximation is obtained by making the mass of the connecting rod or cord very small in comparison with the pendulum bob.

Notice that the displacement x of the bob is not along a straight line but lies along the arc subtended by the angle θ. From methods discussed in Chap. 11, the length of the displacement is simply the product of the angle θ and the length of the

Fig. 14-8

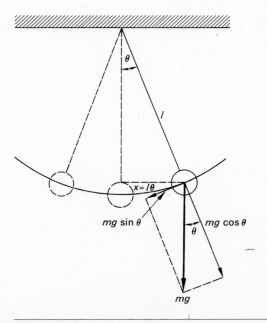

MECHANICS

cord. Thus

$$x = l\theta$$

If the motion of the bob is SHM, the restoring force must be given by

$$F = -kx = -kl\theta \qquad (14\text{-}14)$$

which means that the restoring force should be proportional to θ since the length l is constant. Let us examine the restoring force to see if this is the case. In the back-and-forth movement of the pendulum bob, the necessary restoring force is provided by the tangential component of the weight. From Fig. 14-8, we can write

$$F = -mg \sin \theta \qquad (14\text{-}15)$$

Thus the restoring force is proportional to $\sin \theta$ instead of to θ. The conclusion must be that the bob does not oscillate with SHM. However, if we stipulate that the angle θ is small, $\sin \theta$ will be approximately equal to the angle θ in radians. You can verify this approximation by considering several small angles:

$\sin \theta$	θ (rad)
$\sin 6° = 0.1045$	$6° = 0.1047$
$\sin 12° = 0.208$	$12° = 0.209$
$\sin 27° = 0.454$	$27° = 0.471$

When the approximation $\sin \theta \approx \theta$ is used, Eq. (14-15) becomes

$$F = -mg \sin \theta = -mg\theta$$

Comparing this relation with Eq. (14-14), we obtain

$$F = -kl\theta = -mg\theta$$

from which

$$\frac{m}{k} = \frac{1}{g}$$

Substitution of this proportion into Eq. (14-13) yields an expression for the period of a simple pendulum:

$$\boxed{T = 2\pi\sqrt{\frac{l}{g}}} \qquad (14\text{-}16)$$

Notice that for small amplitudes the period of a simple pendulum is a function of neither the mass of the bob nor the amplitude of vibration. In fact, since the acceleration of gravity is constant, the period is determined solely by the length of the connecting cord or rod.

EXAMPLE 14-3 In a laboratory experiment a student is given a stop watch, a wooden bob, and a piece of cord. He is then asked to determine the acceleration of gravity g. If he constructs a simple pendulum of length 1 m and measures the period to be 2 s, what value will he obtain for g?

Solution Squaring both sides of Eq. (14-16) gives

$$T^2 = 4\pi^2 \frac{l}{g}$$

from which

$$g = \frac{4\pi^2 l}{T^2} = \frac{4\pi^2 (1 \text{ m})}{(2 \text{ s})^2}$$

$$= 9.87 \text{ m/s}^2$$

Fig. 14-9

Another example of SHM is accorded by the torsion pendulum (Fig. 14-9), which consists of a solid disk or cylinder supported at the end of a thin rod. If the disk is twisted through an angle θ, the restoring torque τ is directly proportional to the

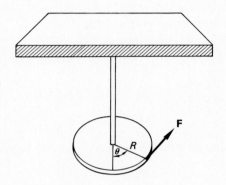

angular displacement. Thus

$$\tau = -k'\theta \tag{14-17}$$

where k' is a constant which depends on the material from which the thin rod is made. (See Prob. 13-11.)

When the disk is released, the restoring torque produces an angular acceleration which is directly proportional to the angular displacement. The period of the simple angular harmonic motion thus produced is given by

$$T = 2\pi \sqrt{\frac{I}{k'}} \tag{14-18}$$

where I is the moment of inertia of the vibrating system and k' is the torsion constant defined in Eq. (14-17).

EXAMPLE 14-4 A solid disk of mass 0.16 kg and radius 0.12 m is twisted through an angle of 1 rad and released. (a) If the torsion constant of the wire supporting the disk is 0.025 N · m/rad, calculate the maximum angular acceleration. (b) What is the period of oscillation?

Solution (a) The moment of inertia of the disk is

$$I = \tfrac{1}{2}mR^2 = \tfrac{1}{2}(0.16 \text{ kg})(0.12 \text{ m})^2$$

$$= 1.15 \times 10^{-3} \text{ kg} \cdot \text{m}^2$$

The angular acceleration is a maximum when the angular displacement is 1 rad. From Eq. (14-17) we have

$$\tau = I\alpha = -k'\theta$$

$$\alpha = -\frac{k'\theta}{I} = \frac{-(0.025 \text{ N} \cdot \text{m/rad})(1 \text{ rad})}{1.15 \times 10^{-3} \text{ kg} \cdot \text{m}^2}$$

$$= -21.7 \text{ rad/s}^2$$

Solution (b) The period from Eq. (14-18) is

$$T = 2\pi\sqrt{\frac{I}{k'}} = 2\pi\sqrt{\frac{1.15 \times 10^{-3} \text{ kg} \cdot \text{m}^2}{0.025 \text{ N} \cdot \text{m/rad}}}$$

$$= 2\pi(0.214)s = 1.35 \text{ s}$$

Notice that the period is not a function of the angular displacement.

SUMMARY

Simple harmonic motion is periodic motion in which the restoring force is proportional to the displacement. Such vibratory motion without friction produces predictable variations in displacement and velocity. The major concepts discussed in this chapter are summarized below:

- Simple harmonic motion is produced by a *restoring force F* given by:

$$\boxed{F = -kx}$$

Restoring Force

- Since $F = ma = -kx$, the acceleration produced by a restoring force is

$$\boxed{a = -\frac{k}{m}x}$$

Acceleration

- A convenient way to study simple harmonic motion is to use the *reference circle*. The variations in displacement x, velocity v, and acceleration a can be seen by reference to Figs. 14-4, 14-5, and 14-6.
- For SHM the displacement, velocity, and acceleration may be expressed in terms of the amplitude A, the time t, and the frequency of vibration f:

$$x = A \cos 2\pi ft \qquad \textit{Displacement}$$
$$v = -2\pi fA \sin 2\pi ft \qquad \textit{Velocity}$$
$$a = -4\pi^2 f^2 x \qquad \textit{Acceleration}$$

- The period T and the frequency f in simple harmonic motion are found from

$$f = \frac{1}{2\pi}\sqrt{-\frac{a}{x}} \quad \text{or} \quad f = \frac{1}{2\pi}\sqrt{\frac{k}{m}} \qquad \textit{Frequency}$$

$$T = 2\pi\sqrt{-\frac{x}{a}} \quad \text{or} \quad T = 2\pi\sqrt{\frac{m}{k}} \qquad \textit{Period}$$

- For a simple pendulum of length l, the period is given by

$$T = 2\pi\sqrt{\frac{l}{g}} \qquad \textit{Period of Simple Pendulum}$$

• A torsion pendulum consists of a solid disk or cylinder of moment of inertia I suspended at the end of a thin rod. If the torsion constant k' is known, the period is given by

$$T = 2\pi\sqrt{\frac{I}{k'}}$$ *Period of Torsion Pendulum*

QUESTIONS

14-1. Define the following terms:
a. Simple harmonic motion
b. Spring constant
c. Restoring force
d. Simple pendulum
e. Torsion constant
f. Displacement
g. Amplitude
h. Frequency
i. Period
j. Hertz

14-2. Give several examples of motion which is SHM.

14-3. What effect will doubling the amplitude A of a body moving with SHM have on (a) the period, (b) the maximum velocity, and (c) the maximum acceleration?

14-4. A 2-kg mass m_1 moves in SHM with a frequency f_1. What mass m_2 will cause the system to vibrate with twice the frequency?

14-5. Explain, with the use of diagrams, why the velocity in SHM is greatest when the acceleration is the least.

14-6. An ice puck is attached to a spring of force constant k and set into vibration of amplitude A, as shown in Fig. 14-2. The spring is then replaced with one with a force constant of $4k$, and the ice puck is set into vibration at the same amplitude. Compare their periods and frequencies of oscillation.

14-7. A pendulum clock runs too slow and loses·time. What adjustment should be made?

14-8. Given a spring of known force constant, a meterstick, and a stopwatch, how can you determine the value of an unknown mass?

14-9. How may the principle of the pendulum be used to compute (a) length, (b) mass, and (c) time?

14-10. Explain clearly why the motion of a pendulum is not simple harmonic when the amplitude is large. Is the period larger or smaller than it should be if the motion were strictly simple harmonic?

PROBLEMS

Periodic Motion and the Reference Circle

14-1. A rubber ball is swinging in a horizontal circle 200 cm in diameter, making 20 revolutions in 1 min. A shadow of the ball is projected on a wall by a distant light. What is the amplitude of the motion of the shadow? What is its frequency? What is its period?

Ans. 100 cm, 0.33 Hz, 3 s

14-2. If the rubber ball in Prob. 14-1 describes a circle of radius 12 in. and moves at 300 rpm, what is the frequency of its projection? What is the amplitude? What is the period of the motion?

14-3. A mass oscillates with simple harmonic motion of frequency 3 Hz and amplitude 6 cm. What are its positions at the times $t = 0$ and $t = 2.4$ s?

Ans. 6 cm, 1.85 cm

14-4. An object oscillates with SHM of frequency 0.25 Hz. How much time after it reaches its maximum displacement will its position become zero? How much time will pass before it reaches one-half of its amplitude?

14-5. A 10-lb weight stretches a spring 8 in. before reaching a position of equilibrium. (a) What is the spring constant? (b) If the weight is displaced an additional 6 in. and released, what is the maximum restoring force?

Ans. (a) 1.25 lb/in., (b) −17.5 lb

Velocity in Simple Harmonic Motion

14-6. A body vibrates with a frequency of 5 Hz and an amplitude of 6 cm. What is the maximum velocity? What is its position when the velocity is zero?

14-7. An ice puck attached to a spring is pulled to the right a distance of 4 cm and then released. In 3 s it returns to the point of release and continues to vibrate with SHM. (a) What is the maximum velocity? (b) What are the position and velocity after 2.55 s?

Ans. (a) 8.38 cm/s; (b) 2.35 cm, 6.78 cm/s

* **14-8.** Show that the velocity of an object in SHM can be written as a function of its amplitude and displacement.

$$v = \pm 2\pi f \sqrt{A^2 - x^2}$$

Use this formula to verify the answers of position and velocity in Prob. 14-7.

14-9. A mass vibrating at a frequency of 0.5 Hz has a velocity of 5 cm/s as it passes the center of oscillation. What is the amplitude and what is the time for one vibration?

Ans. 1.59 cm, 2 s

Acceleration in Simple Harmonic Motion

14-10. A body makes one complete vibration in 0.5 s. What is its acceleration when it is displaced 2 cm from its equilibrium position?

* **14-11.** An object is moving with SHM of amplitude 16 cm and frequency 2 Hz. (a) What are the maximum velocity and the maximum acceleration? (b) What are the velocity and the acceleration after 3.2 s?

Ans. (a) 2.01 m/s, (b) 25.3 m/s^2, (c) −1.18 m/s, 20.5 m/s^2

* **14-12.** A body vibrates with SHM of period 1.5 s and amplitude 6 in. (a) What are its maximum velocity and acceleration? (b) What are its position, velocity, and acceleration after it has been vibrating for 7 s?

* **14-13.** A 200-g mass is suspended from a long spiral spring. When displaced 10 cm, the mass is found to vibrate with a period of 2 s. (a) What is the spring constant? (b) What are its velocity and acceleration as it passes upward through the point 5 cm above its equilibrium position?

Ans. (a) 1.97 N/m; (b) 27.2 cm/s, −49.3 cm/s^2

Period and Frequency

* **14-14.** The prong of a tuning fork vibrates with a frequency of 330 Hz and an amplitude of 2 mm. What is the velocity when the displacement is 1.5 mm?

* **14-15.** A 400-g mass stretches a spring 20 cm. The 400-g mass is then removed and replaced with an unknown mass m. When the mass m is pulled down 5 cm and released, it vibrates with a period of 0.1 s. Compute the mass of the vibrating object.

Ans. 4.96 g

*   **14-16.** A long, thin piece of steel is clamped at its lower end with a 2-kg ball fastened to its top end. When the ball is pulled to one side and released it vibrates with a period of 1.5 s. What is the spring constant for this device?

*   **14-17.** A car and its passengers have a total mass of 1600 kg. The frame of the car is supported by four springs, each having a force constant of 20,000 N/m. Find the frequency of vibration of the car when it drives over a bump in the road.
    Ans. 1.13 Hz

The Simple Pendulum

14-18. What are the period and the frequency of a simple pendulum whose length is 2 m?

14-19. What is the length of a pendulum whose period is 2 s?
Ans. 0.993 m

*   **14-20.** A simple pendulum clock beats seconds every time the bob reaches its maximum amplitude on either side. What is the period of this motion? What should be the length of the pendulum at a point where $g = 9.8$ m/s^2?

*   **14-21.** On the surface of the moon, the acceleration due to gravity is only 1.67 m/s^2. If a pendulum clock adjusted for the earth is taken to the moon, the new length of the pendulum must be what percent of its length on the earth if the clock is to remain accurate?
    Ans. 17%

The Torsion Pendulum

14-22. A torsion pendulum has a maximum angular acceleration of 20 rad/s^2 when its displacement is 70°. What is its frequency of vibration?

*   **14-23.** A disk 20 cm in diameter is suspended in a horizontal position by a wire attached to its center. A force of 20 N applied to the rim of the disk causes it to twist through an angle of 12°. If the period of the angular harmonic motion after release is 0.5 s, what is the moment of inertia of the disk?
    Ans. 0.061 kg · m^2

*   **14-24.** An irregular object is suspended by a wire as a torsion pendulum. A torque of 40 lb · ft causes it to twist through an angle of 15°. When released, the body oscillates with a frequency of 3 Hz. What is the moment of inertia of the irregular body?

Additional Problems

*   **14-25.** A 40-g mass is attached to a horizontal spring of force constant 10 N/m and released with an amplitude of 20 cm. What is the velocity of the mass when it is halfway to the equilibrium position? What is the frequency of the motion?
    Ans. 274 m/s, 2.50 Hz

*   **14-26.** A 50-g mass on the end of a spring of force constant 20 N/m is moving at a speed of 120 cm/s when located a distance of 10 cm from the equilibrium position. What is the amplitude of the vibration?

14-27. A 2-kg mass is hung from a light spring. When displaced, it is found that the mass makes 20 complete oscillations in 25 s. What is the period and what is the spring constant?
Ans. 1.25 s, 50.5 N/m

*   **14-28.** A pendulum clock beats seconds every time the bob passes through its lowest point. What must be the length of the pendulum at a place where $g = 32$ ft/s^2? If

the clock is moved to a point where $g = 31$ ft/s^2, how much time will it lose in 1 day?

* 14-29. An object is moving with SHM of amplitude 20 cm and frequency 1.5 Hz. **(a)** What are the maximum acceleration and maximum velocity? **(b)** What are the position, acceleration, and velocity after a time of 1.4 s?

Ans. (a) 188 cm/s, 17.8 m/s^2, (b) 16.2 cm, 14.4 m/s^2, 111 cm/s

* 14-30. A 500-g mass is connected to a device having a spring constant of 6 N/m. The mass is displaced a distance of 5 cm from its equilibrium position and then released. What are its speed and acceleration when it is located 3.0 cm from its equilibrium position?

15 Fluids at Rest

OBJECTIVES

After completing this chapter, you should be able to:

1. Define and apply the concepts of fluid pressure and buoyant force to the solution of physical problems similar to examples in the text.
2. Write and illustrate with drawings your understanding of the four basic principles of fluid pressure as summarized in Sec. 15-3.
3. Define *absolute pressure, gauge pressure,* and *atmospheric pressure,* and demonstrate by examples your understanding of the relationship between these terms.
4. Write and apply formulas for calculating the mechanical advantage of a hydraulic press in terms of input and output forces or areas.

Liquids and gases are called *fluids* because they flow freely and fill their containers. In this chapter you will learn that fluids may exert forces on the walls of their containers. Such forces acting on definite surface areas create a condition of *pressure*. A hydraulic press utilizes fluid pressure to lift heavy loads. The structure of water basins, dams, and large oil tanks is determined largely by pressure considerations. The design of boats, submarines, and weather balloons must take into account the pressure and density of the surrounding fluid.

15-1
DENSITY

Before discussing the statics and dynamics of fluids, it is important to understand the relation of a body's weight to its volume. For example, we refer to lead or iron as *heavy* whereas wood or cork is considered *light*. What we really mean is that a block of wood is lighter than a block of lead *of similar size*. The terms light and heavy are comparative terms. As illustrated in Fig. 15-1, it is possible for a block of lead to weigh the same as a block of wood if their relative size differs greatly. On the other hand, 1 ft^3 of lead weighs more than 6 times as much as 1 ft^3 of wood.

The quantity which relates a body's weight to its volume is known as its *weight density*.

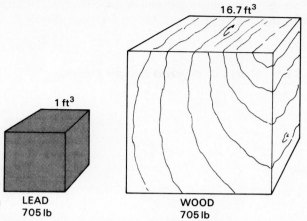

Fig. 15-1 A comparison of weight and volume in lead and wood.

The **weight density** D of a body is defined as the ratio of its weight W to its volume V. Units are the *newton per cubic meter* (N/m^3) and the *pound per cubic foot* (lb/ft^3).

$$D = \frac{W}{V} \qquad W = DV$$

(15-1)

Thus, if a 20-lb object occupies a volume of 4 ft^3, it has a weight density of 5 lb/ft^3.

As mentioned in Chap. 7, the weight of a body is not constant but varies according to location. A more useful relation for density takes advantages of the fact that *mass* is a universal constant, independent of gravity.

The **mass density** ρ of a body is defined as the ratio of its mass m to its volume V.

$$\rho = \frac{m}{V} \qquad m = \rho V$$

(15-2)

The units of mass density are the ratio of a mass unit to a volume unit, i.e., grams per cubic centimeter, kilograms per cubic meter, or slugs per cubit foot.

The relation between weight density and mass density is found by recalling that $W = mg$. Thus

$$D = \frac{mg}{V} = \rho g$$

(15-3)

In USCS units, matter is usually described in terms of its weight. For this reason, weight density is more often used when working with this system of units. In

SI units mass is the more convenient quantity, and the mass density is preferred. Table 15-1 lists the weight densities and mass densities of some common substances.

Table 15-1 Mass Density and Weight Density

Substance	D, lb/ft³	ρ g/cm³	ρ kg/m³
Solids:			
Aluminum	169	2.7	2,700
Brass	540	8.7	8,700
Copper	555	8.89	8,890
Glass	162	2.6	2,600
Gold	1204	19.3	19,300
Ice	57	0.92	920
Iron	490	7.85	7,850
Lead	705	11.3	11,300
Oak	51	0.81	810
Silver	654	10.5	10,500
Steel	487	7.8	7,800
Liquids:			
Alcohol	49	0.79	790
Benzene	54.7	0.88	880
Gasoline	42	0.68	680
Mercury	850	13.6	13,600
Water	62.4	1.0	1,000
Gases (0°C):			
Air	0.0807	0.00129	1.29
Hydrogen	0.0058	0.000090	0.090
Helium	0.0110	0.000178	0.178
Nitrogen	0.0782	0.00126	1.25
Oxygen	0.0892	0.00143	1.43

EXAMPLE 15-1 A cylindrical tank for gasoline is 3 m long and 1.2 m in diameter. How many kilograms of gasoline will the tank hold?

Solution First we find the volume:

$$V = \pi r^2 h = \pi(0.6 \text{ m})^2(3 \text{ m}) = 3.39 \text{ m}^3$$

Substituting the volume and mass density into Eq. (15-1), we obtain

$$m = \rho V = (680 \text{ kg/m}^3)(3.39 \text{ m}^3) = 2310 \text{ kg}$$

Another method of indicating the densities of substances is by their *specific gravity,* which compares their densities to that for water. For example, a substance which is half as dense as water would have a specific gravity of 0.5.

The **specific gravity** of a substance is defined as the ratio of its density to the density of water at 4°C (1000 kg/m³).

A better name for this quantity is *relative density,* but the term specific gravity is much more widely used.

15-2 PRESSURE

The effectiveness of a given force often depends upon the area over which it acts. For example, a woman wearing narrow heels will do much more damage to floors than she would with flat heels. Even though she exerts the same downward force in each case, with the narrow heels her weight is spread over a much smaller surface area. The *normal force per unit area* is called *pressure.* Symbolically, the pressure P is given by

$$P = \frac{F}{A}$$

(15-4)

where A is the area over which the perpendicular force F is applied. The unit of pressure is the ratio of any force unit to a unit of area. Examples are newtons per square meter and pounds per square inch. In SI units, the N/m^2 is renamed the *pascal* (Pa). The *kilopascal* (kPa) is the most appropriate measure for fluid pressure.

$$1 \text{ kPa} = 1000 \text{ N/m}^2 = 0.145 \text{ lb/in.}^2$$

EXAMPLE 15-2 A golf shoe has 10 cleats, each having an area of 0.01 in.^{2} in contact with the floor. Assume that in walking, there is one instant when all 10 cleats support the entire weight of a 180-lb person. What is the pressure exerted by the cleats on the floor? Express the answer in SI units.

Solution The total area in contact with the floor is 0.1 in.^{2} (10 × 0.01 in.^{2}). Substitution into Eq. (15-4) yields

$$P = \frac{F}{A} = \frac{180 \text{ lb}}{0.1 \text{ in.}^2} = 1800 \text{ lb/in.}^2$$

Converting to SI units, we obtain

$$P = (1800 \text{ lb/in.}^2) \left(\frac{1 \text{ kPa}}{0.145 \text{ lb/in.}^2} \right) = 1.24 \times 10^4 \text{ kPa}$$

As the area of the shoe in contact with the floor decreases, the pressure becomes larger. It is easy to see why this factor must be considered in floor construction.

15-3 FLUID PRESSURE

There is a very significant difference between the way a force acts on a fluid and on a solid. Since a solid is a rigid body, it can withstand the application of a force without a significant change in shape. A liquid, on the other hand, can sustain a force only at an enclosed surface or boundary. If a fluid is not restrained, it will flow under a shearing stress instead of being deformed elastically.

> *The force exerted by a fluid on the walls of its container must always act perpendicular to the walls.*

It is this characteristic property of fluids that makes the concept of pressure so useful. Holes bored in the bottom and sides of a barrel of water (Fig. 15-2) demon-

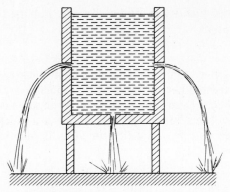

Fig. 15-2 The forces exerted by a fluid on the walls of its container are perpendicular at every point.

strate that the force exerted by the water is everywhere perpendicular to the surface of the barrel.

A moment's reflection will show the student that a liquid also exerts an upward pressure. Anyone who has tried to keep a rubber float under the surface of water is immediately convinced of the existence of an upward pressure. In fact, we find that:

Fluids exert pressure in all directions.

Figure 15-3 shows a liquid under pressure. The forces acting on the face of the piston, the walls of the enclosure, and the surfaces of a suspended object are shown in the figure.

Just as larger volumes of solid objects exert greater forces against their supports, fluids exert greater pressure at increasing depths. The fluid at the bottom of a container is always under a greater pressure than that near the surface. This is due to the weight of the overlying liquid. However, we must point out a distinct difference

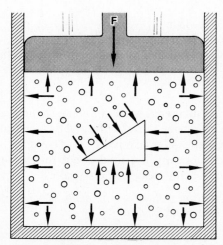

Fig. 15-3 Fluids exert pressure in all directions.

between the pressure exerted by solids and that exerted by liquids. A solid object can exert only a *downward* force due to its weight. At any particular depth in a fluid, the pressure is the same in all directions. If this were not true, the fluid would flow under the influence of a resultant pressure until a new condition of equilibrium was reached.

Since the weight of the overlying fluid is proportional to its density, the pressure at any depth is also proportional to the density of the fluid. This can be seen by considering a rectangular column of water extending from the surface to a depth h, as shown in Fig. 15-4. The weight of the entire column acts on the surface area A at the bottom of the column.

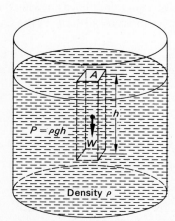

Fig. 15-4 The relationship of pressure, density, and depth.

From Eq. (15-1), we can write the weight of the column as

$$W = DV = DAh$$

where D is the weight density of the fluid. The pressure (weight per unit area) at the depth h is given by

$$P = \frac{W}{A} = Dh$$

or, in terms of mass density,

$$\boxed{P = Dh = \rho gh} \tag{15-5}$$

The fluid pressure at any point is directly proportional to the density of the fluid and to the depth below the surface of the fluid.

EXAMPLE 15-3   The water pressure in a certain house is 160 lb/in.² How high must the water level in a reservoir be above the point of release in the house?

Solution   The weight density of water is 62.4 lb/ft³. The pressure is 160 lb/in². To avoid inconsistency in units, we convert the pressure to units of pounds per square foot.

$$P = (160 \text{ lb/in.}^2) \, \frac{144 \text{ in.}^2}{1 \text{ ft}^2} = 23{,}040 \text{ lb/ft}^2$$

Solving for h in Eq. (15-5), we have

$$h = \frac{P}{D} = \frac{23{,}040 \text{ lb/ft}^2}{62.4 \text{ lb/ft}^2} = 369 \text{ ft}$$

In the above example, no mention was made of the size or shape of the reservoir containing the supply of water. Additionally, no information was given about the path of the water or size of the pipes connecting the reservoir to the home. Can we assume that our answer is correct when it is based only upon the difference in water levels? Doesn't the shape or area of a container have any effect on liquid pressure? In order to answer these questions, we must recall some of the characteristics of fluids already discussed.

Consider a series of vessels of different areas and shapes interconnected, as shown in Fig. 15-5. It would seem at first glance that the greater volume of water in

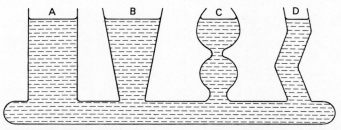

Fig. 15-5 Water seeks its own level, indicating that the pressure is independent of the area or shape of the container.

vessel B should develop a greater pressure at the bottom than vessel D. The effect of such a difference in pressure would then force the liquid to rise higher in vessel D. However, filling the vessels with liquid shows the levels to be the same for each vessel.

Part of the problem in understanding this paradox results from confusing the terms *pressure* and *total force.* Since pressure is measured in terms of a unit area, we do not consider the *total area* when solving problems involving pressure. For example, in vessel A the area of the liquid at the bottom of the vessel is much greater than the area at the bottom of vessel D. This means that the liquid in vessel A will exert a greater *total force* on the bottom than the liquid in vessel D. But the greater force is applied over a larger area so that the pressure is the same in both vessels.

If the bottoms of vessels B, C, and D have the same area, we can say that the total forces are also equal at the bottoms of these containers. (Of course, the pressures are equal at any particular depth.) You may wonder how the total forces can be equal when vessels B and C contain a greater volume of water. The additional water in each case is supported by vertical components of forces exerted by the walls of the container on the fluid. (See Fig. 15-6.) When the walls of a container are vertical, the forces acting on the sides have no upward components. The total force at the bottom of a container is therefore equal to the weight of a straight column of water above the base area.

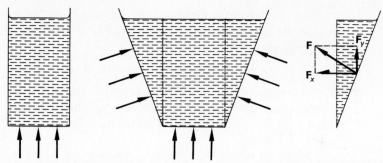

Fig. 15-6 The pressure at the bottom of each vessel is a function only of the depth of the liquid and is the same in all directions. Since the area at the bottom is the same for both vessels, the total force exerted on the bottom of each container is also the same.

EXAMPLE 15-4   Assume that the vessels in Fig. 15-5 are filled with gasoline until the fluid level is 20 cm above the base of each vessel. The areas at the bases of vessels A and B are 20 cm^2 and 10 cm^2, respectively. Compare the pressure and the total force at the base of each container. Compare the pressure and the total force at the base of each container.

Solution   The pressure is the same at the base of either container and is given by

$$P = \rho g h = (680 \text{ kg/m}^3)(9.8 \text{ m/s}^2)(0.20 \text{ m}) = 1330 \text{ Pa}$$

The total force in each case is the product of the pressure and the base area ($F = PA$). Remember that 1 cm^2 = 1 × 10^{-4} m^2.

$$F = PA = (1330 \text{ Pa})(20 \times 10^{-4} \text{ m}^2) = 2.67 \text{ N}$$
$$F = PA = (1330 \text{ Pa})(10 \times 10^{-4} \text{ m}^2) = 1.33 \text{ N}$$

Before considering other applications of fluid pressure, let us summarize the principles discussed in this section for fluids at rest.

1. The forces exerted by a fluid on the walls of its container are always perpendicular.
2. The fluid pressure is directly proportional to the depth of the fluid and to its density.
3. At any particular depth, the fluid pressure is the same in all directions.
4. Fluid pressure is independent of the shape or area of its container.

15-4
MEASURING PRESSURE

The pressure discussed in the previous section is due only to the fluid itself and can be calculated from Eq. (15-5). Unfortunately, this is usually not the case. Any liquid in an open container, for example, is subjected to atmospheric pressure in addition to the pressure of its own weight. Since the liquid is relatively incompressible, the external pressure of the atmosphere is transmitted equally throughout the volume of the liquid. This fact, first stated by the French mathematician Blaise Pascal (1623–1662), is called *Pascal's law*. Generally, it can be stated as follows:

> *An external pressure applied to an enclosed fluid is transmitted uniformly throughout the volume of the liquid.*

Most devices which measure pressure directly actually measure the difference between the *absolute pressure* and *atmospheric pressure*. The result is called the *gauge pressure*.

Absolute pressure = gauge pressure + atmospheric pressure

Atmospheric pressure at sea level is 101.3 kPa, or 14.7 lb/in.[2] Because atmospheric pressure enters into so many calculations, we often use a pressure unit of 1 *atmosphere* (atm), defined as the average pressure exerted by the atmosphere at sea level, that is, 14.7 lb/in.[2]

A common device for measuring gauge pressure is the open-tube *manometer* (muh-nom'-uh-ter), shown in Fig. 15-7. The manometer consists of a U-shaped tube

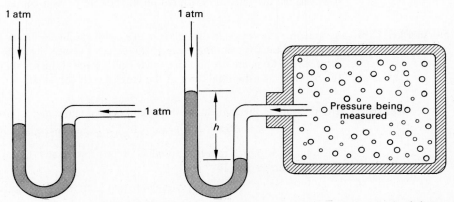

Fig. 15-7 The open-tube manometer. Pressure is measured by the height h of the mercury column.

containing a liquid, usually mercury. When both ends of the tube are open, the mercury seeks its own level because 1 atm of pressure is exerted at each of the open ends. When one end is connected to a pressurized chamber, the mercury will rise in the open tube until the pressures are equalized. The difference between the two levels of mercury is a measure of the gauge pressure, i.e., the difference between the absolute pressure in the chamber and atmospheric pressure at the open end. The manometer is used so often in laboratory situations that atmospheric pressures and other pressures are often expressed in *centimeters of mercury* or *inches of mercury*.

Atmospheric pressure is usually measured in the laboratory with a mercury barometer. The principle of its operation is shown in Fig. 15-8. A glass tube, closed at one end, is filled with mercury. The open end is covered, and the tube is inverted in a bowl of mercury. When the open end is uncovered, the mercury flows out of the tube until the pressure exerted by the column of mercury exactly balances atmospheric pressure acting on the mercury in the bowl. Since the pressure in the tube above the column of mercury is zero, the height of the column above the level of mercury in the bowl indicates the atmospheric pressure. At sea level an atmospheric pressure of 14.7 lb/in.[2] will cause the level of the mercury in the tube to stabilize at a height of 76 cm, or 30 in.

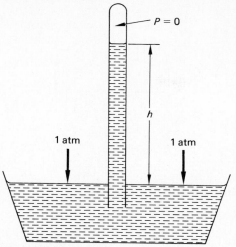

Fig. 15-8 The barometer.

In summary, we can write the following equivalent measures of atmospheric pressure:

$$1 \text{ atm} = 101.3 \text{ kPa} = 14.7 \text{ lb/in.}^2 = 76 \text{ cm of mercury}$$
$$= 30 \text{ in. of mercury} = 2116 \text{ lb/ft}^2$$

EXAMPLE 15-5   The mercury manometer is used to measure the pressure of a gas inside a tank. (Refer to Fig. 15-7.) If the difference between the two mercury levels is 36 cm, what is the absolute pressure inside the tank?

Solution   The gauge pressure is 36 cm of mercury, and atmospheric pressure is 76 cm of mercury. Thus the absolute pressure is found from Eq. (15-5).

$$\text{Absolute pressure} = 36 \text{ cm} + 76 \text{ cm} = 112 \text{ cm of mercury}$$

The pressure in the tank is equivalent to the pressure that would be exerted by a column of mercury 112 cm high.

$$P = Dh = \rho g h$$
$$= (13,600 \text{ kg/m}^3)(9.8 \text{ m/s}^2)(1.12 \text{ m})$$
$$= 1.49 \times 10^5 \text{ N/m}^2 = 149 \text{ kPa}$$

You should verify that this absolute pressure is also 21.6 lb/in.^{2} or 1.47 atm.

15-5

THE HYDRAULIC PRESS

The most universal application of Pascal's law is found with the hydraulic press, shown in Fig. 15-9. According to Pascal's principle, a pressure applied to the liquid in the left column will be transmitted undiminished to the liquid in the column in the right. Thus, if an input force F_i acts upon a piston of area A_i, it will cause an output force F_o to act on a piston of area A_o so that

Input pressure = output pressure

$$\frac{F_i}{A_i} = \frac{F_o}{A_o} \tag{15-6}$$

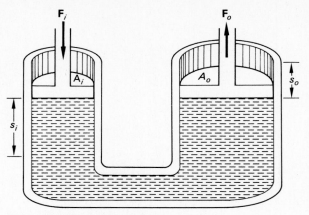

Fig. 15-9 The hydraulic press.

The ideal mechanical advantage of such a device is equal to the ratio of the output force to the input force. Symbolically, we write

$$M_I = \frac{F_o}{F_i} = \frac{A_o}{A_i}$$ (15-7)

A small input force can be multiplied to yield a much larger output force simply by having the output piston much larger in area than the input piston. The output force is given by

$$F_o = F_i \frac{A_o}{A_i}$$ (15-8)

According to the methods developed in Chap. 12 for simple machines, the input work must equal the output work if we neglect friction. If the input force F_i travels through a distance s_i while the output force F_o travels through a distance s_o, we can write

Input work = output work

$$F_i s_i = F_o s_o$$

This relation leads to another useful expression for the ideal mechanical advantage of a hydraulic press.

$$M_I = \frac{F_o}{F_i} = \frac{s_i}{s_o}$$ (15-9)

Notice that the mechanical advantage is gained at the expense of input distance. For this reason, most applications utilize a system of valves to permit the output piston to be raised by a series of short input strokes.

EXAMPLE 15-6 The smaller and larger pistons of a hydraulic press have diameters of 2 and 24 in., respectively. (*a*) What input force is required in order to deliver a total output force of 2000 lb at the larger piston? (*b*) How far must the smaller piston travel in order to lift the larger piston 1 in.?

Solution (a)   The mechanical advantage is

$$M_I = \frac{A_o}{A_i} = \frac{\pi d_o^2/4}{\pi d_i^2/4} = \left(\frac{d_o}{d_i}\right)^2$$

$$= \left(\frac{24 \text{ in.}}{2 \text{ in.}}\right)^2 = (12)^2 = 144$$

The required input force is given by

$$F_i = \frac{F_o}{M_I} = \frac{2000 \text{ lb}}{144} = 13.9 \text{ lb}$$

Solution (b)   Applying Eq. (15-9), we can compute the input distance.

$$s_i = M_I s_o = (144)(1 \text{ in.}) = 144 \text{ in.}$$

The principle of the hydraulic press is found in many engineering and mechanical devices. Power steering, the hydraulic jack, shock absorbers, and automobile braking systems are a few common examples.

15-6 ARCHIMEDES' PRINCIPLE

Anyone familiar with swimming and other water sports has observed that objects seem to lose weight when submerged in water. In fact, the object may even float on the surface because of the upward pressure exerted by the water. An ancient Greek mathematician, Archimedes (287–212 B.C.), first studied the buoyant force exerted by fluids. *Archimedes' principle* can be stated as follows:

> *An object which is completely or partly submerged in a fluid experiences an upward force equal to the weight of the fluid displaced.*

Archimedes' principle can be demonstrated by studying the forces a fluid exerts on a suspended object. Consider a disk of area A and height H which is completely submerged in a fluid, as shown in Fig. 15-10. Recall that the pressure at any depth h in a fluid is given by

$$P = \rho g h$$

where ρ is the mass density of the fluid and g is the acceleration of gravity. Of course, if we wish to represent the absolute pressure within the fluid, we must add the external pressure exerted by the atmosphere. The total downward pressure P_1 on top of the disk in Fig. 15-10 is therefore

$$P_1 = P_a + \rho g h_1 \qquad \text{downward}$$

where P_a is atmospheric pressure and h_1 is the depth at the top of the disk. Similarly, the upward pressure P_2 on the bottom of the disk is

$$P_2 = P_a + \rho g h_2 \qquad \text{upward}$$

where h_2 is the depth at the bottom of the disk. Since h_2 is greater than h_1, the pressure on the bottom of the disk will exceed the pressure at the top, resulting in a net upward force. If we represent the downward force by F_1 and the upward force by F_2, we can write

$$F_1 = P_1 A \qquad F_2 = P_2 A$$

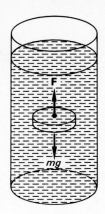 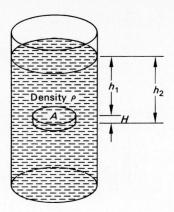

Fig. 15-10 The buoyant force exerted on the disk is equal to the weight of the fluid it displaces.

The net upward force exerted *by* the fluid *on* the disk is called the *buoyant force* and is given by

$$F_B = F_2 - F_1 = A(P_2 - P_1)$$
$$= A(P_a + \rho g h_2 - P_a - \rho g h_1)$$
$$= A\rho g(h_2 - h_1) = A\rho g H$$

where $H = h_2 - h_1$ is the height of the disk. Finally, if we recall that the volume of the disk is $V = AH$, we obtain the important result

$$\boxed{F_B = V\rho g = mg}$$ (15-10)

Buoyant force = weight of displaced fluid

which is Archimedes' principle.

In applying this result, it must be recalled that Eq. (15-10) allows us to compute only the *buoyant force* due to the difference in pressures. It does not represent the resultant force. A submerged body will sink if the weight of the fluid it displaces (the buoyant force) is less than the weight of the body. If the weight of the displaced fluid is exactly equal to the weight of the submerged body, it will neither sink nor rise. In this instance, the body will be in equilibrium. If the weight of displaced fluid exceeds the weight of a submerged body, the body will rise to the surface and float. When the floating body comes to equilibrium at the surface, it will displace its own weight of liquid. Figure 15-11 demonstrates this point with the use of an overflow can and a beaker to catch the fluid displaced by a wooden block.

EXAMPLE 15-7   A cork float has a volume of 4 cm³ and a density of 207 kg/m³. (*a*) What volume of the cork is beneath the surface when the cork floats in water? (*b*) What downward force is needed to submerge the block completely?

Solution (*a*)   The floating cork will displace a volume of water equal to its own weight. Recalling that 1 cm³ $= 1 \times 10^{-6}$ m³, the weight of 4 cm³ of cork is

$$W = \rho g V = (207 \text{ kg/m}^3)(9.8 \text{ m/s}^2)(4 \times 10^{-6} \text{ m}^3)$$
$$= 8.11 \times 10^{-3} \text{ N}$$

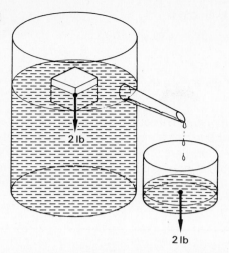

Fig. 15-11 A floating body displaces its own weight of fluid.

2 lb

2 lb

Since $W = \rho g V$, the volume of water displaced is

$$V = \frac{W}{\rho g} = \frac{8.11 \times 10^{-3}\,\text{N}}{(1000\,\text{kg/m}^3)(9.8\,\text{m/s}^2)}$$
$$= 8.28 \times 10^{-7}\,\text{m}^3 \text{ or } 0.828\,\text{cm}^3$$

Thus the volume of the cork under water is also 0.828 cm³.

If the area of the floating surface were known in the previous example, you could calculate how deep the cork would sink into the water. Note that approximately 21 percent of the cork is under water. As an extra example, you should show that the fraction of volume submerged is equal to the specific gravity of an object.

Solution (b)   In order to submerge the cork, a downward force F must be exerted, in addition to the weight W of the cork. The sum of these forces equals the buoyant force F_B:

$$F + W = F_B$$

The required force F is therefore equal to the difference between the buoyant force and the weight of the cork:

$$F = F_B - W$$

The buoyant force is equal to the weight of 4 cm³ of water:

$$F_B = \rho g V = (1000\,\text{kg/m}^3)(9.8\,\text{m/s}^2)(4 \times 10^{-6}\,\text{m}^3)$$
$$= 39.2 \times 10^{-3}\,\text{N}$$

The required force F to submerge the cork is

$$F = 39.2 \times 10^{-3}\,\text{N} - 8.11 \times 10^{-3}\,\text{N} = 31.1 \times 10^{-3}\,\text{N}$$

EXAMPLE 15-8   A weather balloon is to operate at an altitude where the density of air is 0.9 kg/m³. At this altitude, the balloon has a volume of 20 m³ and is filled with hydrogen ($\rho_H = 0.09$ kg/m³). If the balloon bag weighs 118 N, what load can it support at this level?

Solution   The buoyant force is equal to the weight of the displaced air. Thus

$$F_B = \rho g V = (0.9\,\text{kg/m}^3)(9.8\,\text{m/s}^2)(20\,\text{m}^3) = 176\,\text{N}$$

The weight of 20 m³ of hydrogen is

$$W_H = \rho_H \, gV = (0.09 \text{ kg/m}^3)(9.8 \text{ m/s}^2)(20 \text{ m}^3) = 17.6 \text{ N}$$

The load supported is

$$W_L = F_B - W_H - W_B$$
$$= 176 \text{ N} - 17.6 \text{ N} - 118 \text{ N} = 40.4 \text{ N}$$

Large balloons can maintain a condition of equilibrium at any altitude by adjustment of their weight or buoyant force. The weight can be lightened by releasing the ballast provided for that purpose. The buoyant force can be decreased by releasing the gas from the balloon or increased by pumping more gas into the flexible balloon. Hot-air balloons use the lower density of heated air to provide their buoyancy.

SUMMARY

We have discussed the concepts of fluids and pressure. Density, buoyant forces, and other quantities were defined and applied to many physical examples. The key ideas discussed in this chapter are summarized below:

- A very important physical property of matter is its *density*. The weight density D and the mass density ρ are defined as follows:

$$\text{Weight density} = \frac{weight}{volume} \qquad \boxed{D = \frac{W}{V}} \qquad \text{N/m}^3 \text{ or lb/ft}^3$$

$$\text{Mass density} = \frac{mass}{volume} \qquad \boxed{\rho = \frac{m}{V}} \qquad \text{kg/m}^3 \text{ or slugs/ft}^3$$

- Since $W = mg$, the relationship between D and ρ is

$$\boxed{D = \rho g} \qquad \text{Weight density} = \text{mass density} \times \text{gravity}$$

- Important points to remember about fluid pressure:
  a. The forces exerted by a fluid on the walls of its container are always perpendicular.
  b. The fluid pressure is directly proportional to the depth of the fluid and to its density.

$$\boxed{P = \frac{F}{A} \qquad P = Dh \qquad P = \rho gh}$$

  c. At any particular depth, the fluid pressure is the same in all directions.
  d. Fluid pressure is independent of the shape or area of its container.
- Pascal's law states that *an external pressure applied to an enclosed fluid is transmitted uniformly throughout the volume of the liquid.*
- When measuring fluid pressure, it is essential to distinguish between *absolute* pressure and *gauge* pressure:

Absolute pressure = gauge pressure + atmospheric pressure

Atmospheric pressure = 1 atm = 1.013×10^5 N/m^2

$= 1.013 \times 10^5$ Pa = 14.7 lb/in.^{2}

= 76 cm of mercury

- Applying Pascal's law to the hydraulic press gives the following for the ideal advantage:

$$M_I = \frac{F_o}{F_i} = \frac{s_o}{s_i}$$

Ideal Mechanical Advantage for Hydraulic Press

- Archimedes' principle: *An object which is completely or partly submerged in a fluid experiences an upward force equal to the weight of the fluid displaced.*

$$F_B = mg$$ or $$F_B = V\rho g$$ *Buoyant Force*

QUESTIONS

15-1. Define the following terms:
    a. Weight density
    b. Mass density
    c. Pressure
    d. Total force
    e. Pascal's law
    f. Absolute pressure
    g. Gauge pressure
    h. Manometer
    i. Archimedes' principle
    j. Buoyant force

15-2. Make a list of the units for weight density and the similar units for mass density.

15-3. Which is numerically larger: the weight density of an object or its mass density?

15-4. The density of water is given in Table 15-1 as 62.4 lb/ft^3. In performing an experiment with water on the surface of the moon, would you trust this value? Explain.

15-5. Which is heavier, 870 kg of brass or 3.5 ft^3 of copper?

15-6. Why are dams so much thicker at the bottom than at the top? Does the pressure exerted on the dam depend on the length of the reservoir perpendicular to the dam?

15-7. A large block of ice floats in a bucket of water so that the level of the water is at the top of the bucket. Will the water overflow when the ice melts? Explain.

15-8. A tub of water rests on weighing scales which indicate 40 lb total weight. Will the total weight increase when a 5-lb fish is floating on the surface of the water? Discuss.

15-9. Suppose an iron block supported by a string is completely submerged in the tub of Question 15-8. How will the reading on the scales be affected?

15-10. A boy just learning to swim finds that he can float on the surface more easily after inhaling air. He also observes that he can hasten his descent to the bottom of the pool by exhaling air on the way down. Explain his observations.

15-11. A toy sailboat filled with pennies floats in a small tub of water. If the pennies are thrown into the water, what happens to the water level in the tub?

15-12. Is it more difficult to hold a cork float barely under the surface than it is at a depth of 5 ft? Explain.

15-13. Is it possible to construct a barometer using water instead of mercury? How high will the column of water be if the external pressure is 1 atm?

15-14. Discuss the operation of a submarine and a weather balloon. Why will a balloon rise to a definite height and stop? Will a submarine sink to a particular depth and stop if no changes are made after submerging?

PROBLEMS

Density

15-1. What volume does 0.4 kg of alcohol occupy? What is the weight of this volume?
Ans. 5.06×10^{-4} m^3, 3.92 N

15-2. An unknown substance has a volume of 20 ft^3 and weighs 3370 lb. What are the weight density and the mass density?

15-3. What volume of water has the same mass as 100 cm^3 of lead? What is the weight density of lead?
Ans. 1130 cm^3, 1.12×10^5 N/m^3

* 15-4. A 200-mL flask (1 L = 1000 cm^3) is filled with an unknown liquid. An electronic balance indicates that the added liquid has a mass of 176 g. What is the specific gravity of the liquid? Can you guess the identity of the liquid?

Fluid Pressure

15-5. Find the pressure in kilopascals due to a column of mercury 60 cm high. What is this pressure in lb/in.^{2} and in atmospheres?
Ans. 80.0 kPa, 11.6 lb/in.^{2}, 0.79 atm

15-6. A pipe contains water under a gauge pressure of 400 kPa. If you patch a 4-mm-diameter hole in the pipe with a piece of tape, what force must the tape be able to withstand?

15-7. A submarine dives to a depth of 120 ft and levels off. The interior of the submarine is maintained at atmospheric pressure. What are the pressure and the total force applied to a hatch 2 ft wide and 3 ft long? The weight density of sea water is around 64 lb/ft^3. **Ans.** 53.3 lb/in.^{2}, 46,100 lb

15-8. If you constructed a barometer using water as the liquid instead of mercury, what height of water would indicate a pressure of one atmosphere?

15-9. A 20-kg piston rests on a sample of gas in a cylinder 8 cm in diameter. What is the gauge pressure on the gas? What is the absolute pressure?
Ans. 39.0 kPa, 140.3 kPa

* 15-10. An open U-shaped tube such as the one in Fig. 15-12 is 1 cm^2 in cross section.

Fig. 15-12

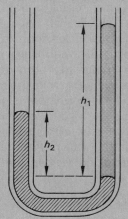

What volume of water must be poured into the right tube in order to cause the mercury in the left tube to rise 1 cm above its original position?

15-11. The gauge pressure in an automobile tire is 28 lb/in.^{2}. If the wheel supports 1000 lb, what area of the tire is in contact with the ground?

Ans. 35.7 in.^{2}

15-12. Two liquids which do not react chemically are placed in a bent tube like the one in Fig. 15-12. Show that the heights of the liquids above their surface of separation are inversely proportional to their densities:

$$\frac{h_1}{h_2} = \frac{\rho_2}{\rho_1}$$

15-13. Assume that the two liquids in the U-shaped tube of Fig. 15-12 are water and oil. Compute the density of the oil if the water stands 19 cm above the interface and the oil stands 24 cm above the interface. Refer to Prob. 15-12.

Ans. 792 kg/m^3

15-14. A water-pressure gauge indicates a pressure of 50 lb/in.^{2} at the foot of a building. What is the maximum height to which the water will rise in the building?

The Hydraulic Press

15-15. The areas of the small and large pistons in a hydraulic press are 0.5 and 25 in.^{2}, respectively. What is the ideal mechanical advantage of the press? What force must be exerted in order to lift a 1-ton (2000-lb) load? Through what distance must the input force act in order to lift this load a distance of 1 in.?

Ans. 50, 40 lb, 50 in

15-16. A force of 400 N is applied to the small piston of a hydraulic press whose diameter is 4 cm. What must be the diameter of the large piston if it is to lift a 200-kg load?

15-17. The inlet pipe which supplies air pressure to operate a hydraulic lift is 2 cm in diameter. The output piston is 32 cm in diameter. What air pressure (gauge pressure) must be used to lift an 1800-kg automobile?

Ans. 219 kPa

15-18. The area of a piston in a force pump is 10 in.^{2}. What force is required to raise water with the piston to a height of 100 ft?

Archimedes' Principle

15-19. A 100-g cube 2 cm on each side is attached to a string and then totally submerged in water. What is the buoyant force and what is the tension in the rope?

Ans. 0.0784 N, 0.902 N

15-20. A solid object weighs 8 N in air. When this object is suspended from a spring scale and submerged in water, the apparent weight is only 6.5 N. What is the density of the object?

15-21. A cube of wood 5.0 cm on each edge floats in water with three-fourths of its volume submerged. (a) What is the weight of the cube? (b) What is the mass of the cube? (c) What is the specific gravity of wood?

Ans. (a) 0.919 N, (b) 93.8 g, (c) 0.75

15-22. A 20-g piece of metal has a density of 4000 kg/m^3. It is hung in a jar of oil (1500 kg/m^3) by a thin thread until it is completely submerged. What is the tension in the thread?

15-23. The mass of a piece of rock is found to be 9.17 g in air. When the rock is submerged in a fluid of density 873 kg/m^3, its apparent mass is only 7.26 g. What is the density of this rock?

Ans. 4191 kg/m^3

15-24. A balloon 40 m in diameter is filled with helium. The mass of the balloon and attached basket is 18 kg. What additional mass can be lifted by this balloon?

15-25. A balloon 40 m in diameter is filled with helium. What force is required to hold the balloon down? The mass of the balloon is 20 kg.

Ans. 365 kN

Additional Problems

15-26. What is the absolute pressure at the bottom of a lake that is 30 m deep?

15-27. Human blood of density 1050 kg/m^3 is held a distance of 60 cm above an arm where it is being administered to a patient. How much higher is the pressure at this position than it would be if it were held at the same level as the arm?

Ans. 6.17 kPa

15-28. A cylindrical tank 50 ft high and 20 ft in diameter is filled with water. (a) What is the water pressure on the bottom of the tank? (b) What is the total force on the bottom? (c) What is the pressure in a water pipe which is located 90 ft below the water level in the tank?

15-29. A block of wood weighs 16 lb in air. A lead sinker, which has an apparent weight of 28 lb in water, is attached to the wood, and both are submerged in water. If their combined apparent weight in water is 18 lb, find the density of the wooden block.

Ans. 38.4 lb/ft^3

15-30. A 100-g block of wood has a volume of 120 cm^3. Will it float in water? Gasoline?

15-31. A vertical test tube has 3 cm of oil (0.8 g/cm^3) floating on 9 cm of water. What is the pressure at the bottom of the tube?

Ans. 1.12 kPa

15-32. What percentage of an iceberg will remain below the surface of seawater (1030 kg/m^3)?

15-33. What is the smallest area of ice 30 cm thick that will support a 90-kg man? The ice is floating in fresh water.

Ans. 3.75 m^2

Fluids in Motion

OBJECTIVES

After completing this chapter, you should be able to:

1. Define the *rate of flow* of a fluid and solve problems which relate the rate of flow to the velocity and cross-sectional area.
2. Write *Bernoulli's equation* in its general form and describe the equation as it would apply to (a) a fluid at rest, (b) fluid flow at constant pressure, and (c) flow through a horizontal pipe.
3. Apply Bernoulli's equation to the solution of problems involving absolute pressure P, density ρ, fluid elevation h, and fluid velocity v.

In Chap. 15 we studied the properties of fluids at rest. The work of Archimedes, Pascal, and Newton all contributed greatly to our present knowledge of fluids. Unfortunately, the mathematical difficulties encountered when dealing with fluids in motion are formidable. However, the fundamental aspects of fluid flow can be analyzed by making certain assumptions and generalizations. This chapter will give you a working knowledge of the mechanics of fluids in motion.

16-1
FLUID FLOW

In studying the dynamics of fluids, we shall assume that all fluids in motion exhibit *streamline flow*.

> Streamline flow is the motion of a fluid in which every particle in the fluid follows the same path (past a particular point) as that followed by previous particles.

Figure 16-1 illustrates the *streamlines* of air flowing past two stationary obstacles. Note that the streamlines break down as air passes over the second obstacle, setting up whirls and eddies. These little whirlpools represent *turbulent flow* and absorb much of the fluid energy, increasing the frictional drag through the fluid.

We shall further consider that fluids are incompressible and have essentially no internal friction. Under these conditions, certain predictions can be made about the rate of fluid flow through a pipe or other container.

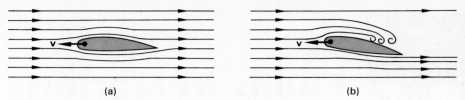

Fig. 16-1 Streamline and turbulent flow of a fluid in its path.

The **rate of flow** is defined as the volume of fluid that passes a certain cross section per unit of time.

In order to express this rate quantitatively, we shall consider a liquid flowing through the pipe of Fig. 16-2 with an average speed v. During a time interval t, each particle

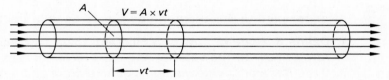

Fig. 16-2 Computing the rate of flow of a fluid through a pipe.

in the stream moves through a distance vt. The volume V flowing through a cross section A is given by

$$V = Avt$$

Thus the rate of flow (volume per unit time) can be calculated from

$$R = \frac{Avt}{t} = vA \qquad (16\text{-}1)$$

Rate of flow = velocity × cross section

The units of R express the ratio of a volume unit to a time unit. Common examples are cubic feet per second, cubic meters per second, liters per second, and gallons per minute.

If the fluid is incompressible and we ignore the effects of internal friction, the rate of flow R will remain constant. This means that a variation in the pipe cross section, as illustrated in Fig. 16-3, will result in a change in speed of the liquid so

Fig. 16-3 In streamline flow the product of the fluid velocity and the cross-sectional area of the pipe is constant at any point.

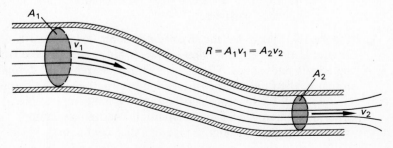

$$R = A_1 v_1 = A_2 v_2$$

that the product vA remains constant. Symbolically, we write

$$R = v_1 A_1 = v_2 A_2 \qquad (16\text{-}2)$$

A liquid will flow faster through a narrow section of pipe and more slowly through the broad sections. It is this principle which causes water to flow more rapidly when the banks of a small stream suddenly come closer together.

EXAMPLE 16-1 Water flows through a rubber hose 2 cm in diameter at a speed of 4 m/s. (*a*) What must be the diameter of the nozzle if water is to emerge at 20 m/s? (*b*) What is the rate of flow in cubic meters per minute?

Solution The rate of flow is constant, so that $A_1 v_1 = A_2 v_2$. Since the area A is proportional to the square of the diameter, we have

$$d_1^2 v_1 = d_2^2 v_2 \qquad \text{or} \qquad d_2^2 = \frac{v_1}{v_2} d_1^2$$

from which

$$d_2^2 = \frac{4 \text{ m/s}}{20 \text{ m/s}} (2 \text{ cm})^2 = 0.8 \text{ cm}^2$$

Taking the square root of each side gives

$$d_2 = 0.894 \text{ cm}$$

In order to find the rate of flow, we must first determine the area of the 2-cm hose.

$$A_1 = \frac{d_1^2}{4} = \frac{(2 \text{ cm})^2}{4} = 3.14 \text{ cm}^2$$

$$= 3.14 \text{ cm}^2 \, \frac{1 \times 10^{-4} \text{ m}^2}{1 \text{ cm}^2} = 3.14 \times 10^{-4} \text{ m}^2$$

The rate of flow is $R = A_1 v_1$, so that

$$R = (3.14 \times 10^{-4} \text{ m}^2)(4 \text{ m/s}) = 1.26 \times 10^{-3} \text{ m}^3/\text{s}$$
$$= (1.26 \times 10^{-3} \text{ m}^3/\text{s})(60 \text{ s/min}) = 0.0754 \text{ m}^3/\text{min}$$

The same rate would be found by considering the product $A_2 v_2$.

16-2 PRESSURE AND VELOCITY

We have noted that a fluid's speed increases when it flows through a constriction. An increase in speed can result only through the presence of an accelerating force. In order to accelerate the liquid as it enters the constriction, the pushing force from the large cross section must be greater than the resisting force from the constriction. In other words, the pressure at points A and C in Fig. 16-4 must be greater than the pressure at B. The tubes inserted into the pipe above these points clearly indicate the difference in pressure. The fluid level in the tube above the restriction is lower than the level in the adjacent areas. If h is the difference in height, the pressure differential is given by

$$P_A - P_B = \rho g h \qquad (16\text{-}3)$$

This assumes that the pipe is horizontal and that no pressure changes are introduced because of a change in potential energy.

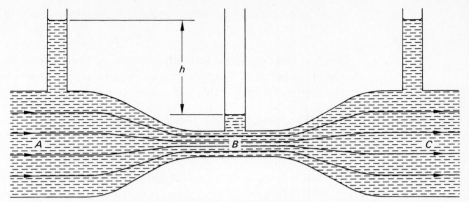

Fig. 16-4 The increased velocity of a fluid flowing through a constriction causes a drop in pressure.

The above example, as illustrated in Fig. 16-4, shows the principle of the *venturi meter*. From a determination of the difference in pressure, this device makes it possible to calculate the velocity of water in a horizontal pipe.

The venturi effect finds many other applications for both liquids and gases. The carburetor in an automobile utilizes the venturi principle to mix gasoline vapor and air. Air passing through a constriction on its way to the cylinders creates a low-pressure area as its speed increases. The decrease in pressure is used to draw fuel into the air column, where it is readily vaporized.

Figure 16-5 shows two methods you can use to demonstrate the decrease in pressure due to an increase in velocity. The simplest example consists of blowing air past the top surface of a sheet of paper, as shown in Fig. 16-5a. The pressure in the airstream above the paper will be reduced. This allows the excess pressure on the bottom to force the paper upward.

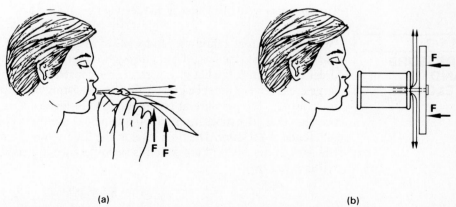

(a)   (b)

Fig. 16-5 Demonstrations of the decrease in pressure resulting from an increase in fluid speeds.

A second demonstration requires a hollow spool, a cardboard disk, and a pin (Fig. 16-5b). The pin is driven through the cardboard disk and placed in one end of the hollow spool, as shown in the figure. If you blow through the open end, you will find that the disk becomes more tightly pressed to the other end. One would expect the cardboard disk to fly off immediately. The explanation is that air blown into the spool must escape through the narrow space between the disk and the end of the spool. This action creates a low-pressure area, allowing the external atmospheric pressure to push the disk tight against the spool.

16-3
BERNOULLI'S
EQUATION

In our discussion of fluids, we have emphasized four quantities: the pressure P, the density ρ, the velocity v, and the height h above some reference level. The relationship between these quantities and their ability to describe fluids in motion was first established by Daniel Bernoulli (1700–1782), a Swiss mathematician. Steps leading to the development of this fundamental relationship can be understood by considering Fig. 16-6.

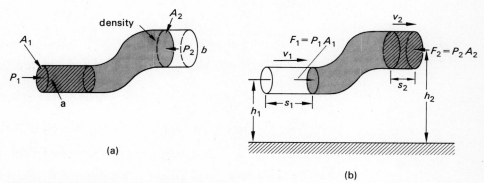

Fig. 16-6 Derivation of Bernoulli's equation.

Since a fluid has mass, it must obey the same conservation laws established earlier for solids. Consequently, the work required to move a certain volume of fluid through a pipe must equal the total change in kinetic and potential energy. Let us consider the work required to move the fluid from point a to point b in Fig. 16-6a. The net work must be the sum of the work done by the input force F_1 and the negative work done by the resisting force F_2.

$$\text{Net work} = F_1 s_1 - F_2 s_2$$

But $F_1 = P_1 A_1$ and $F_2 = P_2 A_2$, so that

$$\text{Net work} = P_1 A_1 s_1 - P_2 A_2 s_2$$

The product of area and distance represents the volume V of the fluid moved through the pipe. Since this volume is the same at the bottom and at the top of the pipe, we can substitute

$$V = A_1 s_1 = A_2 s_2$$

obtaining

$$\text{Net work} = P_1V - P_2V = (P_1 - P_2)V$$

The kinetic energy E_K of a fluid is defined as $\frac{1}{2}mv^2$, where m is the mass of the fluid and v is its velocity. Since the mass remains constant, a change in kinetic energy ΔE_K results only from the difference in fluid speed. In our example, the change in kinetic energy is

$$\Delta E_K = \frac{1}{2}mv_2^2 - \frac{1}{2}mv_1^2$$

The potential energy of a fluid at a height h above some reference point is defined as mgh, where mg represents the weight of the fluid. The volume of fluid moved through the pipe is constant. Therefore, the change in potential energy ΔE_p results from the increases in height of the fluid from h_1 to h_2:

$$\Delta E_p = mgh_2 - mgh_1$$

We are now prepared to apply the principle of conservation of energy. The net work done on the system must equal the sum of the increases in kinetic and potential energy. Thus

$$\text{Net work} = \Delta E_K + \Delta E_p$$
$$(P_1 - P_2)V = (\tfrac{1}{2}mv_2^2 - \tfrac{1}{2}mv_1^2) + (mgh_2 - mgh_1)$$

If the density of the fluid is ρ, we can substitute $V = m/\rho$, giving

$$(P_1 - P_2)\frac{m}{\rho} = \tfrac{1}{2}mv_2^2 - \tfrac{1}{2}mv_1^2 + mgh_2 - mgh_1$$

Multiplying through by ρ/m and rearranging, we obtain Bernoulli's equation:

$$P_1 + \rho gh_1 + \tfrac{1}{2}\rho v_1^2 = P_2 + \rho gh_2 + \tfrac{1}{2}\rho v_2^2 \qquad (16\text{-}4)$$

Since the subscripts 1 and 2 refer to any two points, Bernoulli's equation can be stated more simply as

$$\boxed{P + \rho gh + \tfrac{1}{2}\rho v^2 = \text{constant}} \qquad (16\text{-}5)$$

Bernoulli's equation finds application in almost every aspect of fluid flow. The pressure P must be recognized as the *absolute* pressure and not the *gauge* pressure. Remember that ρ is the mass density and not the weight density of the fluid. Notice that the units of each term in Bernoulli's equation are units of pressure.

16-4 APPLICATIONS OF BERNOULLI'S EQUATION

In many physical situations, the speed, height, or pressure of a fluid is constant. In such cases Bernoulli's equation holds in simpler form. For example, when a liquid is stationary, both v_1 and v_2 are zero. Bernoulli's equation will then show that the difference in pressure is

$$P_2 - P_1 = \rho g(h_1 - h_2) \qquad (16\text{-}6)$$

This equation is identical to the relationship discussed in Chap. 15 for fluids at rest.

Another important result occurs when there is no change in pressure ($P_1 = P_2$). In Fig. 16-7 a liquid emerges from a hole, or orifice, near the bottom of an open tank. Its velocity as it emerges from the orifice can be determined from Bernoulli's equation. We shall assume that the liquid level in the tank falls slowly in comparison with

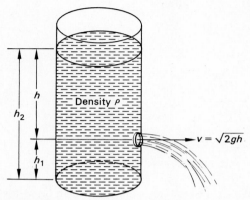

Fig. 16-7 Torricelli's theorem.

the emergent speed so that the speed v_2 at the top can be assumed zero. Additionally, it is noted that the liquid pressure at both the top and at the orifice is equal to atmospheric pressure. Thus $P_1 = P_2$ and $v_2 = 0$, reducing Bernoulli's equation to

$$\rho g h_1 + \tfrac{1}{2}\rho v_1^2 = \rho g h_2$$

or

$$v_1^2 = 2g(h_2 - h_1) = 2gh$$

This relationship is known as *Torricelli's* (toh-ree-chel'-ees) *theorem:*

$$\boxed{v = \sqrt{2gh}}$$

(16-7)

Note that the emergent velocity of a liquid at a depth h is the same as that of an object dropped from rest at a height h.

The rate at which a liquid flows from an orifice is given by vA from Eq. (16-1). Torricelli's relation allows us to express the rate of flow in terms of the height of a liquid above the orifice. Hence

$$R = vA = A\sqrt{2gh}$$

(16-8)

EXAMPLE 16-2 A crack in a water tank has a cross-sectional area of 1 cm². At what rate is water lost from the tank if the water level in the tank is 4 m above the opening?

Solution The area $A = 1 \text{ cm}^2 = 10^{-4} \text{ m}^2$ and the height $h = 4$ m. Direct substitution into Eq. (16-8) gives

$$R = A\sqrt{2gh} = (10^{-4} \text{ m}^2)\sqrt{(2)(9.8 \text{ m/s}^2)(4 \text{ m})}$$
$$= (10^{-4} \text{ m}^2)(8.85 \text{ m/s}) = 8.85 \times 10^{-4} \text{ m}^3/\text{s}$$

An interesting example demonstrating Torricelli's principle is shown in Fig. 16-8. The discharge velocity increases with depth. Note that the maximum range occurs when the opening is at the middle of the water column. Although the dis-

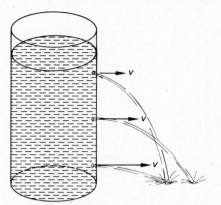

Fig. 16-8 The discharge velocity increases with depth below the surface, but the range is a maximum at the midpoint.

charge velocity increases below the midpoint, the water strikes the floor closer. This is because it strikes the floor sooner. Holes equidistant above and below the midpoint will have the same horizontal range.

As a final application, let us return to the venturi effect, which describes the motion of a fluid through a constriction. If the pipe in Fig. 16-9 is horizontal, we can set $h_1 = h_2$ in Bernoulli's equation, giving

$$P_1 + \tfrac{1}{2}\rho v_1^2 = P_2 + \tfrac{1}{2}\rho v_2^2 \qquad (16\text{-}9)$$

Since v_1 is greater than v_2, it follows that the pressure P_1 must be less than the pressure P_2 in order for Eq. (16-9) to hold. This relationship between velocity and pressure has already been discussed.

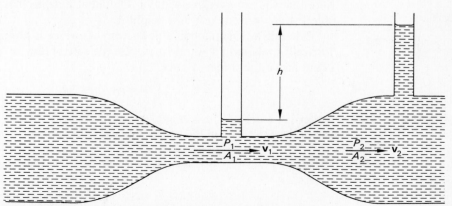

Fig. 16-9 Fluid flow through a constriction in a horizontal pipe.

EXAMPLE 16-3   Water flowing initially at 10 ft/s passes through a venturi tube like the one in Fig. 16-9. If $h = 4$ in., what is the velocity of the water in the constriction?

Solution   The difference in pressure, from Eq. (16-3), is

$$P_2 - P_1 = \rho g h$$

The pressure difference, from Bernoulli's equation, is

$$P_2 - P_1 = \tfrac{1}{2}\rho v_1^2 - \tfrac{1}{2}\rho v_2^2$$

Combining these two relations, we have

$$\rho g h = \tfrac{1}{2}\rho v_1^2 - \tfrac{1}{2}\rho v_2^2$$

Multiplying through by $2/\rho$ gives

$$2gh = v_1^2 - v_2^2$$

Note that this relation is similar to that for a freely falling body with an initial velocity v_2. Solving for v_1^2, we have

$$
\begin{aligned}
v_1^2 &= 2gh + v_2^2 \\
&= (2)(32 \text{ ft/s}^2)(0.333 \text{ ft}) + (10 \text{ ft/s})^2 \\
&= 21.3 \text{ ft}^2/\text{s}^2 + 100 \text{ ft}^2/\text{s}^2 = 121.3 \text{ ft}^2/\text{s}^2
\end{aligned}
$$

Therefore, the velocity through the constriction is

$$v_1 = \sqrt{121.3} = 11 \text{ ft/s}$$

In the above example, the density of the fluid did not enter into our calculations because the fluid in the tubes above the pipe was the same as that flowing through the pipe. When considering the density of the fluid, it must be recognized that the symbol ρ is the mass density and not the weight density.

SUMMARY

In this chapter we have discussed the rate of flow of fluids and how it is related to fluid velocity and cross-sectional area. Bernoulli's equation was derived and applied to problems involving the dynamics of fluids. The essential concepts are summarized below:

- The *rate of flow* is defined as the volume of fluid that passes a certain cross section A per unit of time t. In terms of fluid velocity v, we write

$$R = \frac{V}{t} = vA \qquad \textit{Rate of flow = velocity} \times \textit{cross section}$$

- For an incompressible fluid flowing through pipes in which the cross sections vary, the rate of flow is constant:

$$v_1 A_1 = v_2 A_2 \qquad d_1^2 v_1 = d_2^2 v_2$$

where v is the fluid velocity, A is the cross-sectional area of the pipe, and d is the diameter of the pipe.

- The net work done on a fluid is equal to the changes in kinetic and potential energy of the fluid. Bernoulli's equation expresses this fact in terms of the pressure P, the density ρ, the height of the fluid h, and its velocity v.

$$P + \rho g h + \tfrac{1}{2}\rho v^2 = \text{constant} \qquad \textit{Bernoulli's Equation}$$

If a volume of fluid changes from a state 1 to a state 2, as shown in Fig. 16-6, we can write

$$P_1 + \rho g h_1 + \tfrac{1}{2}\rho v_1^2 = P_2 + \rho g h_2 + \tfrac{1}{2}\rho v_2^2$$

- Special applications of Bernoulli's equation occur when one of the parameters does not change:

For a stationary liquid
$(v_1 = v_2)$
$$P_2 - P_1 = \rho g(h_1 - h_2)$$

If the pressure is constant
$(P_1 = P_2)$
$$v = \sqrt{2gh}$$

For a horizontal pipe
$(h_1 = h_2)$
$$P_1 + \tfrac{1}{2}\rho v_1^2 = P_2 + \tfrac{1}{2}\rho v_2^2$$

QUESTIONS

16-1. Define the following terms:
   a. Streamlining flow
   b. Turbulent flow
   c. Rate of flow
   d. Venturi effect
   e. Bernoulli's equation
   f. Torricelli's theorem

16-2. What assumptions and generalizations are made concerning the study of fluid dynamics?

16-3. Why does the flow of water from a faucet decrease when someone turns on another faucet in the same building?

16-4. Two rowboats moving parallel to each other in the same direction are drawn together. Explain.

16-5. Explain what would happen in a modern jet airliner at high speed if a hijacker fired a bullet through the window or broke open an escape hatch.

16-6. During high-velocity windstorms or hurricanes, the roofs of houses are sometimes blown off without otherwise damaging the homes. Explain with the use of diagrams.

16-7. A small child knocks a balloon over the heating duct in his home and is surprised to find that the balloon remains suspended above the duct, bobbing from one side to the other. Explain.

16-8. What conditions would determine the maximum lift capacity of a streamlined aircraft wing? Draw figures to justify your answer.

16-9. Explain with diagrams how a baseball pitcher throws a rising fast ball, an outside curve ball, and a sinking fast ball. Would a pitcher prefer to throw into the wind or with the wind when delivering the three pitches discussed above?

16-10. Two identical reservoirs are placed on the floor side by side. One is filled with mercury, and the other is filled with water. A hole is bored in each reservoir at the same depth below the surface. Compare the ranges of the emergent fluids.

PROBLEMS

Fluid Flow

16-1. Gasoline flows through a 1-in.-diameter hose at an average speed of 5 ft/s. What is the rate of flow in gallons per minute (1 ft^3 = 7.48 gal)? How much time is required to fill a 20-gal tank?

Ans. 12.2 gal/min, 1.63 min

16-2. Water flows from a terminal 3 cm in diameter and has an average speed of 2 m/s. What is the rate of flow in liters per minute. (1 L = 0.001 m^3.) How much time is required to fill a 40-L container?

16-3. What must be the diameter of a hose if it is to deliver 8 L of oil in 1 min with an exit speed of 3 m/s?

Ans. 7.52 mm

* 16-4. Water flowing from a 2-in. pipe emerges horizontally at the rate of 8 gal/min. What is the emerging velocity? What is the horizontal range of the stream of water if the pipe is 4 ft from the ground?

16-5. Water flowing at 6 m/s through a 6-cm pipe is connected to a 3-cm pipe. What is the velocity in the small pipe? Is the rate of flow greater in the smaller pipe?

Ans. 24 m/s, no

Applications of Bernoulli's Equation

16-6. Consider the situation described by Prob. 16-5. If the centers of each pipe are on the same horizontal line, what is the difference in pressure between the two connecting pipes?

16-7. What is the emergent velocity of water from a crack in its container 6 m below the surface? If the area of the crack is 1.3 cm^2, at what rate of flow does water leave the container?

Ans. 10.8 m/s, 1.41 × 10^{-3} m^3/s

16-8. A 2-cm-diameter hole is in the side of a water tank, and it is located 5 m below the water level in the tank. What is the emergent velocity of the water from the hole? What volume of water will escape from this hole in 1 min?

** 16-9. Water flows through a horizontal pipe at the rate of 82 ft^3/min. A pressure gauge placed on a 6-in. cross section of this pipe reads 16 lb/in.^{2}. What is the gauge pressure in a section of pipe where the diameter is 3 in.?

Ans. 11.1 lb/in.^{2}

* 16-10. Water flows at the rate of 6 gal/min through an opening in the bottom of a cylindrical tank. The water in the tank is 16 ft deep. What is the rate of escape if an added pressure of 9 lb/in.^{2} is applied to the source of the water?

* 16-11. Water moves through a pipe at 4 m/s under an absolute pressure of 200 kPa. The pipe narrows to one-half of its original diameter. What is the absolute pressure in the narrow part of the pipe?

Ans. 80.0 kPa

16-12. Water flows steadily through a horizontal pipe. At a point where the absolute pressure is 300 kPa, the velocity is 2 m/s. The pipe suddenly narrows, causing the absolute pressure to drop to 100 kPa. What will be the velocity of the water in this constriction?

Additional Problems

16-13. A fluid is forced out of a 6-mm-diameter tube so that 200 mL emerges in 32 s. What is the average velocity of the fluid in the tube?

Ans. 0.221 m/s

16-14. A pump of 2 kW output power discharges water from a cellar into a street 6 m above. At what rate in liters per second is the cellar emptied?

16-15. A horizontal pipe of diameter 120 mm has a constriction of diameter 40 mm. The velocity of water in the pipe is 60 cm/s and the pressure is 150 kPa. **(a)** What is the velocity in the constriction? **(b)** What is the pressure in the constriction?

Ans. (a) 540 cm/s, (b) 136 kPa

16-16. The water column in the container shown in Fig. 16-8 stands at a height H above the base of the container. Show that the depth h required to give the horizontal range x is given by

$$h = \frac{H}{2} \pm \frac{\sqrt{H^2 - x^2}}{2}$$

How does this equation show that the holes equidistant above and below the midpoint will have the same horizontal range?

16-17. A column of water stands 16 ft above the base of its container. What are two hole depths at which the emergent water will have a horizontal range of 8 ft?

Ans. 1.07 ft, 14.9 ft

16-18. Refer to Fig. 16-8 and Prob. 16-16. Show that the horizontal range is given by

$$x = 2\sqrt{h(H - h)}$$

Use this relation to show that the maximum range is equal to the height H of the water column.

16-19. Water flows through a horizontal pipe at the rate of 60 gal/min (1 ft^3 = 7.48 gal). What is the velocity in a section of pipe which narrows to a diameter of 1 in.?

Ans. 24.5 ft/s

16-20. An aircraft wing 25 ft long and 5 ft wide experiences a lifting force of 800 lb. What is the difference in pressure between the upper and lower surfaces of the wing?

16-21. Assume that air ($\rho = 1.3$ kg/m^3) flows past the top surface of an aircraft wing at 36 m/s. The air moving past the lower surface of the wing has a speed of 27 m/s. If the wing has a weight of 3700 N and an area of 3.5 m^2, what is the buoyant force on the wing?

Ans. 1290 N

16-22. What must be the gauge pressure in a fire hose if the nozzle is to force water to a height of 20 m?

16-23. Water flows through the pipe shown in Fig. 16-10 at the rate of 30 liters per second. The absolute pressure at point A is 200 kPa, and the point B is 8 m

Fig. 16-10

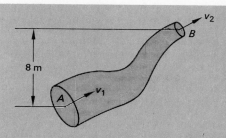

higher than point A. The lower section of pipe has a diameter of 16 cm and the upper section narrows to a diameter of 10 cm. (a) Find the velocities of the stream at points A and B. (b) What is the absolute pressure at point B?

Ans. (a) 1.49 m/s, 3.82 m/s; (b) 115 kPa

** 16-24. Seawater ($D = 64$ lb/ft^3) is pumped through the system of pipes shown in Fig. 16-11 at the rate of 4 ft^3/min. The pipe diameters at the lower and upper sec-

Fig. 16-11

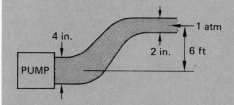

tions are 4 and 2 in., respectively. The water is discharged into the atmosphere at the upper end a distance of 6 ft higher than the lower section. (a) What are the speeds of flow in the upper and lower pipes? (b) What are the pressures in the lower and upper sections?

PART TWO
Heat, Light, and Sound

17 Temperature and Expansion

OBJECTIVES: After completing this chapter, you should be able to:

1. Demonstrate your understanding of the Celsius, Fahrenheit, Kelvin, and Rankine temperature scales by converting from specific temperatures on one scale to corresponding temperatures on another scale.
2. Distinguish between specific temperatures and temperature intervals and convert an interval on one scale to the equivalent interval on another scale.
3. Write formulas for linear expansion, area expansion, and volume expansion and be able to apply them to the solution of problems similar to those given in this chapter.

We have discussed the behavior of systems at rest and in motion. The fundamental quantities of mass, length, and time were introduced to describe the state of a given mechanical system.

Consider, for example, a 10-kg block moving with a constant speed of 20 m/s. The quantities of mass, length, and time are all present, and we find them sufficient to describe the motion. We can speak of the weight of the block, its kinetic energy, or its momentum, but a complete description of a system requires more than a simple statement of these quantities.

This becomes apparent when our 10-kg block encounters frictional forces. As the block slides to a stop, its energy seems to disappear, but the block and its supporting surface are slightly warmer. If energy is to be conserved, we must assume that the lost energy reappears in some form not yet considered. When energy disappears from the visible motion of objects and does not reappear in the form of visible potential energy, we frequently notice a rise in temperature. In this chapter we introduce the concept of temperature as a fourth fundamental quantity.

Until now we have been concerned only with the causes and the effects of *external* motion. A block resting on a table is in translational and rotational equilibrium insofar as its surroundings are concerned. However, a much closer study of the block reveals that it is active internally. Figure 17-1 shows a simple model of a solid. Individual molecules are held together by elastic forces analogous to the springs in

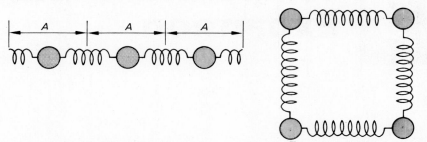

Fig. 17-1 A simplified model of a solid in which the individual molecules are held together by elastic forces.

the figure. These molecules oscillate about their equilibrium positions with a particular frequency and amplitude. Thus both potential energy and kinetic energy are associated with the molecular motion. Since this internal energy is related to the hotness or coldness of a body, it is often referred to as thermal energy.

> **Thermal energy** represents the total internal energy of an object, i.e.,
> the sum of its molecular kinetic and potential energies.

When two objects with different temperatures are placed in contact, energy is transferred from one to the other. For example, suppose hot coals are dropped into a container of water, as shown in Fig. 17-2. Thermal energy will be transferred from the coals to the water until the system reaches a stable condition, called *thermal equilibrium*. When touched, the coals and the water produce similar sensations and there is no more transfer of thermal energy.

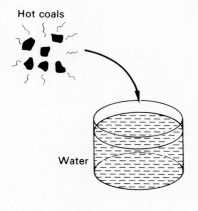
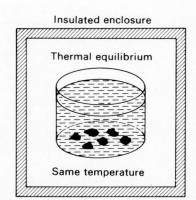

Fig. 17-2 Thermal equilibrium.

Such changes in thermal energy states cannot be satisfactorily explained in terms of classical mechanics alone. Therefore, all objects must have a new fundamental property which determines whether they will be in thermal equilibrium with other objects. This property is called *temperature*. In our example, the coals and the water are said to have the same temperature when the transfer of energy is zero.

> *Two objects are said to be in **thermal equilibrium** if and only if they are at the same temperature.*

Once we establish a means of measuring temperature, we have a necessary and sufficient condition for thermal equilibrium. The transfer of thermal energy which is due only to a difference in temperature is defined as *heat*.

> *Heat is defined as the transfer of thermal energy which is due to a difference of temperature.*

Before discussing the measurement of temperature, we should distinguish clearly between temperature and thermal energy. It is possible for two objects to be in thermal equilibrium (same temperature) with very different thermal energies. For example, consider a pitcher of water and a small cup of water, each having a temperature of 90°C. If they are mixed together, there will be no transfer of energy, but the thermal energy is much greater in the pitcher because it contains many more molecules. Remember that thermal energy represents the *sum* of the kinetic and potential energies of all the molecules. If we pour the water from each container onto separate blocks of ice, as shown in Fig. 17-3, more ice will be melted by the larger volume, indicating that it had more thermal energy.

Fig. 17-3 The distinction between thermal energy and temperature.

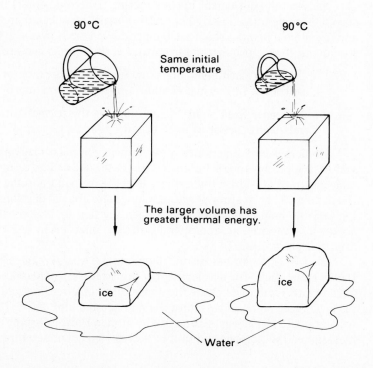

90°C   90°C

Same initial temperature

The larger volume has greater thermal energy.

ice

ice

Water

17-2

**THE
MEASURE-
MENT OF
TEMPERA-
TURE**

Temperature is usually determined by measuring some mechanical, optical, or electrical quantity which varies with temperature. For example, most substances expand as their temperature increases. If a change in any dimension can be shown to have a one-to-one correspondence with changes in temperature, the variation can be calibrated to measure temperature. A device calibrated in this way is called a *thermometer*. The temperature of another object can then be measured by placing the thermometer in close contact with it and allowing the two to reach thermal equilibrium. The temperature indicated by a number on the graduated thermometer also corresponds to the temperature of the surrounding objects.

A **thermometer** is a device which, through marked scales, can give an indication of its own temperature.

Two things are necessary in constructing a thermometer. First, we must have a confirmation that some thermometric property X varies with temperature T. If the variation is linear, we can write

$$T = kX$$

where k is the proportionality constant. The thermometric property should be one that is easily measured, e.g., the expansion of a liquid, the pressure in a gas, or the resistance in an electric circuit. Other quantities which vary with temperature are radiated energy, the color of emitted light, vapor pressure, and magnetic susceptibility. Thermometers have been constructed for each of these thermometric properties. The choice is dictated by the range of temperatures over which the thermometer is linear and by the mechanics of its use.

The second requirement in constructing a thermometer is the establishment of standard temperatures. Early temperature scales were based on the choice of upper and lower *fixed points*, which are temperatures readily available for laboratory measurements. Two convenient temperatures easily reproduced are:

The **lower fixed point** (ice point) is the temperature at which water and ice coexist in thermal equilibrium under a pressure of 1 atm.

The **upper fixed point** (steam point) is the temperature at which water and steam coexist in equilibrium under a pressure of 1 atm.

A widely used temperature measurement for scientific work originated with a scale developed by Anders Celsius (1701–1744), a Swedish astronomer. The *Celsius* scale arbitrarily assigned the number 0 to the ice point and the number 100 to the steam point. Thus at atmospheric pressure there are 100 divisions between the freezing and the boiling point of water. Each division or unit on the scale is called a *degree* (°). For example, room temperature is often taken as 20°C, which is read as *twenty degrees Celsius*.

Another scale for measuring temperature was developed in 1714 by Gabriel Daniel Fahrenheit. The development of this scale was based on the choice of different fixed points. Fahrenheit chose the temperature of a freezing solution of salt water as his lower fixed point and assigned the number and unit of 0°F. The upper fixed point was chosen to be the temperature of the human body. For some unexplained reason, he assigned the number and unit of 96°F for body temperature. The fact that body temperature is really 98.6°F indicates an experimental error in establishing the scale. Relating the Fahrenheit scale to the universally accepted fixed points on the Celsius scale, we note that 0 and 100°C correspond to 32 and 212°F.

We can compare the two scales by calibrating ordinary mercury-in-glass thermometers. This type of thermometer makes use of the fact that liquid mercury expands with increasing temperature. It consists of an evacuated glass capillary tube with a reservoir of mercury at the bottom and a closed top. Since mercury expands more than the glass tube, the mercury column rises in the tube until the mercury, glass, and its surroundings are in equilibrium.

Suppose we make two ungraduated thermometers and place them in a mixture of ice and water, as in Fig. 17-4. After allowing the mercury columns to stabilize, we mark 0°C on one thermometer and 32°F on the other. Next we place the two thermometers directly above boiling water, allowing the mercury columns to stabilize at the steam point. Again we mark the two thermometers, inscribing 100°C and 212°F adjacent to the mercury level above the marks corresponding to the ice point. The level of mercury is the same in each thermometer. Thus the only difference between the two thermometers will be in how they are graduated. There are 100 divisions, or Celsius degrees (C°), between the ice point and steam point on the Celsius thermometer, and there are 180 divisions, or Fahrenheit degrees (F°), on the Fahrenheit thermometer. Thus 100 Celsius degrees represents the same temperature interval as 180 Fahrenheit degrees. Symbolically,

$$100 \text{ C}° = 180 \text{ F}° \qquad \text{or} \qquad 5 \text{ C}° = 9 \text{ F}° \qquad\qquad (17\text{-}1)$$

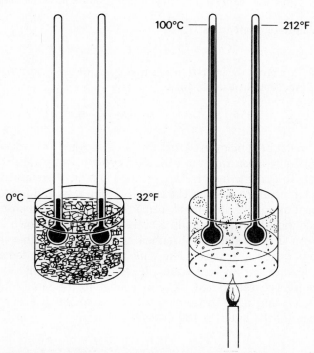

Fig. 17-4 Calibration of Celsius and Fahrenheit thermometers.

The degree mark (°) is placed after the C or F to emphasize that the numbers correspond to temperature intervals and not to specific temperatures. In other words, 20 F° is read "twenty Fahrenheit degrees" and corresponds to a *difference* between two temperatures on the Fahrenheit scale. The symbol 20°F, on the other hand, refers to a specific mark on the Fahrenheit thermometer. For example, suppose a pan of hot food cools from 98 to 76°F. These numbers correspond to specific temperatures, as indicated by the height of a mercury column. However, they represent a temperature interval of

$$\Delta t = 98°F - 76°F = 22 \ F°$$

Δt is used to denote a change in temperature.

The physics which treats the transfer of thermal energy is nearly always concerned with changes in temperature. Thus it often becomes necessary to convert a temperature interval from one scale to the corresponding interval on the other scale. This can best be accomplished by recalling from Eq. (17-1) that an interval of 5 C° is equivalent to an interval of 9 F°. The appropriate conversion factors can be written as

$$\frac{5 \ C°}{9 \ F°} = 1 = \frac{9 \ F°}{5 \ C°} \tag{17-2}$$

When converting F° to C°, the factor on the left should be used; when converting C° to F°, the factor on the right should be used.

EXAMPLE 17-1   During a 24-h period, a steel rail varies in temperature from 20°F at night to 70°F in the middle of the day. Express this range of temperature in Celsius degrees.

Solution   The temperature interval is

$$\Delta t = 70°F - 20°F = 50 \ F°$$

In order to convert the interval to Celsius degrees, we choose the conversion factor which will cancel the Fahrenheit units. Thus

$$\Delta t = 50 \ \cancel{F°} \times \frac{5 \ C°}{9 \ \cancel{F°}} = 27.8 \ C°$$

It must be remembered that Eq. (17-2) applies for temperature intervals. It can be used only when working with *differences* in temperatures. It is another matter entirely to find the temperature on the Fahrenheit scale which corresponds to the same temperature on the Celsius scale. Using ratio and proportion, we can develop an equation which will convert specific temperatures. Suppose, for example, we place two identical thermometers in a beaker of water, as shown in Fig. 17-5. One thermometer is graduated in Fahrenheit degrees and the other in Celsius degrees. The symbols t_C and t_F represent the same temperature (the temperature of the water), but they are on different scales. It should be apparent from the figure that the difference between t_C and 0°C corresponds to the same interval as the difference between t_F and 32°F. The ratio of the former to 100 divisions should be the same as the ratio of the latter to 180 divisions. Hence,

$$\frac{t_C - 0}{100} = \frac{t_F - 32}{180}$$

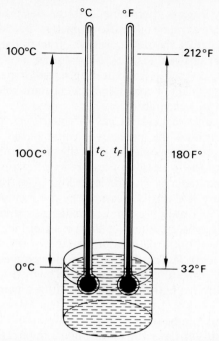

Fig. 17-5 Comparison of the Celsius and Fahrenheit scales.

Simplifying and solving for t_C, we obtain

$$t_C = \tfrac{5}{9}(t_F - 32) \qquad (17\text{-}3)$$

or, solving for t_F,

$$t_F = \tfrac{9}{5}t_C + 32 \qquad (17\text{-}4)$$

EXAMPLE 17-2   The melting point of lead is 330°C. What is the corresponding temperature on the Fahrenheit scale?

Solution   Substitution into Eq. (17-4) yields

$$t_F = \tfrac{9}{5}t_C + 32 = \tfrac{9}{5}(330) + 32$$
$$= 594 + 32 = 626°\text{F}$$

It is important to recognize that the t_F and t_C of Eqs. (17-3) and (17-4) represent identical temperatures. The numbers are different because the origin of each scale was at a different point and the degrees are of different size. What these equations tell us is the relationship between the *numbers* which are assigned to specific temperatures on two *different* scales.

Although the mercury-in-glass thermometer is the best known and most widely used, it is not as accurate as many other thermometers. In addition, mercury freezes at around −40°C, restricting the range over which it can be used. A very accurate thermometer with an extensive measuring range can be constructed by utilizing the properties of a gas. All gases subjected to heating expand in nearly the same manner. If their expansion is prevented by maintaining a constant volume, the pressure will increase in proportion to temperature.

In general, there are two kinds of gas thermometers. One kind maintains a constant pressure and utilizes the increases in volume as an indicator. This type is called a *constant-pressure* thermometer. The other kind, called a *constant-volume* thermometer, measures the increase in pressure as a function of temperature. The constant-volume thermometer is illustrated in Fig. 17-6. The gas is contained in bulb *B*, and the pressure it exerts is measured by the mercury manometer. As the temperature of the gas increases, it expands, forcing the mercury down in the closed tube and up in the open tube. In order to maintain a constant volume of the gas, the open tube must be raised until the level of mercury in the closed tube is brought back to the reference mark *R*. The difference in the two mercury levels is then an indication of the gas pressure at constant volume. The instrument can be calibrated for temperature measurements through the use of fixed points, as in the previous section.

The same apparatus can be used as a constant-pressure thermometer, as illustrated in Fig. 17-7. In this instance the volume of gas in bulb *B* is allowed to increase at constant pressure. The pressure exerted on the gas is maintained constant at 1 atm by lowering or raising the open tube until the mercury levels are the same in both tubes. The change in volume with temperature can then be indicated by the mercury level in the closed tube. Calibration consists of marking the mercury level at the ice point and again at the steam point.

Gas thermometers are very useful because of their almost unlimited range. For this reason and because they are so accurate, they are commonly used in research laboratories and in bureaus of standards. However, they are also large and bulky, rendering them useless for many minute technical measurements.

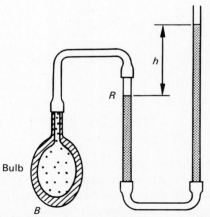

Fig. 17-6 The constant-volume thermometer.

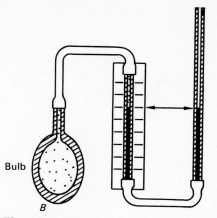

Bulb

B

Fig. 17-7 The constant-pressure thermometer.

It has probably occurred to the reader that Celsius and Fahrenheit scales have a very serious limitation. Neither 0°C nor 0°F represents a true zero of temperature. Consequently, for temperatures much lower than the ice point a negative temperature results. Even more serious is the fact that a formula involving temperature as a variable will not work with the existing scales. For example, we have discussed the expansion of a gas with an increase in temperature. We can state this proportionality as

$$V = kt$$

where k is the proportionality constant and t is the temperature. Certainly, the volume of a gas is not zero at 0°C or negative at negative temperatures, and yet these conclusions might be drawn from the above relationship.

This example provides a clue for establishing an *absolute* scale. If we can determine the temperature at which the volume of a gas under constant pressure becomes zero, we can establish a true zero of temperature. Suppose we use a constant-pressure gas thermometer, like the one in Fig. 17-7. The volume of the gas in the bulb can be measured carefully, first at the ice point and then at the steam point. These two points can be plotted on a graph, as in Fig. 17-8, with the volume as the ordinate and the temperature as the abscissa. The points A and B correspond to the gas volume at temperatures of 0 and 100°C, respectively. A straight line through these two points, extended both to the left and to the right, provides a mathematical description of the change in volume as a function of temperature. Note that the line can be extended indefinitely to the right, indicating that there is no upper limit to temperature. However, we cannot extend the line indefinitely to the left because it will eventually intercept the temperature axis. At this theoretical point, the gas would have zero volume. Further extension of the line would indicate a negative volume, which is meaningless. Therefore, the point at which the line intercepts the temperature axis is called the *absolute zero* of temperature. (Actually, any real gas would liquefy before reaching this point.)

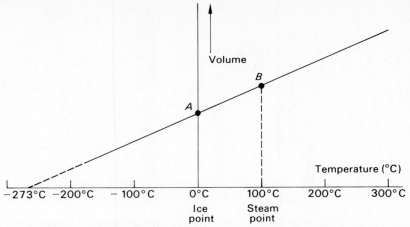

Fig. 17-8 The variation of the volume of a gas as a function of temperature. Absolute zero can be defined by extrapolation to zero volume.

If the above experiment is performed for several different gases, the slope of the curves will vary slightly. But the temperature intercept will always be the same and near −273°C. Ingenious theoretical and experimental procedures have established that the absolute zero of temperature is −273.15°C. In this text we shall assume that it is −273°C without fear of significant error. Conversion to the Fahrenheit scale shows that absolute zero is −460°F on that scale.

An absolute temperature scale has as its zero point the absolute zero of temperature. One such scale was devised by Lord Kelvin (1824–1907). The standard interval on this scale, the *kelvin*, has been adopted by the international (SI) metric system as the base unit for temperature measurement. The interval on the *Kelvin scale* represents the same change in temperature as the Celsius degree. Thus, an interval of 5 K (read "five kelvins") is exactly the same as 5 C°.

The Kelvin scale is related to the Celsius scale by the formula

$$T_K = t_C + 273$$

(17-5)

For example, 0°C would correspond to 273 K, and 100°C would correspond to 373 K. (See Fig. 17-9.) Hereafter we will reserve the symbol T for absolute temperature and the symbol t for other temperatures.

Problems with the reproducibility of accurate measurements for the ice and steam points of water caused the International Committee on Weights and Measures to establish a new standard in 1954. This standard is based on the *triple point of water*, which is the single temperature and pressure at which water, water vapor, and ice can coexist in thermal equilibrium. This convenient event occurs at a temperature of around 0.01°C and at a pressure of 4.58 mm of mercury. In order to remain somewhat consistent with earlier measurements, the temperature for the triple point of water was set at exactly 273.16 K. Thus, *the kelvin is now defined as the fraction 1/273.16 of the temperature of the triple point of water*. The SI temperature is now fixed by this definition, and all other scales must be redefined based on this single standard temperature.

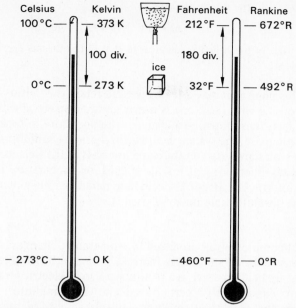

Fig. 17-9 A comparison of the four common temperature scales.

A second absolute scale, called the *Rankine scale*, remains in limited use despite efforts by various organizations to eliminate its use entirely. The Rankine degree is included here only for the sake of completeness. It has its absolute zero point at −460°F, and the degree intervals are identical to the Fahrenheit degree interval. The relationship between the temperature in degrees Rankine (°R) and the corresponding temperature in degrees Fahrenheit is

$$\boxed{T_R = t_F + 460} \tag{17-6}$$

For example, 0°F corresponds to 460°R, and 212°F corresponds to 672°R.

Remember that Eqs. (17-5) and (17-6) apply for specific temperatures. If we are concerned with a change in temperature or a difference in temperature, the absolute change or difference is the same in kelvins as it is in Celsius degrees. It is helpful to recall that

$$\boxed{1\ K = 1\ C°} \qquad \boxed{1\ R° = 1\ F°} \tag{17-7}$$

EXAMPLE 17-3   A mercury-in-glass thermometer may not be used at temperatures below −40°C. This is because mercury freezes at this temperature. (*a*) What is the freezing point of mercury on the Kelvin scale? (*b*) What is the difference between this temperature and the freezing point of water? Express the answer in kelvins.

Solution (a)   Direct substitution of −40°C into Eq. (17-5) yields

$$T_K = -40°C + 273 = 233\ K$$

Solution (b)   The difference in the freezing points is

$$\Delta t = 0°C - (-40°C) = 40\ C°$$

Since the size of the kelvin is identical to that of the Celsius degree, the difference is also 40 kelvins.

At this point you may ask why we still retain the Celsius and Fahrenheit scales. When working with heat, one is nearly always concerned with changes in temperature. In fact, there must be a change in temperature in order for heat to be transferred. Otherwise, the system would be in thermal equilibrium. Since the Kelvin and Rankine scales are based on the same intervals as the Celsius and Fahrenheit scales, it makes no difference which scale is used for temperature intervals. On the other hand, if a formula calls for a specific temperature rather than a temperature difference, the absolute scale must be used.

17-5 LINEAR EXPANSION

The most common effect produced by temperature changes is a change in size. With a few exceptions, all substances increase in size with rising temperature. The atoms in a solid are held together in a regular pattern by electric forces. At any temperature the atoms vibrate with a certain frequency and amplitude. As the temperature is increased, the amplitude (maximum displacement) of the atomic vibrations increases. This results in an overall change in the dimensions of the solid.

A change in any *one* dimension of a solid is called *linear expansion*. It is found experimentally that an increase in a single dimension, for example the length of a rod, is dependent on the original dimension and the change in temperature. Consider, for example, the rod in Fig. 17-10. The initial length is L_0 and the initial temperature is t_0. When heated to a temperature t, the rod's new length is denoted by L. Thus a change in temperature, $\Delta t = t - t_0$, has resulted in a change in length, $\Delta L = L - L_0$. The proportional change in length is given by

$$\Delta L = \alpha L_0\,\Delta t \tag{17-8}$$

where α is the proportionality constant called the *coefficient of linear expansion*. Since an increase in temperature does not produce the same increase in length for all

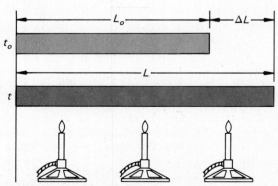

Fig. 17-10 Linear expansion.

materials, the coefficient α is a property of the material. Solving Eq. (17-8) for α, we obtain

$$\alpha = \frac{\Delta L}{L_0 \, \Delta t} \tag{17-9}$$

The coefficient of linear expansion of a substance can be defined as *the change in length per unit length per degree change in temperature*. Since the ratio $\Delta L/L_0$ has no dimensions, the units of α are in inverse degrees, that is, $1/C°$ of $1/F°$. The expansion coefficients for many common materials are given in Table 17-1.

Table 17-1 Linear Expansion Coefficients

Substance	$10^{-5}/C°$	$10^{-5}/F°$
Aluminum	2.4	1.3
Brass	1.8	1.0
Concrete	0.7–1.2	0.4–0.7
Copper	1.7	0.94
Glass, Pyrex	0.3	0.17
Iron	1.2	0.66
Lead	3.0	1.7
Silver	2.0	1.1
Steel	1.2	0.66
Zinc	2.6	1.44

EXAMPLE 17-4 An iron pipe is 300 m long at room temperature (20°C). If the pipe is to be used as a steam pipe, how much allowance must be made for expansion, and what will the new length of the pipe be?

Solution The temperature of steam is 100°C and $\alpha_{iron} = 1.2 \times 10^{-5}/C°$. Thus the increase in length is

$$\Delta L = \alpha L_0 \, \Delta t = (1.2 \times 10^{-5}/C°)(300 \text{ m})(100°C - 20°C)$$
$$= (1.2 \times 10^{-5}/C°)(300 \text{ m})(80 \text{ C}°) = 0.288 \text{ m}$$

Therefore, the length of the pipe at 100°C is

$$L = L_0 + \Delta L = 300.29 \text{ m}$$

We can see from this example that the new length may be calculated by the following relation:

$$\boxed{L = L_0 + \alpha L_0 \, \Delta t} \tag{17-10}$$

Remember, when calculating ΔL, that the units of α must be consistent with the units of Δt.

Linear expansion has both useful and destructive properties when applied to physical situations. The destructive effects require engineers to use expansion joints or rollers to make allowances for expansion. The predictable expansion of some materials, on the other hand, can be used to open or close switches at certain temperatures. Such devices are called *thermostats*.

Probably the most common application of the principle of linear expansion is the bimetallic strip. This device, shown in Fig. 17-11, consists of two flat strips of different metal welded or riveted together. The strips are fused together so that they are

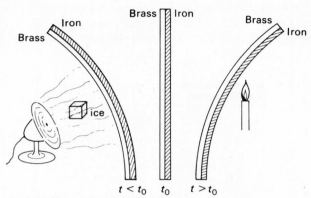

Fig. 17-11 The bimetallic strip.

the same length at a chosen temperature t_0. If we heat the strip, causing a rise in temperature, the material with the larger expansion coefficient will expand more. For example, a brass-iron strip will bend in an arc toward the iron side. When the source of heat is removed, the strip will gradually return to its original position. Cooling the strip below the initial temperature will cause the strip to bend in the other direction. This results because the material with the higher coefficient of expansion also *decreases* in length at a faster rate. The bimetallic strip has many useful applications, from thermostatic control systems to blinking lights. Since the expansion is in direct proportion to an increase in temperature, the bimetallic strip can also be used as a thermometer.

17-6
AREA
EXPANSION

Linear expansion is by no means restricted to the length of a solid. Any line drawn through the solid will increase in length per unit length at the rate given by its expansion coefficient α. For example, in a solid cylinder the length, diameter, and a diagonal drawn through the solid will all increase their dimensions in the same proportion. In fact, the expansion of a surface is exactly analogous to a photographic enlargement, as illustrated in Fig. 17-12. Notice also that if the material contains a hole, the area of the hole expands at the same rate it would if it were filled with material.

Let us consider the area expansion of the rectangular surface in Fig. 17-13. Both the length and width of the material will expand at the rate given by Eq. (17-10). Thus the new length and width are given, in factored form, by

$$L = L_0(1 + \alpha \, \Delta t)$$
$$W = W_0(1 + \alpha \, \Delta t)$$

HEAT, LIGHT, AND SOUND

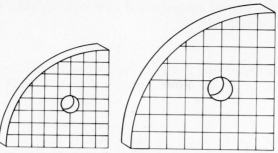

Fig. 17-12 Thermal expansion is analogous to a photographic enlargement. Note that the hole gets larger in the same proportion as the material.

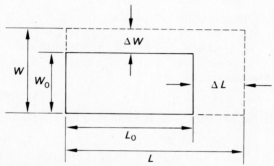

Fig. 17-13 Area expansion.

We can now derive an expression for area expansion by finding the product of these two equations.

$$LW = L_0 W_0(1 + \alpha \, \Delta t)^2$$
$$= L_0 W_0(1 + 2\alpha \, \Delta t + \alpha^2 \, \Delta t^2)$$

Since the magnitude of α is of the order of 10^{-5}, we may certainly neglect the term containing α^2. Hence we can write

$$LW = L_0 W_0(1 + 2\alpha \, \Delta t)$$

or

$$A = A_0(1 + 2\alpha \, \Delta t)$$

where $A = LW$ represents the new area and $A_0 = L_0 W_0$ represents the original area. Rearranging terms, we obtain

$$A - A_0 = 2\alpha A_0 \, \Delta t$$

or

$$\Delta A = 2\alpha A_0 \, \Delta t \qquad (17\text{-}11)$$

The coefficient of area expansion γ *is approximately twice* the coefficient of linear expansion. Symbolically,

$$\gamma = 2\alpha \qquad (17\text{-}12)$$

where γ (gamma) is the change in area per unit initial area per degree change in temperature. Using this definition, we may write the following formulas for area expansion.

$$\Delta A = \gamma A_0 \, \Delta t \qquad (17\text{-}13)$$
$$A = A_0 + \gamma A_0 \, \Delta t \qquad (17\text{-}14)$$

EXAMPLE 17-5 A brass disk has a hole 80 mm in diameter punched in its center at 70°F. If the disk is placed in boiling water, what will be the new area of the hole?

Solution We first compute the area of the hole at 70°F.

$$A_0 = \frac{\pi D^2}{4} = \frac{\pi (80 \text{ mm})^2}{4} = 5027 \text{ mm}^2$$

Now the area expansion coefficient is

$$\gamma = 2\alpha = (2)(1.0 \times 10^{-5}/\text{F°}) = 2 \times 10^{-5}/\text{F°}$$

The increase in the area of the hole is found from Eq. (17-13) as follows:

$$\Delta A = \gamma A_0 \, \Delta t$$
$$= (2 \times 10^{-5}/\text{F°})(5027 \text{ mm}^2)(212°\text{F} - 70°\text{F})$$
$$= 14.3 \text{ mm}^2$$

The new area is found by adding this increase to the original area (see Eq. (17-14)).

$$A = A_0 + \Delta A$$
$$= 5027 \text{ mm}^2 + 14.3 \text{ mm}^2 = 5041.3 \text{ mm}^2$$

17-7
VOLUME EXPANSION

The expansion of heated material is the same in all directions. Therefore, the volume of a liquid, gas, or solid will have a predictable increase in volume with a rise in temperature. Reasoning similar to that of the previous sections will give us the following formulas for volume expansion.

$$\Delta V = \beta V_0 \, \Delta t \qquad (17\text{-}15)$$
$$V = V_0 + \beta V_0 \, \Delta t \qquad (17\text{-}16)$$

The symbol β (beta) is the *volume expansion coefficient*. It represents the *change in volume per unit volume per degree change in temperature*. For solid materials it is approximately three times the linear expansion coefficient.

$$\beta = 3\alpha \qquad (17\text{-}17)$$

When working with solids, we can compute β from the table of linear expansion coefficients (Table 17-1). For different liquids, the volume expansion coefficients are listed in Table 17-2. The molecular separation in gases is so great that they all expand at approximately the same rate. Volumetric expansion of gases will be discussed later.

Table 17-2   Volume Expansion Coefficients

Liquid	β	
	$10^{-4}/C°$	$10^{-4}/F°$
Alcohol, ethyl	11	6.1
Benzene	12.4	6.9
Glycerin	5.1	2.8
Mercury	1.8	1.0
Water	2.1	1.2

EXAMPLE 17-6   A Pyrex glass bulb is filled with 50 cm³ of mercury at 20°C. What volume will overflow if the system is heated uniformly to a temperature of 60°C? Refer to Fig. 17-14.

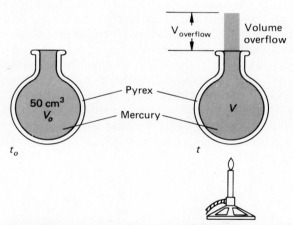

Fig. 17-14 The volume overflow is found by subtracting the change in volume of the glass from the change in volume of the liquid.

Solution   The inside volume of the glass bulb is 50 cm³ initially and will increase according to Eq. (17-15). Remember that $\beta_G = 3\alpha_G$. At the same time, the mercury will increase in volume according to the value of β_M. The overflow will, therefore, be the difference between the two expansions.

$$\left(\begin{array}{c}\text{Volume}\\\text{overflow}\end{array}\right) = \left(\begin{array}{c}\text{volume increase}\\\text{in mercury}\end{array}\right) - \left(\begin{array}{c}\text{volume increase}\\\text{in glass}\end{array}\right)$$

$$V_{\text{overflow}} = \Delta V_M - \Delta V_G$$

$$= \beta_M V_M \Delta t - \beta_G V_G \Delta t$$

We will compute the volume increases separately.

$$\Delta V_M = \beta_M V_M \Delta t = (1.8 \times 10^{-4}/C°)(50 \text{ cm}^3)(40 \text{ C}°) = 0.36 \text{ cm}^3$$

$$\Delta V_G = 3\alpha_G V_G \Delta t = 3(0.3 \times 10^{-5}/C°)(50 \text{ cm}^3)(40 \text{ C}°) = 0.018 \text{ cm}^3$$

Thus the volume overflow is

$$V_{\text{overflow}} = \Delta V_M - \Delta V_G = 0.36 \text{ cm}^3 - 0.018 \text{ cm}^3$$
$$= 0.342 \text{ cm}^3$$

The volume overflow is 0.342 cm³.

17-8
THE UNUSUAL EXPANSION OF WATER

Suppose we fill the bulb of the tube in Fig. 17-15 with water at 0°C so that the narrow neck is partially filled. Expansion or contraction of the water can easily be measured by observing the water level in the narrow tube. As the temperature of the water increases, the water in the tube will gradually sink, indicating a contraction. The contraction continues until the temperatures of the bulb and the water are 4°C. As the temperature increases above 4°C, the water reverses its direction and rises continuously, indicating the normal expansion with an increase in temperature. This means that water has its minimum volume and its maximum density at 4°C.

The variation in the density of water with temperature is shown graphically in Fig. 17-16. If we study the graph from the high-temperature side, we note that the density gradually increases to a maximum of 1.0 g/cm³ at 4°C. The density then decreases gradually until the water reaches the ice point. Ice occupies a greater volume than water, and its formation sometimes results in cracked water pipes if proper precautions are not taken.

The greater volume in the ice results from the way groups of its molecules are bonded in its crystalline structure. As the ice melts, the water formed still contains groups of molecules bonded in this open crystal structure. As these structures begin to collapse, the molecules move closer together, increasing the density. This is the dominant process until the water reaches a temperature of 4°C. From that point to higher temperatures, the increased amplitude of molecular vibrations takes over and the water expands.

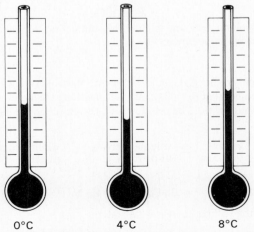

Fig. 17-15 The irregular expansion of water. As the temperature of water is increased from 0 to 8°C, it first contracts and then expands.

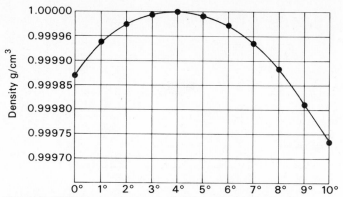

Fig. 17-16 Variation in the density of water near 4°C.

Once again the beginning student may be tempted to marvel that science can be so exact. The very thought that the density of water at 4°C "turns out to be exactly 1.00 g/cm³" must certainly be a remarkable coincidence. However, like the temperatures of the ice point and steam point, this result is also a consequence of a definition. The scientists who originally devised the metric system defined the kilogram as the mass of 1000 cm³ of water at 4°C. Later the kilogram was redefined in terms of a platinum-iridium cylinder which serves as a standard.

SUMMARY

We have seen that because of the existence of four commonly used temperature scales, temperature conversions are important. You have also studied one very important effect of changes in temperature of materials, a change in the physical dimensions. The major concepts are summarized below:

- There are four temperature scales with which you should be thoroughly familiar. These scales are compared in Fig. 17-5, giving values for the steam point, the ice point, and absolute zero on each scale. It is very important for you to distinguish between a temperature interval Δt and a specific temperature t. For temperature intervals:

$$\frac{5\ C°}{9\ F°} = 1 = \frac{9\ F°}{5\ C°} \qquad 1\ K = 1\ C° \qquad 1\ R° = 1\ F°$$

Temperature Intervals

- For specific temperatures you must correct for the interval difference, but you must also correct for the fact that different numbers are assigned for the same temperatures:

$$t_C = \frac{5}{9}(t_F - 32) \qquad t_F = \frac{9}{5}t_C + 32$$

Specific Temperatures

$$T_K = t_C + 273 \qquad T_R = t_F + 460$$

Absolute Temperatures

- The following relations apply for thermal expansion of solids:

$$\Delta L = \alpha L_0\, \Delta t$$ $$L = L_0 + \alpha L_0\, \Delta t$$ *Linear Expansion*

$$\Delta A = \gamma A_0\, \Delta t$$ $$A = A_0 + \gamma A_0\, \Delta t$$ $$\gamma = 2\alpha$$ *Area Expansion*

$$\Delta V = \beta V_0\, \Delta t$$ $$V = V_0 + \beta V_0\, \Delta t$$ $$\beta = 3\alpha$$ *Volume Expansion*

- The volume expansion of a liquid uses the same relation as for a solid except, of course, that there is no linear expansion coefficient α for a liquid. Only β is needed.

QUESTIONS

17-1. Define the following terms:
   a. Thermal energy
   b. Temperature
   c. Thermal equilibrium
   d. Thermometer
   e. Ice point
   f. Steam point
   g. Celsius scale
   h. Fahrenheit scale
   i. Absolute zero
   j. Kelvin scale
   k. Rankine scale
   l. Coefficient of linear expansion

17-2. Two lumps of hot iron ore are dropped into a container of water. The system is insulated and allowed to reach thermal equilibrium. Is it necessarily true that the iron ore and the water have the same thermal energy? Is it necessarily true that they have the same temperature? Discuss.

17-3. Distinguish clearly between thermal energy and temperature.

17-4. If a flame is placed underneath a mercury-in-glass thermometer, the mercury column first drops and then rises. Explain.

17-5. What factors must be considered in the design of a sensitive thermometer?

17-6. How good is our sense of touch as a means of judging temperature? Does the "hotter" object always have the higher temperature?

17-7. Given an unmarked thermometer, how would you proceed to graduate it in Celsius degrees?

17-8. A 6-in. ruler expands 0.0014 in. when the temperature is increased 1 C°. How much would a 6-cm ruler expand during the same temperature interval if it is made of the same material?

17-9. A brass rod connects the opposite sides of a brass ring. If the system is heated uniformly, will the ring remain circular?

17-10. A brass nut is used with a steel bolt. How is the closeness of fit affected when the bolt alone is heated? If the nut alone is heated? If they are both heated equally?

17-11. An aluminum cap is screwed tightly to the top of a pickle jar at room temperature. After the pickle jar has been stored in the refrigerator for a day or two, the cap cannot easily be removed. Explain. Suggest a way to remove the cap with very little effort. How might this problem be solved by the manufacturer?

17-12. Describe the expansion of water near 4°C. Why does a lake freeze at the surface first? What temperature is likely to result at the bottom of the lake if its surface is frozen?

17-13. Follow reasoning similar to that for area expansion to derive Eqs. (17-15) and (17-16). In the text it was stated that γ is only approximately equal to twice α. Why is it not exactly twice α? Is the error larger in Eq. (17-13) or in Eq. (17-15)?

PROBLEMS

The Measurement of Temperature

17-1. Body temperature is normal at 98.6°F. What is the corresponding temperature on the Celsius scale?

Ans. 37.0°C

17-2. The boiling point of sulfur is 444.5°C. What is the corresponding temperature on the Fahrenheit scale?

17-3. A steel rail cools from 70 to 30°C in 1 h. What is the change in temperature in Fahrenheit degrees for the same time period?

Ans. 72 F°

* 17-4. At what temperature will the Celsius and Fahrenheit scales have the same reading?

17-5. A piece of charcoal initially at 180°F experiences a decrease in temperature of 120 F°. Express this change of temperature in Celsius degrees. What is its final temperature on the Celsius scale?

Ans. 66.7 C°, 15.5°C

The Absolute Temperature Scale

17-6. Acetone boils at 56.5°C; liquid nitrogen boils at −196°C. Express these specific temperatures on the Kelvin scale. What is the difference in these temperatures in kelvins?

17-7. The boiling point of oxygen is −297.35°F. Express this temperature in kelvins. If oxygen cools from 120 to 70°F, what is the change of temperature in kelvins?

Ans. 90.0 K, 27.8 K

17-8. Gold melts at 1336 K. What is the corresponding temperature in degrees Celsius and in degrees Fahrenheit?

17-9. A wall of firebrick has an inside temperature of 313°F and an outside temperature of 73°F. Express the difference of temperature in Celsius degrees and in kelvins.

Ans. 133 C°, 133 K

17-10. A sample of gas cools from −120 to −180°C. Express the change of temperature in kelvins.

Linear Expansion

17-11. A piece of copper tubing is 6 m long at 20°C. How much will it increase in length when heated to a temperature of 80°C?

Ans. 6.12 mm

17-12. A silver bar is 1 ft long at 70°F. How much will it increase in length when it is placed in boiling water (212°F)?

17-13. The diameter of a hole in a steel plate is 9 cm when the temperature is 20°C. What will be the diameter of the hole at 200°C?

Ans. 9.02 cm

17-14. A brass rod is 2.00 m long at 15°C. To what temperature must the rod be heated in order that its new length is 2.01 m?

Area Expansion

17-15. A square copper plate 4 cm on a side at 20°C is heated to 120°C. What is the increase in the area of the copper plate temperature?

Ans. 0.0544 cm²

17-16. A circular hole in a steel plate has a diameter of 20 cm at 27°C. To what temperature must the plate be cooled if the desired area of the opening is to be 314.00 cm²?

Volume Expansion

17-17. What is the increase in volume of 16 liters of ethyl alcohol when it is heated from 20 to 50°C?

Ans. 0.528 L

17-18. A Pyrex beaker has an inside volume of 600 mL at 20°C. At what temperature will the inside volume be 603 mL?

17-19. If 200 cm³ of benzene exactly fills an aluminum cup at 40°C, and if the system is then cooled to 18°C, how much benzene (at 18°C) can be added to the cup without overflowing?

Ans. 5.14 cm³

17-20. A Pyrex glass beaker is filled to the top with 200 cm³ of mercury at 20°C. How much mercury will overflow if the temperature of the system is increased to 68°C?

Additional Problems

17-21. The laboratory apparatus for measuring the coefficient of linear expansion is illustrated in Fig. 17-17. The temperature of a metal rod is increased by passing

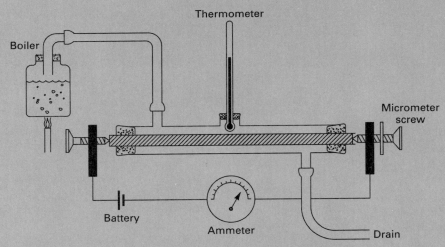

Fig. 17-17 Apparatus for measuring coefficient of linear expansion.

steam through an enclosed jacket. The resulting increase in length is measured with the micrometer screw at one end. Since the original length and temperature are known, the expansion coefficient can be calculated from Eq. (17-8). The following data were recorded during an experiment with a rod of unknown metal:

$$L_0 = 600 \text{ mm} \qquad t_0 = 23°C$$
$$\Delta L = 1.04 \text{ mm} \qquad t_f = 98°C$$

What is the coefficient of linear expansion for this metal? What do you think the metal is?

Ans. $2.3 \times 10^{-5}/C°$, Al

* 17-22. Assume that the end points of a rod are rigidly fixed between two walls to prevent expansion with increasing temperature. From the definition of Young's modulus (Chap. 13) and your knowledge of linear expansion, show that the compressive force F exerted by the walls will be given by

$$F = \alpha A Y \, \Delta t$$

where A = cross section of rod
Y = Young's modulus of rod
Δt = increase in temperature of rod

17-23. The cross section of a steel rod is 2 in.^{2}. What force is needed to prevent it from expanding if the temperature is increased from 70 to 120°F?

Ans. 1.98×10^4 lb

17-24. A rectangular sheet of aluminum measures 6 by 8 cm at 28°C. What is its area at 0°C?

* 17-25. A round brass plug has a diameter of 8.001 cm at 28°C. To what temperature must the plug be cooled if it is to fit snugly into an 8.000-cm hole?

Ans. 21.1°C

* 17-26. A steel ring has an inside diameter of 4.000 cm at 20°C. The ring is to be slipped over a copper shaft that has a diameter of 4.003 cm at 20°C. (a) To what temperature must the ring be heated? (b) If the ring and the shaft are uniformly cooled, at what temperature will the ring just slip off the shaft?

* 17-27. A 100-ft steel tape correctly measures distance when the temperature is 20°C. What is the true measurement if this tape indicates a distance of 94.62 ft on a day when the temperature is 36°C?

Ans. 94.64 ft

* 17-28. Prove that the density of a material changes with temperature so that the new density ρ is given by

$$\rho = \frac{\rho_0}{1 + \beta \, \Delta t}$$

where ρ_0 = original density
β = volume expansion coefficient
t = change in temperature

17-29. The density of mercury at 0°C is 13.6 g/cm^3. What is its density at 60°C?

Ans. 13.5 g/cm^3

Quantity of Heat

OBJECTIVES

After completing this chapter, you should be able to:

1. Define quantity of heat in terms of the *calorie,* the *kilocalorie,* the *joule,* and the *British thermal unit* (Btu).
2. Write a formula for the *specific heat capacity* of a material and apply it to the solution of problems involving the loss and gain of heat.
3. Write formulas for calculating the *latent heats of fusion* and *vaporization* and apply them to the solution of problems in which heat produces a change in phase of a substance.
4. Define the *heat of combustion* and apply it to problems involving the production of heat.

Thermal energy is the energy associated with random molecular motion, but it is not possible to measure the position and velocity of every molecule in a substance in order to determine its thermal energy. However, we can measure *changes* in thermal energy by relating it to a change in temperature.

For example, when two systems at different temperatures are placed together, they eventually reach a common intermediate temperature. From this observation it is safe to say that the system at the higher temperature has lost thermal energy to the system at the lower temperature. The thermal energy lost or gained by objects is called *heat.* This chapter is concerned with the quantitative measurement of heat.

18-1
THE MEANING OF HEAT

It was originally believed that two systems reach thermal equilibrium through the transfer of a substance called *caloric.* It was postulated that all bodies contain an amount of caloric in proportion to their temperature. Thus, when two objects were placed in contact, the object of higher temperature transferred caloric to the object of lower temperature until they reached the same temperature. The idea that a substance is transferred carries with it the implication that there is a limit to the amount of heat energy that can be withdrawn from a body. It was this point that eventually led to the downfall of the caloric theory.

Count Rumford of Bavaria was the first to shed doubt on the caloric theory. He made his discovery in 1798 while supervising the boring of a cannon. The bore of the cannon was kept full of water to prevent overheating. As the water boiled away, it was replenished. According to the existing theory, caloric had to be supplied to boil the water. The apparent production of caloric was explained by supposing that when matter is finely divided, it loses some of its ability to retain caloric. Rumford devised an experiment to show that even a dull boring tool which did not cut the gun metal at all produced enough caloric to boil the water. In fact, it seemed that as long as mechanical work was supplied, the tool was an inexhaustible source of caloric. Rumford ruled out the caloric theory on the basis of his experiments and suggested that the explanation must be related to motion. Hence the idea that mechanical work is responsible for the creation of heat was introduced. The equivalence of heat and work as two forms of energy was established later by Sir James Prescott Joule.

18-2
THE QUANTITY OF HEAT

The idea of heat as a substance must be discarded. It is not something that an object *has* but something that it *gives up* or *absorbs*. Heat is simply another form of energy that can be measured only in terms of the effect it produces. The SI unit of energy, the *joule,* is also the preferred unit for heat since heat is a form of energy. However, there are three older units which remain in use today, and they will also be treated in this text. These earlier units were based on the thermal energy required to produce a standard change. They are the *calorie,* the *kilocalorie,* and the *British thermal unit.*

> One **calorie** (cal) is the quantity of heat required to change the temperature of one gram of water through one Celsius degree.

> One **kilocalorie** (kcal) is the quantity of heat required to change the temperature of one *kilogram* of water through one Celsius degree. (1 kcal = 1000 cal.)

> One **British thermal unit** (Btu) is the quantity of heat required to change the temperature of one *standard pound* (lb) through one *Fahrenheit* degree.

Aside from the fact that these older units imply that heat energy cannot be related to other forms of energy, there are other problems. The heat required to change the temperature of water from 92 to 93°C is not exactly the same as that needed to increase the temperature from 8 to 9°C. Thus, it is necessary to specify the interval of temperature in strict applications for the calorie and for the British thermal unit. The chosen intervals were 14.5 to 15.5°C and 63 to 64°F. Additionally, the pound unit which appears in the definition of the Btu must be recognized as the *mass of the standard pound.* This represents a departure from USCS units, in which the unit pound was reserved for weight. Therefore, in this chapter, when we refer to 1 lb of water, we shall be referring to a *mass* of water equivalent to about 1/32 slug. This distinction is necessary because the pound of water must represent a constant quantity of matter, independent of its location. By definition the pound mass is related to the gram and kilogram as follows:

$$1 \text{ lb} = 454 \text{ g} = 0.454 \text{ kg}$$

The difference between these older units for heat results from the difference in masses and the difference in temperature scales. It is left as an exercise for you to show that

$$1 \text{ Btu} = 252 \text{ cal} = 0.252 \text{ kcal} \qquad (18\text{-}1)$$

The first quantitative relationship between these older units and the traditional units for mechanical energy was established by Joule in 1843. Although Joule devised many different experiments to demonstrate the equivalence of heat units and energy units, the apparatus most often remembered is illustrated in Fig. 18-1. Falling

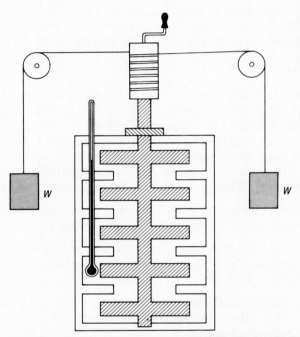

Fig. 18-1 Joule's experiment for determining the mechanical equivalent of heat. The falling weights do work in stirring the water and raising its temperature.

weights supplied the mechanical energy, which rotated a set of paddles in a water can. The quantity of heat absorbed by the water was measured from the known mass and the measured increase in temperature of the water.

In modern times the mechanical equivalent of heat has been accurately established by a variety of techniques. The accepted results are:

$$1 \text{ cal} = 4.186 \text{ J}$$
$$1 \text{ kcal} = 4186 \text{ J}$$
$$1 \text{ Btu} = 778 \text{ ft} \cdot \text{lb}$$

Thus, 4.186 J of heat is needed to raise the temperature of one gram of water from 14.5 to 15.5°C. Since each of the above units remains in common use, it will often become necessary to compare units or to convert from one unit to another.

Now that we have defined units for the quantitative measurement of heat, the distinction between quantity of heat and temperature should be clear. For example, suppose we pour 200 g of water into one beaker and 800 g of water into another beaker, as shown in Fig. 18-2. The initial temperature of the water in each beaker is measured to be 20°C. A flame is placed under each beaker for the same length of time, delivering 8000 J of heat energy to the water in each beaker. The temperature of the 800-g quantity of water increases by a little more than 2 C°, but the temperature of the 200-g quantity increases by almost 10 C°. Yet the same quantity of heat was applied to the water in each beaker.

18-3

SPECIFIC HEAT CAPACITY

We have defined a quantity of heat as the thermal energy required to raise the temperature of a given mass. However, the amount of thermal energy required to raise the temperature of a substance varies for different materials. For example, suppose we apply heat to five balls, all of the same size but made of different materi-

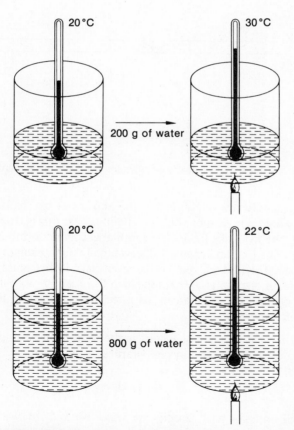

Fig. 18-2 The same quantity of heat is applied to different masses of water. The larger mass experiences a smaller rise in temperature.

als, as shown in Fig. 18-3a. If we want to raise the temperature of each ball to 100°C, we shall find that some of the balls must be heated longer than the others. For illustration, let us assume that each ball has a volume of 1 cm³ and an initial temperature of 0°C. Each ball is heated with a burner capable of delivering thermal energy at the rate of 1 cal/s. The approximate times required to obtain a temperature of 100°C for each ball are given in Fig. 18-3. Notice that the lead ball reaches the final temperature in only 37 s whereas the iron ball requires 90 s of continuous heating. The glass, aluminum, and copper balls require intermediate times.

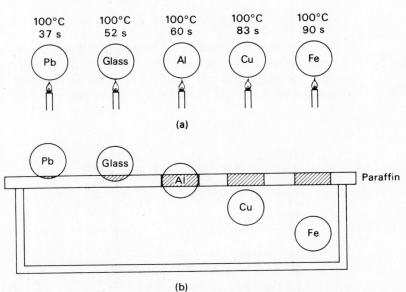

Fig. 18-3 A comparison of heat capacities for five balls of different materials.

Since more heat was absorbed by the iron and copper balls, it might be expected that they would release more heat on cooling. That this is true can be demonstrated by placing the five balls (at 100°C) simultaneously on a thin strip of paraffin, as in Fig. 18-3b. The iron and copper balls eventually melt through the paraffin and drop into the pan. The lead and glass balls never make it through. Clearly, there must be some property of a material that relates to the quantity of heat absorbed or released during a change in temperature. As a step in establishing this property, we first define *heat capacity*.

> The **heat capacity** of a body is the ratio of heat supplied to the corresponding rise in temperature of the body.

$$\text{Heat capacity} = \frac{Q}{\Delta t} \tag{18-2}$$

The SI units for heat capacity are *joules per kelvin* (J/K), but since the Celsius interval is the same as the kelvin and more commonly used, the *joule per Celsius degree* (J/C°) is used in this text. Other units are *calories per Celsius degree* (cal/C°), *kilocalories per Celsius degree* (kcal/C°), and *Btu per Fahrenheit degree* (Btu/F°). In the above example, 89.4 cal of heat was required to raise the temperature of the iron ball by 100 C°. Thus the heat capacity of this particular iron ball is 0.894 cal/C°.

The mass of an object is not included in the definition of heat capacity. Therefore, heat capacity is a property of the object. To make it a property of the material, the *heat capacity per unit mass* is defined. We call this property the *specific heat capacity,* denoted *c*.

> The **specific heat capacity** of a material is the quantity of heat required to raise the temperature of a unit mass through one degree.

$$ c = \frac{Q}{m\, \Delta t} \qquad Q = mc\, \Delta t \tag{18-3} $$

The SI unit of specific heat assigns the *joule* for heat, the *kilogram* for mass, and the *kelvin* for temperature. If we again replace the kelvin with the Celsius degree, the units for *c* are J/kg · C°. In industry most temperature measurements are made in C° or F°, and the calorie and the Btu are still used more than are the SI units. We will therefore continue to speak of specific heat in units of cal/g · C° and Btu/lb · F°, but we will also use the preferred SI units in some cases. In the example of the iron ball, the mass is found to be 7.85 g. The specific heat of iron is therefore

$$ c = \frac{Q}{m\, \Delta t} = \frac{89.4 \text{ cal}}{(7.85 \text{ g})(100 \text{ C°})} = 0.114 \text{ cal/g} \cdot \text{C°} $$

Notice that we speak of the heat capacity of the *ball* and the *specific* heat capacity of *iron.* The former relates to the object itself whereas the latter relates to the material from which the object is made. In our experiment with the balls, we noted only the quantity of heat required to raise their temperature 100 C°. The density of the materials was not considered. If the sizes of the balls were adjusted so that each had the same mass, we would observe different results. Since the specific heat of aluminum is highest, more heat would be required for the aluminum ball than for the others. Similarly, the aluminum ball would release more heat in cooling.

For most practical applications, the specific heat of water can be taken as

$$ \begin{array}{ll} 4186 \text{ J/kg} \cdot \text{C°} & 4.186 \text{ J/g} \cdot \text{C°} \\ 1 \text{ cal/g} \cdot \text{C°} \quad \text{or} & 1 \text{ Btu/lb} \cdot \text{F°} \end{array} $$

Notice that the numerical values are the same for the specific heat expressed in cal/g · C° and in Btu/lb · F°. This is a consequence of their definitions and can be demonstrated by unit conversion:

$$ 1\, \frac{\text{Btu}}{\text{lb} \cdot \text{F°}} \times \frac{9 \text{ F°}}{5 \text{ C°}} \times \frac{1 \text{ lb}}{454 \text{ g}} \times \frac{252 \text{ cal}}{1 \text{ Btu}} = 1 \text{ cal/g} \cdot \text{C°} $$

The specific heats for many common substances are listed in Table 18-1.

Table 18-1   Specific Heat Capacities

Substance	J/kg · C°	cal/g · C° or Btu/lb · F°
Aluminum	920	0.22
Brass	390	0.094
Copper	390	0.093
Ethyl alcohol	2500	0.60
Glass	840	0.20
Gold	130	0.03
Ice	2300	0.5
Iron	470	0.113
Lead	130	0.031
Mercury	140	0.033
Silver	230	0.056
Steam	2000	0.48
Steel	480	0.114
Turpentine	1800	0.42
Zinc	390	0.092

Once the specific heats of a large number of materials have been established, the thermal energy released or absorbed in many experiments can be determined. For example, the quantity of heat Q required to raise the temperature of a mass m through an interval t, from Eq. (18-3), is

$$Q = mc\, \Delta t \qquad\qquad (18\text{-}4)$$

where c is the specific heat of the mass.

EXAMPLE 18-1   How much heat is required to raise the temperature of 200 g of mercury from 20 to 100°C?

Solution   Substitution into Eq. (18-4) yields

$$Q = mc\, \Delta t = (0.2 \text{ kg})(140 \text{ J/kg} \cdot \text{C°})(80 \text{ C°})$$
$$= 2200 \text{ J}$$

You should show that this is approximately 530 cal.

18-4

THE MEASURE-MENT OF HEAT

We have often emphasized the distinction between thermal energy and temperature. The term *heat* has now been introduced as the thermal energy *absorbed* or *released* during a temperature change. The quantitative relationship between heat and temperature is best described by the concept of specific heat as it appears in Eq. (18-4). The physical relationships between all these terms are now beginning to fall into place.

The principle of thermal equilibrium tells us that whenever objects are placed together in an insulated enclosure, they will eventually reach the same temperature. This is the result of a transfer of thermal energy from the warmer bodies to the cooler bodies. If energy is to be conserved, we say that *the heat lost by the warm bodies must equal the heat gained by the cool bodies.* That is,

$$\boxed{Heat\ lost = heat\ gained} \tag{18-5}$$

This equation expresses the net result of heat transfer within a system.

The heat lost or gained by an object is not related in a simple way to the molecular energies of the objects. Whenever thermal energy is supplied to an object, it can absorb the energy in many different ways. The concept of specific heat is needed to measure the abilities of different materials to utilize thermal energy to increase their temperatures. The same amount of applied thermal energy does not result in the same temperature increase for all materials. For this reason, we say that temperature is a *fundamental* quantity. Its measurement is necessary in order to determine the quantity of heat lost or gained in a given process.

In applying the general equation for the conservation of thermal energy [Eq. (18-4)], the quantity of heat gained or lost by each item is calculated from the equation

$$Q = mc\ \Delta t \tag{18-4}$$

The term Δt represents the absolute change in temperature when applied to the conservation equation. The procedure is best demonstrated by an example.

EXAMPLE 18-2 A handful of copper shot is heated to 90°C and then dropped into 80 g of water at 10°C. The final temperature of the mixture is 18°C. What was the mass of the shot?

Solution Applying Eq. (18-5), we write

$$Heat\ lost\ by\ shot = heat\ gained\ by\ water$$
$$m_s c_s\ \Delta t_s = m_w c_w\ \Delta t_w$$
$$m_s c_s(t_s - t_e) = m_w c_w(t_e - t_w)$$

The change in temperature of the shot is computed by subtracting the equilibrium temperature t_e from the initial temperature of the shot t_s. On the other hand, the change in temperature of the water is computed by subtracting the initial temperature of the water t_w from the equilibrium temperature. This does not represent an error in sign because the quantity on the left represents a heat *loss* and the quantity on the right represents a heat *gain*. Obtaining the required specific heat from Table 18-1 and substituting other known quantities, we have

$$m_s(0.093\ cal/g \cdot C°)(90°C - 18°C) = (80\ g)(1\ cal/g \cdot C°)(18°C - 10°C)$$
$$m_s(0.093\ cal/g \cdot C°)(72\ C°) = (80\ g)(1\ cal/g \cdot C°)(8\ C°)$$
$$m_s = 95.6\ g$$

In this simple example we have neglected two important facts: (1) the water must have a container, which will also absorb heat from the shot; (2) the entire system must be insulated from external temperatures. Otherwise, the equilibrium temperature will always be room temperature. A laboratory device called a *calorimeter* (Fig. 18-4) is used to control these difficulties. The calorimeter consists of a thin metallic vessel K, generally aluminum, held centrally within an outer jacket A by a nonconducting rubber support H. Loss of heat is minimized in three ways: (1) the rubber gasket prevents loss by conduction; (2) the dead air space between the container walls prevents heat loss by air currents; and (3) highly polished metal vessels reduce the loss of heat by radiation. These three methods of heat transfer will be

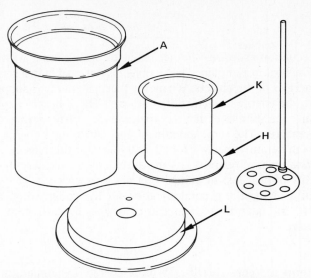

Fig. 18-4 The laboratory calorimeter. *(Central Scientific Co.)*

discussed in the following chapter. The wooden cover *L* has holes in the top for insertion of a thermometer and an aluminum stirrer.

EXAMPLE 18-3   In a laboratory experiment it is desired to use a calorimeter to find the specific heat of iron. Eighty grams of dry iron shot is placed in a cup and heated to a temperature of 95°C. The mass of the inner aluminum cup and of the aluminum stirrer is 60 g. The calorimeter is partially filled with 150 g of water at 18°C. The hot shot is quickly poured into the cup, and the calorimeter is sealed, as shown in Fig. 18-5. After the system has reached thermal equilibrium, the final temperature is 22°C. Compute the specific heat of iron.

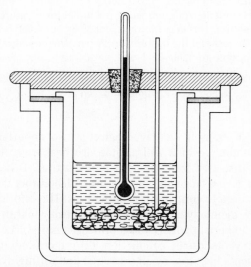

Fig. 18-5 A calorimeter can be used to compute the specific heat of a substance.

Solution   The heat lost by the iron shot must equal the heat gained by the water plus the heat gained by the aluminum cup and stirrer. We can assume that the initial temperature of the cup is the same as that of the water and stirrer (18°C). We will calculate the heat gained by the water and by the aluminum separately.

$$Q_{water} = mc \, \Delta t = (150 \text{ g})(1 \text{ cal/g} \cdot \text{C°})(22\text{°C} - 18\text{°C})$$
$$= (150 \text{ g})(1 \text{ cal/g} \cdot \text{C°})(4 \text{ C°}) = 600 \text{ cal}$$
$$Q_{Al} = mc \, \Delta t = (60 \text{ g})(0.22 \text{ cal/g} \cdot \text{C°})(22\text{°C} - 18\text{°C})$$
$$= (60 \text{ g})(0.22 \text{ cal/g} \cdot \text{C°})(4 \text{ C°}) = 52.8 \text{ cal}$$

Now the total heat gained is the sum of these values.

$$\text{Heat gained} = 600 \text{ cal} + 52.8 \text{ cal} = 652.8 \text{ cal}$$

This amount must equal the heat lost by the iron shot:

$$\text{Heat lost} = Q_s = mc_s \, \Delta t = (80 \text{ g})c_s(95\text{°C} - 22\text{°C})$$

Setting the heat lost equal to the heat gained gives

$$(80 \text{ g})c_s(73 \text{ C°}) = 652.8 \text{ cal}$$

Solving for c_s, we obtain

$$c_s = \frac{652.8 \text{ cal}}{(80 \text{ g})(73 \text{ C°})} = 0.11 \text{ cal/g} \cdot \text{C°}$$

In this experiment the heat gained by the thermometer was neglected. In an actual experiment, the portion of the thermometer inside the calorimeter would absorb about the same amount of heat as an extra 0.5 g of water. This quantity, called the *water equivalent* of the thermometer, should be added to the mass of water in an accurate experiment.

18-5
CHANGE OF PHASE

When a substance absorbs a given amount of heat, the speed of its molecules usually increases and its temperature rises. Depending on the specific heat of the substance, the rise in temperature is directly proportional to the quantity of heat supplied and inversely proportional to the mass of the substance. However, a curious thing happens when a solid melts or when a liquid boils. In these cases, the temperature remains constant until all the solid melts or until all the liquid boils.

To understand what happens to the applied energy, let us consider a simple model, as illustrated in Fig. 18-6. Under the proper conditions of temperature and pressure, all substances can exist in three *phases,* solid, liquid, or gas. In the solid phase, the molecules are held together in a rigid, crystalline structure so that the substance has a definite shape and volume. As heat is supplied, the energies of the particles in the solid gradually increase and its temperature rises. Eventually, the kinetic energy becomes so great that some of the particles overcome the elastic forces that hold them in fixed positions. The increased separation gives them the freedom of motion which we associate with the liquid phase. At this point, the energy absorbed by the substance is used in separating the molecules more than in the solid phase. The temperature does not increase during such a change of phase. The change of phase from a solid to a liquid is called *fusion,* and the temperature at which this change occurs is called the *melting point.*

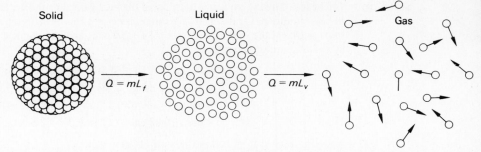

Fig. 18-6 A simplified model showing the relative molecular separations in the solid, liquid, and gaseous phases. During a change of phase, the temperature remains constant.

The quantity of heat required to melt a unit mass of a substance at its melting point is called the *latent heat of fusion* for that substance.

> The **latent heat of fusion** L_f of a substance is the heat per unit mass required to change the substance from the solid to the liquid phase at its melting temperature.

$$L_f = \frac{Q}{m} \qquad Q = mL_f$$

(18-6)

The latent heat of fusion L_f is expressed in *joules per kilogram* (J/kg), *calories per gram* (cal/g), or *Btu per pound* (Btu/lb). At 0°C, 1 kg of ice will absorb about 334,000 J of heat in forming 1 kg of water at 0°C. Thus the latent heat of fusion for water is 334,000 J/kg. The term *latent* arises from the fact that the temperature remains constant during the melting process. The heat of fusion for water is any of the following:

$$3.34 \times 10^5 \text{ J/kg} \qquad 334 \text{ J/g}$$
$$80 \text{ cal/g} \quad \text{or} \quad 144 \text{ Btu/lb}$$

After all the solid melts, the kinetic energy of the particles in the resulting liquid increases in accordance with its specific heat, and the temperature rises again. Eventually the temperature will level off as the thermal energy is used to change the molecular structure, forming a gas or vapor. The change of phase from a liquid to a vapor is called *vaporization,* and the temperature associated with this change is called the *boiling point* of the substance.

The quantity of heat required to vaporize a unit mass is called the *latent heat of vaporization.*

> The **latent heat of vaporization** L_v of a substance is the heat per unit mass required to change the substance from a liquid to a vapor at its boiling temperature.

$$L_v = \frac{Q}{m} \qquad Q = mL_v$$

(18-7)

The latent heat of vaporization L_v is expressed in units of joules per kilogram, calories per gram, and Btu per pound. It is found that 1 kg of water at 100°C absorbs 2,260,000 J of heat in forming 1 kg of steam at the same temperature. The heat of fusion for water is

$$2.26 \times 10^6 \text{ J/kg} \qquad 2260 \text{ J/g}$$
$$540 \text{ cal/g} \quad \text{or} \quad 970 \text{ Btu/lb}$$

Values for the heat of fusion and the heat of vaporization of many substances are given in Table 18-2. They are given in SI units and in calories per gram. It should be noted that the Btu per pound (Btu/lb) equivalents can be obtained by multiplying the value in cal/g by (9/5). They differ only because of the difference in the temperature scales. The industrial use of the SI units of J/kg for both L_f and L_v is strongly encouraged, but few American companies have made these conversions.

Table 18-2 Heats of Fusion and Heats of Vaporization for Various Substances

Substance	Melting point, °C	Heat of fusion		Boiling point, °C	Heat of vaporization	
		J/kg	cal/g		J/kg	cal/g
Alcohol, ethyl	−117.3	104×10^3	24.9	78.5	854×10^3	204
Ammonia	−75	452×10^3	108.1	−33.3	1370×10^3	327
Copper	1080	134×10^3	32	2870	4730×10^3	1130
Helium	−269.6	5.23×10^3	1.25	−268.9	20.9×10^3	5
Lead	327.3	24.5×10^3	5.86	1620	871×10^3	208
Mercury	−39	11.5×10^3	2.8	358	296×10^3	71
Oxygen	−218.8	13.9×10^3	3.3	−183	213×10^3	51
Silver	960.8	88.3×10^3	21	2193	2340×10^3	558
Water	0	334×10^3	80	100	2256×10^3	540
Zinc	420	100×10^3	24	918	1990×10^3	475

It is often helpful in studying the changes of phase of a substance to plot a graph showing how the temperature of the substance varies as thermal energy is applied. Such a graph is shown in Fig. 18-7 for water. If a quantity of ice is taken from a freezer at −20°C and heated, its temperature will increase gradually until the ice begins to melt at 0°C. For each degree rise in temperature, each gram of ice will absorb 0.5 cal of heat energy. During the melting process, the temperature remains constant, and each gram of ice will absorb 80 cal of heat energy in forming 1 g of water.

Once all the ice has melted, the temperature begins to rise again at a uniform rate until the water begins to boil at 100°C. For each degree increase in temperature, each gram will absorb 1 cal of heat energy. During the vaporization process, the temperature remains constant. Each gram of water absorbs 540 cal of heat energy in forming 1 g of water vapor at 100°C. If the resulting water vapor is contained and the heating is continued until all the water is gone, the temperature will again start to rise. The specific heat of steam is 0.48 cal/g · C°.

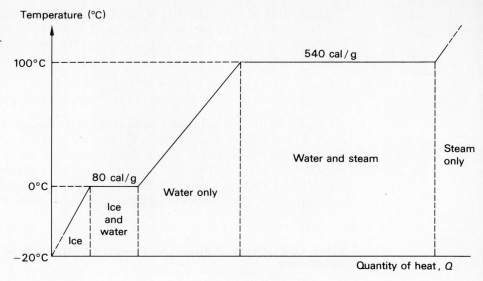

Fig. 18-7 The variation of temperature with a change in thermal energy for water.

Temperature (°C)

100°C ---- 540 cal/g

Steam only

Water and steam

0°C ---- 80 cal/g

Water only

Ice and water

Ice

−20°C

Quantity of heat, Q

EXAMPLE 18-4   What quantity of heat is required to change 20 lb of ice at 12°F to steam at 212°F?

Solution   The heat necessary to raise the temperature of the ice to its melting point is

$$Q_1 = mc\,\Delta t = (20\text{ lb})(0.5\text{ Btu/lb}\cdot\text{F}°)(32°\text{F} - 12°\text{F})$$
$$= 200\text{ Btu}$$

The heat required to melt the ice is given from Eq. (18-9).

$$Q_2 = mL_f = (20\text{ lb})(144\text{ Btu/lb}) = 2880\text{ Btu}$$

The heat necessary to raise the temperature of the resulting water to 212°F is

$$Q_3 = mc\,\Delta t = (20\text{ lb})(1\text{ Btu/lb}\cdot\text{F}°)(212°\text{F} - 32°\text{F})$$
$$= 3600\text{ Btu}$$

The heat required to vaporize the water is, from Eq. (18-7),

$$Q_4 = mL_v = (20\text{ lb})(970\text{ Btu/lb}) = 19{,}400\text{ Btu}$$

The total heat required is

$$Q = Q_1 + Q_2 + Q_3 + Q_4$$
$$= (200 + 2880 + 3600 + 19{,}400)\text{ Btu}$$
$$= 26{,}080\text{ Btu}$$

When heat is removed from a gas, its temperature drops until it reaches the temperature at which it boiled. As more heat is removed, the vapor returns to the liquid phase. This process is referred to as *condensation*. In condensing, a vapor gives up an amount of heat equivalent to the heat required to vaporize it. Thus the *heat of condensation* is equivalent to the heat of vaporization. The difference lies only in the direction of heat transfer.

Similarly, when heat is removed from a liquid, its temperature will drop until it reaches the temperature at which it melted. As more heat is removed, the liquid returns to its solid phase. This process is called *freezing* or *solidification*. The heat of solidification is exactly equal to the heat of fusion. Thus the only distinction between freezing and melting lies in whether heat is being released or absorbed.

Under the proper conditions of temperature and pressure, it is possible for a substance to change from the solid phase directly to the gaseous phase without passing through the liquid phase. This process is referred to as *sublimation*. Solid carbon dioxide (dry ice), iodine, and camphor (mothballs) are examples of substances which are known to sublime at normal temperatures. The quantity of heat absorbed per unit mass in changing from a solid to a vapor is called the *heat of sublimation*.

Before we leave the subject of fusion and vaporization, it will be instructive to offer examples of their measurement. In any given mixture, the quantity of heat absorbed must equal the quantity of heat released. This principle holds even if a change of phase occurs. The procedure is demonstrated in the examples below.

EXAMPLE 18-5   After 12 g of crushed ice at $-10°C$ is dropped into a 50-g aluminum calorimeter cup containing 100 g of water at 50°C, the system is sealed and allowed to reach thermal equilibrium. What is the resulting temperature?

Solution   The heat lost by the calorimeter and water must equal the heat gained by the ice, including any changes of phase that take place. Let us assume that all the ice melts, leaving only water at the equilibrium temperature t_e.

$$Total\ heat\ lost = heat\ lost\ by\ calorimeter + heat\ lost\ by\ water$$
$$= m_c c_c(50°C - t_e) + m_w c_w(50°C - t_e)$$
$$= (50)(0.22)(50°C - t_e) + (100)(1)(50°C - t_e)$$
$$= 550°C - 11t_e + 5000°C - 100t_e$$
$$= 5550°C - 111t_e$$

$$Total\ heat\ gained = heat\ gained\ by\ ice + heat\ to\ melt\ ice + heat\ to\ bring\ water\ to\ t_e$$
$$= m_i c_i\,\Delta t_i + m_i L_f + m_i c_w(t_e - 0°C)$$
$$= (12)(0.5)(10) + (12)(80) + (12)(1)t_e$$
$$= 1020 + 12t_e$$

$$Total\ heat\ lost = total\ heat\ gained$$
$$5550 - 111t_e = 1020 + 12t_e$$
$$123t_e = 4530$$
$$t_e = 36.8°C$$

EXAMPLE 18-6   If 10 g of steam at 100°C is introduced into a mixture of 200 g of water and 120 g of ice, find the final temperature and composition of the mixture.

Solution   The rather small amount of steam in comparison with ice and water makes us wonder whether enough heat can be released by the steam to melt all the ice. To check this suspicion, we shall compute the heat required to melt the 120 g of ice completely at 0°C.

$$Q_1 = m_i L_f = (120\ g)(80\ cal/g) = 9600\ cal$$

The maximum heat we can expect the steam to give up is

$$Q_2 = m_s L_v + m_s c_w(100°C - 0°C)$$
$$= (10)(540) + (10)(1)(100) = 6400\ cal$$

Since 9600 cal was needed to melt all the ice and only 6400 cal can be delivered by the steam, the final mixture must consist of ice and water at 0°C.

To determine the final composition of the mixture, note that 3200 additional calories would be required to melt the remaining ice. Hence

$$m_i L_f = 3200 \text{ cal}$$

$$m_i = \frac{3200 \text{ cal}}{80 \text{ cal/g}} = 40 \text{ g}$$

Therefore, there must be 40 g of ice in the final mixture. The amount of water remaining is

Water remaining = initial water + melted ice + condensed steam

$$= 200 \text{ g} + 80 \text{ g} + 10 \text{ g} = 290 \text{ g}$$

The final composition consists of a mixture of 40 g of ice in 290 g of water at 0°C.

Suppose we had assumed in the above example that all the ice melted, attempting to solve for t_e as in Example 18-5. In this case, we would have obtained a value for the equilibrium temperature which was below the freezing point (0°C). Clearly, this kind of answer could result only from a false assumption.

An alternative procedure for solving Example 18-6 would be to solve directly for the number of grams of ice which must have melted in order to balance the 6400 cal of heat energy released by the steam. It is left as an exercise for you to show that the same results are obtained.

18-6
HEAT OF COMBUSTION

Whenever a substance is burned, it releases a definite quantity of heat. The quantity of heat per unit mass, or per unit volume, when the substance is completely burned is called the *heat of combustion*. Commonly used units are Btu per pound mass, Btu per cubic foot, calories per gram, and kilocalories per cubic meter. For example, the heat of combustion of coal is approximately 13,000 Btu/lb$_m$. This means that each pound of coal, when completely burned, should release 13,000 Btu of heat energy.

SUMMARY

In this chapter you have studied the quantity of heat as a measurable quantity that is based on a standard change. The British thermal unit and the calorie are measures of the heat required to raise the temperature of a unit mass of water by a unit degree. By applying these standard units to experiments with a variety of materials, we have learned to predict heat losses or heat gains in a constructive fashion. The essential concepts presented in this chapter are as follows:

- The **British thermal unit (Btu)** is the heat required to change the temperature of one pound-mass of water one Fahrenheit degree.
- The **calorie** is the heat required to raise the temperature of one gram of water by one Celsius degree.
- Several conversion factors may be useful for problems involving thermal energy:

1 Btu = 252 cal = 0.252 kcal	1 cal = 4.186 J
1 Btu = 778 ft · lb	1 kcal = 4186 J

- The specific heat capacity c is used to determine the quantity of heat Q absorbed or released by a unit mass m as the temperature changes by an interval Δt.

$$c = \frac{Q}{m\,\Delta t} \qquad Q = mc\,\Delta t \qquad \textit{Specific Heat Capacity}$$

- Conservation of thermal energy requires that in any exchange of thermal energy the heat lost must equal the heat gained.

$$\textit{Heat lost} = \textit{heat gained} \qquad \Sigma\,(mc\,\Delta t)\text{loss} = \Sigma\,(mc\,\Delta t)\text{gain}$$

As an example, suppose body 1 transfers heat to bodies 2 and 3 as the system reaches an equilibrium temperature t_e:

$$m_1 c_1 (t_1 - t_e) = m_2 c_2 (t_e - t_2) + m_3 c_3 (t_e - t_3)$$

- The latent heat of fusion L_f and the latent heat of vaporization L_v are heat losses or gains by a unit mass m during a phase change. There is no change in temperature.

$$L_f = \frac{Q}{m} \qquad Q = mL_f \qquad \textit{Latent Heat of Fusion}$$

QUESTIONS

18-1. Define the following terms:
  a. Heat
  b. Temperature
  c. Calorie
  d. British thermal unit
  e. Mechanical equivalent of heat
  f. Heat capacity
  g. Specific heat capacity
  h. Conservation of heat energy
  i. Calorimeter
  j. Water equivalent
  k. Fusion
  l. Melting point
  m. Heat of fusion
  n. Vaporization
  o. Boiling point
  p. Heat of vaporization
  q. Condensation
  r. Freezing
  s. Sublimation
  t. Heat of combustion

18-2. Discuss the caloric theory of heat. In what ways is it successful in explaining heat phenomena? Where does it fail?

18-3. Blocks of five different metals—aluminum, copper, zinc, iron, and lead—are constructed with the same mass and the same cross-sectional area. Each block is heated to a temperature of 100°C and placed on a block of ice. Which will melt the ice to the greatest depth? List the remaining four metals in the order of decreasing penetration depths.

18-4. On a winter day the snow is observed to melt from the concrete sidewalk before it melts from the road. Which has the higher heat capacity?

18-5. If two objects have the same heat capacity, are they necessarily constructed of the same material? What if they have the same specific heat capacities?

18-6. Why is temperature considered a *fundamental quantity*?

18-7. A mechanical analogy to the concept of thermal equilibrium is given in Fig.

Fig. 18-8 A mechanical analogy to the equalization of temperature.

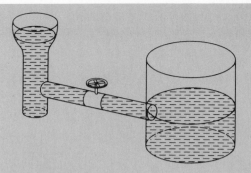

18-8. When the valve is opened, the water will flow until it has the same level in each tube. What are the analogies to temperature and thermal energy?

18-8. The mechanical equivalence of heat is established so that heat and work can be expressed in the same units. How then can we distinguish between the terms *work* and *heat*?

18-9. Discuss the change of phase from solid to liquid to vapor in terms of the molecular theory of matter.

18-10. In a mixture of ice and water, the temperature of both the ice and the water is 0°C. Why then does the ice feel colder to the touch?

18-11. Why does steam at 100°C produce a far worse burn than water at 100°C?

18-12. The temperature of 1 g of iron is raised by 1 C°. How much more heat would be required to raise the temperature of 1 lb$_m$ of iron by 1 F°?

NOTE: *Refer to Tables 18-1 and 18-2 for accepted values for specific heat, heat of fusion, and heat of vaporization for the substances in the problems below.*

Quantity of Heat and Specific Heat Capacity

18-1. What quantity of heat is required to change the temperature of 200 g of lead from 20 to 100°C? Express your answer in joules, in calories, and in Btu.

Ans. 2080 J, 496 cal, 1.97 Btu

18-2. What quantity of heat will be released when 40 lb of copper cools from 78 to 32°F?

18-3. A lawn mower engine does work at the rate of 7 hp (1 hp = 550 ft · lb/s). What equivalent amount of heat energy will the engine give off in 1 h?

Ans. 17,800 Btu

18-4. The mechanical output of an electric motor is 2 kW. This represents 80 percent of the input electric energy; the remainder is lost to heat. Express this rate of heat loss in joules per second and in calories per second.

18-5. Hot coffee is poured into a 0.5-kg ceramic cup with a specific heat of 880 J/kg · C°. How much heat is absorbed by the cup if its temperature increases from 20 to 80°C?

Ans. 26.4 kJ

18-6. In a heat-treating operation, a hot copper part is cooled quickly in water (quenched). If the temperature of the part drops from 400 to 30°C and the part loses 80 kcal of heat, what was the mass of the copper part in grams?

18-7. A 4-lb copper sleeve must be heated from 70 to 250°F so that it will expand enough to slip over a shaft. How much heat is needed?

Ans. 67.0 Btu

18-8. The specific heat for 4 kg of a given metal is 320 J/kg · C° and it is initially at 300°C. What will be its final temperature if it loses 50 kJ of heat?

CONSERVATION OF ENERGY; CALORIMETRY

18-9. A 0.4-kg copper part initially at 200°C is dropped into a container filled with 3 kg of water at 20°C. If other heat exchanges are neglected, what is the equilibrium temperature of the mixture?

Ans. 22.2°C

18-10. What mass of water initially at 20°C must be mixed with 2 kg of iron in order to bring the iron from 250°C to an equilibrium temperature of 25°C?

18-11. A 450-g metal cylinder is heated to 100°C and dropped into a 50-g copper calorimeter. The calorimeter contains 100 g of water initially at 10°C. Other heat exchanges are negligible, and the equilibrium temperature is 21.1°C. What is the specific heat of the unknown metal?

Ans. 0.033 cal/g · C°

18-12. How much iron at 212°F must be mixed with 10 lb of water at 68°F to bring the equilibrium temperature to 100°F?

18-13. A workman needs to know the temperature of the inside of an oven. He removes a 2-kg iron bar from the oven and places it in a 1-kg aluminum container partially filled with 2 kg of water. If the temperature of the water rises from 21 to 50°C, what was the oven temperature? Neglect other heat exchanges.

Ans. 337°C

18-14. Fifty grams of brass shot is heated to 200°C and then dropped into a 50-g aluminum calorimeter cup containing 160 g of water. What is the equilibrium temperature, if the temperature of the cup and water is initially 20°C?

Heat of Fusion and Heat of Vaporization

18-15. A foundry has an electric furnace that can completely melt 540 kg of copper. If the temperature of the copper was initially 20°C, how much heat is required?

Ans. 75,600 kcal

18-16. How much heat is required to completely melt 20 g of silver at its melting temperature?

18-17. What quantity of heat is needed to convert 2 kg of ice at −25°C to steam at 100°C?

Ans. 1465 kcal

18-18. What total heat is released when 0.5 lb of steam at 212°F changes to ice at 10°F?

18-19. In an experiment to determine the latent heat of vaporization for water, a student measures the mass of an aluminum calorimeter cup to be 50 g. After a quantity of water is added, the combined mass of the water and cup is 120 g. The initial temperature of the cup and water is 18°C. A quantity of steam at 100°C is passed into the calorimeter, and the system is observed to reach equilibrium at 47.4°C. The total mass of the final mixture is 124 g. What value will the student obtain for the heat of vaporization?

Ans. 543 cal/g

*   18-20. When 100 g of ice at 0°C is mixed with 600 g of water at 25°C, what will be the equilibrium temperature of the mixture?

Additional Problems

18-21. If 1600 J of heat is applied to a brass ball, its temperature rises from 20 to 70°C. What is the mass of the ball?

Ans. 82.0 g

18-22. How much heat does an electric freezer absorb in lowering the temperature of 2 kg of water from 80 to 20°C?

18-23. A heating element supplies an output power of 12 kW. How much time is needed to completely melt a 2-kg silver block? Assume that none of the output power is wasted.

Ans. 14.7 s

*   18-24. How much heat is developed by the brakes of a 4000-lb truck in order to bring it to a rest from a speed of 60 mi/h?

*   18-25. Suppose that 200 g of copper at 300°C is dropped into a 310-g copper calorimeter cup filled with 300 g of water. If the initial temperature of the mixture was 15°C, what is the equilibrium temperature?

Ans. 30.3°C

** 18-26. If 4 g of steam at 100°C is mixed with 20 g of ice at −5°C, find the final temperature and composition of the mixture.

** 18-27. If 10 g of ice at −5°C is mixed with 6 g of steam at 100°C, find the final temperature and composition of the mixture.

Ans. 100°C, 13.38 g of water, 2.62 g of steam

*   18-28. What equilibrium temperature is reached when 2 lb of ice at 0°F is dropped into a 3-lb aluminum calorimeter cup filled with 7.5 lb of water at 200°F?

18-29. How many pounds of coal must be burned to melt completely 50 lb of ice in a heater that is 60 percent efficient?

Ans. 0.923 lb

** 18-30. If 100 g of water at 20°C is mixed with 100 g of ice at 0°C and 4 g of steam at 100°C, find the equilibrium temperature and the composition of the mixture.

18-31. How much fuel oil (15,000 Btu/lb) is needed to raise the temperature of 120 lb of steel from 75 to 900°F?

Ans. 0.752 lb

Transfer of Heat

19

After completing this chapter, you should be able to:

1. Demonstrate by definition and example your understanding of *thermal conductivity, convection,* and *radiation*.
2. Solve *thermal conductivity* problems involving such parameters as quantity of heat *Q*, surface area *A*, surface temperature *t*, time τ, and material thickness *L*.
3. Solve problems involving the transfer of heat by *convection* and discuss the meaning of the *convection coefficient*.
4. Define the *rate of radiation* and *emissivity* and use these concepts to solve problems involving thermal radiation.

We have referred to heat as a form of energy in transit. Whenever there is a difference of temperature between two bodies or between two portions of the same body, heat is said to *flow* in a direction from higher to lower temperature. There are three principal methods by which this heat exchange occurs: *conduction, convection,* and *radiation*. Examples of all three are shown in Fig. 19-1.

19-1
METHODS OF HEAT TRANSFER

Most of our discussion has involved heat transfer by conduction, i.e., by molecular collisions between neighboring molecules. For example, if we hold one end of an iron rod in a fire, the heat will eventually reach our hand through the process of conduction. The increased molecular activity at the heated end is passed on from molecule to molecule until it reaches the hand. The process will continue as long as there is a difference in temperature along the rod.

> **Conduction** is the process in which heat energy is transferred by adjacent molecular collisions throughout a material medium. The medium itself does not move.

The most common application of the principle of conduction is probably cooking.

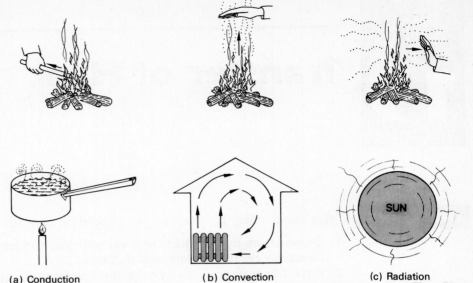

(a) Conduction   (b) Convection   (c) Radiation

Fig. 19-1 The three methods of heat transfer: *(a)* conduction, *(b)* convection, and *(c)* radiation.

On the other hand, if we place our hand above a fire, as shown in Fig. 19-1*b*, we can feel the transfer of heat by the hot rising air. This process, called convection, differs from conduction in that the material medium is moving. Heat is transferred by moving masses instead of being passed along by neighboring molecules.

> **Convection** is the process by which heat is transferred by the actual mass motion of a fluid.

Convection currents form the basis for heating and cooling most houses.

When we hold our hand to the side of a fire, the primary source of heat is through *thermal radiation*. Radiation involves the emission or absorption of electromagnetic waves originating at the atomic level. These waves travel at the speed of light (3×10^8 m/s) and require no material medium for their passage.

> **Radiation** is the process by which heat is transferred by electromagnetic waves.

The most obvious source of radiant energy is our own sun. Neither conduction nor convection can play a role in transferring heat energy through space to the earth. The enormous heat energy received on the earth is transferred by electromagnetic radiation. Where a material medium is involved, however, the transfer of heat due to radiation is usually very small in comparison with that transferred by conduction and convection.

Unfortunately, there are many factors which affect the transfer of heat energy by all three methods. The computation of the quantity of thermal energy transferred in a given process is very complicated. The relationships discussed in the following sections are based on empirical observations and depend on ideal conditions. The extent to which these conditions can be met usually determines the accuracy of their predictions.

When two parts of a material are maintained at different temperatures, energy is transferred by molecular collisions from higher to lower temperature. This process of conduction is also aided by the motion of free electrons within the substance. These electrons have become disassociated from their parent atoms and are free to move about when stimulated either thermally or electrically. Most metals are good conductors of heat because they have a number of free electrons which can distribute heat in addition to that propagated by molecular agitation. In general, a good conductor of electricity is also a good conductor of heat.

The fundamental law of heat conduction is a generalization of experimental results concerning the flow of heat through a slab of material. Consider the slab of thickness L and area A in Fig. 19-2. One face is maintained at a temperature t and the

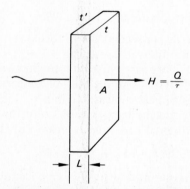

Fig. 19-2 Measurement of thermal conductivity.

other at a temperature t'. The quantity of heat Q that flows perpendicular to the face during a time τ is measured. Repeating the experiment for many different materials of different thickness and face areas, we are able to make some general observations concerning the conduction of heat:

1. The quantity of heat transferred per unit of time is directly proportional to the temperature difference ($\Delta t = t' - t$) between the two faces.
2. The quantity of heat transferred per unit of time is directly proportional to the area A of the slab.
3. The quantity of heat transferred per unit of time is inversely proportional to the thickness L of the slab.

These results can be expressed in equation form by introducing the proportionality constant k. We write

$$H = \frac{Q}{\tau} = kA\frac{\Delta t}{L}$$ (19-1)

where H represents the rate at which heat is transferred. Although the equation was introduced for a slab of material, it also holds for a rod of cross section A and length L.

The proportionality constant k is a property of the material called its *thermal conductivity*. From the defining equation, one can see that substances with a large thermal conductivity are good conductors of heat whereas substances of low conductivity are poor conductors, or *insulators*.

The **thermal conductivity** of a substance is a measure of its ability to conduct heat and is defined by the relation

$$k = \frac{QL}{A\tau\,\Delta t} \tag{19-2}$$

The numerical value for the thermal conductivity depends upon the units chosen for heat, thickness, area, time, and temperature. A substitution of SI units for each of these quantities yields the following accepted units:

$$\text{SI units: } \text{J/s} \cdot \text{m} \cdot \text{C}° \quad \text{or} \quad \text{W/m} \cdot \text{K}$$

You will remember that the joule per second (J/s) is the power in watts (W) and that kelvin and Celsius temperature intervals are equal.

Unfortunately, the SI units are rarely used for conductivity in industry today. The choice of units is more often made on the basis of the convenience of measurement. For example, in the USCS, heat is measured in Btu, thickness in inches, area in square feet, time in hours, and the temperature interval in Fahrenheit degrees. Consequently, the units for thermal conductivity, from Eq. (19-2), are

$$\text{USCS: } k = \text{Btu} \cdot \text{in./ft}^2 \cdot \text{h} \cdot \text{F}°$$

In the metric system, heat transfer in calories is often encountered more than heat transfer expressed in joules. Therefore, the following units are frequently used:

$$\text{Metric units: } k = \text{kcal/m} \cdot \text{s} \cdot \text{C}°$$

The following conversion factors will be helpful:

$$1 \text{ kcal/s} \cdot \text{m} \cdot \text{C}° = 4186 \text{ W/m} \cdot \text{K}$$

$$1 \text{ Btu} \cdot \text{in./ft}^2 \cdot \text{h} \cdot \text{F}° = 3.445 \times 10^{-5} \text{ kcal/m} \cdot \text{s} \cdot \text{C}°$$

The thermal conductivities for various materials are listed in Table 19-1.

Table 19-1 Thermal Conductivities and R-values

| Substance | Conductivity, k | | | R-value* |
	W/m · K	kcal/m · s · C°	Btu · in./ft² · h · F°	ft² · h · F°/Btu
Aluminum	205	5.0×10^{-2}	1451	0.00069
Brass	109	2.6×10^{-2}	750	0.0013
Copper	385	9.2×10^{-2}	2660	0.00038
Silver	406	9.7×10^{-2}	2870	0.00035
Steel	50.2	1.2×10^{-2}	320	0.0031
Brick	0.7	1.7×10^{-4}	5.0	0.20
Concrete	0.8	1.9×10^{-4}	5.6	0.18

Table 19-1 (Continued)

Substance	Conductivity, k			R-value*
	W/m · K	kcal/m · s · C°	Btu · in./ft² · h · F°	ft² · h · F°/Btu
Corkboard	0.04	1.0×10^{-5}	0.3	3.3
Drywall	0.16	3.8×10^{-5}	1.1	0.9
Fiberglass	0.04	1.0×10^{-5}	0.3	3.3
Glass	0.8	1.9×10^{-4}	5.6	0.18
Polyurethane	0.024	5.7×10^{-6}	0.17	5.9
Sheathing	0.55	1.3×10^{-5}	0.38	2.64
Air	0.024	5.7×10^{-6}	0.17	5.9
Water	0.6	1.4×10^{-4}	4.2	0.24

*R-values based on 1-in. thickness.

EXAMPLE 19-1 The outside wall of a brick barbecue pit is 3 in. thick. The inside surface is at 300°F, and the outside surface is 85°F. How much heat is lost in 1 h through an area of 1 ft²?

Solution Solving for Q in Eq. (19-1), we obtain

$$Q = kA\tau\frac{\Delta t}{L}$$

$$= (5 \text{ Btu} \cdot \text{in./ft}^2 \cdot \text{h} \cdot \text{F°})(1 \text{ ft}^2)(1 \text{ h})\frac{300°F - 85°F}{3 \text{ in.}}$$

$$= 358 \text{ Btu}$$

It is always a good idea to carry the units of each quantity throughout the entire solution of a problem. This practice will save many needless errors. For example, it is sometimes easy to forget that in USCS units the thickness must be expressed in inches and the area in square feet. If the units of thermal conductivity are given with their numerical value in the equation, these errors will not be made.

When two materials of different thermal conductivities and similar cross sections are connected, the rate at which heat is conducted through each material must be constant. If there are no sources or sinks of heat energy within the materials and the end points are maintained at a constant temperature, a steady flow will eventually be reached. The heat cannot "pile up" or "speed up" at any point.

EXAMPLE 19-2 The wall of a freezing plant is composed of 10 cm of corkboard inside 14 cm of solid concrete (Fig. 19-3). The temperature of the inside cork surface is −20°C, and the outside surface of the concrete is at 24°C. (a) Find the temperature at the corkboard-concrete interface. (b) Calculate the rate of heat loss in watts per square meter.

Solution (a) For steady flow, the rate of heat flow through the corkboard is equal to the rate of heat flow through the concrete. We will use the subscript 1 to refer to the corkboard and the subscript 2 to refer to the concrete. Thus, letting t_i be the temperature of the interface, we have

$$\frac{H_1}{A_1}(\text{corkboard}) = \frac{H_2}{A_2}(\text{concrete})$$

$$\frac{k_1[t_1 - (-20°C)]}{0.10 \text{ m}} = \frac{k_2(24°C - t_i)}{0.14 \text{ m}}$$

$$\frac{(0.04 \text{ W/mK})(t_i + 20°C)}{0.10 \text{ m}} = \frac{(0.8 \text{ W/mK})(24°C - t_i)}{0.14 \text{ m}}$$

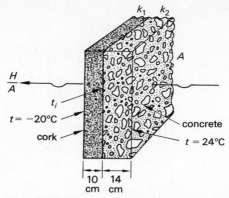

Fig. 19-3 Heat conduction through a compound wall.

Now, we simplify by multiplying each term by 14 to obtain

$$5.6(t_i + 20°C) = 80(24°C - t_i)$$
$$t_i = 21.1°C$$

Solution (b)   The heat flow per unit area per unit time can now be found from Eq. (19-1) applied to either the corkboard or the concrete. For the concrete, we have

$$\frac{H_2}{A_2} = \frac{k_2(24°C - t_i)}{0.14 \text{ m}}$$
$$= \frac{(0.8 \text{ W/mK})(24°C - 21.1°C)}{0.14 \text{ m}}$$
$$= 16.4 \text{ W/m}^2$$

Notice that the kelvin interval cancels with the Celsius interval since they are equal. The same rate would be calculated through the corkboard. However, the difference in temperature between the end points of the corkboard is 41.1 C° whereas the temperature difference in the concrete is only 2.9 C°. The very different temperature intervals result primarily from the difference in the thermal conductivities of the walls.

19-3

INSULATION;
THE *R*-VALUE

Heat losses in homes and in industry are often based on the insulating properties of various composite walls. For example, one might wish to know the effects of replacing dead air spaces in walls with fiberglass insulation. In such cases, the concept of *thermal resistance R* has been introduced into engineering applications. The *R*-value of a material of thickness *L* and of thermal conductivity *k* is defined as

$$R = \frac{L}{k}$$

If we recognize that the steady-state flow of heat through a composite wall is constant (see Example 19-2) and apply the thermal conductivity equation to a number of thicknesses of different materials, it can be shown that

$$\frac{Q}{\tau} = \frac{A \, \Delta t}{\Sigma_i \, (L_i/k_i)} = \frac{A \, \Delta t}{\Sigma_i \, R_i} \tag{19-3}$$

The quantity of heat flowing per unit of time (Q/τ) through a number of thicknesses of different materials is equal to the product of the area A and the temperature difference Δt divided by the sum of the R-values of the various materials. The R-values of common building materials are most often given in USCS units. For example, fiberglass ceiling insulation that is 6 in. thick has an R-value of 18.8 ft^2 · F° · h/Btu. A 4-in. brick has an R-value of 4.00 ft^2 · F° · h/Btu. These materials placed side by side would have a total R-value of 22.8 ft^2 · F° · h/Btu.

19-4
CONVECTION

Convection has been defined as the process in which heat is transferred by the actual mass motion of a material medium. A current of liquid or gas that absorbs energy at one place and then moves to another place, where it releases heat to a cooler portion of the fluid, is called a *convection current*. A laboratory demonstration of a convection current is illustrated in Fig. 19-4. A rectangular section of glass tubing is filled with water and heated at one of the lower corners. The water near the flame is heated and expands, becoming less dense than the cooler water above it. As the heated water rises, it is replaced by cooler water from the lower tube. This process continues until a counterclockwise convection current is circulating throughout the tubing. The existence of such a current is vividly demonstrated by dropping ink into the opening

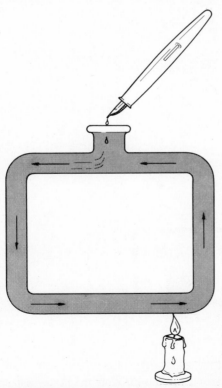

Fig. 19-4 An example of natural convection.

at the top. The ink will be carried along by the convection current until it finally returns to the top from the right section of the tube.

If the motion of a fluid is caused by a difference in density that accompanies a change in temperature, the current produced is referred to as *natural convection*. The water flowing through the glass tubing in the above example represents a natural-convection current. When a fluid is caused to move by the action of a pump or fan, the current produced is referred to as *forced convection*. Many homes are heated by using fans to force hot air from a furnace throughout the rooms.

Both forced- and natural-convection currents occur in the process of heating a room with a radiator. (Refer to Fig. 19-5.) A circulating water pump forces hot water through the pipes to the radiator and back to the heater or furnace. The heat from the water is conducted through walls of the radiator to the air in contact with it. The heated air rises and displaces the cooler air, thus establishing natural-convection currents throughout the room. Although some heating occurs by the process of radiation, conduction and convection play the large heating roles. The name "radiator" is a misnomer.

The calculation of heat transferred by convection is an enormously difficult task. So many physical properties of a fluid depend upon temperature and pressure that we can hope for only an estimate in most situations. We present a working relationship

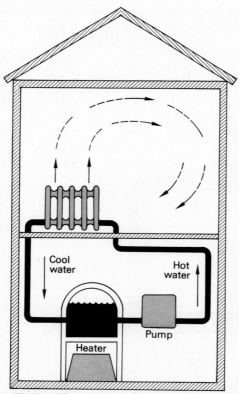

Fig. 19-5 Forced convection currents circulate the heated water and return it to the furnace. The room is heated by natural convection currents in the air.

based on experimental observations. Suppose we consider a conducting slab of material of area A and temperature t_s. If this vertical slab is completely submerged in a cooler fluid at a temperature t_f, natural-convection currents will be set up in the fluid, as illustrated in Fig. 19-6. The fluid which comes in contact with the walls will

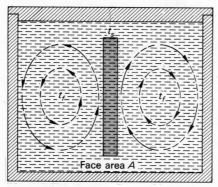

Fig. 19-6 When a heated slab is placed in a cool fluid, convection currents transfer heat away from the slab at a rate proportional to the difference in temperatures and to the area of the slab.

rise and displace the cooler air. Experimental observation shows that the rate H at which heat is transferred by convection is proportional to the area A and to the difference in temperature Δt between the wall and the fluid. We write

$$H = \frac{Q}{\tau} = hA \, \Delta t \qquad (19\text{-}3)$$

where h is the proportionality constant called the *convection coefficient.*

Unlike thermal conductivity, the convection coefficient is not a property of the solid or fluid but depends upon many parameters of the system. It is known to vary with the geometry of the solid and its surface finish, the velocity of the fluid, the density of the fluid, and the thermal conductivity. Differences in the temperature and pressure of the fluid also affect the value for h. Approximate convection coefficients for vertical and horizontal surfaces can be calculated from the formulas given in Table 19-2. The units are in $W/m^2 \cdot K$ or in $kcal/m^2 \cdot s \cdot C°$. When using the SI units, it should be recalled that the difference of temperature in $C°$ is equivalent to the difference in kelvins and that the watt is equal to a joule per second.

Table 19-2 Convection Coefficients

Geometry	$W/m^2 \cdot K$	$kcal/m^2 \cdot s \cdot C°$
Vertical surface	$1.77(\Delta t)^{1/4}$	$(4.24 \times 10^{-4})(\Delta t)^{1/4}$
Horizontal surface:		
Floor (facing up)	$2.49(\Delta t)^{1/4}$	$(5.95 \times 10^{-4})(\Delta t)^{1/4}$
Ceiling (facing down)	$1.31(\Delta t)^{1/4}$	$(3.14 \times 10^{-4})(\Delta t)^{1/4}$

EXAMPLE 19-3 A flat vertical wall 6 m² in area is maintained at a constant temperature of 116°C, and the surrounding air on both sides is at 35°C. How much heat is lost from the wall to both sides in 1 h by natural convection?

Solution We must first compute h for a vertical wall. From Table 19-2 we have

$$h = (4.24 \times 10^{-4})\sqrt[4]{116 - 35}$$
$$= (4.24 \times 10^{-4})\sqrt[4]{81} \text{ kcal/m}^2 \cdot \text{s} \cdot \text{C}°$$
$$= 1.27 \times 10^{-3} \text{ kcal/m}^2 \cdot \text{s} \cdot \text{C}°$$

The quantity of heat transferred from each surface can be found by solving for Q in Eq. (19-3).

$$Q = hA\tau \, \Delta t$$
$$= (1.27 \times 10^{-3} \text{ kcal/m}^2 \cdot \text{s} \cdot \text{C}°)(6 \text{ m}^2)(3600 \text{ s})(81 \text{ C}°)$$
$$= 2.22 \times 10^3 \text{ kcal}$$

Since there are two surfaces, the total heat transferred is

$$Q = (2)(2.22 \times 10^3 \text{ kcal}) = 4.44 \times 10^3 \text{ kcal}$$

19-5 RADIATION

The term radiation refers to the continuous emission of energy in the form of electromagnetic waves originating at the atomic level. Gamma-rays, x-rays, light waves, infrared rays, radio waves, and radar waves are all examples of electromagnetic radiation; they differ only in their wavelength. In this section, we shall be concerned with *thermal radiation.*

> **Thermal radiation** consists of electromagnetic waves emitted by a solid, liquid, or gas by virtue of its temperature.

All objects are continuously emitting radiant energy. At low temperatures the rate of emission is small, and the radiation is predominantly of long wavelengths. As the temperature is increased, the rate of emission increases very rapidly, and the predominant radiation shifts to shorter wavelengths. If an iron rod is heated continuously, it will eventually give off radiation in the visible region; hence the terms *red hot* and *white hot.*

Experimental measurements have shown that the rate at which thermal energy is radiated from a surface *varies directly with the fourth power of the absolute temperature of the radiating body.* Thus, if the temperature of an object is doubled, the rate at which it emits thermal energy will be increased sixteenfold.

An additional factor which must be considered in computing the rate of heat transfer by radiation is the nature of the exposed surfaces. Objects that are good emitters of thermal radiation are also good absorbers of radiation. An object which absorbs all the radiation incident on its surface is called an *ideal absorber.* Such an object will also be an *ideal radiator.* There is no such thing as an *ideal* absorber; but, in general, the blacker a surface is, the better it absorbs thermal energy. For example, a black shirt absorbs more of the sun's radiant energy than a lighter shirt. Since the black shirt is also a good emitter, its external temperature will be higher than our body temperature, making us uncomfortable.

An ideal absorber or an ideal radiator is sometimes referred to as a *blackbody* for the reasons mentioned above. The radiation emitted from a blackbody is called *black-*

body radiation. Although such bodies do not actually exist, the concept is very useful as a standard for comparing the abilities of various surfaces to absorb or emit thermal energy.

Emissivity e is a measure of a body's ability to absorb or emit thermal radiation.

The emissivity is a unitless quantity which has a numerical value between 0 and 1, depending upon the nature of the surface. For a blackbody, the emissivity is equal to unity. For a highly polished silver surface, it is near zero.

The rate of radiation R of a body is formally defined as the radiant energy emitted per unit area per unit time, i.e., the power per unit area. Symbolically,

$$R = \frac{E}{\tau A} = \frac{P}{A} \tag{19-4}$$

If the radiant power P is expressed in watts and the surface area A in square meters, the rate of radiation will be in watts per square meter. As we have discussed earlier, this rate depends on two factors, the absolute temperature T and the emissivity e of the radiating body. The formal statement of this dependence, known as the *Stefan–Boltzmann law,* can be written

$$\boxed{R = \frac{P}{A} = e\sigma T^4} \tag{19-5}$$

The proportionality constant σ is a universal constant completely independent of the nature of the radiation. If the radiant power is expressed in watts and the surface in square meters, σ has the value of $5.67 \times 10^{-8} \, \text{W/m}^2 \cdot \text{K}^4$. The emissivity e has values from 0 to 1, depending upon the nature of the radiating surface. A summary of the symbols and their definitions is given in Table 19-3.

Table 19-3 Definition of Symbols in the Stefan–Boltzmann law ($R = e\sigma T^4$)

Symbol	Definition	Comment
R	Energy radiated per unit time per unit area	$\dfrac{E}{\tau A}$ or $\dfrac{P}{A}$
e	Emissivity of the surface	0–1
σ	Stefan's constant	$5.67 \times 10^{-8} \, \text{W/m}^2 \cdot \text{K}^4$
T^4	The fourth power of the absolute temperature	K^4

EXAMPLE 19-4 What power will be radiated from a spherical silver surface 10 cm in diameter if its temperature is 527°C? The emissivity of the surface is 0.04.

Solution   We must first compute the surface area from the known diameter of the sphere.

$$A = 4\pi R^2 = \pi D^2 = \pi(0.1 \text{ m})^2 = 0.0314 \text{ m}^2$$

The absolute temperature is

$$T = 527 + 273 = 800 \text{ K}$$

Solving for P in Eq. (19-5), we obtain

$$P = e\sigma A T^4$$
$$= (0.04)(5.67 \times 10^{-8} \text{ W/m}^2 \cdot \text{K}^4(0.0314 \text{ m}^2)(800 \text{ K})^4$$
$$= 29.2 \text{ W}$$

We have said that all objects continuously emit radiation, regardless of their temperature. If this is true, why don't the objects eventually run out of fuel? The answer is that they *would* run down if no energy were supplied to them. The filament in an electric light bulb cools rather quickly to room temperature when the supply of electric energy is shut off. It does not cool further because, at this point, the filament is absorbing radiant energy at the same rate that it is emitting radiant energy. The law covering this phenomenon is known as *Prevost's law of heat exchange:*

> *A body at the same temperature as its surroundings radiates and absorbs heat at the same rates.*

Figure 19-7 shows an isolated object in thermal equilibrium with the walls of its container.

The rate at which energy is absorbed by a body is also given by the Stefan–Boltzmann law [Eq. (19-5)]. Thus we can figure the net transfer of radiant energy by an object surrounded by walls at a different temperature. For example, consider a thin wire filament in a lamp which is covered with an envelope, as shown in Fig. 19-8. Let the temperature of the filament be denoted by T_1 and the temperature of the surrounding envelope be denoted by T_2. The emissivity of the filament is e, and

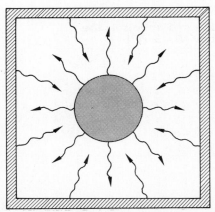

Fig. 19-7 When an object and its surroundings are at the same temperature, the radiant energy emitted is the same as that absorbed.

HEAT, LIGHT, AND SOUND

Fig. 19-8 The net energy emitted by a radiator in surroundings of a different temperature.

only radiative processes are considered. In this example, we note that

Net rate of radiation = rate of energy emission − rate of energy absorption

$$R = e\sigma T_1^4 - e\sigma T_2^4$$

$$\boxed{R = e\sigma(T_1^4 - T_2^4)} \tag{19-6}$$

Equation (19-6) can be applied to any system for computing the net energy emitted by a radiator of temperature T_1 and emissivity e in the presence of surroundings at a temperature T_2.

SUMMARY

Heat is the transfer of thermal energy from one place to another. We have seen that the rate of transfer by conduction, convection, and radiation can be predicted from experimental formulas. The effects of material, surface areas, and temperature differences must be understood for many industrial applications of heat transfer. The major concepts presented in this chapter are:

- In the transfer of heat by conduction, the quantity of heat Q transferred per unit of time τ through a wall or rod of length L is given by

$$H = \frac{Q}{\tau} = \frac{kA\,\Delta t}{L} \qquad \qquad \textit{Conduction}$$

where A is the area and Δt is the difference in surface temperatures. The SI unit for H is the *watt* (W). Other commonly used units are kcal/s and Btu/h. From this relation, the thermal conductivity is

$$k = \frac{QL}{\tau A\,\Delta t} \qquad \qquad \textit{Thermal Conductivity}$$

The SI units for k are W/m · K. Useful conversions can be made from the following definitions:

$$1 \text{ kcal/m} \cdot \text{s} \cdot \text{C}° = 4186 \text{ W/m} \cdot \text{K}$$
$$1 \text{ W/m} \cdot \text{K} = 6.94 \text{ Btu in./ft}^2 \cdot \text{h} \cdot \text{F}°$$

- The R-value is an engineering term to measure the thermal resistance offered to the conduction of heat. It is defined as follows:

$$R = \frac{L}{k} \qquad \qquad R\text{-value}$$

This concept applied to a number of thicknesses of different materials yields the following useful equation:

$$\frac{Q}{\tau} = \frac{A \, \Delta t}{\Sigma_i \, (L_i/k_i)} = \frac{A \, \Delta t}{\Sigma_i \, R_i}$$

The quantity of heat flowing per unit of time (Q/τ) through a number of thicknesses of different materials is equal to the product of the area A and the temperature difference Δt divided by the sum of the R-values of the various materials. In keeping with current engineering practice the units for the R-value are $\text{ft}^2 \cdot \text{F}° \cdot \text{h/Btu}$.

- The heat transferred by convection for a surface A is given by

$$\boxed{H = \frac{Q}{\tau} = hA \, \Delta t} \qquad \qquad \textit{Convection}$$

In this case Δt is the difference in temperature between the surface and the fluid. The convection coefficient for several cases is given in Table 19-2.

- For heat transfer by radiation, we define the rate of radiation as the energy emitted per unit area per unit time (or simply the power per unit area):

$$\boxed{R = \frac{E}{\tau A} = \frac{P}{A}} \qquad \qquad \textit{Rate of Radiation}, \text{ W/m}^2$$

According to *Stefan–Boltzmann's law,* this rate is given by

$$\boxed{R = \frac{P}{A} = e\sigma T^4} \qquad \sigma = 5.67 \times 10^{-8} \text{ W/m}^2 \cdot \text{K}^4$$

- Prevost's law of heat exchange states that *a body at the same temperature as its surroundings radiates and absorbs heat at the same rates.*

19-1. Define the following terms:
  a. Conduction
  b. Thermal conductivity
  c. Natural convection

  g. Blackbody
  h. Emissivity
  i. Stefan–Boltzmann law

d. Forced convection
e. Convection coefficient
f. Thermal radiation

j. Stefan's constant σ
k. Prevost's law

19-2. Discuss the vacuum bottle and explain how it minimizes transfer of heat by conduction, convection, and radiation.

19-3. What determines the *direction* of heat transfer?

19-4. Heat flows both by conduction and by radiation. In what ways are they different? In what ways are they similar?

19-5. A hot chunk of iron is suspended centrally within an evacuated calorimeter. Can we determine the specific heat of iron by the techniques introduced in Chap. 18? Discuss.

19-6. Discuss the analogies that exist between steady state heat flow and the flow of an incompressible fluid.

19-7. A pan of water is placed over a gas burner on a kitchen stove until the water boils vigorously. Discuss the heat transfers which take place. How would you explain the fact that the bubbles forming in the water are carried to the surface in the form of a pyramid instead of rising directly to the surface?

19-8. By placing a flame underneath a paper cup filled with water it is possible to bring the water to a boil without burning the bottom of the cup. Explain.

19-9. When a piece of paper is wrapped around a stick of wood and the system is heated with a flame, the paper will begin to burn. But if the paper is wrapped tight around a copper rod and heated in the same manner, it does not burn. Why?

19-10. On a very cold day, a piece of iron feels colder to the touch than a piece of wood. Explain.

19-11. Copper has about twice the thermal conductivity of aluminum, but its specific heat is a little less than half that of aluminum. A rectangular block is made from each material so that they have identical masses and the same surface area at their bases. Each block is heated to 300°C and placed on the top of a large cube of ice. Which block will stop sinking first? Which will sink deeper?

19-12. Distinguish between thermal conductivity and specific heat as they relate to heat transfer.

19-13. The term *absorptivity* is sometimes used in place of the term *emissivity*. Can you justify this practice?

19-14. Why is more air conditioning required to cool the inside of a navy-blue car than a white car of the same size?

19-15. If a house is to be designed for maximum comfort in both summer and winter, would you prefer a light roof or a dark roof? Explain.

19-16. If you are interested in the number of kilocalories transferred by radiation in a unit of time, the Stefan–Boltzmann law can be written in the form

$$\frac{Q}{\tau} = e\sigma A T^4$$

Show that Stefan's constant σ is equal to 1.35×10^{-11} kcal/m$^2 \cdot$ s $\cdot$ K^4 for this form of the law.

19-17. When a liquid is heated in a glass beaker, a wire gauze is usually placed between the flame and the bottom of the beaker. Why is this a wise practice?

19-18. Does the warm air over a burning fire rise, or is it forced upward by the flames?

19-19. Should a hot-water or steam radiator be painted with a good emitter or a poor one? If it is painted black, will it be more efficient? Why?

19-20. Which is a faster process, conduction or convection? Give an illustration to justify your conclusion.

Thermal Conductivity

19-1. A block of copper has a cross section of 20 cm^2 and a length of 50 cm. The left end is maintained at 0°C and the right end is at 100°C. **(a)** What is the rate of heat flow in watts? **(b)** What is the rate of heat flow if the copper block is replaced with an identical block of glass?

                                                      Ans. **(a)** 154 W, **(b)** 0.32 W

19-2. A pane of window glass is 10 in. wide, 16 in. long, and 1/8 in. thick. If the inside surface of the glass is at 60°F and the outside surface is at 20°F, how many Btu are transferred from the inside to the outside surface in a time of 2 h?

* **19-3.** One end of an iron rod 30 cm long and 4 cm^2 in cross section is placed in a bath of ice and water. The other end is placed in a steam bath. How much time in minutes is required to transfer 1.0 kcal of heat? In what direction does the heat flow?

                                        Ans. 10.4 min, toward the ice bath

19-4. A steel plate has a cross section of 600 cm^2. One side is at 170°C and the other is at 120°C. If the steel is 20 mm thick, what is the rate of heat transfer in watts?

19-5. How much heat will be lost in 12 h by conduction through a 3-in. brick firewall if one side is at 330°F and the other at 78°F? The area of the brick wall is 10 ft^2.

                                        Ans. 50,400 Btu

** **19-6.** A steel rod 30 cm long is rigidly attached to a silver rod 60 cm long. Both rods have a cross-sectional area of 4 cm^2. The free end of the silver rod is maintained at 5°C, and the free end of the steel rod is maintained at 95°C. **(a)** What is the temperature at the interface? **(b)** How much heat is conducted in 1 min?

Insulation; The R-Value

19-7. Find the *R*-value for a composite wall consisting of 3 in. of brick, 0.5 in. of sheathing, a vertical air space 3.5 in. thick, and 0.5 in. of drywall. How many Btu per hour will be conducted through each square foot of this wall if the inside wall temperature is 78°F and the outside temperature is 30°F? The R-value for the air is 0.97.

                                  Ans. 3.34 ft$^2 \cdot$ h $\cdot$ F°/Btu, 14.4 Btu/h

19-8. Replace the dead air space in Prob. 19-7 with 3.5 in. of fiberglass insulation. What are the new *R*-value and the new rate of heat loss through each square foot of wall?

* **19-9.** The thermal conductivity of white pine is 0.11 W/m $\cdot$ K. **(a)** What is the conductivity in Btu $\cdot$ in./ft$^2 \cdot$ h $\cdot$ F°? **(b)** What is the *R*-value for a thickness of 1 in. of white pine? **(c)** What thickness of concrete is needed to provide the same insulation as 1 in. of white pine?

             Ans. **(a)** $0.763 \dfrac{\text{Btu} \cdot \text{in.}}{\text{ft}^2 \cdot \text{h} \cdot \text{F°}}$, **(b)** $1.31 \dfrac{\text{ft}^2 \cdot \text{h} \cdot \text{F°}}{\text{Btu}}$, **(c)** 7.28 in.

Convection

19-10. A vertical wall 2 m tall and 4 m wide is maintained at 101°C while surrounded by air at 20°C. How much heat is lost each minute from either side?

19-11. A flat vertical wall 4 by 6 m in size is maintained at a temperature of 90°C. The air on both sides is at 30°C. How much heat is lost from *both* sides of the wall in 1 h?

Ans. 5.11×10^7 J or 1.22×10^4 kcal

19-12. Assume the wall in Prob. 19-13 is horizontal instead of vertical. How much heat is lost from both top and bottom sides in 1 h?

Thermal Radiation

19-13. What is the rate of radiation for a spherical blackbody at a temperature of 327°C? Will this rate of radiation change if the radius is doubled for the same temperature?

Ans. 7.35 kW/m², no

19-14. Thermal radiation is incident upon a body at the rate of 100 W/m². If the body absorbs 20 percent of the incident radiation, what is its emissivity? What energy in joules will be emitted by this body in 1 min if its surface area is 1 m² and its temperature is 727°C?

19-15. The operating temperature of the filament in a 25-W lamp is 2727°C. If the emissivity of the filament is 0.3, what is its surface area?

Ans. 0.181 cm²

Additional Problems

19-16. A solid wall of concrete is 80 ft high, 100 ft wide, and 6 in. thick. The temperature of one side is 30°F, and the temperature of the other side is 100°F. How many minutes will pass before 400,000 Btu of heat is transferred by conduction?

19-17. The bottom of a metal pan has an area of 86 cm². The pan is filled with boiling water (100°C) and is placed on top of a corkboard 5 mm thick. The formica table top underneath the corkboard maintains a constant temperature of 20°C. How much heat is conducted through the cork in 2 min?

Ans. 0.167 kcal

19-18. A glass window is 1/8 in. thick and has a length of 3 ft and a height of 2 ft. How much heat will be conducted through the glass in 1 day if the surface temperatures are 48 and 45°F?

19-19. The bottom of an aluminum pan is 3 mm thick and has a surface area of 120 cm². How many calories per minute are conducted through the bottom of the pan if the temperature of the outer surface is 114°C and that of the inner surface is 117°C?

Ans. 36,000 cal

19-20. What thickness of copper is required to have the same insulating value as 2 in. of corkboard? What thicknesses of aluminum and of brass are needed?

19-21. The wall of a freezing plant consists of 6 in. of concrete and 4 in. of corkboard. The temperature of the inside cork surface is −15°F, and the temperature of the outside concrete surface is 70°F. (a) What is the temperature at the interface between the cork and concrete? (b) How much heat is conducted through each ft² of wall in 1 h?

Ans. 63.7°F, 5.95 Btu

19-22. A wooden icebox has walls that are 4 cm thick, and the overall effective surface area is 2 m². How many grams of ice will be melted in 1 min if the inside temperature is 4°C and the outside temperature is 20°C? ($k = 0.10$ W/m² · K.)

** 19-23. The air in a room at 26°C is separated from the outside air at −4°C by a vertical glass window 3 mm thick and 10 m² in area. We must expect a small difference in temperature between the inner and outer surfaces of the glass. This is due to the fact that in steady-state heat flow the rate of heat transfer by convection inside, the rate of heat conduction through the glass, and the rate of heat transfer by convection outside must all be equal. For the purposes of calculation, assume that the center of the glass is at the midway point in temperature (11°C). (a) What is the steady-state rate of heat flow? (b) What are the inside and outside temperatures of the glass surfaces?

Ans. (a) 0.125 kcal/s; (b) $t_i = 11.098$°C, $t_o = 10.902$°C.

19-24. A plate-glass window in an office building measures 2 by 6 m and is 1.2 cm thick. When its outer surface is at 23°C and the inner surface is at 25°C, how many joules of heat are transferred through the glass in 1 h?

* 19-25. The filament in a lamp operates at a temperature of 727°C and is surrounded by an envelope at 227°C. If the filament has an emissivity of 0.25 and a surface area of 0.30 cm², what is the operating power of the lamp?

Ans. 0.399 W

1000°

20 Thermal Properties of Matter

After completing this chapter, you should be able to:

1. Write and apply the relationship between the volume and the pressure of a gas at constant temperature (*Boyle's law*).
2. Write and apply the relationship between the volume and the temperature of a gas under conditions of constant pressure (*Charles' law*).
3. Write and apply the relationship between the temperature and pressure of a gas under conditions of constant volume (*Gay-Lussac's law*).
4. Apply the *general gas law* to the solution of problems involving changes in mass, volume, pressure, and temperature of gases.
5. Define *vapor pressure, dew point,* and *relative humidity,* and apply these concepts to the solution of problems.

Now that we have an understanding of the concepts of heat and temperature, we proceed to study the thermal behavior of matter. Four measurable quantities are of interest: the pressure, volume, temperature, and mass of a sample. Together, these variables determine the *state* of a given sample of matter. Depending upon its state, matter may exist in the liquid, solid, or gaseous phase. Thus it is important to distinguish between the terms *state* and *phase*. We begin by studying the thermal behavior of gases.

20-1
IDEAL GASES AND BOYLE'S LAW

In a gas the individual molecules are so far apart that the cohesive forces between them are usually very small. Even though the molecular structure of different gases may vary considerably, their behavior is affected little by the size of the individual molecules. It is usually safe to say that when a large quantity of gas is confined in a rather small volume, the volume occupied by the molecules is still a tiny fraction of the total volume.

One of the most useful generalizations about gases is the concept of an *ideal gas,* whose behavior is completely unaffected by cohesive forces or molecular volumes. Of course, no real gas is *ideal,* but under ordinary conditions of temperature and pressure, the behavior of any gas conforms very closely to the behavior of an ideal gas. Therefore, experimental observations of many real gases can lead to the derivation of general physical laws governing their thermal behavior. The degree to which any real gas obeys these relations is determined by how closely it approximates an ideal gas.

The first experimental measurements of the thermal behavior of gases were made by Robert Boyle (1627–1691). He made an exhaustive study of the changes in the volume of gases as a result of changes in pressure. All other variables, such as mass and temperature, were kept constant. In 1660, Boyle demonstrated that the volume of a gas is inversely proportional to its pressure. In other words, doubling the volume *decreases* the pressure to one-half its original value. This finding is now known as *Boyle's law*.

> *Boyle's law: Provided that the mass and temperature of a sample of gas are held constant, the volume of the gas is inversely proportional to its absolute pressure.*

Another way of stating Boyle's law is to say that the product of the pressure P of a gas and its volume V will be constant as long as the temperature does not change. Consider, for example, a closed cylinder equipped with a movable piston, as shown in Fig. 20-1. In Fig. 20-1a, the initial state of the gas is described by its pressure P_1 and

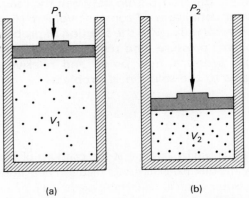

(a)   (b)

Fig. 20-1 When a gas is compressed at constant temperature, the product of its pressure and its volume is always constant, that is, $P_1V_1 = P_2V_2$.

its volume V_1. If the piston is pressed downward until it reaches the new position shown in Fig. 20-1b, its pressure will increase to P_2 while its volume decreases to V_2. This process is shown graphically in Fig. 20-2. If the process occurs without a change in temperature, Boyle's law reveals that

$$P_1V_1 = P_2V_2$$

With Constant
m and T   (20-1)

In other words, the product of pressure and volume in the initial state is equal to the product of pressure and volume in the final state. Equation (20-1) is a mathematical statement of Boyle's law. The pressure P must be the *absolute* pressure and not *gauge* pressure. (See Chap. 15.)

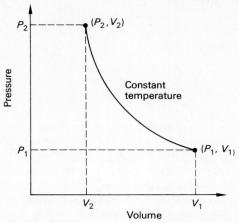

Fig. 20-2 A *P-V* diagram illustrating that the pressure of an ideal gas varies inversely with its volume.

EXAMPLE 20-1   What volume of hydrogen gas at atmospheric pressure is required to fill a 5000-cm³ tank under a gauge pressure of 530 kPa?

Solution   Recall that atmospheric pressure is 101.3 kPa absolute. Thus, the absolute pressures initially and finally are

$$P_1 = 101.3 \text{ kPa} \qquad P_2 = 530 \text{ kPa} + 101.3 \text{ kPa} = 631 \text{ kPa}$$

The final volume V_2 is 5000 cm³. Applying Eq. (20-1), we have

$$P_1V_1 = P_2V_2$$
$$(101.3 \text{ kPa})V_1 = (631 \text{ kPa})(5000 \text{ cm}^3)$$
$$V_1 = 31{,}100 \text{ cm}^3$$

Notice that it was not necessary for the units for pressure to be consistent with the units for volume. Since *P* and *V* appear on each side of the equation, it is only necessary to choose the same units for pressure. The units for volume will then be the units substituted for V_2.

In Chap. 17 we used the fact that the volume of a gas increased directly with its temperature to help us define absolute zero. We found the result ($-273°$C) by extending the line on the graph in Fig. 20-3. Of course, any real gas will become a liquid before its volume reaches zero. But the direct relationship is a valid approximation for most gases which are not subjected to extreme conditions of temperature and pressure.

This direct proportionality between volume and temperature was first experimentally tested by Jacques Charles in 1787. *Charles' law* may be stated as follows:

> *Charles' Law: Provided that the mass and pressure of a gas are held constant, the volume of the gas is directly proportional to its absolute temperature.*

If we use the subscript 1 to refer to an initial state of a gas and the subscript 2 to refer to the final state, a mathematical statement of Charles' law is obtained.

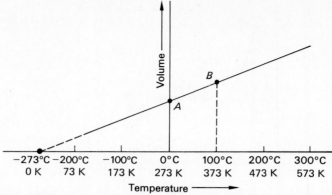

Fig. 20-3 The variation of volume as a function of temperature. When the volume is extrapolated to zero, the temperature of a gas is at absolute zero (0 K).

$$\boxed{\frac{V_1}{T_1} = \frac{V_2}{T_2}}$$

With Constant
m and P (20-2)

In this equation V_1 refers to the volume of a gas at the *absolute* temperature T_1, and V_2 is the later volume of the same sample of gas when its absolute temperature is T_2.

EXAMPLE 20-2   A large balloon filled with air has a volume of 200 liters at 0°C. What will its volume be at 57°C if the pressure is unchanged?

Solution   Since Charles' law applies only for absolute temperatures, we must first convert the given temperatures to kelvins.

$$T_1 = 273 \text{ K} \qquad T_2 = 57 + 273 = 330 \text{ K}$$

Now we may substitute and solve Eq. (20-2) for V_2.

$$\frac{V_1}{T_1} = \frac{V_2}{T_2}$$

$$\frac{200 \text{ liters}}{273 \text{ K}} = \frac{V_2}{330 \text{ K}}$$

$$V_2 = \frac{(200 \text{ liters})(330 \text{ K})}{273 \text{ K}} = 242 \text{ liters}$$

20-2

GAY-LUSSAC'S LAW

The three quantities which determine the state of a given mass of gas are its pressure, its volume, and its temperature. Boyle's law deals with changes in pressure and volume under constant temperature, and Charles' law applies for volume and temperature under constant pressure. The variation in pressure as a function of temperature is described in a law attributed to Gay-Lussac.

> *Gay-Lussac's Law: If the volume of a sample of gas remains constant, the absolute pressure of the gas is directly proportional to its absolute temperature.*

This means that doubling the pressure applied to a gas will cause its absolute temperature to double also. In equation form, Gay-Lussac's law may be written as

$$\boxed{\frac{P_1}{T_1} = \frac{P_2}{T_2}}$$ With Constant m and V (20-3)

EXAMPLE 20-3 An automobile tire is inflated to a gauge pressure of 30 lb/in.^{2} at a time when the surrounding pressure is 14.4 lb/in.^{2} and the temperature is 70°F. After driving, the temperature of the air in the tire increases to 100°F. Assuming that the volume changes only slightly, what will be the new gauge pressure in the tire?

Solution Gay-Lussac's law applies for constant volume, but we must first convert to absolute pressure and to absolute temperature.

$$P_1 = 30 \text{ lb/in.}^2 + 14.4 \text{ lb/in.}^2 = 44.4 \text{ lb/in.}^2$$
$$T_1 = 70 + 460 = 530°\text{R} \qquad T_2 = 100 + 460 = 560°\text{R}$$

Now we calculate the new pressure P_2 from Eq. (20-3).

$$\frac{P_1}{T_1} = \frac{P_2}{T_2}$$

$$\frac{44.4 \text{ lb/in.}^2}{530°\text{R}} = \frac{P_2}{560°\text{R}}$$

$$P_2 = \frac{(44.4 \text{ lb/in.}^2)(560°\text{R})}{530°\text{R}} = 46.9 \text{ lb/in.}^2$$

Note that 46.9 lb/in.^{2} will be the absolute pressure. The gauge pressure will be 14.4 lb/in.^{2} less.

Gauge pressure = 46.9 lb/in.^{2} − 14.4 lb/in.^{2} = 32.5 lb/in.^{2}

20-3
GENERAL GAS LAWS

Thus far, we have discussed three laws which can be used to describe the thermal behavior of gases. Boyle's law, as given by Eq. (20-1) applies for a sample of gas whose temperature is unchanged. Charles' law, as given by Eq. (20-2) applies for a gas sample under a constant pressure. Gay-Lussac's law, in Eq. (20-3), is for a gas sample under constant volume. Unfortunately, none of these conditions is usually satisfied. Normally, a system undergoes changes in volume, temperature, and pressure as a result of a thermal process. A more general relation combines the three laws as follows:

$$\boxed{\frac{P_1 V_1}{T_1} = \frac{P_2 V_2}{T_2}}$$ With Constant m (20-4)

where (P_1, V_1, T_1) may be considered coordinates of the initial state and (P_2, V_2, T_2) the coordinates of the final state. In other words, for a given mass the ratio PV/T is constant for any ideal gas. Equation (20-4) can be remembered by "a private (PV/T) is always a private."

EXAMPLE 20-4 An oxygen tank with internal volume of 20 liters is filled with oxygen under an absolute pressure of 6×10^6 N/m^2 at 20°C. The oxygen is to be used in a high-flying aircraft, where the absolute pressure is 7×10^4 N/m^2 and the temperature is −20°C. What volume of oxygen can be supplied by the tank under these conditions?

Solution After converting the temperatures to the absolute kelvin scale, we apply Eq. (20-4)

$$\frac{P_1 V_1}{T_1} = \frac{P_2 V_2}{T_2}$$

$$\frac{(6 \times 10^6 \text{ N/m}^2)(20 \text{ liters})}{293 \text{ K}} = \frac{(7 \times 10^4 \text{ N/m}^2) V_2}{253 \text{ K}}$$

$$V_2 = 1480 \text{ liters}$$

Let us now consider the effect of a change in mass on the behavior of gases. If the temperature and volume of an enclosed gas are held constant, the addition of more gas will result in a proportional increase in pressure. Similarly, if the pressure and temperature are fixed, an increase in the mass will result in a proportional increase in the volume of the container. We can combine these experimental observations with Eq. (20-4) to obtain the general relation:

$$\boxed{\frac{P_1 V_1}{m_1 T_1} = \frac{P_2 V_2}{m_2 T_2}} \qquad (20\text{-}5)$$

where m_1 is the initial mass and m_2 is the final mass. A study of this relation will reveal that Boyle's law, Charles' law, Gay-Lussac's law, and Eq. (20-4) are all special cases of the more general equation (20-5).

EXAMPLE 20-5 The pressure gauge on a helium storage tank reads 2000 lb/in.^{2} when the temperature is 27°C. The container develops a leak overnight, and the gauge pressure the next morning is found to be 1500 lb/in.^{2} at a temperature of 17°C. What percentage of the original mass of helium remains inside the container?

Solution Since the volume of the container remains constant, $V_1 = V_2$ and Eq. (20-5) gives

$$\frac{P_1}{m_1 T_1} = \frac{P_2}{m_2 T_2}$$

The ratio m_2/m_1 represents the fraction of the helium mass remaining. Hence

$$\frac{m_2}{m_1} = \frac{P_2 T_1}{P_1 T_2}$$

The pressures and temperatures are adjusted to their absolute values as follows:

$$P_1 = 2000 \text{ lb/in.}^2 + 14.7 \text{ lb/in.}^2 = 2014.7 \text{ lb/in.}^2$$
$$P_2 = 1500 \text{ lb/in.}^2 + 14.7 \text{ lb/in.}^2 = 1514.7 \text{ lb/in.}^2$$
$$T_1 = 27 + 273 = 300 \text{ K}$$
$$T_2 = 17 + 273 = 290 \text{ K}$$

Substitution of these values yields

$$\frac{m_2}{m_1} = \frac{(1514.7)(300)}{(2014.7)(290)} = 0.778$$

Therefore, 77.8 percent of the helium still remains inside the container.

Equation (20-5) is very general because it accounts for variance in the pressure, volume, temperature, and mass of a gas. However, the quantity which affects pressure and volume is not the mass of a gas but the number of molecules in the gas. According to the kinetic theory of gases, the pressure is due to molecular collisions with the walls of the container. Increasing the number of molecules will increase the number of particles impacting per second, and the pressure of the gas will become greater. If we are considering a thermal process involving quantities of the same gas, it is safe to apply Eq. (20-5) because the mass is proportional to the number of molecules.

When dealing with different kinds of gas, such as hydrogen compared with oxygen, it is necessary to refer to equal numbers of molecules rather than equal masses. When they are placed in similar containers, 6 g of hydrogen will yield a much greater pressure than 6 g of oxygen. There are many more hydrogen molecules in 6 g of H_2 than there are oxygen molecules in 6 g of O_2. To be more general, we must revise Eq. (20-5) to account for differences in the number of gas molecules instead of the difference in mass. However, first we must develop methods of relating the quantity of a gas to the number of molecules present.

20-4 MOLECULAR MASS AND THE MOLE

Although the mass of individual atoms is difficult to determine because of their size, experimental methods have been successful in measuring atomic mass. For example, we know that one atom of helium has a mass of 6.65×10^{-24} g. When working with macroscopic quantities, such as volume, pressure, and temperature, it is much more convenient to compare the *relative masses* of individual atoms.

The relative atomic masses are based upon the mass of a reference atom known as *carbon 12*. By arbitrarily assigning exactly 12 *atomic mass units* (u) to this atom, we have a standard for comparison of other atomic masses.

> The **atomic mass** of an element is the mass of an atom of that element compared with the mass of an atom of carbon taken as 12 atomic mass units.

On this basis, the atomic mass of hydrogen is approximately 1 u, and the atomic mass of oxygen is approximately 16 u.

A molecule consists of two or more atoms in chemical combination. The definition of molecular mass follows from the definition of relative atomic mass.

> The **molecular mass** M is the sum of the atomic masses of all the atoms making up the molecule.

For example, a molecule of oxygen (O_2) contains two atoms of oxygen. Its molecular mass is 16 u × 2 = 32 u. A molecule of carbon dioxide (CO_2) contains one atom of carbon and two atoms of oxygen. Thus the molecular mass of CO_2 is 44 u:

$$
\begin{aligned}
1\ C &= 1 \times 12 = 12\ u \\
2\ O &= 2 \times 16 = 32\ u \\
\hline
CO_2 &= 44\ u
\end{aligned}
$$

In dealing with gases, we have noted that it is more meaningful to treat the amount of substance present in terms of the number of molecules present. This is accomplished by establishing a new unit of measure called the *mole* (mol).

> A **mole** is that quantity of a substance which contains the same number of particles as there are atoms in 12 g of carbon 12.

On the basis of this definition, 1 mol of carbon must be equal to 12 g by definition. Since the molecular mass of any substance is based on carbon 12 as a standard, it follows that:

> *One mole is the mass in grams equal numerically to the molecular mass of a substance.*

For example, 1 mol of hydrogen (H_2) is 2 g; 1 mol of oxygen (O_2) is 32 g; and 1 mol of carbon dioxide (CO_2) is 44 g. In other words, 2 g of H_2, 32 g of O_2, and 44 g of CO_2 all have the same number of molecules. This number N_A is known as *Avogadro's number*.

The ratio of the number of molecules N to the number of moles n must equal Avogadro's number N_A. Symbolically,

$$N_A = \frac{N}{n} \qquad \qquad \textit{Molecules per Mole} \quad (20\text{-}6)$$

There are several experimental methods of determining Avogadro's number. The accepted value for N_A is

$$N_A = 6.023 \times 10^{23} \text{ molecules per mole} \qquad (20\text{-}7)$$

The easiest way to determine the number of moles n contained in a gas is to divide its mass m in grams by its molecular mass M. Thus

$$\boxed{n = \frac{m}{M}} \qquad \qquad \textit{Number of Moles} \quad (20\text{-}8)$$

EXAMPLE 20-6   (a) How many moles of gas are present in 200 g of CO_2? (b) How many molecules are present?

Solution (a)   The molecular mass of CO_2 is 44 u or 44 g/mol. Therefore,

$$n = \frac{m}{M} = \frac{200 \text{ g}}{44 \text{ g/mol}} = 4.55 \text{ mol}$$

Solution (b)   From Eq. (20-9),

$$n = \frac{N}{N_A} = 4.55 \text{ mol}$$

$$N = (6.023 \times 10^{23} \text{ molecules/mol})(4.55 \text{ mol})$$

$$= 2.74 \times 10^{24} \text{ molecules}$$

Let us now return to our search for a more general gas law. If we substitute the number of moles n for the mass m in Eq. (20-5), we can write

$$\frac{P_1 V_1}{n_1 T_1} = \frac{P_2 V_2}{n_2 T_2} \qquad (20\text{-}9)$$

This equation represents the most useful form of a general gas law when all the parameters of an initial state and of a final state are known except for a single quantity.

An alternative expression of Eq. (20-9) is

$$\frac{PV}{nT} = R \qquad (20\text{-}10)$$

where R is known as the *universal gas constant*. If we can evaluate R under certain known values of P, V, n, and T, Eq. (20-10) can be used directly without any information concerning initial and final states. The numerical value for R, of course, depends on the choice of units for P, V, n, and T. In SI units, the value has been determined as

$$R = 8.314 \text{ J/mol} \cdot \text{K}$$

Other choices of units lead to the following equivalent values:

$$R = 0.0821 \text{ L} \cdot \text{atm/mol} \cdot \text{K}$$
$$= 1.99 \text{ cal/mol} \cdot \text{K}$$

If the pressure is measured in pascals and the volume in cubic meters, one should use 8.314 J/mol · K for the constant R. However, often the pressure is expressed in atmospheres and the volume in liters. Rather than making appropriate conversions, it would probably be simpler to use $R = 0.0821$ L · atm/mol · K.

Equation (20-10) is known as the *ideal-gas law* and is usually written in the form

$$PV = nRT \qquad (20\text{-}11)$$

Another useful form of the ideal-gas law makes use of the fact that $n = m/M$. Thus,

$$PV = \frac{m}{M} RT \qquad (20\text{-}12)$$

EXAMPLE 20-7   Find the volume of 1 mole of any ideal gas at a condition of standard temperature (0°C = 273 K) and pressure (1 atm = 101.3 kPa).

Solution   Solving Eq. (20-11) for V, we obtain

$$V = \frac{nRT}{P} = \frac{(1 \text{ mol})(0.0821 \text{ L} \cdot \text{atm/mol} \cdot \text{K})(273 \text{ K})}{1 \text{ atm}}$$
$$= 22.4 \text{ L or } 0.0224 \text{ m}^3$$

EXAMPLE 20-8   How many grams of oxygen will occupy a volume of 1.6 m³ at a pressure of 200 kPa and a temperature of 27°C?

Solution   Solving for m in Eq. (20-12), we obtain

$$m = \frac{MPV}{RT} = \frac{(32 \text{ g/mol})(200{,}000 \text{ Pa})(1.6 \text{ m}^3)}{(8.314 \text{ J/mol} \cdot \text{K})(300 \text{ K})}$$
$$= 4110 \text{ g or } 4.11 \text{ kg}$$

20-6
**LIQUEFAC-
TION OF
A GAS**

We have defined an ideal gas as one whose thermal behavior is completely unaffected by cohesive forces or molecular volume. Such a gas compressed at a constant temperature will remain a gas no matter how great a pressure is supplied to it. In other words, it will obey Boyle's law at any temperature. The binding forces necessary for liquefaction are never present.

All real gases experience intermolecular forces. However, at rather low pressures and high temperatures, real gases behave very much like an ideal gas. Boyle's law applies because the intermolecular forces under these conditions are practically negligible. A real gas at high temperatures can be compressed in a cylinder, as in Fig. 20-4, to relatively high pressures without liquefying. By plotting the increase in pressure as a function of the volume, the curve A_1B_1 is obtained. Note the similarity between this curve and that for an ideal gas, as shown in Fig. 20-2.

If the same gas is compressed at a much lower temperature, it will begin to condense at a particular pressure and volume. Further compression will continue to liquefy the gas at essentially constant pressure until all the gas has condensed. At

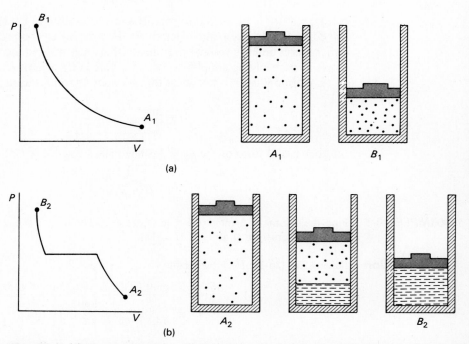

Fig. 20-4 (*a*) Compression of an ideal gas at any temperature or a real gas at high temperature. (*b*) Liquefaction of a real gas when it is compressed at lower temperatures.

HEAT, LIGHT, AND SOUND

that point, a sharp rise in pressure occurs with a slight decrease in volume. The entire process is shown graphically as the curve A_2B_2 in Fig. 20-4.

Let us now begin with the high-temperature compression and perform the experiment at lower and lower constant temperatures. Eventually, a temperature will be reached at which the gas will just begin to liquefy under compression. The highest temperature at which this liquefication occurs is called the *critical temperature*.

> The **critical temperature** of a gas is that temperature above which the gas will not liquefy regardless of the amount of pressure applied to it.

If any gas is to be liquefied, it must first be cooled below its critical temperature. Before this concept was understood, scientists attempted to liquefy oxygen by subjecting it to extreme pressures. Their attempts failed because the critical temperature of oxygen is −119°C. After cooling the gas below this temperature, it can be easily liquefied by compression.

20-7
VAPORIZA-
TION

In Chap. 18 we discussed at length the process of vaporization in which a definite quantity of heat is required to change from the liquid phase to the vapor phase. There are three ways in which such a change may occur: (1) evaporation, (2) boiling, and (3) sublimation. During evaporation, vaporization occurs at the surface of a liquid as the more energetic molecules leave the surface. In the process of boiling, vaporization occurs within the body of the liquid. Sublimation occurs when a solid vaporizes without passing through the liquid phase. In each case, an amount of energy equal to the latent heat of vaporization or sublimation must be lost by the liquid or solid.

The molecular theory of matter assumes that a liquid consists of molecules crowded fairly close together. These molecules have an average kinetic energy which is related to the temperature of the liquid. However, because of random collisions or vibratory motion, not all the molecules move at the same rate of speed; some move faster than others.

Because the molecules are so close together, the forces between them are relatively large. As a molecule approaches the surface of a liquid, as in Fig. 20-5, it experiences a resultant downward force. The net force results from the fact that there are no liquid molecules above the surface to offset the downward attraction of those below the surface. Only the *faster-moving* particles can approach the surface with sufficient energy to overcome the retarding forces. These molecules are said to *evaporate* because, upon leaving the liquid, they become typical gas particles. They have

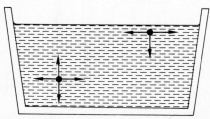

Fig. 20-5 A molecule near the surface of a liquid experiences a net downward force. Only the more energetic molecules are able to overcome this force and leave the liquid.

not changed chemically; the only difference between a liquid and its vapor is the distance between molecules.

Since only the most energetic molecules are able to break away from the surface, the average kinetic energy of the molecules remaining in the liquid is reduced. Hence *evaporation is a cooling process*. (If you place a few drops of alcohol on the back of your hand, you will feel a cooling sensation.) The rate of evaporation is affected by the temperature of the liquid, the number of molecules above the liquid (the pressure), the exposed surface area, and the extent of ventilation.

20-8 VAPOR PRESSURE

A jar is partially filled with water, as shown in Fig. 20-6. The pressure exerted by the molecules above the surface of the water is measured by an open-tube mercury manometer. In Fig. 20-6a there are as many molecules of air inside the jar as are contained in an equal volume of air outside the jar. In other words, the pressure inside the jar is equal to 1 atm, as indicated by the equal levels of mercury in the manometer.

When a high-energy liquid molecule breaks away from the surface, it becomes a vapor molecule and mixes with the air molecules above the liquid. These vapor molecules collide with air molecules, other vapor molecules, and the walls of the jar. The additional vapor molecules cause a rise in pressure inside of the jar, as indicated in Fig. 20-6b. The vapor molecules may also rebound back into the liquid, where they are held as liquid molecules. This process is called *condensation*. Eventually the rate of evaporation will equal the rate of condensation, and a condition of equilibrium will exist, as shown in Fig. 20-6c. Under these conditions, the space above the liquid is said to be *saturated*. The pressure exerted by the saturated vapor against the walls of the jar, over and above that exerted by the air molecules, is called the *saturated vapor pressure*. It is characteristic of the substance and the temperature but independent of the volume of the vapor.

> The **saturated vapor pressure** of a substance is the additional pressure exerted by vapor molecules on the substance and its surroundings under a condition of saturation.

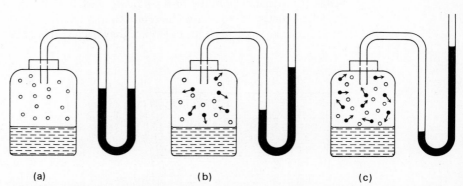

Fig. 20-6 Measuring the vapor pressure of a liquid. (*a*) Air pressure only. (*b*) Partial vapor pressure. (*c*) Saturated vapor pressure.

HEAT, LIGHT, AND SOUND

Once a condition of saturation is obtained for a substance and its vapor at a particular temperature, the vapor pressure remains essentially constant. If the temperature is increased, the molecules in the liquid will acquire more energy and evaporation will occur more rapidly. The condition of equilibrium remains upset until once again the rate of condensation has caught up with the rate of evaporation. The saturated vapor pressure of a substance therefore increases with a rise in temperature.

The saturated-vapor-pressure curve for water is plotted in Fig. 20-7. Note that the vapor pressure increases rapidly with temperature. At room temperature (20°C), it is around 17.5 mm of mercury; at 50°C it has increased to 92.5 mm; and at 100°C it is equal to 760 mm, or 1 atm. The latter point is of importance in distinguishing between *evaporation* and *boiling*.

When a liquid boils, bubbles of its vapor can be seen rising toward the surface from within the liquid. The fact that these bubbles are stable and do not collapse indicates that the pressure inside the bubble is equal to the pressure outside the bubble. The pressure inside the bubble is the vapor pressure at that temperature; the pressure on the outside is the pressure at that depth in the liquid. Under this condition of equilibrium, vaporization occurs freely throughout the liquid, causing the liquid to become agitated.

> **Boiling** is defined as vaporization within the body of a liquid when its vapor pressure equals the pressure in the liquid.

If the pressure on the liquid surface is 1 atm, as it would be in an open container, the temperature at which boiling occurs is called the *normal boiling point* for that liquid. The normal boiling point for water is 100°C because that is the temperature at which the vapor pressure of water is 1 atm (760 mm of mercury). If the pressure on any liquid surface is lower than 1 atm, boiling will occur at a temperature lower than the normal boiling point. If the external pressure is greater than 1 atm, boiling will occur at a higher temperature.

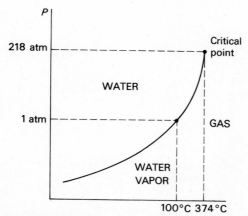

Fig. 20-7 The vaporization curve for water. Any point on the curve represents a condition of pressure and temperature which allows water to boil. The curve ends abruptly at the critical temperature because water can exist only as a gas beyond that point.

We have discussed in detail the process of vaporization, and in Fig. 20-7 we constructed a vaporization curve for water. This curve is represented by the line *AB* in the general phase diagram of Fig. 20-8. Any point on this curve represents a temperature and pressure at which water and its vapor can coexist in equilibrium.

A similar curve can be plotted for the temperatures and pressures at which a substance in the solid phase can coexist with its liquid phase. Such a curve is called a *fusion curve*. The fusion curve for water is represented by the line *AC* in the phase diagram. At any point on this curve, the rate at which ice is melting is equal to the rate at which water is freezing. Note that as the pressure increases, the melting temperature (or freezing temperature) is lowered.

A third graph, called the *sublimation curve,* can be plotted to show the temperatures and pressures at which a solid may coexist with its vapor. The sublimation curve for water is represented by line *AD* of Fig. 20-8.

Let us now study the phase diagram for water more closely to illustrate the usefulness of such a graph for any substance. The coordinates of any point on the graph represent a particular pressure *P* and a particular temperature *T*. The volume must be considered constant for any thermal change indicated by the graph. For any point which falls in the fork between the vaporization and fusion curves, water will exist in its liquid phase. The vapor and the solid regions are also indicated on the diagram. The point *A*, at which all three curves intersect, is called the *triple point* for water. This is the temperature and pressure at which ice, liquid water, and water vapor coexist in equilibrium. Careful measurements have shown that the triple point for water is at 0.01°C and 4.62 mm of mercury (Hg).

The air in our atmosphere consists largely of nitrogen and oxygen with small amounts of water vapor and other gases. It is often useful to describe the water-vapor content of the atmosphere.

The **absolute humidity** is defined as the mass of water vapor per unit volume of air.

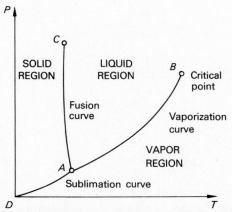

Fig. 20-8 Triple-point phase diagram for water or other substance which expands on freezing.

For example, if 7 g of water vapor is contained in every cubic meter of air, the absolute humidity is 7 g/m³. Other units of absolute humidity are pounds per cubic foot and grains per cubic foot (7000 grains = 1 lb).

A more useful method of expressing the water-vapor content in air is to compare the actual vapor pressure at a particular temperature with the saturated vapor pressure at that temperature. When the atmosphere is holding all the water possible for a certain temperature, it is saturated. The addition of more vapor molecules will simply result in an equal amount of condensation.

The **relative humidity** is defined as the ratio of the actual vapor pressure in the air to be saturated vapor pressure at that temperature.

$$Relative\ humidity = \frac{actual\ vapor\ pressure}{saturated\ vapor\ pressure} \qquad (20\text{-}13)$$

The relative humidity is usually expressed as a percentage.

If the air in a room is not already saturated, it can be made so either by adding more water vapor to the air or by lowering the room temperature until the vapor already present is sufficient. *The temperature to which the air must be cooled at constant pressure in order to produce saturation is called the dew point.* Thus, if ice is placed in a glass of water, moisture will eventually collect on the outside walls of the glass when its temperature reaches the dew point. Given the temperature and the dew point, the relative humidity can be computed from saturated-vapor-pressure tables. Table 20-1 lists the saturated vapor pressure for water at various temperatures.

Table 20-1 Saturated Vapor Pressure for Water

Temperature		Pressure,	Temperature		Pressure,
°C	°F	mmHg	°C	°F	mmHg
0	32	4.62	50	122	92.5
5	41	6.5	60	140	149.4
10	50	9.2	70	158	233.7
15	59	12.8	80	176	355.1
17	62.6	14.5	85	185	433.6
19	66.2	16.5	90	194	525.8
20	68	17.5	95	203	633.9
22	71.6	19.8	98	208.4	707.3
24	75.2	22.4	100	212	760.0
26	78.8	25.2	103	217.4	845.1
28	82.4	28.3	105	221	906.1
30	86	31.8	110	230	1074.6
35	95	42.2	120	248	1489.1
40	104	55.3	150	302	3570.5

EXAMPLE 20-9   On a clear day, the air temperature is 86°F, and the dew point is 50°F. What is the relative humidity?

Solution The pressure of saturated vapor at 50°F is 9.2 mm, from Table 20-1. The pressure of saturated vapor at 86°F is 31.8 mm. From Eq. (20-13),

$$\text{Relative humidity} = \frac{9.2}{31.8} = 0.29$$

Thus the relative humidity is 29 percent.

SUMMARY

The thermal properties of matter must be understood if we are to apply the many laws discussed in this chapter. The relationships between mass, temperature, volume, and pressure allow us to explain and predict the behavior of gases. The major concepts discussed in this chapter are summarized below:

- A useful form of the general gas law which does not involve the use of moles is written on the basis that PV/mT is constant. When a gas in state 1 changes to another state 2, we may write

$$\frac{P_1 V_1}{m_1 T_1} = \frac{P_2 V_2}{m_2 T_2} \qquad \begin{aligned} P &= \text{pressure} & V &= \text{volume} \\ m &= \text{mass} & T &= \text{absolute temperature} \end{aligned}$$

When one or more of the parameters m, P, T, or V is constant that factor disappears from both sides of the above equation. *Boyle's law*, *Charles' law*, and *Gay-Lussac's law* are the following special cases:

$$P_1 V_1 = P_2 V_2 \qquad \frac{V_1}{T_1} = \frac{V_2}{T_2} \qquad \frac{P_1}{T_1} = \frac{P_2}{T_2}$$

- When applying the general gas law in any of its forms, it must be remembered that the pressure is *absolute pressure* and the temperature is *absolute temperature*.

$$\textit{Absolute pressure = gauge pressure + atmospheric pressure}$$
$$T_K = t_C + 273 \qquad T_R = t_F + 460$$

For example, a pressure measured in an auto tire is 30 lb/in.^{2} at 37°C. These values must be adjusted before substitution into the gas laws:

$$P = 30 \text{ lb/in.}^2 + 14.7 \text{ lb/in.}^2 = 44.7 \text{ lb/in.}^2 \qquad \text{(absolute)}$$
$$T = 37 + 273 = 310 \text{ K}$$

- A more general form of the gas law is obtained by using the concepts of molecular mass M, and the number of *moles* n for a gas. The number of molecules in 1 mol is Avogadro's number N_A.

$$N_A = \frac{N}{n} \qquad N_A = 6.023 \times 10^{23} \text{ molecules/mol} \qquad \begin{aligned} &\textit{Avogadro's} \\ &\textit{Number} \end{aligned}$$

The number of moles is found by dividing the mass of a gas (in grams) by its molecular mass M:

$$n = \frac{m}{M}$$

Number of Moles

Often it is desired to determine the mass, pressure, volume, or temperature of a gas in a single state. The ideal-gas law uses the molar concept to arrive at a more specific equation:

$$PV = nRT \qquad R = 8.314 \text{ J/mol} \cdot \text{K}$$

It should be noted that use of the constant given above restricts the units of P, V, T, and n to those which are in the constant.

- The *relative humidity* can be computed from saturated-vapor-pressure tables according to the following definition:

$$\text{Relative humidity} = \frac{\text{actual vapor pressure}}{\text{saturated vapor pressure}}$$

Remember that the *actual* vapor pressure at a particular temperature is the same as the *saturated* vapor pressure for the dew-point temperature. Refer to the example in the text.

QUESTIONS

20-1. Define the following terms:
 a. Ideal gas
 b. Boyle's law
 c. Charles' law
 d. Atomic mass
 e. Mole
 f. Evaporation
 g. Boiling
 h. Sublimation
 i. Saturation
 j. Molecular mass
 k. Avogadro's number
 l. Ideal-gas law
 m. Critical temperature
 n. Vapor pressure
 o. Triple point
 p. Absolute humidity
 q. Relative humidity
 r. Dew point

20-2. Distinguish between *state* and *phase*.

20-3. Explain Boyle's law in terms of the molecular theory of matter.

20-4. Explain Charles' law in terms of the molecular theory of matter.

20-5. Why must the absolute temperature be used in Charles' law?

20-6. A closed steel tank is filled with an ideal gas and heated. What happens to the (a) mass, (b) volume, (c) density, and (d) pressure of the enclosed gas?

20-7. Prove the accuracy of the following equations which involve the density ρ of an ideal gas.

(a) $\dfrac{P_1}{\rho_1 T_1} = \dfrac{P_2}{\rho_2 T_2}$

(b) $\rho = \dfrac{PM}{RT}$

20-8. Suppose we wish to express the pressure of an ideal gas in millimeters of mercury and the volume in cubic centimeters. Show that the universal gas constant will be equal to 6.23×10^4 mm · cm³/mol · K.

20-9. A mole of any gas occupies 22.4 liters at STP. Can we also say that the same mass of any gas will occupy the same volume? Explain.

20-10. Distinguish between evaporation, boiling, and sublimation.

20-11. From your experience, would you expect alcohol to have a higher vapor pressure than water? Why?

20-12. Explain the principle of operation for the pressure cooker and the vacuum pan in cooking.

20-13. Can a solid have a vapor pressure? Explain.

20-14. If evaporation is a cooling process, is condensation a heating process? Explain.

20-15. Explain the cooling effects of evaporation in terms of the latent heat of vaporization.

20-16. Distinguish between a vapor and a gas by discussing critical temperature.

20-17. Will it take longer to cook an egg by boiling it in water on Mt. Everest or at the seashore? Why?

20-18. Water is brought to a vigorous boil in an open flask. When the flask is removed from the flame and tightly stoppered, the boiling stops. Why? The stoppered flask is then inverted and held under a stream of cold running water. The boiling begins again. Explain. As soon as the flask is removed from the water, boiling stops. If the flask is cooled, boiling begins again. How long can the process of making the water boil by cooling be continued?

20-19. On a cool day, the relative humidity inside a house is the same as the relative humidity outside the house. Are the dew points necessarily the same? Explain.

20-20. Explain what is meant by *critical pressure*.

20-21. Is it possible for ice to exist in equilibrium with boiling water? Explain.

20-22. The formation of moisture on the walls and windows inside a home can cause considerable damage. What causes this moisture? Discuss several ways of reducing or preventing the formation of moisture.

20-23. Discuss the formation of fog and clouds. Why are fog conditions usually worse in the fall and early spring?

PROBLEMS

General Gas Laws: Initial and Final States

20-1. A flexible container is filled with 2 m³ of gas at an absolute pressure of 400 kPa. If the volume is slowly increased to 5 m³ at constant temperature with no change in mass, what is the new absolute pressure?

Ans. 160 kPa

* 20-2. The inside of an automobile tire is under a gauge pressure of 30 lb/in.² on a morning when the air temperature is 4°C. After driving a few hours, the air temperature in the tires rises to 50°C. Assume that the volume is essentially constant and that no air is removed from the tires. What is the new gauge pressure?

* 20-3. A tank with a capacity of 14 L contains helium gas at 24°C under a gauge pressure of 2700 kPa. (a) What will be the volume of a balloon filled with this gas if the helium expands to an internal absolute pressure of 1 atm and the

temperature drops to −35°C? **(b)** Now suppose the system returns to its original temperature (24°C). What is the final volume of the balloon?

Ans. (a) 310 L, (b) 387 L

20-4. A 6-L tank holds a sample of gas under an absolute pressure of 600 kPa and a temperature of 57°C. What will be the new pressure if the same sample of gas is placed into a 3-L container at 7°C?

20-5. An air compressor takes in 2 m^3 of air at 20°C and atmospheric pressure (101.3 kPa). If the compressor discharges into a 0.3-m^3 tank at an absolute pressure of 1500 kPa, what is the temperature of the discharged air?

Ans. 651 K

20-6. If 0.8 L of a gas at 10°C is heated to 90°C at constant pressure, what will be the new volume?

* 20-7. A steel tank is filled with oxygen. One evening when the temperature is 27°C, the gauge at the top of the tank indicates a pressure of 400 kPa. During the night a leak develops in the tank. The next morning a technician notices that the gauge pressure is only 300 kPa and the temperature is 15°C. What percent of the original gas still remains in the tank?

Ans. 83.4%

20-8. Five liters of a gas at an absolute pressure of 200 kPa and a temperature of 25°C is heated to 60°C and the absolute pressure is reduced to 120 kPa. What volume will the gas occupy under these conditions?

Molecular Mass and the Mole

20-9. How many moles of gas are there in 400 g of nitrogen gas? The molecular mass of nitrogen is 28 g/mol. How many molecules are there in this sample?

Ans. 14.3 mol, 8.60 × 10^{24} molecules

20-10. How many moles are there in 0.3 kg of water (H_2O)? How many molecules?

* 20-11. The molecular mass of a CO_2 molecule is 44 g/mole. What is the mass of a single molecule of CO_2?

Ans. 7.31 × 10^{-26} kg

* 20-12. What is the mass of a 4-mol sample of air ($M = 29$ g/mol)?

The Ideal Gas Law

20-13. Three moles of an ideal gas under a pressure of 300 kPa are confined to a volume of 0.026 m^3. What is the temperature of the gas in degrees Celsius?

Ans. 39.7°C

20-14. A 16-L tank contains 0.2 kg of air ($M = 29$ g/mol) at 27°C. **(a)** How many moles of air are in the tank? **(b)** What is the absolute pressure in kilopascals?

20-15. How many grams of nitrogen gas ($M = 28$ g/mol) will occupy a volume of 2000 L at an absolute pressure of 202 kPa and a temperature of 80°C?

Ans. 3.85 kg

20-16. What volume is occupied by 8 g of oxygen ($M = 32$ g/mol) at standard temperature and pressure (STP)?

** 20-17. A 2-L flask is filled with nitrogen ($M = 28$ g/mol) at 27°C and at 1 atm of absolute pressure. A stopcock at the top of the flask is opened to the air and the system is heated to a temperature of 127°C. The stopcock is then closed and the system is allowed to return to 27°C. (a) What mass of nitrogen is in the flask? (b) What is the final absolute pressure?

Ans. (a) 1.71 g, (b) 0.75 atm

Humidity

20-18. The relative humidity in a room is 65 percent at 26°C. What is the relative humidity if the temperature drops to 22°C?

20-19. If the air temperature is 20°C and the dew point is 12°C, what is the relative humidity?

Ans. 60.8%

20-20. When room temperature is 28°C, the relative humidity is 77 percent. What is the dew point?

20-21. What is the pressure of water vapor in the air on a day when the temperature is 86°F and the relative humidity is 80 percent?

Ans. 25.4 mm of Hg

Additional Problems

20-22. A given mass of gas occupies 12 L at 7°C and an absolute pressure of 102 kPa. Calculate its temperature when its volume is reduced to 10 L and the pressure is increased to 230 kPa.

20-23. A tractor tire contains 2.8 ft^3 of air at a gauge pressure of 70 lb/in.^{2}. What volume of atmospheric air is required to fill this tire assuming there is no change in its volume or temperature?

Ans. 16.1 ft^3

* 20-24. The density of nitrogen gas at STP is 1.25 kg/m^3. What is the density of nitrogen at 18 atm and 60°C?

20-25. How many moles of helium gas are in a 6-L tank when the pressure is 2×10^5 Pa and the temperature is 27°C? What is the mass of the helium?

Ans. 0.481 mol, 1.92 g

* 20-26. What is the volume of 8 g of sulfur dioxide (M = 64 g/mol) if it has an absolute pressure of 10 atm and a temperature of 300 K? If 10^{20} molecules leak from this volume every second, how long will it take to reduce the pressure by one-half?

20-27. What mass of oxygen will fill a 16-L tank at a pressure of 20 atm and a temperature of 27°C?

Ans. 0.416 kg

* 20-28. A 5000-cm^3 tank is filled with carbon dioxide at 1 atm of pressure and at a temperature of 300 K. How many grams of CO_2 can be added to the tank if the maximum absolute pressure is 60 atm and there is no change in temperature?

* 20-29. A flask contains 2 g of helium at a pressure of 12 atm and a temperature of 57°C. What is the volume of the flask? The system is checked at a later time when the temperature is 17°C and the pressure is found to be only 7 atm. How many grams of helium have leaked out of the flask?

Ans. 1.13 L, 0.672 g

20-30. The molecular mass of air is 29 g/mol. How many grams of air must be pumped into an automobile tire if it is to have a gauge pressure of 31 lb/in.^{2}? Consider the volume of the tire is a constant 5000 cm^3 and the temperature remains at 27°C.

Thermodynamics

After completing this chapter, you should be able to:

1. Demonstrate by definition and examples your understanding of the *first* and *second laws of thermodynamics.*
2. Define and give illustrated examples of *adiabatic, isochoric,* and *isothermal* processes.
3. Write and apply a relationship for determining the *ideal efficiency* of a heat engine.
4. Define the *coefficient of performance* for a refrigerator and solve refrigeration problems similar to those discussed in the text.

Thermodynamics treats the transformation of heat energy into mechanical energy and the reverse process, the conversion of work into heat. Since almost all the energy available from raw materials is liberated in the form of heat, it is easy to see why thermodynamics plays such an important role in science and technology.

In this chapter, we shall study two basic laws which must be obeyed when heat energy is used to accomplish work. The first law is simply a restatement of the principle of conservation of energy. The second law places restrictions on the efficient use of the available energy.

21-1
HEAT AND WORK

The equivalence of heat and work as two forms of energy has been clearly established. Rumford destroyed the caloric theory of heat by showing that it is possible to remove heat indefinitely from a system so long as external work is supplied. Joule then sealed the case by demonstrating the mechanical equivalence of heat.

Work, like heat, involves a transfer of energy, but there is a very important distinction between the two terms. In mechanics, we define work as a scalar quantity, equal in magnitude to the product of a force and a displacement. Temperature plays no role in this definition. Heat, on the other hand, is energy that flows from one body to another because of a difference in temperature. A temperature difference is a necessary condition for the transfer of heat. Displacement is the necessary condition for the performance of work.

The important point in this discussion is to recognize that both heat and work represent changes which occur in a given process. Usually these changes are accompanied by a change in internal energy. Consider the two situations illustrated in Fig. 21-1. In Fig. 21-1a the internal energy of the water is increased by the performance of mechanical work. In Fig. 21-1b the internal energy of the water is increased through the flow of heat.

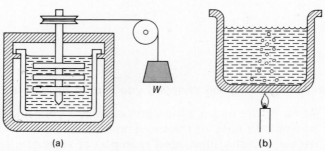

(a)   (b)

Fig. 21-1 Increasing the internal energy of a system *(a)* By the performance of work and *(b)* by supplying heat to the system.

<table>
<tr><td>

21-2

THE INTERNAL-ENERGY FUNCTION

</td><td>

A system is said to be in *thermodynamic equilibrium* if there is no resultant force on the system and if the temperature of the system is the same as its surroundings. This condition requires that no work be done on or by the system and that there be no exchange of heat between the system and its surroundings. Under these conditions, the system has a definite internal energy U. Its *thermodynamic state* can be described by three coordinates: (1) its pressure P, (2) its volume V, and (3) its temperature T. Whenever energy is absorbed or released by such a system, either in the form of heat or work, it will reach a new state of equilibrium in such a way that energy is conserved.

</td></tr>
</table>

In Fig. 21-2, let us consider a general thermodynamic process in which a system is caused to change from an equilibrium state 1 to an equilibrium state 2. In Fig. 21-2a the system is in thermodynamic equilibrium, with an initial internal energy U_1 and thermodynamic coordinates (P_1, V_1, T_1). In Fig. 21-2b the system reacts with its surroundings. Heat Q may be absorbed by the system and/or released to its environment. The heat transfer is considered positive for heat input and negative for heat output. The net heat *absorbed* by the system is represented by ΔQ. Work W may be done *by* the system and/or *on* the system. Output work is considered positive, and input work is considered negative. Thus ΔW represents the net work done *by* the system (output work). In Fig. 21-2c the system has reached its final state 2 and is again in equilibrium, with a final internal energy U_2. Its new thermodynamic coordinates are (P_2, V_2, T_2).

If energy is to be conserved, the change in internal energy

$$\Delta U = U_2 - U_1$$

Fig. 21-2 A schematic diagram of a thermodynamic process.

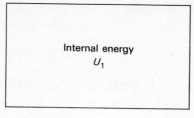

(a) Initial state of system (P_1, V_1, T_1)

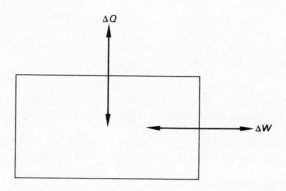

(b) System undergoing thermodynamic process

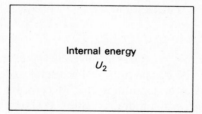

(c) Final state of system (P_2, V_2, T_2)

must represent the difference between the net heat ΔQ absorbed by the system and the net work ΔW done by the system on its surroundings.

$$\Delta U = \Delta Q - \Delta W \tag{21-1}$$

Thus the change in internal energy is uniquely defined in terms of the measurable quantities heat and work. Equation (21-1) states the existence of an *internal energy function* which is determined by the thermodynamic coordinates of a system. Its value at the final state minus its value at the initial state represents the change in energy of the system.

21-3
THE FIRST LAW OF THERMO-DYNAMICS

The *first law of thermodynamics* is simply a restatement of the principle of *conservation of energy:*

> *Energy cannot be created or destroyed but can change from one form to another.*

Applying this law to a thermodynamic process, we note from Eq. (21-2) that

$$\Delta Q = \Delta W + \Delta U \qquad (21\text{-}2)$$

This equation represents a mathematical statement of the *first law of thermodynamics* which can be stated as follows:

> *In any thermodynamic process, the net heat absorbed by a system is equal to the sum of the thermal equivalent of the work done by the system and the change in the internal energy of the system.*

EXAMPLE 21-1
In a certain process, a system absorbs 400 cal of heat and at the same time does 80 J of work on its surroundings. What is the increase in the internal energy of the system?

Solution
Applying the first law, we have

$$\Delta U = \Delta Q - \Delta W$$

$$= 400 \text{ cal} - 80 \text{ J} \, \frac{1 \text{ cal}}{4.186 \text{ J}}$$

$$= 400 \text{ cal} - 19.1 \text{ cal} = 380.9 \text{ cal}$$

Thus the 400 cal of input thermal energy is used to perform 19.1 cal of work while the internal energy of the system is increased by 380.9 cal. Energy is conserved.

21-4
THE *P-V* DIAGRAM

Many thermodynamic processes involve energy changes which occur to gases enclosed in cylinders. At this point, it will be useful to derive an expression for computing the work done by an expanding gas. We consider a system consisting of a gas in a cylinder equipped with a movable piston, as shown in Fig. 21-3a. The piston has a cross-sectional area A and rests on a column of gas under a pressure P. Heat can flow in or out of the gas through the cylinder walls. Work can be done on or by the gas by pushing the piston down or by allowing it to expand upward.

Let us first consider the work done by the gas when it expands at a constant pressure P. The force exerted by the gas on the piston will be equal to PA. If the piston moves upward through a distance Δx, the work ΔW of this force will be given by

$$\Delta W = F \, \Delta x = PA \, \Delta x$$

But $A \, \Delta x = \Delta V$, where ΔV represents the change in volume of the gas. Hence

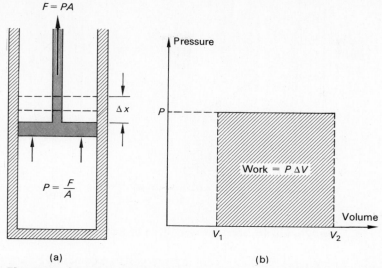

Fig. 21-3 *(a)* Calculating the work done by a gas expanding at constant pressure. *(b)* The work is done to the area under the curve on a *P-V* diagram.

$$\boxed{\Delta W = P\,\Delta V} \tag{21-3}$$

In other words, the work done by a gas expanding at constant pressure is equal to the product of the pressure and the change in volume of the gas.

The process can be shown graphically by plotting the increase in volume as a function of pressure (Fig. 21-3*b*). This representation, called a *P-V diagram*, is extremely useful in thermodynamics. In the above example, the pressure was constant, so that the graph is a straight line. Note that the area under the line, indicated by the shaded portion in the figure, is

$$\text{Area} = P(V_2 - V_1) = P\,\Delta V$$

which is equal to the work ΔW, from Eq. (21-3). This leads us to a very important principle:

> *When a thermodynamic process involves changes in volume and/or pressure, the work done by the system is equal to the area under the curve on a P-V diagram.*

In general, the pressure will not be constant during a piston displacement. For example, in the power stroke of a gasoline engine, the fluid is ignited under high pressure, and the pressure decreases as the piston is displaced downward. The *P-V* diagram in this case is a sloping curve, as shown in Fig. 21-4*a*. The volume increases from V_1 to V_2 while the pressure decreases from P_1 to P_2. To compute the work in such a process, we must resort to a graphical analysis and measure the area under the *P-V* curve.

That the area under the curve is equal to the work when the pressure is not constant can be demonstrated in Fig. 21-4. The area of the narrow shaded rectangle

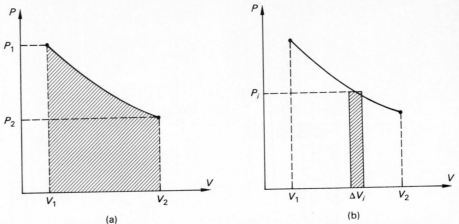

Fig. 21-4 Calculating the work done by a gas expanding under a varying pressure.

represents the work done by the gas in expanding by an increment ΔV_i under a constant pressure P_i. If the area under the entire curve is split up into many of these rectangles, we can sum all the products $P_i \, \Delta V_i$ to obtain the total work. Thus one can see that the total work is simply the area under the $P\text{-}V$ diagram between the points V_1 and V_2 on the volume axis.

<table>
<tr><td>

21-5

THE GENERAL CASE FOR THE FIRST LAW

</td><td>

The first law of thermodynamics stipulates that energy must be conserved in any thermodynamic process. In the mathematical formulation

$$\Delta Q = \Delta W + \Delta U$$

There are three quantities which may undergo changes. The most general process is one in which all three quantities are involved. For example, the fluid in Fig. 21-5 is expanded while in contact with a hot flame. Considering the gas as a system, there is a net heat transfer ΔQ imparted to the gas. This energy is used in two ways: (1) the internal energy ΔU of the gas is increased by a portion of the input thermal energy, and (2) the gas does an amount of work ΔW on the piston which is equivalent to the remainder of the available energy.

Special cases of the first law arise when one or more of the three quantities— ΔQ, ΔW, or ΔU—do not undergo change. In these instances, the first law is considerably simplified. In the following sections we consider several of these special processes.

</td></tr>
<tr><td>

21-6

ADIABATIC PROCESSES

</td><td>

Suppose a system is completely isolated from its surroundings so that there can be no exchange of thermal energy Q. Any process which occurs completely inside such an isolated chamber is called an *adiabatic process*, and the system is said to be surrounded by *adiabatic* walls.

> An adiabatic process is one in which there is no exchange of thermal energy ΔQ between a system and its surroundings.

</td></tr>
</table>

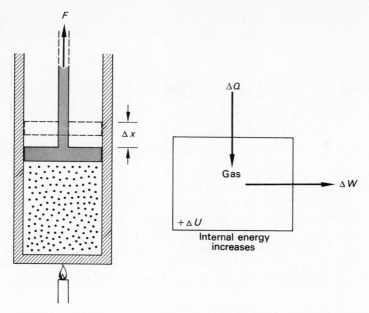

Fig. 21-5 Part of the energy ΔQ supplied to the gas by the flame results in external work ΔW. The remainder increases the internal energy ΔU of the gas.

Applying the first law to a process in which $\Delta Q = 0$, we obtain

$$\Delta W = -\Delta U \qquad \qquad \textit{Adiabatic} \quad (21\text{-}4)$$

Equation (21-4) tells us that in an adiabatic process the work is done at the *expense* of internal energy. The decrease in thermal energy is usually accompanied by a decrease in temperature.

As an example of an adiabatic process, consider Fig. 21-6, in which a piston is lifted by an expanding gas. If the walls of the cylinder are insulated and the expansion occurs rapidly, the process will be approximately adiabatic. As the gas expands, it does work on the piston but loses internal energy and experiences a drop in tempera-

Fig. 21-6 In an adiabatic process there is no transfer of heat, and work is done at the *expense* of internal energy.

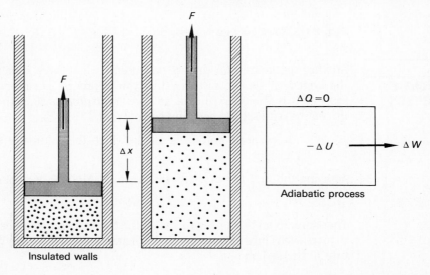

ture. If the process is reversed by forcing the piston back down, work is done *on the* gas ($-\Delta W$) and there will be an increase in internal energy (ΔU) such that

$$-\Delta W = +\Delta U$$

In this instance the temperature will rise.

Another example of an adiabatic process which finds very useful applications in industrial refrigeration is referred to as a *throttling process.*

> A **throttling process** is one in which a fluid at high pressure seeps adiabatically through a porous wall or narrow opening into a region of low pressure.

Consider a gas forced by a pump to circulate through the apparatus in Fig. 21-7. Gas from the high-pressure side of the pump is forced through the narrow constriction, called the *throttling valve,* to the low-pressure side. The valve is heavily insulated, so that the process is adiabatic and $\Delta Q = 0$. According to the first law, $\Delta W = -\Delta U$, and the net work done by the gas in passing through the valve is accomplished at the expense of internal energy. In refrigeration, a liquid coolant undergoes a drop in temperature and partial vaporization as a result of the throttling process.

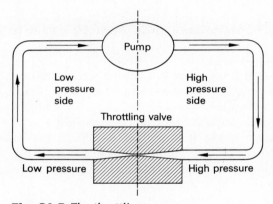

Fig. 21-7 The throttling process.

21-7

ISOCHORIC PROCESSES

Another special case for the first law occurs when there is no work done, either *by* the system or *on* the system. This type of process is referred to as an *isochoric process.* It is also referred to as an *isovolumic process* since there can be no change in volume without the performance of work.

> An **isochoric process** is one in which the volume of the system remains constant.

Applying the first law to a process in which $\Delta W = 0$, we obtain

$$\Delta Q = \Delta U \qquad\qquad\qquad Isochoric \quad (21\text{-}5)$$

Therefore, in an isochoric process all the thermal energy absorbed by a system goes to increase its internal energy. In this instance there is usually a rise in the temperature of the system.

An isochoric process occurs when water is heated in a container of fixed volume, as shown in Fig. 21-8. As heat is supplied, the increase in internal energy results in a rise in the temperature of the water until it begins to boil. Further increases in internal energy go into the process of vaporization. However, the volume of the system, consisting of the water and its vapor, remains constant, and no external work is done.

When the flame is removed, the process is reversed as heat leaves the system through the bottom of the cylinder. The water vapor will condense, and the temperature of the resulting water will eventually drop to room temperature. This process represents a loss of heat and a corresponding decrease in internal energy, but, once again, no work is done.

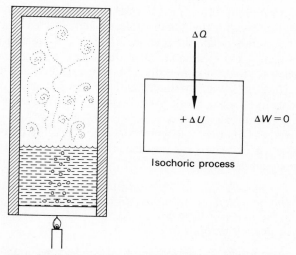

ΔQ

$+\Delta U$ $\Delta W = 0$

Isochoric process

Fig. 21-8 In an isochoric process, the volume of the system (water and vapor) remains constant.

<table>
<tr><td>21-8</td></tr>
</table>

ISOTHERMAL PROCESSES

It is possible for the pressure and volume of a gas to vary without a change in temperature. In Chap. 20 we introduced Boyle's law to describe volume and pressure changes during such a process. A gas can be compressed in a cylinder so slowly that it will essentially remain in thermal equilibrium with its surroundings. The pressure increases as the volume decreases, but the temperature is essentially constant.

> An **isothermal** process is one in which the temperature of the system remains constant.

If there is no change of phase, a constant temperature indicates that there is no change in the internal energy of the system. Applying the first law to a process in which $\Delta U = 0$, we obtain

$$\Delta Q = \Delta W \qquad\qquad \textit{Isothermal} \quad (21\text{-}6)$$

Thus, in an isothermal process all the energy absorbed by a system is converted into output work.

THERMODYNAMICS

When we rub our hands together vigorously, the work done against friction increases the internal energy and causes a rise in temperature. The surrounding air forms a large reservoir at a lower temperature, and the heat energy is transferred to the air without changing its temperature appreciably. When we stop rubbing, our hands return to the same state as before. According to the first law, mechanical energy has been transformed into heat with 100 percent efficiency.

$$\Delta W = \Delta Q$$

Such a transformation can be continued indefinitely as long as work is supplied.

Let us now consider the reverse process. Is it possible to convert heat energy into work with 100 percent efficiency? In the above example, is it possible to capture all the heat transferred to the air and return it to our hands, causing them to rub together indefinitely of their own accord? On a cold winter day, this process would be a boon to hunters with cold hands. Unfortunately, such a process cannot occur even though it does not violate the first law. Neither is it possible to retrieve all the heat lost in braking a car in order to start the wheels rolling again.

We shall see that the conversion of heat energy into mechanical work is a losing process. The first law of thermodynamics tells us that we cannot win in such an experiment. In other words, it is impossible to get more work out of a system than the heat put into the system. It does not, however, preclude us from breaking even. Clearly, we need another rule which states that the 100 percent conversion of heat energy into useful work is not possible. This rule forms the basis for the *second law of thermodynamics*.

> *The Second Law of Thermodynamics: It is impossible to construct an engine which, operating continuously, produces no effect other than the extraction of heat from a reservoir and the performance of an equivalent amount of work.*

To give more insight and application to this principle, suppose we study the operation and efficiency of heat engines. A particular system might be a gasoline engine, a jet engine, a steam engine, or even the human body. The operation of a heat engine is best described by a diagram similar to that shown in Fig. 21-9. During the operating of such a general engine, three processes occur:

1. A quantity of heat Q_{in} is supplied to the engine from a reservoir at a high temperature T_{in}.
2. Mechanical work W_{out} is done by the engine through the use of a portion of the heat input.
3. A quantity of heat Q_{out} is released to a reservoir at a low temperature T_{out}.

Since the system is periodically returned to its initial state, the net change in internal energy is zero. Thus, the first law tells us that

Work output = heat input − heat output

$$W_{out} = Q_{in} - Q_{out} \tag{21-7}$$

The efficiency of a heat engine is defined as the ratio of the useful work done by the engine to the heat put into the engine, and it is usually expressed as a percentage.

Fig. 21-9 A schematic diagram for a heat engine.

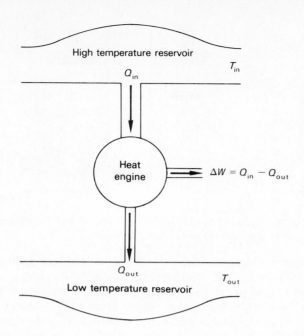

High temperature reservoir

T_{in}

Q_{in}

Heat engine

$\Delta W = Q_{in} - Q_{out}$

Q_{out}

T_{out}

Low temperature reservoir

$$\text{Efficiency} = \frac{\text{work output}}{\text{heat input}}$$

$$E = \frac{Q_{in} - Q_{out}}{Q_{in}} \qquad (21\text{-}8)$$

For example, an engine that is 25 percent efficient ($E = 0.25$) might absorb 1000 Btu, perform 250 Btu of work, and reject 750 Btu as wasted heat. A 100 percent efficient engine is one in which all the input heat is converted to useful work. In this case, no heat would be rejected to the environment ($Q_{out} = 0$). Although such a process would conserve energy, it violates the second law of thermodynamics. The most efficient engine is the one which rejects the *least* possible heat to the environment.

21-10 THE CARNOT CYCLE

All heat engines are subject to many practical difficulties. Friction and the loss of heat through conduction and radiation prevent actual engines from obtaining their maximum efficiency. An idealized engine free of such problems was suggested by Sadi Carnot in 1824. The *Carnot engine* has the maximum possible efficiency for an engine which absorbs heat from a high-temperature reservoir, performs external work, and deposits heat to a low-temperature reservoir. The effectiveness of a given engine can therefore be determined by comparing it with the Carnot engine operating between the same temperatures.

The Carnot cycle is illustrated in Fig. 21-10. A gas contained in a cylinder equipped with a movable piston is placed in contact with a reservoir at a high temperature T_{in}. A quantity of heat Q_{in} is absorbed by the gas which expands isothermally

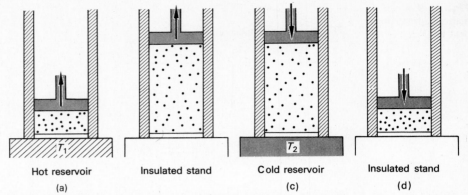

Hot reservoir   Insulated stand   Cold reservoir   Insulated stand

    (a)   (c)   (d)

Fig. 21-10 The Carnot cycle: *(a)* isothermal expansion, *(b)* adiabatic expansion, *(c)* isothermal compression, and *(d)* adiabatic compression.

as the pressure decreases. This first stage of a Carnot cycle is shown graphically as the curve *AB* in the *P-V* diagram (Fig. 21-11). The cylinder is next placed on an insulated stand, where it continues to expand adiabatically as the pressure drops to its lowest value. This stage is represented graphically by the curve *BC*. In the third stage the cylinder is removed from the insulated pad and placed on a reservoir at a low temperature T_{out}. A quantity of heat Q_{out} is exhausted from the gas as it is compressed isothermally from point *C* to point *D* in the *P-V* diagram. Finally, the cylinder is again placed on the insulated pad, where it is compressed adiabatically to its original state along the path *DA*. The engine does external work during the expansion processes and returns to its initial state during the compression processes.

21-11

THE EFFICIENCY OF AN IDEAL ENGINE

The efficiency of a real engine is difficult to predict from Eq. (21-8) because the quantities Q_{in} and Q_{out} are difficult to calculate. Frictional and heat losses through the cylinder walls and around the piston, incomplete burning of the fuel, and even the physical properties of different fuels all frustrate our attempts to measure the efficiency of such engines. However, we can imagine an *ideal engine*, one which is not restricted by these practical difficulties. The efficiency of such an engine depends

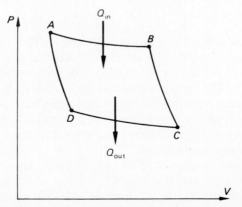

Fig. 21-11 A *P-V* diagram of the ideal Carnot cycle.

only on the quantities of heat absorbed and rejected between two well-defined heat reservoirs. It does not depend on the thermal properties of the working fuel. In other words, regardless of the internal changes in pressure, volume, length, or other factors, all ideal engines have the same efficiency when they are operating between the same two temperatures (T_{in} and T_{out}).

> An *ideal engine* is one which has the highest possible efficiency for the temperature limits within which it operates.

If we can define the efficiency of an engine in terms of input and output temperatures instead of in terms of the input and output of heat, we will have a more useful formula. For an ideal engine it can be shown that the ratio of Q_{in}/Q_{out} is the same as the ratio of T_{in}/T_{out}. The actual proof is beyond the scope of this text. The efficiency of an ideal engine can, therefore, be expressed as a function of the absolute temperatures of the input and output reservoirs. Equation (21-8), for an ideal engine, becomes

$$E = \frac{T_{in} - T_{out}}{T_{in}} \qquad (21\text{-}9)$$

It can be shown that no engine operating between the same two temperatures can be more efficient than would be indicated by Eq. (21-9). This ideal efficiency thus represents an upper limit to the efficiency of any practical engine. The greater the difference in temperature between two reservoirs, the greater the efficiency of any engine.

EXAMPLE 21-2 *(a)* What is the efficiency of an ideal engine operating between two heat reservoirs at 400 and 300 K? *(b)* How much work is done by the engine in one complete cycle if 800 cal of heat is absorbed from the high-temperature reservoir? *(c)* How much heat is delivered to the low-temperature reservoir?

Solution (a) The ideal efficiency is found from Eq. (21-9).

$$E = \frac{T_{in} - T_{out}}{T_{in}} = \frac{400 \text{ K} - 300 \text{ K}}{400 \text{ K}} = 0.25$$

Thus, the ideal efficiency is 25 percent.

Solution (b) The efficiency is the ratio of W_{out}/W_{in}, so that

$$\frac{W_{out}}{Q_{in}} = 0.25 \qquad \text{or} \qquad W_{out} = (0.25)Q_{in}$$

$$W_{out} = (0.25)(800 \text{ cal}) = 200 \text{ cal}$$

A 25 percent efficient engine delivers one-fourth of the input heat to useful work. The rest must be lost (Q_{out}).

Solution (c) The first law of thermodynamics requires that

$$W_{out} = Q_{in} - Q_{out}$$

Solving for Q_{out} we obtain

$$Q_{out} = Q_{in} - W_{out} = 800 \text{ cal} - 200 \text{ cal} = 600 \text{ cal}$$

THERMODYNAMICS

The work output is usually expressed in joules. Conversion to these units gives

$$W_{out} = (200 \text{ cal})(4.186 \text{ J/cal}) = 837 \text{ J}$$

21-12
INTERNAL COMBUSTION ENGINES

An internal combustion engine generates the input heat within the engine itself. The most common engine of this variety is the four-stroke gasoline engine, in which a mixture of gasoline and air is ignited by a spark plug in each cylinder. The thermal energy released is converted into useful work by the pressure exerted on a piston by the expanding gases. The four-stroke process is illustrated in Fig. 21-12. During the

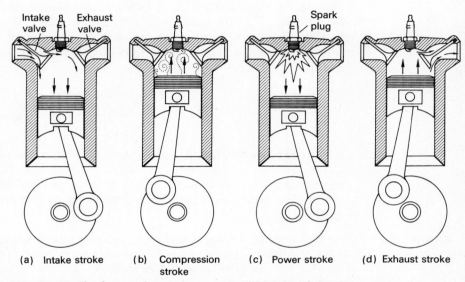

(a) Intake stroke (b) Compression stroke (c) Power stroke (d) Exhaust stroke

Fig. 21-12 The four-stroke gasoline engine: *(a)* intake stroke, *(b)* compression stroke, *(c)* power stroke, and *(d)* exhaust stroke.

intake stroke (Fig. 21-12*a*), a mixture of air and gasoline vapor enters the cylinder through the intake valve. Both valves are closed during the *compression stroke* (Fig. 21-12*b*) as the piston moves upward, causing a rise in pressure. Just before the piston reaches the top, the mixture is ignited, causing a sharp increase in temperature and pressure. In the power stroke (Fig. 21-12*c*) the expanding gases force the piston downward, performing external work. The fourth stroke (Fig. 21-12*d*) pushes the burned gases out of the cylinder through the exhaust valve. The entire cycle is then repeated for as long as the combustible fuel is supplied to the cylinder.

The ideal cycle used by the engineer to perfect the gasoline engine is shown in Fig. 21-13. It is named the *Otto cycle* after its inventor. The compression stroke is represented by the curve *ab*. The pressure increases adiabatically as the volume is reduced. At point *b* the mixture is ignited, supplying a quantity of heat Q_{in} to the system. This causes the sharp rise in pressure indicated by the line *bc*. In the power stroke (*cd*), the gases expand adiabatically, performing external work. The system then cools at constant volume to point *a*, releasing a quantity of heat Q_{out}. The

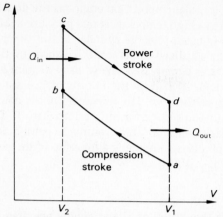

Fig. 21-13 The Otto cycle for a four-stroke gasoline engine.

burned gases are exhausted on the next upward stroke, and more fuel is drawn in on the next downward stroke. Then the cycle begins all over again. The volume ratio V_1/V_2, as indicated in the P-V diagram, is called the *compression* ratio and is about 8 for most automobile engines.

The efficiency of the ideal Otto cycle can be shown to be

$$E = 1 - \frac{1}{(V_1/V_2)^{\gamma-1}} \tag{21-10}$$

where γ is the adiabatic constant for the working substance. The adiabatic constant is defined by

$$\gamma = \frac{c_p}{c_v}$$

where c_p is the specific heat of the gas at constant pressure and c_v is the specific heat at constant volume. For monatomic gases $\gamma = 1.67$, and for diatomic gases $\gamma = 1.4$. In the gasoline engine the working substance is mostly air, for which $\gamma = 1.4$. In the ideal case, Eq. (21-10) shows that the higher compression ratios yield higher efficiencies since γ is always greater than 1.

EXAMPLE 21-3   Compute the efficiency of a gasoline engine for which the compression ratio is 8 and $\gamma = 1.4$.

Solution   From the given information, we note that

$$\frac{V_1}{V_2} = 8 \quad \text{and} \quad \gamma - 1 = 1.4 - 1 = 0.4$$

Thus, from Eq. (21-3),

$$E = 1 - \frac{1}{8^{0.4}} = 1 - \frac{1}{2.3} = 57\%$$

In the above example, 57 percent represents the maximum possible efficiency of a gasoline engine with the parameters given. Actually, the efficiency of such an engine is normally around 30 percent because of uncontrolled heat losses.

A second type of internal combustion engine is the diesel engine. In this engine the air is compressed to a high temperature and pressure near the top of the cylinder. Diesel fuel, which is injected into the cylinder at this point, ignites and forces the piston downward. The idealized diesel cycle is shown by the P-V diagram in Fig. 21-14. Starting at a, air is compressed adiabatically to point b where the diesel fuel is injected. The diesel fuel, ignited by the hot air, delivers a quantity of heat Q_{in} at nearly constant pressure (line bc). The remainder of the power stroke consists of an adiabatic expansion to point d, performing external work. During the exhaust and intake strokes, the gas cools at constant volume to point a, losing a quantity of heat Q_{out}. The efficiency of a diesel engine is a function of the compression ratio (V_1/V_2) and the *expansion ratio* (V_1/V_3).

<table>
<tr><td>

21-13

**REFRIGER-
ATION**

</td><td>

A refrigerator can be thought of as a heat engine operated in reverse. A schematic diagram of a refrigerator is shown in Fig. 21-15. During every cycle, a compressor or similar device supplies mechanical work W to the system, extracting a quantity of heat Q_{cold} from a cold reservoir and depositing a quantity of heat Q_{hot} to a hot reservoir. According to the first law, the input work is given by

</td></tr>
</table>

$$W = Q_{hot} - Q_{cold}$$

The effectiveness of any refrigerator is determined by the amount of heat Q_{cold} extracted for the least expenditure of mechanical work W. The ratio Q_{cold}/W is therefore a measure of the cooling efficiency of a refrigerator and is called its *coefficient of performance η*. Symbolically,

$$\eta = \frac{Q_{cold}}{W} = \frac{Q_{cold}}{Q_{hot} - Q_{cold}} \qquad (21\text{-}11)$$

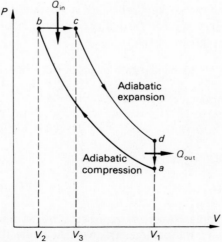

Fig. 21-14 The ideal diesel cycle.

Fig. 21-15 A schematic diagram for a refrigerator.

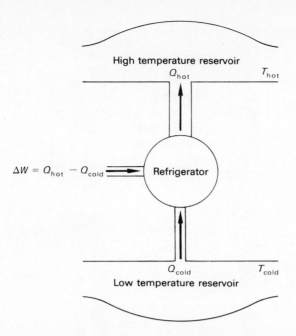

The maximum efficiency can be expressed in terms of absolute temperatures:

$$\eta = \frac{T_{\text{cold}}}{T_{\text{hot}} - T_{\text{cold}}} \tag{21-12}$$

To understand the refrigeration process better let us consider the general schematic in Fig. 21-16. This diagram may refer to a number of refrigeration devices, from a commercial plant to a household refrigerator. The working substance, called the *refrigerant*, is a gas which is easily liquefied by an increase in pressure or a drop in temperature. In the liquid phase it can be vaporized readily by passing it through a throttling process (see Sec. 21-6) near room temperature. Common refrigerants are ammonia, Freon 12, methyl chloride, and sulfur dioxide. Ammonia, the most common industrial refrigerant, boils at −28°F under a pressure of 1 atm. Freon 12, the most common household refrigerant, boils at −22°F at atmospheric pressure. Variation in pressure radically affects the condensation and evaporation temperatures of all refrigerants.

As shown in the schematic, a typical refrigeration system consists of a *compressor*, a *condenser*, a *liquid storage tank*, a *throttling valve*, and an *evaporator*. The compressor provides the necessary input work to move the refrigerant through the system. As the piston moves to the right, it sucks in the refrigerant through the intake valve at a little above atmospheric pressure and near room temperature. During the power stroke, the intake valve closes and the discharge valve opens. The emergent refrigerant, at high temperature and pressure, passes into the condenser where it is cooled until it liquefies. The condenser may be cooled by running water or by an electric fan. It is during this phase that a quantity of heat Q_{hot} is rejected from the system. The condensed liquid refrigerant, still under a condition of high pressure and temperature, is collected in a liquid reservoir. Then the liquid refrigerant is

THERMODYNAMICS **393**

Fig. 21-16 The basic components of a refrigeration system.

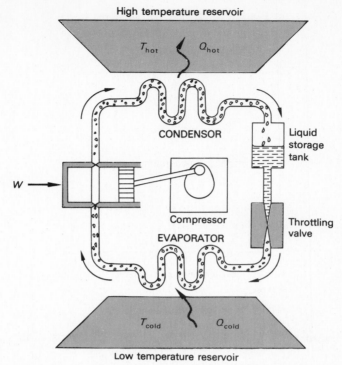

drawn from the storage tank through a throttling valve, causing a sudden drop in temperature and pressure. As the cold liquid refrigerant flows through the evaporator coils, it absorbs a quantity of heat Q_{cold} from the space and products being cooled. This heat boils the liquid refrigerant and is carried away by the gaseous refrigerant as latent heat of vaporization. This phase is the "payoff" for the entire operation, and all components just contribute to the effective transfer of heat to the evaporator. Finally, the refrigerant vapor leaves the evaporator and is sucked into the compressor to begin another cycle.

SUMMARY

Thermodynamics is the science which treats the conversion of heat into work or the reverse process, the conversion of work into heat. We have seen that not only must energy be conserved in such processes, but also there are limits on the efficiency. The major concepts presented in this chapter are summarized below:

- The *first law of thermodynamics* is a restatement of the conservation-of-energy principle. It says that the net heat ΔQ put into a system is equal to the net work ΔW done by the system plus the net change in internal energy ΔU of the system. Symbolically.

$$\Delta Q = \Delta W + \Delta U$$

First Law of Thermodynamics

- In thermodynamics, work ΔW is often done on a gas. In such cases the work is often represented in terms of pressure and volume. A *P-V* diagram is also useful for measuring ΔW. If the pressure is constant,

$$\boxed{\Delta W = P\,\Delta V} \qquad \Delta W = \text{area under } P\text{-}V \text{ curve}$$

- Special cases of the first law occur when one of the quantities doesn't undergo a change.

    (a) *Adiabatic process:* $\Delta Q = 0$ $\Delta W = -\Delta U$

    (b) *Isochoric process:* $\Delta V = 0$ $\Delta W = 0$ $\Delta Q = \Delta U$

    (c) *Isothermal process:* $\Delta T = 0$ $\Delta U = 0$ $\Delta Q = \Delta W$

- The *second law of thermodynamics* places restrictions on the possibility of satisfying the first. In short, it points out that in every process there is some loss of energy due to frictional forces or other dissipative forces. A 100 percent efficient engine, one which converts all input heat to useful output work, is not possible.

- A heat engine is represented generally by Fig. 21-9. The meaning of the symbols used in the equations below can be taken from that figure. The work done by the engine is the difference between input heat and output heat.

$$\boxed{W = Q_{in} - Q_{out}} \qquad\qquad\qquad \textit{Work} \text{ (kcal } or \text{ J)}$$

- The *efficiency E* of an engine is the ratio of the work output to the heat input. It can be calculated for an ideal engine from either of the following relations:

$$\boxed{E = \frac{Q_{in} - Q_{out}}{Q_{in}}} \qquad \boxed{E = \frac{T_{in} - T_{out}}{T_{in}}} \qquad \textit{Efficiency}$$

- A *refrigerator* is a heat engine operated in reverse. A measure of the performance of such a device is the amount of cooling you get for the work you must put into the system. Cooling occurs as a result of the extraction of heat Q_{cold} from the cold reservoir. The coefficient of performance η is given by either

$$\boxed{\eta = \frac{Q_{cold}}{Q_{hot} - Q_{cold}}} \qquad \text{or} \qquad \boxed{\eta = \frac{T_{cold}}{T_{hot} - T_{cold}}}$$

QUESTIONS

21-1. Define the following terms:

    a. Thermodynamics
    b. *P-V* diagram
    c. Adiabatic process
    d. Isochoric process
    e. Heat engine
    f. Refrigerator
    g. Refrigerant
    h. Compressor
    i. Condenser
    j. Evaporator

    k. Internal-energy function
    l. First law of thermodynamics
    m. Throttling process
    n. Isothermal process
    o. Second law of thermodynamics
    p. Thermal efficiency
    q. Carnot cycle
    r. Carnot efficiency
    s. Coefficient of performance

21-2. The latent heat of vaporization of water is 540 cal/g. However, when 1 g of water is completely vaporized at constant pressure, the internal energy of the system

---

increases by only 500 cal. What happened to the remaining 40 cal? Is this an isochoric process? Is it an isothermal process?

21-3. If both heat and work can be expressed in the same units, why is it necessary to distinguish between them?

21-4. Is it necessary to use the concept of molecular energy in order to describe and use the internal-energy function? Explain.

21-5. A gas undergoes an adiabatic expansion. Does it perform external work? If so, what is the source of energy?

21-6. What happens to the internal energy of a gas undergoing (a) adiabatic compression, (b) isothermal expansion, and (c) a throttling process?

21-7. A gas performs external work during an isothermal expansion. What is the source of energy?

21-8. In the text, only one statement was given for the second law of thermodynamics. Discuss each of the following statements, showing them to be equivalent to that given in the text:
(a) It is impossible to construct a refrigerator which, working continuously, will extract heat from a cold body and exhaust it to a hot body without the performance of work on the system.
(b) The natural direction of heat flow is from a body at high temperature to a body at a low temperature, regardless of the size of each reservoir.
(c) All natural spontaneous processes are irreversible.
(d) Natural events always proceed in the direction from order to disorder.

21-9. It is energetically possible to extract the heat energy contained in the ocean and use it to power a steamship across the sea. What objections can you offer?

21-10. In an electric refrigerator, heat is transferred from the cool interior to warmer surroundings. Why isn't this in violation of the second law of thermodynamics?

21-11. Consider the performance of external work by the isothermal expansion of an ideal gas. Why isn't this process of converting heat into work in violation of the second law of thermodynamics?

21-12. If natural processes tend to decrease order in the universe, how can you explain the evolution of biological systems to a highly organized state? Does this violate the second law of thermodynamics?

21-13. Will keeping the door of an electric refrigerator open warm or cool a room? Explain.

21-14. What temperature must the cold reservoir have if a Carnot engine is to be 100 percent efficient? Can this ever happen? If it is impossible for a Carnot engine to have a 100 percent efficiency, why is it called the *ideal* engine?

21-15. What determines the efficiency of heat engines? Why is it generally so low?

PROBLEMS

The First Law of Thermodynamics

21-1. In an industrial chemical process, 600 cal of heat is supplied to a system while 200 J of work is done *by* the system. What is the increase in the internal energy of the system?

Ans. 2.31 kJ

21-2. A piston does 3000 ft · lb of work on a gas, which then expands, performing 2500 ft · lb of work on its surroundings. What is the change in internal energy of the gas in Btu?

21-3. A system absorbs 200 J of heat and at the same time performs 50 J of work. What is the change in internal energy?

Ans. 150 J

* **21-4.** A 10-kg block slides down a plane through a vertical distance of 10 m. If the block is moving with a velocity of 10 m/s when it reaches the bottom, how many calories of heat were lost because of friction?

* **21-5.** At a constant pressure of 101.3 kPa, 1 g of water (1 cm³) is completely vaporized and has a final volume of 1671 cm³ in its vapor form. (a) Compute the external work done by the system in expanding against its surroundings. (b) What is the increase in internal energy of the system?

Ans. (a) 169 J, (b) 2090 J

Thermodynamic Processes

21-6. During the isothermal expansion of an ideal gas, 3 Btu of heat energy is absorbed. The piston weighs 2000 lb. How high will it rise above its initial position?

21-7. The work done on a gas during an adiabatic compression is 140 J. Calculate the increase in internal energy in calories.

Ans. 33.4 cal

21-8. A gas contracts from a volume of 300 cm³ to a volume of 100 cm³ during an isobaric process. If the pressure is 2×10^5 Pa, what is the net work done *by* the gas?

* **21-9.** Consider the *P-V* diagram shown in Fig. 21-17 where the pressure and volume are indicated for each of the points *A*, *B*, *C*, and *D*. Starting at point *A*, a 2-L sample of gas absorbs 800 J of heat, causing the pressure to increase from 1×10^5 to 2×10^5 Pa. Next the gas expands from *B* to *C*, absorbing an additional 200 J of heat while its volume increases to 5 L. (a) Find the net work done and the change in internal energy for each of the processes *AB* and *BC*. (b) What are the net work and total change in internal energy for the process *ABC*? (c) What kind of processes are illustrated by *AB* and *BC*?

Ans. (a) 0, 800 J; 600 J, −400 J; (b) 600 J, +400 J; (c) isochoric, isobaric

Fig. 21-17

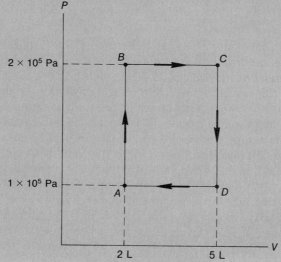

21-10. The cycle begun in Prob. 21-9 now continues from C to D and back to A. (a) What is the net work for the entire cycle? (b) How much heat was lost during the entire cycle?

21-11. A gas expands against a movable piston, lifting it through 2 in. at constant speed. (a) How much work is done by the gas if the piston weighs 200 lb and has a cross-sectional area of 12 in.^{2}? (b) If the expansion is adiabatic, what is the change in internal energy in Btu? (c) Does ΔU represent an increase or decrease in internal energy?

Ans. (a) 33.3 ft · lb, (b) 0.043 Btu, (c) decrease

The Second Law of Thermodynamics

21-12. During a complete thermodynamic cycle, a system absorbs 600 cal of heat and rejects 200 cal to the environment. (a) How many joules of work is done? (b) What is the efficiency of the cycle?

21-13. A steam engine takes steam from a boiler at 200°C and exhausts it directly into the air at 100°C. What is its ideal efficiency?

Ans. 21.1%

21-14. In a Carnot cycle, the isothermal expansion of a gas takes place at 400 K and 500 cal of heat is absorbed by the gas. (a) How much heat is rejected by the system during the isothermal compression if the process occurs at 300 K? (b) How many calories are rejected to the low-temperature reservoir? (c) How much net work is done?

21-15. A Carnot engine absorbs heat from a reservoir at 500 K and rejects heat to a reservoir at 300 K. In each cycle the engine receives 1200 cal of heat from the high-temperature reservoir. (a) What is the Carnot efficiency? (b) How many calories are rejected to the low-temperature reservoir? (c) How much external work is done in joules?

Ans. (a) 40%, (b) 720 cal, (c) 2010 J

21-16. If the engine in Prob. 21-11 is operated in reverse (as a refrigerator) and extracts 1200 cal of heat from the low-temperature reservoir, how many calories will be delivered to the high-temperature reservoir? How much mechanical work is required?

21-17. A Carnot refrigerator has a coefficient of performance of 2.33. If 600 J of input work is done in each cycle, how many joules of heat are extracted from the cold reservoir? How many joules are exhausted to the high-temperature reservoir?

Ans. 1400 J, 2000 J

Additional Problems

21-18. In a thermodynamic process, 200 Btu are supplied to produce an isobaric expansion under a pressure of 100 lb/in.^{2}. The internal energy of the system does not change. What is the increase in volume of the gas?

21-19. A 100-cm^3 sample of gas at a pressure of 1×10^5 Pa is heated isochorically from point A to point B until its pressure reaches 3×10^5 Pa. It then is caused to expand isobarically to point C, where its volume is 400 cm^3. The pressure then returns to 1×10^5 Pa at point D with no change in volume before returning to its original state at point A. Draw the P-V diagram for this cycle. What net work is done for the entire cycle?

Ans. 60 J

21-20. Find the net work done by a gas as it is carried around the cycle shown in Fig. 21-18.

Fig. 21-18

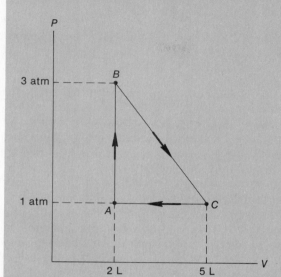

21-21. What is the net work done for the process *ABCD* as described by Fig. 21-19?

Ans. 85 J

Fig. 21-19

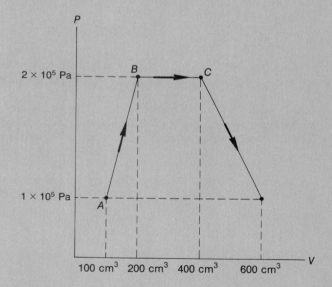

21-22. A real engine operates between a heat source at 327°C and a heat sink at 0°C, and it has an output power of 8 kW. **(a)** What is the ideal efficiency for this engine? **(b)** How much power is wasted if the actual efficiency is only 25 percent?

21-23. The Otto efficiency for a gasoline engine is 50 percent, and the adiabatic constant is 1.4. Compute the compression ratio.

Ans. 5.66

21-24. A heat pump takes heat from a water reservoir at 41°F and delivers it to a system of pipes in a house at 78°F. The energy required to operate the heat pump is about twice that required to operate a Carnot pump. How much mechanical work must be supplied by the pump in order to deliver 1×10^6 Btu of heat energy to the house?

21-25. A Carnot engine has an efficiency of 48 percent. If the working substance enters the system at 400°C, what is the exhaust temperature?

Ans. 77°C

21-26. During the compression stroke of an automobile engine, the volume of the combustible mixture decreases from 18 to 2 in.^{3}. If the adiabatic constant is 1.4, what is the maximum possible efficiency for the engine?

21-27. How many joules of work must be done by the compressor in a refrigerator in order to change 1 kg of water at 20°C to ice at -10°C? The coefficient of performance is 3.5.

Ans. 1.26×10^5 J

21-28. In a mechanical refrigerator the low-temperature coils of the evaporator are at -30°C, and the condenser has a temperature of 60°C. What is the maximum possible coefficient of performance?

21-29. Consider a specific mass of gas which is forced through an adiabatic throttling process. Before entering the valve, it has internal energy U_1, pressure P_1, and volume V_1. After passing through the valve, it has internal energy U_1, pressure P_2, and volume V_2. The net work is the work done *by* the gas minus the work done *on* the gas. Show that

$$U_1 + P_1V_1 = U_2 + P_2V_2$$

The quantity $U + PV$, called the *enthalpy*, is conserved during a throttling process.

21-30. An ideal compression refrigeration system has a coefficient of performance of 4.0. If the evaporating temperature is -12°C, what is the condensing temperature?

21-31. An engine has a thermal efficiency of 27 percent and an exhaust temperature of 230°C. What is the lowest possible input temperature?

Ans. 416°C

22 Wave Motion

OBJECTIVES

After completing this chapter, you should be able to:

1. Demonstrate by definition and example your understanding of *transverse* and *longitudinal* wave motion.
2. Define, relate, and apply the meaning of the terms *frequency*, *wavelength*, and *speed* for wave motion.
3. Solve problems involving the *mass, length, tension,* and *wave velocity* for transverse waves in a string.
4. Write and apply an expression for determining the *characteristic frequencies* for a vibrating string with fixed end points.

Energy can be transferred from one place to another by several means. In driving a nail, the bulk kinetic energy of a hammer is converted into useful work on the nail. Wind, projectiles, and most simple machines also perform work at the expense of material motion. Even the conduction of heat and electricity involves the motion of elementary particles called electrons. In this chapter we study the transfer of energy from one point to another without the physical transfer of material between the points.

22-1 MECHANICAL WAVES

When a stone is dropped into a pool of water, it creates a disturbance which spreads out in concentric circles, eventually reaching all parts of the pool. A small stick, floating on the surface of the water, bobs up and down as the disturbance passes. Energy has been transferred from the point of impact of the stone into the water to the floating stick some distance away. This energy is passed along by the agitation of neighboring water particles. Only the disturbance moves through the water. The actual motion of any particular water particle is comparatively small. Energy propagation by means of a disturbance in a medium instead of the motion of the medium itself is called *wave motion*.

The example above is referred to as a *mechanical* wave because its existence depends upon a mechanical source and a material medium.

A **mechanical wave** is a physical disturbance in an elastic medium.

It is important to recognize that all disturbances are not necessarily mechanical. For example, light waves, radio waves, and heat radiation propagate their energy by means of electric and magnetic disturbances. No physical medium is necessary for the transmission of electromagnetic waves. However, many of the basic ideas presented in this chapter for mechanical waves are also applicable to electromagnetic waves.

22-2
TYPES OF WAVES

Waves are classified according to the motion of a local part of a medium with respect to the direction of wave propagation.

In a **transverse wave** the vibration of the individual particles of the medium is perpendicular to the direction of wave propagation.

For example, suppose we fasten one end of a rope to a post and grasp the other end with the hand, as shown in Fig. 22-1. By moving the free end up and down quickly, we send a single disturbance called a *pulse* down the length of rope. Three equally spaced knots at points *a, b,* and *c* demonstrate that the individual particles move up and down while the disturbance moves to the right with velocity v.

Another type of wave, which may occur in a coiled spring, is illustrated in Fig. 22-2. The coils near the left end are pinched closely together to form a *condensation*. When the distorting force is removed, a condensation pulse is propagated throughout the length of the spring. No part of the spring moves very far from its equilibrium position, but the pulse continues to travel along the spring. Such a wave is called a *longitudinal wave* because the spring particles are displaced along the same direction in which the disturbance is traveling.

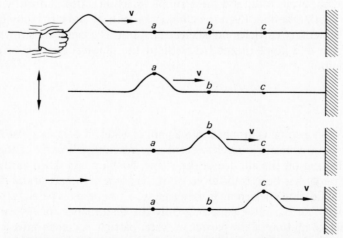

Fig. 22-1 In a transverse wave the individual particles move perpendicular to the direction of wave propagation.

HEAT, LIGHT, AND SOUND

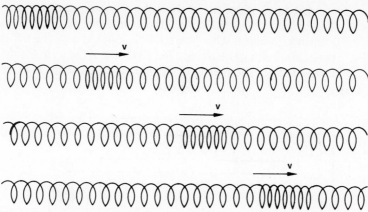

Fig. 22-2 In a longitudinal wave the motion of the individual particles is parallel to the direction of wave propagation. The illustration demonstrates the motion of a condensation pulse.

In a **longitudinal wave** the vibration of the individual particles is parallel to the direction of wave propagation.

If the coils of the spring in our example were forced apart at the left, a *rarefaction* would be formed, as shown in Fig. 22-3. Upon removal of the disturbing force, a longitudinal rarefaction pulse would be propagated along the spring. In general, a longitudinal wave consists of a series of condensations and rarefactions moving in a determined direction.

22-3
CALCULATING WAVE SPEED

The speed with which a pulse moves through a medium depends upon the elasticity of the medium and the inertia of its particles. The more elastic materials yield greater restoring forces when distorted. The less dense materials offer less resistance to motion. In either case, the ability of particles to pass on a disturbance to neighboring particles is improved, and a pulse will travel at a greater speed.

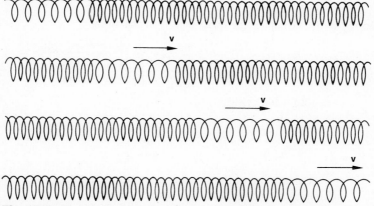

Fig. 22-3 Longitudinal motion of a rarefaction pulse in a coiled spring.

Let us consider the motion of a transverse pulse down the string in Fig. 22-4. The string of mass m and length l is maintained under a constant tension F by the suspended weight. When the string is plucked near the left end, a transverse pulse is propagated along the string. The elasticity of the string is measured by the tension F in the string. The inertia of the individual particles is determined by the *mass per*

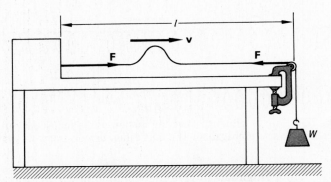

Fig. 22-4 Computing the speed of a transverse pulse in a string.

unit length μ of the string. It can be shown that the speed of a transverse pulse in a string is given by

$$v = \sqrt{\frac{F}{\mu}} = \sqrt{\frac{F}{m/l}} \qquad (22\text{-}1)$$

The mass per unit length μ is usually referred to as the *linear density* of the string. If F is expressed in newtons and μ in kilograms per meter, the speed is in meters per second.

EXAMPLE 22-1   The length l of the string in Fig. 22-4 is 2 m, and the string has a mass of 0.3 g. Calculate the speed of a transverse pulse in the string if it is under a tension of 20 N.

Solution   We first compute the linear density of the string.

$$\mu = \frac{m}{l} = \frac{0.3 \times 10^{-3} \text{ kg}}{2 \text{ m}}$$
$$= 1.5 - 10^{-4} \text{ kg/m}$$

Then Eq. (22-1) yields

$$v = \sqrt{\frac{F}{\mu}} = \sqrt{\frac{20 \text{ N}}{1.5 \times 10^{-4} \text{ kg/m}}}$$
$$= 365 \text{ m/s}$$

Calculation of the speed of a longitudinal pulse will be reserved for the following chapter, in connection with sound waves.

So far we have been considering single, nonrepeated disturbances called pulses. What happens if similar disturbances are repeated periodically?

Suppose we attach the left end of a string to the end point of an electromagnetic vibrator, as shown in Fig. 22-5. The end of the metal vibrator is driven with harmonic motion by an oscillating magnetic field. Since the string is attached to the end of the vibrator, a series of periodic transverse pulses is sent down the string. The resulting waves consist of many crests and troughs, which move down the string at a constant rate of speed. The distance between any two adjacent crests or troughs in such a wave train is called the *wavelength,* denoted by λ.

(a)   (b)

Fig. 22-5 *(a)* Production and propagation of a periodic transverse wave. *(b)* The wavelength λ is the distance between any two particles in phase, such as those at adjacent crests or at points A and B.

As a wave travels along the string, each particle of the string vibrates about its equilibrium position with the same frequency and amplitude as the vibrating source. However, the particles of the string are not in corresponding positions at the same times. Two particles are said to be *in phase* if they have the same displacement and if they are moving in the same direction. The particles A and B of Fig. 22-5b are in phase. Since particles at the crests of a given wave train are also in phase, we can provide a more general definition for the wavelength.

> The wavelength λ of a periodic wave train is the distance between any two adjacent particles which are in phase.

Each time the end point P of the vibrator makes a complete oscillation, the wave will move through a distance of one wavelength. The time required to cover this distance is therefore equal to the period T of the vibrating source. Hence the wave speed v can be related to the wavelength λ and period T by the equation

$$v = \frac{\lambda}{T} \tag{22-2}$$

The frequency f of a wave is the number of waves that pass a particular point in a unit of time. It is the same as the frequency of the vibrating source and is therefore equal to the reciprocal of the period ($f = 1/T$). The units of frequency may be expressed in waves per second, oscillations per second, or cycles per second. The SI unit for frequency is the *hertz* (Hz), which is defined as a cycle per second.

$$1 \text{ Hz} = 1 \text{ cycle/s} = \frac{1}{s}$$

Thus, if 40 waves pass a point every second, the frequency is 40 Hz.

The speed of a wave is more often expressed in terms of its frequency rather than its period. Thus Eq. (22-2) can be written

$$v = f\lambda \qquad (22\text{-}3)$$

Equation (22-3) represents an important physical relationship between the speed, frequency, and wavelength of *any* periodic wave. An illustration of each quantity is given in Fig. 22-6 for a periodic transverse wave.

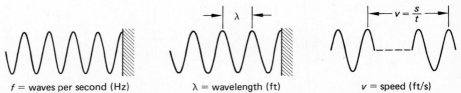

f = waves per second (Hz) λ = wavelength (ft) v = speed (ft/s)

Fig. 22-6 The relationship between the frequency, the wavelength, and the speed of a transverse wave.

EXAMPLE 22-2   A man sits near the end of a fishing dock and counts the water waves as they strike a supporting post. In 1 min he counts 80 waves. If a particular crest travels 20 m in 8 s, what is the wavelength of the waves?

Solution   The frequency and velocity of the waves are calculated as follows:

$$f = \frac{80 \text{ waves}}{60 \text{ s}} = 1.33 \text{ Hz}$$

$$v = \frac{20 \text{ m}}{8 \text{ s}} = 2.5 \text{ m/s}$$

From Eq. (22-3), the wavelength is

$$\lambda = \frac{v}{f} = \frac{2.5 \text{ m/s}}{1.33 \text{ Hz}} = 1.88 \text{ m}$$

A longitudinal periodic wave can be generated by the apparatus shown in Fig. 22-7. The left end of a coiled spring is connected to a metal ball supported at the end of a clamped hacksaw blade. When the metal ball is displaced to the left and released, it vibrates with harmonic motion. The resulting condensations and rarefactions are passed along the spring, producing a periodic longitudinal wave. Each particle of the

Fig. 22-7 Production and propagation of a periodic longitudinal wave.

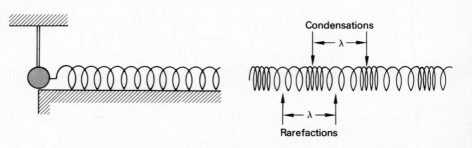

HEAT, LIGHT, AND SOUND

coiled spring oscillates back and forth horizontally with the same frequency and amplitude as the metal ball. The distance between any two adjacent particles which are in phase is the wavelength. As indicated in Fig. 22-7, the distance between adjacent condensations or adjacent rarefactions is a convenient measure of the wavelength. Equation (22-3) also applies for a periodic longitudinal wave.

22-5
ENERGY OF A PERIODIC WAVE

We have seen that each particle in a periodic wave oscillates with simple harmonic motion determined by the source of the wave. The energy content of a wave can be analyzed by considering the harmonic motion of the individual particles. For example, consider a periodic transverse wave in a string at the instant shown in Fig. 22-8. Particle *a* has reached its maximum amplitude; its velocity is zero, and it is experi-

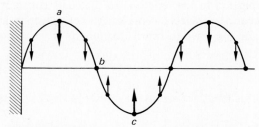

Fig. 22-8 The restoring forces which act on particles of a vibrating string.

encing its maximum restoring force. Particle *b* is passing through its equilibrium position, where the restoring force is zero. At this instant, particle *b* has its greatest speed and hence its maximum energy. Particle *c* is at its maximum displacement in the negative direction. As the periodic wave passes along the string, each particle oscillates back and forth through its own equilibrium position.

In Chap. 14 on harmonic motion, we found that the maximum velocity of a particle oscillating with frequency *f* and amplitude *A* is given by

$$v_{max} = 2\pi f A$$

When a particle has this speed, it is passing through its equilibrium position, where its potential energy is zero and its kinetic energy is a maximum. Thus the total energy of the particle is

$$E = E_p + E_k = (E_k)_{max}$$
$$= \tfrac{1}{2}mv_{max}^2 = \tfrac{1}{2}m(2\pi f A)^2$$
$$= 2\pi^2 f^2 A^2 m \tag{22-4}$$

As a periodic wave passes through a medium, each element of the medium is continuously doing work on adjacent elements. Therefore, the energy transmitted along the length of a vibrating string is not confined to one position. Let us apply the result obtained for a single particle to the entire length of a vibrating string. The energy content of the entire string is the sum of the individual energies of its constituent particles. If we let *m* refer to the entire mass of the string instead of the mass of an individual particle, Eq. (22-4) represents the total wave energy in the string. In a string of length l, the wave energy per unit length is given by

$$\frac{E}{l} = 2\pi^2 f^2 A^2 \frac{m}{l}$$

Substituting μ for the mass per unit length, we can write

$$\frac{E}{l} = 2\pi^2 f^2 A^2 \mu \qquad (22\text{-}5)$$

The wave energy is proportional to the square of the frequency f, to the square of the amplitude A, and to the linear density μ of the string. It must be recognized that the linear density is not a function of the length of string. This is true because the mass increases in proportion to the length l, so that μ is constant for any length.

Suppose that a wave travels down a length l of a given string with a speed v. The time t required for the wave to travel this length is

$$t = \frac{l}{v}$$

If the energy in this length of string is represented by E, the power P of the wave is given by

$$P = \frac{E}{t} = \frac{E}{l/v} = \frac{E}{l} v \qquad (22\text{-}6)$$

This represents the *rate* at which energy is propagated down the string. Substitution from Eq. (22-5) yields

$$P = 2\pi^2 f^2 A^2 \mu v \qquad (22\text{-}7)$$

The wave power is directly proportional to the energy per unit length and to the wave speed.

The dependence of wave energy and wave power on f^2 and A^2, as found in Eqs. (22-5) and (22-7), is a general conclusion for all kinds of waves. The same ideas will be applied in the following chapter when the energy of a sound wave is discussed.

22-6
THE SUPER-POSITION PRINCIPLE

Until now we have been considering the motion of a single train of pulses passing through a medium. We now consider what happens when two or more wave trains pass simultaneously through the same medium. Let us consider transverse waves in a vibrating string. The speed of a transverse wave is determined by the tension of the string and its linear density. Since these parameters are a function of the medium and not the source, any transverse wave will have the same speed for a given string under constant tension. However, the frequency and amplitude may vary considerably.

> When two or more wave trains exist simultaneously in the same medium, each wave travels through the medium as though the other were not present.

The resultant wave is a superposition of the component waves. In other words, the resultant displacement of a particular particle on the vibrating string is the algebraic sum of the displacements each wave would produce independently of the other. This is the *superposition principle:*

> *When two or more waves exist simultaneously in the same medium, the resultant displacement at any point and time is the algebraic sum of the displacements of each wave.*

It should be noted that the superposition principle, as stated here, applies only for *linear* media, i.e., those for which the response is directly proportional to the cause. Also, the sum of displacements is *algebraic* only if the waves have the same plane of polarization. For our purposes, a vibrating string will be assumed to satisfy both these conditions.

The application of this principle is shown graphically in Fig. 22-9. Two waves, indicated by the solid and dashed lines, superpose to form the resultant wave indicated by the heavy line. In Fig. 22-9*a* the superposition results in a wave of larger amplitude. These waves are said to interfere *constructively. Destructive interference* occurs when the resulting amplitude is smaller, as in Fig. 22-9*b*.

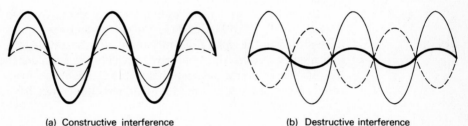

(a) Constructive interference   (b) Destructive interference

Fig. 22-9 The superposition principle.

22-7
STANDING WAVES

Let us consider the reflection of a transverse pulse, as shown for the string in Fig. 22-10. When the end of the string is rigidly fixed to a support, the arriving pulse strikes the support, exerting an upward force on it. The reaction force exerted by the support kicks downward on the string, setting up a reflected pulse. Both the displacement and velocity are reversed in the reflected pulse. In other words, a pulse which arrives as a crest is reflected as a trough with the same speed traveling in the opposite direction, and vice versa.

Suppose we consider the wave set up by a vibrating string whose end points are fixed, as in Fig. 22-11. We can use the superposition principle to analyze the resultant waveform at any instant. In Fig. 22-11*a*, we consider the incident and reflected waves at a particular time $t = 0$. The incident wave, traveling to the right, is indicated by a light solid line. The reflected wave, traveling to the left, is indicated by a dotted line. The two waves have the same speed and wavelength, but they are oppositely directed. At this instant all particles of the string lie in a straight horizontal line, as shown by the heavy line. Note that the heavy line is a superposition of the incident and reflected waves at a moment when their displacements add to zero. A snapshot of the string an instant later will show that, with a few exceptions, the particles have all

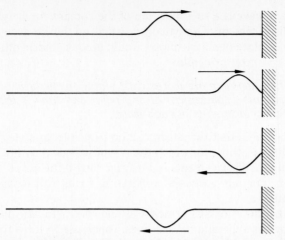

Fig. 22-10 The reflection of a transverse pulse at a fixed boundary.

changed positions. This is because the component waves have moved a finite distance.

Let us now consider the resultant wave at time t equal to one-fourth of a period later ($t = \frac{1}{4}T$), as in Fig. 22-11*b*. The component wave indicated by a solid line will have moved to the *right* a distance of one-fourth wavelength. The component wave indicated by a dashed line will have moved to the *left* a distance of one-fourth wavelength. The resultant wave and hence the shape of the string at this time are indicated on the heavy solid line. Constructive interference has resulted in a wave with

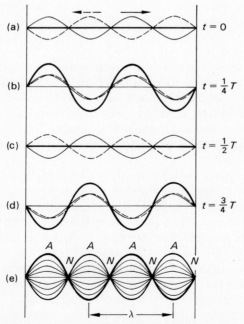

Fig. 22-11 Production of a standing wave.

twice the amplitude of either of the component waves. When the time t is one-half a period ($t = \frac{1}{2}T$), total destructive interference occurs, and once again the shape of the string is a straight line, as in Fig. 22-11c. At $t = \frac{3}{4}T$ the shape of the string reaches its maximum amplitude in the opposite direction. This constructive interference is shown by the heavy line in Fig. 22-11d.

A series of snapshots of the vibrating string at closely spaced time intervals would reveal a number of loops, as shown in Fig. 22-11e. Such a wave is called a *standing wave*. Notice that there are certain points along the string which remain at rest. These positions, called *nodes*, are labeled N in the figure. A flea resting on the vibrating string at these points would not be moved up and down by the wave motion.

Between the nodal points, the particles of the string move up and down with simple harmonic motion. The points of maximum amplitude occur midway between the nodes and are called *antinodes*. A flea resting on the string at any of these points, labeled A, would experience maximum speeds and displacements in the upward and downward oscillation of the string.

> *The distance between alternate nodes or alternate antinodes in a standing wave is a measure of the wavelength of the component waves.*

Longitudinal standing waves may also occur by the continuous reflection of condensation and rarefaction pulses. In this case the nodes exist where the particles of the medium are stationary, and the antinodes occur where the particles of the medium oscillate with maximum amplitude in the direction of propagation. Longitudinal standing waves will be discussed in the following chapter in connection with sound waves.

22-8 CHARACTERISTIC FREQUENCIES

Let us now consider the possible standing waves which can be set up in a string of length l whose ends are fixed, as in Fig. 22-12. When the string is set into vibration, the incident and reflected wave trains travel in opposite directions with the same wavelength. The fixed end points represent *boundary conditions* which restrict the

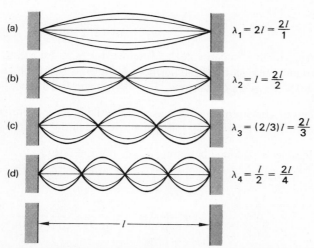

(a) $\lambda_1 = 2l = \dfrac{2l}{1}$

(b) $\lambda_2 = l = \dfrac{2l}{2}$

(c) $\lambda_3 = (2/3)l = \dfrac{2l}{3}$

(d) $\lambda_4 = \dfrac{l}{2} = \dfrac{2l}{4}$

Fig. 22-12 Possible standing-wave patterns in a vibrating string.

possible wavelengths that will produce standing waves. These end points must be displacement nodes for any resulting wave pattern.

The simplest possible standing wave occurs when the wavelengths of incident and reflected waves are equal to twice the length of the string. The standing wave consists of a single loop with nodal points at each end, as shown in Fig. 22-12a. This pattern of vibration is referred to as the *fundamental mode of oscillation*. The *higher modes* of oscillation occur for shorter and shorter wavelengths. From the figure, it is noted that the allowable wavelengths are

$$\frac{2l}{1}, \frac{2l}{2}, \frac{2l}{3}, \frac{2l}{4}, \ldots$$

or, in equation form,

$$\lambda_n = \frac{2l}{n} \qquad n = 1, 2, 3, \ldots \tag{22-8}$$

The corresponding frequencies of vibration are, from $v = f\lambda$,

$$f_n = \frac{nv}{2l} = n\frac{v}{2l} \qquad n = 1, 2, 3, \ldots \tag{22-9}$$

where v is the speed of the transverse waves. This speed is the same for all wavelengths because it depends only upon the characteristics of the vibrating medium. The frequencies given by Eq. (22-9) are called the *characteristic frequencies of vibration*. In terms of string tension F and linear density μ, the characteristic frequencies are

$$\boxed{f_n = \frac{n}{2l}\sqrt{\frac{F}{\mu}} \qquad n = 1, 2, 3, \ldots} \tag{22-10}$$

The lowest possible frequency ($v/2l$) is called the *fundamental frequency* f_1. The others, which are integral multiples of the fundamental, are known as the *overtones*. The entire series,

$$f_n = nf_1 \qquad n = 1, 2, 3, \ldots \tag{22-11}$$

consisting of the fundamental and its overtones, is known as the *harmonic series*. The fundamental is the first harmonic, the first overtone ($f_2 = 2f_1$) is the second harmonic, the second overtone ($f_3 = 3f_1$) is the third harmonic, and so on.

EXAMPLE 22-3   A steel piano wire 50 cm long has a mass of 5 g and is under a tension of 400 N. What are the frequencies of its fundamental mode of vibration and the first two overtones?

Solution   The fundamental is found by setting $n = 1$ in Eq. (22-10).

$$f_1 = \frac{1}{2l}\sqrt{\frac{F}{\mu}} = \frac{1}{2l}\sqrt{\frac{F}{m/l}}$$

$$= \frac{1}{(2)(0.5 \text{ m})}\sqrt{\frac{400 \text{ N}}{(0.005 \text{ kg})/(0.5 \text{ m})}}$$

$$= 200 \text{ Hz}$$

The first and second overtones are

$$f_2 = 2f_1 = 400 \text{ Hz}$$
$$f_3 = 3f_1 = 600 \text{ Hz}$$

We shall see in the following chapter that as a string vibrates in one or more of its possible modes, energy is transmitted to the surrounding air in the form of sound waves. These longitudinal waves consist of condensations and rarefactions of the same frequency as the vibrating strings. The human ear interprets these waves as sound.

SUMMARY

We have investigated mechanical wave motion in which energy is transferred by a physical disturbance in an elastic medium. The fundamental laws developed in this chapter are very important because they also apply for many other types of waves which will be studied later. The essential concepts are summarized below:

- The velocity of a transverse wave in a string of mass m and length l is given by

$$v = \sqrt{\frac{F}{\mu}} \qquad \mu = \frac{m}{l} \qquad v = \sqrt{\frac{Fl}{m}} \qquad \textit{Wave Speed}$$

	Force F	Mass m	Length l	Speed v
SI units	N	kg	m	m/s
USCS units	lb	slug	ft	ft/s

- For any wave of period T or frequency f, the speed v can be expressed in terms of the wavelengths λ as follows:

$$v = \frac{\lambda}{T} \qquad\qquad v = f\lambda \qquad \textit{Frequency is in } \text{Hz} = 1/\text{s}$$

- The *energy per unit length* and the *power* of wave propagation can be found from

$$\frac{E}{l} = 2\pi^2 f^2 A^2 \mu \qquad\qquad P = 2\pi^2 f^2 A^2 \mu v$$

- The characteristic frequencies for the possible modes of vibration is in a stretched string are found from:

$$f_n = \frac{n}{2l} \sqrt{\frac{F}{\mu}} \qquad n = 1, 2, 3, \ldots \qquad \textit{Characteristic Frequencies}$$

- The series $f_n = nf_1$ is called the *harmonics*. They are integral multiples of the fundamental f_1. These are mathematical values and all harmonics may not exist. The actual possibilities beyond the fundamental are called *overtones*. Since all harmonics are possible for the vibrating string, the first overtone is the second harmonic, the second overtone is the third harmonic, and so on.

22-1. Define the following terms:
   a. Wave motion
   b. Wave speed
   c. Wavelength
   d. Frequency
   e. Phase
   f. Nodes
   g. Antinodes
   h. Fundamental
   i. Overtone
   j. Harmonic
   k. Mechanical wave
   l. Transverse wave
   m. Longitudinal wave
   n. Linear density
   o. Amplitude
   p. Superposition principle
   q. Constructive interference
   r. Destructive interference
   s. Standing waves
   t. Characteristic frequencies

22-2. Explain how a water wave is both transverse and longitudinal.

22-3. Describe an experiment to demonstrate that energy is associated with wave motion.

22-4. In a *torsional wave* the individual particles of the medium vibrate with angular harmonic motion about the axis of propagation. Give a mechanical example of such a wave.

22-5. Discuss the interference of waves. Is there a loss of energy when waves interfere? Explain.

22-6. A transverse pulse is sent down a string of mass m and length l under a tension F. How will the speed of the pulse be affected if (a) the mass of the string is quadrupled, (b) the length of the string is quadrupled, and (c) the tension is reduced by one-fourth?

22-7. Draw graphs of a periodic transverse wave and a periodic longitudinal wave. Indicate on the figures the wavelength and amplitude of each wave.

22-8. Which harmonic is indicated by Fig. 22-12d? Which overtone is present?

22-9. We have seen that boundary conditions determine possible modes of vibration. Draw a diagram of the fundamental and of the first two overtones for a vibrating rod (a) clamped at one end and (b) clamped at its midpoint.

22-10. A vibrating string has a fundamental frequency of 200 Hz. If the length is reduced by one-fourth, what will the new fundamental frequency be? Has the wave speed been altered by shortening the string? Assume constant tension.

22-11. Show graphically the superposition of two waves traveling in the same direction. The second wave has three-fourths the amplitude and one-half the wavelength of the first wave.

22-12. In an experiment with the vibrating string, one end of the string is attached to the tip of a vibrator and the other end passes over a pulley. Suspended weights are used to produce the fundamental and the first three overtones. What effect will stretching the string have on frequency calculations?

Calculating Wave Speed

22-1. A metal wire of mass 500 g and length 50 cm is under a tension of 80 N. What is the speed of a transverse wave in the wire? If the length is reduced by one-half, what will be the new mass of the wire? Show that the speed of a wave in the wire is unchanged.

Ans. 8.94 m/s, 250 g

22-2. A 1.2-kg rope is stretched over a distance of 5.2 m and placed under a tension of 120 N. Compute the speed of a transverse wave in the rope.

22-3. A 30-m cord under a tension of 200 N sustains a transverse wave whose speed is 72 m/s. What is the mass of the cord?

Ans. 1.16 kg

22-4. A string 200 cm long has a mass of 500 g. What string tension is required to produce a wave speed of 120 cm/s?

22-5. A longitudinal wave has a frequency of 200 Hz and a wavelength of 4.2 m. What is the speed of the waves?

Ans. 840 m/s

22-6. A wooden float at the end of a fishing line makes eight complete oscillations in 10 s. If it takes 3.6 s for a single wave to travel 11 m, what is the wavelength of the water waves?

Energy of a Periodic Wave

22-7. One end of a long horizontal rope oscillates with a frequency of 2 Hz and an amplitude of 50 mm. A 2-m length of the string has a mass of 0.3 kg. If the rope is under a tension of 48 N, how much energy per second must be supplied to the rope?

Ans. 0.530 W.

* **22-8.** A source of a transverse wave train has a frequency of 8 Hz and an amplitude of 4 cm. If the string is 40 m long and has a mass of 80 g, how much energy per unit length passes along the string? If the wavelength of the transverse wave is 1.6 m, what power must be supplied by the source?

Standing Waves and Characteristic Frequencies

22-9. A string 4 m long has a mass of 10 g and a tension of 64 N. What is the frequency of its fundamental mode of vibration? What are the frequencies of the first and second overtones?

Ans. 20, 40, and 60 Hz

22-10. The second harmonic of a vibrating string is 200 Hz. If the length of the string is 3 m and its tension is 200 N, compute the linear density of the string.

* **22-11.** A 4.3-m string has a tension of 300 N and a mass of 0.5 g. If it is fixed at each end and vibrates in three segments, what is the frequency of the standing waves?

Ans. 560 Hz

* **22-12.** A string vibrates with standing waves in five segments when the frequency is 600 Hz. What frequency will cause the string to vibrate in two segments?

Additional Problems

* **22-13.** The steel guy wire supporting a pole is 18.9 m long and 9.5 mm in diameter. It has a linear density of 0.474 kg/m. It is struck at one end by a hammer, and the pulse returns in 0.3 s. What is the tension in the wire?

Ans. 7520 N

* **22-14.** A 30-m cable weighing 400 N is stretched between two telephone poles with a tension of 1800 N. How much time is required for a pulse to make a round trip if the bale is struck at one end? What time is needed if the tension in the wire is doubled?

22-15. A bass guitar string 750 mm long is stretched with sufficient tension to produce a fundamental vibration of 220 Hz. What is the velocity of the transverse waves in this string?

Ans. 330 m/s

22-16. Transverse waves have a speed of 20 m/s on a string whose tension is 8 N. What tension is required to give a wave speed of 30 m/s in the same string?

22-17. In a laboratory experiment, an electromagnetic vibrator is used as a source of standing waves in a string. The linear density of the string is 0.006 g/cm. One end of the string is connected to the tip of the vibrator. The other end passes over a pulley 1 m away and is attached to a weight hanger. It is found that a mass of 392 g hanging from the free end causes the string to vibrate in three segments. What is the frequency of the vibrator?

Ans. 120 Hz

22-18. Refer to Prob. 22-17. What mass would be required to cause the string to vibrate in four segments?

23 Sound

After completing this chapter, you should be able to:

1. Define *sound* and solve problems involving its velocity in metal, in a liquid, and in a gas.
2. Use boundary conditions to derive and apply relationships for calculating the *characteristic frequencies* for an open pipe and for a closed pipe.
3. Compute the intensity level in *decibels* for a sound whose intensity is given in *watts per square meter*.
4. Use your understanding of the *Doppler effect* to predict the apparent change in sound frequency which occurs as a result of relative motion between a source and an observer.

When a periodic disturbance takes place in air, longitudinal *sound* waves travel out from it. For example, if a tuning fork is struck with a hammer, the vibrating prongs send out longitudinal waves, as shown on Fig. 23-1. An ear, acting as a receiver for these periodic waves, interprets them as sound.

Fig. 23-1 A tuning fork acts as a source of longitudinal sound waves.

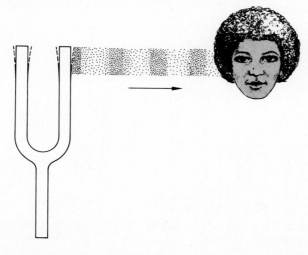

Is the ear necessary for sound to exist? If the tuning fork were struck in the atmosphere of a distant planet, would there be sound even though no ear could interpret the disturbance? The answer depends upon the definition of sound.

The term *sound* is used in two different ways. Physiologists define sound in terms of the auditory sensations produced by longitudinal disturbances in air. For them, sound does not exist on a distant planet. In physics, on the other hand, we refer to the disturbances themselves rather than the sensations produced.

> **Sound** is a longitudinal mechanical wave which travels through an elastic medium.

In this case, sound does exist on the planet. In this chapter *sound* will be used in its physical sense.

23-1
PRODUCTION OF A SOUND WAVE

Two things must exist in order to produce a sound wave. There must be a source of mechanical vibration, and there must be an elastic medium through which the disturbance can travel. The source may be a tuning fork, a vibrating string, or a vibrating air column in an organ pipe. *Sounds are produced by vibrating matter.* The requirement of an elastic medium can be demonstrated by placing an electric bell inside an evacuable flask, as shown in Fig. 23-2. With the bell connected to a battery so that it rings continuously, the flask is slowly evacuated. As more and more of the air is pumped from the flask, the sound of the bell becomes fainter and fainter until finally it cannot be heard at all. When air is allowed to reenter the flask, the sound of the bell returns. Thus, air is necessary to transmit sound.

Let us now examine more closely the longitudinal sound waves in air as they proceed from a vibrating source. A thin strip of metal clamped tight at its base is pulled to one side and released. As the free end oscillates to and fro with simple harmonic motion, a series of periodic, longitudinal sound waves spreads through the air away from the source. The air molecules in the vicinity of the metal strip are alternately compressed and expanded, sending out a wave like that illustrated in Fig. 23-3a. The dense regions where many molecules are packed tightly together are called *compressions*. They are exactly analogous to the *condensations* discussed for

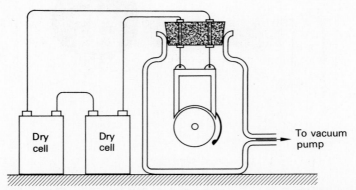

Fig. 23-2 A bell ringing in a vacuum cannot be heard. A material medium is necessary for the production of sound.

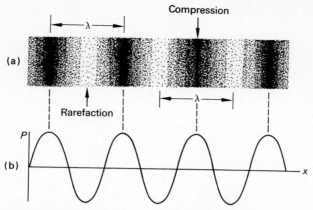

Fig. 23-3 *(a)* Compressions and rarefactions in a sound wave in air at a particular instant. *(b)* The sinusoidal variation in pressure as a function of displacement.

longitudinal waves in a coiled spring. The regions with relatively few molecules are referred to as *rarefactions*. The compressions and rarefactions alternate throughout the medium as the individual air particles oscillate to and fro in the direction of wave propagation. Since a compression corresponds to a high-pressure region and a rarefaction corresponds to a low-pressure region, a sound wave can also be represented by plotting the change in pressure *P* as a function of the distance *x*. (See Fig. 23-3*b*.) The distance between two successive compressions or rarefactions is the wavelength.

23-2

THE SPEED OF SOUND

Anyone who has seen a weapon being fired at a distance has observed the smoke from the weapon before hearing the report. Similarly, we observe the flash of lightning before hearing the thunder. Even though both light and sound travel with finite speeds, the speed of light is so much greater in comparison that it can be considered instantaneous. The speed of sound can be measured directly by observing the time required for the waves to move through a known distance. In air at 0°C, sound travels at a speed of 331 m/s or 1087 ft/s.

In Chap. 22 we established the idea that wave speed depends upon the elasticity of the medium and the inertia of its particles. The more elastic materials sustain greater wave speeds whereas the denser materials retard wave motion. The following empirical relationships are based on these proportionalities.

For longitudinal sound waves in a wire or rod, the wave speed is given by

$$v = \sqrt{\frac{Y}{\rho}} \qquad \qquad Rod \quad (23\text{-}1)$$

where *Y* is Young's modulus for the solid and ρ is its density. This relation is valid only for rods whose diameters are small in comparison with the longitudinal wavelengths of sound passing through them.

In an *extended* solid, the longitudinal wave speed is a function of the shear modulus *S*, the bulk modulus *B*, and the density ρ of the medium. The wave speed can be calculated from

$$v = \sqrt{\frac{B + \frac{4}{3}S}{\rho}}$$ *Extended Solid*   (23-2)

For longitudinal waves in a fluid, the wave speed is found from

$$v = \sqrt{\frac{B}{\rho}}$$ *Fluid*   (23-3)

where B is the bulk modulus for the fluid and ρ is its density.

In computing the speed of sound in a gas, the bulk modulus is given by

$$B = \gamma P$$

where γ is the adiabatic constant ($\gamma = 1.4$ for air and diatomic gases) and P is the pressure of the gas. Thus the speed of longitudinal waves in a gas, from Eq. (23-3), is given by

$$v = \sqrt{\frac{B}{\rho}} = \sqrt{\frac{\gamma P}{\rho}}$$ (23-4)

But for an ideal gas

$$\frac{P}{\rho} = \frac{RT}{M}$$ (23-5)

where R = universal gas constant
T = absolute temperature of gas
M = molecular mass of gas

Substitution of Eq. (23-5) into Eq. (23-4) yields

$$v = \sqrt{\frac{\gamma P}{\rho}} = \sqrt{\frac{\gamma RT}{M}}$$ *Gas*   (23-6)

EXAMPLE 23-1   Compute the speed of sound in an aluminum rod.

Solution   Young's modulus and the density for aluminum are

$$Y = 68,900 \text{ MPa} = 6.89 \times 10^{10} \text{ N/m}^2$$
$$\rho = 2.7 \text{ g/cm}^3 = 2.7 \times 10^3 \text{ kg/m}^3$$

From Eq. (23-1)

$$v = \sqrt{\frac{Y}{\rho}} = \sqrt{\frac{6.89 \times 10^{10} \text{ N/m}^2}{2.7 \times 10^3 \text{ kg/m}^3}}$$
$$= \sqrt{2.55 \times 10^7 \text{ m}^2/\text{s}^2} = 5050 \text{ m/s}$$

This speed is approximately 15 times the speed of sound in air.

EXAMPLE 23-2   Compute the speed of sound in air on a day when the temperature is 27°C. The molecular mass of air is 29.0, and the adiabatic constant is 1.4.

Solution   From Eq. (23-6)

$$v = \sqrt{\frac{\gamma R T}{M}} = \sqrt{\frac{(1.4)(8.31 \text{ J/mol} \cdot \text{K})(300 \text{ K})}{29 \times 10^{-3} \text{ kg/mol}}}$$

$$= 347 \text{ m/s}$$

The speed of sound is significantly greater at 27°C than at 0°C. At standard temperature and pressure (0°C, 1 atm) the speed of sound is 331 m/s. For each Celsius degree rise in temperature (above 0°C) the speed of sound in air increases by approximately 0.6 m/s. Hence the speed v of sound can be approximated by

$$\boxed{v = 331 \text{ m/s} + \left(0.6 \frac{\text{m/s}}{\text{C}°}\right) t} \tag{23-7}$$

where t is the Celsius temperature of the air.

EXAMPLE 23-2   What is the approximate speed of sound in air at room temperature (20°C)?

Solution   From Eq. (23-7),

$$v = 331 \text{ m/s} + \left(0.6 \frac{\text{m/s}}{\text{C}°}\right)(20°\text{C}) = 343 \text{ m/s}$$

23-3 VIBRATING AIR COLUMNS

In the previous chapter we described the possible modes of vibration for a string fixed at both ends. The frequency of the sound waves set up in the air surrounding the string is identical with the frequency of the vibrating string. Thus the possible frequencies, or the *harmonics,* of sound waves produced by a vibrating string are given by

$$\boxed{f_n = \frac{nv}{2l} \qquad n = 1, 2, 3, \ldots} \tag{23-8}$$

where v is the velocity of transverse waves in the string.

Sound can also be produced by the longitudinal vibrations of an air column in an open or closed pipe. As in the vibrating string, the possible modes of vibrations are determined by the boundary conditions. The possible modes of vibration for the air in a closed pipe are illustrated in Fig. 23-4. When a compressional wave is set up in the pipe, the displacement of the air particles at the closed end must be zero.

The closed end of a pipe must be a displacement node.

The air at the open end of a pipe has the greatest freedom of motion, and so the displacement is free at an open end.

The open end of a pipe must be a displacement antinode.

Fig. 23-4 Possible standing waves in a closed pipe.

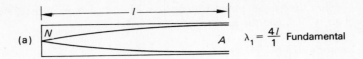

(a) $\lambda_1 = \dfrac{4l}{1}$ Fundamental

(b) $\lambda_3 = \dfrac{4l}{3}$ First overtone

(c) $\lambda_5 = \dfrac{4l}{5}$ Second overtone

(d) $\lambda_7 = \dfrac{4l}{7}$ Third overtone

The sinusoidal curves in Fig. 23-4 represent maximum displacements.

The fundamental mode of oscillation for an air column in a closed pipe has a node at the closed end and an antinode at the open end. Thus the wavelength of the fundamental is 4 times the length l of the pipe (Fig. 23-4a). The next possible mode, which is the first overtone, occurs when there are two nodes and two antinodes, as shown in Fig. 23-4b. The wavelength of the first overtone is therefore equal to $4l/3$. Similar reasoning will show that the second and third overtones occur for wavelengths equal to $4l/5$ and $4l/7$, respectively. In summary, the possible wavelengths are

$$\lambda_n = \frac{4l}{n} \qquad n = 1, 3, 5, \ldots \tag{23-9}$$

The speed of the sound waves is given by $v = f\lambda$, so that the possible frequencies for a *closed pipe* are

$$\boxed{f_n = \frac{nv}{4l} \qquad n = 1, 3, 5, \ldots} \qquad \textit{Closed Pipe} \quad (23\text{-}10)$$

Notice that only the *odd harmonics* are allowed for a closed pipe. The first overtone is the third harmonic, the second overtone is the fifth harmonic, and so on.

An air column vibrating in a pipe open at *both* ends must be bounded by displacement antinodes. Figure 23-5 shows the fundamental and first three overtones for an open pipe. Note that the fundamental wavelength is twice the length l of the pipe. When the number of nodes is increased one at a time, it is seen that the possible wavelengths in an open pipe are

$$\lambda_n = \frac{2l}{n} \qquad n = 1, 2, 3, \ldots \tag{23-11}$$

The possible frequencies are therefore

$$f_n = \frac{nv}{2l} \qquad n = 1, 2, 3, \ldots \qquad \textit{Open Pipe} \quad (23\text{-}12)$$

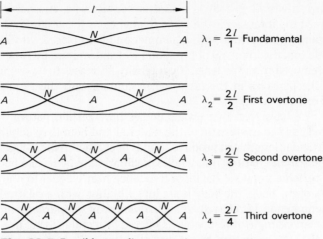

$\lambda_1 = \dfrac{2l}{1}$ Fundamental

$\lambda_2 = \dfrac{2l}{2}$ First overtone

$\lambda_3 = \dfrac{2l}{3}$ Second overtone

$\lambda_4 = \dfrac{2l}{4}$ Third overtone

Fig. 23-5 Possible standing waves in an open pipe.

where v is the velocity of the sound waves. Thus all the harmonics are possible for a vibrating air column in an open pipe. Open pipes of varying lengths are used in many musical instruments, e.g., organs, flutes, and trumpets.

EXAMPLE 23-4   What are the frequencies of the fundamental and first two overtones for a 12-cm closed pipe? Air temperature is 30°C.

Solution   The velocity of the sound is

$$v = 331 \text{ m/s} + \left(0.6 \, \frac{\text{m/s}}{\text{C}^\circ}\right)(30°\text{C}) = 349 \text{ m/s}$$

Thus, from Eq. (23-10),

$$f_1 = \frac{1v}{4l} = \frac{349 \text{ m/s}}{(4)(0.12 \text{ m})} = 727 \text{ Hz}$$

The first and second overtones are the third and fifth harmonics. Hence

$$\text{First overtone} = 3f_1 = 2181 \text{ Hz}$$
$$\text{Second overtone} = 5f_1 = 3635 \text{ Hz}$$

EXAMPLE 23-5   What length of open pipe will have a frequency of 1200 Hz as its first overtone? Take the speed of sound as 340 m/s.

Solution   The first overtone in an open pipe is equal to the second harmonic. Thus we can set $n = 2$ in Eq. (23-12)

$$f_2 = \frac{2v}{2l} = \frac{v}{l}$$

$$l = \frac{v}{f_2} = \frac{340 \text{ m/s}}{1200 \text{ Hz}} = 0.283 \text{ m}$$

23-4
FORCED VIBRATION AND RESONANCE

When a vibrating body is placed in contact with another body, the second body is forced to vibrate with the same frequency as the original vibrator. For example, if a tuning fork is struck with a hammer and then placed with its base against a wooden table top, the intensity of the sound will suddenly be increased. When the tuning fork is removed from the table, the intensity decreases to its original level. The vibrations of the particles in the table top in contact with the tuning fork are called *forced vibrations.*

We have seen that elastic bodies have certain natural frequencies of vibration which are characteristic of the material and boundary conditions. A taut string of a particular length can produce sounds of characteristic frequencies. An open or closed pipe also has natural frequencies of vibration. Whenever a body is acted on by a series of periodic impulses having a frequency nearly equal to one of the natural frequencies of the body, the body is set into vibration with a relatively large amplitude. This phenomenon is referred to as *resonance* or *sympathetic vibration.*

An example of resonance is offered by a child sitting in a swing. Experience tells us that the swing can be set into vibration with large amplitude by a series of small pushes at just the right intervals. Such resonance occurs only when the pushes are in phase with the natural frequency of vibration for the swing. A slight variation of the input pulses would result in little or no vibration.

Reinforcement of sound by resonance has many useful applications as well as many unpleasant consequences. The resonance of an air column in an organ pipe amplifies the weak sound of a vibrating air jet. Many musical instruments are designed with resonant cavities to produce varying sounds. Electrical resonance in radio receivers enables the listener to hear weak signals clearly. When tuned to the frequency of a desired station, the signal is amplified by electrical resonance. In poorly designed auditoriums or long hallways, music and voices may have a hollow sound which is unpleasant to the ear. Bridges have been known to collapse because of sympathetic vibrations set up by gusts of wind.

23-5
AUDIBLE SOUND WAVES

We have defined sound as a *longitudinal mechanical wave traveling through an elastic medium.* This is a very broad definition that makes no restriction whatsoever on the frequencies of sound. The physiologist is concerned primarily with sound waves which are capable of affecting the sense of hearing. Thus it is useful to divide the spectrum of sound according to the following definitions:

> **Audible sound** refers to sound waves in the frequency range from 20 to 20,000 Hz.
>
> Sound waves having frequencies below the audible range are termed **infrasonic.**
>
> Sound waves having frequencies above the audible range are termed **ultrasonic.**

When studying audible sound, the physiologist uses the terms *loudness, pitch,* and *quality* to describe the sensations produced. Unfortunately, these terms represent sensory magnitudes and are therefore subjective. What is *loud* to one person is moderate to another. What one person perceives as quality, another considers inferior. As always, the physicist must deal with explicit measurable definitions. The physicist therefore attempts to correlate the sensory effects with the physical properties of waves. These correlations can be summarized as follows:

Sensory effects		Physical property
Loudness	↔	Intensity
Pitch	↔	Frequency
Qualty	↔	Waveform

The meaning of the terms on the left may vary considerably among individuals. The terms on the right are measurable and objective.

Sound waves constitute a flow of energy through matter. The intensity of a given sound wave is a measure of the rate at which energy is propagated through a given volume of space. A convenient method of specifying sound intensity is in terms of the rate at which energy is transferred through a unit area normal to the direction of wave propagation (see Fig. 23-6). Since the rate at which energy flows is the *power* of

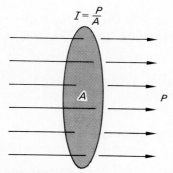

Fig. 23-6 The intensity of a sound wave is a measure of the power transmitted per unit of area perpendicular to the direction of wave propagation.

a wave, the intensity can be related to the power per unit area passing a given point.

Sound **intensity** is the power transferred by a sound wave through a unit of area normal to the direction of the propagation.

$$I = \frac{P}{A}$$

(23-13)

The units for intensity are the ratio of a power unit to an area unit. In SI units, the intensity is in W/m^2, and that is the unit we will use in this text. However, the rate of energy flow in sound waves is very small, and industry still uses the μW/cm^2 in many applications. The conversion factor is

$$1 \ \mu W/cm^2 = 1 \times 10^{-2} \ W/m^2$$

It can be shown by methods similar to those utilized for a vibrating string that the sound intensity varies directly with the square of the frequency f and with the square of the amplitude A of a given sound wave. Symbolically, the intensity I is given by

$$I = 2\pi^2 f^2 A^2 \rho v \tag{23-14}$$

where v is the sound velocity in a medium of density ρ. The symbol A in Eq. (23-14) refers to the amplitude of the sound wave and not the unit area, as in Eq. (23-13).

The intensity I_0 of the faintest audible sound is of the order of 10^{-12} W/m². This intensity, which is referred to as the *hearing threshold,* has been adopted by acoustical experts as the minimum intensity for audible sound.

> The **hearing threshold** represents the standard minimum of intensity for audible sound. Its value at a frequency of 1000 Hz is

$$I_0 = 1 \times 10^{-12} \text{ W/m}^2 = 1 \times 10^{-10} \ \mu\text{W/cm}^2 \tag{23-15}$$

The range of intensities over which the human ear is sensitive is enormous. It extends from the hearing threshold I_0 to an intensity 10^{12} times as great. The upper extreme represents the point at which the intensity is intolerable for the human ear. The sensation becomes one of feeling or pain instead of simply hearing.

> The **pain threshold** represents the maximum intensity that the average ear can record without feeling or pain. Its value is

$$I_p = 1 \text{ W/m}^2 = 100 \ \mu\text{W/cm}^2 \tag{23-16}$$

In view of the wide range of intensities over which the ear is sensitive, it is more convenient to set up a logarithmic scale for the measurement of sound intensities. Such a scale is established by the following rule:

> *When the intensity I_1 of one sound is 10 times as great as the intensity I_2 of another, the ratio of intensities is said to be 1 bel* (B).

Thus, when comparing the intensities of two sounds, we refer to a difference in intensity levels given by

$$B = \log \frac{I_1}{I_2} \quad \text{bels (B)} \tag{23-17}$$

where I_1 is the intensity of one sound and I_2 is the intensity of the other.

EXAMPLE 23-6 Two sounds have intensities of 2.5×10^{-8} W/m² and 1.2 W/m². Compute the difference in intensity levels in bels.

Solution
$$B = \log \frac{I_1}{I_2} = \log \frac{1.2 \text{ W/m}^2}{2.5 \times 10^{-8} \text{ W/m}^2}$$
$$= \log 4.8 \times 10^7 = 7.68 \text{ B}$$

In practice, the unit of 1 B is too large. To obtain a more useful unit, we define a *decibel* (dB) as one-tenth of a bel. Thus the answer to Example 23-6 can also be expressed as 76.8 dB.

By using the standard intensity I_0 as a comparison for all intensities, a general scale has been devised for rating any sound. The intensity level in decibels of any sound of intensity I can be found from the general relation

$$\beta = 10 \log \frac{I}{I_0} \quad \textit{decibels (dB)} \qquad (23\text{-}18)$$

where I_0 is the intensity at the hearing threshold (1×10^{-12} W/m^2). The intensity level for I_0 is zero decibels.

EXAMPLE 23-7 Compute the intensity level of a sound whose intensity is 1×10^{-4} W/m^2.

Solution

$$\beta = 10 \log \frac{I}{I_0} = 10 \log \left(\frac{10^{-4} \text{ W/m}^2}{10^{-12} \text{ W/m}^2} \right)$$
$$= 10 \log 10^8 = 10(8) = 80 \text{ dB}$$

Through the logarithmic decibel notation, we have reduced the wide range of intensities to intensity levels from 0 to 120 dB. However, we must remember that the scale is not linear but logarithmic. A 40-dB sound is much more than twice as intense as a 20-dB sound. A sound which is 100 times as intense as another is only 20 dB larger. Several examples of the intensity levels for common sounds are given in Table 23-1.

Table 23-1 Intensity Levels for Common Sounds

Sound	Intensity level, dB
Hearing threshold	0
Rustling leaves	10
Whisper	20
Quiet radio	40
Normal conversation	65
Busy street corner	80
Subway car	100
Pain threshold	120
Jet engine	140–160

23-6

PITCH AND QUALITY

The effect of intensity on the human ear manifests itself as *loudness*. In general, sound waves which are more intense are also louder, but the ear is not equally sensitive to sounds of all frequencies. Therefore, a high-frequency sound may not seem as loud as one of lower frequency which has the same intensity.

The frequency of a sound determines what the ear judges as the *pitch* of the sound. Musicians designate pitch by letters corresponding to key notes on the piano. For example the C note, D note, and F note each refer to a specific pitch, or frequency. A siren disk, shown in Fig. 23-7, can be used to demonstrate how the pitch is determined by the frequency of a sound. A stream of air is directed against a row of evenly spaced holes. By varying the rate of rotation of the disk the pitch of the resulting sound is increased or decreased.

Two sounds of the same pitch can easily be distinguished. For example, suppose we sound a C note (256 Hz) successively on a piano, a flute, a trumpet, and a violin. Even though each sound has the same pitch, there is a marked difference in the tones. This distinction is said to result from a difference in the *quality* of sound.

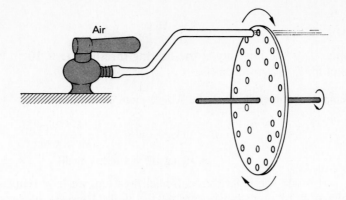

Fig. 23-7 Demonstrating the relationship between pitch and frequency.

Air

Regardless of the source of vibration in musical instruments, several modes of oscillation are usually excited simultaneously. Therefore, the sound produced consists not only of the fundamental but also many of the overtones. *The quality of a sound is determined by the number and relative intensities of the overtones present.* The difference in quality between two sounds can be observed objectively by analyzing the complex *waveforms* resulting from each sound. In general, the more complex the wave, the greater the number of harmonics that contribute to it.

<table>
<tr><td>**23-7**</td></tr>
<tr><td>**INTERFERENCE AND BEATS**</td></tr>
</table>

In Chap. 22 we discussed the superposition principle as a method for studying interference in transverse waves. Interference also occurs in longitudinal sound waves, and the superposition principle can be applied for them also. A common example of the interference in sound waves occurs when two tuning forks (or other single-frequency sound sources) whose frequencies differ only slightly are struck simultaneously. The sound produced fluctuates in intensity, alternating between loud tones and virtual silence. These regular pulsations are referred to as *beats*. The *vibrato* effect obtained on some organs is an application of this principle. Every vibrato note is produced by two pipes tuned to slightly different frequencies.

To understand the origin of beats, let us examine the interference set up between sound waves proceeding from two tuning forks of slightly different frequency, as shown in Fig. 23-8. The superposition of waves A and B illustrates the origin of beats. The loud tones occur when the waves interfere constructively and the quiet tones when the waves interfere destructively. Observation and calculation show that the two waves interfere constructively $f - f'$ times per second. Thus we can write

$$\text{Number of beats per second} = |f - f'| \qquad (23\text{-}19)$$

For example, if tuning forks of frequencies 256 and 259 Hz are struck simultaneously, the resulting sound will pulsate three times every second.

<table>
<tr><td>**23-8**</td></tr>
<tr><td>**THE DOPPLER EFFECT**</td></tr>
</table>

Whenever a source of sound is moving relative to an observer, the pitch of the sound, as heard by the observer, may not be the same as that perceived when the source is at rest. For example, if we stand near a railway track as a train blowing its whistle approaches, we notice that the pitch of the whistle is *higher* than the normal one

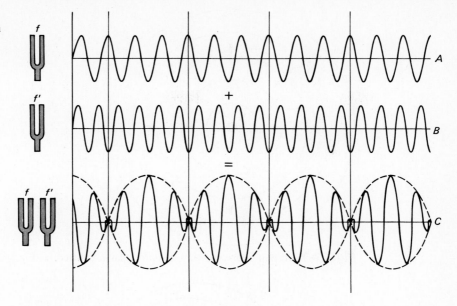

Fig. 23-8 Diagram illustrating the origin of beats. The wave C is a superposition of waves A and B.

when the train is stationary. As the train recedes, the pitch is observed to be *lower* than normal. Similarly, at the race tracks the sound of cars driving toward the stands is considerably higher in pitch than the sound of cars driving away from the stands.

The phenomenon is not restricted to the motion of the source. If the source of sound is stationary, a listener moving toward the source will observe a similar increase in pitch. A listener leaving the source of sound will hear a lower-pitched sound. The change in frequency of sound resulting from relative motion between a source and an observer is called the *Doppler effect*.

> The **Doppler effect** refers to the apparent change in frequency of a source of sound when there is relative motion of the source and the listener.

The origin of the Doppler effect can be demonstrated graphically by representing the periodic waves emitted by a source as concentric circles moving radially outward, as in Fig. 23-9. The distance between any two circles represents the wavelength λ of the sound traveling with a velocity V. The frequency with which these waves strike the ear determines the pitch of sound heard.

Let us first consider that the source is moving to the right toward a stationary observer A, as in Fig. 23-10. As the moving source emits sound waves, it tends to overtake waves traveling in the same direction as the source. Each successive wave is emitted from a point closer to the observer than its predecessor. The result is that the distance between successive waves, or the wavelength, is smaller than usual. A smaller wavelength results in a higher frequency of waves, which increases the pitch of the sound heard by observer A. Similar reasoning will show that an *increase* in the length of waves reaching observer B will cause B to hear a *lower*-frequency sound.

We can now derive a relationship for predicting the change in observed frequency. During one complete vibration of a stationary source (a time equal to the

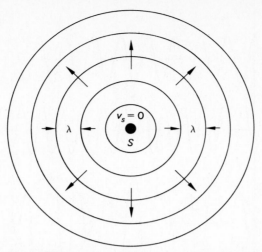

Fig. 23-9 Graphic representation of sound waves emitted from a stationary source.

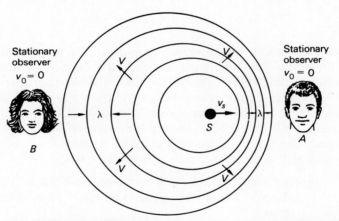

Fig. 23-10 Illustration of the Doppler effect. The waves in front of a moving source are closer together than the waves behind a moving source.

period T), each wave will move through a distance of one wavelength. This distance is represented by λ in Fig. 23-11a and is given by

$$\lambda = VT = \frac{V}{f_s} \qquad \textit{Stationary Source}$$

where V is the velocity of sound and f_s is the frequency of the source. If the source is moving to the right with a velocity v_s, as in Fig. 23-11b, the new wave-length λ' in front of the source will be given by

$$\lambda' = VT - v_s T = (V - v_s)T$$

But $T = 1/f_s$, so that we write

$$\lambda' = \frac{V - v_s}{f_s} \qquad \textit{Moving Source} \quad (23\text{-}20)$$

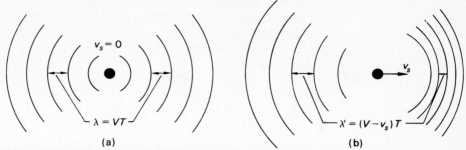

Fig. 23-11 Computing the magnitude of sound emitted from a moving source. The source velocity v_s is considered positive for speeds of approach and negative for speeds of recession.

This equation will also apply for the wavelength on the left of the moving source if we follow the convention that speeds of approach are considered positive and speeds of recession are considered negative. Thus, if we were computing λ' on the left of our moving source, the negative value would be substituted for v_s, resulting in a larger wavelength.

The velocity of sound in a medium is a function of the properties of the medium and does not depend upon the motion of the source. Thus the frequency f_o heard by a stationary observer from a moving source of frequency f_s is given by

$$f_o = \frac{V}{\lambda'} = \frac{V f_s}{V - v_s}$$ *Moving Source* (23-21)

where V is the speed of sound and v_s is the speed of the source. *The speed v_s is reckoned as positive for speeds of approach and negative for speeds of recession.*

EXAMPLE 23-8 A train whistle emits sound at a frequency of 400 Hz. (*a*) What is the pitch of the sound heard when the train is moving toward a stationary observer at a speed of 20 m/s? (*b*) What is the pitch heard when the train is moving away from the observer at this speed? Assume that the speed of sound is 340 m/s.

Solution (*a*) Since the train is approaching the observer, its speed v_s is positive. Substitution into Eq. (23-21) yields

$$f_o = \frac{V f_s}{V - v_s} = \frac{(340 \text{ m/s})(400 \text{ Hz})}{340 \text{ m/s} - 20 \text{ m/s}}$$

$$= \frac{1.36 \times 10^5}{320} = 425 \text{ Hz}$$

Solution (*b*) Now v_s represents a speed of recession, so that -20 m/s should be substituted in Eq. (23-21).

$$f_o = \frac{V f_s}{V - v_s} = \frac{(340 \text{ m/s})(400 \text{ Hz})}{340 \text{ m/s} - (-20 \text{ m/s})}$$

$$= \frac{1.36 \times 10^5}{360} = 378 \text{ Hz}$$

SOUND

Now let us examine the situation in which a source is stationary and the observer moves toward the source with a velocity v_o. In this case, the wavelength of the sound received does not change, but the number of waves encountered per unit of time (the frequency) increases as a result of the observer's speed v_o. Hence the observer will hear the frequency

$$f_o = \frac{f_s(V + v_o)}{V}$$

Moving Observer (23-22)

Here *the speed v_o of the observer should be reckoned as positive for speeds of approach and negative for speeds of recession.*

EXAMPLE 23-9 A stationary source of sound has a frequency of 800 Hz on a day when the speed of sound is 340 m/s. What pitch is heard by a person who is moving from the source at a speed of 30 m/s?

Solution Since v_o represents a speed of recession, $v_o = -30$ m/s must be used in Eq. (23-21). Therefore,

$$f_o = \frac{f_s(V + v_o)}{V} = \frac{(800 \text{ Hz})(340 \text{ m/s} - 30 \text{ m/s})}{340 \text{ m/s}}$$

$$= \frac{(800)(310)}{340} = 729 \text{ Hz}$$

If both the observer and the source are moving Eqs. (23-21) and (23-22) are combined to yield

$$f_o = f_s \frac{V + v_o}{V - v_s}$$

General Motion (23-23)

The same sign convention must be used for this general equation as that developed earlier. Note that Eq. (23-23) will reduce to Eq. (23-21) or to Eq. (23-22) if the observer is stationary ($v_o = 0$) or if the source is stationary ($v_s = 0$).

SUMMARY

We have defined sound as a longitudinal mechanical wave in an elastic medium. Thus the elasticity and density of a medium will affect the speed of sound as it travels in that medium. Under certain conditions we have seen that standing sound waves can produce characteristic frequencies which we observe as the pitch of the sound. The intensity of sound and the Doppler effect were also discussed in this chapter. The major concepts are summarized below:

- Sound is a longitudinal wave traveling through an elastic medium. Its speed in air at 0°C is 331 m/s or 1087 ft/s. At other temperatures the speed of sound is approximated by

$$v = 331 \text{ m/s} + \left(0.6 \frac{\text{m/s}}{\text{C}°}\right) t_C$$

Speed of S~ ¹ in Air

- The speed of sound in other media can be found from the following:

$$v = \sqrt{\frac{Y}{\rho}}$$ *Rod* $$v = \sqrt{\frac{\gamma P}{\rho}} = \sqrt{\frac{\gamma RT}{M}}$$ *Gas*

$$v = \sqrt{\frac{B}{\rho}}$$ *Fluid* $$v = \sqrt{\frac{B + \frac{4}{3}S}{\rho}}$$ *Extended Solid*

- Standing longitudinal sound waves may be set up in a vibrating air column for a pipe that is open at both ends or for one that is closed at one end. The characteristic frequencies are

$$f_n = \frac{nv}{2l}$$ $n = 1, 2, 3, \ldots$ *Open Pipe of Length l*

$$f_n = \frac{nv}{4l}$$ $n = 1, 3, 5, \ldots$ *Closed Pipe of Length l*

Note that *only the odd harmonics are possible for a closed pipe.* In this case, the first overtone is the third harmonic, the second overtone is the fifth harmonic, and so on.
- The intensity of a sound is the power P unit area A perpendicular to the direction of propagation.

$$I = \frac{P}{A} = 2\pi^2 f^2 A^2 \rho v$$ *Intensity*, W/m^2

- The intensity level in decibels is given by

$$\beta = 10 \log \frac{I}{I_0}$$ $I_0 = 1 \times 10^{-12}$ W/m^2 *Intensity level*

- Whenever two waves are nearly the same frequency and exist simultaneously in the same medium, beats are set up such that

$$\text{Number of beats per second} = f - f'$$

- The general equation for the Doppler effect is

$$f_o = f_s \frac{V + v_o}{V - v_s}$$ *Doppler Effect*

where f_o = observed frequency
f_s = frequency of source
V = velocity of sound
v_o = velocity of observer
v_s = velocity of source

Note: Speeds are reckoned as positive for approach and negative for recession.

23-1. Define the following terms:

a. Sound
b. Compression
c. Condensation
d. Resonance
e. Infrasonic
f. Ultrasonic
g. Loudness
h. Intensity
i. Decibels
j. Pitch

k. Forced vibration
l. Audible sound
m. Hearing threshold
n. Pain threshold
o. Intensity level
p. Frequency
q. Quality
r. Waveform
s. Beats
t. Doppler effect

23-2. What is the physiological definition of sound? What is the meaning of sound in physics?

23-3. Why must astronauts on the surface of the moon communicate by radio? Can they hear another spacecraft as it lands nearby? Can they hear by touching helmets?

23-4. How is the sound of a person's voice affected by inhaling helium gas? Is the effect one of pitch, loudness, or quality?

23-5. Vocal sounds originate with the vibration of vocal cords. The mouth and nasal openings act as a resonant cavity to amplify and distinguish sounds. Suppose you hum at a constant pitch equal to the C note on a piano. By opening and closing your mouth, what physiological property of the sound is affected?

23-6. The distance in miles to a thunderstorm can be estimated by counting the number of seconds elapsing between the flash of lightning and the arrival of a clap of thunder and dividing the result by 5. Explain why this is a reasonable approximation.

23-7. A store window is broken by an explosion several miles away. A glass of thin crystal shatters when a high note is reached on a violin. Are 'ie causes of damage similar? What physical property of sound was principally responsible in each case?

23-8. Compare the speeds of sound in solids, liquids, and gases. Explain the reason for differences in speed.

23-9. Perform a unit analysis of Eq. (23-1) showing that $\sqrt{Y/\rho}$ will yield units of speed.

23-10. How will the speed of sound in a gas be affected if the temperature of the gas is quadrupled?

23-11. An electric bell operates inside an evacuated flask. No sound is heard because of the absence of a medium. Explain what happens if the flask is tilted until the bell touches the walls of the flask.

23-12. Draw diagrams to demonstrate the differences between a progressive longitudinal wave and a standing longitudinal wave.

23-13. A standing wave is set up in a vibrating string. How are the harmonics of the possible sounds related to the number of loops in the string? How are the harmonics related to the number of nodes?

23-14. What effect will closing one end of an open pipe have on the frequency of a vibrating air column?

23-15. Compare the quality of sound produced by a violin with that produced by tuning fork.

23-16. If the average ear cannot hear sounds of frequencies much in excess of 15,000 Hz, what is the advantage of building stereo music systems which have frequency responses much higher than 15,000 Hz?

23-17. A vibrating tuning fork mounted on a resonating box is moved toward a wall and away from an observer. The resulting sound pulsates in intensity. Explain.

23-18. An instructor attempts to explain the Doppler effect by using baseballs and a bicycle. He proceeds as follows: "Suppose I am at rest, and I release one baseball in the same direction every second at constant speed. Consider me as the source of sound waves and the baseballs as advancing wavefronts. The spacing between the balls at any instant will be constant and analogous to the wavelength of sound waves. Now, suppose I ride a bicycle in the forward direction at a constant speed and continue to release balls in the forward direction and in the backward direction at the same rate and at the same speed. The spacing of the balls in front of me will be closer together because each time I release a ball in that direction I will have also moved in that direction. Similarly, the balls released in the backward direction will be spaced further apart than normal." Give a careful analysis of his explanation. In what ways is his analogy correct? In what very important aspect does his analogy fail? Why would an equation similar to Eq. (23-20) fail as a means of predicting the spacing of the baseballs? Why does it work for sound waves?

PROBLEMS

The Speed of Sound Waves

23-1. Compute the speed of sound in a copper rod of density 8800 kg/m^3. Young's modulus for copper is 11×10^{10} Pa.

Ans. 3540 m/s

23-2. The speed of longitudinal waves in a certain metal rod of density 7850 kg/m^3 is measured to be 3380 m/s. (a) What is Young's modulus for the metal? (b) If the frequency of the waves is 312 Hz, what is the wavelength?

23-3. Compare the theoretical speeds of sound in hydrogen ($\gamma = 1.4$) and helium ($\gamma = 1.66$) at 0°C. The molecular masses for hydrogen and helium are $M_H = 2.0$ and $M_{He} = 4.0$.

Ans. $V_{He} = 0.77 v_H$

* 23-4. A sound wave is sent from a ship to the ocean floor, where it is reflected and returned. If the round trip takes 0.6 s, how deep is the ocean floor? Consider the bulk modulus for seawater to be 2.1×10^9 Pa and its density to be 1030 kg/m^3.

Vibrating Air Columns

23-5. Find the fundamental frequency and the first three overtones for a 20-cm pipe at 20°C (a) if the pipe is open at both ends and (b) if the pipe is closed at one end.
Ans. (a) 858, 1720, 2570, and 3430 Hz; (b) 429, 1290, 2140, and 3000 Hz

23-6. What length of closed pipe is needed to produce a fundamental frequency of 256 Hz? What length of open pipe will give this same fundamental frequency? Assume the temperature to be 0°C.

23-7. The laboratory apparatus shown in Fig. 23-12 is used to measure the speed of sound in air by the resonance method. A vibrating tuning fork of frequency f is held over the open end of a tube partly filled with water. The length of the air

Fig. 23-12 Laboratory apparatus for computing the velocity of sound by resonance methods.

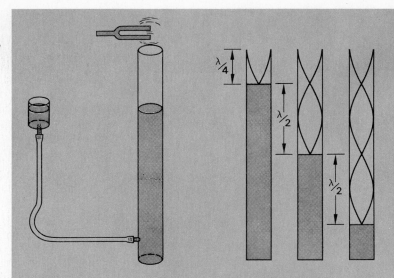

column can be varied by changing the water level. As the water level is gradually lowered from the top of the tube, the sound intensity reaches a maximum at the three levels shown in the figure. The maxima occur whenever the air column resonates with the tuning fork. Thus the distance between successive resonance positions is the distance between adjacent nodes for the standing waves in the air column. The frequency of the fork is 512 Hz, and the resonance positions occur at 17, 51, and 85 cm from the top of the tube. (a) What is the velocity of sound in the air? (b) What is the approximate temperature of the room?

Ans. (a) 348 m/s, (b) 28.3°C

* 23-8. In a resonance experiment, the air in a closed tube of variable length is found to resonate with a tuning fork when the air column is first 6 cm and then 18 cm long. Assuming that the temperature of the air is 10°C, find the frequency of the tuning fork.

* 23-9. A closed pipe and an open pipe are each 3 m long. Compute the wavelength of the fourth overtone for each pipe.

Ans. 1.33 m, 1.2 m

Sound Intensity and Intensity Level

23-10. What is the intensity level in decibels of a sound whose intensity is 4×10^{-6} W/m²?

23-11. Compute the intensity levels in decibels of sounds of the following intensities: (a) 1×10^{-7} W/m², (b) 10 W/m², (c) 2×10^{-3} W/m².

Ans. (a) 50 dB, (b) 130 dB, (c) 93 dB

* 23-12. What is the intensity of a sound whose intensity level is 40 dB?

* 23-13. What is the intensity of a 30-dB sound? If the sound level is doubled to 60 dB, what is the new intensity of the sound?

Ans. 1×10^{-9} W/m², 1×10^{-6} W/m²

* 23-14. How much more intense is a sound that is 20 dB higher than the hearing threshold?

The Doppler Effect

23-15. The fundamental frequency of a train whistle is 300 Hz, and the speed of the train is 20 m/s. On a day when the temperature is 20°C, what frequencies will be heard by a stationary observer as the train passes with its whistle blowing?

Ans. 319, 300, and 283 Hz

23-16. A person in a stranded car blows a 400-Hz horn. What frequencies are heard by the driver of a car passing at a speed of 60 km/h? The speed of sound is 343 m/s.

* **23-17.** An ambulance moves from right to left at 15 m/s. Its siren has a frequency of 600 Hz at rest. A car heads toward the ambulance at 20 m/s. What frequency is heard by the driver of the car **(a)** before they pass and **(b)** after they pass? The velocity of sound is 343 m/s.

Ans. **(a)** 664 and **(b)** 541 Hz

Additional Problems

* **23-18.** On a day when the air temperature is 27°C, a stone is dropped down a mine shaft 200 m deep. How much later will its impact on the bottom be heard? (Allow for time to drop.)

* **23-19.** What is the length of a closed pipe if the frequency of its second overtone is 900 Hz on a day when the temperature is 20°C?

Ans. 47.6 cm

23-20. How many beats per second are heard when two tuning forks of 256 and 259 Hz are sounded together?

* **23-21.** A 60-cm steel rod is first clamped at one end and then it is clamped at its midpoint as shown in Fig. 23-13. Sketch the fundamental and first overtone for these boundary conditions. What are the wavelengths in each case?

Ans. **(a)** 240 and 80 cm; **(b)** 60 and 30 cm

Fig. 23-13

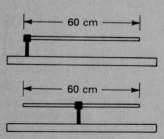

* **23-22.** What is the velocity of sound in a steel rod ($Y = 2 \times 10^{11}$ Pa and =7800 kg/m^3). What is the length of the steel rod shown in Fig. 23-13a if the fundamental frequency of vibration for the rod is 3000 Hz?

* **23-23.** A certain loudspeaker has a circular opening with an area of 6 cm^2. The power being radiated by this speaker is 6×10^{-7} W. What is the intensity of the sound at the opening? What is the intensity level?

Ans. 1×10^{-3} W/m^2, 90 dB

* **23-24.** Find the ratio of the intensities of two sounds if one is 12 dB higher than the other.

23-25. What is the difference in decibels between two sounds whose intensities are 2×10^{-5} W/m^2 and 0.90 W/m^2?

Ans. 46.5 dB

23-26. The noon whistle at a textile mill has a frequency of 360 Hz. What are the frequencies heard by the driver of a car passing the mill at 25 m/s on a day when the air temperature is 20°C?

* 23-27. A tuning fork of frequency 512 Hz is moved away from an observer and toward a flat wall with a speed of 3 m/s. The speed of sound in the air is 340 m/s. **(a)** What is the apparent frequency of the unreflected sound waves? **(b)** What is the apparent frequency of the reflected sound waves? **(c)** How many beats are heard each second?

Ans. **(a)** 508 Hz, **(b)** 516 Hz, **(c)** 9 beats/s

24

Light and Illumination

After completing this chapter, you should be able to:

1. Discuss the historical investigation into the nature of light and explain how light sometimes behaves as a wave and sometimes as particles.
2. Describe the broad classifications in the electromagnetic spectrum on the basis of frequency, wavelength, or energy.
3. Write and apply formulas for the relationship between velocity, wavelength, and frequency, and between energy and frequency for electromagnetic radiation.
4. Describe experiments which will result in a reasonable estimation of the speed of light.
5. Illustrate with drawings your understanding of the formation of shadows, labeling the *umbra* and *penumbra*.
6. Demonstrate your understanding of the concepts of *luminous flux, luminous intensity,* and *illumination,* and solve problems similar to those in the text.

An iron bar resting on a table is in thermal equilibrium with its surroundings. From its outward appearance one would never guess that it is very active internally. All objects are continuously emitting radiant heat energy which is related to their temperature. The bar is in thermal equilibrium only because it is radiating and absorbing energy at the same rates. If the balance is upset by placing one end of the bar in a hot flame, the bar becomes more active internally and emits heat energy at a greater rate. As the heating continues to around 600°C, some of the radiation emitted from the bar becomes *visible*. That is, it affects our sense of sight. The color of the bar becomes a dull red, which turns brighter as more heat is supplied.

The radiant energy emitted by the object before this effect is visible consists of electromagnetic waves of longer wavelengths than red light. Such waves are referred to as *infrared rays,* meaning "beyond the red." If the temperature of the bar is increased to around 3000°C, it becomes white hot, indicating a further extension of the radiant energy into the visible region.

This example sets the stage for our discussion of light. The nature of light is no different fundamentally from the nature of other electromagnetic radiations, e.g., heat, radio waves, or ultraviolet radiation. The characteristic which distinguishes light from the other radiations is its energy.

> **Light** is electromagnetic radiation which is capable of affecting the sense of sight.

The energy content of visible light varies from about 2.8×10^{-19} J to around 5.0×10^{-19} J.

24-1
WHAT IS LIGHT?

The answer to that question has been extremely elusive throughout the history of science. The long search for an answer provides an inspiring example of the scientific approach to the solution of a problem. Every hypothesis put forth to explain the nature of light was tested both by logic and by experimentation. The ancient philosophers' contention that visual rays are emitted from the eye to the seen object failed the test of both logic and experience.

By the latter part of the seventeenth century, two theories were being advanced to explain the nature of light, the particle (corpuscular) theory and the wave theory. The principal advocate of the corpuscular theory was Sir Isaac Newton. The wave theory was upheld by Christiaan Huygens (1629–1695), Dutch mathematician and scientist, 13 years older than Newton. Each theory set out to explain the characteristics of light observed at that time. Three important characteristics can be summarized as follows:

1. *Rectilinear propagation:* Light travels in straight lines.
2. *Reflection:* When light is incident on a smooth surface, it turns back into the original medium.
3. *Refraction:* The path of light changes when it enters a transparent medium.

According to the corpuscular theory, tiny particles of unsubstantial mass were emitted by light sources such as the sun or a flame. These particles traveled outward from the source in straight lines at enormous speeds. When the particles entered the eye, the sense of sight was stimulated. Rectilinear propagation was easily explained in terms of particles. In fact, one of the strongest arguments for the corpuscular theory was based on this property. It was reasoned that particles cast sharp shadows, as illustrated in Fig. 24-1a, whereas waves can bend around edges. Such bending of waves, as shown in Fig. 24-1b, is called *diffraction*.

Fig. 24-1 A strong argument for the particle theory of matter is the formation of sharp shadows. Waves were known to bend around obstacles in their path.

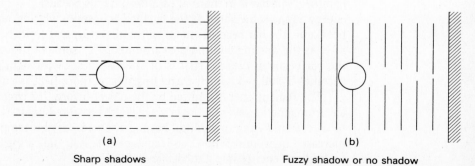

(a)

Sharp shadows

(b)

Fuzzy shadow or no shadow

The sharp shadows formed by light beams indicated to Newton that light must consist of particles. Huygens, on the other hand, explained that the bending of water waves and sound waves around obstacles is easily noticed because of the long wavelengths. He reasoned that if light were a wave with a very short wavelength it would appear to cast a sharp shadow because the amount of bending would be very small.

It was also difficult to explain why particles traveling in straight lines from many directions could cross without impeding one another. In a paper published in 1690, Huygens wrote:

> If, furthermore, we pay attention to, and weigh up, the extraordinary speed with which light spreads in all directions, and also the fact that coming, as it does, from quite different, indeed from opposite, directions, the rays interpenetrate without impeding one another, then we may well understand that whenever we see a luminous object, this cannot be due to the transmission of matter which reaches us from the object, as for instance a projectile or an arrow flies through the air.

Huygens explained the propagation of light in terms of the motion of a disturbance through the distance between a source and the eye. He based his argument on a simple principle which is still useful today in describing the propagation of light. Suppose we drop a stone into a quiet pool of water. A disturbance is created which moves outward from the point of impact in a series of concentric waves. The disturbance continues even after the stone has struck the bottom of the pool. Such an example prompted Huygens to postulate that disturbances existing at all points along a moving wavefront at one instant can be considered sources for the wavefront at the next instant. Huygens' principle states:

> *Every point on an advancing wavefront can be considered a source of secondary waves called wavelets. The new position of the wavefront is the envelope of the wavelets emitted from all points of the wavefront in its previous position.*

The application of this principle is illustrated in Fig. 24-2 for the common cases of a plane wave and a circular wave.

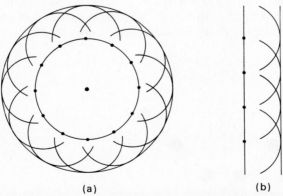

(a)   (b)

Fig. 24-2 Illustration of Huygens' principle *(a)* for a spherical wave and *(b)* for a plane wave.

Huygens' principle was particularly successful in explaining reflection and refraction. Figure 24-3 shows how the principle can be used to explain the bending of light as it passes from air to water. When the plane waves strike the water surface at an angle, points *A, C,* and *E* become the sources of new wavelets. The envelope of these secondary wavelets indicates a change in direction. A similar construction can be made to explain reflection.

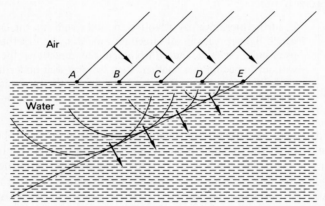

Fig. 24-3 Huygens' explanation of refraction in terms of the wave theory.

Reflection and refraction were also easily explained in terms of the particle theory. Figures 24-4 and 24-5 illustrate models which can be used to explain reflection and refraction on the basis of tiny corpuscles. Perfectly elastic particles of unsubstantial mass rebounding from an elastic surface could explain the regular reflection of light from smooth surfaces. Refraction could be analogous to the change in direction of a rolling ball as it encounters an incline. This explanation required that the particles of light travel faster in the refracted medium, whereas the wave theory required light to travel more slowly in the refracted medium. Newton recognized that if it could ever be shown that light travels more slowly in a material medium than it does in air, he would have to abandon the particle theory. It was not until the middle of the nineteenth century that Jean Foucault successfully demonstrated that light travels more slowly in water than it does in air.

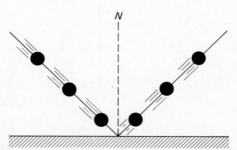

Fig. 24-4 Explanation of reflection in terms of the particle theory of light.

HEAT, LIGHT, AND SOUND

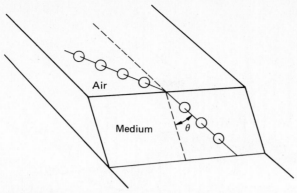

Fig. 24-5 Refraction of light as it passes from air to another medium was explained by this mechanical example.

24-2

THE PROPAGATION OF LIGHT

The discovery of interference and diffraction in 1801 and 1816 swung the debate solidly toward Huygens' wave theory. Clearly, interference and diffraction could be explained only in terms of a wave theory. However, there still remained one problem. All wave phenomena were thought to require the existence of a medium. How, for example, could light waves travel through a vacuum if there were nothing to "vibrate"? Indeed, how could light reach the earth from the sun or other stars through millions of miles of empty space? In order to avoid this contradiction, physicists postulated the existence of a "light-carrying ether." This all-penetrating universal medium was thought to fill all the space between and within all material bodies. But what was the nature of this ether? Certainly it could not be a gas, solid, or liquid that obeyed the physical laws known at that time. And yet the wave theory could not be denied in light of the evidence of interference and diffraction. No other choice seemed possible except for the definition of ether as "that which carries light."

In 1865 a Scottish physicist, James Clerk Maxwell, set out to determine the properties of a medium which would carry light and also account for the transmission of heat and electric energy. His work demonstrated that *an accelerated charge can radiate electromagnetic waves into space.* Maxwell explained that the energy in an electromagnetic wave is equally divided between electric and magnetic fields which are mutually perpendicular. Both fields oscillate perpendicular to the direction of wave propagation, as shown in Fig. 24-6. Thus a light wave would not have to depend on the vibration of matter. It could be propagated by oscillating transverse fields. Such a wave could "break off" from the region around an accelerating charge and fly off into space with the velocity of light. Maxwell's equations predicted that heat and electric action, as well as light, were propagated at the speed of light as electromagnetic disturbances.

Experimental confirmation of Maxwell's theory was achieved in 1885 by H. R. Hertz, who proved that radiation of electromagnetic energy can occur *at any frequency.* In other words, light, heat radiation, and radio waves are of the same nature, and they all travel at the speed of light (3×10^8 m/s). All types of radiation can be reflected, focused by lenses, polarized, and so forth. It seemed that the wave nature of light could no longer be doubted.

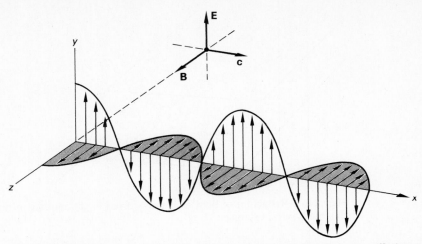

Fig. 24-6 Electromagnetic theory holds that light is propagated as oscillating transverse fields. The energy is equally divided between electric E and magnetic B fields, which are mutually perpendicular.

Confirmation of the electromagnetic theory paved the way for the eventual downfall of the "light-carrying ether" postulate. In 1887, A. A. Michelson, an American physicist, showed conclusively that the velocity of light is a constant, independent of the motion of the source. He could not establish any difference between the speed of light traveling in the direction of the earth's motion and traveling opposite to the earth's motion. Those who are interested in a thorough discussion of the subject should look up the *Michelson–Morley experiment*. Einstein later interpreted Michelson's results to mean that the notion of ether must be abandoned in favor of a completely empty space.

24-3

THE ELECTRO-MAGNETIC SPECTRUM

Today, the electromagnetic spectrum is known to spread over a tremendous range of frequencies. A chart of the electromagnetic spectrum is presented in Fig. 24-7. The wavelength λ of electromagnetic radiation is related to its frequency f by the general equation

$$c = f\lambda \tag{24-1}$$

where c is the velocity of light (3×10^8 m/s). In terms of wavelengths, the tiny segment of the electromagnetic spectrum referred to as the *visible region* lies between 0.00004 and 0.00007 cm.

Because of the small wavelengths of light radiation, it is more convenient to define smaller units of measure. The SI unit is the *nanometer* (nm).

One **nanometer** (1 nm) is defined as one-billionth of a meter.

$$1 \text{ nm} = 10^{-9} \text{ m} = 10^{-7} \text{ cm}$$

Other older units are the *millimicron* ($m\mu$), which is the same as a nanometer, and the *angstrom* (Å), which is 0.1 nm.

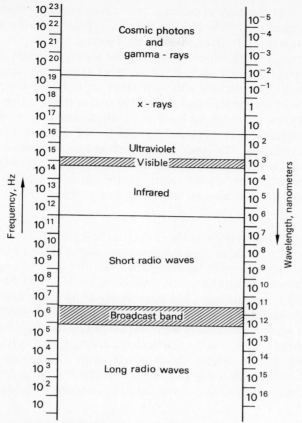

Fig. 24-7 Electromagnetic spectrum.

The visible region of the electromagnetic spectrum extends from 400 nm for violet light to approximately 700 nm for red light.

EXAMPLE 24-1   The wavelength of yellow light from a sodium flame is 589 nm. Compute its frequency.

Solution   The frequency is found from Eq. (24-1).

$$f = \frac{c}{\lambda} = \frac{3 \times 10^8 \text{ m/s}}{589 \times 10^{-9} \text{ m}}$$
$$= 5.09 \times 10^{14} \text{ Hz}$$

Newton was the first to make detailed studies of the visible region by dispersing "white light" through a prism. In order of increasing wavelength, the spectral colors are violet (450 nm), blue (480 nm), green (520 nm), yellow (580 nm), orange (600 nm), and red (640 nm). Anyone who has seen a rainbow has seen the effects that different wavelengths of light have on the human eye.

The electromagnetic spectrum is continuous; there are no gaps between one form of radiation and another. The boundaries we set are purely arbitrary, depending upon our ability to sense one small portion directly and to discover and measure those portions outside the visible region.

The first discovery of radiation of wavelengths longer than those of red light was accomplished in 1800 by William Herschel. These waves are now known as thermal radiation and are referred to as *infrared waves.*

Shortly after the discovery of infrared waves, radiation of wavelengths shorter than visible light were noticed. These waves, now known as *ultraviolet waves,* were discovered in connection with their effect on certain chemical reactions.

How far the infrared region might extend in the direction of longer wavelengths was not known throughout most of the nineteenth century. Fortunately, Maxwell's electromagnetic theory opened the door for the discovery of many other classifications of radiation. The spectrum of electromagnetic waves is now conveniently divided into the eight major regions shown in Fig. 24-7: (1) long radio waves, (2) short radio waves, (3) the infrared region, (4) the visible region, (5) the ultraviolet region, (6) x-rays, (7) gamma-rays, and (8) cosmic photons.

<table>
<tr><td>24-4</td></tr>
<tr><td>THE
QUANTUM
THEORY</td></tr>
</table>

The work of Maxwell and of Hertz in establishing the electromagnetic nature of light waves was truly one of the most important events in the history of science. Not only was the wave nature of light explained, but the door was opened to an enormous range of electromagnetic waves. Remarkably, only 2 years after Hertz' verification of Maxwell's wave equations, the wave theory of light was again challenged. In 1887, Hertz noticed that an electric spark would jump more readily between two charged spheres when their surfaces were illuminated by the light from another spark. This phenomenon, known as the *photoelectric effect,* is demonstrated by the apparatus in Fig. 24-8. A beam of light strikes a metal surface *A* in an evacuated tube. Electrons ejected by the light are drawn to the collector *B* by external batteries. The flow of electrons is indicated by a device called an *ammeter.* The photoelectric effect defied

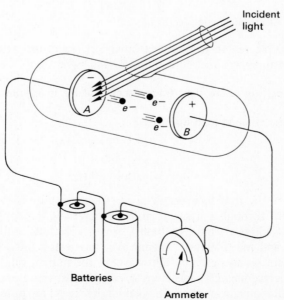

Fig. 24-8 The photoelectric effect.

explanation in terms of the wave theory. In fact, the ejection of electrons could be accounted for more easily in terms of the old particle theory. Still, there could be no doubt of the wave properties either. Science faced a remarkable paradox.

The photoelectric effect, along with several other experiments involving the emission and absorption of radiant energy, could not be accounted for purely in terms of Maxwell's electromagnetic wave theory. In an attempt to bring experimental observation into agreement with theory, Max Planck, a German physicist, published his *quantum* hypothesis in 1901. He found that the problems with the radiation theory lay with the assumption that energy is radiated continuously. It was postulated that electromagnetic energy is absorbed or emitted in discrete packets, or *quanta*. The energy content of these quanta, or *photons* as they were called, is proportional to the frequency of the radiation. Planck's equation can be written

$$E = hf \qquad (24\text{-}2)$$

where E = energy of photon
$\quad f$ = frequency of photon
$\quad h$ = proportionality factor called Planck's constant (6.625×10^{-34} J/Hz)

In 1905, Einstein extended the idea proposed by Planck and postulated that the energy in a light beam does not spread continuously through space. By assuming that light energy is concentrated in small packets (photons) whose energy content is given by Planck's equation, Einstein was able to predict the photoelectric effect mathematically. At last, theory was reconciled with experimental observation.

Thus it appears that light is *dualistic*. The wave theory is retained by considering the photon to have a frequency and an energy proportional to the frequency. The modern practice is to use the wave theory when studying the propagation of light. The corpuscular theory, on the other hand, is necessary to describe the interaction of light with matter. We may think of light as radiant energy transported in photons carried along by a wave field.

The origin of light photons was not understood until Niels Bohr in 1913 devised a model of the atom based on quantum ideas. Bohr postulated that electrons can move about the nucleus of an atom only in certain orbits or *discrete energy levels,* as shown in Fig. 24-9. Atoms were said to be *quantized*. If the atoms are somehow energized, as by heat, the orbital electrons may be caused to jump into a *higher* orbit. At some later time, these excited electrons will fall back to their original level, releasing as photons the energy which was originally absorbed. Although Bohr's model was not strictly correct, it provided a basis for understanding the emission and absorption of electromagnetic radiation in quantum units.

24-5
THE SPEED OF LIGHT

The speed of light is probably the most important constant used in physics, and its accurate determination represents one of the most precise measurements accomplished by man. Its magnitude is so great (about 186,000 mi/s) that experimental measurements were entirely unsuccessful until the latter part of the seventeenth century. It was generally believed that the transmission of light must be instantaneous.

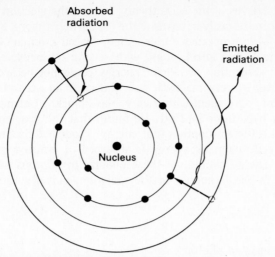

Fig. 24-9 Bohr's theory of the atom.

The first attempts to measure the speed of light experimentally were made in 1667 by Galileo. Briefly, his method consisted of stationing two observers in towers separated by a known distance. Signals were to be made at night with lanterns. The second experimenter was to uncover his lantern as soon as he received a light signal from the first experimenter. The speed of light could then be determined from the time required for the light to travel between the two towers. The experiment was inconclusive, and Galileo merely added to the prevailing opinion that light transmission was instantaneous. Such an experiment seems humorous to us only because of our present understanding of the true magnitude of the velocity of light.

Eight years later, a Danish astronomer, Olaus Roemer, made the first measurement of the speed of light. He based his calculations on irregularities in the predicted eclipses of one of the moons of the planet Jupiter. Roemer had been measuring the time interval between successive eclipses for several years. He noticed a consistent variation in his computations when measurements were made from different portions of the earth's orbit about the sun. Roemer correctly concluded that the irregularities in his measurements were due to the distance which light traveled to reach the earth. On the basis of his rather inaccurate measurements, Roemer calculated that the speed of light was 140,000 mi/s.

The first successful terrestrial measurement of the speed of light was made by A. H. L. Fizeau, a French scientist. In 1849, his experiment involved the simple computation of the time required for light to traverse a known distance. A schematic diagram of the apparatus is shown in Fig. 24-10. A rotating toothed wheel W allows light coming from the source to be interrupted, producing a series of short flashes. Let us assume first that the wheel is stationary and that light passes through one of the openings between the teeth. Light from the source S is reflected by the half-silvered plate glass P through the opening in the wheel and onto the plane mirror M located a few miles away. The reflected light returns through the opening and passes through the glass P and to the observer O.

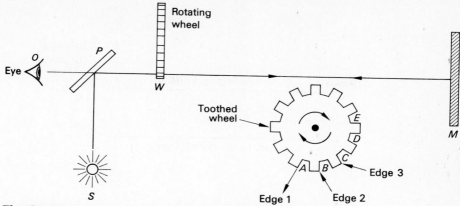

Fig. 24-10 Fizeau's apparatus for measuring the speed of light.

Now suppose the toothed wheel is rotated at a constant frequency. Light from the source will pass through the opening A to the distant mirror and back to the wheel. The rotational frequency of the wheel is adjusted so that the returning light will just pass through the opening B. Under these conditions light from the source passes by edge 1 onto the mirror and returns past edge 2. Thus the time required for the light to travel the known distance is the same as the time for the wheel to rotate through the combined width of one opening and one tooth. The speed of light is therefore a function of the known frequency of rotation of the wheel and can be calculated. If the frequency is gradually increased, the light will eventually be blocked again until it reappears past edge 3. The speed of light calculated by Fizeau was 3.13×10^8 m/s. The error is attributed to inaccurate measurements.

In 1850, Foucault refined the apparatus developed by Fizeau by replacing the toothed wheel with a rotating mirror. He is remembered primarily for his measurements of the velocity of light in water. It was the first conclusive evidence that light travels more slowly in water than in air.

Probably no other scientist is more remembered for his work on measuring the speed of light than Albert A. Michelson (1852–1931). Using the Foucault method, he was able to obtain measurements which were extremely accurate. A schematic diagram of his apparatus is shown in Fig. 24-11. The light source, the eight-sided rotating mirror, and the telescope were located on Mt. Wilson, California. The reflecting mirrors were located approximately 22 miles away on the top of Mt. San Antonio. From the known distances and the time required for light to complete the trip, Michelson determined that the speed of light in air is 2.997×10^8 m/s. This value is accepted as one of the most precise measurements made using the Foucault principle.

Research concerning the velocity of light has continued until recent times when laser techniques have produced the value

$$c = 2.99792457 \times 10^8 \text{ m/s}$$

which is accurate to about 12 parts in 100 million. However, this figure is based on the older definition of the meter in terms of the krypton wavelength. The second can

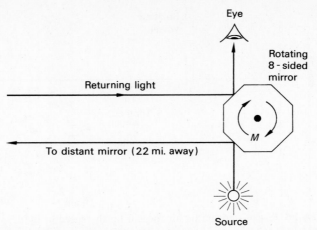

Fig. 24-11 Michelson's method for measuring the speed of light in air.

now be measured with an uncertainty of only 1 part in 10 trillion. As discussed in the chapter on measurements, the meter was redefined in 1983 in terms of the unit of time. More precisely, the meter is established as the distance traveled by light in a time of 1/299,792,458 s as measured by a cesium clock. This decision established the velocity of light, by definition, to be

$$c = 2.99792458 \times 10^8 \text{ m/s} \qquad \text{(exactly)} \qquad (24\text{-}3)$$

Useful approximations are 3×10^8 m/s and 186,000 mi/s. These values may be used in most physical calculations without fear of significant error.

24-6
LIGHT RAYS AND SHADOWS

One of the first properties of light to be studied was rectilinear propagation and the formation of shadows. Instinctively, we rely quite heavily on this property for estimating distances, directions, and shapes. The formation of sharp shadows on a sundial is used to estimate time. In this section we discuss how we can predict the formation of shadows.

According to Huygens' principle, every point on a moving wavefront can be considered a source for secondary wavelets. The wavefront at any time is the envelope of these wavelets. Thus the light emitted in all directions by the point source of light in Fig. 24-12 can be represented by a series of spherical wavefronts moving away from the source at the speed of light. For our purposes, a *point source* of light is one whose dimensions are small in comparison with the distances studied. Note that the spherical wavefronts become essentially plane wavefronts in any specific direction at a long distance away from the source. An imaginary straight line drawn perpendicular to the wavefronts in the direction of the moving wavefronts is called a *ray*. There are, of course, an infinite number of rays starting from the point source.

Any dark-colored object absorbs light, but a black one absorbs nearly all the light it receives. Light that is not absorbed upon striking an object is either reflected or transmitted. If all the light incident upon an object is reflected or absorbed, the object is said to be *opaque*. Since light cannot pass through an opaque body, a

HEAT, LIGHT, AND SOUND

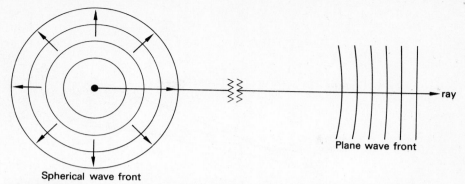

Spherical wave front

Plane wave front

Fig. 24-12 A ray is an imaginary line drawn perpendicular to advancing wavefronts, which indicates the direction of light propagation.

shadow will be produced in the space behind the object. The shadow formed by a point source of light is illustrated in Fig. 24-13. Since light is propagated in straight lines, rays drawn from the source past the edges of the opaque object form a sharp shadow proportional to the shape of the object. Such a region in which no light has entered is called an *umbra.*

If the source of light is an extended one rather than a point, the shadow will consist of two portions, as shown in Fig. 24-14. The inner portion receives no light from the source and is therefore the umbra. The outer portion is called the *penumbra.* An observer within the penumbra would see a portion of the source but not all the source. An observer located outside both regions would see all the source. Solar and lunar eclipses can be studied by similar construction of shadows.

<table>
<tr><td>**24-7**
LUMINOUS
FLUX</td><td>Most sources of light emit electromagnetic energy distributed over many wavelengths. Electric power is supplied to a lamp, and radiation is emitted. The radiant energy emitted per unit of time by the lamp is called the radiant power or the *radiant flux.* Only a small portion of this radiant power is in the visible region, i.e., in the region between 400 and 700 nm. The sensation of sight depends only on the visible, or *luminous,* energy radiated per unit of time.</td></tr>
</table>

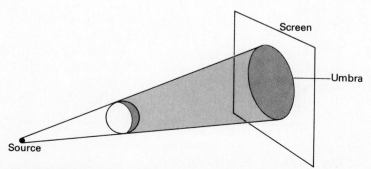

Fig. 24-13 Shadow formed by a point source of light.

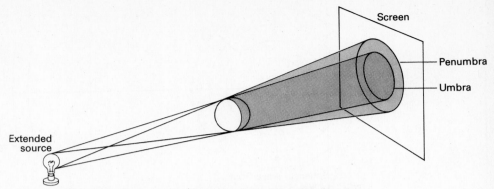

Fig. 24-14 Shadows formed by an extended source of light.

The **luminous flux** F is that part of the total radiant power emitted from a light source which is capable of affecting the sense of sight.

In a common incandescent light bulb, only about 10 percent of the energy radiated is luminous flux. The bulk of the radiant power is nonluminous.

The human eye is not equally sensitive to all colors. In other words, equal radiant power of different wavelengths does not produce equal brightness. A 40-W green light bulb appears brighter than a 40-W blue light bulb. A graph portraying the response of the eye to various wavelengths is shown in Fig. 24-15. Note that the sensitivity curve is bell-shaped around the center of the visible spectrum. Under normal conditions, the eye is most sensitive to yellow-green light of wavelength 555 nm. The sensitivity falls off rapidly for longer and shorter wavelengths.

If the unit chosen for luminous flux is to correspond to the sensual response of the human eye, a new unit must be defined. The watt (W) is not sufficient because the visual sensations are not the same for different colors. What is needed is a unit

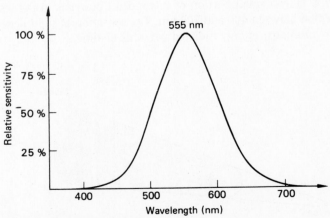

Fig. 24-15 Sensitivity curve.

which measures *brightness*. Such a unit is the *lumen* (lm), which is determined by comparison with a standard source.

To understand the definition of a lumen in terms of the standard source, we must first develop the concept of a solid angle. A solid angle in steradians (sr) is defined the same way a plane angle is defined in radians. In Fig. 24-16 the angle θ in radians is

$$\theta = \frac{S}{R} \quad \text{rad} \tag{24-4}$$

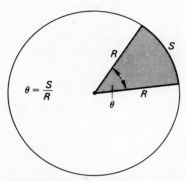

Fig. 24-16 Definition of a plane angle Θ expressed in radians.

where S is the arc length and R is the radius. The solid angle Ω is similarly defined by Fig. 24-17. It may be thought of as the opening from the tip of a cone which is subtended by a segment of area on the spherical surface.

> One **steradian** (sr) is the solid angle subtended at the center of a sphere by an area A on its surface that is equal to the square of its radius R.

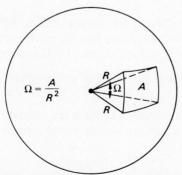

Fig. 24-17 Definition of a solid angle Ω in steradians.

In general, the solid angle in steradians is given by

$$\Omega = \frac{A}{R^2} \quad \text{sr}$$

(24-5)

The steradian, like the radian, is a unitless quantity.

Just as there are 2π rad in a complete circle, it can be shown by Eq. (24-5) that there are 4π sr in a complete sphere.

$$\Omega = \frac{A}{R^2} = \frac{4\pi R^2}{R^2} = 4\pi \quad \text{sr}$$

Notice that the solid angle is independent of the distance from the source. There are 4π sr in a sphere regardless of the length of its radius.

EXAMPLE 24-2   What solid angle is subtended at the center of an 8-m-diameter sphere by a 1.5-m² area on its surface?

Solution   Equation (24-5) yields

$$\Omega = \frac{1.5 \text{ m}^2}{(4 \text{ m})^2} = 0.0938 \text{ sr}$$

We are now in a position to clarify the definition of a unit which measures luminous flux. The *lumen* is defined by comparison with an internationally recognized standard source.

> One **lumen** (lm) is the luminous flux (or visible radiant power) emitted from a $\frac{1}{60}$-cm² opening in a standard source and included within a solid angle of 1 sr.

The standard source consists of a hollow enclosure maintained at the temperature of solidification of platinum, about 1773°C. In practice it is more convenient to use standard incandescent lamps which have been rated by comparison with the standard.

Another convenient definition of the lumen utilizes the sensitivity curve (Fig. 24-15) as a basis for establishing luminous flux. By referring to the standard source, 1 lm is defined in terms of the radiant power of yellow-green light.

> *One lumen is equivalent to* $\frac{1}{680}$ W *of yellow-green light of wavelength 555* nm.

To determine the luminous flux emitted by light of a different wavelength, the luminosity curve must be used to compensate for visual sensitivity.

EXAMPLE 24-3   A source of monochromatic red light (600 nm) produces a visible radiant power of 4 W. What is the luminous flux in lumens?

Solution If the light were yellow-green (555 nm) instead of red, it would have a luminous flux F given by

$$F = (680 \text{ lm/W})(4 \text{ W}) = 2720 \text{ lm}$$

From the sensitivity curve, red light of wavelength 600 nm evokes about 59 percent of the response obtained with yellow-green light. Hence the luminous flux issuing from the red light source is

$$F = (0.59)(2720 \text{ lm}) = 1600 \text{ lm}$$

The luminous flux is often calculated in the laboratory by determining the illumination it produces on a known surface area.

24-8 LUMINOUS INTENSITY

Light travels radially outward in straight lines from a source which is small in comparison with its surroundings. For such a source of light, the luminous flux included in a solid angle Ω remains the same at all distances from the source. Therefore, it is frequently more useful to speak of the *flux per unit solid angle* than simply to express the total flux. The physical quantity which expresses this relationship is called the *luminous intensity*.

The **luminous intensity** I of a source of light is the luminous flux F emitted per unit solid angle Ω.

$$I = \frac{F}{\Omega} \qquad (24\text{-}6)$$

The unit for intensity is the *lumen per steradian* (lm/sr), called a *candela* (cd). The candela, or *candle* as it was sometimes called, originated when the international standard was defined in terms of the quantity of light emitted by the flame of a certain make of candle. This standard was found unsatisfactory, and eventually it was replaced by the platinum standard.

EXAMPLE 24-4 Most light sources have different luminous intensities in different directions. An *isotropic* source is one which emits light uniformly in all directions. What is the total luminous flux emitted by an isotropic source of intensity I?

Solution From Eq. (24-6), the flux is given by

$$F = \Omega I$$

The total solid angle Ω for an isotropic source is 4π sr. Thus

$$F = 4\pi I \qquad \text{\textit{Isotropic Source}} \quad (24\text{-}7)$$

EXAMPLE 24-5 A spotlight is equipped with a 40-cd bulb which concentrates a beam on a vertical wall. The beam covers an area of 9 m^2 on the wall, and the spotlight is located 20 m from the wall. Calculate the luminous intensity of the spotlight.

Solution The total flux emitted by the 40-cd bulb is

$$F = 4\pi I = (4\pi)(40 \text{ cd}) = 160\pi \text{ lm}$$

This total flux is concentrated by reflectors and lens into a solid angle given by

$$\Omega = \frac{A}{R^2} = \frac{9 \text{ m}^2}{(20 \text{ m})^2} = 0.0225 \text{ sr}$$

The intensity of the beam is now found from Eq. (25-2).

$$I = \frac{F}{\Omega} = \frac{160\pi \text{ lm}}{0.0225 \text{ sr}} = 2.23 \times 10^4 \text{ cd}$$

Note that the units of intensity (cd) and the units of flux (lm) are the same dimensionally. This is true because the solid angle in steradians is dimensionless.

<table>
<tr><td>

24-9

ILLUMINATION

</td><td>

If the intensity of a source is increased, the luminous flux transmitted to each unit of surface area in the vicinity of the source is also increased. The surface appears brighter. In the measurement of light efficiency, the engineer is concerned with the density of luminous flux falling on a surface. We are therefore led to a discussion of the *illumination* of a surface.

</td></tr>
</table>

> The **illumination** E of a surface A is defined as the luminous flux F per unit area.

$$\boxed{E = \frac{F}{A}} \tag{24-8}$$

When the flux F is measured in lumens and the area A in square meters, the illumination E has the units of *lumens per square meter* or *lux* (lx). When A is expressed in square feet, E is expressed in *lumens per square foot*. The lumen per square foot is sometimes loosely referred to as the footcandle.

Direct application of Eq. (24-8) requires a knowledge of the luminous flux falling on a given surface. Unfortunately, the flux of common light sources is difficult to determine. For this reason, Eq. (24-8) is most often used to calculate the flux when A is known and E is computed from the measured intensity.

To see the relationship between intensity and illumination, let us consider a surface A at a distance R from a point source of intensity I, as shown in Fig. 24-18. The solid angle Ω subtended by the surface at the source is

$$\Omega = \frac{A}{R^2}$$

where the area A is perpendicular to the emitted light. If the luminous flux makes an angle θ with the normal to the surface, as shown in Fig. 24-19, we must consider the projected area $A \cos \theta$. This represents the effective area that the flux "sees." Therefore, the solid angle, in general, can be found from

$$\Omega = \frac{A \cos \theta}{R^2}$$

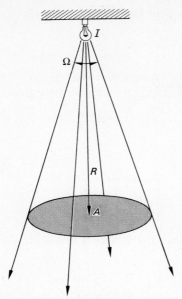

Fig. 24-18 Computing the illumination of a surface perpendicular to the incident flux.

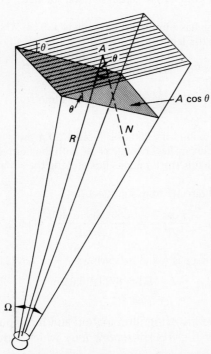

Fig. 24-19 When a surface makes an angle with the incident flux, the illumination is proportional to the component of the surface perpendicular to the flux.

Solving for the luminous flux F in Eq. (24-6), we obtain

$$F = I\Omega = \frac{IA \cos \theta}{R^2} \qquad (24\text{-}9)$$

We are now ready to express the illumination as a function of intensity. Substituting Eq. (24-9) into the defining equation for illumination gives

$$E = \frac{F}{A} = \frac{IA \cos \theta}{AR^2}$$

or

$$\boxed{E = \frac{I \cos \theta}{R^2}} \qquad (24\text{-}10)$$

For the special case in which the surface is normal to the flux, $\theta = 0°$, and Eq. (24-10) is simplified to

$$\boxed{E = \frac{I}{R^2}} \qquad \textit{Normal Surface} \quad (24\text{-}11)$$

You should verify that the units of *candela per square meter* are equivalent dimensionally to the units of *lumens per square meter* or *lux*.

EXAMPLE 24-6 A 100-W incandescent lamp has a luminous intensity of 125 cd. What is the illumination of a surface located 3 ft below the lamp?

Solution By direct substitution into Eq. (24-11) we obtain

$$E = \frac{I}{R^2} = \frac{125 \text{ cd}}{(3 \text{ ft})^2} = 13.9 \text{ lm/ft}^2$$

EXAMPLE 24-7 A tungsten-filament lamp of intensity 300 cd is located 2.0 m away from a surface whose area is 0.25 m^2. The luminous flux makes an angle of 30° with the normal to the surface. (*a*) What is the illumination? (*b*) What is the luminous flux striking the surface? (Refer to Fig. 24-19.)

Solution (a) The illumination is found directly from Eq. (24-10).

$$E = \frac{I \cos \theta}{R^2} = \frac{(300 \text{ cd})(\cos 30°)}{(2 \text{ m})^2} = 65 \text{ lx}$$

Solution (b) The flux falling on the surface is found by solving for F in Eq. (24-7). Thus

$$F = EA = (65 \text{ lx})(0.25 \text{ m}^2)$$
$$= 16.2 \text{ lm}$$

The equations above involving illumination and luminous intensity are mathematical formulations of the *inverse-square law*, which can be stated as follows:

> *The illumination of a surface is proportional to the luminous intensity of a point light source and is inversely proportional to the square of the distance.*

If a light which is illuminating a surface is raised to twice the original height, the illumination will be only one-fourth as great. If the lamp distance is tripled, the illumination is reduced by one-ninth. This inverse-square relationship is illustrated in Fig. 24-20.

Fig. 24-20 Illumination of a surface varies inversely with the square of the distance from a point source.

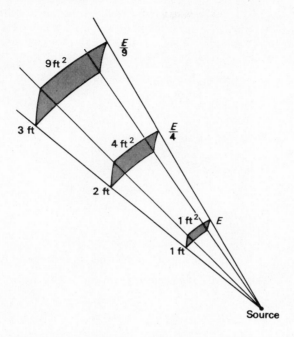

SUMMARY

The investigation of the nature of light is still continuing, but experiments show that it sometimes behaves as particles and sometimes as a wave. Modern theory holds that light is electromagnetic radiation, and that its radiant energy is transported in photons carried along by a wave field. The main ideas and formulas presented in this chapter are summarized below:

- The wavelength λ of electromagnetic radiation is related to its frequency f by the general equation

$$c = f\lambda \qquad c = 3 \times 10^8 \text{ m/s}$$

- The range of wavelengths for visible light goes from 400 nm for violet to 700 nm for red.

$$1 \text{ nm} = 10^{-9} \text{ m}$$

The Nanometer is Used for Wavelengths.

- The energy of light photons is proportional to the frequency.

$$E = hf \qquad E = \frac{hc}{\lambda} \qquad h = 6.626 \times 10^{-34} \text{ J/Hz}$$

The constant h is *Planck's constant.*

Luminous flux has been defined as that portion of the total radiant power emitted from a light source which is capable of affecting the sense of sight. Since visual brightness and sensitivity varies for different individuals, it was necessary to define luminous intensity in terms of a standard source and a well-defined solid angle (the steradian). By comparison to such standards, we are able to treat the illumination of surfaces, which is so important for the design of industrial workplaces. The basic concepts are summarized below:

- The *luminous intensity* of a light source is the luminous flux F per unit solid angle Ω. *Luminous flux* is the radiant power in the visible region. It is measured in *lumens.*

$$1 \text{ lm} = \frac{1}{680} \text{ W}$$ for 555 nm light *The Lumen*

$$1 \, \Omega = \frac{A}{R^2}$$ *Solid Angle in Steradians*

$$I = \frac{F}{\Omega}$$ *Luminous Intensity* (1 cd = 1 lm/sr)

- For an isotropic source, one emitting light in all directions, the luminous flux is

$$F = 4\pi I$$ *Isotropic Source*

- The *illumination* E of a surface A is defined as the luminous flux per unit area.

$$E = \frac{F}{A}$$ $$E = \frac{I \cos \theta}{R^2}$$ *Illumination* (lm/m^2), lx

QUESTIONS

24-1. Define the following terms:
a. light
b. electromagnetic waves
c. visible region
d. nanometer
e. quantum theory
f. photons
g. light ray
h. umbra
i. penumbra
j. luminous flux
k. luminous intensity
l. steradian
m. illumination
n. isotropic source

24-2. What is meant by the dualistic nature of light? In what ways does light behave as particles? In what ways does light behave as a wave?

24-3. Explain how the energy of an electromagnetic wave depends on its frequency and how it depends on the wavelength.

24-4. When light enters glass from air, its energy in glass is the same as its energy in air. Is its frequency the same? What about its wavelength? Explain.

24-5. Microwave ovens, television, and radar utilize electromagnetic waves between infrared and radio waves. Compare the energy, frequency, and wavelengths of these waves with the energy, frequency, and wavelengths of visible radiation.

24-6. Review the definition of a radian, and discuss how the steradian for solid angles is similar to the radian for plane angles. How many radians are in a complete circle? How many steradians are in a complete sphere?

24-7. Draw a diagram illustrating a solar eclipse, labeling the umbra and penumbra regions. If you view a partial eclipse of the sun, are you standing in the umbra or the penumbra region?

24-8. Can you justify the following definition for a lumen? *A lumen is equal to the luminous flux falling on a surface of one square meter, all points of which are one meter from a uniform point source of one candela?*

24-9. An older unit for the illumination was the *footcandle*, that illumination E received by a surface 1 ft^2 located at a distance of 1 ft from a 1-cd source of light. Explain how this definition is the same as that given in this text.

24-10. Describe the distribution of luminous flux from an incandescent lamp. Why is the lamp not an isotropic source?

24-11. Discuss the factors which will affect the illumination of a table in a machine shop.

24-12. Illumination is sometimes referred to as *flux density*. Explain why such a term might be appropriate.

24-13. Photometry is the science of measuring light. The intensity of a light source can be determined with the photometer, shown in Fig. 24-21. The luminous intensity I_x of an unknown source is found by visually comparing the unknown with a standard source of known intensity I_s. If the distances from each source are adjusted so that the grease spot is equally illuminated by each source, the unknown intensity I_x can be calculated from the inverse-square law. Derive the photometry equation

$$\boxed{\frac{I_x}{r_x^2} = \frac{I_s}{r_s^2}}$$

Photometry Equation (24-12)

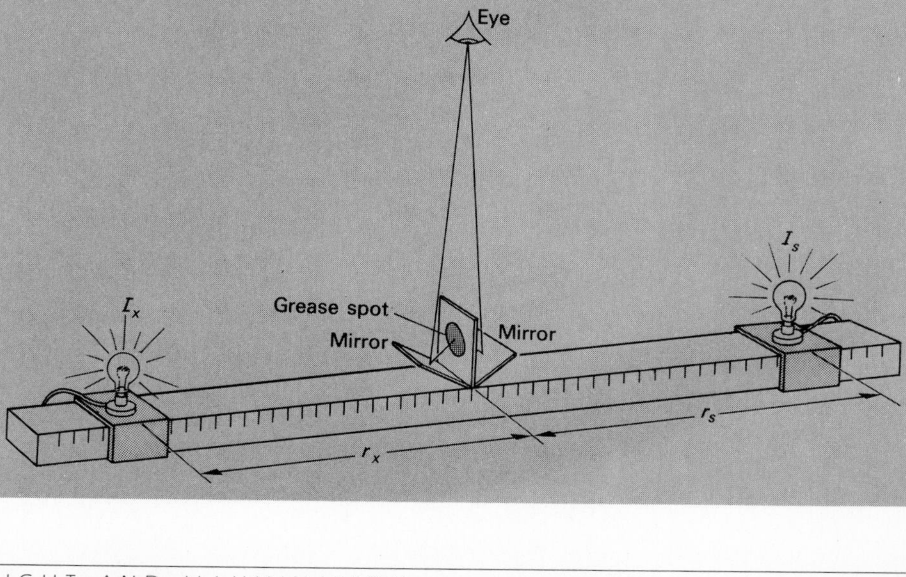

Fig. 24-21 Grease-spot photometer used to measure light intensity.

where r_s is the distance to the standard and r_x is the distance to the unknown source.

24-14. If two 40-W light bulbs are compared by using the photometer, will they necessarily be located at equal distances from the grease spot?

PROBLEMS

Light and the Electromagnetic Spectrum

24-1. An infrared spectrophotometer scans the wavelengths from 1 to 16 μm. Express this range in terms of the frequencies of the infrared rays.

Ans. 30×10^{13} to 1.88×10^{13} Hz

24-2. What is the frequency of violet light of wavelength 410 nm?

24-3. A microwave radiator used in measuring automobile speeds emits radiation of frequency 1.2×10^9 Hz. What is the wavelength in nanometers? In angstroms?

Ans. 2.5×10^8 nm, 2.5×10^9 Å

24-4. What is the range of frequencies for visible light?

24-5. If Planck's constant is $h = 6.625 \times 10^{-34}$ J $\cdot$ s, what is the energy of yellow light (600 nm)?

Ans. 3.31×10^{-19} J

24-6. What is the frequency of light whose energy is 5×10^{-19} J?

The Velocity of Light

24-7. The sun is approximately 93 million miles from the earth. How much time is required for the light emitted by the sun to reach us on earth?

Ans. 8.33 min

24-8. If two experimenters in Galileo's experiment were separated by a distance of 5 km, how much time would have passed from the instant the lantern was opened until the light was observed?

24-9. The light reaching us from the nearest star, Alpha Centauri, requires 4.3 years to reach us. How far is this in miles? In kilometers?

Ans. 2.53×10^{13} mi, 4.07×10^{13} km

24-10. A spacecraft circling the moon at a distance of 384,000 km from the earth communicates by radio with a base on earth. (a) How much time elapses between the sending and receiving of a signal? (b) How long would be required for a TV signal to reach the earth from Venus when it is 4×10^7 km from the earth?

Light Rays and Shadows

24-11. A point source of light is placed 15 cm from an upright 6-cm ruler. Calculate the length of the shadow formed by the ruler on a wall 40 cm from the ruler.

Ans. 22 cm

24-12. How far must an 80-mm-diameter plate be placed in front of a point source of light if it is to form a shadow 400 mm in diameter at a distance of 2 m from the light source?

* 24-13. A lamp is covered with a box, and a 20-mm-long narrow slit is cut in the box so that light shines through. An object 30 mm tall blocks the light from the slit at a distance of 500 mm. Calculate the length of the umbra and penumbra formed on a screen located 1.5 m from the slit.

Ans. 50 mm, 130 mm

24-14. A source of light 40 mm in diameter shines through a pinhole in the tip of a cardboard box 2 m from the source. What is the diameter of the image formed on the bottom of the box if the height of the box is 60 mm?

Illumination of Surfaces

24-15. What is the solid angle subtended at the center of a 3.2-m-diameter sphere by a 0.5-m^2 area on its surface?

Ans. 0.195 sr

24-16. A point source of light is placed at the center of a sphere 70 mm in diameter. A hole is cut in the surface of the sphere allowing the flux to pass through a solid angle of 0.12 sr. What is the diameter of the opening?

24-17. An 8⅛ by 11 cm sheet of metal is illuminated by a source of light located 1.3 m directly above it. (a) Compute the luminous flux falling on the metal if the source has an intensity of 200 cd. (b) What is the total luminous flux emitted by the light source?

Ans. 1.11 lm, 2510 lm

24-18. A 40-W monochromatic source of yellow-green light (555 nm) illuminates a 0.5-m^2 surface at a distance of 1 m. (a) What is the luminous intensity of the source? (b) How many lumens fall on the surface?

24-19. What is the illumination produced by a 200-cd source on a small surface 4 m away?

Ans. 12.5 lx

24-20. A lamp 2 m from a small surface produces an illumination of 100 lx on the surface. What is the intensity of the source?

24-21. A table top 1 m wide and 2 m long is located 4 m from a lamp. If 40 lm of flux fall on this surface, what is the illumination E of the surface? What should be the location of the lamp in order to produce twice the illumination?

Ans. 20 lx, 2.83 m

Additional Problems

24-22. When light of wavelength 550 nm passes from air into a thin glass plate and out again into the air, the frequency remains constant, but the speed through the glass is reduced to 2×10^8 m/s. What is the wavelength inside the glass?

24-23. The wavelength changes from 660 to 455 nm as light passes through a material medium. What is the velocity of light in the medium if the frequency is unaltered?

Ans. 2.07×10^8 m/s.

24-24. In Michelson's measurements of the speed of light, as shown in Fig. 24-11, he obtained a value of 2.997×10^8 m/s. If the total light path was 35 km, what was the rotational frequency of the eight-sided mirror?

24-25. In Fizeau's experiment, the plane mirror was located at a distance of 8630 m. He used a wheel containing 720 teeth (and 720 voids). Every time the rotational speed of the wheel was increased by 24.2 rev/s, the light came through to his eye. What value did he obtain for the speed of light?

Ans. 3.01×10^8 m/s

24-26. A 40-W monochromatic source of yellow-green light (555 nm) illuminates a 0.5-m^2 surface at a distance of 1 m. What is the solid angle subtended at the source? What is the luminous intensity of the source?

24-27. A 30-cd standard light source is compared with a lamp of unknown intensity using a grease-spot photometer (refer to Fig. 24-17). The two light sources are placed 1 m apart, and the grease spot is moved toward the standard light. When the grease spot is 25 cm from the standard light source, the illumination is equal on both sides. Compute the unknown intensity.

Ans. 270 cd

* **24-28.** Where should the grease spot in Prob. 24-23 be placed for the illumination by the unknown light source to be exactly twice the illumination of the standard source?

* **24-29.** At what distance from a wall will a 35-cd lamp provide the same illumination as an 80-cd lamp located 4 m from the wall?

Ans. 2.65 m

24-30. The illumination of a given surface is 80 lx when it is 3 m away from the light source. At what distance will the illumination be 20 lx?

* **24-31.** How much must a small lamp be lowered to double the illumination on an object which is 80 cm directly under it?

Ans. 23.4 cm.

24-32. A light is suspended 9 m above a street and provides an illumination of 36 lx at a point directly below it. Determine the luminous intensity of the light.

* **24-33.** Compute the illumination of a surface 140 cm from a 74-cd light source if the normal to the surface makes an angle of 38° with the flux.

Ans. 29.8 lx

* **24-34.** What angle θ between the flux and a line drawn normal to a surface will cause the illumination of the surface to be reduced by one-half when the distance to the surface has not changed?

* **24-35.** A circular table top is located 4 m below and 3 m to the left of a lamp which emits 1800 lm. What illumination is provided on the surface of the table? What is the area of the table top if 3 lm of flux falls on its surface?

Ans. 4.58 lx, 0.655 m^2

25

Reflection and Mirrors

OBJECTIVES

After completing this chapter, you should be able to:

1. Define and illustrate with drawings your understanding of the following terms: *virtual images, real images, converging mirror, diverging mirror, magnification, focal length,* and *spherical aberration.*
2. Use ray-tracing techniques to construct images formed by spherical mirrors.
3. Predict mathematically the nature, size, and location images formed by spherical mirrors.
4. Determine the *magnification* and/or the *focal length* of spherical mirrors by mathematical and experimental methods.

The eye responds to light. Every object viewed is seen with light—either the light emitted by the object or light that is reflected from it. We now have a general understanding of the nature of light, and we have studied luminous objects and methods of measuring the light emitted from them.

Although all light can be traced to sources of energy, e.g., the sun, an electric light bulb, or a burning candle, most of what we see in the physical world is a result of reflected light. In this chapter we treat the laws describing how light is turned back into its original medium as a result of striking a surface. Although this phenomenon, called *reflection,* can be interpreted in terms of Maxwell's electromagnetic wave theory, it is much simpler to describe it by tracing *rays.*

The ray treatment, generally referred to as *geometrical optics,* is based on the application of Huygens' principle. Remember that light rays are imaginary lines drawn perpendicular to advancing wavefronts in the direction of light propagation.

25-1
THE LAWS OF REFLECTION

When light strikes the boundary between two media, such as air and glass, one or more of three things can happen. As illustrated in Fig. 25-1, some of the light incident on a glass surface is reflected, and some passes into the glass. The light that

---

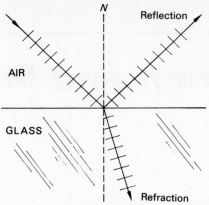

Fig. 25-1 When light strikes the boundary between two media, it may be reflected, refracted, or absorbed.

enters the glass is partially absorbed and partially transmitted. The transmitted light usually undergoes a change in direction, called *refraction*. In this chapter we shall concern ourselves only with the phenomenon of reflection.

The reflection of light obeys the same general law of mechanics that governs other bouncing phenomena; i.e., the angle of incidence equals the angle of reflection. For example, let us consider the pool table in Fig. 25-2a. In order to hit the black ball on the right, we must aim at a point on the rail such that the incident angle θ_i is equal to the reflected angle θ_r. Similarly, light reflected from a smooth

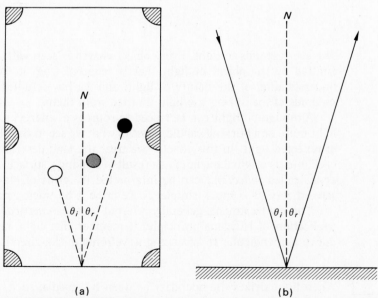

(a)   (b)

Fig. 25-2 Reflection of light follows the same path as that expected for a bouncing billiard ball. The angle of incidence is equal to the angle of reflection.

surface, as in Fig. 25-2*b*, has equal angles of incidence and reflection. The angles θ_i and θ_r are measured with respect to the normal to the surface. Two basic laws of reflection can be stated:

The angle of incidence is equal to the angle of reflection.

The incident ray, the reflected ray, and the normal to the surface all lie in the same plane.

Light reflection from a smooth surface, in Fig. 25-3*a*, is called *regular* or *specular* reflection. Light striking the surface of a mirror or glass is specularly reflected. If all the incident light which strikes a surface were reflected in this manner, we could not see the surface. We would see only images of other objects. It is *diffuse* reflection (Fig. 25-3*b*) that enables us to see a surface. A rough or irregular surface will spread out and scatter the incident light, resulting in illumination of the surface. Reflection of light from brick, concrete, or newsprint provides example of diffuse reflection.

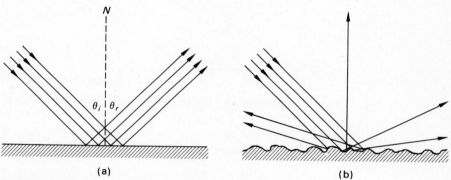

Fig. 25-3 *(a)* Specular reflection; *(b)* diffuse reflection.

25-2
PLANE
MIRRORS

A highly polished surface which forms images by specular reflection of light is called a *mirror.* The mirrors hanging on the walls of our homes are usually flat, or *plane,* and we all have a certain familiarity with the images framed by plane mirrors. In every case, the image appears to be as far behind the mirror as the actual object is in front of it. As shown in Fig. 25-4, the images also appear right–left reversed. Anyone who has learned to tie a necktie by looking in a mirror is well aware of these effects.

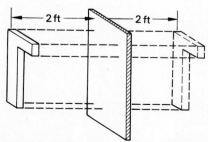

Fig. 25-4 Images formed by mirrors are right-left reversed.

In order to understand the formation of images by a plane mirror, let us first consider image *I* formed by rays emitted from a point *O* in Fig. 25-5. Four light rays are traced from the point source of light. The light ray *OV* is reflected back on itself at the mirror. Since the reflected light appears to have traveled the same distance as the incident light, the image will appear to be an equal distance behind the mirror when viewed along the normal to the reflecting surface. When the reflected light is viewed at an angle to the mirror, the same conclusion results; i.e., the image distance *q* is

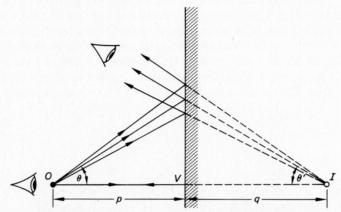

Fig. 25-5 Constructing the image of a point object formed by a plane mirror.

equal to the object distance *p*. This is true because the angle θ is equal to the angle θ' in the figure. Thus we can say that:

> *The object distance is equal in magnitude to the image distance for a plane mirror.*

$$p = q$$

Plane Mirror (25-1)

We now consider the image formed by an extended object, as shown in Fig. 25-6. An extended object may be thought of as consisting of many point objects arranged

Fig. 25-6 Image of an extended object.

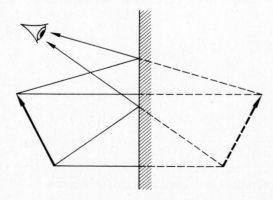

HEAT, LIGHT, AND SOUND

according to the shape and size of the object. Each point on the object will have an image point located an equal distance behind the mirror. It follows that the image will have the same size and shape as the object. However, right and left will be reversed, as discussed earlier.

Notice that the images formed by a plane mirror are, in truth, reflections of real objects. The images themselves are not real because no light passes through them. These images which *appear* to the eye to be formed by rays of light but which in truth do not exist are called *virtual* images. A *real* image is an image formed by actual light rays.

> A **virtual image** is one which seems to be formed by light coming from the image, but no light rays actually pass through it.

> A **real image** is formed by actual light rays which pass through it. Real images can be projected on a screen.

Since virtual images are not formed by real light rays, they cannot be projected on a screen.

Real images cannot be formed by a plane mirror because the light reflected at a plane surface diverges. But if a plane mirror forms virtual images which do not physically exist, how can we see them? The full answer to this question must wait for a discussion of refraction and lenses. A preliminary answer is illustrated by Fig. 25-7, which also serves to demonstrate the two types of images. The eye makes use of the principle of refraction to reconverge the reflected light which *seems* to come from the virtual image. A *real* image is therefore projected on the retina of the eye. This image, which is formed by real, reflected light rays, is interpreted by the brain to have originated from a point behind the mirror. The brain is conditioned to the rectilinear propagation of light. It is fooled when light is somehow caused to change directions. Men who do not believe the brain can be conditioned to interpret images should try to tie someone's necktie without looking in a mirror. In this case, the real object seems less natural than its virtual image.

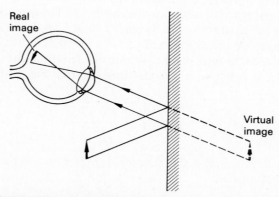

Fig. 25-7 Image formed by a plane mirror is virtual. Such images appear to the eye to be located behind the mirror.

The same geometrical methods applied for the reflection of light from a plane mirror can be used for a curved mirror. The angle of incidence is still equal to the angle of reflection, but the normal to the surface changes at every point along the surface. A complicated relation between the object and its image results.

Most curved mirrors used in practical application are spherical. A *spherical mirror* is a mirror which may be thought of as a portion of a reflecting sphere. The two kinds of spherical mirrors are illustrated in Fig. 25-8. If the inside of the spherical surface is the reflecting surface, the mirror is said to be *concave*. If the outside portion is the reflecting surface, the mirror is *convex*. In either case, R is the radius of curvature, and C is the *center of curvature* for the mirrors. The segment AB, often useful in optical problems, is called the *linear aperture* of the mirror. The dashed line CV which passes through the center of curvature and the topographical center, or *vertex,* of the mirror is known as the *axis* of the mirror.

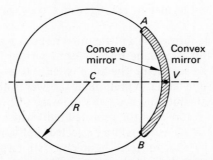

Fig. 25-8 Definition of terms for spherical mirrors.

Let us now examine the reflection of light from a spherical surface. As a simple case, suppose a beam of parallel light rays to be incident on a concave surface, as shown in Fig. 25-9. Since the mirror is perpendicular to the axis at its vertex V, a light ray CV is reflected back on itself. In fact, any light ray which proceeds along a radius of the mirror will be reflected back along itself. The parallel light ray MN is reflected so that the angle of incidence θ_i is equal to the angle of reflection θ_r. Both angles are measured with respect to the radius CN. The geometry of the reflection is such that the reflected ray passes through a point F on the axis halfway between the center of curvature C and the vertex V. The point F, to which parallel light rays converge, is called the *focal point* of the mirror. The distance from F to V is called the *focal length f.* It is left as an exercise to show from Fig. 25-9a that

$$f = \frac{R}{2}$$

(25-2)

The focal length f of a concave mirror is one-half of its radius of curvature R.

All light rays from a distant object, such as the sun, will converge at the focal point F, as shown in Fig. 25-9b. For this reason, concave mirrors are frequently called *con-*

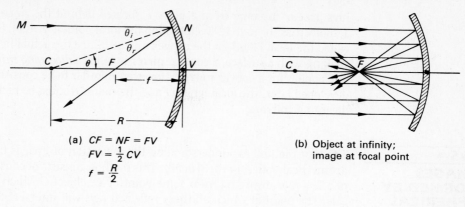

(a) $CF = NF = FV$
$FV = \frac{1}{2}CV$
$f = \frac{R}{2}$

(b) Object at infinity;
image at focal point

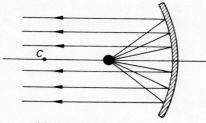

(c) Object at focal point;
image at infinity

Fig. 25-9 Focal point of a concave mirror: *(a)* The focal length is one-half the radius of curvature; *(b)* the object at infinity and the image at the focal point; *(c)* the object at the focal point and the image at infinity.

verging mirrors. The focal point can be found experimentally by converging sunlight to a point on a piece of paper. The point along the axis of the mirror where the image formed on the paper is brightest will correspond to the focal point of the mirror.

Since light rays are reversible, a source of light placed at the focal point of a converging mirror will form its image at infinity. In other words, the emerging light beam will be parallel to the axis of the mirror, as shown in Fig. 25-9c.

A similar discussion holds for a convex mirror, as illustrated in Fig. 25-10. Note that a parallel light beam incident on a convex surface diverges. The reflected light

Fig. 25-10 Focal point of a convex mirror.

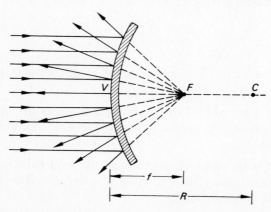

rays *appear* to come from a point F located behind the mirror, but no light rays actually pass through it. Even though the focal point is virtual, the distance VF is still called the focal length of the convex mirror. Since the actual light rays diverge when striking such a surface, a convex mirror is called a *diverging mirror.* Equation (25-2) also applies for a convex mirror. However, in order to be consistent with theory to be developed later, the focal length f and the radius R must be reckoned as negative for diverging mirrors.

<table>
<tr><td>

25-4

IMAGES FORMED BY SPHERICAL MIRRORS

</td><td>

The best method for understanding the formation of images by mirrors is through geometrical optics, or *ray tracing.* This method consists of considering the reflection of a few rays diverging from some point of an object O which is *not* on the mirror axis. The point at which all these reflected rays will intersect determines the image location. We shall discuss three rays whose path can easily be traced. Each ray is illustrated for a converging (concave) mirror in Fig. 25-11 and for a diverging (convex) mirror in Fig. 25-12.

</td></tr>
</table>

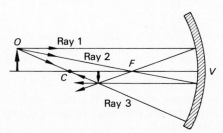

Fig. 25-11 Principal rays for constructing images formed by concave mirrors.

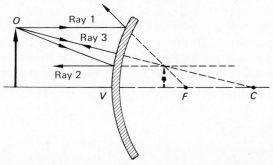

Fig. 25-12 Principal rays for constructing the images formed by convex mirrors.

Ray 1 A ray parallel to the mirror axis passes through the focal point of a concave mirror or seems to come from the focal point of a convex mirror.

Ray 2 A ray which passes through the focal point of a concave mirror or proceeds toward the focal point of a convex mirror is reflected parallel to the mirror axis.

Ray 3 A ray which proceeds along a radius of the mirror is reflected back along its original path.

In any given situation, only two of these rays are necessary to locate the image of a point. By choosing rays from an extreme point of the object, the remainder of the image can usually be filled in by symmetry. In the figures, dashed lines are used to identify virtual rays and virtual images.

To illustrate the graphical method and at the same time to visualize some of the possible images, let us consider several images formed by a concave mirror. Figure 25-13*a* illustrates the image formed by an object *O* which is located outside the center of curvature of the mirror. Note that the image is between the focal point *F* and the center of curvature *C*. The image is *real, inverted,* and *smaller* than the object.

In Fig. 25-13*b* the object *O* is located at the center of curvature *C*. The concave mirror forms an image at the center of curvature which is *real, inverted,* and the *same size* as the object.

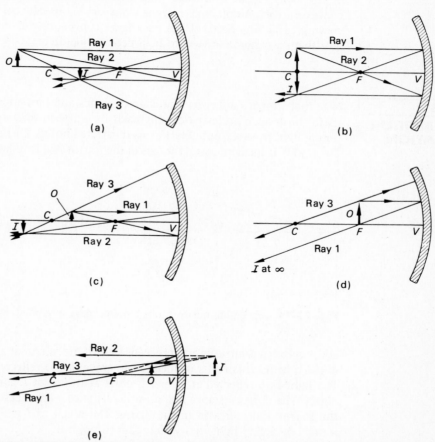

Fig. 25-13 Images formed by a converging mirror for the following object distances: *(a)* beyond the center of curvature *C*; *(b)* at *C*; *(c)* between *C* and the focal length *F*; *(d)* at *F*; and *(e)* inside *F*.

In Fig. 25-13c the object O is located between C and F. Ray tracing shows that the image is located beyond the center of curvature. It is *real, inverted,* and *larger* than the object.

When the object is at the focal point F, all reflected rays are parallel (see Fig. 25-13d). Since the reflected rays will never intersect when extended in either direction, no image will be formed. (Some prefer to say that the image distance is infinite.)

When the object is located inside the focal point F, as shown in Fig. 25-13e, the image *appears* to be behind the mirror. This can be seen by extending the reflected rays to a point behind the mirror. Thus the image is *virtual*. Notice also that the image is *enlarged* and erect (right side up). The magnification in this instance is the principle behind shaving mirrors and other mirrors which form enlarged virtual images.

On the other hand, all images formed by *convex* mirrors have the same characteristics. As was illustrated in Fig. 25-12, such images are *virtual, erect,* and *reduced in size*. This results in a wider field of view and accounts for many of the uses of convex mirrors. Automobile rear-view mirrors are usually convex to give maximum viewing capability. Some stores have large convex mirrors conveniently located to give them a panoramic view to help detect shoplifters.

25-5
THE MIRROR EQUATION

Now that we have a feel for the characteristics and formation of images, it will be useful to develop an analytical approach to image formation. Consider the reflection of light from a point object O, as illustrated in Fig. 25-14 for a concave mirror. The ray OV is incident along the axis of the mirror and is reflected back on itself. Ray

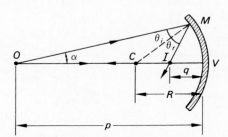

Fig. 25-14 Converging mirror forms a point image of a point object.

OM is selected arbitrarily and proceeds toward the mirror at an angle α with the axis of the mirror. This ray is incident at an angle θ_i and reflected at an equal angle θ_r. The light rays reflected at M and V cross at the point I, forming an image of the object. The object distance p and image distance q are measured from the vertex of the mirror and indicated in the figure. The image at I is a *real* image since it is formed by actual light rays which pass through it.

Now, let us consider the image formed by an extended object OA, as shown in Fig. 25-15. The image of the point O is found to be I, as before. By tracing rays from the top of the arrow, we are able to draw the image of A at B. The ray AM passes through the center of curvature and is reflected back on itself. A ray AV which strikes

HEAT, LIGHT, AND SOUND

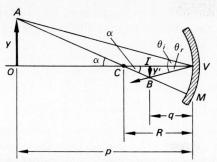

Fig. 25-15 Deriving the mirror equation.

the vertex of the mirror forms equal angles θ_i and θ_r. Rays VB and AM cross at B, forming an image of the top of the arrow at that point. The rest of the image IB can be constructed by tracing similar rays for corresponding points on the object OA. Notice that the image is *real* and *inverted*.

The following quantities are identified in Fig. 25-15.

$$\text{Object distance} = OV = p$$
$$\text{Image distance} = IV = q$$
$$\text{Radius of curvature} = CV = R$$
$$\text{Object size} = OA = y$$
$$\text{Image size} = IB = y'$$

We now attempt to relate these quantities. From the figure, it is noted that angles OCA and VCM are equal. Labeling this angle by α, we can write

$$\tan \alpha = \frac{y}{p - R} = \frac{-y'}{R - q}$$

from which

$$\frac{-y'}{y} = \frac{R - q}{p - R} \qquad (25\text{-}3)$$

The image size y' is negative because it is inverted in the figure. Similarly, angles θ_i and θ_r in the figure are equal, so that

$$\tan \theta_i = \tan \theta_r \qquad \frac{y}{p} = \frac{-y'}{q} \qquad (25\text{-}4)$$

Combining Eqs. (25-3) and (25-4), we have

$$\frac{-y'}{y} = \frac{q}{p} = \frac{R - q}{p - R} \qquad (25\text{-}5)$$

Rearranging terms, we obtain the important relation

$$\boxed{\frac{1}{p} + \frac{1}{q} = \frac{2}{R}} \qquad (25\text{-}6)$$

This relationship is known as the *mirror equation*. It is more often written in terms of the focal length f of the mirror, instead of the radius of curvature. Recalling that $f = R/2$, we can rewrite Eq. (25-6) as

$$\frac{1}{p} + \frac{1}{q} = \frac{1}{f} \qquad (25\text{-}7)$$

A similar derivation can be made for a convex mirror, and the same equations apply if the proper sign convention is adopted. Object and image distances, p and q, must be considered *positive* for real objects and images and *negative* for virtual objects and images. The radius of curvature R and focal length f must be considered *positive* for converging (concave) mirrors and *negative* for diverging (convex) mirrors.

EXAMPLE 25-1   What is the focal length of a converging mirror whose radius of curvature is 20 cm? What are the nature and location of an image formed by the mirror if an object is placed 15 cm from the vertex of the mirror?

Solution   The focal length is one-half of the radius of curvature, and the radius is positive for a converging mirror.

$$f = \frac{R}{2} = \frac{+20 \text{ cm}}{2} = +10 \text{ cm}$$

The location of the image is found from the mirror equation.

$$\frac{1}{p} + \frac{1}{q} = \frac{1}{f}$$

Solving explicitly for q yields

$$q = \frac{pf}{p - f}$$

from which

$$q = \frac{(15 \text{ cm})(10 \text{ cm})}{15 \text{ cm} - 10 \text{ cm}} = +30 \text{ cm}$$

Therefore, the image is real and located 30 cm from the mirror. Ray tracing similar to that of Fig. 25-13c shows that the image will also be inverted.

It is usually easier to solve the mirror equation explicitly for the unknown quantity than to substitute directly. The student will find the following forms useful in most mirror problems:

$$p = \frac{qf}{q - f} \qquad q = \frac{pf}{p - f} \qquad f = \frac{pq}{p + q} \qquad (25\text{-}7)$$

The sign convention is summarized below:

1. The object distance p is positive for real objects and negative for virtual objects.
2. The image distance q is positive for real images and negative for virtual images.

3. The radius of curvature R and focal length f are positive for converging mirrors and negative for diverging mirrors.

This convention applies only to the *numerical* values substituted into Eq. (25-7). The quantities q, p, and f should maintain their signs unchanged until the substitution is made.

EXAMPLE 25-2   Find the position of the image if an object is located 4 cm from a convex mirror whose focal length is 6 cm.

Solution   In this case, $p = 4$ cm and $f = -6$ cm. The minus sign is necessary because a convex mirror is a diverging mirror.

$$q = \frac{pf}{p - f} = \frac{(4 \text{ cm})(-6 \text{ cm})}{4 \text{ cm} - (-6 \text{ cm})}$$

$$= \frac{-24 \text{ cm}^2}{4 \text{ cm} + 6 \text{ cm}} = -2.4 \text{ cm}$$

The image distance is negative, indicating that the image is virtual.

25-6
MAGNIFI-CATION

The images formed by spherical mirrors may be larger, smaller, or equal in size to the objects. The ratio of the image size to the object size is the *magnification M* of the mirror.

$$\text{Magnification} = \frac{\text{image size}}{\text{object size}} = \frac{y'}{y} \tag{25-8}$$

The *size* refers to any linear dimension, e.g., height or width. Referring to Eq. (25-5) and to Fig. 25-15, we obtain the useful relation

$$M = \frac{y'}{y} = \frac{-q}{p} \tag{25-9}$$

where q is the image distance and p is the object distance. A convenient feature of Eq. (25-9) is that *an inverted image will always have a negative magnification, and an erect image will have a positive magnification.*

EXAMPLE 25-3   A source of light 6 cm high is located 60 cm from a concave mirror whose focal length is 20 cm. Find the position, nature, and size of the image.

Solution   We first find the image distance q, as follows:

$$q = \frac{pf}{p - f} = \frac{(60 \text{ cm})(20 \text{ cm})}{60 \text{ cm} - 20 \text{ cm}}$$

$$= \frac{1200 \text{ cm}^2}{40 \text{ cm}} = 30 \text{ cm}$$

Since q is positive, the image is real. The image size is found from Eq. (26-9).

$$M = \frac{y'}{y} = -\frac{q}{p}$$

$$y' = -\frac{qy}{p} = -\frac{(30\ \text{cm})(6\ \text{cm})}{60\ \text{cm}}$$

$$= -3\ \text{cm}$$

The negative sign indicates that the image is inverted. Note that the magnification is $-\frac{1}{2}$.

EXAMPLE 25-4   How far should a pencil be held from a convex mirror to form an image one-half the size of the pencil? The radius of the mirror is 40 cm.

Solution   The focal length of the mirror is

$$f = \frac{R}{2} = \frac{-40\ \text{cm}}{2} = -20\ \text{cm}$$

The minus sign occurs because of the diverging mirror. Such a mirror always forms an erect image, reduced in size. (See Fig. 25-12.) The magnification in this case is $+\frac{1}{2}$. Thus

$$M = -\frac{q}{p} = +\frac{1}{2}$$

$$q = -\frac{p}{2}$$

From the mirror equation, q is also

$$q = \frac{pf}{p - f}$$

Combining the two equations for q, we have

$$\frac{pf}{p - f} = -\frac{p}{2}$$

Dividing by p gives

$$\frac{f}{p - f} = -\frac{1}{2}$$

$$2f = -p + f$$

$$p = -f = -(-20\ \text{cm}) = 20\ \text{cm}$$

Thus when an object is held at a distance equal to the focal length from a convex mirror, the image size is one-half of the object size.

25-7
SPHERICAL ABERRATION

In practice, spherical mirrors form reasonably sharp images as long as their apertures are small compared with their focal lengths. However, when large mirrors are used, some of the rays from objects strike near the outer edges and are focused to different points on the axis. This focusing defect, illustrated in Fig. 25-16, is known as *spherical aberration*.

A parabolic mirror does not exhibit this defect. Theoretically, parallel light rays incident on a parabolic reflector will focus at a single point on the mirror axis. (See Fig. 25-17.) A small source of light located at the focal point of a parabolic reflector is

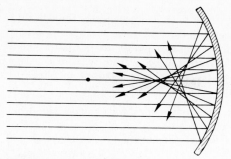

Fig. 25-16 Spherical aberration.

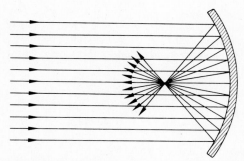

Fig. 25-17 Parabolic reflector focuses all incident parallel light to the same point.

the principle used in many spotlights and search lights. The beam emitted from such a device is parallel to the axis of the reflector.

SUMMARY

In this chapter we have studied the reflective properties of converging and diverging spherical mirrors. The focal length and radius of curvature of such mirrors determine the nature and size of the images they form. Application of the formulas and ideas given in this chapter are necessary for understanding the operation and use of many technical instruments. The main concepts are summarized below:

- The formation of images by spherical mirrors can be visualized more easily with ray-tracing techniques. The three principal rays are listed below. You should refer to Fig. 25-18a for converging mirrors and to Fig. 25-18b for diverging mirrors.

Fig. 25-18 *(a)* Ray tracing for a converging mirror.

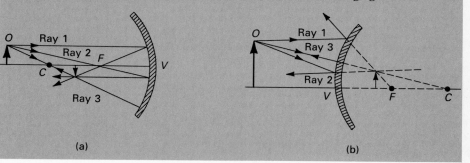

(a)

(b)

Ray 1 A ray parallel to the mirror axis passes through the focal point of a concave mirror or seems to come from the focal point of a convex mirror.

Ray 2 A ray which passes through the focal point of a concave mirror or proceeds toward the focal point of a convex mirror is reflected parallel to the mirror axis.

Ray 3 A ray which proceeds along a radius of the mirror is reflected back along its original path.

- Before listing the mirror equations, you should review what the symbols mean and the sign conventions.

R = radius of curvature	$+$ for converging, $-$ for diverging
f = focal length	$+$ for converging, $-$ for diverging
p = object distance	$+$ for real object, $-$ for virtual
q = image distance	$+$ for real images, $-$ for virtual
y = object size	$+$ if erect, $-$ if inverted
y' = image size	$+$ if erect, $-$ if inverted
M = magnification	$+$ if both erect or both inverted

- The mirror equations can be applied to either converging (concave) or diverging (convex) spherical mirrors:

$$f = \frac{R}{2} \qquad R = 2f \qquad M = \frac{y'}{y} \qquad \frac{1}{p} + \frac{1}{q} = \frac{1}{f}$$

Mirror Equations

- Alternative forms for the last equation are:

$$p = \frac{qf}{q - f} \qquad q = \frac{pf}{p - f} \qquad f = \frac{pq}{p + q}$$

QUESTIONS

1. Define the following terms:
   a. Plane mirror
   b. Spherical mirror
   c. Linear aperture
   d. Focal length
   e. Real image
   f. Virtual image
   g. Diverging mirror
   h. Magnification
   i. Specular reflection
   j. Diffuse reflection
   k. Geometrical optics
   l. Radius of curvature
   m. Converging mirror
   n. Mirror equation
   o. Spherical aberration
   p. Parabolic mirror

2. Discuss the statement: One cannot "see" the surface of a perfect mirror.

3. Prove by a diagram that rays diverging from a point source of light appear to diverge from a virtual point after reflection from a plane surface.

4. Can an image of a real object be projected on a screen by a plane mirror? By a convex mirror? By a concave mirror?

5. State the laws of reflection and show how they may be demonstrated in a laboratory.

6. Use the mirror equation to show that the image of an infinitely distant object is formed at the focal point of a spherical mirror.

7. Use the mirror equation to show that the image of an object placed at the focal point of a concave mirror is located at infinity.

8. Use the mirror equation to show that for a plane mirror the image distance is equal in magnitude to the object distance. What is the magnification of a plane mirror?

9. In a concave shaving mirror, will greater magnification be achieved when the object is closer to the focal point or when it is closer to the vertex? Use diagrams to verify your conclusion.

10. Do objects moving closer to the vertex of a convex mirror form larger or smaller virtual images? Explain with diagrams.

11. Without looking at Fig. 25-13 in the text, construct the images formed by a concave mirror when the object is (a) beyond C, (b) at C, (c) between C and F, (d) at F, and (e) between F and V. Discuss the nature and relative size of each image.

12. Several small spherical mirrors are lying on a laboratory table. Describe how you would distinguish the diverging mirrors from the converging mirrors without touching them.

13. For real objects, is it possible to construct an inverted image by using a diverging mirror? What can you say about the magnification of diverging mirrors?

14. You wish to choose a shaving mirror which will give maximum magnification with an erect image. Does the focal length of the mirror play a part in determining its magnification? Explain.

15. Two concave spherical mirrors have the same focal length, but one has a larger linear aperture. Which forms the sharper image? Why?

16. Show how it is possible for a plane mirror to form a real image if light from the object is first converged by a concave mirror.

PROBLEMS

Reflection from Plane Mirrors

25-1. A man 1.8 m tall stands 1.2 m from a large plane mirror. (a) How tall is his image? (b) How far is he from his image? (c) What is the shortest mirror length required to enable him to see his entire image?

Ans. (a) 1.8 m, (b) 2.4 m, (c) 0.9 m

25-2. A woman 5 ft., 8 in. tall wants to buy a plane mirror long enough to reflect her full-length image. What length of mirror in inches is required? Does it matter what distance she stands from the mirror? Explain using diagrams.

* 25-3. A plane mirror moves at a speed of 30 km/h away from a stationary person. How fast does this person's image appear to be moving away from him?

Ans. 60 km/h

* 25-4. The *optical lever* is a very sensitive measuring device which utilizes minute rotations of a plane mirror to measure small deflections. The device is illustrated in Fig. 25-19. When the mirror is in position 1, the light ray follows the path IVR_1. If the mirror is rotated through an angle θ to position 2, the ray will follow the path IVR_2. Show that the reflected beam turns through an angle 2θ, which is twice the angle through which the mirror itself turns.

Fig. 25-19 The optical lever.

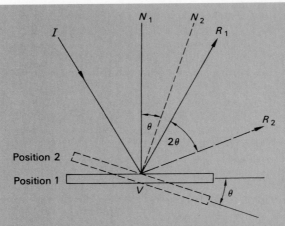

Images Formed by Spherical Mirrors

25-5. A light bulb 3 cm high is placed 20 cm in front of a concave mirror with a radius of curvature of 15 cm. Determine the nature, size, and location of the image formed. Sketch the ray-tracing diagram.

> **Ans.** Real, $y' = -1.8$ cm, $q = +12$ cm

25-6. An object 5 cm tall is placed halfway between the focal point and the center of curvature of a concave spherical mirror. If the radius of the mirror is 30 cm, determine the nature, size, and location of the image formed. Sketch the ray-tracing diagram.

25-7. A 4-cm-high source of light is placed in front of a spherical concave mirror whose radius is 40 cm. Determine the nature, location, and height of images formed for the following object distances: (a) 60 cm, (b) 40 cm, (c) 30 cm, (d) 20 cm, and (e) 10 cm. Draw the ray-tracing diagram for each case.

> **Ans.** (a) $q = 30$ cm, $y' = -2$ cm, real, inverted, diminished;
> (b) $q = 40$ cm, $y' = -4$ cm, real, inverted, same size;
> (c) $q = 60$ cm, $y' = -8$ cm, real, inverted, enlarged;
> (d) $q =$ infinity, no image is formed;
> (e) $q = -20$ cm, $y' = 8$ cm, virtual, erect, enlarged

25-8. A 60-mm-high source of light is placed in front of a spherical concave mirror whose radius is 80 mm. Determine the nature, size, and location of the images formed for the following object distances: (a) 100 mm, (b) 80 mm, (c) 60 mm, (d) 40 mm, and (e) 20 mm.

Magnification

25-9. A Christmas tree ornament has a silvered surface and a diameter of 3 in. What is the magnification of an object placed 6 in. from the surface of this ornament?

> **Ans.** +0.111

25-10. What type of mirror is required to form an image on a screen 2 m away from the mirror when an object is placed 12 cm in front of the mirror? What is the magnification?

25-11. A concave shaving mirror has a focal length of 520 mm. How far away from it should an object be placed for the image to be erect and twice its actual size?

> **Ans.** 260 mm

25-12. If a magnification of +3 is desired, how far should the mirror of Prob. 25-11 be placed from the face?

25-13. In an experiment to determine the radius of curvature, a student notes that an erect image, one-third the size of the object, is formed when the object is placed 12 cm from the surface. What is the radius of the mirror?

Ans. -6 cm

25-14. The diameter of the moon is 3480 km, and it is 3.84×10^5 km from the earth. A telescope on the earth utilizes a spherical concave mirror, whose radius is 8 m, to form an image of the moon. What is the diameter of the image formed. What is the magnification of the mirror?

Additional Problems

25-15. An object 6 cm tall is placed in front of a convex spherical mirror of radius -40 cm. Determine the nature, size, and location of the images formed for the following object distances: **(a)** 60 cm, **(b)** 40 cm, **(c)** 20 cm.

Ans. All images are virtual, erect, and diminished.
(a) $q = -15$ cm, $y = 1.5$ cm;
(b) $q = -13.3$ cm, $y = 2.0$ cm;
(c) $q = -10$ cm, $y = 3.0$ cm

25-16. An object is placed 200 mm from the vertex of a convex spherical mirror whose radius is 400 mm. What is the magnification of the mirror?

25-17. An image 60 mm long is formed on a wall located 2.3 m away from a source of light 20 mm high. What is the focal length of this mirror? Is it diverging or converging?

Ans. $+862$ mm, converging

25-18. A convex spherical mirror has a radius of -60 cm. How far away should an object be held if the image is to be one-third the size of the object?

25-19. What should be the radius of curvature of a convex spherical mirror in order to produce an image one-fourth as large as the object located 40 in. from the mirror?

Ans. -26.7 in.

25-20. An object 80 mm tall is placed in front of a converging mirror of radius -600 mm. Determine the nature, size, and location of the images at object distances of **(a)** 1 m, **(b)** 600 mm, and **(c)** 400 mm.

25-21. A silver ball is 4 cm in diameter. Locate the image of a 6-cm object located 9 cm from the surface of the ball. What is the magnification?

Ans. -9 mm, $+0.1$

25-22. A convex mirror has a focal length of -500 mm. If an object is placed 400 mm from the vertex, what is the magnification?

25-23. What should be the radius of curvature of a concave spherical mirror in order to produce an image one-fourth as large as an object 50 cm away?

Ans. 10 cm

25-24. A spherical mirror forms a *real* image 18 cm from the surface. The image is twice as large as the object. Find the position of the object and the focal length of the mirror.

25-25. A certain mirror placed 2 m from an object produces an erect image enlarged 3 times. **(a)** Is the mirror a diverging or converging mirror? **(b)** What is the radius of the mirror?

Ans. (a) Converging, (b) +3 m

25-26. The magnification of a mirror is $-1/3$. **(a)** Is it a diverging mirror or a converging mirror? **(b)** Where is the object located if its image is formed on a card 540 mm from the mirror? **(c)** What is the focal length of the mirror?

25-27. A concave mirror of radius 800 mm is placed 600 mm from a plane mirror which faces it. A source of light placed midway between the mirrors is shielded so that the light is first reflected from the concave surface. What are the position and magnification of the image formed after reflection from the plane mirror? (Treat the image formed by the first mirror as the object for the second mirror.)

Ans. 1.8 m behind plane mirror, +4

25-28. Derive an expression for calculating the focal length of a mirror in terms of the object distance p and the magnification M.

26 Refraction

After completing this chapter, you should be able to:

1. Define the *index of refraction* and state three laws which describe the behavior of refracted light.
2. Apply Snell's law to the solution of problems involving the transmission of light in two or more media.
3. Determine the change in velocity or wavelength of light as it moves from one medium into another.
4. Explain the concepts of *total internal reflection* and the *critical angle* and use these ideas to solve problems similar to those in the text.

Light travels in straight lines at a constant speed in a uniform medium. If the medium changes, the speed will also change and the light will travel in a straight line along a new path. The bending of a light ray as it passes obliquely from one medium to another is known as *refraction*. The principle of refraction is illustrated in Fig. 26-1 for a light wave entering water from the air. The angle θ_i that the incident beam

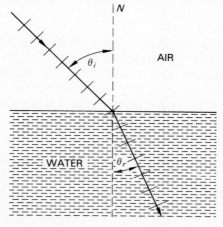

Fig. 26-1 Refraction of a wavefront at the boundary between two media.

makes with the normal to the surface is referred to as the *angle of incidence*. The angle θ_r between the refracted beam and the normal is called the *angle of refraction*.

Refraction explains such familiar phenomena as the apparent distortion of objects partially submerged in water. The stick appears to be bent at the surface of the water in Fig. 26-2a, and the fish in Fig. 26-2b appears to be closer to the surface than it really is. In this chapter we study the properties of refractive media and develop equations to predict their effect on incident light rays.

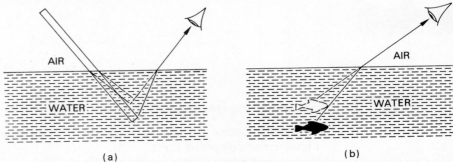

Fig. 26-2 Refraction is responsible for the distortion of images. *(a)* The stick appears to be bent; *(b)* the fish seems to be closer to the surface than it actually is.

26-1
INDEX OF REFRACTION

The velocity of light in a material substance is generally less than the free-space velocity of 3×10^8 m/s. In water the speed of light is almost 2.25×10^8 m/s, which is just about three-fourths of its velocity in air. Light travels about two-thirds as fast in glass, or around 2×10^8 m/s. The ratio of the velocity c of light in a vacuum to the velocity v of light in a particular medium is called the *index of refraction n* for that material.

> The **index of refraction** n of a particular material is the ratio of the free-space velocity of light to the velocity of light through the material.

$$n = \frac{c}{v}$$

Index of Refraction (26-1)

The index of refraction is a unitless quantity, which is generally greater than unity. For water, $n = 1.33$, and for glass, $n = 1.5$. Table 26-1 lists the index of refraction for several common substances. Note that the values given apply for yellow light of wavelength 589 nm. The velocity of light in material substances is different for different wavelengths. This effect, known as *dispersion,* will be discussed in a later section. When the wavelength of light is not specified, the index is usually assumed to correspond to that for yellow light.

EXAMPLE 26-1   Compute the velocity of yellow light in a diamond whose refractive index is 2.42.

Solution   Solving for v in Eq. (26-1) gives

$$v = \frac{c}{n} = \frac{3 \times 10^8 \text{ m/s}}{2.42} = 1.24 \times 10^8 \text{ m/s}$$

Table 26-1   Index of Refraction for Yellow Light of Wavelength 589 nm

Substance	n	Substance	n
Benzene	1.50	Glycerin	1.47
Carbon disulfide	1.63	Ice	1.31
Diamond	2.42	Quartz	1.54
Ethyl alcohol	1.36	Rock salt	1.54
Fluorite	1.43	Water	1.33
Glass		Zircon	1.92
Crown	1.52		
Flint	1.63		

It is the exceptionally large refractive index which provides one of the most positive tests for the identification of diamonds.

26-2
THE LAWS OF REFRACTION

Two basic laws of refraction have been known and observed since ancient times. These laws are stated as follows (refer to Fig. 26-3):

The incident ray, the refraction ray, and the normal to the surface all lie in the same plane.

The path of a ray refracted at the interface between two media is exactly reversible.

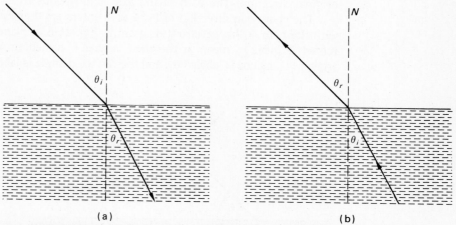

Fig. 26-3 *(a)* The incident ray, the refracted ray, and the normal to the surface are in the same plane. *(b)* Refracted rays are reversible.

These two laws are easily demonstrated by observation and experiment. However, it is of much more importance, in a practical sense, to understand and predict the *degree* of bending which occurs.

In order to understand how a change in the velocity of light can alter its path through a medium, let us consider the mechanical analogy shown in Fig. 26-4. In Fig. 26-4*a* light incident on the glass plate is first bent toward the normal as it passes through the denser medium, and then it is bent away from the normal as it returns

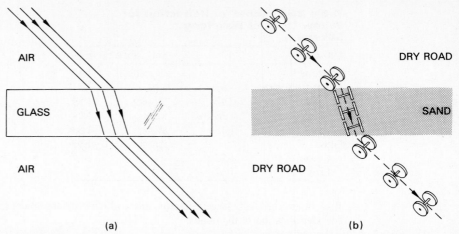

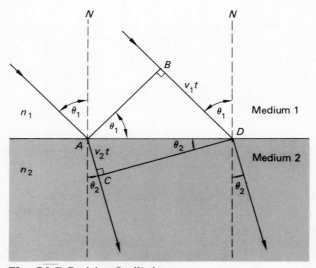

Fig. 26-4 *(a)* The lateral displacement of light passing through glass. *(b)* A mechanical analogy.

to the air. In Fig. 26-4*b* the action of wheels encountering a patch of sand resembles the behavior of light. As they approach the sand, one wheel strikes first and slows down. The other wheel continues at the same speed, causing the axle to assume a new angle. When both wheels are in the sand, the wheels again move in a straight line with uniform speed. The first wheel to enter the sand is also the first to leave, and its speeds up as it leaves the patch of sand. Thus the axle swings around to its original direction. The path of the axle is analogous to the path of a wavefront.

The change in direction of light as it enters another medium can be analyzed with the help of the wavefront diagram in Fig. 26-5. A plane wave in a medium of refractive index n_1 meets at the plane surface of a medium whose index is n_2. The angle of incidence is labeled θ_1, and the refracted angle is labeled θ_2. In the figure, it

Fig. 26-5 Deriving Snell's law.

is assumed that the second medium has a greater optical density than the first $(n_2 > n_1)$. An example is light passing from air $(n_1 = 1)$ to water $(n_2 = 1.33)$. The line AB represents the wavefront at time $t = 0$ when it just comes into contact with medium 2. The line CD represents the same wavefront after the time t required to enter the second medium completely. Light travels from B to D in medium 1 in the same time t required for light to travel from A to C in medium 2. Assuming the velocity v_2 in the second medium is smaller than the velocity v_1 in the first medium, the distance AC will be shorter than the distance BD. These lengths are given by

$$AC = v_2 t \qquad BD = v_1 t$$

It can be shown from geometry that angle BAD is equal to θ_1 and that angle ADC is equal to θ_2, as indicated in Fig. 26-5. The line AD forms a hypotenuse that is common to the two triangles ADB and ADC. From the figure,

$$\sin \theta_1 = \frac{v_1 t}{AD} \qquad \sin \theta_2 = \frac{v_2 t}{AD}$$

Dividing the first equation by the second, we obtain

$$\boxed{\frac{\sin \theta_1}{\sin \theta_2} = \frac{v_1}{v_2}} \qquad (26\text{-}2)$$

The ratio of the sine of the angle of incidence to the sine of the angle of refraction is equal to the ratio of the velocity of light in the incident medium to the velocity of light in the refracted medium.

This rule was first discovered by the seventeenth-century Dutch astronomer Willebrord Snell and is called *Snell's law* in his honor. An alternative form for the law can be obtained by expressing the velocities v_1 and v_2 in terms of the indexes of refraction for the two media. Recall that

$$v_1 = \frac{c}{n_1} \qquad \text{and} \qquad v_2 = \frac{c}{n_2}$$

Utilizing these relations in Eq. (26-2), we write

$$\boxed{n_1 \sin \theta_1 = n_2 \sin \theta_2} \qquad (26\text{-}3)$$

Since the sine of an angle increases as the angle increases, we see that an increase in the index of refraction results in a decrease in the angle and vice versa.

EXAMPLE 26-2 Light passes at an angle of incidence of 35° from water into the air. What is the angle of refraction if the index of refraction for water is 1.33?

Solution The angle θ_a can be found from Snell's law.

$$n_w \sin \theta_w = n_a \sin \theta_a$$
$$1.33 \sin 35° = 1.0 \sin \theta_a$$
$$\sin \theta_a = 1.33 \sin 35° = 0.763$$
$$\theta_a = 49.7°$$

The index of refraction *decreased* from 1.33 to 1.0, and so the angle *increased*.

EXAMPLE 26-3   A ray of light in water ($n_w = 1.33$) is incident upon a plate of glass ($n_g = 1.5$) at an angle of 40°. What is the angle of refraction into the glass? Refer to Fig. 26-6.

Fig. 26-6

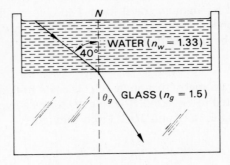

Solution   Applying Snell's law at the interface, we obtain

$$n_w \sin \theta_w = n_g \sin \theta_g$$
$$1.33 \sin 40° = 1.5 \sin \theta_g$$
$$\sin \theta_g = \frac{1.33}{1.5} \sin 40° = 0.57$$
$$\theta_g = 34.7°$$

This time n increased, and so θ decreased.

As an additional exercise, you should show that the ray is reversible.

26-3

WAVE-LENGTH AND REFRACTION

We have seen that light slows down when passing into a medium of greater optical density. What happens to the wavelength of light entering a new medium? In Fig. 26-7, light traveling in air at a velocity c encounters a medium through which it travels at the reduced speed v_m. Upon returning to the air, it again travels at the speed c of light in air. This does not violate the conservation of energy because the

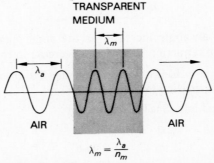

Fig. 26-7 The wavelength of light is reduced when it enters a medium of greater optical density.

---

energy of a light wave is proportional to its frequency (see Sec. 24-4). The frequency f is the same inside the medium as it is outside the medium. That this is true will be realized if you consider that the frequency is the number of waves passing any point per unit of time. The same number of waves leaves the medium in a second as enters the medium in a second. Thus the frequency inside the medium cannot change. The velocity is related to the frequency and wavelength by

$$c = f\lambda_a \qquad \text{and} \qquad v_m = f\lambda_m \qquad\qquad (26\text{-}4)$$

where c and v_m are the speeds in air and inside the medium and λ_a and λ_m are the respective wavelengths. Since the velocity decreases inside the medium, the wavelength inside the medium must decrease proportionately for the frequency to remain constant. Dividing the first equation by the second in Eq. (26-4) yields

$$\frac{c}{v_m} = \frac{f\lambda_a}{f\lambda_m} = \frac{\lambda_a}{\lambda_m}$$

If we substitute $v_m = c/n_m$, we obtain

$$n_m = \frac{\lambda_a}{\lambda_m}$$

Therefore, the wavelength λ_m inside the medium is reduced by

$$\lambda_m = \frac{\lambda_a}{n_m} \qquad\qquad (26\text{-}5)$$

where n_m is the index of refraction of the medium and λ_a is the wavelength of the light in air.

EXAMPLE 26-4   Monochromatic red light of wavelength 640 nm passes from air into a glass plate of refractive index 1.5. What is the wavelength of the light inside the medium?

Solution   Direct substitution into Eq. (26-5) yields

$$\lambda_m = \frac{\lambda_a}{n_m} = \frac{640 \text{ nm}}{1.5} = 427 \text{ nm}$$

As a summary of the relations discussed so far, we can write

$$\boxed{\frac{\sin \theta_1}{\sin \theta_2} = \frac{v_1}{v_2} = \frac{n_2}{n_1} = \frac{\lambda_1}{\lambda_2}} \qquad\qquad (26\text{-}6)$$

where the subscripts 1 and 2 refer to different media. Here we see the relationship between all the important quantities affected by refraction.

26-4
DISPERSION

We have already mentioned that the velocity of light in different substances varies with different wavelengths. We defined the index of refraction as the ratio of the free-space velocity c to the velocity inside a medium.

$$n = \frac{c}{v_m}$$

The values given in Table 26-1 are strictly valid for monochromatic yellow light (589 nm). A different wavelength of light, such as blue light or red light, results in a slightly different index of refraction. Red light travels faster through a particular medium than blue light. This can be shown by passing white light through a glass prism, as in Fig. 26-8. Because of the different speeds inside the medium, the beam is *dispersed* into its component colors.

Dispersion is the separation of light into its component wavelengths.

From such an experiment we conclude that white light is actually a mixture of light, consisting of several colors. The projection of a dispersed beam is called a *spectrum*.

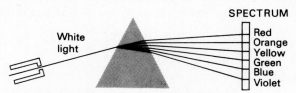

Fig. 26-8 Dispersion of light by a prism.

26-5

TOTAL INTERNAL REFLECTION

A fascinating phenomenon, known as *total internal reflection,* can occur when light passes obliquely from one medium to a medium with a lower optical density. To understand this phenomenon, consider a source of light submerged in medium 1, as illustrated in Fig. 26-9. Consider the four rays A, B, C, and D, which diverge from the submerged source. Ray A passes into medium 2 normal to the interface. The angle of incidence and the angle of refraction are both zero for this special case. Ray B is incident at an angle θ_1 and refracted away from the normal at an angle θ_2. The angle θ_2 is greater than θ_1 because the index of refraction for medium 1 is greater than that for medium 2 ($n_1 > n_2$). As the angle of incidence θ_1 increases, the angle of refraction θ_2 also increases until the refracted ray C emerges tangent to the surface. The angle of incidence θ_c for which this occurs is known as the *critical angle.*

The **critical angle** θ_c is the limiting angle of incidence in a denser medium which results in an angle of refraction of 90°.

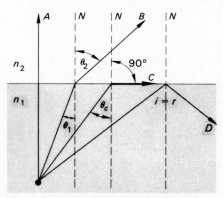

Fig. 26-9 Critical angle of incidence.

A ray approaching the surface at an angle greater than the critical angle is reflected back inside medium 1. The ray D in Fig. 26-9 does not pass into the upper medium at all but is *totally internally reflected* at the interface. This type of reflection follows the same laws as any other reflection; i.e., the angle of incidence is equal to the angle of reflection. Total internal reflection can occur only when light is incident from a denser medium $(n_1 > n_2)$.

The critical angle for two given media can be calculated from Snell's law.

$$n_1 \sin \theta_c = n_2 \sin \theta_2$$

where θ_c is the critical angle and $\theta_2 = 90°$. Simplifying, we write

or

$$n_1 \sin \theta_c = n_2(1)$$

$$\boxed{\sin \theta_c = \frac{n_2}{n_1}}$$ *Critical Angle* (26-7)

Since $\sin \theta_c$ can never be greater than 1, n_1 must be greater than n_2.

EXAMPLE 26-5 What is the critical angle for a glass-to-air surface if the refractive index of the glass is 1.5?

Solution Direct substitution yields

$$\sin \theta_c = \frac{n_a}{n_g} = \frac{1.0}{1.5} = 0.667$$
$$\theta_c = 42°$$

The fact that the critical angle for glass is 42° makes it possible to use 45° prisms in many optical instruments. Two such uses are illustrated in Fig. 26-10. In Fig. 26-10a a 90° reflection can be obtained with very little loss of intensity. In Fig. 26-10b a 180° deflection is obtained. In each case, total internal reflection occurs because the angles of incidence are all 45° and therefore greater than the critical angle.

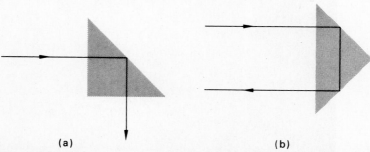

(a)   (b)

Fig. 26-10 Right-angle prisms make use of the principle of total internal reflection to deviate the path of light.

Because we are accustomed to light traveling in straight lines, refraction and total internal reflection often present us with pictures we do not believe. Atmospheric refraction accounts for many illusions which are referred to as *mirages*. Figure 26-11 provides two examples of such occurrences. In Fig. 26-11*a* a layer of hot air in contact with the heated ground is less dense than the cool layers of air above it. Consequently, light from distant objects is refracted upward, making them appear inverted.

Cold air

Warm air

(a)

Warm air

Cold air

(b)

Fig. 26-11 Atmospheric refraction accounts for the mirage in *(a)* and explains the phenomenon of *looming* in *(b)*.

At night the situation is sometimes reversed; i.e., the cool layer of air is beneath warmer layers. The headlights of the car in Fig. 26-11*b* appear to be *looming* in the air. Many scientists believe that some of the unidentified flying objects (UFOs) reported for centuries can be explained in terms of atmospheric refraction.

Combinations of refraction and total internal reflection account for the bizarre photograph in Fig. 26-12. The picture was taken by an underwater camera looking

Fig. 26-12 A photograph taken by an underwater camera presents a bizarre picture of a girl sitting at the edge of a swimming pool. *(From the film Introduction to Optics, Education Development Center.)*

upward at a girl sitting on the edge of a pool with her legs dangling in the water. An explanation is given in Fig. 26-13. The upper portion of the picture results from refraction at the surface. The inverted legs in the middle of the picture are due to total internal reflection at the surface of the water. The bottom of the picture represents the only undistorted image since the legs are viewed directly and in the same medium.

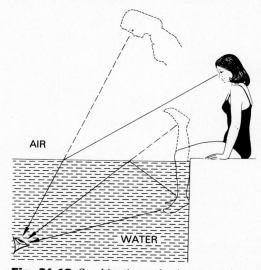

Fig. 26-13 Combinations of refraction, reflection, and total internal reflection serve to deceive the eye.

26-7
APPARENT DEPTH

Refraction causes an object submerged in a liquid of higher index of refraction to appear closer to the surface than it actually is. This shallowing effect is illustrated in Fig. 26-14. The object O appears to be at I because of the refraction of light from the

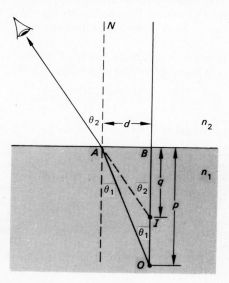

Fig. 26-14 Relation between apparent depth and actual depth.

object. The apparent depth is denoted by q, and the actual depth is denoted by p. Snell's law applied at the surface gives

$$\frac{\sin \theta_1}{\sin \theta_2} = \frac{n_2}{n_1} \qquad (26\text{-}8)$$

If we can relate the ratio of the indexes of refraction to actual and apparent depths, a useful relation can be obtained for predicting the apparent depths of submerged objects. From Fig. 26-14, it is noted that

$$\angle AOB = \theta_1 \qquad \text{and} \qquad \angle AIB = \theta_2$$

and that

$$\sin \theta_1 = \frac{d}{OA} \qquad \sin \theta_2 = \frac{d}{IA}$$

Using this information in Eq. (27-8), we obtain

$$\frac{\sin \theta_1}{\sin \theta_2} = \frac{d/OA}{d/IA} = \frac{IA}{OA} \qquad (26\text{-}9)$$

If we restrict ourselves to rays that are nearly vertical, the angles θ_1 and θ_2 will be small, so that the following approximations apply:

$$OA \approx p \qquad \text{and} \qquad IA \approx q$$

Applying these approximations to Eqs. (26-8) and (26-9), we can write

$$\frac{\sin \theta_1}{\sin \theta_2} = \frac{n_2}{n_1} = \frac{q}{p}$$

$$\boxed{\frac{\text{Apparent depth } q}{\text{Actual depth } p} = \frac{n_2}{n_1}} \qquad (26\text{-}10)$$

EXAMPLE 26-6 A coin rests on the bottom of a container filled with water ($n_w = 1.33$). The apparent distance of the coin from the surface is 9 cm. How deep is the container?

Solution In this problem, $n_1 = 1.33$ and $n_2 = 1.0$ are the indexes of refraction of water and air, respectively. The actual depth p is found by solving for p in Eq. (26-10).

$$p = \frac{qn_1}{n_2} = \frac{(9 \text{ cm})(1.33)}{1.0} = 12 \text{ cm}$$

The apparent depth is approximately three-fourths of the actual depth.

SUMMARY

Refraction has been defined as the bending of a light ray as it passes obliquely from one medium to another. We have seen that the degree of bending can be predicted based on either the change in velocity or the known index of refraction for each medium. The concepts of refraction, critical angle, dispersion, and internal reflection play important roles in the operation of many instruments. The major concepts covered in this chapter are summarized below:

- The index of refraction of a particular material is the ratio of the free-space velocity of light c to the velocity v of light through the medium.

$$n = \frac{c}{v} \qquad c = 3 \times 10^8 \text{ m/s} \qquad \textit{Index of Refraction}$$

- When light enters from medium 1 and is refracted into medium 2, Snell's law can be written in the following two forms (see Fig. 26-15):

$$n_1 \sin \theta_1 = n_2 \sin \theta_2 \qquad \frac{v_1}{v_2} = \frac{\sin \theta_1}{\sin \theta_2} \qquad \textit{Snell's Law}$$

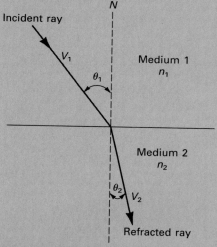

Fig. 26-15 Snell's law.

- When light enters medium 2 from medium 1, its wavelength is changed by the fact that the index of refraction is different.

$$\frac{\lambda_1}{\lambda_2} = \frac{n_2}{n_1} \qquad \lambda_2 = \frac{n_1 \lambda_1}{n_2}$$

- The critical angle θ_c is the maximum angle of incidence from one medium which will still produce refraction (at 90°) into a bordering medium. From the definition, we obtain

$$\sin \theta_c = \frac{n_2}{n_1} \qquad \textit{Critical Angle}$$

- Refraction causes an object in one medium to be observed at a different depth when viewed from above in another medium.

$$\frac{\text{Apparent depth } q}{\text{Actual depth } p} = \frac{n_2}{n_1}$$

QUESTIONS

26-1. Define the following terms:
- a. Refraction
- b. Optical density
- c. Snell's law
- d. Dispersion
- e. Index of refraction
- f. Total internal reflection
- g. Critical angle
- h. Apparent depth

26-2. State three laws of refraction and show how they can be demonstrated in the laboratory.

26-3. Is the index of refraction a constant for a particular medium? Explain.

26-4. Explain how the day is lengthened by atmospheric refraction.

26-5. A coin is placed on the bottom of a bucket so that it is just out of sight when viewed at an angle from the top. Show by the use of diagrams why the coin becomes visible if the bucket is filled with water.

26-6. Will objects of higher optical density have greater or smaller critical angles when surrounded by air?

26-7. On the basis of topics discussed in this chapter, explain why a diamond is much more brilliant than a glass replica.

26-8. Explain why right-angle prisms are more efficient reflectors than mirrored surfaces.

26-9. Why are colors observed in the light from a diamond?

26-10. A child stands waist deep in a swimming pool whose depth is uniform throughout. Why does it appear to him that he is standing in the deepest part of the pool?

26-11. The wavelength λ of a certain source of radiation is increased to 2λ. If the index of refraction was originally measured to be 1.5, what will it be when the wavelength is doubled?

26-12. What is the critical angle for an irregular-shaped piece of glass submerged in a liquid of the same index of refraction? Why would the glass be invisible in this case?

PROBLEMS

The Index of Refraction

26-1. The speed of light through a certain medium is measured to be 1.6×10^8 m/s. What is the index of refraction for the medium?

Ans. 1.88

26-2. If the speed of light is to be reduced by one-third, what must be the index of refraction for the medium through which the light travels?

26-3. Compute the speed of light in (a) crown glass, (b) diamond, (c) water, and (d) ethyl alcohol.

Ans. (a) 1.97×10^8, (b) 1.24×10^8, (c) 2.26×10^8, (d) 2.21×10^8 m/s

26-4. If light travels at 2.10×10^8 m/s in a transparent medium, what is the index of refraction in that medium?

The Laws of Refraction

26-5. Light is incident at an angle of 37° from air to flint glass ($n = 1.6$). What is the angle of refraction into the glass? What is the speed of light in the glass?

Ans. 22.1°, 1.88×10^8 m/s

26-6. A beam of light makes an angle of 60° with the surface of water. What is the angle of refraction into the water?

26-7. Light strikes from medium A into medium B at an angle of 35° with the horizontal boundary between the two media. If the angle of refraction is also 35°, what is the *relative* index of refraction between the two media?

Ans. 1.43

26-8. Light incident from air at 45° is refracted into a transparent medium at an angle of 34°. What is the index of refraction for the material?

26-9. A ray of light originating in air (Fig. 26-16) is incident on water ($n = 1.33$) at an angle of 60°. It then passes through the water, entering glass ($n = 1.5$) and finally emerging back into air again. Compute the angle of emergence.

Ans. 60°

Fig. 26-16

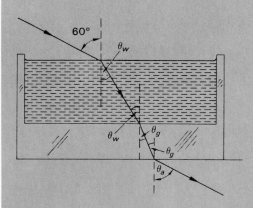

26-10. Prove that, no matter how many parallel layers of different media are traversed by light, the entrance angle and the emergent angle will be equal when the initial and final media are the same.

Wavelength and Refraction

26-11. The index of refraction of a certain glass is 1.5 for light whose wavelength is 600 nm in air. What is the wavelength of this light as it passes through the glass?

Ans. 400 nm

26-12. Red light (650 nm) changes to blue light (478 nm) when it passes into a liquid. **(a)** What is the index of refraction for the liquid? **(b)** What is the velocity in the liquid? **(c)** From Table 26-1, what might the liquid be?

26-13. A ray of monochromatic light of wavelength 400 nm in medium 1 is incident at 30° at the boundary of another medium 2. If the ray is refracted at an angle of 50°, what is its wavelength in medium 2?

Ans. 613 nm

Total Internal Reflection

26-14. If the critical angle for a liquid-to-air surface is 46°, what is the index of refraction for the liquid?

26-15. What is the critical angle for **(a)** diamond, **(b)** water, and **(c)** ethyl alcohol if the exterior surface is air?

Ans. (a) 24.4°, (b) 48.8°, (c) 47.3°

26-16. What is the critical angle for flint glass immersed in ethyl alcohol?

* **26-17.** A right-angle prism like the one shown in Fig. 26-10a is submerged in water. What is the minimum index of refraction for the material in order to achieve total internal reflection?

Ans. 1.88

Additional Problems

26-18. A ray of light strikes a pane of glass at an angle of incidence of 60°. If the angle of refraction is 30°, what is the index of refraction for the glass?

26-19. A beam of light is incident on a plane surface separating two media of indexes 1.6 and 1.4. The angle of incidence is 30° in the medium of higher index. What is the angle of refraction?

Ans. 34.8°

26-20. In going from glass ($n = 1.5$) to water ($n = 1.33$), what is the critical angle for total internal reflection?

* **26-21.** Light of wavelength 650 nm in a particular glass has a speed of 1.7×10^8 m/s. What is the index of refraction for this glass? What is the wavelength of this light in air?

Ans. 1.76, 1146 nm

26-22. The critical angle for a certain substance is 38° when it is surrounded by air. What is the index of refraction of the substance?

** **26-23.** Light passing through a plate of transparent material of thickness t suffers a lateral displacement d, as shown in Fig. 26-17. Compute the lateral displacement if the light passes through glass surrounded by air. The angle of incidence θ_1 is 40° and the glass ($n = 1.5$) is 2 cm thick.

Ans. 5.59 mm

Fig. 26-17

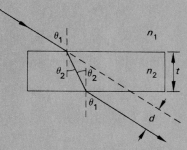

** **26-24.** Prove that the lateral displacement in Fig. 26-17 can be calculated from

$$d = t \sin \theta_1 \left(1 - \frac{n_1 \cos \theta_1}{n_2 \cos \theta_2} \right)$$

Use this relation to verify the answer in Prob. 26-23.

26-25. The water in a swimming pool is 2 m deep. How deep does it appear to a person looking vertically down?

Ans. 1.5 m

26-26. A plate of glass ($n = 1.5$) is placed over a coin on a table. The coin appears to be 3 cm below the top of the glass plate. What is the thickness of the glass plate?

27 Lenses and Optical Instruments

After completing this chapter, you should be able to:

1. Determine mathematically or experimentally the focal length of a lens and state whether it is converging or diverging.
2. Apply the lensmaker's equation to solve for unknown parameters related to the construction of lenses.
3. Use ray-tracing techniques to construct images formed by diverging and converging lenses for various object locations.
4. Predict mathematically or determine experimentally the nature, size, and location of images formed by converging and diverging lenses.

A *lens* is a transparent object that alters the shape of a wavefront passing through it. Lenses are usually constructed of glass and shaped so that refracted light will form images similar to those discussed for mirrors. Anyone who has examined objects through a magnifying glass, observed distant objects through a telescope, or experimented in photography knows something of the effects lenses have on light. In this chapter we study the images formed by lenses and discuss their application.

27-1
SIMPLE LENSES

The simplest way of understanding how a lens works is to consider the refraction of light by prisms, as illustrated in Fig. 27-1. When Snell's law is applied to each surface of a prism, it will be seen that light is bent toward the normal when entering a prism

Fig. 27-1 Parallel rays of light are bent toward the base of a prism and remain parallel.

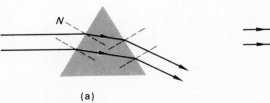

(a)

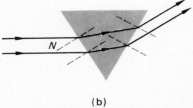

(b)

and away from the normal on leaving. The effect, in either case, is to cause the light beam to be deviated toward the base of the prism. The light rays remain parallel because both the entrance and emergent surfaces are planes forming equal angles with all rays passing the prism. Thus a prism merely alters the direction of a wavefront.

Suppose we place two prisms base to base, as shown in Fig. 27-2a. Light incident from the left will converge, but it will not come to a focus. In order to focus the light rays to a point, the extreme rays must be deviated more than the central rays. This is accomplished by grinding the surfaces so that they have a uniformly curved cross section, as indicated in Fig. 27-2b. A lens which brings a parallel beam of light to a point focus in this fashion is called a *converging lens*.

A **converging lens** is one which refracts and converges parallel light to a point focus beyond the lens.

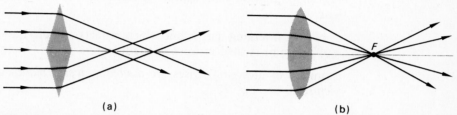

(a)   (b)

Fig. 27-2 *(a)* Two prisms placed base to base will converge rays but will not bring all rays to a common focus. *(b)* A converging lens can be constructed by curving the surfaces uniformly.

The curved surfaces of lenses may be of any regular shape, such as spherical, cylindrical, or parabolic. Since spherical surfaces are easier to make, most lenses are constructed with two spherical surfaces. The line joining the centers of the two spheres is known as the *axis* of the lens. Three examples of converging lenses are shown in Fig. 27-3, *double convex, plano-convex,* and *converging meniscus.* Note that converging lenses are thicker in the middle than at the edge.

A second type of lens can be constructed by making the edges thicker than the middle, as shown in Fig. 27-4. Parallel light rays passing through such a lens bend

CONVERGING LENSES

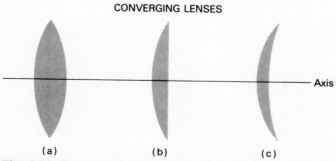

(a)   (b)   (c)

Fig. 27-3 Examples of converging lenses: *(a)* double convex, *(b)* plano-convex, and *(c)* converging meniscus.

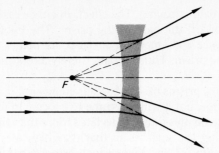

Fig. 27-4 A diverging lens refracts light so that it appears to come from a point on the same side of the lens as the incident light.

toward the thicker part, causing the beam to diverge. Projection of the refracted light rays shows that the light appears to come from a virtual focal point in front of the lens.

A **diverging lens** is one which refracts and diverges parallel light from a point located in front of the lens.

Examples of diverging lenses are *double concave, plano-concave,* and *diverging meniscus.* See Fig. 27-5.

27-2
FOCAL LENGTH AND THE LENS-MAKER'S EQUATION

A lens is regarded as "thin" if the thickness is small in comparison with the other dimensions involved. As with mirrors, image formation by thin lenses is a function of the focal length. However, there are very important differences. One obvious difference is that light may pass *through* a lens in two directions. This results in two focal points for each lens, as shown in Fig. 27-6 for a converging lens and in Fig. 27-7 for a diverging lens. The former has a *real focus F,* and the latter has a *virtual focus F'.* The distance between the optical center of a lens and the focus on either side of the lens is the *focal length f.*

The **focal length** *f* of a lens is reckoned as the distance from the optical center of the lens to either focus.

DIVERGING LENSES

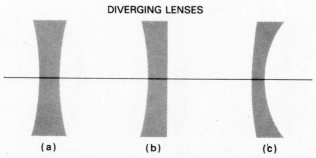

(a)   (b)   (c)

Fig. 27-5 Examples of diverging lenses: *(a)* double concave, *(b)* plano-concave, and *(c)* diverging meniscus.

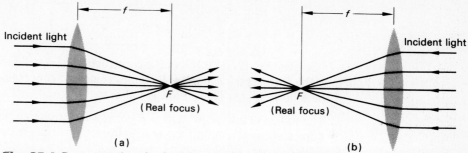

Fig. 27-6 Demonstrating the focal length of a converging lens. The focal point is real because actual light rays pass through it.

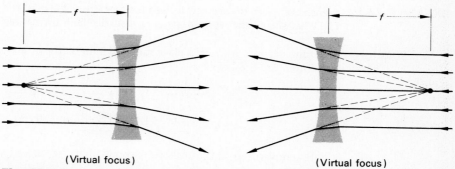

Fig. 27-7 Demonstrating the virtual focal points of a diverging lens.

Since light rays are reversible, a source of light placed at either focus of a converging lens results in a parallel light beam. This can be seen by reversing the direction of the rays illustrated in Fig. 27-6.

The focal length f of a lens is not equal to one-half the radius of curvature, as for spherical mirrors; it depends on the index of refraction n of the material from which it is made. It also is determined by the radii of curvature R_1 and R_2 of its surfaces, as defined in Fig. 27-8a. For thin lenses, these quantities are related by the equation

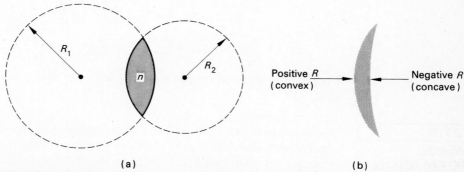

Fig. 27-8 (a) The focal point of a lens is determined by the radii of its surfaces and the index of refractions. (b) The sign convention for the radius of a lens surface.

$$\frac{1}{f} = (n - 1) \left(\frac{1}{R_1} + \frac{1}{R_2} \right) \qquad (27\text{-}1)$$

Because Eq. (27-1) involves the construction parameters for a lens, it is referred to as the *lensmaker's equation*. It applies equally to converging and diverging lenses if the following sign convention is followed:

1. The radius of curvature (either R_1 or R_2) is considered positive if the surface is curved outward (convex) and as negative if the surface is curved inward (concave). (Refer to Fig. 27-8*b*.)
2. The focal length f of a converging lens is considered positive, and the focal length of a diverging lens is considered negative.

EXAMPLE 27-1 A lensmaker plans to construct a plano-concave lens out of glass with an index of refraction of 1.5. What should the radius of its curved surface be if the desired focal length is -30 cm?

Solution The radius of curvature R_1 for a plane surface is infinity. The radius R_2 of the concave surface is found from Eq. (27-1):

$$\frac{1}{f} = (n - 1) \left(\frac{1}{\infty} + \frac{1}{R_2} \right) = (n - 1) \frac{1}{R_2}$$
$$R_2 = (n - 1) f = (1.5 - 1.0)(-30 \text{ cm})$$
$$= -15 \text{ cm}$$

By convention, the minus sign shows that the curved surface is concave.

EXAMPLE 27-2 A meniscus lens has a convex surface whose radius of curvature is 10 cm and a concave surface whose radius is -15 cm. If the lens is constructed from glass with an index of refraction of 1.52, what will its focal length be?

Solution Substitution into the lensmaker's equation yields

$$\frac{1}{f} = (n - 1) \left(\frac{1}{R_1} + \frac{1}{R_2} \right)$$
$$= (1.52 - 1) \left(\frac{1}{10 \text{ cm}} - \frac{1}{15 \text{ cm}} \right)$$
$$= 0.52 \left(\frac{3 - 2}{30 \text{ cm}} \right) = \frac{0.52}{30 \text{ cm}}$$
$$f = \frac{30 \text{ cm}}{0.52} = 57.7 \text{ cm}$$

The fact that the focal length is positive indicates that it is a *converging* meniscus lens.

27-3
IMAGE FORMATION BY THIN LENSES

To understand how images are formed by lenses, we now introduce ray-tracing methods similar to those discussed for spherical mirrors. The method consists of tracing two or more rays from a chosen point on the object and using the point of intersection as the image of that point. The entire deviation of a ray passing through a thin lens can be considered to take place at a plane through the center of the lens. In a previous section it was noted that a lens has two focal points. We define the *first focal point* F_1 as the one located on the same side of the lens as the incident light. The *second focal point* F_2 is located on the opposite, or far, side of the lens. With these definitions in mind, there are three principal rays which can easily be traced through a lens. These rays are illustrated in Fig. 27-9 for a converging lens and in Fig. 27-10 for a diverging lens:

> **Ray 1** A ray parallel to the axis passes through the second focal point F_2 of a converging lens or appears to come from the first focal point F_1 of a diverging lens.
>
> **Ray 2** A ray that passes through the first focal point F_1 of a converging lens or proceeds toward the second focal point F_2 of a diverging lens is refracted parallel to the lens axis.
>
> **Ray 3** A ray that passes through the geometrical center of a lens will not be deviated.

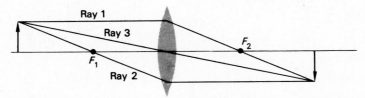

Fig. 27-9 Principal rays for image construction using a converging lens.

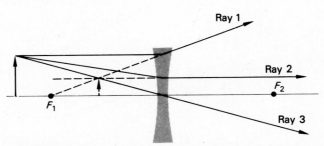

Fig. 27-10 Principal rays for constructing images formed by diverging lenses.

The intersection of any two of these rays (or their extensions) from an object point represents the image of that point. Since a real image produced by a lens is formed by rays of light which actually pass through the lens, *a real image is always formed on the side of the lens opposite the object. A virtual image will appear to be on the same side of the lens as the object.*

To illustrate the graphical method and, at the same time, to understand the various images formed by lenses, we shall consider several examples. Images formed by a converging lens are shown in Fig. 27-11 for the following object locations:

(*a*) Object located at a distance beyond twice the focal length. A real, inverted, and diminished image is formed between F_2 and $2F_2$ on the opposite side of the lens.

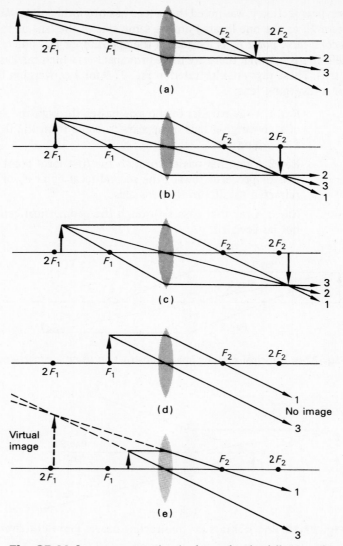

Fig. 27-11 Image construction is shown for the following object distances: *(a)* beyond $2F_1$; *(b)* at $2F_1$, *(c)* between $2F_1$ and F_1; *(d)* at F_1; and *(e)* inside F_1.

(b) Object at a distance equal to twice the focal length. A real, inverted image the same size as the object is located at $2F_2$ on the opposite side of the lens.

(c) Object located at a distance between one and two focal lengths from the lens. A real, inverted, and enlarged image is formed beyond $2F_2$ on the opposite side of the lens.

(d) Object at the first focal point F_1. No image is formed. The refracted rays are parallel.

(e) Object located inside the first focal point. A virtual, erect, and enlarged image is formed on the same side of the lens as the object.

Notice that the images formed by a *convex* lens are similar to those formed by concave mirrors. This is true because they both converge light. Since concave lenses diverge light, we would expect them to form images similar to those formed by a diverging mirror (convex mirror). Figure 27-12 demonstrates this similarity.

> *Images of real objects formed by diverging lenses are always virtual, erect, and diminished in size.*

To avoid confusion, one should identify both lenses and mirrors as either *converging* or *diverging*. Diverging lenses are often used to reduce or neutralize the effect of converging lenses.

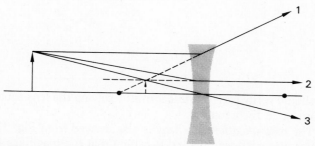

Fig. 27-12 Images formed by diverging lenses are always virtual, erect, and diminished in size.

27-4
THE LENS EQUATION AND MAGNIFICA-TION

The characteristics, size, and location of images can also be determined analytically from the *lens equation*. This important relation can be deduced by applying plane geometry to Fig. 27-13. The derivation is similar to the one used to derive the mirror equation, and the final form is exactly the same. The lens equation can be written

$$\frac{1}{p} + \frac{1}{q} = \frac{1}{f}$$

(27-2)

where p = object distance
q = image distance
f = focal length of lens

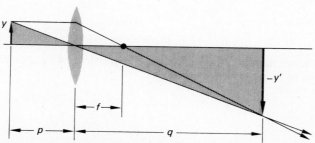

Fig. 27-13 Deriving the lens equation and the magnification.

The same sign conventions established for mirrors can be used in the lens equation if converging and diverging lenses are compared with converging and diverging mirrors. This convention is summarized as follows:

1. **The object distance p and the image distance q are considered positive for real objects and images and negative for virtual objects and images.**
2. **The focal length f is considered positive for converging lenses and negative for diverging lenses.**

The following alternative forms of the lens equation are useful in solving optical problems:

$$p = \frac{fq}{q - f} \qquad q = \frac{fp}{p - f}$$

$$f = \frac{qp}{p + q}$$

(27-2)

You should verify each of these forms by solving the lens equation explicitly for each parameter in the equation.

The *magnification* of a lens is also derived from Fig. 27-13 and has the same form as discussed for mirrors. Recalling that the magnification M is defined as the ratio of image size y' to the object size y, we can write

$$M = \frac{y'}{y} = -\frac{q}{p}$$

(27-3)

where q is the image distance and p is the object distance. *A positive magnification indicates that the image is erect whereas a negative magnification occurs only when the image is inverted.*

EXAMPLE 27-3 An object 4 cm high is located 10 cm from a thin converging lens having a focal length of 20 cm. What are the nature, size, and location of the image?

Solution This situation corresponds to that illustrated in Fig. 27-11e. The image distance is found from Eq. (27-2).

$$q = \frac{pf}{p - f} = \frac{(10 \text{ cm})(20 \text{ cm})}{10 \text{ cm} - 20 \text{ cm}}$$

$$= \frac{200 \text{ cm}^2}{-10 \text{ cm}} = -20 \text{ cm}$$

The minus sign indicates that the image is virtual. The magnification relation [Eq. (27-3)] now allows us to compute the image size.

$$M = \frac{y'}{y} = -\frac{q}{p}$$

$$y' = \frac{-qy}{p} = \frac{-(-20 \text{ cm})(4 \text{ cm})}{10 \text{ cm}}$$

$$= +8 \text{ cm}$$

The positive sign indicates that the image is erect. This example illustrates the principle of a magnifying glass. A converging lens held closer to an object than its focal point produces an erect and enlarged virtual image.

EXAMPLE 27-4 A diverging meniscus lens has a focal length of -16 cm. If the lens is held 10 cm from an object, where is the image located? What is the magnification of the lens?

Solution Direct substitution yields

$$q = \frac{pf}{p - f} = \frac{(10 \text{ cm})(-16 \text{ cm})}{10 \text{ cm} - (-16 \text{ cm})}$$

$$= \frac{-160}{10 + 16} = -6.15 \text{ cm}$$

The minus sign again indicates that the image is virtual. The magnification is

$$M = -\frac{q}{p} = \frac{-(-6.15 \text{ cm})}{10 \text{ cm}} = +0.615$$

The positive magnification means that the image is erect.

27-5
COMBINA-TIONS OF LENSES

When light passes through two or more lenses, the combined action can be determined by considering the image that would be formed by the first lens as the object for the second lens, and so on. Consider, for example, the lens arrangement illustrated in Fig. 27-14. Lens 1 forms a real, inverted image I_1 of the object O. By

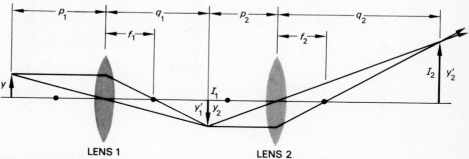

LENS 1   **LENS 2**
Fig. 27-14 The microscope.

considering this intermediate image as a real object for lens 2, the final image I_2 is seen to be real, erect, and enlarged. The lens equation can be applied successively to the two lenses to determine the location of the final image analytically.

The total magnification produced by a system of lenses is the product of the magnifications produced by each lens in the system. That this is true can be seen from Fig. 27-14. The magnifications in this case are

$$M_1 = \frac{y_1'}{y_1} \qquad M_2 = \frac{y_2'}{y_2}$$

Since $y_1' = y_2$, the product M_1M_2 yields

$$\frac{y_1'}{y_1}\frac{y_2'}{y_2} = \frac{y_2'}{y_1}$$

But y_2'/y_1 is the overall magnification M. In general, we can write

$$\boxed{M = M_1M_2}$$

(27-4)

Applications of the above principles are found for the microscope, the telescope, and other optical instruments.

27-6
THE COMPOUND MICROSCOPE

A compound microscope consists of two converging lenses, arranged as shown in Fig. 27-15. The left lens is of short focal length and is called the *objective lens*. This lens has a large magnification and forms a real, inverted image of the object being studied. The image is further enlarged by the *eyepiece,* which forms a virtual final image. The total magnification achieved is the product of the magnifications of the objective lens and the eyepiece.

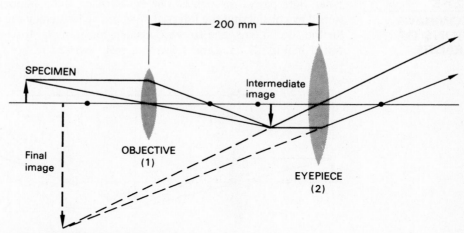

Fig. 27-15 Combinations of lenses.

EXAMPLE 27-5   In a compound microscope, the objective lens has a focal length of 8 mm, and the eyepiece has a focal length of 40 mm. The distance between the two lenses is 200 mm, and the final image appears to be at a distance of 250 mm from the eyepiece. *(a)* How far is the object from the objective lens? *(b)* What is the total magnification? (Refer to Fig. 27-15.)

Solution (a)   We begin by labeling the objective lens as 1 and the eyepiece as 2. Since more information is given about parameters that affect the second lens, we shall compute the position of the intermediate image first. From the lens equation, the distance p_2 of this image from the second lens is

$$p_2 = \frac{f_2 q_2}{q_2 - f_2} = \frac{(40 \text{ mm})(-250 \text{ mm})}{-250 \text{ mm} - 40 \text{ mm}}$$
$$= 34.5 \text{ mm}$$

The minus sign was used for the image distance q_2 because it was measured to a virtual image. Now that p_2 is known, we can compute the image distance q_1 for the first image.

$$q_1 = 200 \text{ mm} - 34.5 \text{ mm} = 165.5 \text{ mm}$$

Thus the object distance p_1 must be

$$p_1 = \frac{q_1 f_1}{q_1 - f_1} = \frac{(165.5 \text{ mm})(8 \text{ mm})}{165.5 \text{ mm} - 8 \text{ mm}}$$
$$= 8.41 \text{ mm}$$

Solution (b) The total magnification is the product of the individual magnifications:

$$M_1 = \frac{-q_1}{p_1} \qquad M_2 = \frac{-q_2}{p_2}$$

Applying Eq. (27-4), we obtain

$$M = M_1 M_2 = \frac{+q_1 q_2}{p_1 p_2}$$
$$= \frac{(165.5 \text{ mm})(-250 \text{ mm})}{(8.41 \text{ mm})(34.5 \text{ mm})} = -143$$

The negative magnification indicates that the final image is inverted. This microscope would be rated 143×, and the object under study should be placed 8.41 mm from the objective lens.

27-7
TELESCOPE

The optical system of a refracting telescope is essentially the same as that of a microscope. Both instruments use an eyepiece, or *ocular,* to enlarge the image produced by an objective lens, but a telescope is used to examine large distant objects and a microscope is used for small nearby objects.

The refracting telescope is illustrated in Fig. 27-16. The objective lens forms a real, inverted, and diminished image of the distant object. As with the microscope, the eyepiece forms an enlarged and virtual final image of the distant object.

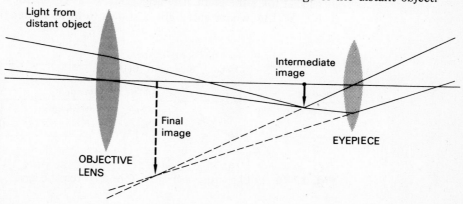

Fig. 27-16 The refracting telescope.

A telescope image is usually smaller than the object being observed. Linear magnification is therefore not a very meaningful way to describe the effectiveness of a given telescope. A better measure would be to compare the size of the final image with the size of the object observed without the telescope. If the image seen by the eye is larger than it would be without the telescope, the effect will be to make the object appear closer to the eye than it actually is.

Spherical lenses often fail to produce perfect images because of defects inherent in their construction. Two of the most common defects are known as *spherical aberration* and *chromatic aberration.* Spherical aberration, as discussed earlier for mirrors, is the inability of a lens to focus all parallel rays to the same point. (See Fig. 27-17.)

> **Spherical aberration** is a lens defect in which the extreme rays are brought to a focus nearer the lens than rays entering near the optical center of the lens.

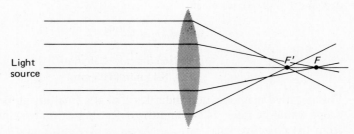

Fig. 27-17 Spherical aberration.

This effect can be minimized by placing a *diaphragm* in front of the lens. The diaphragm blocks off the extreme rays, producing a sharper image along with a reduction in light intensity.

In Chapter 26 we discussed the fact that the index of refraction for a given transparent material varies with the wavelength of the light passing through it. Thus, if white light is incident upon a lens, the rays of the component colors are not focused at the same point. The defect, known as chromatic aberration, is illustrated in Fig. 27-18*a*, where blue light is shown to focus nearer the lens than red light.

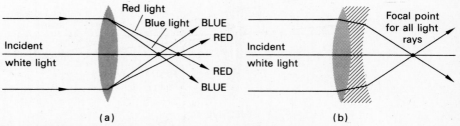

(a)   (b)

Fig. 27-18 *(a)* Chromatic aberration. *(b)* An achromatic lens.

Chromatic aberration is a lens defect which reflects its inability to focus light of different colors to the same point.

The remedy for this defect is the *achromatic lens,* illustrated in Fig. 27-18*b*. Such a lens can be constructed by combining a converging lens of crown glass ($n = 1.52$) with a diverging lens of flint glass ($n = 1.63$). These lenses are chosen and constructed so that the dispersion of one is equal and opposite to that of the other.

SUMMARY

A lens is a transparent device which converges or diverges light to or from a focal point. Lenses are used extensively in the design of many industrial instruments, and an understanding of how they form images is valuable. A summary of the major concepts discussed in this chapter is provided below:

- Image formation by thin lenses can be understood more easily through raytracing techniques as shown in Fig. 27-9 for converging lenses and in Fig. 27-10 for diverging lenses. Remember that the first focal point F_1 is the one on the same side of the lens as the incident light. The second focal point F_2 is on the far side.

  Ray 1 A ray parallel to the axis passes through the second focal point F_2 of a converging lens or appears to come from the first focal point F_1 of a diverging lens.

  Ray 2 A ray that passes through F_1 of a converging lens or proceeds toward F_2 of a diverging lens is refracted parallel to the lens axis.

  Ray 3 A ray which passes through the geometrical center of a lens will not be deviated.

- The *lensmaker's equation* is a relationship between the focal length, the radii of the two lens surfaces, and the index of refraction of the lens material. The meaning of these parameters is seen from Fig. 27-8.

$$\frac{1}{f} = (n - 1)\left(\frac{1}{R_1} + \frac{1}{R_2}\right)$$ *Lensmaker's Equation*

R_1 or R_2 is positive if outside surface is convex, negative if concave.
f is considered positive for converging lens and negative for diverging.

- The equations for object and image locations and for the magnification are the same as for the mirror equations.

$$\frac{1}{p} + \frac{1}{q} = \frac{1}{f}$$ $$p = \frac{qf}{q - f}$$ $$q = \frac{pf}{p - f}$$ $$f = \frac{pq}{p + q}$$

$$\text{Magnification} = \frac{\text{image size}}{\text{object size}}$$ $$M = \frac{y'}{y} = \frac{-q}{p}$$

p or q is positive for real and negative for virtual
y or y' is positive if erect and negative if inverted

27-1. Define the following terms:
  a. Lens
  b. Meniscus lens
  c. Lens equation
  d. Magnification
  e. Microscope
  f. Objective lens
  g. Eyepiece
  h. Telescope
  i. Converging lens
  j. Diverging lens
  k. Lensmaker's equation
  l. Spherical aberration
  m. Chromatic aberration
  n. Achromatic lens
  o. Diaphragm
  p. Virtual focus

27-2. Illustrate by diagrams the effect of a converging lens on a plane wavefront passing through it. What would the effect of a diverging lens be?

27-3. Explain why olives appear to be larger when viewed through their cylindrical glass container. Is the magnification due to the liquid, the glass, or both?

27-4. What happens to the focal length of a converging lens when it is immersed in water? What happens to the focal length of a diverging lens?

27-5. Distinguish between a *real* focus and a *virtual* focus. Which is on the same side of the lens as the incident light?

27-6. Distinguish between the first focal point and the second focal point as defined in the text.

27-7. The focal length of a converging lens is 20 cm. Choose a suitable scale and determine by graphical construction the nature, location, and magnification for the following object distances: (a) 15 cm, (b) 20 cm, (c) 30 cm, (d) 40 cm, (e) 60 cm, (f) infinity.

27-8. An object is moved from the surface of a lens to the focal point of the lens. Explain what happens to the image if the lens is (a) converging; (b) diverging.

27-9. Describe what happens to the magnification and location of an image as the object moves from infinity to the surface of (a) a converging lens; (b) a diverging lens.

27-10. Discuss the similarities and the differences between lenses and mirrors.

27-11. A camera uses a diaphragm to control the amount of light reaching the film. On a bright day, the diaphragm is almost closed, whereas on a dark day it must be opened wide to expose the film properly. Discuss the quality of the images produced in each case if the lens is not corrected for aberrations.

27-12. In a microscope the objective lens has a short focal length, whereas in a telescope the objective lens has a long focal length. Explain the reason for the difference in focal lengths.

27-13. Derive the lens equation with the help of Fig. 27-13.

27-14. Derive the magnification relation [Eq. (27-3)] with the help of Fig. 27-13.

27-15. Describe two methods you might use to determine the focal length of a double-concave lens.

27-16. Describe an experiment to determine the focal length of a double-concave lens.

27-17. Without referring to the text, write down the various sign conventions which must be applied when working with thin lenses.

27-18. According to convention, the object distance is considered negative when measured to a *virtual object*. Suggest some examples of virtual objects.

Focal Length and the Lensmaker's Equation

27-1. A lens such as that shown in Fig. 27-8a is constructed from glass ($n = 1.5$). The radius of the first surface is 15 cm and the radius of the second surface is 10 cm. What is the focal length of this lens?

Ans. 12 cm

27-2. A meniscus lens has a convex surface whose radius of curvature is 20 cm and a concave surface whose radius is −30 cm. What is the focal length if the refractive index is 1.54?

27-3. A plano-convex lens is ground from crown glass ($n = 1.52$). What should be the radius of the curved surface if the desired focal length is to be 400 mm?

Ans. 208 mm

27-4. The concave surface of a glass lens ($n = 1.5$) has a radius of 200 mm. The convex surface has a radius of 600 m. What is the focal length? Is it diverging or converging?

27-5. A plastic lens ($n = 1.54$) has a convex surface of radius 25 cm and a concave surface of 70 cm. What is the focal length? Is it diverging or converging?

Ans. 72 cm, converging

Images Formed by Thin Lenses

27-6. An object 5 cm high is located 12 cm from a thin converging lens of focal length 24 cm. What are the nature, size, and location of the image?

27-7. A light source is 600 mm from a converging lens of focal length 180 mm. Construct the image using ray diagrams. What is the image distance? Is the image real or virtual?

Ans. 257 mm, real

27-8. A plano-convex lens is held 40 mm from an object. What are the nature and location of the image formed if the focal length is 60 mm?

27-9. An object 6 cm high is held 4 cm from a diverging meniscus lens of focal length −24 cm. What are the nature, size, and location of the image?

Ans. $y' = 5.14$ cm, $q = -3.43$ cm

27-10. An object 450 mm from a converging lens forms a real image 900 mm from the lens. (a) What is the focal length of the lens? (b) What is the size of the image if the height of the object is 30 mm?

27-11. The focal length of a converging lens is 200 mm. An object 60 mm high is mounted on a movable track so that the distance from the lens can be varied. Calculate the nature, size, and location of the image formed for the following object distances: (a) 150 mm, (b) 200 mm, (c) 300 mm, (d) 400 mm, and (e) 600 mm.

Ans. (a) $q = -600$ mm, $y = 240$ mm, virtual, erect, enlarged; (b) $q = $ infinity, no image formed; (c) $q = 600$ mm, $y = -120$ mm, real, inverted, enlarged; (d) $q = 400$ mm, $y = -60$ mm, real, inverted, same size; (e) $q = 300$ mm, $y = -30$ mm, real, inverted, diminished

Magnification

27-12. An object is located 20 cm from a converging lens. If the magnification is −2, what is the image distance?

27-13. A pencil is held 20 cm from a diverging lens of focal length −10 cm. What is the magnification?

Ans. +⅓

27-14. A magnifying glass has a focal length of 27 cm. How close must this glass be held to an object to produce an erect image three times the size of the object?

Ans. 18 cm

27-15. A magnifying glass is held 40 mm from a specimen. What must be the focal length of the lens in order to produce an erect image that is twice as large as the specimen?

Ans. $f = +80$ mm

27-16. What is the magnification of a lens if the focal length is 40 cm and the object distance is 65 cm?

Optical Instruments

27-17. A camera consists of a converging lens of focal length 50 mm mounted in front of a light-sensitive film as shown in Fig. 27-19. **(a)** when photographing infinite objects, how far should the lens be from the film? **(b)** How far from the film should the lens be to photograph a flower 500 mm from the lens? **(c)** What is the magnification?

Ans. **(a)** 50 mm, **(b)** 55.5 mm, **(c)** −0.111

Fig. 27-19 The camera.

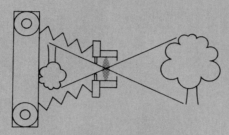

27-18. An object is placed 30 cm from a screen. At what points between the object and the screen can a lens of focal length 5 cm be placed to obtain an image on the screen?

27-19. A simple projector is illustrated in Fig. 27-20. The condenser provides even illumination of the film by the light source. The frame size of regular 8-mm film is 5×4 mm. It is desired to project an image 600×480 mm on a screen located 6 m from the projection lens. **(a)** What should be the focal length of the projection lens? **(b)** How far should the film be from the lens?

Ans. **(a)** $f = 49.6$ mm, **(b)** 50 mm

Fig. 27-20 The projector.

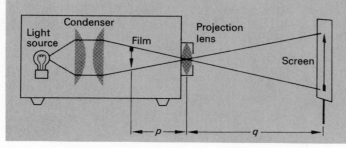

27-20. A telescope has an objective lens of focal length 900 mm and an eyepiece of focal length 50 mm. The telescope is used to examine a rabbit 30 cm high at a distance of 60 m. **(a)** What is the distance between the lenses if the final image is 25 cm in front of the eyepiece? **(b)** What is the apparent height of the rabbit as seen through the telescope?

Additional Problems

27-21. A converging lens has a focal length of 20 cm. An object is placed 15 cm from the lens. Find the image distance and the nature of the image.

Ans. -60 cm, virtual

27-22. How far from a source of light must a lens be placed if it is to form an image 800 mm from the lens? The focal length is 200 mm.

27-23. A source of light 36 cm from a lens projects an image on a screen 18 cm from the lens. What is the focal length of the lens? Is it converging or diverging?

Ans. 12 cm, converging

27-24. What is the minimum film size needed to project the image of a student who is 2 m tall? Assume that the student is located 2.5 mm from the camera lens whose focal length is 55 mm.

27-25. When parallel light strikes a lens the light diverges, apparently coming from a point 80 mm behind the lens. How far from an object should this lens be held to form an image one-fourth the size of the object?

Ans. 240 mm

27-26. It is desired to construct a symmetrical, double convex lens out of crown glass ($n = 1.52$). Assume that each surface has the same radius of curvature. What must be this radius if the resulting lens is to have a focal length of 25 cm?

27-27. The first surface of a thin lens has a convex radius of 20 cm. What should be the radius of the second surface in order to produce a converging lens of focal length 8 cm?

Ans. $+5$ cm

27-28. Two thin converging lenses are placed 60 cm apart and have the same axis. The first lens has a focal length of 10 cm, and the second has a focal length of 15 cm. If an object 6 cm high is placed 20 cm in front of the first lens, what are the location and size of the final image? Is it real or virtual?

27-29. A converging lens of focal length 25 cm is placed 50 cm in front of a diverging lens whose focal length is -25 cm. If an object is placed 75 cm in front of the converging lens, what is the location of the final image? What is the total magnification? Is the image real or virtual?

Ans. $q_2 = -8.33$ cm in front of diverging lens, -0.333, virtual

27-30. The Galilean telescope consists of a diverging lens as the eyepiece and a converging lens as the objective. The focal length of the objective is 30 cm, and the focal length of the eyepiece is -2.5 cm. An object 40 m away from the objective has a final image located 25 cm in front of the diverging lens. What is the separation of the lenses? What is the total magnification?

27-31. The focal length of the eyepiece of a particular microscope is 3 cm, and the focal length of the objective lens is 19 mm. The separation of the two lenses is 26.5 cm, and the final image formed by the eyepiece is at infinity. How far should the objective lens be placed from the specimen being studied?

Ans. 20.7 mm

28 Interference, Diffraction, and Polarization

OBJECTIVES

After completing this chapter, you should be able to:

1. Demonstrate by definition and drawings your understanding of the terms *constructive interference, destructive interference, diffraction, polarization,* and *resolving power.*
2. Describe Young's experiment and be able to use the results to predict the location of bright and dark fringes.
3. Discuss the use of a diffraction grating, derive the grating equation, and apply it to the solution of optical problems.

Light is dualistic in nature, sometimes exhibiting the properties of particles and sometimes those of waves. The demonstrative proof of the wave nature of light came with the discovery of interference and diffraction. Then polarization studies showed that, unlike sound waves, light waves are transverse instead of longitudinal.

In this chapter we study these phenomena and their significance in physical optics. We shall find that the neat geometrical-ray approach, which was so helpful in studying mirrors and lenses, must be discarded in favor of a more rigorous wave analysis.

28-1
DIFFRACTION

When light waves pass through an aperture or past the edge of an obstacle, they always bend to some degree into the region not directly exposed to the light source. This phenomenon is called *diffraction.*

> **Diffraction** is the ability of waves to bend around obstacles placed in their path.

To understand this bending of waves, let us consider what happens when water waves strike a narrow opening. A plane-wave generator can be used in a ripple tank, as shown in Fig. 28-1. The vibrating strip of metal serves as a source of waves at one end of a tray of water. The plane waves strike the barrier, spreading out into the region

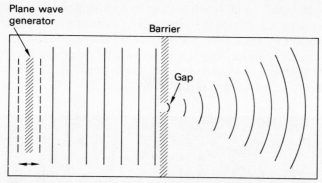

Fig. 28-1 Schematic diagram illustrating the diffraction of plane water waves through a narrow gap.

behind the gap. The diffracted waves appear to originate at the gap in accordance with Huygens' wave principle: *each point on a wavefront can be regarded as a new source of secondary waves.*

A similar experiment can be performed with light, but for the diffraction to be observable, the slit in the barrier must be very narrow. In fact, diffraction is pronounced only when the dimensions of an opening or an obstacle are comparable to the wavelength of the waves striking it. This explains why diffraction of water waves and sound waves is often observed in nature and the diffraction of light is not.

28-2

YOUNG'S EXPERIMENT: INTERFERENCE

The first convincing evidence of diffraction was demonstrated by Thomas Young in 1801. A schematic diagram of Young's apparatus is shown in Fig. 28-2. Light from a monochromatic source falls on a slit A, which acts as a source of secondary waves. Two more slits S_1 and S_2 are parallel to A and equidistant from it. Light from A passes

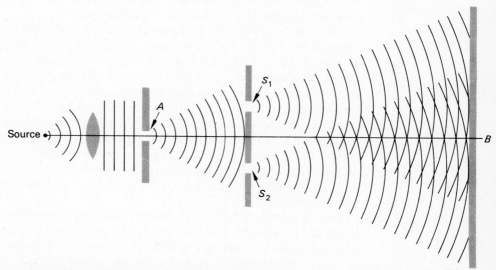

Fig. 28-2 Illustration of Young's experiment.

through both S_1 and S_2 and then on to a viewing screen. If light were not diffracted, the viewing screen would be completely dark. Instead, the screen is illuminated, as shown in Fig. 28-3. Even the point B located behind the barrier in direct line with slit A is illuminated. It is easy to see why this experiment caused early physicists to doubt that light consisted of particles traveling in straight lines. The results could be explained only in terms of the wave theory.

Fig. 28-3 Photograph of an interference pattern in Young's experiment. *(From F. A. Jenkins and H. E. White, "Fundamentals of Optics," 4th Ed., McGraw-Hill Book Company, New York, 1976, Reprinted by permission.)*

The illumination of the screen in alternate bright and dark lines can also be explained in terms of the wave theory. To understand their origin, we recall the *superposition principle,* introduced in Chap. 22 for the study of interference in waves:

> *When two or more waves exist simultaneously in the same medium, the resultant amplitude at any point is the sum of the amplitudes of the composite waves at that point.*

Two waves are said to interfere *constructively* when the amplitude of the resultant wave is greater than the amplitudes of either component wave. *Destructive interference* occurs when the resultant amplitude is smaller.

In Young's experiment, light waves from slit A reach slits S_1 and S_2 at the same time and originate from a single source of one wavelength. Therefore, the secondary wavelets leaving slits S_1 and S_2 are *in phase.* The sources are said to be *coherent.* Figure 28-4 shows how the light and dark bands are produced. The bright bands occur whenever the waves arriving from the two slits interfere constructively. Dark bands occur when destructive interference takes place. At point B in the center of the screen, light travels the same distances p_1 and p_2 from each slit. The difference in path length $\Delta p = 0$, and constructive interference results in a bright central band. At point C the difference in path lengths Δp causes the waves to interfere destructively. Another bright band occurs at point D when the waves from each slit again reinforce each other. The overall interference pattern is similar to that produced by the water waves in Fig. 28-5.

Let us now consider the theoretical conditions necessary for the production of bright and dark bands. Consider light reaching the point D at a distance y from the central axis AB, as shown in Fig. 28-6. The separation of the two slits is represented by d, and the screen is located at a distance x from the slits. The point D on the screen makes an angle θ with the axis of the system. The line S_2C is drawn so that the

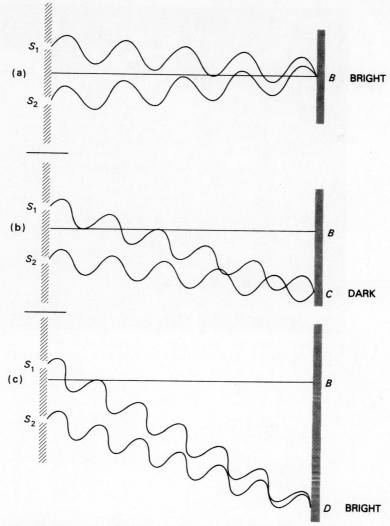

Fig. 28-4 Origin of light and dark bands in an interference pattern.

distances CD and p_2 are equal. As long as the distance x to the screen is much greater than the slit separation d, we may consider that p_1, p_2, and DA are all approximately perpendicular to the line S_2C. Therefore, the angle S_1S_2C is equal to the angle θ, and the difference Δp in path lengths of light coming from S_1 and S_2 is given by

$$\Delta p = p_1 - p_2 = d \sin \theta$$

Constructive interference will occur at D when this difference in path length is equal to

$$0, \lambda, 2\lambda, 3\lambda, \ldots, n\lambda$$

where λ is the wavelength of the light. Therefore, the conditions for bright fringes

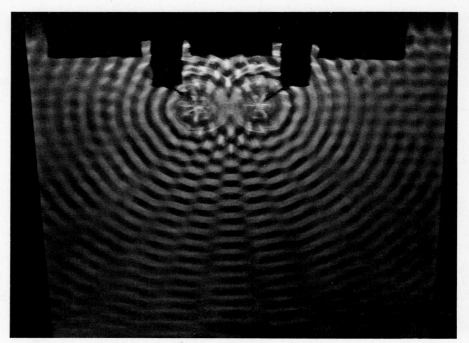

Fig. 28-5 Interference pattern set up in water waves by two coherent sources. *(From "PSSC Physics," D. C. Heath and Company, Lexington, Mass., 1965.)*

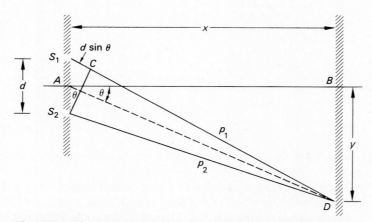

Fig. 28-6 Theoretical interpretation of the double-slit experiment.

are given by

$$d \sin \theta = n\lambda \qquad n = 0, 1, 2, 3, \ldots \qquad (28\text{-}1)$$

where d is the slit separation and θ is the angle the fringe makes with the axis.

The conditions required for the formation of dark fringes at D will be satisfied when the path difference is

$$\frac{\lambda}{2}, \frac{3\lambda}{2}, \frac{5\lambda}{2}, \ldots$$

Under these conditions, destructive interference will cancel the waves. Thus dark fringes will occur when

$$d \sin \theta = n\frac{\lambda}{2} \qquad n = 1, 3, 5, \ldots \qquad (28\text{-}2)$$

The above equations can be put into a more useful form by expressing them in terms of the measurable distances x and y. For small angles

$$\sin \theta \approx \tan \theta = \frac{y}{x}$$

Substitution of y/x for $\sin \theta$ in Eqs. (28-1) and (28-2) yields

Bright fringes:
$$\frac{yd}{x} = n\lambda \qquad n = 0, 1, 2, \ldots \qquad (28\text{-}3)$$

Dark fringes:
$$\frac{yd}{x} = n\frac{\lambda}{2} \qquad n = 1, 3, 5, \ldots \qquad (28\text{-}4)$$

In experiments designed to measure the wavelength of light, x and d are known initially. The distance y to any particular fringe can be measured and used to determine the wavelength.

EXAMPLE 28-1 In Young's experiment, the two slits are 0.04 mm apart, and the screen is located 2 m away from the slits. The third bright fringe from the center is displaced 8.3 cm from the central fringe. (a) Determine the wavelength of the incident light. (b) Where will the second dark fringe appear?

Solution (a) For the third bright fringe, $n = 3$ in Eq. (28-3). Thus

$$\frac{yd}{x} = 3\lambda$$

$$\lambda = \frac{yd}{3x} = \frac{(8.3 \times 10^{-2} \text{ m})(4 \times 10^{-5} \text{ m})}{3(2 \text{ m})}$$

$$= 5.53 \times 10^{-7} \text{ m} = 553 \text{ nm}$$

Solution (b) The displacement of the second dark fringe is found by setting $n = 3$ in Eq. (28-4).

$$\frac{yd}{x} = \frac{3\lambda}{2}$$

$$y = \frac{3\lambda x}{2d} = \frac{(3)(5.53 \times 10^{-7} \text{ m})(2 \text{ m})}{2(4 \times 10^{-5} \text{ m})}$$

$$= 4.15 \times 10^{-2} \text{ m}$$

28-3

THE DIFFRACTION GRATING

If many parallel slits similar to those in Young's experiment are regularly spaced and of the same width, a brighter and sharper diffraction pattern can be obtained. Such an arrangement is known as a *diffraction grating*. Gratings are made by ruling thousands of parallel grooves on a glass plate with a diamond point. The grooves act

---

as opaque barriers to light, and the clear spaces form the slits. Most laboratory gratings are ruled from 10,000 to 30,000 lines/in.

A parallel beam of monochromatic light striking a diffraction grating, as shown in Fig. 28-7, is diffracted in a manner similar to that in Young's experiment. Only a few slits are shown in the figure, each separated by a distance d. Each slit acts as a source of secondary Huygens' wavelets, producing an interference pattern. A lens is used to focus the light from the slits on a screen.

The diffracted rays leave the slits in many directions. In the forward direction, the paths of all rays will be of the same length, setting up a condition for constructive interference. A central bright image of the source is formed on the screen. In certain other directions, the diffracted rays will also be in phase, giving rise to other bright images. One of these directions θ is illustrated in Fig. 28-7. The first bright line formed on either side of the central image is called the *first-order* image. The second bright line on either side of the central maximum is called the *second-order* fringe, and so forth. The condition for the formation of these bright fringes is the same as that derived for Young's experiment. (Refer to Fig. 28-8.) Therefore, we can write the grating equation as

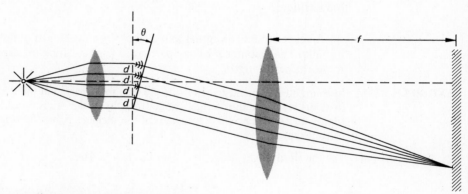

Fig. 28-7 The diffraction grating.

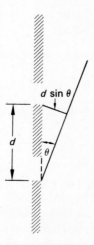

Fig. 28-8

$$d \sin \theta_n = n\lambda \qquad n = 1, 2, 3, \ldots \tag{28-5}$$

where d = spacing of slits

λ = wavelength of incident light

θ_n = deviation angle for nth bright fringe

The first-order bright fringe occurs when $n = 1$. As illustrated in Fig. 28-9a, this image occurs when the paths of diffracted rays from each slit differ by an amount equal to one wavelength. The second-order image occurs when the paths differ by two wavelengths (Fig. 28-9b).

Fig. 28-9

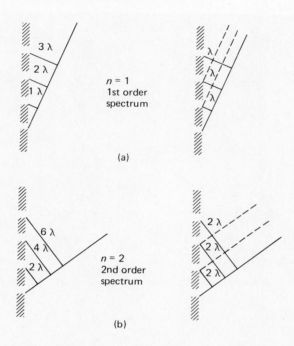

(a)

(b)

EXAMPLE 28-2   A diffraction grating having 20,000 lines/in. is illuminated by parallel light of wavelength 589 nm. What are the angles at which the first- and second-order bright fringes are formed?

Solution   The slit spacing d that corresponds to 20,000 lines/in. is

$$d = \frac{2.54 \text{ cm/in.}}{20,000 \text{ lines/in.}} = 1.27 \times 10^{-4} \text{ cm}$$

The first-order bright fringe occurs for $n = 1$, and the angle θ_1 is found from Eq. (28-5).

$$\sin \theta_1 = 1\frac{\lambda}{d} = \frac{5.89 \times 10^{-5} \text{ cm}}{1.27 \times 10^{-4} \text{ cm}} = 0.464$$

$$\theta_1 = 27.6°$$

The second-order fringe occurs when $n = 2$. Thus,

$$\sin \theta_2 = (2)(0.464) = 0.928$$

$$\theta_2 = 68.1°$$

The third-order fringe will not be formed because $\sin \theta_n$ cannot exceed 1.00. (The beam will not be deviated by an angle greater than 90°.)

28-4

RESOLVING POWER OF INSTRUMENTS

We have learned that light passing through a small opening or past an obstacle is diffracted so that the images formed are fuzzy. Interference fringes near the edges of images sometimes make it difficult to determine the exact shape of the source. Figure 28-10 shows the diffraction pattern formed by passing light through a small

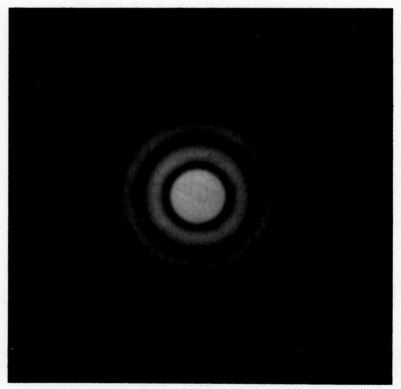

Fig. 28-10 Photograph of the interference pattern formed by passing light through a small circular opening. *(Reprinted from Cagnet et al., "Atlas of Optical Phenomena," with permission from Springer-Verlag.)*

circular opening. Notice the large central maximum surrounded by dark and bright interference bands. This diffraction is of extreme importance in optical instruments because it sets the ultimate limit on the possible magnification.

To understand this limitation, consider light from the two sources in Fig. 28-11, passing through a small circular opening in an opaque barrier. In Fig. 28-11a the images of the sources A and B are distinguished as separate images. The sources are said to be *resolved*. However, if they are brought closer together, as in Fig. 28-11b, their images overlap, resulting in a confused image. When the sources are so close

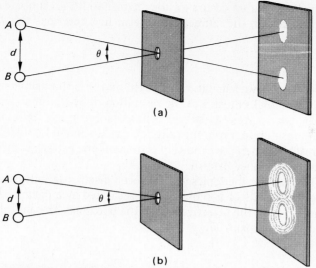

(a)

(b)

Fig. 28-11 *(a)* The images of the sources *A* and *B* are easily distinguished. *(b)* As the sources are brought closer together, the images overlap, resulting in a confused image.

together (or the opening is so small) that the separate images can no longer be distinguished, the sources are said to be *unresolved*.

> The **resolving power** of an instrument is a measure of its ability to produce well-defined separate images.

A useful method of expressing the resolving power of an instrument is in terms of the angle θ subtended at the opening by the objects being resolved. (See Fig. 28-11.) The smallest angle θ_0 for which the images can be distinguished separately is a measure of the resolving power.

No matter how perfectly a lens is constructed, the image of a point source of light will not be focused at a point. It will appear as only a tiny bright dot with light and dark fringes around it. Resolution improves as the diameter of a lens is increased. It can be shown that for a given lens of diameter *D* (see Fig. 28-12) the

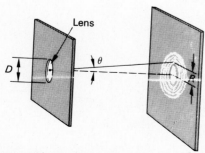

Fig. 28-12 The limit of resolution.

optimum resolution occurs for the angular width θ_0, subtended by the radius of the central image in the diffraction pattern. In a telescope this limiting angle is

$$\theta_0 = \frac{1.22\lambda}{D} \tag{28-6}$$

where λ is the wavelength of the light and D is the diameter of the objective lens.

Figure 28-13 illustrates the resolution of two point sources of light. In Fig. 28-13a the two images can still be distinguished; in Fig. 28-13b the images are at the limit of resolution. From the criterion used to obtain Eq. (28-6), two such images are just resolved when the central maximum of one pattern coincides with the first dark fringe of the other pattern.

To illustrate the topics we have been discussing in this section, consider a telescope. In Fig. 28-14 two objects are separated by a distance s_0 and are located at a distance p from the objective lens in the telescope. The angle between the objects, in radians, is approximately

$$\theta_0 = \frac{s_0}{p} \tag{28-7}$$

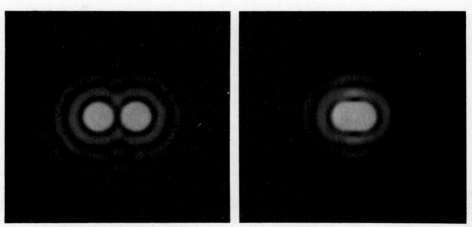

Fig. 28-13 *(a)* Separate images formed from two point sources. *(b)* The images of two sources at the limit of resolution. *(From Cagnet et al., "Atlas of Optical Phenomena," Springer-Verlag, Reprinted by permission.)*

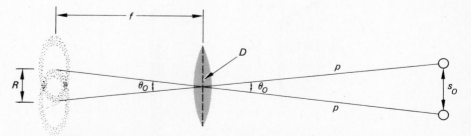

Fig. 28-14 Resolution of two distant objects by a spherical lens.

The subscript 0 is used to show that θ_0 and s_0 represent the minimum conditions for resolution. Thus we can rewrite Eq. (28-6) as

$$\boxed{\theta_0 = 1.22\frac{\lambda}{D} = \frac{s_0}{p}}$$

(28-8)

Since s_0 represents the minimum separation of objects which can be resolved, this distance is also used to indicate the resolving power of an instrument.

EXAMPLE 28-3 One of the largest refracting telescopes in the world is the 40-in.-diameter instrument at Yerkes Observatory in Wisconsin; its objective lens has a focal length of 19.8 m (65 ft). (a) What is the minimum separation of two features of the moon's surface such that they are just resolved by this telescope? (b) What is the radius of the central maximum in the diffraction pattern set up by the objective lens? (For white light, the central wavelength of 500 nm can be used for computing the resolution. The moon is 3.84×10^5 km from the earth.)

Solution (a) We must first convert all distances to meters. Thus

$$p = 3.84 \times 10^5 \text{ km} = 3.84 \times 10^8 \text{ m}$$
$$\lambda = 500 \text{ nm} = 5 \times 10^{-7} \text{ m}$$
$$D = 40 \text{ in.}(2.54 \times 10^{-2} \text{ m/in.}) = 1.02 \text{ m}$$

The minimum separation s_0 is found by solving for s_0 in Eq. (28-8).

$$s_0 = 1.22\frac{\lambda p}{D}$$
$$= \frac{(1.22)(5 \times 10^{-7} \text{ m})(3.84 \times 10^8 \text{ m})}{1.02 \text{ m}} = 230 \text{ m}$$

Solution (b) From Eq. (28-8), the radius R of the central maximum is

$$R = f\theta_0 = f\frac{s_0}{p}$$
$$= \frac{(19.8 \text{ m})(230 \text{ m})}{3.84 \times 10^8 \text{ m}} = 1.19 \times 10^{-5} \text{ m}$$

The 200-in. reflecting telescope at Mt. Palomar in California can distinguish features on the moon 46.3 m (or 152 ft) apart.

28-5

POLARIZATION

All the phenomena discussed so far can be explained on the basis of either transverse or longitudinal waves. Interference and diffraction occur in sound waves, which are longitudinal, as well as in water waves, which are transverse. More experimental evidence is needed to determine whether light waves are longitudinal or transverse. In this section we introduce a property of light waves that can be interpreted only in terms of transverse waves.

Let us first consider a mechanical example of transverse waves in a vibrating string. If the source of the wave causes each particle of the rope to vibrate up and down in a single plane, the waves are *plane-polarized*. If the rope is vibrated in such a manner that each particle moves in a random manner at all possible angles, the waves are *unpolarized*.

> **Polarization** is the process by which the transverse oscillations of a wave motion are confined to a definite pattern.

As an illustration of the polarization of a transverse wave, consider the rope passing through slotted frames, as shown in Fig. 28-15. The unpolarized vibrations pass through slot A and emerge polarized in the vertical plane. This frame is called the *polarizer*. Only waves with vertical vibrations can pass through the slot; all other vibrations will be blocked. The slotted frame B is called the *analyzer* because it can be used to test whether the incoming waves are plane-polarized. If the analyzer is rotated so that the slots in B are perpendicular to those in A, all the incoming waves are stopped. This can happen only if the waves reaching B are polarized in a plane perpendicular to the slots in B.

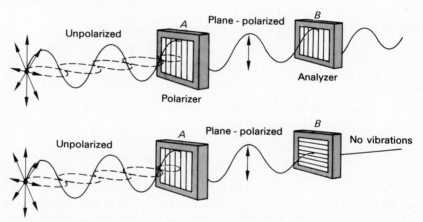

Fig. 28-15 Mechanical analogy explaining the polarization of a transverse wave.

Polarization is a characteristic of *transverse* waves. Should the rope in our example be replaced by a spring, as in Fig. 28-16, the longitudinal waves would pass on through the slots, regardless of their orientation.

Now, let us consider light waves. In Chap. 24 we discussed the electromagnetic nature of light waves. Recall that such a wave consists of an oscillating electric field and an oscillating magnetic field perpendicular to each other and to the direction of propagation. Therefore, light waves consist of *oscillating fields* rather than of vibrating particles, as was the case for the waves in a rope. If it can be shown that these

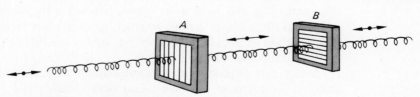

Fig. 28-16 A longitudinal wave cannot be polarized.

oscillations can be polarized, we can state conclusively that the oscillations are transverse.

A number of substances exhibit different indexes of refraction for light with different planes of polarization relative to their crystalline structure. Some examples are calcite, quartz, and tourmaline. Plates can be constructed from these materials that transmit light in only a single plane of oscillation. Thus they can be used as polarizers for incident light whose oscillations are randomly oriented. Analogous to the vibrating rope passing through slots, two polarizing plates can be used to determine the transverse nature of light waves.

As shown in Fig. 28-17, light emitted by most sources is unpolarized. Upon passing through a tourmaline plate (the polarizer), the light beam emerges plane-polarized but with reduced intensity. Another plate serves as the analyzer. As this plate is rotated with respect to the polarizer, the intensity of the light passing through the system is gradually reduced until relatively no light is transmitted. It is therefore demonstrated that light waves are transverse rather than longitudinal.

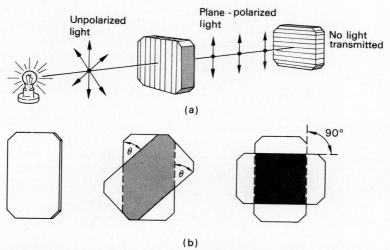

Fig. 28-17 *(a)* Proof that light can be polarized. *(b)* The reduction of transmitted light intensity as the analyzer is rotated from 0 to 90°.

Polarization has demonstrated the transverse nature of light in laboratories for many years, but, the useful applications of the principle were not realized until the development of *Polaroid* sheets. These sheets are constructed by sandwiching a thin layer of iodosulfate crystals between two sheets of plastic. The crystals are aligned by a strong electric field. Two such sheets can be used to control the intensity of light. Photographers use Polaroid filters to vary the intensity and to reduce the glare from reflected light. Complex engineering studies can be made by examining stress patterns in certain plastic tool models. The light and dark fringes set up by polarized light give an indication of the areas of varying stress.

In this chapter we have discussed several ways in which light behaves as a wave. The bending of light around obstacles placed in its path is called diffraction. A combination of diffraction and interference of light waves led to Young's experiment and to the diffraction grating. Modern industrial instruments use these concepts for a variety of applications. The major ideas presented in this chapter are summarized below:

- In Young's experiment interference and diffraction account for the production of bright and dark fringes. The location of these fringes is given by the following equations (see Fig. 28-6):

$$\text{Bright fringes: } \frac{yd}{x} = n\lambda \qquad n = 0, 1, 2, 3, \ldots$$

$$\text{Dark fringes: } \frac{yd}{x} = n\frac{\lambda}{2} \qquad n = 1, 3, 5, 7, \ldots$$

- In a diffraction grating of slit separation d, the wavelengths of the nth order fringes are given by

$$\boxed{d \sin \theta_n = n\lambda} \qquad \boxed{n = 1, 2, 3, \ldots}$$

- The resolving power of an instrument is a measure of its ability to produce well-defined separate images. The minimum conditions for resolution are illustrated in Fig. 28-14. For this situation the resolution equation is

$$\boxed{\theta_0 = 1.22\frac{\lambda}{D} = \frac{s_0}{p}} \qquad \textit{Resolving Power}$$

28-1. Define the following terms:
  a. Diffraction
  b. Interference
  c. Coherent
  d. Resolving power
  e. Resolution
  f. Polarization
  g. Polarizer
  h. Analyzer
  i. Young's experiment
  j. Huygens' principle
  k. Superposition principle
  l. Interference fringes
  m. Diffraction grating
  n. Order of an image
  o. Plane-polarized
  p. Polaroid sheets

28-2. Radio waves and light waves are both electromagnetic radiations. Explain why radio waves can be received behind tall buildings whereas light cannot reach these areas.

28-3. Consider plane water waves striking a gap in a barrier, as in Fig. 28-1. Explain how the diffraction varies as (a) the slit width is decreased and (b) the wavelength of incoming waves is reduced.

28-4. Which effects in Young's experiment are due to diffraction and which are due to interference?

28-5. In a diffraction grating, how does the spacing of the lines affect the separation of the fringes in the interference pattern?

28-6. If white light were incident upon a diffraction grating, instead of monochromatic light, what would the resulting interference pattern look like?

28-7. In Young's experiment, what effect will reducing the wavelength of incident light have on the interference pattern?

28-8. What effect does increasing the aperture of a lens have on its resolving power? Will a larger wavelength of light result in increased resolution if other conditions are kept constant?

28-9. The resolving power of some microscopes is increased by illuminating the object with ultraviolet light. Explain.

28-10. A polarizing plate absorbs about 50 percent of the intensity of an unpolarized beam. What accounts for this?

28-11. Assume that a beam of unpolarized light passes through a polarizer and an analyzer, as shown in Fig. 28-17. The analyzer has been rotated by an angle θ from the position for maximum transmission. Only the component of the amplitude A in the plane-polarized beam that lies along the axis is transmitted by the analyzer. This component is given by $A \cos \theta$. The intensity I of light is proportional to the square of the amplitude A. Show that the intensity I of the beam transmitted by the analyzer is given by

$$I = I_0 \cos^2 \theta \qquad (28-9)$$

where θ is the angle at which the analyzer is placed relative to the position for maximum transmitted intensity I_0.

28-12. Which of the following waves can be polarized: (a) x-rays, (b) water waves, (c) sound waves, and (d) radio waves?

28-13. When white light falls on a prism, it is dispersed, forming a spectrum of colors with the red component receiving the least deviation. Compare this spectrum with that produced by a diffraction grating.

28-14. Give a strong argument for the wave theory of light.

28-15. Suppose you are using a diffraction grating with a spacing of 3000 lines/cm. Discuss the usefulness of this grating for examining (a) infrared radiation of wavelength 3 nm; (b) ultraviolet radiation of wavelength 100 nm.

PROBLEMS

Young's Experiment: Interference

28-1. Monochromatic light illuminates two parallel slits 0.2 mm apart. On a screen 1 m from the slits, the first bright fringe is separated from the central fringe by 2.5 mm. What is the wavelength of the light?

Ans. 500 nm

28-2. Monochromatic light from a sodium flame illuminates two slits separated by 1 mm. A viewing screen is 1 m from the slits, and the distance from the central bright fringe to the bright fringe nearest it is 0.589 mm. What is the frequency of the light?

28-3. Two slits 0.05 mm apart are illuminated by green light of wavelength 520 nm. A diffraction pattern is formed on a viewing screen 2 m away. (a) Determine the distance from the center of the screen to the first bright fringe. (b) What is the distance to the third dark fringe?

Ans. (a) 2.08 cm, (b) 5.2 cm

* 28-4. For the situation described in Prob. 28-3, what is the separation between the two first-order bright fringes located on each side of the central band?

The Diffraction Grating

28-5. A small sodium lamp emits light of wavelength 589 nm, which illuminates a diffraction grating having 5000 lines per centimeter. Calculate the angular deviation of the first- and second-order images.

Ans. 17.1°, 36°

28-6. A parallel beam of light illuminates a diffraction grating with 6000 lines per centimeter. The second-order image is located 32 cm from the central image on a screen 50 cm from the grating. Calculate the wavelength of the light.

28-7. The visible-light spectrum ranges in wavelength from 400 to 700 nm. Find the angular width of the first-order spectrum produced by passing white light through a grating having 20,000 lines per inch.

Ans. 15°

28-8. An infrared spectrophotometer uses gratings to disperse infrared light. One grating is ruled with 240 lines per millimeter. What is the maximum range of wavelengths in nanometers that can be studied with this grating?

Resolving Power of Instruments

28-9. What is the minimum separation of two points on the moon's surface that can just be resolved by the Mt. Palomar 200-in. telescope? Assume the reflected light from the moon has a wavelength of 500 nm and that it travels a distance of 384,000 km.

Ans. 46.1 m

28-10. What is the angular limit of resolution of a man's eye when the diameter of the opening is 0.3 cm? Assume that the wavelength of the light is 500 nm. At what distance could these eyes resolve the wires in a door screen which are separated by ¼ cm?

28-11. A certain radiotelescope has a parabolic reflector which is 70 m in diameter. Radio waves from outer space have a wavelength of 21 cm. Calculate the theoretical limit of angular resolution in radians.

Ans. 3.66×10^{-3} rad

Additional Problems

28-12. A Michelson interferometer, as shown in Fig. 28-18, can be used to measure very small distances. The beam splitter partially reflects and partially transmits monochromatic light of wavelength λ from the source S. One mirror M_1 is fixed and another M_2 is movable. The light rays reaching the eye from each mirror differ, causing constructive and destructive interference patterns to move across the scope as the mirror M_2 is moved a distance x. Show that this distance is given by

$$x = m \frac{\lambda}{2}$$

where m is the number of dark fringes that cross an indicator line on the scope as the mirror moves a distance x.

28-13. A Michelson interferometer is used to measure the advance of a small screw. How far has the screw advanced if krypton 86 light ($\lambda = 606$ nm) is used and 4000 fringes move across the field of view as the screw advances?

Ans. 1.21 mm

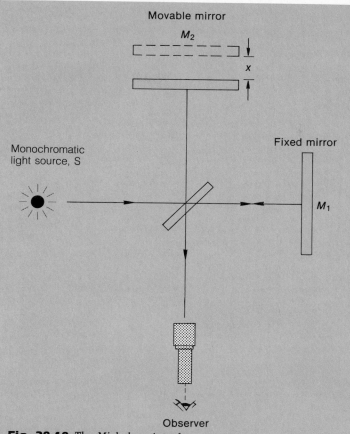

Movable mirror

M_2

x

Fixed mirror

M_1

Monochromatic
light source, S

Observer

Fig. 28-18 The Michelson interferometer.

* 28-14. If the separation of the two slits in Young's experiment is 0.1 mm and the distance to the screen is 50 cm, find the distance between the first dark fringe and the third bright fringe when the slits are illuminated with light of wavelength 600 nm.

28-15. A transmission grating ruled with 6000 lines per centimeter forms a second-order bright fringe at an angle of 53° from the central fringe. What is the wavelength of the incident light?

Ans. 666 nm

* 28-16. Light from a mercury-arc lamp is incident on a diffraction grating ruled with 7000 lines per inch. The spectrum consists of a yellow line (579 nm) and a blue line (436 nm). Compute the angular separation of these lines in the third-order spectrum.

* 28-17. It is planned to use a telescope to resolve two points on a mountain 160 km away. If the separation of the points is 2 m, what is the minimum diameter for the objective lens? Assume that the light has an average wavelength of 500 nm.

Ans. 4.88 cm

* 28-18. The intensity of unpolarized light is reduced by one-half when it passes through a polarizer. In the case of the plane-polarized light reaching the analyzer, the intensity I of the transmitted beam is given by

$$I = I_0 \cos^2\theta$$

where I_0 is the maximum intensity transmitted and θ is the angle through which the analyzer has been rotated. Three Polaroid plates are stacked so that the axis of each is turned 30° with respect to the preceding plate. By what percentage will incident light be reduced in intensity when it passes through all three plates?

PART THREE
Electricity and Magnetism

29 The Electric Force

OBJECTIVES

After completing this chapter, you should be able to:

1. Demonstrate the existence of two kinds of electric charge and verify the *first law of electrostatics* using laboratory materials.
2. Explain and demonstrate the processes of charging by *contact* and by *induction,* and use an *electroscope* to determine the nature of an unknown charge.
3. State *Coulomb's law* and apply it to the solution of problems involving electric forces.
4. Define the *electron,* the *coulomb,* and the *microcoulomb* as units of electric charge.

A hard-rubber comb or a plastic rod acquires a strange ability to attract other objects after it is rubbed on a coat sleeve. An annoying *shock* is sometimes experienced when you touch the handle of a car door after sliding across the seat. Stacked sheets of paper tend to resist separation. All these occurrences are examples of *electrification,* a phenomenon which frequently occurs as a result of rubbing objects together. Long ago, the name *charging* was given to the rubbing process, and the electrified object was said to be charged. This chapter begins our study of *electrostatics,* the science which treats electric charges at rest.

29-1
THE ELECTRIC CHARGE

The best way to begin a study of electrostatics is to experiment with objects which become electrified by rubbing. The materials illustrated in Fig. 29-1 are commonly found in the physics laboratory. In the order of their appearance in the figure, they are a hard-rubber rod resting on a piece of cat's fur, a glass rod resting on a piece of silk, the pith-ball electroscope, suspended pith balls, and the gold-leaf electroscope. A pith ball is a very light sphere of wood pith painted with metallic paint and usually suspended from a silk thread. An *electroscope* is a sensitive laboratory instrument used to detect the presence of an electric charge.

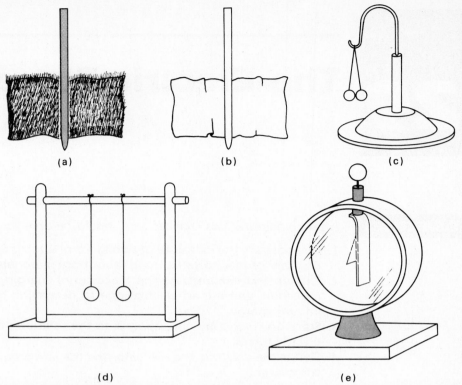

Fig. 29-1 Laboratory materials for studying electrostatics: *(a)* a hard-rubber rod resting on a piece of cat's fur; *(b)* a glass rod resting on a piece of silk; *(c)* the pith-ball electroscope; *(d)* suspended pith balls; and *(e)* the gold-leaf electroscope.

The pith-ball electroscope can be used to study the effects of electrification. Two metallic-coated pith balls are suspended by silk threads from a common point. We begin by vigorously rubbing the rubber rod with cat's fur (or wool). Then if the rubber rod is brought near the electroscope, the suspended pith balls will be attracted to the rod, as shown in Fig. 29-2a. After remaining in contact with the rod for an instant, the balls will be repelled from the rod and from each other. When the rod is removed, the pith balls remain separated, as shown in the figure. The initial attraction (due to a redistribution of charge on the neutral pith balls) will be explained later. The repulsion must be due to some property acquired by the pith balls as a result of their contact with the charged rod. We may reasonably assume that some of the *charge* has been transferred from the rod to the pith balls and that all three objects become similarly charged. From these observations, we can state the conclusion that:

> *A force of repulsion exists between two substances which are electrified in the same way.*

Let us continue our experimentation by picking up the glass rod and rubbing it vigorously on a silk cloth. When the charged rod is brought near the pith balls, the same sequence of events occurs as with the rubber rod. See Fig. 29-2b. Does this

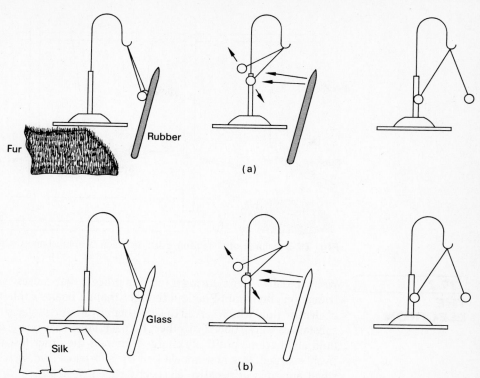

Fig. 29-2 *(a)* Charging the pith-ball electroscope with a rubber rod. *(b)* Charging the pith balls with a glass rod.

mean that the nature of the charge is the same on both rods? Our experiment neither proves nor disproves this assumption. In each case, the rod and balls were electrified in the same way, and so repulsion occurs in each case.

To test whether the two processes are identical, let us charge one pith ball with a glass rod and the other with a rubber rod. As shown in Fig. 29-3, a force of *attraction* exists between the balls charged in this manner. We conclude that the charges produced on the glass and rubber rods are opposite.

Similar experimentation with many different materials demonstrates that all electrified objects can be divided into two groups: (1) those which have a charge like that produced on glass and (2) those which have a charge like that produced on rubber. According to a convention established by Benjamin Franklin, objects in the former group are said to have a *positive* (+) charge, and objects belonging to the latter group are said to have a *negative* (−) charge. These terms have no mathematical significance; they simply denote the two opposite kinds of electric charge.

We are now in a position to state the *first law of electrostatics,* which is based on our experimentation:

> *Like charges repel and unlike charges attract.*

Two negatively charged objects or two positively charged objects repel each other, as demonstrated by Fig. 29-2a and *b,* respectively. Figure 29-3 demonstrates that a positively charged object attracts a negatively charged object.

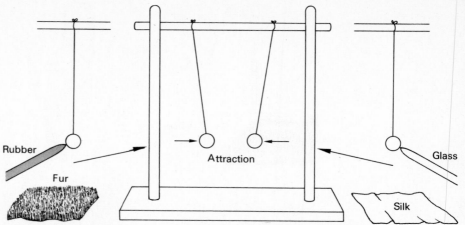

Fig. 29-3 A force of attraction exists between two substances which are oppositely charged.

What actually occurs during a rubbing process that causes the phenomenon of electrification? Benjamin Franklin thought that all bodies contained a specified amount of electric fluid which served to keep them in an uncharged state. When two different substances were rubbed together, he postulated that one accumulated an excess of fluid and became positively charged whereas the other lost fluid and became negatively charged. It is now known that the substance transferred is not a fluid but very small amounts of negative electricity called *electrons*.

The modern atomic theory of matter holds that all substances are made up of atoms and molecules. Each atom has a positively charged central core, called the *nucleus*, which is surrounded by a cloud of negatively charged electrons. The nucleus consists of a number of *protons*, each with a single unit of positive charge, and (except for hydrogen) one or more *neutrons*. As the name suggests, a neutron is a neutral particle. Normally, an atom of matter is in a *neutral* or *uncharged* state because it contains the same number of protons in its nucleus as there are electrons surrounding the nucleus. A schematic diagram of the neon atom is shown in Fig. 29-4. If, for some reason, a neutral atom loses one or more of its outer electrons, the atom has a net positive charge and is referred to as a positive *ion*. A negative ion is an atom which has gained one or more additional charges.

When two particular materials are brought in close contact, some of the loosely held electrons may be transferred from one material to the other. For example, when a hard-rubber rod is rubbed against fur, electrons are transferred from the fur to the rod, leaving an *excess* of electrons on the rod and a *deficiency* of electrons on the fur. Similarly, when a glass rod is rubbed on a silk cloth, electrons are transferred from the glass to the silk. We can now state:

> *An object which has an excess of electrons is negatively charged, and an object which has a deficiency of electrons is positively charged.*

A laboratory demonstration of the transfer of charge is illustrated in Fig. 29-5. A hard-rubber rod is rubbed vigorously on a piece of fur. One pith ball is charged negatively with the rod, and the other is touched with the fur. The resulting attrac-

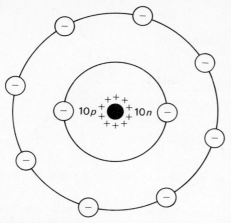

Fig. 29-4 The neon atom consists of a tightly packed nucleus containing 10 protons *(p)* and 10 neutrons *(n)*. The atom is electrically neutral because it is surrounded by 10 electrons.

tion shows that the fur is oppositely charged. The process of rubbing has left a deficiency of electrons on the fur.

29-3
INSULATORS
AND
CONDUCTORS

A solid piece of matter is composed of many atoms arranged in a manner peculiar to that material. Some materials, primarily metals, have a large number of *free electrons,* which can move about through the material. These materials have the ability to transfer charge from one object to another, and they are called *conductors.*

> A **conductor** is a material through which charge can easily be transferred.

Most metals are good conductors. In Fig. 29-6 a copper rod is supported by a glass stand. The pith balls can be charged by touching the right end of the copper with a

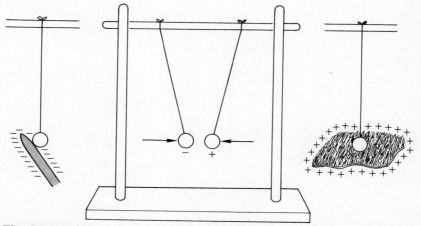

Fig. 29-5 Rubbing a hard-rubber rod on a piece of fur transfers electrons from the fur to the rod.

THE ELECTRIC FORCE

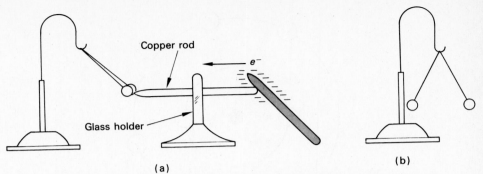

Copper rod

e^-

Glass holder

(a)

(b)

Fig. 29-6 Electrons are conducted by the copper rod to charge the pith balls.

charged rubber rod. The electrons are transferred or *conducted* through the rod to the pith balls. Note that none of the charge is transferred to the glass support or to the silk thread. These materials are poor conductors and are referred to as *insulators*.

An **insulator** is a material which resists the flow of charge.

Other examples of good insulators are rubber, plastic, mica, Bakelite, sulfur, and air.

A **semiconductor** is a material intermediate in its ability to carry charge.

Examples are silicon, germanium, and gallium arsenide. The ease with which a semiconductor carries charge can be greatly varied by the addition of impurities or by a change in temperature.

29-4

THE GOLD-LEAF ELEC-TROSCOPE

The gold-leaf electroscope shown in Fig. 29-7 consists of a strip of gold foil attached to a conducting rod. The rod and the foil are protected from air currents by a cylindrical metal case with glass windows. The rod is fitted with a spherical knob at the top and is insulated from the case by a block of hard rubber or amber. Whenever

Fig. 29-7 The gold-leaf electroscope.

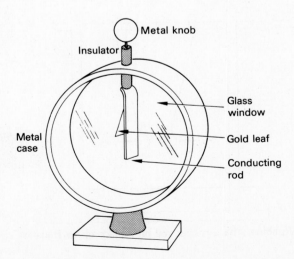

Metal knob

Insulator

Glass window

Gold leaf

Conducting rod

Metal case

the knob is given a charge, the repulsion of like charges on the rod and the gold leaf causes the leaf to diverge away from the rod.

Figure 29-8 illustrates charging the electroscope by *contact.* When the knob is touched with the negatively charged rod, electrons flow from the rod to the leaves, leaving an excess of electrons on the electroscope. When the knob is touched with a positively charged rod, electrons are transferred from the knob to the rod, leaving the electroscope deficient in electrons. Note that the residual charge on the electroscope is of the same sign as that of the charging rod.

Once the electroscope is charged, either negatively or positively, it can be used to detect the presence and nature of other charged objects. (See Fig. 29-9.) For example, consider what happens to the leaf of a negatively charged electroscope as a positively charged rod is brought near the knob. Some electrons are drawn from the leaf and rod up into the knob. As a result, the leaf converges. Bringing the rod closer produces a proportionately greater convergence of the leaf as more and more electrons are attracted to the knob. There appears to be a direct proportion between the number of charges on the leaf and rod and the force of repulsion between them. Moreover, there must be some *inverse* relation between the separation of the charged

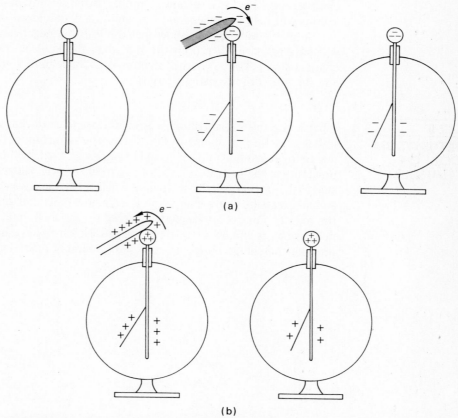

Fig. 29-8 Charging the electroscope by contact with *(a)* a negatively charged rod and *(b)* a positively charged rod.

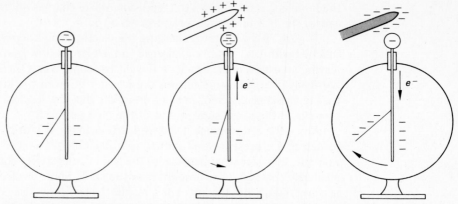

Fig. 29-9 The negatively charged electroscope can be used to detect the presence of other charge.

rod and the knob and the force attracting the electrons from the leaf and rod of the electroscope. This force becomes stronger as the separation decreases. These observations will help us understand Coulomb's law, developed in a later section.

Similar reasoning will show that the leaf of a negatively charged electroscope will be repelled further from the rod when the knob is placed near a negatively charged object. Thus a charged electroscope can be used to indicate both the nature and the presence of nearby charges.

| 29-5 |

**REDISTRIBU-
TION OF
CHARGE**

When a negatively charged rod is brought close to an uncharged pith ball, there is an initial attraction, as shown in Fig. 29-10. The attraction of the uncharged object is due to the separation of positive and negative electricity within the neutral body. The proximity of the negatively charged rod repels loosely held electrons to the opposite side of the uncharged object, leaving a deficiency (positive charge) on the near side and an excess (negative charge) on the far side. Since the unlike charge is nearer to the rod, the force of attraction will exceed the force of repulsion and the electrically neutral object will be attracted to the rod. No charge is gained or lost during this process; the charge on the neutral body is simply redistributed.

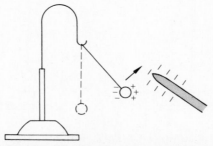

Fig. 29-10 Attraction of a neutral body due to a redistribution of charge.

The redistribution of charge due to the presence of a nearby charged object can be useful in charging objects without contact. This process, called charging by *induction,* can be accomplished without any loss of charge from the charging body. For example, consider the two neutral metal spheres placed in contact as shown in Fig. 29-11. When a negatively charged rod is brought near the left sphere (without touching it), a redistribution of charge occurs. Electrons are forced from the left sphere to the right sphere through the point of contact. Now, if the spheres are separated in the presence of the charging rod, the electrons cannot return to the left sphere. Thus the left sphere will have a deficiency of electrons (a positive charge), and the right sphere will have an excess of electrons (a negative charge).

A charge can also be induced on a single sphere. This process is illustrated with the electroscope in Fig. 29-12. A negatively charged rod is placed near the metal knob, causing a redistribution of charge. The repelled electrons cause the leaf to diverge, leaving a deficiency of electrons on the knob. By touching the knob with a finger or by connecting a wire from the knob to the earth a path is provided for the repelled electrons to leave the electroscope. The body or the ground will acquire a

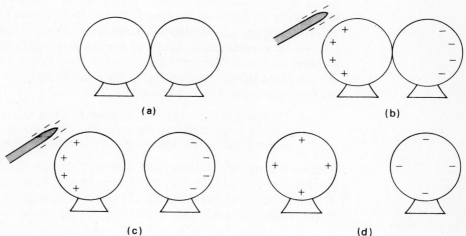

Fig. 29-11 Charging two metal spheres by induction.

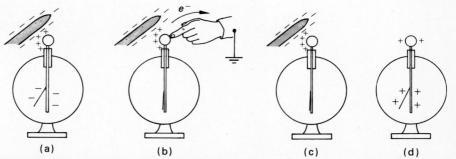

Fig. 29-12 Charging an electroscope by induction. Note that the residual charge is opposite that of the charging body.

negative charge equal to the positive charge (deficiency) left on the electroscope. When the charging rod is removed, the leaf of the electroscope will again diverge, as shown in the figure. Charging by induction always leaves a residual charge which is opposite that of the charging body.

29-7

COULOMB'S LAW

As usual, the task of the physicist is to measure the interactions between charged objects in some quantitative fashion. It is not sufficient to state that an electric force exists; we must be able to predict its magnitude.

The first theoretical investigation of the electric forces between charged bodies was accomplished by Charles Augustin de Coulomb in 1784. His studies were made with a torsion balance to measure the variation in force with separation and quantity of charge. The separation r of two charged objects is reckoned as the straight-line distance between their centers. The quantity of charge q can be thought of as the excess number of electrons or protons in the body.

Coulomb found that the force of attraction or repulsion between two charged objects is inversely proportional to the square of their separation distance. In other words, if the distance between two charged objects is reduced by one-half, the force of attraction or repulsion between them will be increased fourfold.

The concept of a quantity of charge was not clearly understood in Coulomb's time. There was no established unit of charge and no means for measuring it, but his experiments clearly showed that the electric force between two charged objects is directly proportional to the product of the quantity of charge on each object. Today, his conclusions are stated in *Coulomb's law:*

> *The force of attraction or repulsion between two point charges is directly proportional to the product of the two charges and inversely proportional to the square of the distance between them.*

In order to arrive at a mathematical statement of Coulomb's law, let us consider the charges in Fig. 29-13. In Fig. 29-13a the force F of attraction between two like charges in indicated, and in Fig. 29-13b a force of repulsion is shown for like charges. In either case, the magnitude of the force is determined by the magnitudes of the

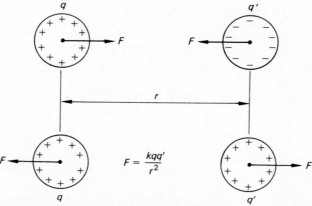

$$F = \frac{kqq'}{r^2}$$

Fig. 29-13 Illustrating Coulomb's law.

charges q and q' and by their separation r. From Coulomb's law, we can write

$$F \propto \frac{qq'}{r^2}$$

or

$$F = \frac{kqq'}{r^2} \tag{29-1}$$

The proportionality constant k takes into account the properties of the medium separating the charged bodies and has the dimensions dictated by Coulomb's law.

In SI units, the practical system for the study of electricity, the unit of charge is expressed in *coulombs* (C). In this case, the quantity of charge is not defined by Coulomb's law but is related to the flow of charge through a conductor. We shall find later that this rate of flow is measured in *amperes*. A formal definition of the coulomb is as follows:

> One **coulomb** is the charge transferred through any cross section of a conductor in one second by a constant current of one ampere.

Since current theory is not a part of this chapter, it will suffice to compare the coulomb with the charge of an electron.

$$1 \text{ C} = 6.25 \times 10^{18} \text{ electrons}$$

Obviously the coulomb is an enormously large unit from the standpoint of most problems in electrostatics. The charge of one electron expressed in coulombs is

$$e^- = -1.6 \times 10^{-19} \text{ C} \tag{29-2}$$

where e^- is the symbol for the electron and the minus sign denotes the nature of the charge.

A more convenient unit for electrostatics is the microcoulomb (μC), defined by

$$1 \ \mu\text{C} = 10^{-6} \text{ C} \tag{29-3}$$

Since the SI units of force, charge, and distance do not depend on Coulomb's law, the proportionality constant k must be determined by experiment. A large number of experiments have shown that when the force is in newtons, the distance is in meters, and the charge is in coulombs the proportionality constant is approximately

$$k = 9 \times 10^9 \text{ N} \cdot \text{m}^2/\text{C}^2 \tag{29-4}$$

When applying Coulomb's law in SI units, one must substitute this value for k in Eq. (29-1):

$$F = \frac{(9 \times 10^9 \text{ N} \cdot \text{m}^2/\text{C}^2)qq'}{r^2} \tag{29-5}$$

It must be remembered that **F** represents the force on a charged particle and is, therefore, a vector quantity. The direction of the force is determined solely by the nature of the charges q and q'. Therefore, it is easier to use the absolute values of q and q' in Coulomb's law, Eq. (29-5). Then remember that like charges repel and opposite charges attract to get the direction of the force. When more than one force acts on a charge, the resultant force is the vector sum of the separate forces.

EXAMPLE 29-1   A -3-μC charge is placed 100 mm from a $+3$-μC charge. Calculate the force between the two charges.

Solution   First we convert to appropriate units.

$$3 \ \mu C = 3 \times 10^{-6} \ C \qquad 100 \ mm = 100 \times 10^{-3} \ m = 0.1 \ m$$

Then we use the absolute values, so that both q and q' are equal to 3×10^{-6} C. Applying Coulomb's law, we obtain

$$F = \frac{kqq'}{r^2} = \frac{(9 \times 10^9 \ N \cdot m^2/C^2)(3 \times 10^{-6} \ C)(3 \times 10^{-6} \ C)}{(0.1 \ m)^2}$$

$$F = 8.1 \ N \qquad\qquad\qquad\qquad\qquad\qquad\qquad\qquad\qquad \textit{Attraction}$$

This is a force of *attraction* because the charges had opposite signs.

EXAMPLE 29-2   Two charges $q_1 = -8 \ \mu C$ and $q_2 = +12 \ \mu C$ are placed 120 mm apart in the air. What is the resultant force on a third charge $q_3 = -4 \ \mu C$ placed midway between the other two charges?

Solution   We convert the charges to coulombs ($1 \ \mu C = 1 \times 10^{-6} \ C$), use their absolute values, and convert the distance to meters ($120 \ mm = 0.12 \ m$). One-half of 0.12 m is 0.06 m. A sketch is drawn in Fig. 29-14 to help visualize the forces. The force on q_3 due to q_1 is directed to the right and is calculated from Coulomb's law.

Fig. 29-14 Computing the resultant force on a charge placed midway between two other charges.

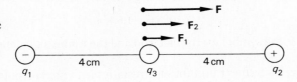

$$F_1 = \frac{kq_1q_3}{r^2} = \frac{(9 \times 10^9 \ N \cdot m^2/C^2)(8 \times 10^{-6} \ C)(4 \times 10^{-6} \ C)}{(0.06 \ m)^2}$$

$$F_1 = \frac{0.288 \ N \cdot m^2}{0.0036 \ m^2} = 80 \ N \qquad\qquad\qquad \textit{Repulsion, to the Right}$$

Similarly, the force F_2 on q_3 due to q_2 is equal to

$$F_2 = \frac{kq_2q_3}{r^2} = \frac{(9 \times 10^9 \ N \cdot m^2/C^2)(12 \times 10^{-6} \ C)(4 \times 10^{-6} \ C)}{(0.06 \ m)^2}$$

$$F_2 = 120 \ N \qquad\qquad\qquad\qquad\qquad \textit{Attraction, Also to the Right}$$

The resultant force F is the vector sum of F_1 and F_2. Thus

$$F = 80 \ N + 120 \ N$$

$$= 200 \ N \qquad\qquad\qquad\qquad\qquad\qquad \textit{Directed to the Right}$$

Note that the signs of the charges were used only to determine the direction of the forces; they were not substituted into the calculations.

EXAMPLE 29-3 Three charges $q_1 = +4 \times 10^{-9}$ C, $q_2 = -6 \times 10^{-9}$ C, and $q_3 = -8 \times 10^{-9}$ C are arranged as shown in Fig. 29-15. What is the resultant force on q_3 due to the other two charges?

Fig. 29-15

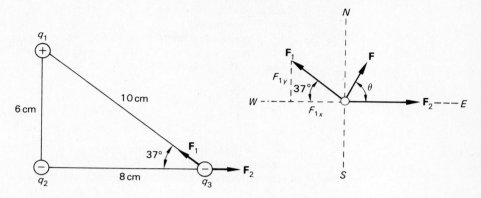

Solution Let $\mathbf{F}_1$ be the force on q_3 due to q_1, and let $\mathbf{F}_2$ be the force on q_3 due to q_2. $\mathbf{F}_1$ is a force of attraction and $\mathbf{F}_2$ is a force of repulsion, as shown in the figure. Their magnitudes are found from Coulomb's law.

$$F_1 = \frac{(9 \times 10^9 \text{ N} \cdot \text{m}^2/\text{C}^2)(4 \times 10^{-9} \text{ C})(8 \times 10^{-9} \text{ C})}{(0.1 \text{ m})^2}$$

$$= 2.88 \times 10^{-5} \text{ N} \qquad (37° \text{ north of west})$$

$$F_2 = \frac{(9 \times 10^9 \text{ N} \cdot \text{m}^2/\text{C}^2)(6 \times 10^{-9} \text{ C})(8 \times 10^{-9} \text{ C})}{(8 \times 10^{-2} \text{ m})^2}$$

$$= 6.75 \times 10^{-5} \text{ N} \qquad \text{east}$$

We must now find the resultant $\mathbf{F}$ of the forces $\mathbf{F}_1$ and $\mathbf{F}_2$. From the free-body diagram, we note that

$$F_x = F_2 - F_{1x} = F_2 - F_1 \cos 37°$$
$$= (6.75 \times 10^{-5} \text{ N}) - (2.88 \times 10^{-5} \text{ N})(0.8)$$
$$= 4.45 \times 10^{-5} \text{ N}$$
$$F_y = F_1 \sin 37° = 1.73 \times 10^{-5} \text{ N}$$

From the components we find

$$\tan \theta = \frac{F_y}{F_x} = \frac{1.73 \times 10^{-5} \text{ N}}{4.45 \times 10^{-5} \text{ N}} = 0.389$$

$$\theta = 21°$$

$$F = \frac{F_y}{\sin \theta} = \frac{1.73 \times 10^{-5} \text{ N}}{\sin 21°} = 4.80 \times 10^{-5} \text{ N}$$

Therefore, the resultant force on q_3 is 4.8×10^{-5} N directed 21° north of east.

SUMMARY

Electrostatics is the science which treats charges at rest. We have seen that there are two kinds of charge that exist in nature. If an object has an excess of electrons it is said to be *negatively* charged; if it has a deficiency of electrons it is *positively* charged.

Coulomb's law was introduced to provide a quantitative measure of electrical forces between such charges. The major concepts are:

- The first law of electrostatics states that *like charges repel each other and unlike charges attract each other.*
- Coulomb's law states that *the force of attraction or repulsion between two point charges is directly proportional to the product of the two charges and inversely proportional to the separation of the two charges.*

$$F = \frac{kqq'}{r^2}$$

$$k = 9 \times 10^9 \text{ N} \cdot \text{m}^2/\text{C}^2$$

Coulomb's Law

The force F is in newtons (N) when the separation r is in meters (m) and the charge q is measured in coulombs (C).

- When solving the problems in this chapter, it is important to use the *sign* of the charges to *determine* the direction of forces and Coulomb's law to determine their *magnitudes.* The resultant force on a particular charge is then found by the methods of vector mechanics.

QUESTIONS

29-1. Define the following terms:
  - a. Electrostatics
  - b. Charging
  - c. Electron
  - d. Negative charge
  - e. Positive charge
  - f. Ion
  - g. Induced charge
  - h. Conductor
  - i. Insulator
  - j. Semiconductor
  - k. Coulomb's law
  - l. Coulomb
  - m. Microcoulomb
  - n. Electroscope

29-2. Discuss several examples of static electricity in addition to those mentioned in the text.

29-3. In the process of rubbing a glass rod with a silk cloth, is charge *created?* Explain.

29-4. What is the nature of the charge on the silk cloth in Question 29-3?

29-5. An insulated stand supports a charged metal ball in the laboratory. Describe several procedures of determining the nature of the charge on the ball.

29-6. During an experiment in the laboratory, two bodies are seen to attract each other. Is this conclusive proof that they are both charged? Explain.

29-7. Two bodies are found to repel each other with an electric force. Is this conclusive proof that they are both charged? Explain.

29-8. One of the fundamental principles of physics is the principle of the conservation of charge which states that *the total quantity of electric charge in the universe does not change.* Can you offer reasons for accepting this law?

29-9. Describe what happens to the leaf of a positively charged electroscope as **(a)** a negatively charged rod is brought closer and closer to the knob without touching it; **(b)** a positively charged rod is brought closer and closer to the knob.

29-10. When charging the leaf electroscope by induction, should the finger be removed before the charging rod is taken away? Explain.

29-11. List the units for each parameter in Coulomb's law for SI units.

29-12. Coulomb's law is valid only when the separation r is large in comparison with the radii of the charge. What accounts for this limitation?
29-13. How many electrons would be required to give a metal sphere a negative charge of (a) 1 C, (b) 1 μC?

PROBLEMS

Coulomb's Law

29-1. Two balls each having a charge of 3 μC are separated by 20 mm. What is the force of repulsion between them?

Ans. 202 N

29-2. Two point charges of -3 and $+4$ μC are 12 mm apart in a vacuum. What is the electrostatic force between them?

29-3. An alpha particle consists of two protons and two neutrons. What is the repulsive force between two alpha particles separated by a distance of 2 nm?

Ans. 2.30×10^{-10} N

29-4. The radius of the electron's orbit around the proton in a hydrogen atom is approximately 52 pm (1 pm $= 1 \times 10^{-12}$ m). What is the electrostatic force of attraction?

29-5. What is the separation of two -4-μC charges if the force of repulsion between them is 200 N?

Ans. 26.8 mm

* 29-6. Two charges attract each other with a force of 6×10^{-3} N when they are a certain distance apart in a vacuum. If their separation is decreased to one-third of the original distance, what is the new force of attraction?

The Resultant Electrostatic Force

29-7. A $+60$-μC charge is placed 60 mm to the left of a $+20$-μC charge. What is the resultant force on a -35-μC charge placed midway between the two charges?

Ans. 1.4×10^4 N, leftward

29-8. A point charge of $+36$ μC is placed 80 mm from a second point charge of -22 μC. **(a)** What force is exerted on each charge? **(b)** What is the resultant force on a third charge of $+12$ μC placed between the other charges and located 60 mm from the $+36$-μC charge?

* 29-9. A 64-μC charge is located 30 mm to the left of a 16-μC charge. What is the resultant force on a -12-μC charge positioned exactly 50 mm below the 16-μC charge?

Ans. 2650 N, 113.3°

* 29-10. A charge of $+60$ nC is located 80 mm above a -40-nC charge. What is the resultant force on a -50-nC charge located 45 mm horizontally to the right of the -40-nC charge?

* 29-11. Three point charges $q_1 = +8$ μC, $q_2 = -4$ μC, and $q_3 = +2$ μC are placed at the corners of an equilateral triangle. Each side of the triangle is 80 mm long. What are the magnitude and direction of the resultant force on the $+8$-μC charge? Assume that the base of the triangle is formed by a line joining the 8- and 4-μC charges.

Ans. 39 N at 330°

THE ELECTRIC FORCE

Additional Problems

29-12. What should be the separation of two $+5$-μC charges in order that the force of repulsion is 4 N?

29-13. The repulsive force between two pith balls is found to be 60 μN. If each pith ball carries a charge of 8 nC, what is their separation?

Ans. 98 mm

29-14. Two identical unknown charges experience a 48-N repulsive force when separated by 60 mm. What is the magnitude of each charge?

29-15. A small metal sphere is given a charge of $+40$ μC, and a second sphere located 8 cm away is given a charge of -12 μC. (a) What is the force of attraction between them? (b) If the two spheres are allowed to touch and are then again placed 8 cm apart, what new electric force exists between them?

Ans. (a) 675 N, attraction; (b) 276 N, repulsion

29-16. What is the resultant force on a $+2$-μC charge at a distance of 60 mm from each of two -4-μC charges which are 80 mm apart in air?

29-17. Two charges of $+25$ and $+16$ μC are 80 mm apart in air. A third charge of $+60$ μC is placed 30 mm from the $+25$-μC charge and between the two charges. What is the magnitude of the resultant force on the third charge?

Ans. 1.15×10^4 N

29-18. A charge of $+8$ nC is placed 40 mm to the left of a -14-nC charge. Where should a third charge of $+4$ nC be placed if the resultant force on that charge is to be zero?

29-19. Two charges of $+16$ and $+9$ μC are 80 mm apart in air. Where should a third charge be placed in order that the resultant force on it be zero? Why is it not necessary to specify either the magnitude or the sign of the third charge?

Ans. 34.3 mm from the 9-μC charge

29-20. A 0.02-g pith ball is suspended freely. The ball is given a charge of $+20$ μC and placed 0.6 m from a charge of $+50$ μC. What will be the initial acceleration of the pith ball?

29-21. Two 8-g pith balls are suspended from silk threads 60 cm long and attached to a common point. When the spheres are given equal amounts of negative charge, the balls come to rest 30 cm apart. Calculate the magnitude of the charge on each pith ball.

Ans. 450 nC

29-22. If it were possible to place 1 C of charge on each of two spheres separated by a distance of 1 m, what would be the repulsive force in newtons? How many electrons must be added to each of two metal spheres (1 m apart in air) if the repulsive force is to be 1 N?

29-23. Two 3-g spheres are suspended from the ceiling by light silk threads of negligible mass. The threads are each 80 mm long and are attached at a common point on the ceiling. What charge must be placed on each sphere if the resulting horizontal separation is to be 50 mm?

Ans. 51.8 nC

29-24. A -40-nC charge is placed 40 mm to the left of a $+6$-nC charge. What is the resultant force on a -12-nC charge placed 8 mm to the right of the $+6$-nC charge?

29-25. Two charges q_1 and q_2 are separated by a distance r. They experience a force F at this distance. If the initial separation is decreased by only 40 mm, the force between the two charges is doubled. What was the initial separation?

Ans. 137 mm

30 The Electric Field

OBJECTIVES
After completing this chapter, you should be able to:

1. Define the electric field and explain what determines its magnitude and direction.
2. Write and apply an expression which relates the electric field intensity at a point to the distance(s) from the known charge(s).
3. Explain and illustrate the concept of electric field lines, and discuss the two rules which must be followed in the construction of such lines.
4. Explain the concept of the *permittivity* of a medium, and how it affects the field intensity and the construction of field lines.
5. Write and apply Gauss's law as it applies to the electric fields surrounding surfaces of known charge density.

In our study of mechanics we discussed force and motion at great length. Newton's laws of motion were normally used to describe the application and consequences of *contact* forces. A moment's reflection on the universe as a whole convinces us of the enormous number of objects which are *not* in contact.

A projectile experiences a downward force which cannot be explained in terms of its interaction with air particles; planets revolve continuously through the void surrounding the sun; and the sun is pulled along an elliptical path by forces which do not touch it. Even at the atomic level, there are no "strings" to hold the electrons in their orbits about the nucleus.

If we are really to understand our universe, we must develop laws to predict the magnitude and direction of forces which are not transmitted by contact. Two such laws have already been discussed:

1. Newton's law of universal gravitation:

$$F_g = G\frac{m_1 m_2}{r^2} \tag{30-1}$$

2. Coulomb's law for electrostatic forces:

$$F_e = k \frac{q_1 q_2}{r^2} \tag{30-2}$$

Newton's law predicts the force which exists between two masses separated by a distance r; Coulomb's law deals with the electrostatic force, as discussed in Chap. 29. In applying such laws, we find it useful to develop certain properties of the space surrounding masses or charges.

30-1

THE CONCEPT OF A FIELD

Both the electric force and the gravitational force are examples of *action-at-a-distance forces,* which are extremely difficult to visualize. To overcome this fact, early physicists postulated the existence of an invisible material, called *ether,* which was thought to pervade all space. The gravitational force of attraction could then be due to strains in the ether caused by the presence of various masses. Certain optical experiments have now shown the ether theory to be untenable (see Sec. 24-2), and we are forced to consider whether space itself possesses properties of interest to the physicist.

It may be postulated that the mere presence of a mass alters the space surrounding it so as to produce a gravitational force on another nearby mass. We describe this alteration in space by introducing the concept of a *gravitational field* which surrounds all masses. Such a field may be said to exist in any region of space where a test mass will experience a gravitational force. The strength of the field at any point would be proportional to the force a given mass experiences at that point. For example, at every point in the vicinity of the earth, the gravitational field could be represented quantitatively by

$$g = \frac{F}{m} \tag{30-3}$$

where g = acceleration due to gravity
F = gravitational force
m = test mass (see Fig. 30-1)

If g is known at every point above the earth, the force F which will act on a given mass m placed at that point can be determined from Eq. (30-3).

The concept of a field can also be applied to electrically charged objects. The space surrounding a charged object is altered by the presence of the charge. We may postulate the existence of an *electric field* in this space.

> *An electric field is said to exist in a region of space in which an electric charge will experience an electric force.*

This definition provides a test for the existence of an electric field. Simply place a charge at the point in question. If an electric force is observed, an electric field exists at that point.

Just as the force per unit mass provides a quantitative definition of a gravitational field, the strength of an electric field can be represented by the force per unit charge. We define the electric field intensity E at a point in terms of the force F

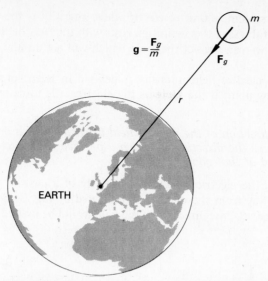

Fig. 30-1 The gravitational field at any point above the earth can be represented by the acceleration g that a small mass m would experience if it were placed at that point.

experienced by small positive charge $+q$ when it is placed at that point (see Fig. 30-2). The magnitude of the electric field intensity is given by

$$E = \frac{F}{q} \qquad (30\text{-}4)$$

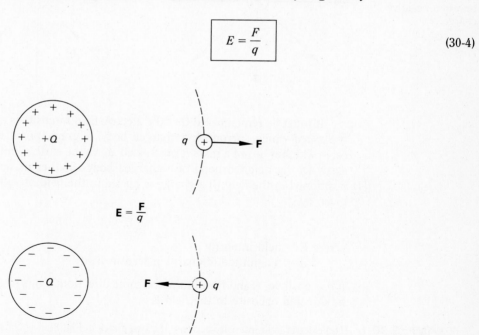

Fig. 30-2 The direction of the electric field intensity at a point is the same as the direction in which a positive charge $+q$ would move when placed at that point. Its magnitude is the force per unit charge (F/q).

In the metric system, a unit of electric field intensity is the newton per coulomb (N/C). The usefulness of this definition rests with the fact that if the field is known at a given point, we can predict the force which will act on any charge placed at that point.

Since the electric field intensity is defined in terms of a *positive* charge, its direction at any point is the same as the electrostatic force on a positive charge at that point.

> *The direction of the electric field intensity* E *at a point in space is the same as the direction in which a positive charge would move if it were placed at that point.*

On this basis, the electric field in the vicinity of a positive charge $+Q$ would be outward, or away from the charge, as indicated by Fig. 30-3a. In the vicinity of a negative charge $-Q$, the direction of the field would be inward, or toward the charge (Fig. 30-3b).

Fig. 30-3 *(a)* The field in the vicinity of a positive charge is directed radially outward at every point. *(b)* The field is directed inward or toward a negative charge.

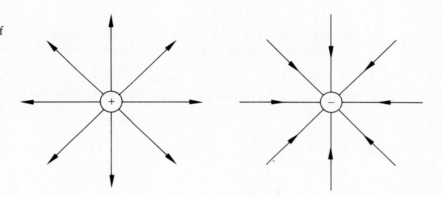

It must be remembered that the electric field intensity is a property assigned to the *space* which surrounds a charged body. A gravitational field exists above the earth whether or not a mass is positioned above the earth. Similarly, an electric field exists in the neighborhood of a charged body whether or not a second charge is positioned in the field. If a charge *is* placed in the field, it will experience a force F given by

$$F = q\text{E} \tag{30-5}$$

where E = field intensity
$\quad q$ = magnitude of charge placed in field

If q is positive, E and F will have the same direction; if q is negative, the force F will be directed opposite to the field E.

EXAMPLE 30-1   The electric field intensity between the two plates in Fig. 30-4 is constant and directed downward. The magnitude of the electric field intensity is 6×10^4 N/C. What are the magnitude and the direction of the electric force exerted on an electron projected horizontally between the two plates?

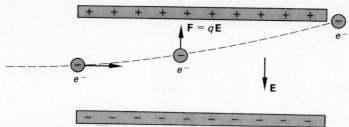

Fig. 30-4 An electron projected into an electric field of constant intensity.

Solution   Since the direction of the field intensity E is defined in terms of a positive charge, the force on an electron will be upward, or opposite to the field direction. The charge on an electron is -1.6×10^{-19} C. Thus the electric force is given by Eq. (30-5).

$$F = qE = (1.6 \times 10^{-19} \text{ C})(6 \times 10^4 \text{ N/C})$$
$$= 9.6 \times 10^{-15} \text{ N} \qquad \text{upward}$$

Remember that the absolute value of the charge is used. The directions of **F** and **E** are the same for positive charges and opposite for negative charges.

EXAMPLE 30-2   Show that the gravitational force on the electron in Example 30-1 may be neglected. (The mass of an electron is 9.1×10^{-31} kg.)

Solution   The downward force due to the weight of the electron is

$$F_g = mg = (9.1 \times 10^{-31} \text{ kg})(9.8 \text{ m/s}^2)$$
$$= 8.92 \times 10^{-30} \text{ N}$$

The electric force is larger than the gravitational force by a factor of 1.08×10^{15}.

30-2
COMPUTING THE ELECTRIC INTENSITY

We have discussed one method of measuring the magnitude of the electric field intensity at a point in space. A known charge is placed at the point, and the resultant force is measured. The force per unit charge is then a measure of the electric intensity at that point. The disadvantage of this method is that it bears no obvious relationship to the charge Q which creates the field. Experimentation will quickly show that the magnitude of the electric field surrounding a charged body is directly proportional to the quantity of charge on the body. It can also be demonstrated that at points farther and farther away from a charge Q a test charge q will experience smaller and smaller forces. The exact relationship is derived from Coulomb's law.

Suppose we wish to calculate the field intensity E at a distance r from a single charge Q, as shown in Fig. 30-5. The force F that Q exerts on a test charge q at the point in question is, from Coulomb's law,

$$F = \frac{kQq}{r^2} \tag{30-6}$$

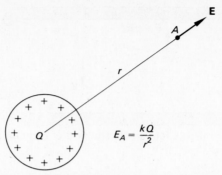

$$E_A = \frac{kQ}{r^2}$$

Fig. 30-5 Calculating the electric field intensity at a distance r from the center of a single charge Q.

Substituting this value for F into Eq. (30-4), we obtain

$$E = \frac{F}{q} = \frac{kQq/r^2}{q}$$

$$\boxed{E = \frac{kQ}{r^2}} \qquad (30\text{-}7)$$

where k is equal to 9×10^9 N · m²/C². The direction of the field is away from Q if Q is positive and toward Q if Q is negative. We now have a relation which allows us to compute the field intensity at a point without having to place a second charge at the point.

EXAMPLE 30-3   What is the electric field intensity at a distance of 2 m from a charge of -12 μC?

Solution   Since the charge Q is negative, the field intensity will be directed toward Q. Its magnitude, from Eq. (30-7), is

$$E = \frac{kQ}{r^2} = \frac{(9 \times 10^9 \text{ N} \cdot \text{m}^2/\text{C}^2)(12 \times 10^{-6} \text{ C})}{(2 \text{ m})^2}$$

$$= 27 \times 10^3 \text{ N/C} \qquad \text{toward } Q$$

When more than one charge contributes to the field, as in Fig. 30-6, the resultant field is the vector sum of the contributions from each charge.

$$\mathbf{E} = \mathbf{E}_1 + \mathbf{E}_2 + \mathbf{E}_3 + \cdots$$

$$E = \frac{kQ_1}{r_1^2} + \frac{kQ_2}{r_2^2} + \frac{kQ_3}{r_3^3} + \cdots \qquad \textit{Vector Sum}$$

In an abbreviated form, the resultant field at a point in the vicinity of a number of charges is given by

$$\boxed{E = \sum \frac{kQ}{r^2}} \qquad \textit{Vector Sum} \quad (30\text{-}8)$$

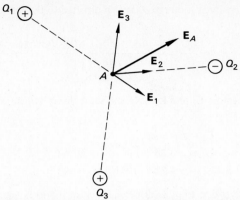

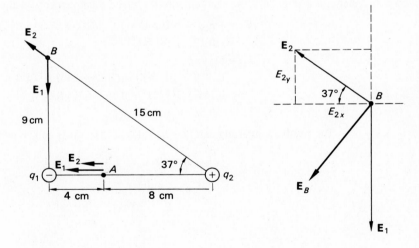

Fig. 30-6 The field in the vicinity of a number of charges is equal to the vector sum of the fields due to the individual charges.

It must be remembered that this is a *vector sum* and not an algebraic sum.

EXAMPLE 30-4   Two point charges, $q_1 = -6$ nC and $q_2 = +6$ nC, are 12 cm apart, as shown in Fig. 30-7. Determine the electric field *(a)* at point A and *(b)* at point B.

Fig. 30-7

Solution (a)   By convention, let us consider any vector directed to the right or up as positive and any vector directed to the left or down as negative. The field at A due to q_1 is directed to the left since q_1 is negative, and its magnitude is

$$E_1 = \frac{kq_1}{r_1^2} = \frac{-(9 \times 10^9 \text{ N} \cdot \text{m}^2/\text{C}^2)(6 \times 10^{-9} \text{ C})}{(4 \times 10^{-2} \text{ m})^2}$$

$$= -3.38 \times 10^4 \text{ N/C} \qquad \text{(left)}$$

The electric intensity at A due to q_2 is also directed to the left since q_2 is positive, and it is given by

$$E_2 = \frac{kq_2}{r_2^2} = \frac{-(9 \times 10^9 \text{ N} \cdot \text{m}^2/\text{C}^2)(6 \times 10^{-9} \text{ C})}{(8 \times 10^{-2} \text{ m})^2}$$

$$= -0.844 \times 10^4 \text{ N/C}$$

Since the two vectors $\mathbf{E}_1$ and $\mathbf{E}_2$ have the same direction, the resultant intensity at A is simply

$$E_A = E_1 + E_2$$

$$= -3.38 \times 10^4 \text{ N/C} - 0.844 \times 10^4 \text{ N/C}$$

$$= -4.22 \times 10^4 \text{ N/C} \qquad \text{(left)}$$

Solution (b) The field intensity at B due to q_1 is directed downward and is equal to

$$E_1 = \frac{kq_1}{r_1^2} = \frac{-(9 \times 10^9 \text{ N} \cdot \text{m}^2/\text{C}^2)(6 \times 10^{-9} \text{ C})}{(9 \times 10^{-2} \text{ m})^2}$$

$$= -0.667 \times 10^4 \text{ N/C}$$

Similarly, the field due to q_2 is

$$E_2 = \frac{kq_2}{r_2^2} = \frac{(9 \times 10^9 \text{ N} \cdot \text{m}^2/\text{C}^2)(6 \times 10^{-9} \text{ C})}{(15 \times 10^{-2} \text{ m})^2}$$

$$= 0.240 \times 10^4 \text{ N/C} \qquad \text{at } 37°$$

The resultant intensity at point B is the vector sum of $\mathbf{E}_1$ and $\mathbf{E}_2$. Referring to the vector diagram in Fig. 30-7, we can compute the x and y components of the resultant.

$$E_x = -E_{2x} = -(0.240 \times 10^4 \text{ N/C}) \cos 37°$$

$$= -0.192 \times 10^4 \text{ N/C}$$

$$E_y = E_{2y} - E_1$$

$$= (0.240 \times 10^4 \text{ N/C}) \sin 37° - 0.667 \times 10^4 \text{ N/C}$$

$$= 0.144 \times 10^4 \text{ N/C} - 0.667 \times 10^4 \text{ N/C}$$

$$= -0.523 \times 10^4 \text{ N/C}$$

The resultant intensity can now be calculated from its components:

$$\tan \phi = \frac{E_y}{E_x} = \frac{-0.523 \times 10^4 \text{ N/C}}{-0.192 \times 10^4 \text{ N/C}}$$

$$\phi = 69.8°$$

$$E = \frac{E_y}{\sin \phi} = \frac{0.523 \times 10^4 \text{ N/C}}{\sin 69.8°}$$

$$= 0.557 \times 10^4 \text{ N/C}$$

Therefore, the resultant field intensity at B is 0.557×10^4 N/C directed 69.8° downward and to the left.

30-3

ELECTRIC FIELD LINES

An ingenious aid to the visualization of electric fields was introduced by Michael Faraday (1791–1867) in his early work in electromagnetism. The method consists of representing both the strength and the direction of an electric field by imaginary lines called *electric field lines*.

Electric field lines are imaginary lines drawn in such a manner that their direction at any point is the same as the direction of the electric field at that point.

For example, the lines drawn radially outward from the positive charge in Fig. 30-3a represent the direction of the field at any point on the line. The electric lines in the vicinity of a negative charge would be radially inward and directed toward the charge, as in Fig. 30-3b. We shall see later that the density of these lines in any region of space is a measure of the *magnitude* of the field intensity in that region.

In general, the direction of the electric field in a region of space varies from place to place. Thus the electric lines are normally curved. For instance, let us consider the construction of an electric field line in the region between a positive charge and a negative charge, as illustrated in Fig. 30-8.

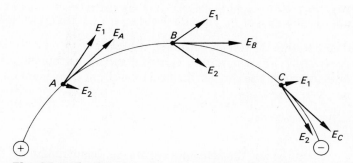

Fig. 30-8 The direction of an electric field line at any point is the same as the direction of the resultant electric field intensity at that point.

The direction of the electric field line at any point is the same as the direction of the resultant electric field vector at that point. Two rules must be followed when constructing electric field lines:

1. The direction of the field line at any point is the same as the direction in which a positive charge would move if placed at that point.
2. The spacing of the field lines must be such that they are close together where the field is strong and far apart where the field is weak.

Following these very general rules, one can construct the electric field lines for the two common cases shown in Fig. 30-9. As a consequence of how electric lines are drawn, *they will always leave positive charges and enter negative charges.* No lines can originate or terminate in space although one end of an electric line may proceed to infinity.

30-4

GAUSS'S LAW

For any given charge distribution, we can draw an infinite number of electric lines. Clearly, if the spacing of the lines is to be a standardized indication of field strength, we must set a limit on the number of lines drawn in any situation. For example, let us consider the field lines directed radially outward from a positive point charge.

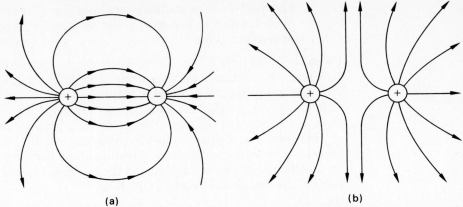

Fig. 30-9 *(a)* A graphical illustration of the electric field lines in the region surrounding two opposite charges. *(b)* The field lines between two positive charges.

(Refer to Fig. 30-10.) We shall use the letter N to represent the number of lines drawn. Now let us imagine a spherical surface surrounding the point charge at a distance r from the charge. The field intensity at every point on such a sphere would be given by

$$E = \frac{kq}{r^2} \qquad (30\text{-}9)$$

From the way the field lines are drawn, we might also say that the field at a tiny element of surface area ΔA is proportional to the number of lines ΔN penetrating that area. In other words, the density of lines (lines per unit area) is directly proportional to the field strength. Symbolically,

$$\frac{\Delta N}{\Delta A} \propto E_n \qquad (30\text{-}10)$$

The subscript n indicates that the field is everywhere normal to the surface area. This proportionality is true regardless of the total number of lines N which may be drawn.

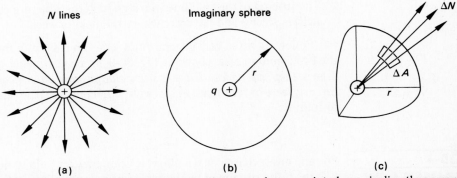

Fig. 30-10 Electric field intensity at a distance r from a point charge is directly proportional to the number of lines ΔN penetrating a unit area ΔA of an imaginary spherical surface constructed at that distance.

However, once we choose a proportionality constant for Eq. (30-10), we automatically set a limit to the number of lines drawn for any given situation. It has been found that the most convenient choice for this spacing constant is ϵ_0. It is called the *permittivity of free space* and is defined by

$$\epsilon_0 = \frac{1}{4\pi k} = 8.85 \times 10^{-12} \text{ C}^2/\text{N} \cdot \text{m}^2$$

(30-11)

where $k = 9 \times 10^9$ N $\cdot$ m^2/C^2 from Coulomb's law. Hence Eq. (30-10) can be written

$$\frac{\Delta N}{\Delta A} = \epsilon_0 E_n$$

(30-12)

or

$$\Delta N = \epsilon_0 E_n \Delta A$$

(30-13)

When E_n is constant over the entire surface, the total number of lines radiating outward from the enclosed charge is

$$N = \epsilon_0 E_n A$$

(30-14)

It can be seen that the choice of ϵ_0 is a convenient one by substituting Eq. (30-11) into Eq. (30-9):

$$E_n = \frac{1}{4\pi\epsilon_0} \frac{q}{r^2}$$

Substituting this expression into Eq. (30-14) and recalling that the area of a spherical surface is $A = 4\pi r^2$, we obtain

$$N = \epsilon_0 E_n A$$

$$= \frac{\epsilon_0}{4\pi\epsilon_0} \frac{q}{r^2} 4\pi r^2 = q$$

The choice of ϵ_0 as the proportionality constant has resulted in the fact that *the total number of lines passing normally through a surface is numerically equal to the charge contained within the surface.* Although this result was obtained using a spherical surface, it will apply to any other surface. The more general statement of the result is known as *Gauss's law*:

> The net number of electric lines of force crossing any closed surface in an outward direction is numerically equal to the net total charge within that surface.

$$N = \sum \epsilon_0 E_n A = \sum q$$

(30-15)

Gauss's law can be used to compute the field intensity near surfaces of charge. This represents a distinct advantage over methods developed so far because the previous equations apply only to point charges. The best way to understand the application of Gauss's law is through examples.

30-5
APPLICATIONS OF GAUSS'S LAW

Since most charged conductors have large quantities of charge on them, it is not practical to treat the charges individually. Generally, we speak of the charge density σ, defined as the charge per unit area of surface.

$$\sigma = \frac{q}{A} \qquad q = \sigma A \tag{30-16}$$

EXAMPLE 30-5 Calculate the electric field intensity at a distance r from an infinite sheet of positive charge, as shown in Fig. 30-11.

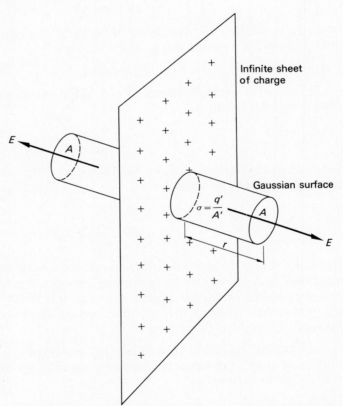

Fig. 30-11 Computing the field outside an infinite sheet of positive charge.

Solution The solution of problems using Gauss's law usually necessitates the construction of an imaginary surface of simple geometric form, e.g., a sphere or a cylinder. These are referred to as *gaussian surfaces*. In this example, a cylindrical surface is imagined to penetrate the sheet of positive charge so that it projects a distance r on either side of the thin sheet. The area A of each end of the cylinder is the same as the area cut on the sheet of charge. Thus the total charge contained inside the cylinder is

$$\sum q = \sigma A$$

where σ represents the charge density. Because of symmetry, the resultant field intensity E must be directed perpendicular to the sheet of charge at any point near the sheet. This means that no field lines will penetrate the curved sides of the cylinder, and the two ends of area A will represent the total area penetrated by the field lines. From Gauss's law,

$$\sum \epsilon_0 E_n A = \sum q$$
$$\epsilon_0 E A + \epsilon_0 E A = \sigma A$$
$$2\epsilon_0 E A = \sigma A$$
$$E = \frac{\sigma}{2\epsilon_0} \qquad (30\text{-}17)$$

Note that the field intensity E is independent of the distance r from the sheet.

Before the reader assumes that the example of an infinite sheet of charge is impractical, it should be pointed out that "infinity" in a practical sense implies only that the dimensions of the sheet are beyond the point of electrical interaction. In other words, Eq. (30-17) applies when the length and width of the sheet are very large in comparison with the distance r from the sheet.

EXAMPLE 30-6   Show by using Gauss's law that all the excess charge resides on the surface of a charged conductor.

Solution   Let us first consider the charged solid conductor in Fig. 30-12. Within such a conductor, charges are free to move if a resultant force is applied to them. After a time, we could safely

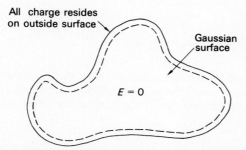

All charge resides
on outside surface

Gaussian
surface

$E = 0$

Fig. 30-12 Gauss's law demonstrates that all the charge resides on the surface of a conductor.

assume that all charges in the conductor are at rest. Under this condition, the electric field intensity within the conductor must be zero. Otherwise, the charges would be moving. Now, if we construct a gaussian surface just inside the surface of the conductor, as shown in the figure, we can write (for this surface)

$$\sum \epsilon_0 E A = \sum q$$

Substituting $E = 0$, we also find that $\sum q = 0$ or that no charge is enclosed by the surface. Since the gaussian surface can be drawn as close to the outside of the conductor as we wish, we can conclude that all the charge resides on the surface of the conductor. This conclusion holds even if the conductor is hollow on the inside.

An interesting experiment was devised by Michael Faraday to demonstrate that charge resides on the surface of a hollow conductor. In this experiment, known as the *ice-pail experiment,* a positively charged ball supported by a silk thread is lowered into a hollow metallic conductor. As shown in Fig. 30-13, a redistribution of charge occurs on the walls of the conductor, drawing the electrons to the inner surface. When the ball makes contact with the bottom of the conductor, the induced charge is neutralized, leaving a net positive charge on the outer surface. Probes with an electroscope will show that no charge resides inside the conductor and that a net positive charge remains on the outer surface.

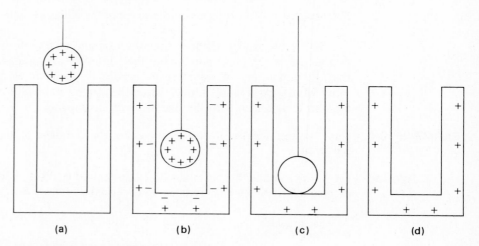

(a)   (b)   (c)   (d)

Fig. 30-13 Faraday's ice-pail experiment.

EXAMPLE 30-7   A capacitor is an electrostatic device consisting of two metallic conductors of area A separated by a distance d. If equal and opposite charge densities σ are placed on the conductors, an electric field E will exist between them. Determine the charge density on either plate. (Refer to Fig. 30-14.)

Solution   A gaussian cylinder is constructed as shown in Fig. 30-14 for the inner surface of either plate. There is no field inside the conducting plate, and the only area penetrated by the field lines is the surface A' projecting into the space between the plates. Applying Gauss's law to the left plate, we have

$$\sum \epsilon_0 EA = \sum q$$

$$\epsilon_0 EA' = \sigma A'$$

$$E = \frac{\sigma}{\epsilon_0} \tag{30-18}$$

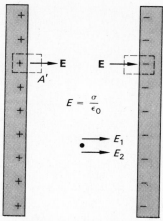

$$E = \frac{\sigma}{\epsilon_0}$$

E_1

E_2

Fig. 30-14 Electric field in the region between two oppositely charged plates is equal to the ratio of the charge density σ to the permittivity ϵ_0.

The same result would be obtained if the gaussian cylinder on the right were used. However, the lines would be directed inward, indicating an enclosed negative charge.

Notice that the field between the two plates in this example is exactly twice the field due to a thin sheet of charge, as given by Eq. (30-17). One can understand this relationship by considering the field **E** between the plates as a superposition of the fields due to two oppositely charged sheets.

$$E = E_1 + E_2 = \frac{\sigma}{2\epsilon_0} + \frac{\sigma}{2\epsilon_0} = \frac{\sigma}{\epsilon_0}$$

The field $\mathbf{E}_1$ due to the positive sheet of charge is in the same direction as the field $\mathbf{E}_2$ due to the negative sheet of charge. To the left and right of the two plates $\mathbf{E}_1$ and $\mathbf{E}_2$ are oppositely directed and cancel out.

SUMMARY

The concept of an electric field has been introduced to describe the region surrounding an electric charge. Its magnitude is determined by the force a unit charge will experience at a given location, and its direction is the same as the force on a positive charge at that point. Electric field lines were postulated to give a visual picture of electric fields, and the density of such field lines is an indication of the intensity of the electric field. The major concepts to be remembered are:

- An *electric field* is said to exist in a region of space in which an electric charge will experience an electric force. The *magnitude* of the electric field intensity E is given by the force F per unit of charge q.

$$E = \frac{F}{q} \qquad E = \frac{(9 \times 10^9 \text{ N} \cdot \text{m}^2/\text{C}^2)Q}{r^2}$$

The metric unit for the electric field intensity is the newton per coulomb (N/C). In the above equation, r is the distance from the charge Q to the point in question.

- The resultant field intensity at a point in the vicinity of a number of charges is the *vector* sum of the contributions from each charge.

$$E = E + E_2 + E_3 + \cdots \qquad \boxed{E = \sum \frac{kQ}{r^2}} \qquad \textit{Vector Sum}$$

It must be emphasized that this is a vector sum and not an algebraic sum. Once the magnitude and direction of each vector is determined, the resultant can be found from vector mechanics.

- The permittivity of free space ϵ_0 is a fundamental constant defined as

$$\boxed{\epsilon_0 = \frac{1}{4\pi k} = 8.85 \times 10^{-12} \ \text{C}^2/\text{N} \cdot \text{m}^2} \qquad \textit{Permittivity}$$

- Gauss's law states that the net number of electric field lines crossing any closed surface in an outward direction is numerically equal to the net total charge within that surface.

$$\boxed{N = \sum \epsilon_0 E_n A = \sum q} \qquad \textit{Gauss's law}$$

- In applications of Gauss's law the concept of charge density σ as the charge q per unit area A of surface is often utilized:

$$\boxed{\sigma = \frac{q}{A}} \qquad \boxed{q = \sigma A} \qquad \textit{Charge Density}$$

QUESTIONS

30-1. Define the following terms:
   a. Electric field
   b. Electric field intensity
   c. Electric field lines
   d. Permittivity ϵ_0
   e. Charge density
   f. Gauss's law
   g. Gaussian surface
   h. Faraday's ice pail

30-2. Some texts refer to electric field lines as "lines of force." Discuss the advisability of this description.

30-3. Can an electric field exist in a region of space in which an electric charge would not experience a force? Explain.

30-4. Is it necessary that a charge be placed at a point in order to have an electric field at that point? Explain.

30-5. Using a procedure similar to that for electric fields, show that the gravitational acceleration can be calculated from

$$g = \frac{GM}{r^2}$$

where M = mass of earth

r = distance from center of earth

30-6. Discuss the similarities between electric fields and gravitational fields. In what ways do they differ?

30-7. In Gauss's law the constant ϵ_0 was chosen as the proportionality factor between line density and field intensity. In a theoretical sense, this is a wise choice because it leads to the conclusion that the total number of lines is equal to the enclosed charge. Is such a choice practical for graphically illustrating field lines? According to Gauss's relation, how many field lines would emanate from a charge of 1 C?

30-8. Justify the following statement: the electric field intensity on the surface of any charged conductor must be directed perpendicular to the surface.

30-9. Electric field lines will never intersect. Explain.

30-10. Suppose you connect an electroscope to the outside surface of Faraday's ice pail. Show graphically what will happen to the gold leaf during each of the steps illustrated in Fig. 30-13.

30-11. Can an electric field line begin and end on the same conductor? Discuss.

30-12. What form would Gauss's law take if we had chosen k for the proportionality constant instead of the permittivity ϵ_0?

30-13. In Gauss's law, demonstrate that the units of $\epsilon_0 EA$ are dimensionally equivalent to the units of charge.

30-14. Show that the field in the region outside the two parallel plates in Fig. 30-14 is equal to zero.

30-15. Why is the field intensity constant in the region between two oppositely charged plates? Draw a vector diagram of the field due to each plate at various points between the plates.

PROBLEMS

The Electric Field Intensity

30-1. A charge of $+2\ \mu C$ placed at a point P in an electric field experiences a downward force of 8×10^{-4} N. What is the electric field intensity at point P?

Ans. 400 N/C, downward

30-2. A charge of -3 nC placed at point A experiences a downward force of 6×10^{-5} N. What is the electric field intensity at point A?

30-3. What are the magnitude and direction of the force that would act on an electron $(-1.6 \times 10^{-19}$ C$)$ if it were placed at **(a)** point P in Prob. 30-1? **(b)** point A in Prob. 30-2?

Ans. (a) 6.40×10^{-17} N, up; (b) 3.20×10^{-15}, down

30-4. What must be the magnitude and direction of the electric field intensity between two horizontal plates if it is desired to produce an upward force of 6×10^{-4} N on a $+60$-μC charge?

30-5. The uniform electric field between two horizontal plates is 8×10^4 N/C. The top plate is positively charged and the lower plate is negatively charged. What are the magnitude and direction of the force exerted on an electron as it passes horizontally through these plates?

Ans. 1.28×10^{-14} N, upward

30-6. Find the electric field intensity at a point P, located 6 mm to the left of a point charge of $8\ \mu C$. What are the magnitude and direction of the force on a -2-nC charge placed at point P?

30-7. Determine the electric field intensity at a point P, located 4.0 cm above a -12-μC charge. What are the magnitude and direction of the force on a $+3$-nC charge placed at point P?

Ans. 6.75×10^7 N/C, down; 0.202 N, down

Calculating the Resultant Electric Field Intensity

30-8. Determine the electric field intensity at the midpoint between a -60-μC charge and a $+40$-μC charge. The charges are 70 mm apart in air.

30-9. An 8-nC charge is located 80 mm to the right of a $+4$-nC charge. Determine the field intensity at the midpoint of a line joining the two charges.

Ans. 2.25×10^4 N/C, left

30-10. Find the electric field intensity at a point 30 mm to the right of a 16-nC charge and 40 mm to the left of a 9-nC charge.

* **30-11.** A charge of -20 μC is placed 50 mm to the right of a 49-μC charge. What is the resultant field intensity at a point located 24 mm directly above the -20-nC charge?

Ans. 2.82×10^8 N/C, 296.7°

* **30-12.** Two charges of $+12$ nC and $+18$ nC are separated horizontally by a distance of 28 mm. What is the resultant field intensity at a point 20 mm from each charge and above a line joining the two charges?

Applications of Gauss's Law

30-13. Two parallel plates are each 2 cm wide and 4 cm long. The electric field intensity between the two plates is 10,000 N/C directed upward. What is the charge on each plate?

Ans. 7.08×10^{-11} C

* **30-14.** Use Gauss's law to show that the field outside a solid charged sphere at a distance r from its center is given by

$$E = \frac{1}{4\pi\epsilon_0} \frac{Q}{r^2}$$

where Q is the total charge on the sphere.

* **30-15.** A sphere of 8 cm in diameter has 4 μC of charge placed on the surface. What is the electric field intensity (a) at the surface, (b) 2 cm outside the surface, (c) inside the sphere?

Ans. (a) 2.25×10^{-7} N/C, (b) 1.00×10^{-7} N/C, (c) 0

Additional Problems

30-16. How far from a point charge of 80 nC will the field intensity be 500 N/C?

30-17. The electric field intensity at a point in space is found to be 5×10^5 N/C, directed due west. What are the magnitude and direction of the force on a -4-μC charge placed at that point?

Ans. 2 N, east

30-18. An alpha particle consists of two protons and two neutrons, and it has a net positive charge of $+2e$. What are the magnitude and direction of the force on an alpha particle as it passes into an upward electric field of intensity 8×10^4 N/C?

Fig. 30-15

* 30-19. A 20-mg particle is placed in a uniform downward field of 2000 N/C. How many excess electrons must be placed on the particle in order for the electric and gravitational forces to balance?

Ans. 6.12×10^{11} electrons

30-20. What is the acceleration of an electron placed in a constant field of 4×10^5 N/C?

* 30-21. Charges of -2 and $+4\ \mu$C are placed at the base corners of an equilateral triangle with 10-cm sides. **(a)** What is the magnitude of the electric field intensity at the top corner? **(b)** What are the magnitude and direction of the force which would act on a charge of $-2\ \mu$C placed at that corner?

Ans. **(a)** 3.12×10^6 N/C, 150°; **(b)** 6.24 N, 330°

** 30-22. The electric field intensity between the plates in Fig. 30-15 is 4000 N/C. What is the magnitude of the charge on the suspended pith ball whose mass is 3 mg?

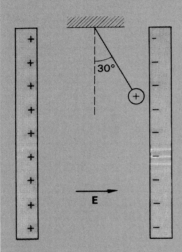

** 30-23. Two charges of $+16\ \mu$C and $+8\ \mu$C are 200 mm apart in air. At what point between the two charges is the field intensity equal to zero?

Ans. 82.8 mm from the 8-μC charge

** 30-24. Two charges of $+8$ nC and -5 nC are 40 mm apart in air. At what point on a line joining the two charges will the electric field intensity be zero?

** 30-25. The electric field intensity between the two plates in Fig. 30-4 is 2000 N/C. The length of the plates is 4 cm, and their separation is 1 cm. An electron is projected into the field from the left with horizontal velocity of 2×10^7 m/s. What is the upward deflection of the electron at the instant it leaves the plates?

Ans. 0.7 mm

* 30-26. Use Gauss's law to show that the electric field intensity at a distance R from an infinite line of charge is given by

$$E = \frac{\lambda}{2\pi\epsilon_0 R}$$

where λ is the charge per unit length. Construct a gaussian surface, as shown in Fig. 30-16.

Fig. 30-16

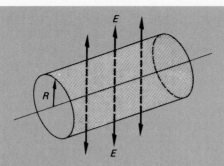

* 30-27. A uniformly charged conducting sphere has a radius of 24 cm and a surface charge density of $+16$ μC/m^2. What is the total number of electric field lines leaving the sphere?

    Ans. 1.16×10^{-5} lines

* 30-28. Use Gauss's law to show that the field just outside any solid conductor is given by

$$E = \frac{\sigma}{\epsilon_0}$$

* 30-29. What is the electric field intensity 2 m from the surface of a sphere 20 cm in diameter having a surface charge density of $+8$ nC/m^2?

    Ans. 2.05 N/C

Electric Potential

OBJECTIVES

After completing this chapter, you should be able to:

1. Demonstrate by definition and example your understanding of *electric potential energy, electric potential,* and *electric potential difference.*
2. Compute the potential energy of a known charge at a given distance from other known charges, and state whether the energy is negative or positive.
3. Compute the absolute potential at any point in the vicinity of a number of known charges.
4. Use your knowledge of potential difference to calculate the work required to move a known charge from any point *A* to another point *B* in an electric field created by one or more point charges.
5. Write and apply a relationship between the electric field intensity, the potential difference, and the plate separation for parallel plates of equal and opposite charge.

In our study of mechanics, many problems were simplified by introducing the concepts of energy. The conservation of mechanical energy allowed us to say certain things about the initial and final states of systems without having to analyze the motion between states. The concept of a change of potential energy into kinetic energy avoids the problem of varying forces.

In electricity, many practical problems can be solved by considering the changes in energy experienced by a moving charge. For example, if a certain quantity of work is required to move a charge against electric forces, the charge should have a *potential* for giving up an equivalent amount of energy when it is released. In this chapter we develop the idea of electric potential energy.

31-1
ELECTRIC POTENTIAL ENERGY

One of the best ways to understand the concept of electric potential energy is by comparing it with gravitational potential energy. In the gravitational case, consider that a mass *m* is moved from level *A* in Fig. 31-1 to level *B*. An external force *F* equal to the weight *mg* must be applied to move the mass against gravity. The work done

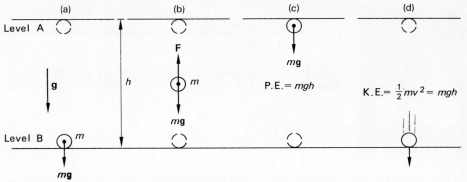

Fig. 31-1 A mass m lifted against a gravitational field g results in a potential energy of mgh at level B. When released, this energy will be transformed entirely into kinetic energy as it falls to level A.

by this force is the product of mg and h. When the mass m reaches level B, it has a potential for doing work relative to level A. The system has a *potential energy* (P.E.) which is equal to the work done against gravity.

$$\text{P.E.} = mg \cdot h$$

This expression represents the potential for doing work after the mass m is released at level B and falls the distance h. Therefore, the magnitude of the potential energy at B does not depend on the path taken to reach that level.

Now, let us consider a positive charge $+q$ resting at point A in a uniform electric field E between two oppositely charged plates. (See Fig. 31-2.) An electric force qE acts downward on the charge. The work done against the electric field in moving the charge from A to B is equal to the product of the force qE and the distance d. Hence, the electric potential energy at point B relative to point A is

$$\text{P.E.} = qE \cdot d \tag{31-1}$$

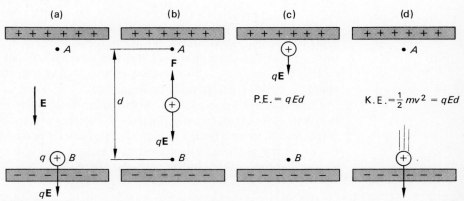

Fig. 31-2 A positive charge $+q$ is moved against a constant electric field E through a distance d. At point B the potential energy will be gEd relative to point A. When released, the charge will gain an equivalent amount of kinetic energy.

Before we proceed, we should point out an important difference between gravitational potential energy and electric potential energy. In the case of gravity, there is only one kind of mass, and the forces involved are always forces of attraction. Therefore, a mass at higher elevations always has greater potential energy relative to the earth. This is not true in the electrical case because of the existence of negative charge. In the example, a positive charge, as in Fig. 31-2, has a greater potential energy at point B than at point A. This is true regardless of the reference point chosen for measuring the energy because work has been done *against* the electric field. (Refer to Fig. 31-3.) On the other hand, if a negative charge were moved from point A to point B, work would be done *by* the field. A negative charge would have a *lower* potential energy at B, which is exactly opposite to the situation for a positive charge.

> *Whenever a positive charge is moved against an electric field, the potential energy increases; whenever a negative charge moves against an electric field, the potential energy decreases.*

The above rule is a direct consequence of the fact that the direction of the electric field is defined in terms of a positive charge.

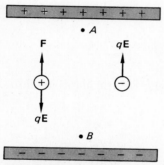

Fig. 31-3 A positive charge increases its potential energy when moved from A to B; a negative charge *loses* potential energy when it moves from A to B.

31-2
CALCULATING POTENTIAL ENERGY

In considering the space between two oppositely charged plates, work computations are fairly simple because the electric field is uniform. The electric force that a charge experiences is constant as long as it remains between the plates. In general, however, the field will not be constant, and we must make allowances for a varying force. For example, consider the electric field in the vicinity of a positive charge Q, as illustrated in Fig. 31-4. The field is directed radially outward, and its intensity falls off inversely with the square of the distance from the center of the charge. The field at points A and B is

$$E_A = \frac{kQ}{r_A^2} \qquad E_B = \frac{kQ}{r_B^2}$$

Where r_A and r_B are the respective distances to points A and B.

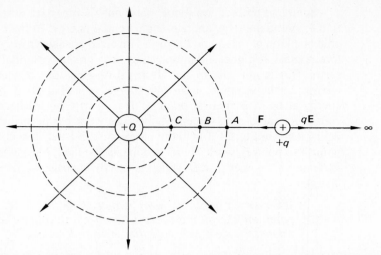

Fig. 31-4 The potential energy due to a charge placed in an electric field is equal to the work done *against* electric forces in bringing the charge from infinity to the point in question.

The average electric force experienced by a charge $+q$ when it is moved from point A to point B is

$$F = \frac{kQq}{r_A r_B} \qquad (31\text{-}2)$$

Thus the work done against the electric field in moving through the distance $r_A - r_B$ is equal to

$$\text{Work}_{A \to B} = \frac{kQq}{r_A r_B}(r_A - r_B)$$

$$= kQq\left(\frac{1}{r_B} - \frac{1}{r_A}\right) \qquad (31\text{-}3)$$

Note that the work is a function of the distances r_A and r_B. The path traveled is unimportant. The same work would be done against the field in moving a charge from any point on the dashed circle passing through A to any point on the circle passing through B.

Suppose now we compute the work done against electric forces in moving a positive charge from infinity to a point a distance r from the charge Q. From Eq. (31-3), the work is given by

$$\text{Work}_{\infty \to r} = kQq\left(\frac{1}{r} - \frac{1}{\infty}\right)$$

$$= \frac{kQq}{r} \qquad (31\text{-}4)$$

Since we have shown that the work done against the electric field equals the increase in potential energy, Eq. (31-4) is the potential energy at r relative to infinity.

Often potential energy is taken as zero at infinity so that the potential energy of a system composed of a charge q and another charge Q separated by a distance r is

$$\text{P.E.} = \frac{kQq}{r} \qquad\qquad (31\text{-}5)$$

The **potential energy** of the system is equal to the work done against the electric forces in moving the charge $+q$ from infinity to that point.

EXAMPLE 31-1   A charge of $+2$ nC is 20 cm away from another charge of $+4$ μC. (a) What is the potential energy of the system? (b) What is the change in potential energy if the 2-μC charge is moved to a distance of 8 cm from the $+4$-nC charge?

Solution (a)   The potential energy at 20 cm is given by Eq. (31-5).

$$
\begin{aligned}
\text{P.E.} &= \frac{kQq}{r} \\
&= \frac{(9 \times 10^9 \text{ N} \cdot \text{m}^2/\text{C}^2)(4 \times 10^{-6} \text{ C})(2 \times 10^{-9} \text{ C})}{0.2 \text{ m}} \\
&= 36 \times 10^{-5} \text{ J}
\end{aligned}
$$

Solution (b)   The potential energy at a distance of 8 cm is

$$
\begin{aligned}
\text{P.E.} &= \frac{kQq}{r} \\
&= \frac{(9 \times 10^9 \text{ N} \cdot \text{m}^2/\text{C}^2)(4 \times 10^{-6} \text{ C})(2 \times 10^{-9} \text{ C})}{0.08 \text{ m}} \\
&= 90 \times 10^{-5} \text{ J}
\end{aligned}
$$

The change in potential energy is

$$90 \times 10^{-5} \text{ J} - 36 \times 10^{-5} \text{ J} = 54 \times 10^{-5} \text{ J}$$

Notice that the difference is positive, indicating an increase in potential energy. If the charge Q were negative and all other parameters were unchanged, the potential energy would have decreased by this same amount.

31-3
POTENTIAL

When we first introduced the concept of an electric field as force per unit charge, we pointed out that the primary advantage of such a concept was that it allowed us to assign an electrical property to space. If the field intensity is known at some point, the force on a charge placed at that point can be predicted. It is equally convenient to assign another property to the space surrounding a charge that would allow us to predict the potential energy due to another charge placed at any point. This property of space is called *potential* and is defined as follows:

The **potential** V at a point a distance r from a charge Q is equal to the work per unit charge done against electric forces in bringing a positive charge $+q$ from infinity to that point.

In other words, the potential at some point A, as shown in Fig. 31-5, is equal to *the potential energy per unit charge.* The units of potential are expressed in *joules per coulomb,* defined as a *volt* (V).

$$V_A(V) = \frac{\text{P.E.(J)}}{q(C)}$$

(31-6)

Thus, a potential of *one volt* at point A means that *if a charge of one coulomb were to be placed at A, the potential energy would be one joule.* In general when the potential is known at a point A, the potential energy due to a charge q at that point can be found from

$$\boxed{\text{P.E.} = qV_A}$$

(31-7)

Substitution of Eq. (31-5) into Eq. (31-6) yields an expression for directly computing the potential.

$$V_A = \frac{\text{P.E.}}{q} = \frac{kQq/r}{q}$$

$$\boxed{V_A = \frac{kQ}{r}}$$

(31-8)

The symbol V_A refers to the potential at point A located at a distance r from the charge Q.

It is noted now that the potential is the same at equal distances from a spherical charge. For this reason, the dashed lines in Figs. 31-4 and 31-5 are called *equipotential lines.* Note that the lines of equal potential are always perpendicular to the electric field lines. If this were not true, work would be done by a resultant force

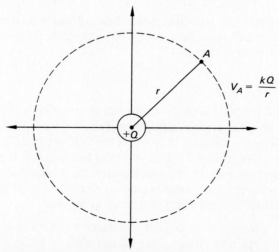

Fig. 31-5 Calculating the potential at a distance r from a charge $+Q$.

when a charge moved along an equipotential line. Such work would increase or decrease the potential.

Equipotential lines are always perpendicular to electric field lines.

Before offering an example, it must be pointed out that the potential at a point is defined in terms of a positive charge. This means that the potential will be negative at a point in space surrounding a negative charge. As a rule, we must remember that:

The potential due to a positive charge is positive, and the potential due to a negative charge is negative.

Using a negative sign for a negative charge Q in Eq. (31-8) results in a negative value for the potential.

EXAMPLE 31-2   (*a*) Calculate the potential at a point A which is 30 cm distant from a charge of $-2\ \mu\text{C}$. (*b*) What is the potential energy if a $+4$-nC charge is placed at A?

Solution (a)   From Eq. (31-8)

$$V_A = \frac{kQ}{r} = \frac{(9 \times 10^9\ \text{N} \cdot \text{m}^2/\text{C}^2)(-2 \times 10^{-6}\ \text{C})}{0.3\ \text{m}}$$

$$= -6 \times 10^4\ \text{V}$$

Solution (b)   The potential energy due to the $+4$-nC charge is given by direct substitution into Eq. (31-7).

$$\text{P.E.} = qV_A = (4 \times 10^{-9}\ \text{C})(-6 \times 10^4\ \text{V})$$

$$= -24 \times 10^{-5}\ \text{J}$$

A negative potential energy means that work must be done *against* the electric field in moving the charges apart. In the above example, 24×10^{-5} J of work would have to be supplied by an external force in order to remove the charge to infinity.

Now let us consider the more general case, which deals with the potential in the neighborhood of a number of charges. As illustrated in Fig. 31-6:

The potential in the vicinity of a number of charges is equal to the algebraic sum of the potentials due to each charge.

$$V_A = V_1 + V_2 + V_3$$

$$= \frac{kQ_1}{r_1} + \frac{kQ_2}{r_2} + \frac{kQ_3}{r_3}$$

Fig. 31-6 The potential in the vicinity of a number of charges.

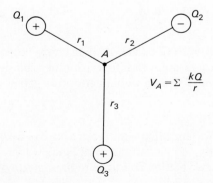

$$V_A = \Sigma\ \frac{kQ}{r}$$

Whenever the charge is negative, as with Q_2 in Fig. 31-6, the sign of the charge is inserted into the calculations. In general,

$$V = \sum \frac{kQ}{r}$$ (31-9)

This equation is an algebraic sum since potential is a scalar quantity and not a vector quantity.

EXAMPLE 31-3 Two charges, $Q_1 = +6\ \mu C$ and $Q_2 = -6\ \mu C$, are separated by 12 cm, as shown in Fig. 31-7. Calculate the potential (a) at point A and (b) at point B.

Fig. 31-7

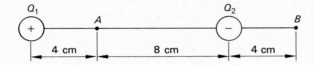

Solution (a) The potential at A is found from Eq. (31-9).

$$V_A = \frac{kQ_1}{r_1} + \frac{kQ_2}{r_2}$$
$$= \frac{(9 \times 10^9\ \text{N} \cdot \text{m}^2/\text{C}^2)(6 \times 10^{-6}\ \text{C})}{4 \times 10^{-2}\ \text{m}} + \frac{(9 \times 10^9\ \text{N} \cdot \text{m}^2/\text{C}^2)(-6 \times 10^{-6}\ \text{C})}{8 \times 10^{-2}\ \text{m}}$$
$$= 13.5 \times 10^5\ \text{V} - 6.75 \times 10^5\ \text{V}$$
$$= 6.75 \times 10^5\ \text{V}$$

This means that the electric field will do 6.75×10^5 J of work on each coulomb of positive charge that it moves from A to infinity.

Solution (b) The potential at B is

$$V_B = \frac{kQ_1}{r_1} + \frac{kQ_2}{r_2}$$
$$= \frac{(9 \times 10^9\ \text{N} \cdot \text{m}^2/\text{C}^2)(6 \times 10^{-6}\ \text{C})}{16 \times 10^{-2}\ \text{m}} + \frac{(9 \times 10^9\ \text{N} \cdot \text{m}^2/\text{C}^2)(-6 \times 10^{-6}\ \text{C})}{4 \times 10^{-2}\ \text{m}}$$
$$= 3.38 \times 10^5\ \text{V} - 13.5 \times 10^5\ \text{V}$$
$$= -10.1 \times 10^5\ \text{V}$$

The negative value indicates that the field will hold onto a positive charge. In order to move 1 C of positive charge from A to infinity, another source of energy must perform 10.1×10^5 J of work. The field would perform negative work in this amount.

31-4

POTENTIAL DIFFERENCE

In practical electricity, we are seldom interested in the work per unit charge to remove a charge to infinity. More often, we want to know the work requirements for moving charges between two points. This leads to the concept of *potential difference*.

The **potential difference** between two points is the work per unit positive charge done by electric forces in moving a small test charge from the point of higher potential to the point of lower potential.

Another way of stating this would be to say that the potential difference between two points is the difference in the potentials at those points. For example, if the potential at some point A is 100 V and the potential at another point B is 40 V, the potential difference is

$$V_A - V_B = 100 \text{ V} - 40 \text{ V} = 60 \text{ V}$$

This means that 60 J of work will be done by the field on each coulomb of positive charge moved from A to B. In general, the work done by the electric field in moving a charge q from point A to point B can be found from

$$\boxed{\text{Work}_{A \rightarrow B} = q(V_A - V_B)} \qquad (31\text{-}10)$$

EXAMPLE 31-4   What is the potential difference between points A and B in Fig. 31-7. How much work is done by the electric field in moving a -2-nC charge from point A to point B?

Solution   The potentials at points A and B were calculated in the previous example. They are

$$V_A = 6.75 \times 10^5 \text{ V} \qquad V_B = -10.1 \times 10^5 \text{ V}$$

Therefore the potential difference between points A and B is

$$V_A - V_B = 6.75 \times 10^5 \text{ V} - (-10.1 \times 10^5 \text{ V})$$
$$= 16.9 \times 10^5 \text{ V}$$

Since A is at a higher potential than B, positive work would be done by the field when a *positive* charge is moved from A to B. If a *negative* charge is moved, the work done by the field in moving it from A to B will be negative. In this example, the work is

$$\text{Work}_{A \rightarrow B} = q(V_A - V_B)$$
$$= (-2 \times 10^{-9} \text{ C})(16.9 \times 10^5 \text{ V})$$
$$= -3.37 \times 10^{-3} \text{ J}$$

Since the work done by the field is negative, another source of energy must supply the work to move the charge.

Let us now return to the example of a uniform electric field **E** between two oppositely charged plates, as in Fig. 31-8. We shall assume that the plates are separated by a distance d. A charge q placed in the region between the plates A and B will experience a force given by

$$\mathbf{F} = q\mathbf{E}$$

The work done by this force in moving the charge q from plate A to plate B is given by

$$Fd = (qE)d$$

But this work is also equal to the product of the charge q and the potential difference $V_A - V_B$ between the two plates, so that we can write

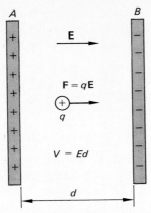

Fig. 31-8 The potential between two oppositely charged plates.

$$q(V_A - V_B) = qEd$$

If we divide through by q and represent the potential difference by the single symbol V, we obtain

$$\boxed{V = Ed} \qquad (31\text{-}11)$$

The potential difference between two oppositely charged plates is equal to the product of the field intensity and the plate separation.

EXAMPLE 31-5 The potential difference between two plates 5 mm apart is 10 kV. Determine the electric field intensity between the plates.

Solution Solving for E in Eq. (31-11) gives

$$E = \frac{V}{d} = \frac{10 \times 10^3 \text{ V}}{5 \times 10^{-3} \text{ m}} = 2 \times 10^6 \text{ V/m}$$

As an exercise, the student should show that the *volt per meter* is equivalent to the *newton per coulomb*. The electric field expressed in volts per meter is sometimes referred to as the *potential gradient*.

31-5

MILLIKAN'S OIL-DROP EXPERIMENT

Now that we have developed the concepts of the electric field and potential difference, we are ready to describe a classic experiment designed to determine the smallest unit of charge. Robert A. Millikan, an American physicist, devised a series of experiments in the early 1900s. A schematic diagram of his apparatus is shown in Fig. 31-9. Very tiny oil droplets are sprayed into the region between the two metallic plates. Electrons are freed from air molecules by passing ionizing x-rays through the medium. These electrons attach themselves to the oil droplets, giving them a net negative charge.

The downward motion of the oil droplets can be observed by a microscope as they fall slowly under the influence of their weight and the upward viscous force of

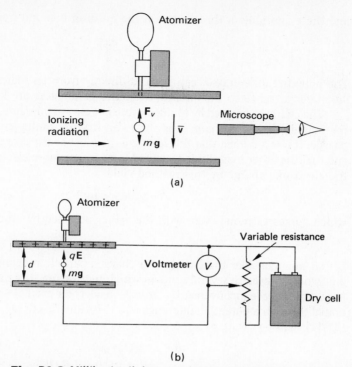

Fig. 31-9 Millikan's oil-drop experiment: (*a*) The mass *m* of the droplet is determined from its rate of fall against the viscous force of air resistance. (*b*) The magnitude of the charge is computed from the equilibrium conditions which suspend the charge between oppositely charged plates.

air resistance. (Refer to Fig. 31-9*a*.) The laws of hydrostatics can be used to calculate the mass *m* of a particular oil drop from its measured rate of fall.

After the necessary data for determining the mass *m* have been recorded, and external battery is connected to establish a uniform electric field **E** between the oppositely charged plates. (See Fig. 31-9*b*.) The magnitude of the field intensity can be controlled by a variable resistor in the electric circuit. The field is adjusted until the upward electric force on the drop is equal to the downward gravitational force, so that the oil drop stops moving. Under these conditions

$$qE = mg$$

where q = net charge of oil drop
m = mass of oil drop
g = acceleration of gravity

The field intensity **E**, as determined from Eq. (31-11), is a function of the applied voltage V and the plate separation d. Therefore, Eq. (31-12) becomes

$$q\frac{V}{d} = mg$$

and the magnitude of the charge on the oil drop is found from

$$q = \frac{mgd}{V} \qquad (31\text{-}12)$$

The potential difference V can be read directly from an indicating device called a *voltmeter* attached to the circuit. The other parameters are known.

The charges observed by Millikan were not always the same, but he observed that the magnitude of the charge was always an integral multiple of a basic quantity of charge. It was assumed that this *least* charge must be the charge of a single electron and that the other magnitudes resulted from two or more electrons. Computation of the electronic charge by this method yields

$$e = 1.6065 \times 10^{-19} \text{ C}$$

which agrees extremely well with the values obtained by other methods.

31-6
THE ELECTRON-VOLT

Let us now consider the energy of a charged particle moving through a potential difference. There are several units in which to measure this energy, but most of the familiar units are inconveniently large. Consider, for example, a charge of 1 C accelerated through a potential difference of 1 V. Its kinetic energy, from Eqs. (31-2) and (31-11), will be

$$\text{K.E.} = qEd = qV$$
$$= (1 \text{ C})(1 \text{ V}) = 1 \text{ C} \cdot \text{V}$$

The coulomb-volt, of course, is a joule. But 1 C of charge is inconveniently large when applied to single particles, and the corresponding unit of energy (the joule) is also large. The most convenient unit of energy in atomic and nuclear physics is the *electronvolt* (eV).

> The electronvolt is a unit of energy equivalent to the energy acquired by an electron which is accelerated through a potential difference of one volt.

The electronvolt differs from the coulomb-volt by the same degree as the difference in the charge of an electron and the charge of 1 C. To compare the two units, suppose we compute the energy in joules acquired by an electron which has been accelerated through a potential difference of 1 V:

$$\text{K.E.} = qV$$
$$= (1.6 \times 10^{-19} \text{ C})(1 \text{ V})$$
$$= 1.6 \times 10^{-19} \text{ J}$$

Thus 1 eV is equivalent to an energy of 1.6×10^{-19} J.

SUMMARY

The concepts of potential energy, potential, and potential difference have been extended to electrical phenomena. The many problems dealing with electrostatic potential have been designed to provide a base for direct current electricity which will be discussed later. The essential elements of this chapter are summarized below:

- When a charge q is moved against a constant electric force for a distance d, the potential energy of the system is

$$\text{P.E.} = qEd$$

where E is the constant electric field intensity. If the charge is released, it will acquire a kinetic energy

$$\text{K.E.} = \tfrac{1}{2}mv^2 = qEd$$

as it returns for the same distance.
- Due to the existence of positive and negative charges and the opposite effects produced by the same field, we must remember that:

  The potential energy increases as a positive charge is moved against the electric field, and the potential energy decreases as a negative charge is moved against the same field
- In general, the potential energy due to a charge q placed at a distance r from another charge Q is equal to the work done against electric forces in moving the charge $+q$ from infinity.

$$V = \frac{kQq}{r} \qquad \text{\textit{Electric Potential Energy}}$$

Note that the distance r is not squared, as it was for the electric field intensity.
- The electric *potential* V at a point a distance r from a charge Q is equal to the work per unit charge done against electric forces in bringing a positive charge $+q$ from infinity.

$$V = \frac{kQ}{r} \qquad \text{\textit{Electric Potential}}$$

- The unit of electric potential is the joule per coulomb (J/C), which is renamed the volt (V).

$$1\,\text{V} = \frac{1\,\text{J}}{1\,\text{C}}$$

- The potential at a point in the vicinity of a number of charges is equal to the algebraic sum of the potentials due to each charge:

$$V = \sum \frac{kQ}{r} = \frac{kQ_1}{r_1} + \frac{kQ_2}{r_2} + \frac{kQ_3}{r_3} + \cdots \qquad \text{\textit{Algebraic Sum}}$$

- The potential difference between two points A and B is the difference in the potentials at those points.

$$V_{AB} = V_A - V_B \qquad \text{\textit{Potential Difference}}$$

- The work done by an electric field in moving a charge q from point A to point B can be found from

$$\text{Work}_{AB} = q(V_A - V_B)$$

Work and Potential Difference

- The potential difference between two oppositely charged plates is equal to the product of the field intensity and the plate separation.

$$V = Ed$$

$$E = \frac{V}{d}$$

QUESTIONS

31-1. Define the following terms:
a. Electric potential energy
b. Electric work
c. Potential
d. Volt
e. Equipotential lines
f. Potential difference
g. Potential gradient
h. Electronvolt

31-2. Distinguish clearly between positive and negative work. Distinguish between positive and negative potential energy.

31-3. Is it possible for a mass m to increase the potential energy by moving it to a lower elevation? Is it possible for a charged object to increase the potential energy as it is moved to a position of lower potential? Explain.

31-4. Give an example in which the electric potential is zero at some point where the electric field intensity is not zero.

31-5. The electric field inside an electrostatic conductor is zero. Is the electric potential inside the conductor zero also? Explain.

31-6. If the electric field intensity is known at some point, can one determine the electric potential at that point? What information is needed?

31-7. The surface of any conductor is an equipotential surface. Justify this statement.

31-8. Is the direction of the electric field intensity from higher to lower potential? Illustrate.

31-9. Apply the potential concept to the gravitational field to obtain an expression similar to Eq. (31-9) for computing the potential energy per unit mass. Discuss the applications of such a formula.

31-10. Show that the volt per meter is dimensionally equivalent to the newton per coulomb.

31-11. Distinguish between potential difference and a difference in potential energy.

31-12. A potential difference of 220 V is maintained between the ends of a long high-resistance wire. If the center of the wire is grounded ($V = 0$), what is the potential difference between the center and the end points?

31-13. The potential due to a negative charge is negative, and the potential due to a positive charge is positive. Why? Is it also true that the potential energy due to a negative charge is negative? Explain.

31-14. Is potential a property assigned to space or to a charge? What is potential energy assigned to?

Work and Electric Potential Energy

31-1. Two metal plates are separated by 30 mm and are oppositely charged so that a constant electric field of 6×10^4 N/C exists between them. How much work must be done *against* the electric field in order to move a $+4$-μC charge from the negative plate to the positive plate? How much work is done *by* the electric field? What is the potential energy when the charge is at the positive plate?

Ans. 7.2 mJ, -7.2 mJ, 7.2 mJ

31-2. The electric field intensity between two parallel plates, separated by 25 mm of air, is 8000 N/C. How much work is done *by* the electric field in moving a -2-μC charge from the negative plate to the positive plate? What is the work done *by* the electric field in moving the same charge back to the positive plate? What is the potential energy when the charge is at the positive plate? What is the potential energy when the charge is at the negative plate?

31-3. A charge of $+6$ μC is 30 mm away from another charge of 16 μC. (a) What is the potential energy of the system? (b) What is the *change* in potential energy if the 6-μC charge is moved to a distance of only 5 mm? Is this an increase or decrease in potential energy?

Ans. (a) 28.8 J; (b) 144 J, increase

31-4. A -3-μC charge is placed 6 mm away from a $+9$-μC charge. What is the potential energy? Is it negative or positive? What is the potential energy if the separation is increased to a distance of 24 mm? Is this an increase or a decrease in potential energy?

31-5. At what distance from a -7-μC charge must a charge of -12 nC be placed if the potential energy is to be 9×10^{-5} J?

Ans. 8.40 m

31-6. The potential energy of a system consisting of two identical charges is 4.50 mJ when their separation is 38 mm. What is the magnitude of each charge?

Electric Potential and Potential Difference

31-7. Calculate the potential at a point A which is 50 mm distant from a charge of -40 μC. What is the potential energy if a $+3$-μC charge is placed at point A?

Ans. -7.20 MV, -21.6 J

31-8. What is the potential at a point B located 60 mm away from a charge of $+15$ μC? What is the potential energy of a $+2$-nC charge placed at point B?

31-9. Point A is located 40 mm from a 6-μC charge; point B is located 25 mm from the same charge. Calculate the potential difference between points A and B. How much work is required by an external force if a $+5$-μC charge is moved from point A to point B?

Ans. 810 kV, $+4.05$ J

31-10. Point A is 68 mm away from a 90-μC charge and point B is only 26 mm away from the same charge. How much work is done *by* the electric field in moving a -5-nC charge from a point A to a point B?

31-11. A $+45$-nC charge is 68 mm to the left of a -9-nC charge. What is the potential at a point located 40 mm to the left of the -9-nC charge?

Ans. 12.4 kV

31-12. What is the potential at the midpoint of a line joining a -12-μC charge with a $+3$-μC charge? The separation of the two charges is 80 mm.

31-13. Point A is located 90 mm to the right of a -40-μC charge and 30 mm to the left of a $+55$-μC charge. Point B is located 15 mm to the left of the -40-μC charge. What is the potential at A? What is the potential at B? What is the potential difference? How much work is done *by* the electric field in moving a $+4$-nC charge from point A to point B?

> **Ans.** $+12.5$ MV, -20.3 MV, $+32.8$ MV, $+0.131$ J

31-14. A -60-μC charge is located 45 mm to the right of a $+20$-μC charge. Point A is midway between the two charges. Point B is located a distance of 35 mm directly above the $+20$-μC charge. What is the potential difference between A and B? How much work is done when a -5-μC charge is moved from point A to point B? Will this work be done *by* the field or must it be accomplished by an external force?

Additional Problems

31-15. The potential at a certain distance from a point charge is 1200 V, and the electric field intensity at that point is 400 N/C. What is the distance to the charge, and what is the magnitude of the charge?

> **Ans.** 3 m, 400 nC

31-16. What is the difference in potential between two points 30 and 60 cm away from a -50-μC charge?

31-17. The electric field intensity between two parallel plates 4 mm apart is 6000 V/m. What is the potential difference between the plates?

> **Ans.** 24.0 V

31-18. The electric field intensity between two plates separated by a distance of 50 mm is 6×10^5 V/m. What is the potential difference between the plates?

31-19. What must be the separation of two parallel plates if the field intensity is 5×10^4 V/m and the potential difference is 400 V?

> **Ans.** 8 mm

31-20. The potential difference between two parallel plates is 600 V. A 6-μC charge is accelerated through the entire potential difference. What is the kinetic energy given to the charge?

31-21. Determine the kinetic energy of an alpha particle ($+2e$) which is accelerated through a potential difference of 800 kV. Give the answer in both electron-volts and joules.

> **Ans.** 1.60 MeV, 2.56×10^{-13} J

31-22. A linear accelerator accelerates an electron through a potential difference of 4 MV. What is the energy of an emergent electron in electron-volts? In joules?

31-23. An electron acquires an energy of 2.8×10^{-15} J as it passes from point A to point B. What is the potential difference between these points in volts?

> **Ans.** 17.5 kV

31-24. Two charges of $+12$ and -6 μC are separated by 160 mm. What is the potential at the midpoint of a line joining the two charges? If an electron is accelerated from the -6-μC charge, what will be its kinetic energy in electron-volts when it arrives at the $+12$-μC charge?

31-25. Points A, B, and C are at the corners of an equilateral triangle which is 100 mm on each side. At the base AB of the triangle, a $+8$-μC charge is 100 mm to the left of a -8-μC charge. What is the potential at C? What is the potential at a point D which is 20 mm to the left of the -8-μC charge? How much work is done *by* the electric field in moving a $+2$-μC charge from point C to point D?

Ans. 0, -2.70 MV, 5.40 J

* 31-26. Two large plates are 80 mm apart and have a potential difference of 800 kV. What is the magnitude of the force which would act on an electron placed at the midpoint between the two plates? What would be the kinetic energy of the electron if it moved from one plate to the other?

** 31-27. The horizontal plates in Millikan's oil-drop experiment are 20 mm apart. The diameter of a particular drop of oil is 4 μm, and the density of oil is 900 kg/m^3. Assuming that two electrons attach themselves to the droplet, what potential difference must exist between the plates in order to establish equilibrium?

Ans. 18.5 kV.

32 Capacitance

OBJECTIVES

After completing this chapter, you should be able to:

1. Define *capacitance* and apply a relationship between *capacitance*, applied *voltage* and total *charge.*
2. Compute the capacitance of a *parallel-plate capacitor* when the area of the plates and their separation in a medium of known dielectric constant are given.
3. Write and apply expressions for calculating the *dielectric constant* as a function of the voltage, the electric field, or the capacitance before and after insertion of a dielectric.
4. Calculate the equivalent capacitance of a number of capacitors connected in *series* and in *parallel.*
5. Determine the *energy* of a charged capacitor, given the appropriate information.

Any charged conductor may be viewed as a reservoir, or source, of electric charge. If a conducting wire is connected to such a reservoir, electric charge can be transferred to perform useful work. In many applications of electricity, large quantities of charge are stored upon a conductor or group of conductors. Any device designed to store electric charge is called a *capacitor*. In this chapter we discuss the nature and application of these devices.

32-1
LIMITATIONS ON CHARGING A CONDUCTOR

How much electric charge can be placed on a conductor? Are there practical limits to the number of electrons that can be transferred to or from a conductor? Suppose we connect a large reservoir of positive and negative charges, such as the earth, to a conducting object, as illustrated in Fig. 32-1a. The energy necessary to transfer electrons from the earth to the conductor can be provided by an electrical device called a *battery*. Charging the conductor is analogous to pumping air into a hollow steel tank. (Refer to Fig. 32-1b.) As more air is pumped into the tank, the pressure

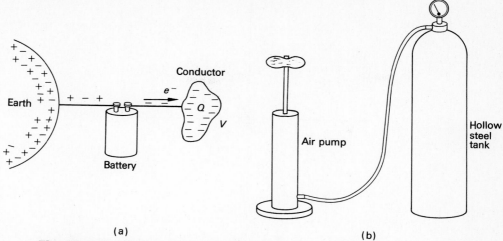

Fig. 32-1 Charging a conductor is analogous to pumping air into a hollow steel tank.

opposing the flow of additional air becomes greater. Similarly, as more charge Q is transferred to the conductor, the potential V of the conductor becomes higher, making it increasingly difficult to transfer more charge. We can say that the increase in potential V is directly proportional to the charge Q placed on the conductor. Symbolically,

$$V \propto Q$$

Therefore, the ratio of the quantity of charge Q to the potential V produced will be a constant for a given conductor. This ratio reflects the ability of a conductor to store charge and is called its *capacitance C.*

$$C = \frac{Q}{V}$$

(32-1)

The unit of capacitance is the *coulomb per volt* which is redefined as a *farad* (F). Thus, *if a conductor has a capacitance of one farad, a transfer of one coulomb of charge to the conductor will raise its potential by one volt.*

Let us return to our original question about the limitations placed on charging a conductor. We have said that every conductor has a capacitance C for storing charge. The value of C for a given conductor is not a function of either the charge placed on a conductor or the potential produced. In principle, the ratio Q/V will remain constant as charge is added indefinitely, but the capacitance depends on the *size* and *shape* of a conductor as well as on the nature of the *surrounding medium.*

Suppose we try to place an indefinite quantity of charge Q on a spherical conductor of radius r, as illustrated in Fig. 32-2. The air surrounding the conductor is an insulator, sometimes called a *dielectric*, which contains very few charges free to

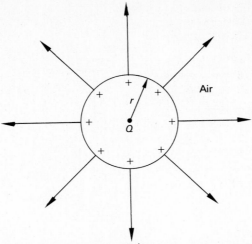

Fig. 32-2 The amount of charge which can be placed on a conductor is limited by the dielectric strength of the surrounding medium.

move. The electric field intensity E and the potential V at the surface of the sphere are given by

$$E = \frac{kQ}{r^2} \quad and \quad V = \frac{kQ}{r}$$

Since the radius r is constant, both the field intensity and the potential at the surface of the sphere increase in direct proportion to the charge Q. However, there is a limit to the field intensity which can exist on a conductor without ionizing the surrounding air. When this occurs, the air essentially becomes a conductor, and any additional charge placed on the sphere will "leak off" to the air. This limiting value of electric field intensity for which a material loses its insulation properties is called the dielectric strength of that material.

> The **dielectric strength** for a given material is that electric field intensity for which the material ceases to be an insulator and becomes a conductor.

The dielectric strength for dry air at 1 atm pressure is around 3 MN/C. Since the dielectric strength of a material varies considerably with environmental conditions, such as pressure, humidity, etc., it is difficult to compute accurate values.

EXAMPLE 32-1   What is the maximum charge that may be placed on a spherical conductor of radius 50 cm?

Solution   The field intensity at the surface of the sphere, at the point of breakdown in the air, is given by

$$E = \frac{kQ}{r^2} = 3 \text{ MN/C}$$

where 3 MN/C is assumed to be the dielectric strength of air. Solving for Q gives

---

596

$$Q = \frac{(3 \times 10^6 \text{ N/C})r^2}{k}$$

$$= \frac{(3 \times 10^6 \text{ N/C})(0.5 \text{ m})^2}{9 \times 10^9 \text{ N} \cdot \text{m}^2/\text{C}^2}$$

$$= 8.33 \times 10^{-5} \text{ C} = 83.3 \text{ } \mu\text{C}$$

This example illustrates the enormous magnitude of the coulomb when it is applied as a unit of electrostatic charge.

Note that the amount of charge which can be placed on a spherical conductor decreases with the radius of the sphere. Thus smaller conductors usually can hold less charge. But the shape of a conductor also influences its ability to retain charge. Consider the charged conductors illustrated in Fig. 32-3. If these conductors are

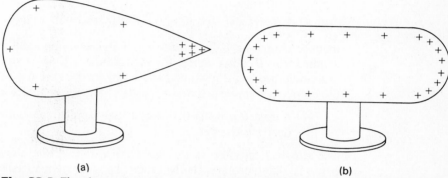

(a)   (b)

Fig. 32-3 The charge density on a conductor is greatest at regions of greatest curvature.

tested with an electroscope, it will be discovered that the charge on the surface of a conductor is concentrated at points of greatest curvature. Because of the greater charge density in these regions, the electric field intensity is also greater in regions of larger curvature. If the surface is reshaped to a sharp point, the field intensity may become great enough to ionize the surrounding air. A slow leakage of charge sometimes occurs at these locations, producing a *corona discharge,* which is often observed as a faint violet glow in the vicinity of the sharply pointed conductor. It is important to remove all sharp edges from electrical equipment in order to minimize this leakage of charge.

32-2
THE
CAPACITOR

When a number of conductors are placed near one another, the potential of each is affected by the presence of the other. For example, suppose we connect a negatively charged plate A to an electroscope, as in Fig. 32-4. The divergence of the gold leaf in the electroscope provides a measure of the potential of the conductor. Now let us suppose that another conductor B is placed parallel to A a short distance away. When the second conductor is grounded, a positive charge will be induced on it as the electrons are forced into the ground. Immediately, the gold leaf will collapse slightly, indicating a drop in the potential of conductor A. Because of the presence of the

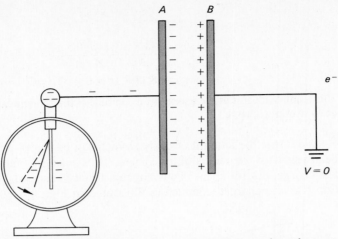

Fig. 32-4 A capacitor consists of two closely spaced conductors.

induced charge on *B,* less work is required to bring additional units of charge to conductor *A.* In other words, the capacitance of the system for holding charge has been increased by the proximity of the two conductors. Two such conductors in close proximity, carrying equal and opposite charges, constitute a *capacitor.*

> A **capacitor** consists of two closely spaced conductors carrying equal and opposite charges.

The simplest capacitor is the *parallel-plate capacitor,* illustrated in Fig. 32-4. A potential difference between two such plates can be realized by connecting a battery to them, as shown in Fig. 32-5. Electrons are transferred from plate *A* to plate *B,*

Fig. 32-5 Charging a capacitor by transferring charge from one plate to the other.

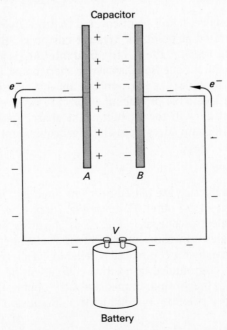

producing an equal and opposite charge on the plates. The capacitance of this arrangement is defined as follows:

The **capacitance** between two conductors having equal and opposite charges is the ratio of the magnitude of the charge on either conductor to the resulting potential difference between the two conductors.

The equation for the capacitance of a capacitor is the same as Eq. (32-1) for a single conductor, except that the symbol V now applies to the *potential difference* and the symbol Q refers to the charge on *either* conductor.

$$C = \frac{Q}{V}$$

(32-2)

$$1 \text{ F} = \frac{1 \text{ C}}{1 \text{ V}}$$

Because of the enormous size of the coulomb as a unit of charge, the farad as a unit of capacitance is usually too large for practical application. Consequently, the following submultiples are commonly used:

$$1 \text{ microfarad } (\mu\text{F}) = 10^{-6} \text{ F}$$
$$1 \text{ picofarad } (\text{pF}) = 10^{-12} \text{ F}$$

Capacitances as low as a few picofarads are not uncommon in some electrical communication applications.

EXAMPLE 32-2   A capacitor having a capacitance of 4 μF is connected to a 60-V battery. What is the charge on the capacitor?

Solution   The charge *on* a capacitor refers to the magnitude of the charge on either plate of the capacitor. From Eq. (32-2),

$$Q = CV = (4 \text{ }\mu\text{F})(60 \text{ V}) = 240 \text{ }\mu\text{C}$$

32-2

COMPUTING THE CAPACITANCE

In general, a larger conductor can hold a greater quantity of charge, and a capacitor can store more charge than a single conductor because of the inductive effect of two closely spaced conductors. The closer the spacing of these conductors, the greater the inductive effect and hence the easier it becomes to transfer additional charge from one conductor to the other. On the basis of these observations, one might suspect that *the capacitance of a given capacitor will be directly proportional to the area of the plates and inversely proportional to their separation.* The exact relationship can be determined by considering the electric field intensity between the capacitor plates.

The electric field intensity between the plates of the charged capacitor in Fig. 32-6 can be found from

$$E = \frac{V}{d}$$

(32-3)

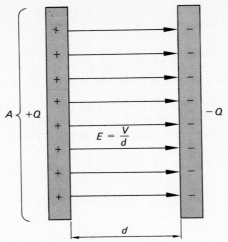

$E = \dfrac{V}{d}$

Fig. 32-6 The capacitance is directly proportional to the area of either plate and inversely proportional to the plate separation.

where V = potential difference between plates, V
$\quad\quad d$ = separation of plates, m

An alternative equation for computing the electric field intensity was derived in Chapter 30, using Gauss's law. It relates the field intensity E to the charge density σ as follows:

$$E = \frac{\sigma}{\epsilon_0} = \frac{Q}{A\epsilon_0}$$ (32-4)

where Q = charge on either plate
$\quad\quad A$ = area of either plate
$\quad\quad \epsilon_0$ = permittivity of vacuum (8.85×10^{-2} C²/N · m²)

For a capacitor with a vacuum between its plates, we combine Eqs. (32-3) and (32-4) and get

$$\frac{V}{d} = \frac{Q}{A\epsilon_0}$$

Realizing that the capacitance C is the ratio of charge to voltage, we can rearrange terms and obtain

$$\boxed{C_0 = \frac{Q}{V} = \epsilon_0 \frac{A}{d}}$$ (32-5)

The subscript 0 is used to indicate that a vacuum exists between the plates of the capacitor. To a very close approximation, Eq. (32-5) can also be used when air is between the capacitor plates.

EXAMPLE 32-3   The plates of a parallel-plate capacitor are 3 mm apart in air. If the area of each plate is 0.2 m², what is the capacitance?

Solution   Direct substitution into Eq. (32-5) gives

$$C_0 = \epsilon_0 \frac{A}{d} = \frac{(8.85 \times 10^{-12} \text{ C}^2/\text{N} \cdot \text{m}^2)(0.2 \text{ m}^2)}{3 \times 10^{-3} \text{ m}}$$

$$= 590 \times 10^{-12} \text{ F} = 590 \text{ pF}$$

Parallel-plate capacitors are frequently made with a stack of several plates by connecting alternate plates as shown in Fig. 32-7. By making one of the sets of plates movable a variable capacitor can be constructed. Rotating one set of plates relative to the other set varies the effective area of the capacitor plates, causing a variation in the capacitance. Variable capacitors are often used in the tuning circuits of radios.

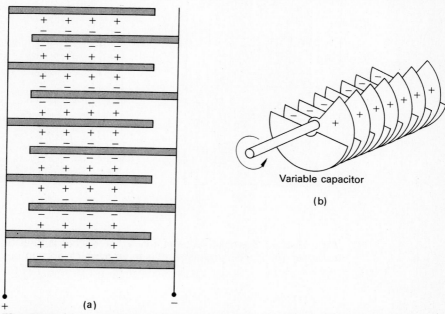

Variable capacitor

(b)

(a)

Fig. 32-7 *(a)* A capacitor consisting of a number of stacked plates, alternating with positive and negative charges. *(b)* A variable capacitor allows one set of plates to be rotated relative to the other, causing a variation in effective area.

32-4

DIELECTRIC CONSTANT; PERMITTIVITY

The amount of charge which can be put on a conductor is determined to a large degree by the dielectric strength of the surrounding medium. Similarly, the dielectric strength of the material between the plates of a capacitor limits its ability to store charge. Most capacitors have a nonconducting material, called a *dielectric,* between the plates to provide a dielectric strength greater than that for air. The following advantages are realized:

1. A dielectric material provides for a small plate separation without contact.
2. A dielectric increases the capacitance of a capacitor.

CAPACITANCE

3. Higher voltages can be used without danger of dielectric breakdown.
4. A dielectric often provides greater mechanical strength.

Common dielectric materials are mica, paraffined paper, ceramics, and plastics. Alternating sheets of metal foil and paraffin-coated paper can be rolled up to provide a compact capacitor with a capacitance of several microfarads.

In order to understand the effect of a dielectric, let us consider the insulating material of Fig. 32-8 placed between capacitor plates having a potential difference V.

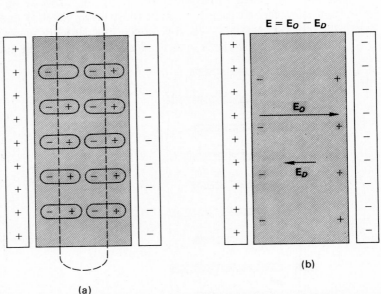

Fig. 32-8 *(a)* The polarization of a dielectric when it is inserted between the plates of a capacitor. *(b)* The polarization results in an overall reduction in the electric field intensity.

The electrons in the dielectric are not free to leave their parent atoms, but they do shift toward the positive plate. The protons and electrons of each atom align themselves as shown in the figure. The material is said to become polarized, and the atoms form *dipoles*. All the positive and negative charges inside the dotted ellipse in Fig. 32-8a neutralize each other. However, a layer of negative charge on one surface and a layer of positive charge on the other are not neutralized. An electric field E_D is set up in the dielectric *opposing* the field E_0 which would exist without the dielectric. The resulting electric field intensity is

$$E = E_0 - E_D \tag{32-6}$$

Therefore, insertion of a dielectric results in a reduction of the field intensity between the capacitor plates.

Since the potential difference V between the plates is proportional to the electric field intensity, $V = Ed$, a reduction in the intensity will cause a drop in potential difference. This fact is illustrated by the example shown in Fig. 32-9. The insertion of a dielectric causes a divergence of the gold leaf on the electroscope.

It can be seen from the definition of capacitance, $C = Q/V$, that a drop in voltage will result in an increased capacitance. If we represent the capacitance before insertion of a dielectric by C_0 and the capacitance after insertion by C, the ratio C/C_0 will denote the relative increase in capacitance. Although this ratio varies from material to material, it is constant for a particular dielectric.

> The dielectric constant K for a particular material is defined as the ratio of the capacitance C of a capacitor with the material between its plates to the capacitance C_0 for a vacuum.

$$K = \frac{C}{C_0}$$

(32-7)

The dielectric constant for various dielectric materials is given in Table 32-1 along with the dielectric strength for the materials. Note that for air K is approximately 1.

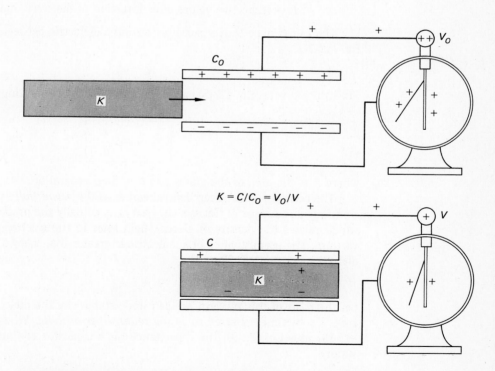

Fig. 32-9 The insertion of a dielectric between the plates of a capacitor causes a drop in potential difference resulting in an increased capacitance.

Table 32-1 Dielectric Constant and Dielectric Strength

Material	Average dielectric constant	Average dielectric strength, MV/m
Air, dry at 1 atm	1.006	3
Bakelite	7.0	16
Glass	7.5	118
Mica	5.0	200
Nitrocellulose plastics	9.0	250
Paper, paraffined	2.0	51
Rubber	3.0	28
Teflon	2.0	59
Transformer oil	4.0	16

On the basis of proportionalities, it can be shown that the dielectric constant is also given by

$$K = \frac{V_0}{V} = \frac{E_0}{E}$$ (32-8)

where V_0, E_0 = voltage and electric field with vacuum between capacitor plates
V, E = respective values after insertion of dielectric material

The capacitance C of a capacitor having a dielectric between its plates is, from Eq. (32-7),

$$C = KC_0$$

Substituting from Eq. (32-5), we have a relation for computing C directly:

$$\boxed{C = K\epsilon_0 \frac{A}{d}}$$ (32-9)

where A is the area of the plates and d is their separation.

The constant ϵ_0 has been defined earlier as the *permittivity* of a vacuum. Recall from our discussions of Gauss's law that ϵ_0 is actually the proportionality constant which relates the density of electric field lines to the electric field intensity in a vacuum. The permittivity ϵ of a dielectric is greater than ϵ_0 by a factor equal to the dielectric constant K. Thus

$$\epsilon = K\epsilon_0$$ (32-10)

On the basis of this relation, we can understand why the dielectric constant, $K = \epsilon/\epsilon_0$, is sometimes referred to as the *relative permittivity*. When we substitute Eq. (32-10) into Eq. (32-9), the capacitance for a capacitor containing a dielectric is simply

$$C = \epsilon \frac{A}{d}$$ (32-11)

This relation is the most general equation for computing capacitance. When a vacuum or air is between the capacitor plates, $\epsilon = \epsilon_0$ and Eq. (32-11) reduces to Eq. (32-5).

EXAMPLE 32-4 A certain capacitor has a capacitance of 4 μF when its plates are separated by 0.2 mm of vacant space. A battery is used to charge the plates to a potential difference of 500 V and is then disconnected from the system. (a) What will be the potential difference across the plates if a sheet of mica 0.2 mm thick is inserted between the plates? (b) What will the capacitance be after the dielectric is inserted? (c) What is the permittivity of mica?

Solution (a) The dielectric constant of mica is 5. Thus Eq. (32-8) gives

$$V = \frac{V_0}{K} = \frac{500 \text{ V}}{5} = 100 \text{ V}$$

Solution (b) From Eq. (32-7),

$$C = KC = (5)(4\mu\text{F}) = 20 \ \mu\text{F}$$

Solution (c) The permittivity is found from Eq. (32-10):

$$\epsilon = K\epsilon_0 = (5)(8.85 \times 10^{-12} \text{ C}^2/\text{N} \cdot \text{m}^2)$$
$$= 44.2 \times 10^{-12} \text{ C}^2/\text{N} \cdot \text{m}^2$$

It should be noted that the charge on the capacitor is the same before and after insertion since the voltage source did not stay connected to the capacitor.

EXAMPLE 32-5 Assume that the source of voltage stays connected to the capacitor in Example 32-4. What will be the increase in charge as a result of insertion of the mica sheet?

Solution In this instance, the voltage stays at 500 V when the dielectric is inserted. Since the capacitance is increased by the dielectric, an increase in charge must result. The charge across the capacitor initially was

$$Q_0 = C_0 V_0 = (4 \ \mu\text{F})(500 \text{ V}) = 2000 \ \mu\text{C}$$

The charge after insertion of the mica is determined by the new capacitance of 20 μF:

$$Q = CV = (20 \ \mu\text{F})(500 \text{ V}) = 10,000 \ \mu\text{C}$$

Thus the increase in charge is

$$\Delta Q = Q - Q_0 = 10,000 \ \mu\text{C} - 2000 \ \mu\text{C} = 8000 \ \mu\text{C}$$

This 8000 μC was provided by the voltage source.

32-5
CAPACITORS IN PARALLEL AND IN SERIES

Electric circuits often contain two or more capacitors grouped together. In considering the effect of such a grouping, it is convenient to resort to the *circuit diagram,* in which electrical devices are represented by symbols. Four symbols commonly used with capacitors are defined in Fig. 32-10. The high-potential side of a battery is denoted by the longer line. The high-potential side of a capacitor may be represented as a straight line, with a curved line representing the low-potential side. An arrow indicates a variable capacitor. A *ground* is an electrical connection between the

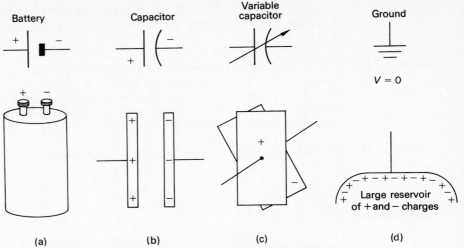

$V = 0$

Large reservoir of $+$ and $-$ charges

(a)   (b)   (c)   (d)

Fig. 32-10 Definition of symbols frequently used with capacitors.

wiring of an apparatus and its metal framework or any other large reservoir of positive and negative charges.

First we consider the effect of a group of capacitors connected along a single path, as shown in Fig. 32-11. Such a connection, in which the positive plate of one capacitor is connected to the negative plate of another, is called a *series connection*. The battery maintains a potential difference V between the positive plate of C_1 and the negative plate of C_3, transferring electrons from one to the other. The charge cannot pass between the plates of a capacitor. Therefore, all the charge inside the dotted circle in Fig. 32-11a is induced charge. For this reason, the charge on each capacitor is identical. We write

$$Q = Q_1 = Q_2 = Q_3$$

where Q is the effective charge transferred by the battery.

All three capacitors can be replaced by an equivalent capacitance C_e without changing the external effect. We now derive an expression for calculating this equivalent capacitance for the series connection. Since the potential difference between A

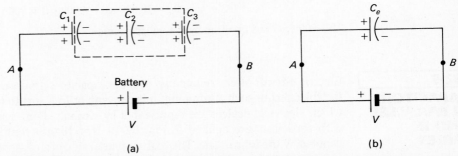

(a)   (b)

Fig. 32-11 Computing the equivalent capacitance of a group of capacitors connected in series.

and B is independent of the path, the battery voltage must equal the sum of the potential drops across each capacitor.

$$V = V_1 + V_2 + V_3 \qquad (32\text{-}12)$$

If we recall that the capacitance C is defined by the ratio Q/V, Eq. (32-12) becomes

$$\frac{Q}{C_e} = \frac{Q_1}{C_1} + \frac{Q_3}{C_2} + \frac{Q_3}{C_3}$$

For a series connection, $Q = Q_1 = Q_2 = Q_3$, so that we can divide out the charge, yielding

$$\boxed{\frac{1}{C_e} = \frac{1}{C_1} + \frac{1}{C_2} + \frac{1}{C_3}} \qquad \begin{array}{c}\textit{Series} \\ \textit{Connection}\end{array} \quad (32\text{-}13)$$

The total effective capacitance for *two* capacitors in series is

$$C_e = \frac{C_1 C_2}{C_1 + C_2} \qquad (32\text{-}14)$$

The derivation of Eq. (32-14) is left as an exercise.

Now, let us consider a group of capacitors connected so that charge can be shared between two or more conductors. When several capacitors are all connected directly to the same source of potential, as in Fig. 32-12, they are said to be connected in *parallel*. From the definition of capacitance, the charge on each parallel capacitor is

$$Q_1 = C_1 V_1 \qquad Q_2 = C_2 V_2 \qquad Q_3 = C_3 V_3$$

The total charge Q is equal to the sum of the individual charges.

$$Q = Q_1 = Q_2 + Q_3 \qquad (32\text{-}15)$$

The equivalent capacitance of the entire circuit is $Q = CV$, so that Eq. (32-15) becomes

$$CV = C_1 V_1 + C_2 V_2 + C_3 V_3 \qquad (32\text{-}16)$$

For a parallel connection,

$$V = V_1 = V_2 = V_3$$

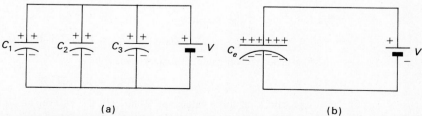

(a)   (b)

Fig. 32-12 Equivalent capacitance of a group of capacitors connected in parallel.

since all capacitors are connected to the same potential difference. Hence the voltages divide out of Eq. (32-16), giving

$$C = C_1 + C_2 + C_3$$

Parallel Connection (32-17)

EXAMPLE 32-6   (a) Find the equivalent capacitance of the circuit illustrated in Fig. 32-13a. (b) Determine the charge on each capacitor. (c) What is the voltage across the 4-μF capacitor?

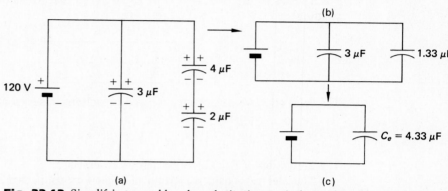

Fig. 32-13 Simplifying a problem by substituting equivalent values for capacitance.

Solution (a)   The 4- and 2-μF capacitors are in series. Their combined capacitance is found from Eq. (32-14).

$$C_{2,4} = \frac{C_2 C_4}{C_2 + C_4} = \frac{(2\ \mu F)(4\ \mu F)}{2\mu F + 4\ \mu F}$$
$$= 1.33\ \mu F$$

These two capacitors can be replaced by their equivalent capacitance, as shown in Fig. 32-13b. The two remaining capacitors are in parallel. Thus the equivalent capacitance is

$$C_e = C_3 + C_{2,4} = 3\ \mu F + 1.33\ \mu F$$
$$= 4.33\ \mu F$$

Solution (b)   The total charge in the network is

$$Q = C_e V = (4.33\ \mu F)(120\ V) = 520\ \mu C$$

The charge Q_3 on the 3-μF capacitor is

$$Q_3 = C_3 V = (3\ \mu F)(120\ V) = 360\ \mu C$$

The remainder of the charge,

$$Q - Q_3 = 520\ \mu C - 360\ \mu C = 160\ \mu C$$

must be deposited on the series capacitors. Hence

$$Q_2 = Q_4 = 160\ \mu C$$

As a check on these values for Q_2 and Q_4, the equivalent capacitance of the two series capacitors can be multiplied by the voltage drop across it:

$$Q_{2,4} = C_{2,4}\ V = (1.33\ \mu F)(120\ V) = 160\ \mu C$$

Solution (c)   The voltage drop across the 4-μF capacitor is

$$V_4 = \frac{Q_4}{C_4} = \frac{160\ \mu C}{4\ \mu F} = 40\ V$$

The remaining 80 V is across the 2-μF capacitor.

32-6
ENERGY OF A CHARGED CAPACITOR

Consider a capacitor that is initially uncharged. When a source of potential difference is connected to the capacitor, the potential difference between the plates increases as charge is transferred. As more and more charge builds up on the capacitor, it becomes increasingly difficult to transfer additional charge. Suppose we represent the total charge transferred by Q and the final potential difference by V. The *average* potential difference through which the charge is moved is given by

$$V_{av} = \frac{V_{final} + V_{initial}}{2} = \frac{V + 0}{2} = \frac{1}{2}V$$

Since the total charge transferred is Q, the total work done against electric forces is equal to the product of Q and the average potential difference V_{av}. Thus

$$\text{Work} = Q(\tfrac{1}{2}V) = \tfrac{1}{2}QV$$

This work is equivalent to the electrostatic potential energy of a charged capacitor. From the definition of capacitance ($Q = CV$), this potential energy can be written in alternative forms:

$$\begin{aligned} \text{P.E.} &= \tfrac{1}{2}QV \\ &= \tfrac{1}{2}CV^2 \\ &= \frac{Q^2}{2C} \end{aligned} \tag{32-18}$$

When C is in farads, V is in volts, and Q is in coulombs, the potential energy will be in joules. These equations apply equally to all capacitors regardless of their construction.

SUMMARY

Stored electric charge is a necessity if large quantities of electrical energy are to be delivered on demand to a modern industrial world. We have studied in this chapter the very basic principles which determine the amount of charge that can be stored on capacitors. Further we have discussed the insertion of capacitors into electric circuits and the factors affecting the distribution of charge in such circuits. The fundamental concepts are summarized below:

• Capacitance is the ratio of charge Q to the potential V for a given conductor. For two oppositely charged plates, the Q refers to the charge on either plate and the V refers to the potential difference between the plates.

$$C = \frac{Q}{V} \qquad 1\ \text{farad (F)} = \frac{1\ \text{coulomb (C)}}{1\ \text{volt (V)}} \qquad \textit{Capacitance}$$

CAPACITANCE

- The dielectric strength is that value for E for which a given material ceases to be an insulator and becomes a conductor. For air this value is

$$E = \frac{kQ}{r^2} = 3 \times 10^6 \text{ N/C}$$ *Dielectric Strength, Air*

- For a parallel-plate capacitor, the material between the plates is called the dielectric. The insertion of such a material has an effect on the electric field and the potential between the plates. Consequently, it changes the capacitance. The dielectric constant K for a particular material is the ratio of the capacitance with the dielectric C to the capacitance for a vacuum C_0.

$$K = \frac{C}{C_0}$$ $$K = \frac{V_0}{V}$$ $$K = \frac{E}{E_0}$$ *Dielectric Constant*

- The permittivity of a dielectric is greater than the permittivity of a vacuum by a factor equal to the dielectric constant. For this reason, K is sometimes referred to as the *relative permittivity.*

$$K = \frac{\epsilon}{\epsilon_0}$$ $$\epsilon = K\epsilon_0$$ $$\epsilon_0 = 8.85 \times 10^{-12} \text{ C}^2/\text{N} \cdot \text{m}^2$$

- The capacitance for a parallel-plate capacitor depends on the surface area A of each plate, the plate separation d, and the permittivity or dielectric constant. The general equation is

$$C = \epsilon \frac{A}{d}$$ $$C = K\epsilon_0 \frac{A}{d}$$ *Capacitance*

For a vacuum, $K = 1$, in the above relationship.
- Capacitors may be connected in series as shown in Fig. 32-11 or in parallel as shown in Fig. 32-12.
  (a) For *series connections*, the charge on each capacitor is the same as the total charge, the potential difference across the battery is equal to the sum of the drops across each capacitor, and the net capacitance is found from

$$Q_T = Q_1 = Q_2 = Q_3$$ $$V_T = V_1 + V_2 + V_3$$

$$\frac{1}{C_e} = \frac{1}{C_1} + \frac{1}{C_2} + \frac{1}{C_3}$$ *Series Connections*

  (b) For *parallel connections*, the total charge is equal to the sum of the charges across each capacitor, the voltage drop across each capacitor is the same as the drop across the battery, and the effective capacitance is equal to the sum of the individual capacitances.

$$Q_T = Q_1 + Q_2 + Q_3 \qquad V_B = V_1 = V_2 = V_3$$

$$C_e = C_1 + C_2 + C_3 \qquad \textit{Parallel Connections}$$

- The potential energy stored in a charged capacitor can be found from any of the following relationships:

$$\text{P.E.} = \tfrac{1}{2}QV \qquad \text{P.E.} = \tfrac{1}{2}CV^2 \qquad \text{P.E.} = \frac{Q^2}{2C}$$

When C is in *farads*, V is in *volts*, and Q is in *coulombs*, the potential energy will be in *joules*.

QUESTIONS

32-1. Define the following terms:
   a. Capacitor
   b. Capacitance
   c. Dielectric
   d. Permittivity
   e. Farad
   f. Corona discharge
   g. Variable capacitor
   h. Dielectric strength
   i. Dielectric constant
   j. Parallel connection
   k. Series connection

32-2. Discuss several factors which limit the ability of a conductor to store charge.

32-3. Air is pumped from one metal tank to another, creating a partial vacuum in one tank and a high-pressure in the other. When the pump is removed, potential energy is stored. The energy is released if the two tanks are reconnected and the pressure in each tank becomes equal. In what ways is this mechanical example analogous to charging and discharging a capacitor?

32-4. Large sparks are often seen jumping from the leather belts driving machinery. Explain.

32-5. The Leyden jar is a capacitor which consists of a glass jar coated inside and out with tinfoil, as shown in Fig. 32-14. Contact with the inside coating is made with a metal chain connected to the central metal rod. From the figure, explain how the capacitor becomes charged. What is the function of the ground wire? What purpose does the glass serve?

32-6. May lightning be considered a capacitor discharge? Explain.

32-7. Discuss the following statement: the permittivity is a measure of how easily a dielectric will permit the establishment of electric field lines within the dielectric.

32-8. A dielectric with a larger permittivity allows for the storage of greater quantities of charge. Explain.

32-9. Distinguish the dielectric strength of a material from its dielectric constant. What part does each play in the design of a capacitor?

32-10. The term *breakdown voltage* is often used in electronics for capacitors. How would you define such a term? In what ways does it differ from the dielectric strength?

32-11. If two point charges are surrounded by a dielectric, will the force each exerts on the other be reduced or increased?

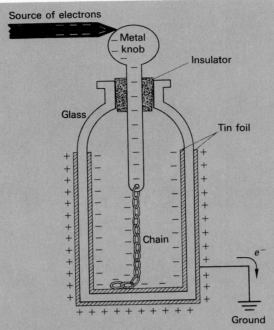

Source of electrons

Metal knob

Insulator

Glass

Tin foil

Chain

e^-

Ground

Fig. 32-14 The Leyden jar.

32-12. The unit of permittivity is the $C^2/N \cdot m^2$. Show that the permittivity can be expressed as farads per meter.

32-13. Prove that each of the expressions for potential energy, as given in Eq. (32-18), will yield a proper unit of energy (the joule).

PROBLEMS

Capacitance

32-1. What is the maximum charge that can be placed on a metal sphere 30 mm in diameter and surrounded by air?

Ans. 75 nC

32-2. How much charge can be placed on a metal sphere of radius 40 mm if it is immersed in transformer oil?

32-3. What would be the radius of a metal sphere such that it could theoretically hold a charge of one coulomb in air?

Ans. 54.8 m

* 32-4. Write an equation for the potential at the surface of a sphere of radius r in terms of the permittivity of the surrounding medium. Show that the capacitance of such a sphere is given by $C = 4\pi\epsilon r$.

32-5. A parallel-plate capacitor has a capacitance of 28 μF. How much charge will be stored by this capacitor when it is connected to a 120-V source of potential difference?

Ans. 3360 μC

32-6. Find the capacitance of a parallel-plate capacitor if 1600 μC of charge is on each plate when the potential difference is 80 V.

32-7. A potential difference of 110 V is applied across the plates of a capacitor. If the charge on each plate is 1200 μC, what is the capacitance?

Ans. 10.9 μF

32-8. What potential difference is required to store a charge of 800 μC on a 40-μF capacitor?

Calculating the Capacitance

32-9. The plates of a certain capacitor are 3 mm apart and have an area of 0.04 m^2. For a dielectric of air, find (a) the capacitance, (b) the electric field intensity between the plates, and (c) the charge on each plate if 200 V is applied to the capacitor.

Ans. (a) 118 pF, (b) 66.7 kN/C, (c) 23.6 nC.

32-10. Answer the questions of Prob. 32-9 if mica ($K = 5.0$) replaces air as the dielectric.

32-11. Find the capacitance of a parallel-plate capacitor if the area of each plate is 0.08 m^2 and the separation of the plates is 4 mm if (a) the dielectric is air, (b) the dielectric is paraffined paper ($K = 2.0$).

Ans. (a) 177 pF, (b) 354 pF

32-12. Two parallel plates of a capacitor are 4.6 mm apart and each plate has an area of 0.03 m^2. The capacitor has a dielectric of glass ($K = 7.5$). (a) What is the capacitance? (b) What is the field intensity between the plates? (c) What is the charge on each plate? The plate voltage is 800 V.

* 32-13. A certain capacitor has a capacitance of 12 μF when its plates are separated by 0.3 mm of vacant space. A battery is used to charge the plates to a potential difference of 400 V and is then disconnected from the system. (a) What will be the potential difference across the plates if a sheet of Bakelite ($K = 7.0$) fills the space between the plates? (b) What is the capacitance after the Bakelite is inserted? (c) What is the permittivity of Bakelite?

Ans. (a) 57.1 V, (b) 84.0 μF, (c) 6.20×10^{-11} C/N $\cdot$ m^2

Capacitors in Series and in Parallel

32-14. Find the equivalent capacitance of a 6-μF capacitor and a 12-μF capacitor connected (a) in series, (b) in parallel.

32-15. A 6-μF capacitor is connected in series with a 15-μF capacitor. What is the effective capacitance (a) for a series connection, (b) for a parallel connection?

Ans. (a) 4.29 μF, (b) 21 μF

* 32-16. Three capacitors A, B, and C have capacitances of 4, 7, and 12 μF, respectively. What is the equivalent capacitance if they are connected in series? In parallel?

* 32-17. Find the equivalent capacitance of a 6-μF capacitor if it is connected in series with two parallel capacitors whose capacitances are 5 and 4 μF.

Ans. 3.6 μF

* 32-18. Find the equivalent capacitance for the entire circuit shown in Fig. 32-15.

* 32-19. A 6-μF capacitor and a 3-μF capacitor are connected to a 24-V battery. If they are connected in series, what are the charge and voltage across each capacitor?

Ans. $V_3 = 16$ V, $Q_3 = 48$ μC, $Q_6 = 48$ μC, $V_6 = 8$ V

Fig. 32-15

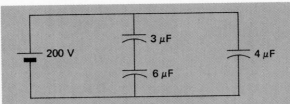

* 32-20. If the capacitors in Prob. 32-19 are reconnected in parallel, what will be charge and voltage across each capacitor?

* 32-21. Compute the equivalent capacitance for the circuit shown is Fig. 32-16. What is the total charge on the equivalent capacitance?

Ans. 1.74 μF, 20.9 μC

Fig. 32-16

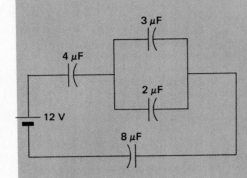

* 32-22. What are the charge and voltage across each capacitor of Fig. 32-16?

The Energy of a Charged Capacitor

32-23. What is the potential energy stored in the electric field of a 200-μF capacitor when it is charge to a voltage of 2400 V?

Ans. 576 J

32-24. What is the energy stored on a 25-μF capacitor when the charge on each plate is 2400 μC? What is the voltage across the capacitor?

32-25. How much work is required to charge a capacitor to a potential difference of 30 kV if 800 μC are on each plate?

Ans. 12 J

Additional Problems

* 32-26. Four capacitors A, B, C, and D have capacitances of 12, 16, 20, and 26 μF, respectively. Capacitors A and B are connected in parallel. The combination is then connected in series with C and D. What is the effective capacitance?

* 32-27. Consider the circuit drawn in Fig. 32-17. (a) What is the equivalent capacitance of the circuit? (b) What are the charge and voltage across the 2-μF capacitor?

Ans. (a) 6.0 μF; (b) 18 μC, 9.0 V

Fig. 32-17

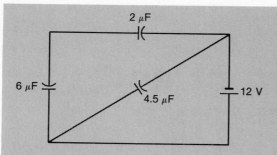

* 32-28. Two identical, 20-μF air capacitors A and B are connected in parallel with a 12-V battery. What is the charge on each capacitor if a sheet of porcelain ($K = 6$) is inserted between the plates of capacitor B and the battery remains connected?

* 32-29. A 4-μF air capacitor is connected to a 500-V source of potential difference. The capacitor is then disconnected from the source and a sheet of mica ($K = 5.0$) is inserted between its plates. What is the new voltage on the capacitor? Now suppose the 500-V source is reconnected and equilibrium is reached. What will be the final charge on the capacitor? What is its potential energy?

  Ans. 100 V; 10,000 μC; 2.5 J

32-30. What voltage will cause a capacitor with a dielectric of glass ($K = 7.5$) to break down if the plate separation is 4 mm?

* 32-31. Three capacitors A, B, and C have respective capacitances of 2, 4, and 6 μF. Compute the equivalent capacitance if they are connected in series with an 800-V source. What are the charge and voltage across the 4-μF capacitor?

  Ans. 1.09 μF, 873 μC, 218 V

* 32-32. Suppose the capacitors of Prob. 32-31 are reconnected in parallel with the 800-V source. What is the equivalent capacitance? What are the charge and voltage across the 4-μF capacitor?

32-33. What voltage will cause a parallel-plate capacitor with a dielectric of air 3 mm thick to break down? What is the breakdown voltage if the dielectric is changed to nitrocellulose plastics ($K = 9.0$)?

  Ans. 9 kV, 750 kV

* 32-34. Show that the total capacitance of a multiple-plate capacitor containing N plates separated by air is given by

$$C_0 = \frac{(N - 1)\epsilon_0 A}{d}$$

where A is the area of each plate and d is the separation of each plate.

* 32-35. A capacitor is formed from 30 parallel plates 20 $\times$ 20 cm. If each plate is separated by 2 mm of dry air, what is the total capacitance?

  Ans. 5.13 nF

33 Current and Resistance

After completing this chapter, you should be able to:

1. Demonstrate by definition and example your understanding of *electric current* and *electromotive force.*
2. Write and apply *Ohm's law* to the solution of problems involving electric resistance.
3. Compute *power losses* as a function of voltage, current, and resistance.
4. Define the *resistivity* of a material and solve problems similar to those in the text.
5. Define the *temperature coefficient of resistance* and calculate the change in resistance which occurs with a change in temperature.

We now leave electrostatics and enter a discussion of charges in motion. We have been concerned with forces, electric fields, and potential energies as they relate to charged conductors. For example, excess electrons, evenly distributed over an insulated spherical surface, will remain at rest. However, if a wire is connected from the sphere to ground, the electrons will flow through the wire to the ground. The flow of charge constitutes an *electric current*. In this chapter the foundation is laid for a study of direct currents and electric resistance.

33-1
THE MOTION OF ELECTRIC CHARGE

Let us begin our discussion of moving charges by considering the discharge of a capacitor. The potential difference V between the two capacitor plates in Fig. 33-1a is indicated by the electroscope. The total charge Q on either plate is given by

$$Q = CV$$

where C is the capacitance. If a path is provided, electrons on one plate will travel to the other, decreasing the net charge and causing a drop in the potential difference. Thus a drop in potential, as indicated by the collapsing leaf of the electroscope, means that charge has been transferred. Any conductor used to connect the plates of

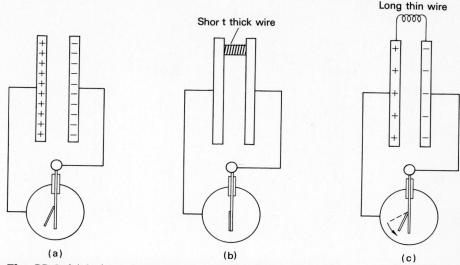

Fig. 33-1 *(a)* A charged capacitor is a source of current. *(b)* If the capacitor plates are joined by a short, thick wire, the capacitor will discharge instantly. *(c)* A long, thin wire allows for a gradual discharge.

a capacitor will cause it to discharge. However, the rate of discharge varies considerably with the size, shape, material, and temperature of the conductor.

If a short, thick wire is connected between the plates of the capacitor, as shown in Fig. 33-1*b*, the electroscope leaf collapses instantly, indicating a very rapid transfer of charge. This current, which exists for a very short time, is called a *transient current*. If we replace the short, thick wire with a long, thin wire of the same material, we shall observe a gradual collapse of the electroscope leaf (Fig. 33-1*c*). Such opposition to the flow of electricity is called electric *resistance*. A quantitative description of electric resistance will be presented in a later section. It is introduced here to illustrate that the rate at which charge flows through a conductor varies. This rate is referred to as the *electric current*.

> The electric current *I* is the rate of flow of charge *Q* past a given point *P* on an electric conductor.

$$I = \frac{Q}{t}$$

(33-1)

The unit of electric current is the *ampere*. One *ampere* (A) represents a flow of charge at the rate of *one coulomb per second* past any point.

$$1 \text{ A} = \frac{1 \text{ C}}{1 \text{ s}}$$

In the example of a discharging capacitor, the current arises from the motion of electrons, as illustrated in Fig. 33-2. The positive charges in a wire are tightly bound and cannot move. The electric field created in the wire because of the potential

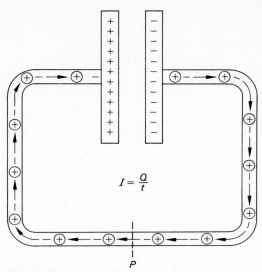

$$I = \frac{Q}{t}$$

Fig. 33-2 Current arises from the motion of electrons and is a measure of the quantity of charge passing a given point in a unit of time.

difference between the plates causes the free electrons in the wire to experience a drift toward the positive plate. The electrons are repeatedly deflected or stopped by processes relating to impurities and thermal motions of the atoms. Consequently, the motion of the electrons is not an accelerated one but a drifting or diffusion process. The average drift velocity of electrons is typically of the order of 4 m/h. This velocity of charge, which is a *distance* per unit of time, should not be confused with current, which is a *quantity* of charge per unit of time.

An analogy to water flowing through a pipe is useful in understanding current flow. The rate of flow of the water in gallons per minute is analogous to the rate of flow of charge in coulombs per second. For a current of 1 A, 6.25×10^{18} electrons (1 C) flow past a given point every second. Just as the size and length of a pipe affect the flow of water, the size and length of a conductor affect the flow of electrons.

EXAMPLE 33-1   How many electrons pass a point in 5 s if a constant current of 8 A is maintained in a conductor?

Solution   From Eq. (33-1),

$$Q = It = (8 \text{ A})(5 \text{ s})$$
$$= (8 \text{ C/s})(5 \text{ s}) = 40 \text{ C}$$
$$= (40 \text{ C})(6.25 \times 10^{18} \text{ electrons/C})$$
$$= 2.50 \times 10^{20} \text{ electrons}$$

33-2

THE DIRECTION OF ELECTRIC CURRENT

Thus far, we have discussed only the magnitude of electric current. The choice of direction is purely arbitrary as long as we apply our definition consistently. The flow of charge caused by an electric field in a gas or a liquid consists of a flow of positive ions in the direction of the field or a flow of electrons opposite to the field direction. As we have seen, the current in a metallic material consists of electrons flowing against the field direction. However, a current that consists of negative particles moving in one direction is electrically the same as a current consisting of positive charges moving in the opposite direction.

There are a number of reasons for preferring the motion of positive charge as an indicator of direction. In the first place, all the concepts introduced in electrostatics were defined in terms of positive charges, e.g., the electric field, potential energy, and potential difference. An electron flows contrary to the electric field and "up a potential hill" from the negative plate to the positive plate. If we define current as a flow of *positive* charge, the loss in energy as charge encounters resistance will be from + to − or "down a potential hill." By convention, we consider all currents as consisting of a flow of positive charge.

The direction of conventional current is always the same as the direction in which positive charges would move, even if the actual current consists of a flow of electrons.

For a metallic conducting wire, both electron flow and conventional current are indicated in Fig. 33-3. The zigzag line is used to indicate the electric resistance R. Note that the conventional current flows from the positive plate of the capacitor, neutralizing negative charge on the other plate. Conventional current is in the same direction as the electric field E producing the current.

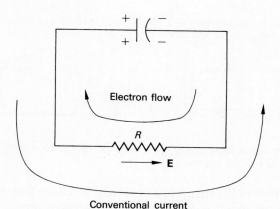

Conventional current

Fig. 33-3 In a metallic conductor, the conventional current is in a direction opposite to the actual flow of electrons.

33-3

ELECTRO-MOTIVE FORCE

The currents discussed in the preceding sections were called *transient currents* because they exist only for a short time. Once the capacitor has been completely discharged, there will no longer be a potential difference to promote the flow of additional charge. If some means were available to keep the capacitor continually charged, a continuous current could be maintained. This would require that electrons be continuously supplied to the negative plate to replace those leaving. In other words, energy must be supplied to replace the energy lost by the charge in the external circuit. In this manner, the potential difference between the plates could be maintained, allowing for a continuous flow of charge. A device with the ability to maintain potential difference between two points is called a *source of electromotive force* (emf).

The most familiar sources of emf are batteries and generators. Batteries convert chemical energy into electric energy, and generators transform mechanical energy into electric energy. The detailed nature and operation of these devices will be discussed in a later chapter.

> A **source of electromotive force (emf)** is a device which converts chemical, mechanical, or other forms of energy into the electric energy necessary to maintain a continuous flow of electric charge.

In an electric circuit, the source of emf is usually represented by the symbol $\mathcal{E}$.

The function of a source of emf in an electric circuit is similar to the function of a water pump in maintaining the continuous flow of water through a system of pipes. In Fig. 33-4*a* the water pump must perform the work on each unit volume of water

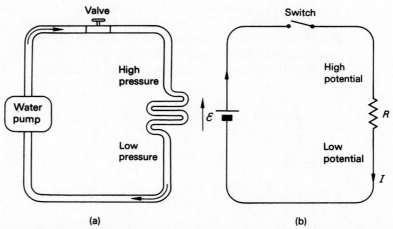

Fig. 33-4 The mechanical analogy of a water pump can be used to explain the function of a source of emf in an electric circuit.

necessary to replace the energy lost by each unit volume flowing through the pipes. In Fig. 33-4*b* the source of emf must do work on each unit of charge which passes through it in order to raise it to a higher potential. This work must be supplied at a rate equal to the rate at which energy is lost in flowing through the circuit.

By convention, we have assumed that the current consists of a flow of positive charge even though in most cases it is negative electrons. Therefore, the charge loses energy in passing through the resistor from a high potential to a low potential. In the hydraulic analogy, water passes from high pressure to low pressure. When the shut-off valve is closed, pressure exists but there is no water flow. Similarly, when the electric switch is *open*, there is voltage but no current.

Since emf is work per unit charge, it is expressed in the same unit as potential difference, i.e., the *joule per coulomb*, or *volt*.

> A source of emf of one volt will perform one joule of work on each coulomb of charge which passes through it.

For example, a 12-V battery performs 12 J of work on each coulomb of charge transferred from the low-potential side (− terminal) to the high-potential side (+ terminal). An arrow (↑) is usually drawn next to the symbol $\mathcal{E}$ for an emf to indicate the direction in which the source, acting alone, would cause a positive charge to move through the external circuit. The conventional current is directed away from the + terminal of a battery, and the hypothetical positive charge flows "downhill" through external resistance to the − terminal of the battery.

In the following sections, circuit diagrams, like that in Fig. 33-4b, will frequently be used to describe electric systems. Many of the symbols we shall use are defined in Fig. 33-5.

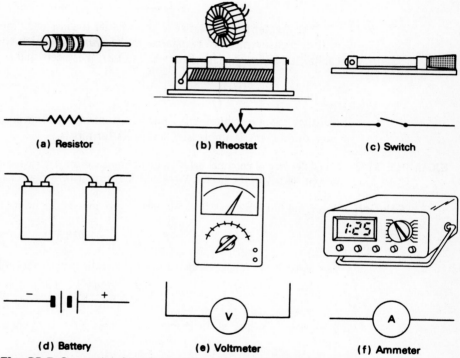

(a) Resistor   (b) Rheostat   (c) Switch

(d) Battery   (e) Voltmeter   (f) Ammeter

Fig. 33-5 Conventional symbols used in electric circuit diagrams.

33-4

**OHM'S LAW;
RESISTANCE**

Resistance (R) is defined as the opposition to the flow of electric charge. Although most metals are good conductors of electricity, all offer some opposition to the surge of electric charge through them. This electric resistance is fixed for many specific materials of known size, shape, and temperature. It is independent of the applied emf and the current passing through it.

The effects of resistance in limiting the flow of charge were first studied quantitatively by Georg Simon Ohm in 1826. He discovered that, *for a given resistor at a particular temperature, the current is directly proportional to the applied voltage.* Just as the rate of flow of water between two points depends on the difference of height between them, the rate of flow of electric charge between two points depends on the difference in potential between them. This proportionality is usually stated as *Ohm's law:*

> *The current produced in a given conductor is directly proportional to the difference of potential between its end points.*

The current I which is observed for a given voltage V is therefore an indication of resistance. Mathematically, the resistance R of a given conductor can be calculated from

$$R = \frac{V}{I} \qquad V = IR \tag{33-2}$$

The greater the resistance R, the smaller the current I for a given voltage V. The unit of measurement of resistance is the *ohm*, for which the symbol is the Greek capital letter *omega* (Ω). From Eq. (33-2), it is seen that

$$1\ \Omega = \frac{1\ \text{V}}{1\ \text{A}}$$

A resistance of *one ohm* will allow a current of *one ampere* when a potential difference of *one volt* is impressed across its terminals.

EXAMPLE 33-2 The difference of potential between the terminals of an electric heater is 80 V when there is a current of 6 A in the heater. What will the current be if the voltage is increased to 120 V?

Solution According to Ohm's law, the resistance of the coils in the heater is

$$R = \frac{V}{I} = \frac{80\ \text{V}}{6\ \text{A}} = 13.3\ \Omega$$

Therefore, if the voltage is increased to 120 V, the new current will be

$$I = \frac{V}{R} = \frac{120\ \text{V}}{13.3\ \Omega} = 9\ \text{A}$$

Here we have neglected any change in resistance due to a rise in temperature of the heating coils.

Four devices commonly used in the laboratory to study Ohm's law are the battery, the voltmeter, the ammeter, and the rheostat. As their names imply, the voltmeter and ammeter are devices to measure voltage and current. The rheostat is simply a variable resistor. A sliding contact changes the number of resistance coils through which charge can flow. A laboratory collection of these electrical devices is illustrated in Fig. 33-6. You should study the circuit diagram in Fig. 33-6*a* and justify the electrical connections shown pictorially in Fig. 33-6*b*. Note that the voltmeter is connected in parallel with the battery whereas the ammeter is connected in series. In general, the positive terminals are color-coded red, and the negative terminals are black.

EXAMPLE 33-3 The apparatus shown in Fig. 33-6 is used to study Ohm's law in the laboratory. The voltage V is determined by the source of emf and remains at 6 V. (*a*) What is the resistance when the

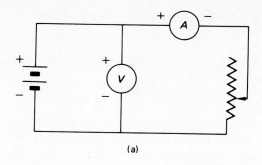

(a)

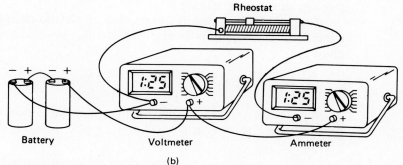

(b)

Fig. 33-6 *(a)* A circuit diagram for studying Ohm's law. *(b)* A pictorial diagram showing how the various elements are connected in the laboratory.

rheostat is varied to indicate a current of 0.4 A? *(b)* If this resistance is doubled, what will the new current be?

Solution (*a*)   From Ohm's law,

$$R = \frac{V}{I} = \frac{6\text{ V}}{0.4\text{ A}} = 15\ \Omega$$

Solution (*b*)   Doubling the resistance to 30 Ω would result in a current given by

$$I = \frac{6\text{ V}}{30\ \Omega} = 0.2\text{ A}$$

Note that doubling the resistance in a circuit reduces the current by one-half.

33-5
ELECTRIC POWER AND HEAT LOSS

We have seen that electric charge gains energy within a generating source of emf and loses energy in passing through external resistance. Inside the source of emf, work is done *by the source* in raising the potential energy of charge. As the charge passes through the external circuit, work is done *by* the charge on the components of the circuit. In the case of a pure resistor, the energy is dissipated in the form of heat. If a motor is attached to the circuit, the energy loss is divided between heat and useful work. In any case, the energy gained in the source of emf must equal the energy lost in the entire circuit.

Let us examine the work accomplished inside a source of emf more closely. By definition, *one joule* of work is accomplished for each *coulomb* of charge moved through a potential difference of *one volt*. Thus

$$\text{Work} = Vq \tag{33-3}$$

where q is the quantity of charge transferred during a time t. But $q = It$, so that Eq. (33-3) becomes

$$\text{Work} = VIt \tag{33-4}$$

where I is the current in *coulombs per second*. This work represents the energy gained by charge in passing through the source of emf during the time t. An equivalent amount of energy will be dissipated in the form of heat as the charge moves through an external resistance.

The rate at which heat is dissipated in an electric circuit is referred to as the *power loss*. When charge is flowing continuously through a circuit, this power loss is given by

$$P = \frac{\text{work}}{t} = \frac{VIt}{t} = VI \tag{33-5}$$

When V is in volts and I is in amperes, the power loss is measured in watts. That the product of voltage and current will give a unit of power is shown as follows:

$$(\text{V})(\text{A}) = \frac{\text{J}}{\text{C}} \frac{\text{C}}{\text{s}} = \frac{\text{J}}{\text{s}} = \text{W}$$

Equation (33-5) can be expressed in alternative forms by using Ohm's law ($V = IR$). Substituting for V, we can write

$$P = VI = I^2R \tag{33-6}$$

Substitution for I in Eq. (33-6) gives another variation:

$$P = VI = \frac{V^2}{R} \tag{33-7}$$

The relation expressed by Eq. (33-6) is so often used in electrical work that heat loss in electrical wiring is often referred to as an "*I*-squared-*R*" loss.

EXAMPLE 33-4   A current of 6 A flows through a resistance of 300 Ω for 1 h. What is the power loss? How much heat is generated in joules?

Solution   From Eq. (33-6),

$$P = I^2R = (6 \text{ A})^2(300 \text{ Ω}) = 10,800 \text{ W}$$

Since the power represents the heat dissipated per unit of time, we obtain

$$\text{Work} = Pt = (10,800 \text{ W})(3600 \text{ s})$$
$$\text{Heat lost} = 3.89 \times 10^7 \text{ J}$$

This represents about 36,900 Btu.

33-6

RESISTIVITY

Just as capacitance is independent of the voltage and quantity of charge, the resistance of a conductor is independent of current and voltage. Both capacitance and resistance are inherent properties of a conductor. The resistance of a wire of uniform cross-sectional area, like the one shown in Fig. 33-7, is determined by the following four factors:

1. The kind of material
2. The length
3. The cross-sectional area
4. The temperature

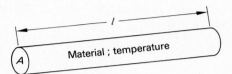

Fig. 33-7 The resistance of a wire depends on the kind of material, the length, the cross-sectional area, and the temperature of the wire.

Ohm, the German physicist who discovered the law that now bears his name, also reported that *the resistance of a conductor at a given temperature is directly proportional to its length, inversely proportional to its cross-sectional area, and dependent upon the material from which it is made.* For a given conductor at a given temperature, the resistance can be computed from

$$R = \rho \frac{l}{A}$$

(33-8)

where R = resistance
l = length
A = area

The proportionality constant ρ is a property of the material called its *resistivity,* given by

$$\rho = \frac{RA}{l}$$

(33-9)

It varies considerably with different materials and is also affected by changes in temperature. When R is in ohms, A is in square meters, and l is in meters, the unit of resistivity is the ohm-meter ($\Omega \cdot$ m):

$$\frac{(\Omega)(m^2)}{m} = \Omega \cdot m$$

Table 33-1 lists the resistivities of several common metals.

EXAMPLE 33-5   What is the resistance of a 20-m length of copper wire with a diameter of 0.8 mm?

Solution   We first compute the cross-sectional area of the wire in square meters.

$$A = \frac{\pi D^2}{4} = \frac{\pi(8 \times 10^{-4} \text{ m})^2}{4} = 5.03 \times 10^{-7} \text{ m}^2$$

The resistivity of copper is $1.72 \times 10^{-8}\ \Omega \cdot m$. Substitution into Eq. (33-8) gives

$$R = \rho \frac{l}{A} = \frac{(1.72 \times 10^{-8}\ \Omega \cdot m)(20\ m)}{5.03 \times 10^{-7}\ m^2}$$

$$= 0.684\ \Omega$$

In many engineering applications, the resistivity is given in mixed units. The length is measured in feet, and the area is measured in circular mils (cmil).

> One **circular mil** (cmil) is defined as the cross-sectional area of a wire 1 mil (0.001 in.) in diameter.

Table 33-1 Resistivities of Various Materials at 20°C

	Resistivity	
	$\Omega \cdot m$	$\Omega \cdot$ cmil/ft
Aluminum	2.8×10^{-8}	17.0
Constantan	49×10^{-8}	295
Copper	1.72×10^{-8}	10.4
Gold	2.2×10^{-8}	13.0
Iron	9.5×10^{-8}	57.0
Nichrome	100×10^{-8}	600
Tungsten	5.5×10^{-8}	33.2
Silver	1.63×10^{-8}	9.6

In order to calculate the area of a wire in circular mils, we first convert its diameter to mils. Since 1 mil is 0.001 in., a diameter can be converted from inches to mils simply by moving the decimal three places to the right. For example,

$$0.128\ in. = 128.0\ mils$$

The area of a wire in *square mils* is found from

$$A = \frac{\pi D^2}{4} \quad \text{square mils} \tag{33-10}$$

However, by definition, a wire having a diameter of 1 mil has an area of 1 cmil. Thus

$$1\ cmil = \frac{\pi}{4} \quad \text{square mils} \tag{33-11}$$

Comparing Eqs. (33-10) and (33-11), it can be seen that *the area in circular mils equals the square of the diameter in mils.* Symbolically,

$$A_{\text{cmils}} = (D_{\text{mils}})^2 \tag{33-12}$$

If the area of a wire is expressed in circular mils and its length in feet, the unit for resistivity from Eq. (33-9) is

$$\rho = \frac{RA}{l} \rightarrow \frac{\Omega \cdot \text{cmil}}{\text{ft}}$$

Table 33-1 also lists the resistivity of materials in the *mil-foot system* of units.

EXAMPLE 33-6   What length of aluminum wire 0.025 in. in diameter is required to construct a 12-Ω resistor?

Solution   The diameter is 25 mils, and the area is found from Eq. (33-12).

$$A_{\text{cmils}} = (D_{\text{mils}})^2 = (25 \text{ mils})^2 = 625 \text{ cmils}$$

From Eq. (33-8), the length required is

$$l = \frac{RA}{\rho} = \frac{(12 \ \Omega)(625 \text{ cmils})}{17 \ \Omega \cdot \text{cmils/ft}} = 441 \text{ ft}$$

The resistivity ρ was taken from Table 33-1.

33-7
TEMPERATURE COEFFICIENT OF RESISTANCE

For most metallic conductors, the resistance tends to increase as the temperature increases. The increased atomic and molecular movement in the conductor hinders the flow of charge. The increase in resistance for most metals is approximately linear when compared with temperature changes. Experiments have shown that the increase in resistance ΔR is proportional to the initial resistance R_0 and the change in temperature Δt. We can write

$$\Delta R = \alpha R_0 \, \Delta t \tag{33-13}$$

The constant α is a characteristic of the material known as the *temperature coefficient of resistance*. The defining equation for α can be found by solving Eq. (33-13):

$$\alpha = \frac{\Delta R}{R_0 \, \Delta t} \tag{33-14}$$

The **temperature coefficient of resistance** is the change in resistance per unit resistance per degree change in temperature.

Since the units of ΔR and R_0 are the same, the unit of the coefficient α is inverse degrees (1/C°).

EXAMPLE 33-7   An iron wire has a resistance of 200 Ω at 20°C. What will its resistance be at 80°C if the temperature coefficient of resistance is 0.006/C°?

Solution   We first compute the change in resistance from Eq. (33-13):

$$\Delta R = \alpha R_0 \, \Delta t$$
$$= (0.006/\text{C}°)(200 \ \Omega)(80°\text{C} - 20°\text{C})$$
$$= 72 \ \Omega$$

Therefore, the resistance at 80°C is

$$R = R_0 + \Delta R = 200 \ \Omega + 72 \ \Omega = 272 \ \Omega$$

The increase in resistance of a conductor with temperature is large enough to be measured easily. This fact is used in resistance thermometers to measure temperatures very accurately. Because of the very high melting point of some metals, resistance thermometers can be used to measure extremely high temperatures.

SUMMARY

In this chapter, we introduced the *ampere* as a unit of electric current, and we discussed the various quantities which affect its magnitude. Ohm's law described mathematically the relationship between current, resistance, and applied voltage. We also learned the factors which affect electric resistance, and applied these concepts to the solution of basic problems in elementary electricity. The major points are summarized below:

- Electric current I is the rate of flow of charge Q past a given point on a conductor:

$$I = \frac{Q}{t}$$
$$1 \text{ ampere (A)} = \frac{1 \text{ coulomb (C)}}{1 \text{ second (s)}}$$

- By convention, the *direction* of electric current is the same as the direction in which *positive* charges would move, even if the actual current consists of a flow of negatively charged electrons.

- Ohm's law states that *the current produced in a given conductor is directly proportional to the difference of potential between its end points:*

$$R = \frac{V}{I} \qquad V = IR \qquad \textit{Ohm's Law}$$

The symbol R represents the resistance in ohms (Ω) defined as

$$1 \text{ ohm } (\Omega) = \frac{1 \text{ ampere (A)}}{1 \text{ volt (V)}}$$

- The electric power in watts is given by any of

$$P = VI \qquad P = I^2R \qquad P = \frac{V^2}{R} \qquad \textit{Power}$$

- The resistance of a wire depends on four factors: (*a*) the kind of *material*, (*b*) the *length*, (*c*) the cross-sectional *area*, and (*d*) the *temperature*. By introducing a property of the material called its *resistivity* ρ, we can write

$$R = \rho \frac{l}{A} \qquad \rho = \frac{RA}{l} \qquad \textit{SI unit for } \rho: \Omega \cdot \text{m}$$

- The *temperature coefficient of resistance* α is the change in resistance per unit resistance per degree change in temperature.

$$\alpha = \frac{\Delta R}{R_0 \, \Delta t} \qquad \Delta R = \alpha R_0 \, \Delta t$$

33-1. Define the following terms:
a. Current
b. Resistance
c. Ampere
d. Emf
e. Ohm
f. Rheostat
g. Ammeter
h. Voltmeter
i. Transient current
j. Source of emf
k. Electric power
l. Ohm's law
m. Resistivity
n. Circular mil
o. Temperature coefficient of resistance

33-2. Distinguish clearly between electron flow and conventional current. What are some reasons for preferring the conventional current?

33-3. Use the mechanical analogy of water flowing through pipes to describe the flow of charge through conductors of various lengths and cross-sectional areas.

33-4. A rheostat is connected across the terminals of a battery. What determines the positive and negative terminals on the rheostat?

33-5. Is the electromotive force really a *force*? What is the function of a source of emf?

33-6. What is wrong with the following statement: the resistivity of a material is directly proportional to its length?

33-7. How many circular mils are equivalent to an area of 1 mil^2?

33-8. Use Ohm's law to verify Eqs. (33-6) and (33-7).

Electric Current and Ohm's Law

33-1. How many electrons pass a point every second in a wire carrying a current of 20 A? How much time is needed to transport 40 C of charge past this point?
Ans. 1.25×10^{10} electrons, 2 s

33-2. If 600 C of charge pass a given point in 3 s, what is the electric current in amperes?

33-3. Find the current in amperes when 690 C of charge pass a given point in 2 min.
Ans. 5.75 A

33-4. If a current of 24 A exists for 50 s, how many coulombs of charge have passed through the wire?

33-5. (a) What is the potential drop across a 4-Ω resistor with a current of 8 A through it? (b) What is the resistance of a rheostat if the drop in potential is 48 V and the current is 4 A? (c) Determine the current through a 5-Ω resistor that has a potential drop a 40 V across it.
Ans. (a) 32 V, (b) 12 Ω, (c) 8 A

33-6. A 2-A fuse is placed in a circuit with a battery having a terminal voltage of 12 V. What is the minimum resistance for a circuit containing this fuse?

33-7. What emf is required to pass 60 mA through a resistance of 20 kΩ? If this same emf is applied to a resistance of 300 Ω, what will be the new current?
Ans. 1200 V, 4 A

Electric Power and Heat Loss

33-8. A soldering iron draws 0.75 A at 120 V. How much energy will it use in 15 min?

33-9. An electric lamp has an 80-Ω filament connected to a 110-V direct-current line. What is the current through the filament? What is the power loss in watts?

Ans. 1.38 A, 151 W

33-10. Assume that the cost of energy in a home is 8 cents per kilowatthour. A family goes on a 2-week vacation leaving a single 80-W light bulb burning. What is the cost?

33-11. A 120-V, direct-current generator delivers 2.4 kW to an electric furnace. What current is supplied? What is the resistance encountered?

Ans. 20 A, 6 Ω

Resistivity

33-12. What length of copper wire ¹⁄₁₆ in. diameter is required to construct a 20-Ω resistor at 20°C?

33-13. Find the resistance of 40 m of copper wire having a diameter of 0.8 mm at 20°C.

Ans. 1.37 Ω

33-14. A nichrome wire has a length of 40 m at 20°C. What is its diameter if the total resistance is 500 Ω?

33-15. What is the resistance of 200 ft of iron wire with a diameter of 0.002 in. at 20°C?

Ans. 2850 Ω

Temperature Coefficient of Resistance

33-16. The resistance of a length of wire is 4.0 Ω at 20°C. What is the resistance at 80°C? (α = 0.04/C°).

33-17. If the resistance of a conductor is 100 Ω at 20°C and 116 Ω at 60°C, what is its temperature coefficient of resistivity?

Ans. 0.004/C°

33-18. A length of copper wire has a resistance of 8 Ω at 20°C. What is the resistance at 90°C? At −30°C?

Additional Problems

33-19. A water turbine delivers 2000 kW to an electric generator. The generator is only 80 percent efficient and has an output terminal voltage of 1200 V. What current is delivered and what is the resistance?

Ans. 1.33 kA, 0.9 Ω

33-20. A 110-V radiant heater draws a current of 6 A. How much heat energy in joules is delivered in 1 h?

33-21. A power line has a total resistance of 4000 Ω. What is the power loss through the wire if the current is reduced to 6 mA?

Ans. 0.144 W

33-22. The fan motor operating a home cooling system is rated at 10 A for a 110-V line. How much energy is required to operate the fan for a 24-h period? At a cost of 8 cents per kilowatthour, what is the cost of operating this fan continuously for 1 month (30 days)? (The current in a home is alternating current instead of direct current, but the same formulas apply.)

33-23. Determine the resistivity of a wire made of an unknown alloy if its diameter is 0.007 in. and 100 ft of the wire is found to have a resistance of 4.0 Ω.

Ans. 1.96 Ω · cmil/ft

33-24. The resistivity of a certain wire is 1.72×10^{-8} Ω · m. A 6-V battery is connected to a 20-m coil of this wire having a diameter of 0.8 mm. What is the current in the wire?

33-25. A certain resistor is used as a thermometer. Its resistance at 20°C is 26.00 Ω, and its resistance at 40°C is 26.20 Ω. What is the temperature coefficient of resistance for this material?

Ans. 3.85×10^{-4}/C°

* **33-26.** What length of copper wire at 20°C has the same resistance as 200 m of iron wire at 20°C? Assume the same area for each wire.

Ans. 1100 m

34 Direct-Current Circuits

OBJECTIVES

After completing the chapter, you should be able to:

1. Determine the effective resistance of a number of resistors connected in *series* and in *parallel*.
2. Write and apply equations involving *voltage, current,* and *resistance* for a circuit containing resistors connected in series and in parallel.
3. Solve problems involving the *emf* of a battery, its *terminal potential difference,* the *internal resistance,* and the *load resistance*.
4. Write and apply *Kirchhoff's laws* for electrical networks similar to those shown in the text.

Two types of current are in use. Direct current (dc) is the continuous flow of charge in only one direction. Alternating current (ac) is a flow of charge continually changing in both magnitude and direction. In this chapter we analyze current, voltage, and resistance for dc circuits. Many of the same methods and procedures can also be applied to ac circuits. The variations required for alternating currents build logically from a strong foundation in dc analysis.

34-1
SIMPLE CIRCUITS; RESISTORS IN SERIES

An electric circuit consists of any number of branches joined together so that at least one closed path is provided for current. The simplest circuit consists of a single source of emf joined to a single external resistance, as shown in Fig. 34-1. If $\mathcal{E}$ represents the emf and R indicates the total resistance, Ohm's law yields

$$\mathcal{E} = IR \qquad (34\text{-}1)$$

where I is the current through the circuit. All the energy gained by charge in passing through the source of emf is lost in flowing through the resistance.

Let us consider the addition of a number of elements to a circuit. Two or more elements are said to be in *series* if they have only *one* point in common that is not connected to some third element. Current can follow only a single path through

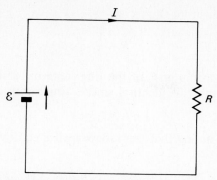

Fig. 34-1 A simple electric circuit.

elements in series. Resistors R_1 and R_2 of Fig. 34-2a are in series because point A is common to both resistors. However, the resistors in Fig. 34-2b are not in series because point B is common to three current branches. Electric current entering such a junction may follow two separate paths.

Suppose that three resistors (R, R_2, and R_3) are connected in series and enclosed in a box, indicated by the shaded portion of Fig. 34-3. The effective resistance R of the three resistors can be determined from the external voltage V and current I, as

Fig. 34-2 *(a)* Resistors connected in series. *(b)* Resistors which are not connected in series.

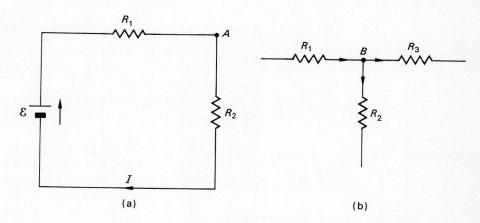

(a)

(b)

Fig. 34-3 The voltmeter-ammeter method of measuring the effective resistance of a number of resistors connected in series.

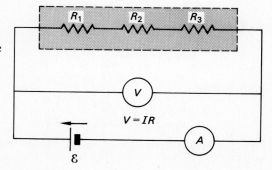

$$V = IR$$

recorded by the meters. From Ohm's law

$$R = \frac{V}{I} \qquad (34\text{-}2)$$

But what is the relationship of R to the three internal resistances? The current through each resistor must be identical since a single path is provided. Thus

$$I = I_1 = I_2 = I_3 \qquad (34\text{-}3)$$

Utilizing this fact and noting that Ohm's law applies equally well to any part of a circuit, we write

$$V = IR \qquad V_1 = IR_1 \qquad V_2 = IR_2 \qquad V_3 = IR_3 \qquad (34\text{-}4)$$

The external voltage V represents the sum of the energies lost per unit of charge in passing through each resistance. Therefore,

$$V = V_1 + V_2 + V_3$$

Finally, if we substitute from Eq. (34-4) and divide out the current, we obtain

$$IR = IR_1 + IR_2 + IR_3$$

$$\boxed{R = R_1 + R_2 + R_3} \qquad \qquad Series \quad (34\text{-}5)$$

To summarize what has been learned about resistors connected in *series:*

1. The current in all parts of a series circuit is the same.
2. The voltage across a number of resistances in series is equal to the sum of the voltages across the individual resistors.
3. The effective resistance of a number of resistors in series is equivalent to the sum of the individual resistances.

EXAMPLE 34-1   The resistances R_1 and R_2 in Fig. 34-2a are 2 and 4 Ω, respectively. If the source of emf maintains a constant potential difference of 12 V, what is the current delivered to the external circuit? What is the potential drop across each resistor?

Solution   The effective resistance is

$$R = R_1 + R_2 = 2\ \Omega + 4\ \Omega = 6\ \Omega$$

The current is then found from Ohm's law:

$$I = \frac{V}{R} = \frac{12\ \text{V}}{6\ \Omega} = 2\ \text{A}$$

The voltage drops are therefore

$$V_1 = IR_1 = (2\ \text{A})(2\ \Omega) = 4\ \text{V}$$
$$V_2 = IR_2 = (2\ \text{A})(4\ \Omega) = 8\ \text{V}$$

Note that the sum of the voltage drops ($V_1 + V_2$) is equal to the applied 12 V.

34-2
RESISTORS IN PARALLEL

There are several limitations in the operation of series circuits. If a single element in a series circuit fails to provide a conducting path, the entire circuit is opened and current ceases. It would be quite annoying if all electrical devices in a home were to cease functioning whenever one lamp burned out. Moreover, each element in a series circuit adds to the total resistance of the circuit, thereby limiting the total current which can be supplied. These objections can be overcome by providing alternative paths for electric current. Such a connection, in which current can be divided between two or more elements, is called a *parallel connection*.

A *parallel circuit* is one in which two or more components are connected to two common points in the circuit. For example, in Fig. 34-4, the resistors R_2 and R_3 are

Fig. 34-4 The resistors R_1 and R_2 are connected in parallel.

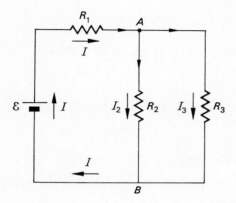

in parallel because they both have points A and B in common. Note that the current I, provided by the source of emf, is divided between resistors R_2 and R_3.

To arrive at an expression for the equivalent resistance R of a number of resistances connected in parallel, we follow a procedure similar to that discussed for a series connection. Assume that three resistors (R_1, R_2, and R_3) are placed inside a box, as shown in Fig. 34-5. The total current I delivered to the box is determined by its effective resistance and the applied voltage:

$$I = \frac{V}{R} \tag{34-6}$$

Fig. 34-5 Computing the equivalent resistance of a number of resistors connected in parallel.

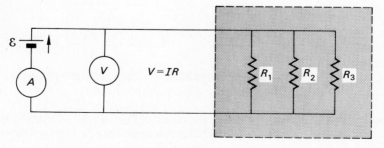

In a parallel connection, the voltage drop across each resistor is the same and equivalent to the total drop in voltage.

$$V = V_1 = V_2 = V_3 \qquad (34\text{-}7)$$

That this is true can be realized when we consider that the same energy must be lost by a unit of charge regardless of the path it travels in the circuit. In this example, charge may flow through any one of the three resistors. Thus the total current delivered is divided between the resistors.

$$I = I_1 + I_2 + I_3 \qquad (34\text{-}8)$$

Applying Ohm's law to Eq. (34-8) yields

$$\frac{V}{R} = \frac{V_1}{R_1} + \frac{V_2}{R_2} + \frac{V_3}{R_3}$$

But the voltages are equal, and we can divide them out.

$$\boxed{\frac{1}{R} = \frac{1}{R_1} + \frac{1}{R_2} + \frac{1}{R_3}} \qquad \textit{Parallel} \quad (34\text{-}9)$$

In summary, for parallel resistors:

1. The total current in a parallel circuit is equal to the sum of the currents in the individual branches.
2. The voltage drops across all branches in a parallel circuit must be of equal magnitude.
3. The reciprocal of the equivalent resistance is equal to the sum of the reciprocals of the individual resistances connected in parallel.

In the case of only two resistors in parallel,

$$\frac{1}{R} = \frac{1}{R_1} + \frac{1}{R_2}$$

Solving this equation algebraically for R, we obtain a simplified formula for computing the equivalent resistance.

$$\boxed{R = \frac{R_1 R_2}{R_1 + R_2}} \qquad (34\text{-}10)$$

The equivalent resistance of two resistors connected in parallel is equal to their product divided by their sum.

EXAMPLE 34-2 The total applied voltage to the circuit in Fig. 34-6 is 12 V, and the resistances R_1, R_2, and R_3 are 4, 3, and 6 Ω, respectively. (a) Determine the equivalent resistance of the circuit. (b) What is the current through each resistor?

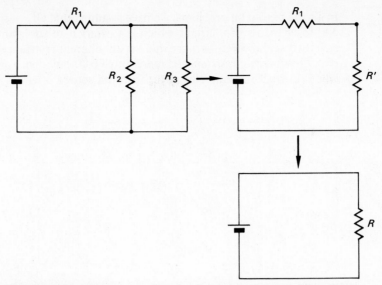

Fig. 34-6 Reducing a complex circuit to a simple equivalent circuit.

Solution (a) The best approach to a problem which contains both series and parallel resistors is to reduce the circuit by steps to its simplest form, as shown in Fig. 34-6. We first find the equivalent resistance R' of the pair of resistors R_2 and R_3.

$$R' = \frac{R_2 R_3}{R_2 + R_3} = \frac{(3\ \Omega)(6\ \Omega)}{(3+6)\Omega} = 2\ \Omega$$

Since the equivalent resistance R' is in series with R_1, the total equivalent resistance is

$$R = R' + R_1 = 2\ \Omega + 4\ \Omega = 6\ \Omega$$

Solution (b) The total current can be found from Ohm's law:

$$I = \frac{V}{R} = \frac{12\ V}{6\ \Omega} = 2\ A$$

The current through R_1 and R' is therefore 2 A since they are in series. To find the currents I_2 and I_3 we must know the voltage drop V' across the equivalent resistance R'.

$$V' = IR' = (2\ A)(2\ \Omega) = 4\ V$$

Thus the potential must drop by 4 V through each of the resistors R_2 and R_3. The currents are found from Ohm's law:

$$I_2 = \frac{V'}{R_2} = \frac{4\ V}{3\ \Omega} = 1.33\ A$$

$$I_3 = \frac{V'}{R_3} = \frac{4\ V}{6\ \Omega} = 0.67\ A$$

Note that $I_2 + I_3 = 2$ A, which is the total current delivered to the circuit.

In all the preceding problems, we have assumed that all resistance to current flow is due to elements of a circuit which are external to the source of emf. This is not strictly true, however, because there is an inherent resistance within every source of emf. This *internal resistance* is represented by the symbol r and is shown schematically as a small resistance in series with the source of emf. (See Fig. 34-7.) When a

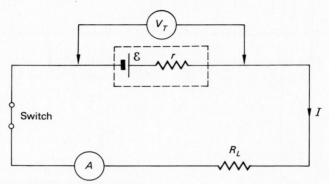

Fig. 34-7 Internal resistance.

current I is flowing through the circuit, there is a loss of energy through the external load R_L and also there is a heat loss due to the internal resistance. Thus the actual terminal voltage V_T across a source of emf $\mathcal{E}$ with an internal resistance r is given by

$$V_T = \mathcal{E} - Ir \qquad (34\text{-}11)$$

The voltage applied to the external load is therefore less than the emf by an amount equal to the internal potential drop. Since $V_T = IR_L$, Eq. (34-11) can be rewritten

$$V_T = IR_L = \mathcal{E} - Ir \qquad (34\text{-}12)$$

Solving Eq. (34-12) for the current I, we have

$$I = \frac{\mathcal{E}}{R_L + r} \qquad (34\text{-}13)$$

The current in a simple circuit containing a single source of emf is equal to the emf $\mathcal{E}$ divided by the total resistance in the circuit (including internal resistance).

EXAMPLE 34-3   A load resistance of 8 Ω is connected to a battery whose internal resistance is 0.2 Ω. (*a*) If the emf of the battery is 12 V, what current is delivered to the load? (*b*) What is the terminal voltage of the battery?

Solution (a)   The current delivered is found from Eq. (34-13).

$$I = \frac{\mathcal{E}}{R_L + r} = \frac{12 \text{ V}}{8 \ \Omega + 0.2 \ \Omega} = 1.46 \text{ A}$$

Solution (b)   The terminal voltage is

$$V_T = \mathcal{E} - Ir = 12\text{ V} - (1.46\text{ A})(0.2\ \Omega)$$
$$= 12\text{ V} - 0.292\text{ V} = 11.7\text{ V}$$

As a check, we can find the voltage drop across the load R_L:

$$V_T = IR_L = (1.46\text{ A})(8\ \Omega) = 11.7\text{ V}$$

34-4
MEASURING INTERNAL RESISTANCE

The internal resistance of a battery can be measured in the laboratory by using a voltmeter, an ammeter, and a known resistance. A voltmeter is an instrument which has an extremely high resistance. When a voltmeter is attached directly to the terminals of a battery, negligible current is drawn from the battery. We can see from Eq. (34.11) that for zero current this terminal voltage is equal to the emf ($V_T = \mathcal{E}$). In fact, the emf of a battery is sometimes referred to as its "open-circuit" potential difference. Thus the emf can be measured with a voltmeter. By connecting a known resistance to the circuit, we can determine the internal resistance by measuring the current delivered to the circuit.

EXAMPLE 34-4   A dry cell gives an open-circuit reading of 1.5 V when a voltmeter is connected to its terminals. When the voltmeter is removed and a load of 3.5 Ω is placed across the terminals of the battery, a current of 0.4 A is measured. What is the internal resistance of the battery?

Solution   Solving for r in Eq. (34-12), we obtain

$$r = \frac{\mathcal{E} - IR_L}{I} = \frac{1.5\text{ V} - (0.4\text{ A})(3.5\ \Omega)}{0.4\text{ A}}$$

$$= \frac{0.10\text{ V}}{0.4\text{ A}} + 0.25\ \Omega$$

As a dry cell ages, its internal resistance increases while its emf remains relatively constant. The increased internal resistance causes a reduction in the current delivered. This fact accounts for the difference in intensity of light between a flashlight using old batteries and one using fresh batteries.

34-5
REVERSING THE CURRENT THROUGH A SOURCE OF EMF

In a battery chemical energy is converted into electric energy in order to maintain current flow in an electric circuit. A generator performs a similar function by converting mechanical energy into electric energy. In either case, the process is reversible. If a source of higher emf is connected in direct opposition to a source of lower emf, the current will pass through the latter from its positive terminal to its negative terminal. Reversing the flow of charge in this manner results in a loss of energy as electric energy is converted into chemical or mechanical energy.

Let us consider the process of charging a battery, as illustrated in Fig. 34-8. As charge flows through the higher source of emf $\mathcal{E}_1$, it gains energy. The terminal voltage for $\mathcal{E}_1$ is given by

$$V_1 = \mathcal{E}_1 - Ir_1$$

in accordance with Eq. (34-12). The output voltage is reduced because of the internal resistance r_1.

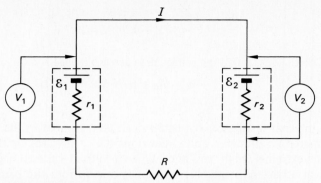

Fig. 34-8 Reversing the current through a source of emf.

Energy is lost in two ways as charge is forced through the battery against its normal output direction:

1. Electric energy in the amount equal to $\mathcal{E}_2$ is stored as chemical energy in the battery.
2. Energy is lost to the internal resistance of the battery.

Therefore, the terminal voltage V_2, which represents the total drop in potential across the battery, is given by

$$V_2 = \mathcal{E}_2 + Ir_2 \qquad (34\text{-}14)$$

where r_2 is the internal resistance. Note that in this case the terminal voltage is *greater* than the emf of the battery. The remainder of the potential supplied by the higher source of emf is lost through the external resistance R.

Throughout the entire circuit, the energy lost must equal the energy gained. Thus we can write

Energy gained per unit charge = energy lost per unit charge

$$\mathcal{E}_1 = \mathcal{E}_2 + Ir_1 + Ir_2 + IR$$

Solving for the current I yields

$$I = \frac{\mathcal{E}_1 - \mathcal{E}_2}{r_1 + r_2 + R}$$

The current supplied to a continuous electric circuit is equal to the net emf divided by the total resistance of the circuit, including internal resistance.

$$\boxed{I = \frac{\Sigma \mathcal{E}}{\Sigma R}} \qquad (34\text{-}15)$$

For the purposes of applying Eq. (34-15), an emf is considered negative when the current flows against its normal output direction.

ELECTRICITY AND MAGNETISM

EXAMPLE 34-5 Assume the following values for the parameters of the circuit in Fig. 34-8: $\mathcal{E}_1 = 12$ V, $\mathcal{E}_2 = 6$ V, $r_1 = 0.2$ Ω, $r_2 = 0.1$ Ω, and $R = 4$ Ω. (a) What is the current in the circuit? (b) What is the terminal voltage across the 6-V battery?

Solution (a) From Eq. (34-15), the current is

$$I = \frac{\mathcal{E}_1 - \mathcal{E}_2}{r_1 + r_2 + R} = \frac{12 \text{ V} - 6 \text{ V}}{0.2 \text{ Ω} + 0.1 \text{ Ω} + 4 \text{ Ω}}$$

$$= \frac{6 \text{ V}}{4.3 \text{ Ω}} = 1.40 \text{ A}$$

Solution (b) The terminal voltage of the battery being charged is, from Eq. (34-14),

$$V_2 = \mathcal{E}_2 + Ir_2$$
$$= 6 \text{ V} + (1.4 \text{ A})(0.1 \text{ Ω})$$
$$= 6.14 \text{ V}$$

34-6
KIRCHHOFF'S LAWS

An electrical network is a complex circuit consisting of a number of current loops or meshes. For complex networks containing several meshes and a number of sources of emf, the application of Ohm's law becomes very difficult. A more straightforward procedure for analyzing such circuits was developed in the nineteenth century by Gustav Kirchhoff, a German scientist. His method involves the use of two laws. *Kirchhoff's first law* is:

> *The sum of the currents entering a junction is equal to the sum of the currents leaving that junction.*

$$\sum I_{\text{entering}} = \sum I_{\text{leaving}}$$

(34-16)

Kirchhoff's second law states:

> *The sum of the emfs around any closed current loop is equal to the sum of all the IR drops around that loop.*

$$\sum \mathcal{E} = \sum IR$$

(34-17)

A junction refers to any point in a circuit where three or more wires come together. The first law simply states that charge must flow continuously; it cannot pile up at a junction. In Fig. 34-9, if 12 C of charge enters the junction every second, then 12 C must leave it every second. The current delivered to each branch is inversely proportional to the resistance of that branch.

The second law is a restatement of the conservation of energy. If we begin at any point in a circuit and travel around any closed current loop, the energy gained by a unit of charge must equal the energy lost by that charge. Energy is gained through the conversion of chemical or mechanical energy into electric energy by a source of

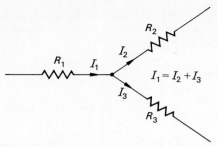

Fig. 34-9 The sum of the currents entering a junction must equal the sum of the currents leaving that junction.

emf. Energy may be lost either in the form of *IR* potential drops or in the process of reversing the current through a source of emf. In the latter case, electric energy is converted into the chemical energy necessary to charge a battery or electric energy is converted to mechanical energy for the operation of a motor.

In applying Kirchhoff's rules, very definite procedures must be followed. The steps in the general procedure will be presented by considering the example offered by Fig. 34-10*a*.

1. Assume a current direction for each loop in the network.

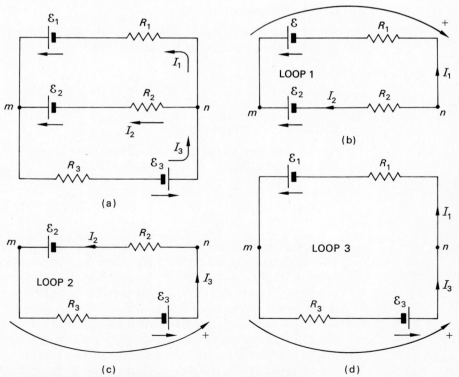

Fig. 34-10 Applying Kirchhoff's laws to a complex circuit.

ELECTRICITY AND MAGNETISM

The three loops which may be considered are those illustrated in Fig. 34.10*b*, *c*, and *d*. The current I_1 is assumed to flow counterclockwise in the top loop, I_2 is assumed to travel to the left in the middle branch, and I_3 is assumed to flow counterclockwise in the lower loop. If we have guessed correctly, the solution to the problem will give a positive value for the current; if we have guessed wrong, a negative value will indicate that the current is actually in the other direction.

2. Apply Kirchhoff's first law to write a current equation for all but one of the junction points.

Writing the current equation for *every* junction would result in a duplicate equation. In our example, there are two junction points which are labeled *m* and *n*, respectively. The current equation for *m* is

$$\sum I_{\text{entering}} = \sum I_{\text{leaving}}$$

$$I_1 + I_2 = I_3 \tag{34-18}$$

The same equation would result if we considered junction *n*, and no new information would be given.

3. Indicate by a small arrow, drawn next to the symbol for each emf, the direction in which the source, acting alone, would cause a positive charge to move through the circuit.

In our example, $\mathcal{E}_1$ and $\mathcal{E}_2$ are directed to the left, and $\mathcal{E}_3$ is directed to the right.

4. Apply Kirchhoff's second law ($\Sigma \mathcal{E} = \Sigma IR$) for one loop at a time. There will be one equation for each loop.

In applying Kirchhoff's second rule, one must begin at a specific point on a loop and *trace* around the loop in a consistent direction back to the starting point. The choice of the *tracing direction* is arbitrary, but, once established, it becomes the positive (+) direction for sign conventions. (Tracing directions for the three loops in our example are labeled in Fig. 34-10.) The following sign conventions apply:

1. When summing the emfs around a loop, the value assigned to the emf is positive if its output (see step 3) is with the tracing direction; it is considered negative if the output is against the tracing direction.
2. An *IR* drop is considered positive when the assumed current is with the tracing direction and negative when the assumed current opposes the tracing direction.

Let us now apply Kirchhoff's second law to each loop in our example:

Loop 1 Starting at point *m* and tracing clockwise, we have

$$-\mathcal{E}_1 + \mathcal{E}_2 = -I_1 R_1 + I_2 R_2 \tag{34-19}$$

Loop 2 Starting at point *m* and tracing counterclockwise, we have

$$\mathcal{E}_3 + \mathcal{E}_2 = I_3 R_3 + I_2 R_2 \tag{34-20}$$

Loop 3 Starting at *m* and tracing counterclockwise we have

$$\mathcal{E}_3 + \mathcal{E}_1 = I_3R_3 + I_1R_1 \qquad (34\text{-}21)$$

If the equation for loop 1 is subtracted from the equation for loop 2, the equation for loop 3 is obtained, showing that the last loop equation gives no new information.

We now have three independent equations involving only three unknowns. They can be solved simultaneously to find the unknowns, and the third loop equation can be used to check the results.

EXAMPLE 34-6   Solve for the unknown currents in Fig. 34-11, using Kirchhoff's laws.

Fig. 34-11

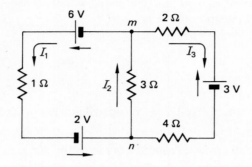

Solution   Current directions are assumed, as indicated in the figure for I_1, I_2, and I_3. If we apply Kirchhoff's first law to junction m, we obtain

$$\sum I \text{ (entering)} = \sum I \text{ (leaving)}$$

$$I_2 = I_1 + I_3 \qquad (34\text{-}22)$$

Next, the direction of positive output is indicated in the figure, adjacent to each source of emf. Since there are three unknowns, we need at least two more equations from the application of Kirchhoff's second law. Starting at m and tracing counterclockwise around the left loop, we write the voltage equation

$$\sum \mathcal{E} = \sum IR$$

$$6\text{ V} + 2\text{ V} = I_1(1\ \Omega) + I_2(3\ \Omega)$$
$$8\text{ V} = (1\ \Omega)I_1 + (3\ \Omega)I_2$$

Dividing through by $1\ \Omega$ and transposing, we obtain

$$I_1 + 3I_2 = 8\text{ A} \qquad (34\text{-}23)$$

The unit ampere arises from the fact that

$$1\text{ V}/\Omega = 1\text{ A}$$

Another voltage equation can be written by starting at m and tracing clockwise around the right loop:

$$-3\text{ V} = I_3(2\ \Omega) + I_3(4\ \Omega) + I_2(3\ \Omega)$$

The negative sign arises from the fact that the output of the source opposes the tracing direction. Simplifying, we have

$$2I_3 + 4I_3 + 3I_2 = -3 \text{ A}$$
$$6I_3 + 3I_2 = -3 \text{ A}$$
$$I_2 + 2I_3 = -1 \text{ A} \qquad (34\text{-}24)$$

The three equations which must be solved simultaneously for I_1, I_2, and I_3 are

$$I_1 - I_2 + I_3 = 0 \qquad (34\text{-}22)$$
$$I_1 + 3I_2 = 8 \text{ A} \qquad (34\text{-}23)$$
$$I_2 + 2I_3 = -1 \text{ A} \qquad (34\text{-}24)$$

From Eq. (34-22), we note that

$$I_1 = I_2 - I_3$$

which, substituted into Eq. (34-23), yields

$$(I_2 - I_3) + 3I_2 = 8 \text{ A}$$
$$4I_2 - I_3 = 8 \text{ A} \qquad (34\text{-}25)$$

Now we can solve Eqs. (34-25) and (34-24) simultaneously by eliminating I_3 from the two equations by addition:

$$
\begin{array}{rl}
(34\text{-}24): & I_2 + 2I_3 = -1 \text{ A} \\
2 \times (34\text{-}25): & \underline{8I_2 - 2I_3 = 16 \text{ A}} \\
& 9I_2 = 15 \text{ A} \\
& I_2 = 1.67 \text{ A}
\end{array}
$$

Substituting $I_2 = 1.67$ A into Eqs. (34-23) and (34-24) gives values for the other currents:

$$I_1 = 3 \text{ A} \qquad I_3 = -1.33 \text{ A}$$

The negative value obtained for I_3 indicates that our assumed current direction was incorrect. Actually, the current flows opposite the assumed direction. However, in working problems, the minus sign should be retained until all unknowns have been determined.

As a check on the above results, we can write one more voltage equation by applying Kirchhoff's second law to the outside loop. Starting at m and tracing counterclockwise, we obtain

$$(6 + 2 + 3) \text{ V} = I_1(1 \text{ }\Omega) - I_3(4 \text{ }\Omega) - I_3(2 \text{ }\Omega)$$
$$I_1 - 6I_3 = 11 \text{ A}$$

Substituting for I_1 and I_3, we obtain

$$3 \text{ A} - (6)(-1.33 \text{ A}) = 11 \text{ A}$$
$$11 \text{ A} = 11 \text{ A} \qquad \qquad \textit{Check}$$

Note again that the negative value for I_3 was used in the mathematics even though it indicates an incorrect assumption.

34-7
THE WHEATSTONE BRIDGE

A very precise laboratory method for measuring an unknown resistance uses a *Wheatstone bridge*. The circuit diagram for the apparatus is shown in Fig. 34-12. The bridge consists of a battery, a galvanometer, and four resistors. The galvanometer is a highly sensitive instrument which indicates the flow of electric charge. In the

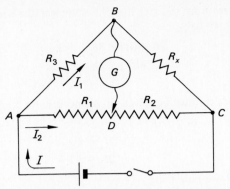

Fig. 34-12 A circuit diagram for the Wheatstone bridge.

diagram, an unknown resistance R_x is balanced with the three known resistances (R_1, R_2, and R_3). Normally, when the switch is closed, the galvanometer will indicate a current between points B and D. By moving the contact point D to the right or left, the resistances R_1 and R_2 can be varied until the galvanometer deflection is zero, indicating zero current between points B and D. Under these conditions, the bridge is said to be balanced, and points B and D must be at the same potential.

$$V_B = V_D \qquad\qquad \textit{Balanced}$$

Thus the potential drop from A to B must equal the potential drop from A to D.

$$V_{AB} = V_{AD} \tag{34-26}$$

Similarly,

$$V_{BC} = V_{DC} \tag{34-27}$$

The current I coming from the source of emf splits at point A, sending the current I_1 through the branch ABC and the current I_2 through the branch ADC. Applying Ohm's law to Eqs. (34-26) and (34-27) gives the relations

$$I_1 R_3 = I_2 R_1 \quad \text{and} \quad I_1 R_x = I_2 R_2$$

Dividing the second equation by the first and solving for R_x, we have

$$\boxed{R_x = R_3 \frac{R_2}{R_1}} \tag{34-28}$$

The laboratory apparatus for the Wheatstone bridge is illustrated in Fig. 34-13. The resistance R_3 is usually a plug-type resistance box, which gives a positive indication of its resistance. Removing plugs will increase the known resistance. The resistances R_1 and R_2 normally consist of a wire of uniform resistance attached to a meterstick. Since the resistance of such a wire is directly proportional to its length, the ratio R_2/R_1 of Eq. (34-28) is equivalent to the ratio of the corresponding wire lengths. If l_1 and l_2 are used to represent the wire segments AD and DC, the unknown resistance can be found from

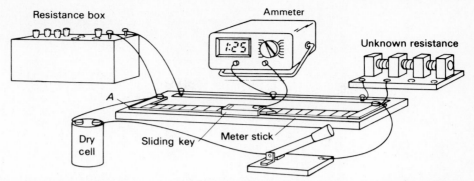

Fig. 34-13 A laboratory apparatus for measuring unknown resistance with the Wheatstone bridge.

$$R_x = R_3 \frac{l_2}{l_1} \qquad (34\text{-}29)$$

Thus it is not necessary to know the actual resistance of R_1 and R_2.

EXAMPLE 34-7 A Wheatstone bridge is used to measure the resistance R_x of a spool of copper wire. It is found that the galvanometer gives a small deflection when 4 Ω is placed on the resistance box. When the contact key is positioned at the 40-cm mark (measured from A), the bridge is balanced. Determine the unknown resistance.

Solution In this example $l_1 = 40$ cm, $l_2 = 60$ cm, and $R_3 = 4\ \Omega$. Substitution into Eq. (34-29) yields

$$R_x = R_3 \frac{l_2}{l_1} = (4\ \Omega) \frac{60\ \text{cm}}{40\ \text{cm}} = 6\ \Omega$$

During preliminary adjustments, when the bridge may be far from balanced, a resistor should be placed in series with the galvanometer to protect this extremely sensitive instrument from large currents. The resistor is removed from the circuit when final adjustments are made.

SUMMARY

An understanding of direct-current circuits is essential as an introduction to electrical technology. Most of the advanced study builds on the ideas presented in this chapter. You should review in detail any sections which are not clear to you. A summary of the major relationships is given below:

- In dc circuits, resistors may be connected in series (refer to Fig. 34-3) or in parallel (refer to Fig. 34-5).
  (a) For *series connections*, the current in all parts of the circuit is the same, the total voltage drop is the sum of the individual drops across each resistor, and the effective resistance is equal to the sum of the individual resistances:

$$\boxed{I_T = I_1 = I_2 = I_3} \qquad \boxed{V_T = V_1 + V_2 + V_3}$$

$$\boxed{R_e = R_1 + R_2 + R_3} \qquad \textit{Series Connections}$$

(b) For *parallel connections,* the total current is the sum of the individual currents, the voltage drops are all equal, and the effective resistance is given by

$$I_T = I_1 + I_2 + I_3 \qquad V_T = V_1 = V_2 = V_3$$

$$\frac{1}{R_e} = \frac{1}{R_1} + \frac{1}{R_2} + \frac{1}{R_3}$$

Parallel Connections

For two resistors connected in parallel, a simpler form is

$$R_e = \frac{R_1 R_2}{R_1 + R_2}$$

Two Resistors in Parallel

- The current supplied to an electric circuit is equal to the *net* emf divided by the total resistance of the circuit, including internal resistances.

$$I = \frac{\sum \mathcal{E}}{\sum R} \qquad \text{for example} \qquad I = \frac{\mathcal{E}_1 - \mathcal{E}_2}{r_1 + r_2 + R_L}$$

The example is for two opposing batteries of internal resistances r_1 and r_2 when the circuit load resistance is R_L.

- According to Kirchhoff's laws, the current entering a junction must equal the current leaving the junction and the net emf around any loop must equal the sum of the *IR* drops. Symbolically,

$$\sum I_{\text{entering}} = \sum I_{\text{leaving}}$$

$$\sum \mathcal{E} = \sum IR$$

Kirchhoff's Laws

- The following steps should be applied in solving circuits with Kirchhoff's laws (see Fig. 34-10):

Step 1 *Assume a current direction for each loop in the network.*

Step 2 *Apply Kirchhoff's first law to write a current equation for all but one of the junction points ($\sum I_{\text{in}} = \sum I_{\text{out}}$).*

Step 3 *Indicate by a small arrow the direction in which each emf acting alone would cause a positive charge to move.*

Step 4 *Apply Kirchhoff's second law ($\sum \mathcal{E} = \sum IR$) to write an equation for all possible current loops. An arbitrary positive tracing direction is chosen. An emf is considered positive if its output direction is the same as your tracing direction. An IR drop is considered positive when the assumed current direction is the same as your tracing direction.*

Step 5 *Solve the equations simultaneously to determine the unknown quantities.*

- A Wheatstone bridge is a device which allows one to determine an unknown resistance R_x by balancing the voltage drops in the circuit. If R_3 is known and the ratio

R_2/R_1 can be determined, we have

$$R_x = R_3 \frac{R_2}{R_1}$$

Wheatstone Bridge

QUESTIONS

34-1. Define the following terms:
   a. Dc circuit
   b. Series connection
   c. Parallel connection
   d. Terminal potential difference
   e. Internal resistance
   f. Kirchhoff's first law
   g. Kirchhoff's second law
   h. Wheatstone bridge

34-2. Defend the following statement: the effective resistance of a group of resistors connected in parallel will be less than any of the individual resistances.

34-3. Discuss the advantages and disadvantages of connecting Christmas-tree lights (a) in series; (b) in parallel.

34-4. What is meant by the "open-circuit" potential difference of a battery?

34-5. Distinguish clearly between terminal potential difference and emf.

34-6. Many electrical devices and appliances are designed to operate at the same voltage. How should such devices be connected in an electric circuit?

34-7. Should elements connected in series be designed to function at a constant current or at a constant voltage?

34-8. In an electric circuit, it is desired to decrease the effective resistance by adding resistors. Should these resistors be connected in parallel or in series?

34-9. Describe a method for measuring the resistance of a spool of wire by using a voltmeter, an ammeter, a rheostat, and a source of emf. Draw the circuit diagram. (The rheostat is used to adjust the current to the range required for the ammeter.)

34-10. In a laboratory experiment with the Wheatstone bridge, the contact key is adjusted for zero galvanometer current. When the key has remained in contact with the wire for a while, a student notes that the bridge becomes unbalanced and there is a slight galvanometer deflection. Suggest a possible explanation.

34-11. Why is it acceptable to use lengths of wire rather than actual resistances when the Wheatstone bridge is balanced?

34-12. Given the emf of a battery, describe a laboratory procedure for determining its internal resistance.

34-13. Compare the formulas for computing equivalent capacitance with the formulas developed in this chapter for resistances in series and in parallel.

34-14. Can the terminal voltage of a battery ever be greater than its emf? Explain.

34-15. Solve Eq. (34-9) explicitly for the equivalent resistance R.

PROBLEMS

Resistors in Series and in Parallel

34-1. An 18-Ω resistor and a 9-Ω resistor are first connected in parallel and then in series with a 24-V battery. (a) What is the effective resistance for each connection? **(b)** What is the current drawn in each case? (Neglect internal battery resistance.)

Ans. (a) 6 Ω, 27 Ω; (b) 4.00 A, 0.889 A

34-2. A 12-Ω resistor and an 8-Ω resistor are connected in parallel with a 28-V source of emf. What is the effective resistance? What current is delivered by the battery? What are the effective resistance and current if the resistors are connected in series?

34-3. An 8-Ω resistor and a 3-Ω resistor are first connected in parallel and then in series with a 12-V source of emf. What are the effective resistance and the current for each case? Draw the circuit diagram for each case.

Ans. 2.18 Ω, 5.5 A, 11 Ω, 1.09 A

34-4. Given three resistors, $R_1 = 80$ Ω, $R_2 = 60$ Ω, and $R_3 = 40$ Ω, find their effective resistance when connected **(a)** in series, **(b)** in parallel.

34-5. Three resistances of 4, 9, and 11 Ω are connected first in series and then in parallel. Find the effective resistance for each connection.

Ans. 24 Ω, 2.21 Ω

* 34-6. A 9-Ω resistor is connected in series with two parallel resistors of 6 and 12 Ω. What is the terminal potential difference if the total current from the battery is 4 A?

* 34-7. Find the equivalent resistance of the circuit shown in Fig. 34-14.

Ans. 8.00 Ω

Fig. 34-14

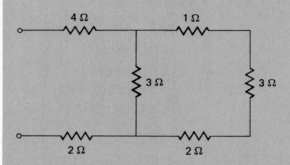

* 34-8. Determine the effective resistance of the circuit illustrated in Fig. 34-15.

Fig. 34-15

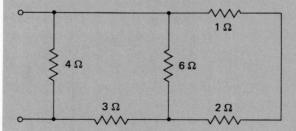

EMF and Terminal Potential Difference

34-9. A resistance of 6 Ω is placed across a 12-V battery whose internal resistance is 0.3 Ω. What is the current delivered to the circuit? What is the terminal voltage?

Ans. 1.90 A, 11.4 V

34-10. Two resistors of 7 and 14 Ω are connected in parallel with a battery whose emf is 16 V. The internal resistance of the battery is found to be 0.25 Ω. What is the load resistance in the circuit? What is the external current delivered? What is the terminal potential difference?

34-11. In an experiment to determine the internal resistance of a battery, its open-circuit potential difference is measured as 6.0 V. The battery is then connected to a 4.0-Ω resistor, and the current is found to be 1.40 A. What is the internal resistance of the battery?

Ans. 0.286 Ω

34-12. A dc motor draws 20 A from a 120-V line. If the internal resistance is 0.2 Ω, what is the emf of the motor? What is the electric power drawn from the line? What portion of this power is dissipated because of heat losses? What is the power delivered by the motor?

* **34-13.** Determine the total current delivered by the source of emf to the circuit in Fig. 34-16. What is the current through each resistor? Assume that $R_1 = 6\ \Omega$, $R_2 = 3\ \Omega$, $R_3 = 1\ \Omega$, $R_4 = 2\ \Omega$, $r = 0.4\ \Omega$, and $\mathcal{E} = 24$ V.

Ans. $I_T = 15$ A, $I_1 = 2$ A, $I_2 = 4$ A, $I_3 = 6$ A, $I_4 = 9$ A

Fig. 34-16

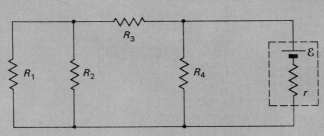

* **34-14.** Find the total current and the current through each resistor for Fig. 34-16 when $\mathcal{E} = 50$ V, $R_1 = 12\ \Omega$, $R_2 = 6\ \Omega$, $R_3 = 6\ \Omega$, $R_4 = 8\ \Omega$, $r = 0.4\ \Omega$.

Kirchhoff's Laws

** **34-15.** Use Kirchhoff's laws to solve for the currents through the circuit illustrated in Fig. 34-17.

Ans. $I_1 = 190$ mA, $I_2 = 23.8$ mA, $I_3 = 214$ mA

Fig. 34-17

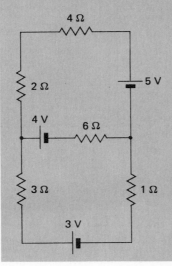

**   34-16. Use Kirchhoff's laws to solve for the currents through the circuit shown in Fig. 34-18.

Fig. 34-18

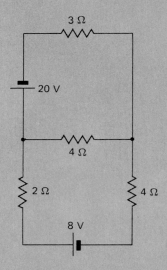

**   34-17. Apply Kirchhoff's laws to the circuit of Fig. 34-19. What are the currents through each branch?

Ans. 536 mA, 732 mA, 439 mA, 634 mA

Fig. 34-19

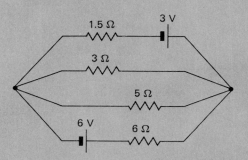

Wheatstone Bridge

34-18. A Wheatstone bridge is used to measure the resistance R_x of a coil of copper wire. The resistance box is adjusted for 6 Ω, and the contact key is positioned at the 45-cm mark when measured from point A of Fig. 34-13. Determine R_x.

34-19. Commercially available Wheatstone bridges are portable and have a self-contained galvanometer. The ratio R_2/R_1 can be set at any integral power of 10 between 0.001 and 1000 by a single dual switch. If this ratio is set at 100 and the known resistance R_3 is adjusted to 46.7 Ω, the galvanometer current drops to zero. What is the unknown resistance?

Ans. 4670 Ω

34-20. In a commercial Wheatstone bridge, R_1 and R_2 have the resistances of 20 and 40 Ω, respectively. If the resistance R_x to be measured is 14 Ω, what must be the known resistance R_3 for zero galvanometer deflection?

Additional Problems

34-21. The circuit illustrated in Fig. 34-7 consists of a 12-V battery, a 4-Ω resistor, and a switch. The internal resistance of the battery is 0.4 Ω. Assume first of all that the switch is left open. (a) What would a voltmeter read when placed across the terminals of the battery? (b) If the switch is closed, what will the voltmeter reading be? (c) With the switch closed, what would the voltmeter read when placed across the 4-Ω resistor?

Ans. (a) 12 V, (b) 10.9 V, (c) 10.9 V

34-22. A 6-Ω resistor R_1 and a 4-Ω resistor R_2 are connected in parallel across a 6-V generator whose internal resistance is 0.3 Ω. (a) Draw a circuit diagram. (b) What is the total current? (c) What is the power *developed* by the generator? (d) What is the power *delivered* by the generator? (e) At what rate is energy lost through R_1? At what rate is energy lost through R_2?

* **34-23.** The generator in Fig. 34-20 develops an emf $\mathcal{E}_1$ of 24 V and has an internal resistance of 0.2 Ω. The generator is used to charge a 12-V battery whose internal resistance is 0.3 Ω. Assume that $R_1 = 4\ \Omega$ and $R_2 = 6\ \Omega$. (a) What current is delivered to the circuit? (b) What is the terminal voltage across the generator? (c) What is the terminal voltage across the battery? (d) Show that the total drop in voltage in the external circuit is equal to the terminal voltage of the generator.

Ans. (a) 1.14 A, (b) 23.8 V, (c) 12.3 V

Fig. 34-20

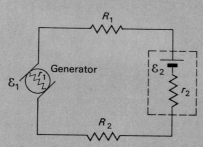

* **34-24.** Assume the following values for the parameters of the circuit illustrated in Fig. 34-8: $\mathcal{E}_1 = 100$ V, $\mathcal{E}_2 = 20$ V, $r_1 = 0.3\ \Omega$, $r_2 = 0.4\ \Omega$, and $R = 4.0\ \Omega$. (a) What are the terminal voltages V_1 and V_2? (b) What is the drop in potential across the external resistor?

** **34-25.** Solve for currents in the branches for Fig. 34-21.

Ans. 1.2 A, 0.6 A, 0.6 A

Fig. 34-21

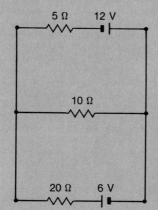

34-26. If the current in the 6-Ω resistor of Fig. 34-22 is 2 A, what is the emf of the battery? Neglect internal resistance. What is the power loss through the 1-Ω resistor?

Fig. 34-22

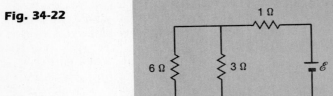

34-27. What is the effective resistance of the external circuit in Fig. 34-23 if internal resistance is negligible? What is the current in the 1-Ω resistor?

Ans. 6.08 Ω, 2.58 A

Fig. 34-23

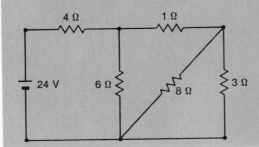

35 Magnetism and the Magnetic Field

OBJECTIVES: After completing this chapter, you should be able to:

1. Demonstrate by definition and example your understanding of *magnetism, induction, retentivity, saturation,* and *permeability.*
2. Write and apply an equation relating the magnetic force on a moving charge to its velocity, its charge, and its direction in a field of known magnetic flux density.
3. Determine the magnetic force on a current-carrying wire placed in a known *B* field.
4. Calculate the magnetic flux density *(a)* at a known distance from a current-carrying wire, *(b)* at the center of a current loop or coil, and *(c)* at the interior of a *solenoid.*

In previous chapters, we saw that electrical charges exert forces on one another. In this chapter, we will study magnetic forces. A magnetic force may be generated by electric charges in motion, and an electric force may be generated by a magnetic field in motion. The operation of electric motors, generators, transformers, circuit breakers, televisions, radios, and most electric meters depends upon the relationship between electric and magnetic forces. We shall begin this chapter by studying the magnetic effects associated with materials and conclude with a discussion of the magnetic effects of charges in motion.

35-1
MAGNETISM

The first magnetic phenomena to be observed were associated with rough fragments of lodestone (an oxide of iron) found near the ancient city of Magnesia some 2000 years ago. These *natural magnets* were observed to attract bits and pieces of unmagnetized iron. This force of attraction is referred to as *magnetism,* and the device which exerts a magnetic force is called a *magnet.*

If a bar magnet is dipped into a pan of iron filings and removed, the tiny pieces of iron are observed to cling most strongly to small areas near the ends (see Fig. 35-1).

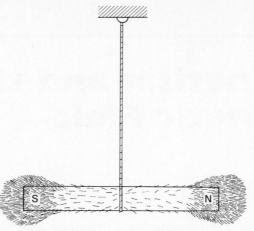

Fig. 35-1 The strength of a magnet is concentrated near its ends.

These regions where the magnet's strength appears to be concentrated are called *magnetic poles.*

When any magnetic material is suspended from a string, it turns about a vertical axis. As illustrated in Fig. 35-2, the magnet aligns itself in a north–south direction. The end pointing toward the north is called the *north-seeking* pole or the north (N) pole of the magnet. The opposite, *south-seeking,* end is referred to as the south (S) pole of the magnet. It is the polarization of magnetic material that accounts for its usefulness as a compass for navigation. The compass consists of a light magnetized needle pivoted on a low-friction support.

That the north and south poles of a magnet are different can easily be demonstrated. When another bar magnet is brought near a suspended magnet, as in Fig. 35-3, two north poles or two south poles repel each other whereas the north pole of

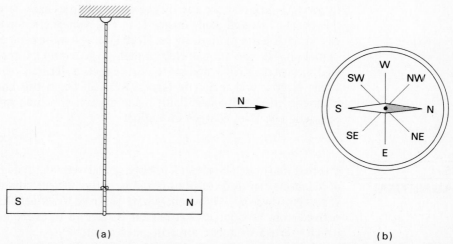

(a)   (b)

Fig. 35-2 *(a)* A suspended bar magnet will come to rest in a north–south direction. *(b)* The top view of a magnetic compass.

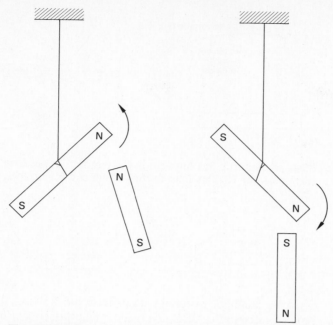

Fig. 35-3 Like magnetic poles repel each other; unlike poles attract each other.

one and the south pole of another attract each other. *The law of magnetic force* states:

> *Like magnetic poles repel each other; unlike magnetic poles attract each other.*

Isolated poles do not exist. No matter how many times a magnet is broken in half, each piece will become a magnet, having both a north and a south pole. We know of no single particle that could create a magnetic field in the same way that a proton or electron can create an electric field.

The attraction of a magnet for unmagnetized iron and the interacting forces between magnetic poles act through all substances. In industry, ferrous materials in trash are separated by magnets for recycling.

35-2
MAGNETIC FIELDS

Every magnet is surrounded by a space in which its magnetic effects are present. Such regions are called *magnetic fields.* Just as electric field lines were useful in describing electric fields, magnetic field lines, called *flux lines,* are useful for visualizing magnetic fields. The direction of a flux line at any point is the same as the direction of the magnetic force on an *imaginary* isolated north pole positioned at that point (see Fig. 35-4a). Accordingly, lines of magnetic flux leave the north pole of a magnet and enter the south pole. Unlike electric field lines, magnetic flux lines do not have beginning or ending points. They form continuous loops, passing through the metallic bar, as shown in Fig. 35-4b. The flux lines in the region between two like or unlike poles are illustrated in Fig. 35-5.

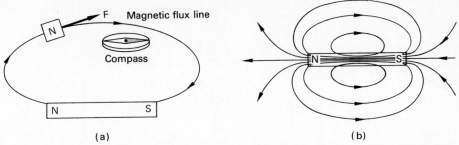

Fig. 35-4 *(a)* Magnetic flux lines are in the direction of the force exerted on an independent north pole. *(b)* The flux lines in the vicinity of a bar magnet.

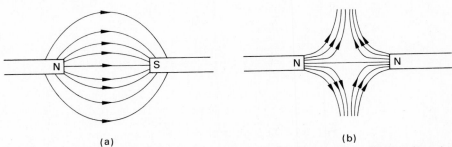

Fig. 35-5 *(a)* Magnetic flux lines between two unlike magnetic poles. *(b)* Flux lines in the space between two like poles.

35-5
THE MODERN THEORY OF MAGNETISM

Magnetism in matter is currently believed to result from the movement of electrons in the atoms of substances. If this is true, magnetism is a property of *charge in motion* and is closely related to electric phenomena. According to classical theory, individual atoms of a magnetic substance are, in effect, tiny magnets with north and south poles. The magnetic polarity of atoms is thought to arise primarily from the spin of electrons and is due only partially to their orbital motions around the nucleus. Figure 35-6 illustrates the two kinds of electron motion. Diagrams such as this should not be taken too seriously because there is still a lot we do not know about the motion of electrons. But we do firmly believe that magnetic fields of all particles must be caused by charge in motion, and such models help us to describe the phenomena.

The atoms in a magnetic material are grouped into microscopic magnetic regions called *domains*. All the atoms within a domain are believed to be magnetically

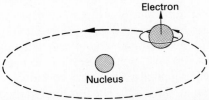

Fig. 35-6 Two kinds of electron motion responsible for magnetic properties.

polarized along a crystal axis. In an unmagnetized material, these domains are oriented in random directions, as indicated by the arrows in Fig. 35-7a. A dot is used to indicate an arrow directed out of the paper, and a cross is used to denote a direction into the paper. If a large number of the domains become oriented in the same direction, as in Fig. 35-7b, the material will exhibit strong magnetic properties.

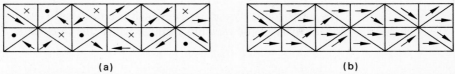

| (a) | (b) |

Fig. 35-7 *(a)* Magnetic domains are randomly oriented in an unmagnetized material. *(b)* The preferred orientation of domains in a magnetized material.

This theory of magnetism is highly plausible in that it offers an explanation for many of the observed magnetic effects of matter. For example, an unmagnetized iron bar can be made into a magnet simply by holding another magnet near it or in contact with it. This process, called *magnetic induction,* is illustrated by Fig. 35-8. The tacks become temporary magnets by induction. Note that the tacks on the right become magnetized even though they do not actually touch the magnet. Magnetic induction is explained by the domain theory. The introduction of a magnetic field aligns the domains, resulting in magnetization.

Induced magnetism is often only temporary, and when the field is removed, the domains gradually become disoriented. If the domains remain aligned to some degree after the field has been removed, the material is said to be *permanently* magnetized. The ability to retain magnetism is referred to as *retentivity.*

Another property of magnetic materials which is easily explained by the domain theory is *magnetic saturation.* There appears to be a limit to the degree of magnetization experienced by a material. Once this limit has been reached, no greater strength of an external field can increase the magnetization. It is believed that all its domains have been aligned.

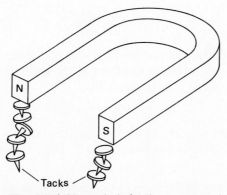

Tacks

Fig. 35-8 Magnetic induction.

In Chap. 30 we stated that electric field lines are drawn so that their spacing at any point will determine the strength of the electric field at that point (refer to Fig. 35-9). The number of lines ΔN drawn through a unit of area ΔA is directly proportional to the electric field intensity E.

$$\frac{\Delta N}{\Delta A} = \epsilon E \tag{35-1}$$

The proportionality constant ϵ, which determines the number of lines drawn, is the permittivity of the medium through which the lines pass.

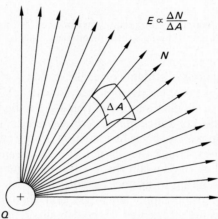

Fig. 35-9 Electric field intensity is proportional to the electric line density.

An analogous description of a magnetic field can be presented by considering the magnetic flux ϕ passing through a unit of perpendicular area A. This ratio B is called the *magnetic flux density.*

> The **magnetic flux density** in a region of a magnetic field is the number of flux lines which pass through a unit perpendicular area in that region.

$$B = \frac{\phi \text{ (flux)}}{A \text{ (area)}} \tag{35-2}$$

The SI unit of magnetic flux is the *weber* (Wb). The unit of flux density would then be webers per square meter, which is redefined as the *tesla* (T). An older unit which remains in use is the *gauss* (G). In summary,

$$1\text{ T} = 1\text{ Wb/m}^2 = 10^4\text{ G} \tag{35-3}$$

EXAMPLE 35-1 A rectangular loop 10 cm wide and 20 cm long makes an angle of 30° with respect to the magnetic flux in Fig. 35-10. If the flux density is 0.3 T, compute the magnetic flux ϕ penetrating the loop.

Solution The effective area penetrated by the flux is that component of the area which is perpendicular to the flux. Thus Eq. (35-2) becomes

---

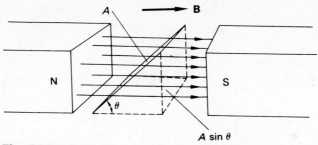

Fig. 35-10 Computing the magnetic flux through a rectangular conductor.

$$B = \frac{\phi}{A \sin \theta} \qquad \text{or} \qquad \phi = BA \sin \theta$$

The magnetic flux in webers is found by substitution into this relation.

$$\phi = (0.3 \text{ T})(0.1 \text{ m} \times 0.2 \text{ m})(\sin 30°)$$
$$= (0.3 \text{ T})(0.02 \text{ m}^2)(0.5)$$
$$= 3 \times 10^{-3} \text{ Wb}$$

The flux density at any point in a magnetic field is strongly affected by the nature of the medium or by the nature of a material placed in the medium. For this reason it is convenient to define a new magnetic field vector, the *magnetic field intensity* **H**, which does not depend on the nature of a medium. In any case, the number of lines established per unit area is directly proportional to the magnetic field intensity **H**. We can write

$$B = \frac{\phi}{A} = \mu H \qquad (35\text{-}4)$$

where the proportionality constant μ is the *permeability* of the medium through which the flux lines pass. Equation (35-4) is exactly analogous to Eq. (35-1), which was developed for electric fields. The permeability of a medium can thus be thought of as a measure of its ability to establish magnetic flux lines. The greater the permeability of a medium, the more flux lines will pass through a unit of area.

The permeability of free space (vacuum) is denoted by μ_0 and has the following magnitude for SI units:

$$\mu_0 = 4\pi \times 10^{-7} \text{ Wb/A} \cdot \text{m} = 4\pi \times 10^{-7} \text{ T} \cdot \text{m/A}$$

The full meaning of the unit webers per ampere-meter will come later. It is determined by the units of ϕ, A, and H of Eq. (35-4), which for a vacuum can be written

$$B = \mu_0 H \qquad \qquad \textit{Vacuum} \quad (35\text{-}5)$$

If a nonmagnetic material, such as glass, is placed in a magnetic field like that in Fig. 35-11, the flux distribution will not vary appreciably from that established for a vacuum. However, when a highly permeable material, such as soft iron, is placed in the field, the flux distribution will be altered considerably. The permeable material becomes magnetized by induction, resulting in a greater field strength for that region. For this reason, the flux density B is also referred to as the *magnetic induction*.

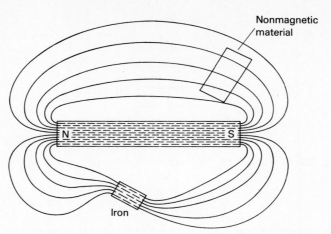

Fig. 35-11 A permeable material becomes magnetized by induction, resulting in a greater flux density in that region.

Magnetic materials are classified according to their permeability compared with that of free space. The ratio of the permeability of a material to that of a vacuum is called its *relative permeability* and is given by

$$\mu_r = \frac{\mu}{\mu_0} \tag{35-6}$$

Consideration of Eqs. (35-5) and (35-6) shows that the relative permeability of a material is a measure of its ability to change the flux density of a field from its value in a vacuum.

Materials with a relative permeability slightly less than unity have the property of being feebly repelled by a strong magnet. Such materials are said to be *diamagnetic,* and the property is referred to as *diamagnetism.* On the other hand, materials with slightly greater permeability than that of a vacuum are said to be *paramagnetic.* These materials are feebly attracted by a strong magnet.

A few materials, e.g., iron, cobalt, nickel, steel, and alloys of these metals, have extremely high permeabilities, ranging from a few hundred to thousands of times that for free space. Such materials are strongly attracted by a magnet and are said to be *ferromagnetic.*

35-5
MAGNETIC FIELD AND ELECTRIC CURRENT

Although the modern theory of magnetism holds that a magnetic field results from the motion of charges, science has not always accepted this proposition. It is fairly easy to show that a powerful magnet exerts no force on a static charge. In the course of a lecture demonstration in 1820, Hans Oersted set up an experiment to show his students that *moving* charges and magnets also do not interact. He placed the magnetic needle of a compass near a conductor, as illustrated in Fig. 35-12. To his surprise, when a current was sent through the wire, a twisting force was exerted on the compass needle until it pointed almost perpendicular to the wire. Further, the magnitude of the force depended upon the relative orientation of the compass needle and the current direction. The maximum twisting force occurred when the wire and

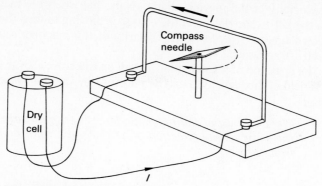

Fig. 35-12 Oersted's experiment.

compass needle were parallel before the current was established. If they were initially perpendicular, no force was experienced. Evidently, a magnetic field is set up by the charge in motion through the conductor.

In the same year that Oersted made his discovery, Ampère found that forces exist between two current-carrying conductors. Two wires with current in the same direction were found to attract each other whereas oppositely directed currents caused a force of repulsion. A few years later, Faraday found that the motion of a magnet toward or away from an electric circuit produces a current in the circuit. The relationship between magnetic and electric phenomena could no longer be doubted. Today, all magnetic phenomena can be explained in terms of electric charges in motion.

35-6
THE FORCE ON A MOVING CHARGE

Let us investigate the effects of a magnetic field by observing the magnetic force exerted on a charge which passes through the field. In studying these effects, it is useful to imagine a positive-ion tube like that in Fig. 35-13. Such a tube allows us to inject a positive ion of constant charge and velocity into a field of magnetic flux density B. By pointing the tube in various directions, we can observe the force exerted on the moving charge. The most striking observation is that the charge experiences a force which is perpendicular to both the magnetic flux density B and to

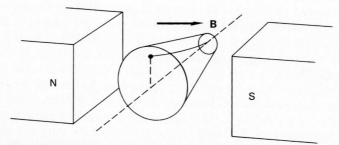

Fig. 35-13 The magnetic force F on a moving charge is perpendicular both to the flux density B and to the charge velocity v.

the velocity **v** of the moving charge. Note that when the magnetic flux is directed from the left to the right and the charge is moving toward the reader, the charge is deflected upward. Reversing the polarity of the magnets causes the charge to be deflected downward.

The direction of the magnetic force **F** on a positive charge moving with a velocity **v** in a field of flux density **B** can be reckoned by the *right-hand-screw rule* (see Fig. 35-14):

> *The direction of the magnetic force* **F** *on a moving positive charge is the same as the direction of advance of a right-hand screw if rotated from* **v** *to* **B**.

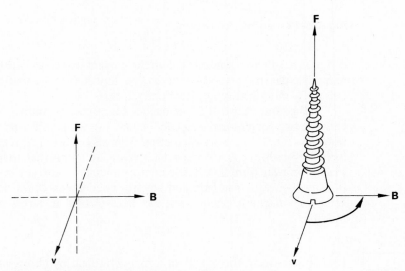

Fig. 35-14 The right-hand-screw rule.

If the moving charge is negative, as for an electron, the magnetic force will be directed *opposite* to the advance of a right-hand screw.

Let us now consider the magnitude of the force on a moving charge. Experimentation has shown that the magnitude of the magnetic force is directly proportional to the magnitude of the charge q and to its velocity v. Greater deflections will be indicated by our positive-ion tube if either of these parameters is increased.

An unexpected variation in the magnetic force will be observed if the ion tube is rotated slowly with respect to the magnetic flux density **B**. As indicated by Fig. 35-15, for a given charge of constant velocity v, the magnitude of the force varies with the angle the tube makes with the field. Particle deflection is a maximum when the charge velocity is perpendicular to the field. As the tube is slowly rotated toward **B**, the particle deflection becomes less and less. Finally, when the charge velocity is directed parallel to **B**, no deflection occurs, indicating that the magnetic force has dropped to zero. Evidently, the magnitude of the force is a function not only of the magnitude of the charge and its velocity but also varies with the angle θ between **v** and **B**. This variation is accounted for by stating that the magnetic force is propor-

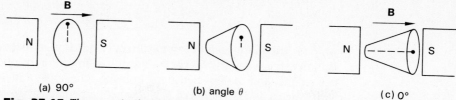

(a) 90°   (b) angle θ (c) 0°

Fig. 35-15 The magnitude of the magnetic force varies with the angle the moving charge makes with the direction of the magnetic field.

tional to the component of the velocity, v sin θ, perpendicular to the field direction. (Refer to Fig. 35-16.)

Fig. 35-16

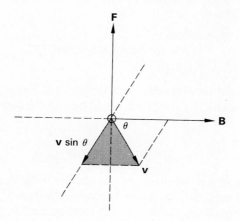

The above observations are summarized by the proportionality

$$F \propto qv \sin \theta \qquad (35\text{-}7)$$

If the proper units are chosen, the proportionality constant can be equated to the magnetic flux density B of the field causing the force. In fact, this proportionality is often used to *define* magnetic flux density as the constant ratio:

$$B = \frac{F}{qv \sin \theta} \qquad (35\text{-}8)$$

A magnetic field having a flux density of one tesla (one weber per square meter) will exert a force of one newton on a charge of one coulomb moving perpendicular to the field with a velocity of one meter per second.

As a consequence of Eq. (35-8), it can be noted that

$$1 \text{ T} = 1 \text{ N}/(\text{C} \cdot \text{m/s}) = 1 \text{ N/A} \cdot \text{m} \qquad (35\text{-}9)$$

These unit relationships are useful in solving problems involving magnetic forces. Solving for the force F in Eq. (35-8), we obtain

$$F = qvB \sin \theta \qquad \text{(35-8)}$$

which is the more useful form for direct calculation of magnetic forces. The force F is in newtons when the charge q is expressed in coulombs; the velocity v is in meters per second; and the flux density is in teslas. The angle θ indicates the direction of **v** with respect to **B**. The force **F** is *always* perpendicular to both **v** and **B**.

EXAMPLE 35-2 An electron is projected from left to right into a magnetic field directed vertically downward. The velocity of the electron is 2×10^6 m/s, and the magnetic flux density of the field is 0.3 T. Find the magnitude and direction of the magnetic force on the electron.

Solution The electron is moving in a direction perpendicular to **B**. Thus, $\sin \theta = 1$ in Eq. (35-8), and we find the force F as follows:

$$\begin{aligned}
F &= qvB \sin \theta \\
&= (1.6 \times 10^{-19}\ \text{C})(2 \times 10^6\ \text{m/s})(0.3\ \text{T})(1) \\
&= 9.6 \times 10^{-14}\ \text{N}
\end{aligned}$$

Application of the right-hand-screw rule will show that the direction of the force is *out of the page,* or toward the reader. (It would be into the page for a positive charge like a proton.)

35-7
FORCE ON A CURRENT-CARRYING WIRE

When an electric current passes through a conductor lying in a magnetic field, each charge q flowing through the conductor experiences a magnetic force. These forces are transmitted to the conductor as a whole, causing each unit of length to experience a force. If a total quantity of charge Q passes through the length l of the wire (Fig. 35-17) with an average velocity $\bar{v}$, perpendicular to a magnetic field B, the net

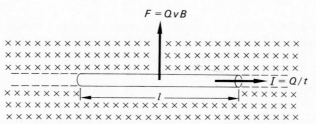

Fig. 35-17 Magnetic force on a current-carrying conductor.

force on that segment of wire is

$$F = Q\bar{v}B$$

The average velocity for each charge passing through the length l in the time t is l/t. Thus the net force on the entire length becomes

$$F = Q\frac{l}{t}B$$

Rearranging and simplifying, we obtain

ELECTRICITY AND MAGNETISM

$$F = \frac{Q}{t} lB = IlB$$

where I represents the current in the wire.

Just as the magnitude of the force on a moving charge varies with velocity direction, the force on a current-carrying conductor depends on the angle the current makes with the flux density. In general, if a wire of length l makes an angle θ with the B field, as illustrated in Fig. 35-18, it will experience a force given by

$$\boxed{F = BIl \sin \theta}$$

(35-10)

where I is the current through the wire. When B is in *teslas*, l is in *meters*, and I is in *amperes*, the force will be in *newtons*.

Fig. 35-18

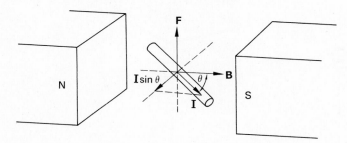

The direction of the magnetic force on a current-carrying conductor can be determined by the right-hand-screw rule in the same way as for a moving charge. When **I** is rotated into **B**, the direction of the force **F** is in the same direction as the advance of a right-hand screw. The force **F** is *always* perpendicular to both **I** and **B**.

EXAMPLE 35-3   The wire in Fig. 35-18 makes an angle of 30° with respect to a B field of 0.2 T. If the length of the wire is 8 cm and a current of 4 A passes through it, determine the magnitude and direction of the resultant force on the wire.

Solution   Direct substitution into Eq. (35-10) yields

$$F = BIl \sin \theta$$
$$= (0.2 \text{ T})(4 \text{ A})(0.08 \text{ m})(\sin 30°)$$
$$= 0.032 \text{ N}$$

Application of the right-hand-screw rule shows the direction of the force to be upward, as indicated in Fig. 35-18. If the current direction were reversed, the force would be directed downward.

35-8

MAGNETIC FIELD OF A LONG STRAIGHT WIRE

Oersted's experiment demonstrated that electric charge in motion, or a current, sets up a magnetic field in the space surrounding it. Up to this point, we have been discussing the force that such a field will exert on a second current-carrying conductor or on a charge moving in the field. We shall now begin to calculate the magnetic fields produced by electric currents.

Let us first examine the flux density surrounding a long straight wire carrying a constant current. If iron filings are sprinkled on the paper surrounding the wire in

Fig. 35-19, they will become aligned in concentric circles around the wire. Similar investigation of the area surrounding the wire with a magnetic compass will confirm that the magnetic field is circular and directed in a clockwise fashion, as viewed along the direction of conventional (positive) current. A convenient method devised by Ampère to determine the direction of the field surrounding a straight wire is called the *right-hand-thumb rule* (refer to Fig. 35-19).

> *If the wire is grasped with the right hand so that the thumb points in the direction of the conventional current, the curled fingers of that hand will point in the direction of the magnetic field.*

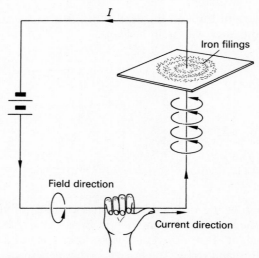

Fig. 35-19 The magnetic field surrounding a straight current-carrying conductor.

The magnetic induction, or flux density, at a perpendicular distance d from a long straight wire carrying a current I, as shown in Fig. 35-20, can be calculated from

$$B = \frac{\mu I}{2\pi d}$$

Long Wire (35-11)

where μ is the permeability of the medium surrounding the wire. In the special cases of a vacuum, air, and nonmagnetic media, the permeability μ_0 is

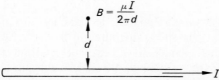

Fig. 35-20 The magnetic field B at a perpendicular distance d from a long current-carrying conductor.

$$\mu_0 = 4\pi \times 10^{-7} \text{ T} \cdot \text{m/A} \qquad (35\text{-}12)$$

The units are determined from Eq. (35-11).

EXAMPLE 35-4 Determine the magnetic induction in the air 5 cm from a long wire carrying a current of 10 A.

Solution From Eq. (35-11)

$$B = \frac{\mu_0 I}{2\pi d} = \frac{(4\pi \times 10^{-7} \text{ T} \cdot \text{m/A})(10 \text{ A})}{(2\pi)(0.05 \text{ m})}$$

$$= 4 \times 10^{-5} \text{ T}$$

The direction of the magnetic induction is determined from the right-hand-thumb rule. For the case illustrated in Fig. 35-20, it would be out of the paper.

For a derivation of Eq. (35-11) and other relations which follow, the reader is referred to the *Biot-Savart law* or to *Ampère's law*. Many conventional physics texts provide a complete analysis, which usually includes the methods of calculus.

35-9
OTHER MAGNETIC FIELDS

If a wire is bent into a circular loop and connected to a source of current, as shown in Fig. 35-21a, a magnetic field very similar to that for a bar magnet will be set up. The right-hand-thumb rule will still serve to give the field direction in a rough manner, but now the flux lines are no longer circular. The magnetic flux density varies considerably from point to point.

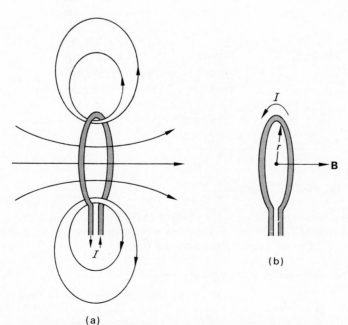

(a)

(b)

Fig. 35-21 The magnetic field at the center of a circular loop.

The magnetic induction at the center of a circular loop of radius r carrying a current I is given by

$$B = \frac{\mu I}{2r}$$

Center of Loop (35-13)

The direction of B is perpendicular to the plane of the loop. If the wire consists of a coil having N turns, Eq. (35-13) becomes

$$B = \frac{\mu NI}{2r}$$

Center of Coil (35-14)

A *solenoid* consists of many circular turns of wire wound in the form of a helix, as illustrated in Fig. 35-22. The magnetic field produced is very similar to that of a bar magnet. The magnetic induction in the interior of a solenoid is given by

$$B = \frac{\mu NI}{L}$$

Solenoid (35-15)

where N = number of turns
    I = current, A
    L = length of the solenoid, m

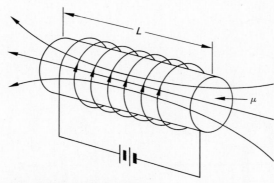

Fig. 35-22 The solenoid.

EXAMPLE 35-5   A solenoid is constructed by winding 400 turns of wire on a 20-cm iron core. The relative permeability of the iron is 13,000. What current is required to produce a magnetic induction of 0.5 T in the center of the solenoid?

Solution   The permeability of the core is

$$\mu = \mu_r \mu_0 = (13,000)(4\pi \times 10^{-7}\,\text{T} \cdot \text{m/A})$$
$$= 1.63 \times 10^{-2}\,\text{T} \cdot \text{m/A}$$

Solving for I in Eq. (35-15) and substituting known values, we obtain

ELECTRICITY AND MAGNETISM

$$I = \frac{BL}{\mu N} = \frac{(0.5 \text{ T})(0.2 \text{ m})}{(1.63 \times 10^{-2} \text{ T} \cdot \text{m/A})(400 \text{ turns})}$$
$$= 0.015 \text{ A}$$

The diameter of the solenoid is not significant provided that it is small relative to the length L.

One type of solenoid, called a *toroid,* is often used in studying magnetic effects. As will be seen in the following section, the toroid consists of a tightly wound coil of wire in the shape of a doughnut. The magnetic flux density in the core of a toroid is also given by Eq. (35-15).

<table>
<tr><td>**35-10**
HYSTERESIS</td><td></td></tr>
</table>

35-10
HYSTERESIS

We have seen that the lines of magnetic flux are more numerous for a solenoid with an iron core than for a solenoid in air. The flux density is related to the permeability μ of the material serving as a core for the solenoid. Recall that the field intensity H and flux density B are related to each other according to the equation

$$B = \mu H$$

Comparison of this relationship with Eq. (35-15) shows that for a solenoid

$$H = \frac{NI}{L} \tag{35-16}$$

Note that the magnetic intensity is independent of the permeability of the core. It is a function only of the number of turns N, the current I, and the solenoid length L. The magnetic intensity is expressed in *amperes per meter.*

We can study the magnetic properties of matter by observing the flux density B produced as a function of the magnetizing current or as a function of the magnetic intensity H. This can be more easily done when a substance is fashioned into the form of a toroid, as in Fig. 35-23. The magnetic field set up by a current in the magnetizing windings is confined wholly to the toroid. Such a device is often called a *Rowland ring* after J. H. Rowland, who used it to study the magnetic properties of many materials.

Fig. 35-23 Rowland's ring.

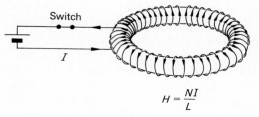

$$H = \frac{NI}{L}$$

Suppose we begin by studying the magnetic properties of a material with an unmagnetized Rowland ring fashioned out of the substance. Initially, $B = 0$ and $H = 0$. The switch is closed, and the magnetizing current I is gradually increased, producing a magnetic intensity given by

$$H = \frac{NI}{L}$$

where L is the circumference of the ring. As the material is subjected to an increasing magnetic intensity H, the flux density B increases until the material is *saturated*. Refer to the curve AB in Fig. 35-24. Now, if the current is gradually reduced to zero, the flux density B throughout the core does not return to zero but lags behind the magnetic intensity, as illustrated by the curve BC. (This essentially refers to residual magnetism.) The lack of retraceability is known as *hysteresis*.

Hysteresis is the lagging of the magnetization behind the magnetic intensity.

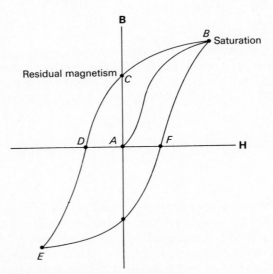

Fig. 35-24 The hysteresis loop.

The only way to bring the flux density B within the ring back to zero is to reverse the direction of the current through the windings. This procedure builds up the magnetic intensity H in the opposite direction, as shown by curve CD. If the magnetization continues to increase in the negative direction, the material will eventually become saturated again with a reversed polarity. Refer to curve DE. Reducing the current to zero again and then increasing it in the positive direction will yield the curve EFB. The entire curve is called the *hysteresis loop*.

The area enclosed by a hysteresis loop is an indication of the quantity of energy that is lost (in the form of heat) by carrying a given material through a complete magnetization cycle. The efficiency of many electromagnetic devices depends on the selection of magnetic materials with low hysteresis. On the other hand, materials which are to remain well magnetized should have large hysteresis.

SUMMARY

We have seen that magnetic fields are created by charge in motion. This very basic principle will underlie much of what follows in your study of electromagnetism. The operation of electric motors, generators, transformers, and an endless variety of industrial instruments requires an understanding of the magnetic field. The major concepts are summarized below:

- The magnetic flux density B in a region of a magnetic field is the number of flux lines which pass through a unit of area perpendicular to the flux.

$$B = \frac{\phi}{A_\perp} = \frac{\phi}{A \sin \theta}$$

Magnetic Flux Density

where ϕ = flux, Wb
A = unit area, m^2
θ = angle that plane of area makes with flux
B = magnetic flux density, T (1 T = 1 Wb/m^2)

- The magnetic flux density B is proportional to the magnetic field intensity H. The constant of proportionality is the permeability of the medium in which the field exists.

$$B = \frac{\phi}{A_\perp} = \mu H$$

For a Vacuum
$\mu_0 = 4\pi \times 10^{-7}$ T $\cdot$ m/A

- The *relative permeability* μ_r is the ratio of μ/μ_0. We can write

$$B = \mu_0 \mu_r H$$ where $$\mu_r = \frac{\mu}{\mu_0}$$

Relative Permeability

- A magnetic field of flux density equal to 1 T will exert a force of 1 N on a charge of 1 C moving perpendicular to the field with a velocity of 1 m/s. The general case is described by Fig. 35-25 in which the charge moves at an angle θ with the field.

$$F = qvB \sin \theta$$ $$B = \frac{F}{qv \sin \theta}$$

Magnetic Force on a Moving Charge

The direction of the magnetic force is given by the right-hand-screw rule, as illustrated in Fig. 35-25.
- The force F on a wire carrying a current I at an angle θ with a flux density B is given by

$$F = BIl \sin \theta$$

Magnetic Force on a Conductor

where l is the length of the conductor.
- Equations for many common magnetic fields are given below:

$$B = \frac{\mu I}{2\pi d}$$ *Long Wire*

$$B = \frac{\mu I}{2r}$$ *Center of Loop*

$$B = \frac{\mu NI}{2r}$$ *Center of Coil*

$$B = \frac{\mu NI}{L}$$ *Solenoid*

Fig. 35-25

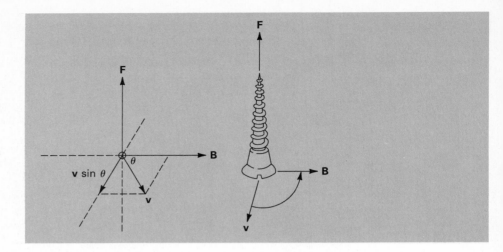

35-1. Define the following terms:
- a. Magnetism
- b. Magnet
- c. Domains
- d. Retentivity
- e. Permeability
- f. Weber
- g. Tesla
- h. Diamagnetic
- i. Paramagnetic
- j. Ferromagnetic
- k. Magnetic poles
- l. Law of magnetic force
- m. Coulomb's law for magnetic forces
- n. Magnetic field
- o. Magnetic flux lines
- p. Magnetic induction
- q. Magnetic saturation
- r. Flux density
- s. Relative permeability
- t. Solenoid
- u. Hysteresis

35-2. How can you positively determine whether a piece of steel is magnetized? How would you determine its polarity if magnetized?

35-3. In general, magnetic materials with high permeability have low retentivity. Why do you think this is true?

35-4. The earth acts like a huge magnet with one pole in the Arctic circle and the other in the Antarctic region. Can you justify the following statement: The geographic North Pole is actually near a magnetic south pole? Explain.

35-5. If an iron bar is placed parallel to a north–south direction and is hammered on one end, the bar becomes a temporary magnet. Explain.

35-6. When a bar magnet is broken into several pieces, each part becomes a magnet with a north and a south pole. Apparently, an isolated pole cannot exist. Explain this, using the domain theory of magnetism.

35-7. Heating magnets or passing electric currents through them will cause a reduction in field strength. Explain.

35-8. The strength of a U magnet will be preserved much longer if an iron plate, called a *keeper*, is placed across the north and south poles. Explain.

35-9. A wire lying along a north–south direction supports an electric current from south to north. What happens to the needle of a compass if the compass is placed (a) above the wire, (b) below the wire, and (c) on the right side of the wire?

35-10. Use the right-hand-thumb rule and the right-hand-screw rule to explain why two adjacent wires experience a force of attraction when the currents are in the same direction. Illustrate your point by drawings.

35-11. Explain with the use of diagrams the repulsion of two adjacent wires carrying oppositely directed currents.

35-12. A circular coil in the plane of the paper supports an electric current. Determine the direction of the magnetic flux near the center of the coil when the current is counterclockwise.

35-13. When an electron beam is projected from left to right into a B field directed into the paper, the beam is deflected into a circular path. Do the electrons travel clockwise or counterclockwise? Why is the path circular? What if the beam consisted of protons?

35-14. A proton passes through a region of space without being deflected. Can we say positively that no magnetic field exists in that region? Discuss.

35-15. Many sensitive electrical instruments are shielded from magnetic effects by surrounding the device with ferromagnetic material. Explain.

35-16. If B is expressed in teslas and μ is in tesla-meters per ampere, what is the SI unit of H?

35-17. Hardened steel has a thick hysteresis loop whereas soft iron has a thin loop. Which should be used to produce a permanent magnet? Which should be used if strong temporary magnetization is desired?

PROBLEMS

Magnetic Fields

35-1. A constant horizontal field of 0.5 T pierces a rectangular loop 120 mm long and 70 mm wide. Determine the magnetic flux through the loop when its plane makes the following angles with the B field: 0°, 30°, 60°, and 90°.

Ans. 0, 2.10 mWb, 3.64 mWb, 4.20 mWb

35-2. A coil of wire 240 mm in diameter is situated so that its plane is perpendicular to a field of density 0.3 T. Determine the magnetic flux through the coil.

35-3. A magnetic flux of 50 μWb passes through a loop of wire having an area of 0.78 m². What is the magnetic flux density?

Ans. 64.1 μT

35-4. A rectangular loop 25 × 15 cm is oriented so that its plane makes an angle θ with a 0.6-T B field. What is the angle θ if the magnetic flux linking the loop is 0.015 Wb?

The Force on a Moving Charge

35-5. A proton ($q = +1.6 \times 10^{-19}$ C) is injected from right to left into a B field of 0.4 T directed vertically upward. If the velocity of the proton is 2×10^6 m/s, what are the magnitude and direction of the magnetic force on the proton?

Ans. 1.28×10^{-13} N, into paper

35-6. An alpha particle ($+2e$) is projected into a 0.12-T magnetic field with a velocity of 3.6×10^6 m/s. What is the magnetic force on the charge at the instant its velocity is directed at an angle of 35° with the magnetic flux?

35-7. An electron moves with a speed of 5×10^5 m/s at an angle of 60° with respect to a magnetic field of density B. If the electron experiences a force of 3.2×10^{-18} N, what is the flux density?

Ans. 46.2 μT

35-8. A particle having a charge q and a mass m is projected into a B field directed into the paper. If the particle has a velocity v, show that it will be deflected into a circular path of radius

$$R = \frac{mv}{qB}$$

Draw a diagram of the motion assuming a positive charge entering the field from left to right. *Hint:* The magnetic force provides the necessary centripetal force for the circular motion.

35-9. A deuteron is a nuclear particle consisting of a proton and a neutron bound together by nuclear forces. The mass of a deuteron is 3.3427×10^{-27} kg, and its charge is $+1.6 \times 10^{-19}$ C. A deuteron projected into a magnetic field of flux density 1.2 T is observed to travel in a circular path of radius 300 mm. What is the velocity of the deuteron? (Refer to Prob. 35-8.)

Ans. 1.72×10^7 m/s

35-10. A proton is moving vertically upward with a speed of 4×10^7 m/s. It passes through a 0.4-T field directed to the right. (a) What are the magnitude and direction of the magnetic force? (b) Suppose an electron replaces the proton but has the same velocity.

Force on a Current-Carrying Conductor

35-11. A long wire carries a current of 6 A in a direction 35° north of an easterly magnetic field of flux density 0.04 T. What are the magnitude and direction of the force on each centimeter of wire?

Ans. 1.38×10^{-3} N, into page

35-12. A wire 1 m long supports a current of 5 A in a direction perpendicular to a magnetic field of flux density 0.034 T. What is the magnetic force on the wire?

35-13. A 12-cm segment of wire carries a current of 4 A and makes an angle of 41° with horizontal magnetic flux. What must be the magnitude of the B field to produce a force of 5 N on this wire segment?

Ans. 15.9 T

35-14. If 80 mm of a straight wire is at an angle of 53° with a B field of 0.23 T, what current is required to give a force of 2 N on this length of wire?

Calculating Magnetic Fields

35-15. What is the magnetic induction B in air at a point 4 cm from a long wire carrying a current of 6 A?

Ans. 30 μT

35-16. Determine the magnetic induction in air 8 mm from a long wire carrying a current of 14 A.

35-17. A circular loop 240 mm in diameter supports a current of 7.8 A. If it is submerged in a medium of relative permeability 2.0, what is the magnetic induction at the center?

Ans. 81.7 μT

35-18. A circular loop of radius 50 mm carries a current of 15 A. If the wire is in air, determine the magnetic induction at the center of the loop. What is the magnetic induction if the loop is submerged in a medium whose relative permeability is 3.0?

35-19. A circular coil having 40 turns of wire in air has a radius of 6 cm. What current must exist in the coil to produce a flux density of 2 mT at its center?

Ans. 4.77 A

35-20. A circular coil having 60 turns has a radius of 75 mm. What current must exist in the coil to produce a flux density of 300 μT at the center of the coil?

35-21. A solenoid of length 30 cm and diameter 4 cm is closely wound with 400 turns of wire around a nonmagnetic material. If the current in the wire is 6 A, determine the magnetic induction at the center of the solenoid.

Ans. 10.1 mT

Additional Problems

35-22. A velocity selector is a device (Fig. 35-26) which utilizes crossed E and B fields to select ions of only one velocity v. Positive ions of charge q are projected into

Fig. 35-26 The velocity selector.

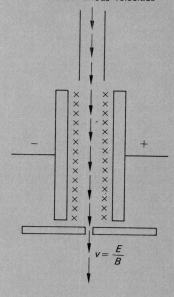

Source of positive ions of various velocities

$$v = \frac{E}{B}$$

the perpendicular fields at varying speeds. Ions with velocities sufficient to make the magnetic force equal and opposite to the electric force pass through the bottom slit undeflected. Show that the speed of these ions can be found from

$$v = \frac{E}{B}$$

35-23. What will be the velocity of protons injected through the velocity selector of Fig. 35-26 if $E = 3 \times 10^5$ V/m and $B = 0.25$ T?

Ans. 1.2×10^6 m/s

35-24. A singly charged Li^7 ion ($+e$) is accelerated through a potential difference of 500 V and then enters at right angles to a magnetic field of 0.4 T. The radius of the resulting circular path is 2.13 cm. What is the mass of the lithium ion?

35-25. A sodium ion ($q = +1.6 \times 10^{-19}$ C, $m = 3.818 \times 10^{-27}$ kg) is moving through a B field with a velocity of 4×10^4 m/s. What must be the magnitude of the field if the ion is to follow a circular path of radius 200 mm?

Ans. 4.77 mT

35-26. Two parallel wires (see Fig. 35-27) carrying currents I_1 and I_2 are separated by a distance d. Show that the force per unit length F/l each wire exerts on the other is given by

$$\frac{F}{l} = \frac{\mu I_1 I_2}{2\pi d}$$

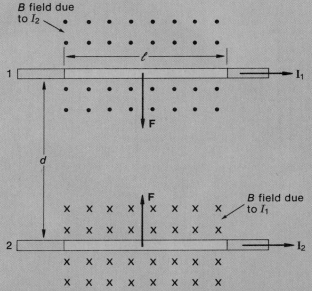

Fig. 35-27 Two parallel wires carrying a steady current exert a force on each other.

35-27. Two wires lying in a horizontal plane carry parallel currents of 15 A each and are 200 mm apart in air. If both currents are directed to the right, what are the magnitude and direction of the flux density at a point midway between the wires? What is the force per unit length that each wire exerts on the other? Is it attraction or repulsion?

Ans. $B(\text{net}) = 0$, 2.25×10^{-4} N/m, attraction

35-28. A solenoid has a length of 20 cm and is wound with 220 turns of wire carrying a current of 5 A. What should be the relative permeability of the core in order to produce a magnetic induction of 0.2 T at the center of the solenoid?

35-29. Two long, fixed, parallel wires A and B are 10 cm apart in air and carry currents of 6 A and 4 A, respectively, in opposite directions. (a) Determine the net flux density on a line midway between these conductors and parallel with them. (b) What is the magnetic force per unit length on a third wire placed midway between A and B and carrying a current of 2 A in the same direction as A?

Ans. (a) 40 μT; (b) 80 μN/m, toward A

35-30. Consider the two wires in Fig. 35-28 where the dot indicates current out of the page and the cross indicates current into the page. What is the resultant flux density at point *A* and at point *B*?

Fig. 35-28

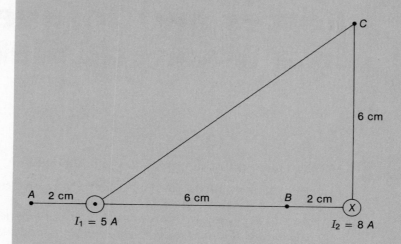

35-31. What are the magnitude and direction of the resultant flux density B at point *C* in Fig. 35-28?

Ans. 22.2 μT, 21.2° N of E

36

Forces and Torques in a Magnetic Field

After completing this chapter, you should be able to:

1. Determine the direction of the magnetic force on a current-carrying conductor in a known B field.
2. Write and apply equations for calculating the *magnetic torque* on a coil or a solenoid of known area, number of turns, and current, when located in a magnetic field of known flux density.
3. Explain with drawings the function of each part of a laboratory *galvanometer*; describe how it may be converted to an ammeter and to a voltmeter.
4. Calculate the *multiplier resistance* necessary to increase the range of a dc voltmeter which contains a galvanometer of fixed sensitivity.
5. Calculate the *shunt resistance* necessary to increase the range of a galvanometer or ammeter of constant sensitivity.
6. Explain the operation of a simple *dc motor*, discussing the function of each of its parts, with particular emphasis on the *split-ring commutator*.

We have seen that a current-carrying conductor placed in a magnetic field will experience a force which is perpendicular both to the current I and to the magnetic induction B. A coil suspended in a magnetic field will experience a *torque* due to equal and opposite magnetic forces on the sides of the coil. Such forces and torques provide the operating principle of many useful devices. In this chapter we discuss the galvanometer, the voltmeter, the ammeter, and the dc motor as applications of electromagnetic forces.

36-1
FORCE AND TORQUE ON A LOOP

A current-carrying conductor suspended in a magnetic field, as illustrated in Fig. 36-1, will experience a magnetic force given by

$$F = BIl \sin \theta = BI_{\perp}l \tag{36-1}$$

where $I_{\perp}$ refers to the current perpendicular to the B field and l is the length of the conductor. The direction of the force is determined from the right-hand-screw rule.

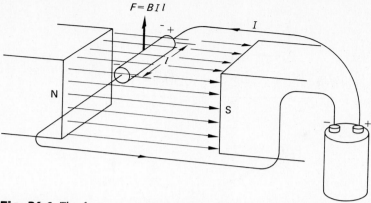

$$F = BIl$$

Fig. 36-1 The force on a current-carrying conductor is directed perpendicular to the magnetic field.

Now let us examine the forces acting on a rectangular current-carrying loop suspended in a magnetic field, shown in Fig. 36-2. The lengths of the sides are a and b, and a current I passes around the loop, as indicated. (The seat of emf and method of leading current into the loop are not shown for simplicity.) Sides mn and op of the loop are each of length a and perpendicular to the magnetic induction B. Thus there are exerted on the sides equal and opposite forces of magnitude

$$F = BIa \qquad (36\text{-}2)$$

The force is directed upward for the segment mn and downward for the segment op.

Similar reasoning will show that equal and opposite forces are also exerted on the other two sides. These forces have a magnitude of

$$F = BIb \sin \alpha$$

where α is the angle that the sides np and mo make with the magnetic field.

Evidently, the loop is in translational equilibrium since the resultant force on the loop is zero. However, the nonconcurrent forces on the sides of length a produce a resultant torque which tends to rotate the coil clockwise. As can be seen from Fig. 36-3, each force produces a torque equal to

$$\tau = BIa \, \frac{b}{2} \cos \alpha$$

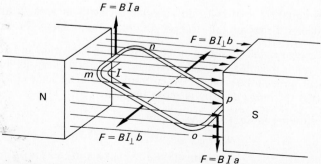

Fig. 36-2 Magnetic forces on a current-carrying loop.

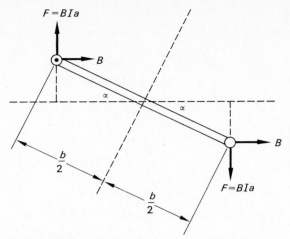

$F = BIa$

B

α

α

B

$\dfrac{b}{2}$

$\dfrac{b}{2}$

$F = BIa$

Fig. 36-3 Calculating the torque on a current loop.

Because the total torque is equal to twice this value, the resultant torque can be found from

$$\tau = BI(a \times b) \cos \alpha \qquad \text{(36-3)}$$

Since $a \times b$ is the area A of the loop, Eq. (36-3) can be written

$$\tau = BIA \cos \alpha \qquad \text{(36-4)}$$

Note that the torque is a maximum when $\alpha = 0°$, that is, when the plane of the loop is parallel with the magnetic field. As the coil turns about its axis, the angle α increases, reducing the rotational effect of the magnetic forces. When the plane of the loop is perpendicular to the field, the angle $\alpha = 90°$ and the resultant torque is zero. The momentum of the coil will cause it to pass this point slightly, but the direction of the magnetic forces ensures that it will oscillate until it reaches equilibrium with the plane of the loop perpendicular to the field.

If the loop is replaced with a closely wound coil having N turns of wire, the general equation for computing the resultant torque is

$$\boxed{\tau = NBIA \cos \alpha} \qquad \text{(36-5)}$$

This equation applies to any complete circuit of area A, and its use need not be restricted to rectangular loops. Any plane loop obeys the same relationship.

EXAMPLE 36-1   A rectangular coil consisting of 100 turns of wire has a width of 16 cm and a length of 20 cm. The coil is mounted in a uniform magnetic field of flux density 8 mT, and a current of 20 A is sent through the windings. When the coil makes an angle of 30° with the magnetic field, what is the torque tending to rotate the coil?

Solution   Substituting in Eq. (36-5), we obtain

$$\tau = NBIA \cos \alpha$$
$$= (100 \text{ turns})(8 \times 10^{-3} \text{ T})(20 \text{ A})(0.16 \text{ m} \times 0.2 \text{ m})(\cos 30°)$$
$$= 0.445 \text{ N} \cdot \text{m}$$

The relationship expressed by Eq. (36-5) applies for computing the torque on a solenoid of area A having N turns of wire. However, in applying the relation we must remember that the angle α is the angle that each turn of wire makes with the field. It is the *complement* of the angle θ between the solenoid axis and the magnetic field (refer to Fig. 36-4). An alternative equation for computing the torque on a solenoid is therefore

$$\tau = NBIA \sin \theta \qquad \qquad \text{Solenoid} \quad (36\text{-}6)$$

You should verify that $\sin \theta$ is equal to $\cos \alpha$ by looking at the figure.

The action of the solenoid in Fig. 36-4 can also be explained in terms of magnetic poles. Applying the right-hand-thumb rule to each turn of wire shows that the solenoid will act as an electromagnet, with north and south poles as indicated.

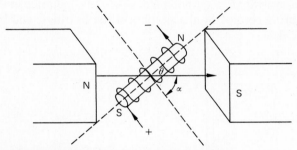

Fig. 36-4 Magnetic torque on a solenoid.

Any device used to detect an electric current is called a *galvanometer*. The operating principle of the majority of such instruments is based on the torque exerted on a coil in a magnetic field. The essential parts are shown in Fig. 36-5. A coil, wrapped around a soft iron core, is pivoted on jeweled bearings between the poles of a perma-

Fig. 36-5 The essential components of a galvanometer.

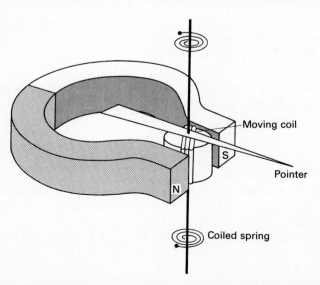

Moving coil

S

Pointer

N

Coiled spring

nent magnet. Its rotational motion is restrained by a pair of spiral springs, which also serve as current leads to the coil. Depending upon the direction of the current being measured, the coil and pointer will rotate either in a clockwise or counterclockwise direction.

For the laboratory galvanometer illustrated in Fig. 36-6 the coil and pointer are shown in the equilibrium position. Note that the permanent magnets are shaped to provide a uniform *radial magnetic field.* This ensures that the pointer deflection will be directly proportional to the current in the coil. If the current through the coil passes into the page on the right and out of the page on the left, magnetic forces in the radial field will produce a clockwise torque. The pointer and coil will move clockwise until the resisting counter-clockwise spring torque equals the magnetic torque produced by the current. Thus the position of the pointer on the marked scale is a measure of the current magnitude. Reversing the current direction will cause an equal pointer deflection in a counterclockwise direction.

The sensitivity of a galvanometer of the type indicated in Fig. 36-6 is determined by the spring torque, by the friction in the bearings, and by the strength of the magnetic field.

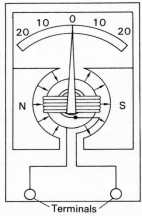

Fig. 36-6 The laboratory galvanometer.

EXAMPLE 36-2   The galvanometer in Fig. 36-6 has a sensitivity of 50 μA per scale division. What current is required to give full deflection of the pointer through 25 scale divisions to the right or left of the equilibrium position?

Solution   The current required is equal to the product of the sensitivity and the number of divisions:

$$I = (50 \ \mu A/div)(25 \ div) = 1250 \ \mu A$$

Galvanometers of this type may be designed to read currents as low as 1 μA (10^{-6} A). Greater sensitivities require a modified design. One such design consists of suspending a coil in a magnetic field by thin thread. The motion of the coil is then observed by mounting a mirror on the thread, utilizing the principle of an optical lever.

Many important dc measuring instruments utilize a galvanometer as an indicating element. Two of the most common are the *voltmeter* and the *ammeter*. The first of these will be discussed in this section as an example of how the galvanometer can be used to measure voltages.

The potential difference across a galvanometer is extremely small even when a large-scale deflection occurs. Thus, if a galvanometer is to be used to measure voltage, it must be converted to a high-resistance instrument. Suppose, for example, it is desired to measure the drop in voltage across the battery in Fig. 36-7. This voltage must be measured without appreciably disturbing the current through the circuit. In other words, the voltmeter must draw negligible current. To accomplish this, a large multiplier resistance R_m is placed in series with the galvanometer as an integral part of a dc voltmeter.

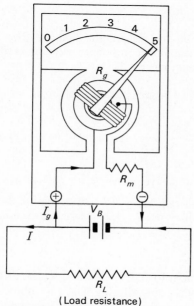

(Load resistance)

Fig. 36-7 The dc voltmeter.

Note that the galvanometer used in a voltmeter is adjusted so that its equilibrium position is to the extreme left on the scale. This allows for a greater range of measurement but unfortunately requires that the current pass through the coils in one direction only. The sensitivity of the galvanometer is determined by the current I_g required for *full-scale deflection* (maximum pointer deflection) as indicated in Fig. 36-7.

Suppose that the galvanometer coil has a resistance R_g and that the meter is designed to yield full-scale deflection for the current I_g. Such a galvanometer, acting alone, could be calibrated to record voltages from zero up to a maximum value given by

$$V_g = I_g R_g \tag{36-7}$$

By properly choosing the multiplier resistance R_m, we can calibrate the meter to read any desired voltage.

Suppose, for example, we want full-scale deflection of the voltmeter for the voltage V_B in Fig. 36-7. The multiplier resistance R_m must be chosen so that only the small current I_g passes through the galvanometer. Under these conditions,

$$V_B = I_g R_g + I_g R_m$$

Solving for R_m, we obtain

$$R_m = \frac{V_B}{I_g} - R_g \qquad (36\text{-}8)$$

Thus we see that the multiplier resistance R_m is equal to the total resistance V_B/I_g less the galvanometer resistance R_g.

EXAMPLE 36-3 A certain galvanometer has an internal resistance of 30 Ω and gives a full-scale deflection for a current of 1 mA. Calculate the multiplier resistance necessary to convert this galvanometer into a voltmeter whose maximum range is 50 V.

Solution The multiplier resistance R_m must be such that the total drop in voltage through R_g and R_m is 50 V. From Eq. (36-8),

$$R_m = \frac{50 \text{ V}}{1 \times 10^{-3} \text{ A}} - 30 \text{ Ω} = 50{,}000 \text{ Ω} - 30 \text{ Ω}$$

$$= 49{,}970 \text{ Ω}$$

Note that the total resistance of the voltmeter $(R_m + R_g)$ is 50 kΩ.

A voltmeter must be connected in *parallel* with the part of the circuit whose potential difference is to be measured. This is necessary so that the large resistance of the voltmeter will not greatly alter the circuit.

36-5
THE DC AMMETER

An ammeter is a device which, through calibrated scales, gives an indication of the electric current without appreciably altering it. A galvanometer is an ammeter, but its range is limited by the extreme sensitivity of the moving coil. The range of the galvanometer can be extended simply by adding a very low resistance, called a *shunt*, in parallel with the galvanometer coil (refer to Fig. 36-8). Placing the shunt in parallel assures that the ammeter as a whole will have a very low resistance, which is necessary if the current is to be essentially unaltered. The major portion of the current will pass through the shunt. Only the small current I_g required for galvanometer deflection will be drawn from the circuit. For example, if 10 A goes through an ammeter, 9.99 A may go through the shunt and only 0.01 A through the coil itself.

Suppose the range of a galvanometer is to be extended to measure a maximum current I of the circuit in Fig. 36-8. A shunt resistance R_s must be chosen such that only the current I_g, required for full-scale deflection, passes through the galvanometer coil. The remainder of the current I_s must pass through the shunt. Since R_g and R_s are in parallel, the *IR* drop across each resistance must be identical:

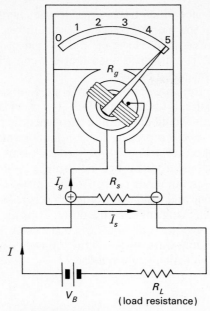

Fig. 36-8 The dc ammeter.

$$I_s R_s = I_g R_g \qquad (36\text{-}9)$$

The shunt current I_s is the difference between the circuit current I and the galvanometer current I_g. Thus Eq. (36-9) becomes

$$(I - I_g)R_s = I_g R_g$$

Solving for the shunt resistance R_s, we obtain the following useful relation:

$$\boxed{R_s = \frac{I_g R_g}{I - I_g}}$$

EXAMPLE 36-4   A certain galvanometer has an internal coil resistance of 46 Ω, and a current of 200 mA is required for full-scale deflection. What shunt resistance must be used to convert the galvanometer into an ammeter whose maximum range is 10 A?

Solution   Equation (36-9) gives

$$R_s = \frac{(0.2 \text{ A})(46 \text{ }\Omega)}{10 \text{ A} - 0.2 \text{ A}} = \frac{9.2 \text{ V}}{9.8 \text{ A}}$$
$$= 0.939 \text{ }\Omega$$

It is important to remember that an ammeter must be connected in *series* with the portion of a circuit through which the current is to be measured. The circuit must be opened at some convenient point and the ammeter inserted. If by mistake the ammeter were placed in parallel, the circuit would be shorted across the ammeter because of its extremely low resistance.

An electric motor is a device which transforms electric energy into mechanical energy. The dc motor, like the moving coil of a galvanometer, consists of a current-carrying coil in a magnetic field. However, the motion of the coil in a motor is unrestrained by springs. The design is such that the coil will rotate continuously under the influence of magnetic torque.

A very simply dc motor, consisting of a single current-carrying loop suspended between two magnetic poles, is illustrated in Fig. 36-9. Normally, the torque exerted on a current-carrying loop would diminish to zero when its plane becomes perpendicular to the field. In order to provide for continuous rotation of the loop, the current in the loop must be automatically reversed each time it turns through 180°.

The current reversal is accomplished by using a *split-ring commutator,* as shown in Fig. 36-9. The commutator consists of two metal half rings fused to each end of the conducting loop and insulated from each other. As the loop rotates, each brush touches first one half ring and then the other. Thus the electrical connections are reversed every half revolution at times when the loop is perpendicular to the magnetic field. In this manner, the torque acting on the loop is always in the same direction, and the loop will rotate continuously.

Fig. 36-9 The dc motor.

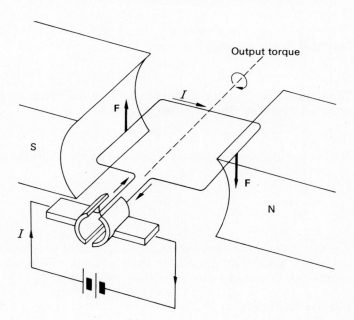

Although actual dc motors operate on the principle described for Fig. 36-9, there are a number of designs which increase the available torque and make it more uniform. One such design is shown in Fig. 36-10. A greater magnetic field is established by replacing the permanent magnets with electromagnets. Additionally, the torque can be increased and made more uniform by adding a number of different coils, each having a large number of turns around a slotted iron core called the *armature.* The commutator is an automatic switching arrangement which maintains the currents in the directions shown in the figure, regardless of the orientation of the armature. More will be said about the dc motor in the following chapter.

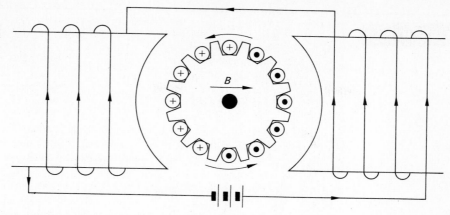

Fig. 36-10 Greater, more uniform torque is possible in commercial motors with many armature coils.

SUMMARY

Magnetic torques on current loops form the basis for so many applications that a strong foundation is essential. The operation of generators, motors, ammeters, voltmeters, and many industrial instruments is directly affected by magnetic forces and torques. The major concepts to be remembered are summarized below:

- The magnetic torque on a current-carrying coil of wire having N turns of wire is given by

$$\tau = NBIA \cos \alpha$$

Magnetic Torque

where N = number of turns of wire
B = flux density, T
I = current, A
A = area of the coil of wire, m^2
α = angle plane of coil makes with field

- The same equation applies for a solenoid, except that the angle α is generally replaced with θ, the angle the solenoid axis makes with the field.

$$\tau = NBIA \sin \theta$$

Torque on a Solenoid

- The multiplier resistance R_m which must be placed in series with a voltmeter to give full-scale deflection for V_B is found from

$$R_m = \frac{V_B - I_g R_g}{I_g}$$

Multiplier Resistance

I_g is the galvanometer current, and R_g is its resistance.
- The shunt resistance R_s which must be placed in parallel with an ammeter to give full-scale deflection for a current I is

$$R_s = \frac{I_g R_g}{I - I_g}$$

Shunt Resistance

36-1. Define the following terms:
   a. Magnetic torque
   b. Galvanometer
   c. Voltmeter
   d. Ammeter
   e. Full-scale deflection
   f. Shunt resistance
   g. Multiplier resistance
   h. Sensitivity
   i. Motor
   j. Commutator
   k. Armature

36-2. The equal and opposite forces acting on a current loop in a magnetic field form what is called a *couple*. Show that the resultant torque on such a couple is the product of one of the forces and the perpendicular distance between their lines of action.

36-3. Why is it necessary to provide a *radial* magnetic field for the coil of a galvanometer?

36-4. A coil of wire is suspended by a thread with the plane of the loop coinciding with the plane of the paper. If the coil is placed in a magnetic field directed from left to right, and if a clockwise current is sent through the coil, describe its motion.

36-5. How are the actions of a galvanometer and a motor similar? How are they different?

36-6. How does the core of a galvanometer coil affect the sensitivity of the instrument?

36-7. Explain the torque exerted on a bar magnet suspended in a magnetic field without referring to magnetic poles. Discuss, from an atomic standpoint, how the observed torque may arise from the same cause as the torque on a current loop.

36-8. Suppose the range of an ammeter is to be increased N-fold. Show that the shunt resistance which must be placed across the terminals of the ammeter is given by

$$R_s = \frac{R_a}{N - 1}$$

where R_a is the ammeter resistance.

36-9. Suppose the range of a given voltmeter is to be increased N-fold. Show that the multiplier resistance which must be placed in series with the voltmeter is given by

$$R_m = (N - 1)R_v$$

where R_v is the voltmeter resistance.

36-10. Show by diagrams how an ammeter and a voltmeter should be connected in a circuit. Compare the resistances of the two devices.

36-11. Discuss the error caused by the insertion of an ammeter into an electric circuit. How is this error minimized?

36-12. Discuss the error caused by the insertion of a voltmeter into a circuit. How is the error minimized?

36-13. A voltmeter is connected to a battery, and the reading is taken. An accurate resistance box is then placed in the circuit and adjusted until the voltmeter reading is one-half of its previous value. Show that the voltmeter resistance must be just equal to the added resistance. (This is called the *half-deflection method* for determining voltmeter resistance.)

36-14. Explain what happens when a voltmeter is erroneously placed in series in a circuit. What happens if an ammeter is mistakenly placed in parallel?

36-15. Prepare a short paper on the following topics:

a. Ballistic galvanometer   c. Dynamometer
b. Ohmmeter   d. Wattmeter

36-16. Plot a graph of the torque as a function of time for a single-loop dc motor.

PROBLEMS

Magnetic Torque on a Loop

36-1. A rectangular loop of wire 6×10 cm is placed with its plane parallel to a magnetic field of flux density 0.08 T. What is the resultant torque on the loop if it carries a current of 14 A?

Ans. 0.00672 N · m

36-2. A rectangular loop of wire has an area of 0.3 m². The plane of the loop makes an angle of 30° with a magnetic field of flux density 0.75 T. What is the resultant torque when the current in the loop is 7 A?

36-3. Calculate the magnetic flux density required to give a loop of 400 turns a torque of 0.5 N · m when its plane is parallel to the field. The dimensions of each turn are 70×120 mm, and the current is 9 A.

Ans. 16.5 mT

36-4. What current is required to produce a maximum torque of 0.8 N · m on a solenoid having 800 turns of area 0.4 m²? The flux density of the field is 3.0 mT. What is the position of the solenoid for maximum torque?

36-5. The axis of a solenoid makes an angle of 34° with a 5-mT magnetic field. If the solenoid has 750 turns, what current is required to produce a torque of 4.0 N · m? The area of each turn is 0.25 m².

Ans. 7.63 A

Galvanometers, Voltmeters, and Ammeters

36-6. A galvanometer coil 50×120 mm is mounted in a constant radial field of flux density 0.2 T. The coil has 600 turns of wire. What current is required to develop a torque of 3.6×10^{-5} N · m?

36-7. A certain voltmeter draws 0.02 mA for full-scale deflection of 50 V. **(a)** What is the resistance of the meter? **(b)** What is the resistance per volt? **(c)** What multiplier resistance must be added to permit the measurement of 150 V full scale?

Ans. (a) 2.5 MΩ, (b) 50,000 Ω/V, (c) 5 MΩ

36-8. A current of only 90 μA will produce full-scale deflection of a voltmeter which is designed to read 50 mV full scale. **(a)** What is the resistance of the meter? **(b)** What multiplier resistance is required to convert this voltmeter to an instrument that reads 100 mV full scale?

36-9. A galvanometer has a sensitivity of 20 μA per scale division. What current is required to give full-scale deflection with 25 divisions on either side of the equilibrium position?

Ans. 500 μA

36-10. A galvanometer has a sensitivity of 15 μA per scale division. How many scale divisions will the galvanometer needle be deflected for a current of 60 μA?

36-11. The coil of an ammeter will burn out if a current of more than 40 mA is sent through it. If the coil resistance is 0.5 Ω, what shunt resistance should be added to permit the measurement of 4 A?

Ans. 0.00505 Ω

36-12. An ammeter which has a resistance of 0.1 Ω is connected in a circuit and indicates a current of 10 A at full scale. A shunt having a resistance of 0.01 Ω is then connected across the terminals of the meter. What new circuit current is needed to produce full-scale deflection of the ammeter?

Additional Problems

36-13. A solenoid consists of 400 turns of wire, each having a radius of 60 mm. What angle does the axis of the solenoid make with the magnetic flux if the current through the wire is 6 A, the flux density is 46 mT, and the resulting torque is 0.80 N · m?

Ans. 39.8°

36-14. A solenoid of 100 turns has a cross-sectional area of 0.25 m². What torque is required to hold the solenoid at an angle of 30° with the field? The current is 10 A and the magnetic field is 42 mT.

36-15. A circular loop consisting of 500 turns carries a current of 10 A in a 0.25-T magnetic field. The area of each turn is 0.2 m². Calculate the torque when the loop makes the following angles with the field: **(a)** 0°, **(b)** 30°, **(c)** 45°, **(d)** 60°, and **(e)** 90°.

Ans. **(a)** 250, **(b)** 217, **(c)** 177, **(d)** 125, **(e)** 0 N · m

36-16. A galvanometer having an internal resistance of 35 Ω requires 1 mA for full-scale deflection. What multiplier resistance is needed to convert this device to a voltmeter which reads a maximum of 30 V?

36-17. A certain galvanometer has an internal resistance of 20 Ω and gives full-scale deflection for a current of 10 mA. Calculate the multiplier resistance required to convert this galvanometer into a voltmeter whose maximum range is 50 V. What is the total resistance of the resulting voltmeter?

Ans. 4980 Ω, 5000 Ω

36-18. What shunt resistance is required to convert the galvanometer of Prob. 36-16 to an ammeter reading 10 mA full scale?

36-19. What shunt resistance is required to convert the galvanometer of Prob. 36-17 to an ammeter which reads a maximum current of 50 mA?

Ans. 5.0 Ω

36-20. A galvanometer has a coil resistance of 50 Ω and a current sensitivity of 1 mA (full scale). What shunt resistance is needed to convert this galvanometer to an ammeter reading 2 A full scale?

36-21. A certain voltmeter reads 150 V full scale. The galvanometer coil has a resistance of 50 Ω and produces a full-scale deflection on 20 mV. Find the multiplier resistance of the voltmeter.

Ans. 374950 Ω

36-22. A commercial 3-V voltmeter requires a current of 0.02 mA to produce full-scale deflection. How can it be converted to an instrument with a range of 150 V?

36-23. A laboratory ammeter has a resistance of 0.01 Ω and reads 5 A full scale. What shunt resistance is needed to increase the range of the ammeter tenfold?

Ans. 0.00111 Ω

36-24. A voltmeter of range 150 V and total resistance 15,000 Ω is connected in series with another voltmeter of range 100 V and total resistance 20,000 Ω. What will each meter read when they are connected across a 120-V battery of negligible internal resistance?

37 Electromagnetic Induction

OBJECTIVES

After completing this chapter, you should be able to:

1. Explain and calculate the current or emf induced by a conductor moving through a magnetic field.
2. Write and apply an equation relating the induced emf in a length of wire moving with a velocity *v* directed at an angle θ with a known *B* field.
3. State *Lenz's law* and use it or the *right-hand rule* to determine the direction of induced emf or current.
4. Explain the operation of simple ac and dc generators; calculate the instantaneous and maximum emf or current generated by a simple ac generator.
5. Demonstrate with diagrams your knowledge of *series-wound* and *shunt-wound* motors and solve for starting current and operating voltage in electrical problems.
6. Explain the operation of a *transformer* and solve problems involving changes in current, voltage, or power.

We have seen that an electric field can produce a magnetic field. In this chapter, you will learn that the reverse is also true: a magnetic field can give rise to an electric field. An electric current is *generated* by a conductor which is caused to move relative to a magnetic field. A rotating coil in a magnetic field *induces* an alternating emf, which produces an *alternating current* (ac). This process is called *electromagnetic induction* and it is the operating principle behind many electrical devices. For example, electric ac generators and transformers use electromagnetic induction to produce and distribute electric power economically.

**37-1
FARADAY'S
LAW**

Faraday discovered that when magnetic flux lines are cut by a conductor, an emf is produced between the end points of the conductor. For example, an electric current is induced in the conductor of Fig. 37-1a as it is moved downward across the flux

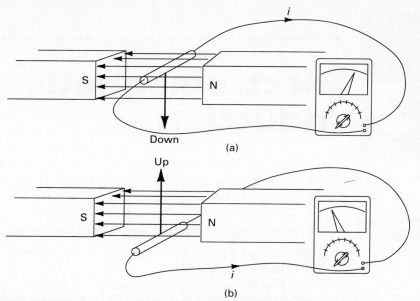

(a)

(b)

Fig. 37-1 When magnetic flux lines are cut by a conductor, an electric current is induced.

lines. (The lower case symbol *i* will be used for induced currents and for varying currents.) The faster the movement, the more pronounced the galvanometer deflection. When the conductor is moved upward across the flux lines, a similar observation is made except that the current is reversed (see Fig. 37-1*b*). If no flux lines are crossed, e.g., the conductor is moved parallel to the field, no current is induced.

Suppose that a number of conductors are moved through a magnetic field, as illustrated by dropping a coil of *N* turns across the flux lines in Fig. 37-2. The magnitude of the induced current is directly proportional to the number of coils and to the rate of motion. Evidently, *an emf is induced by the relative motion between the conductor and the magnetic field.* The same effect is observed when the coil is held stationary and the magnet is moved upward.

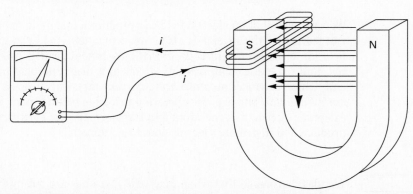

Fig. 37-2 Induced emf in a coil is proportional to the number of turns of wire passing through the field.

Summarizing what we have learned from these experiments, we can state that:

1. *Relative motion between a conductor and a magnetic field induces an emf in the conductor.*
2. *The direction of the induced emf depends upon the direction of motion of the conductor with respect to the field.*
3. *The magnitude of the emf is directly proportional to the rate at which magnetic flux lines are cut by the conductor.*
4. *The magnitude of the emf is directly proportional to the number of turns of the conductor crossing the flux lines.*

A quantitative relationship for computing the induced emf in a coil of N turns is given by

$$\mathcal{E} = -N\frac{\Delta\phi}{\Delta t} \qquad\qquad (37\text{-}1)$$

where $\mathcal{E}$ = average induced emf

$\Delta\phi$ = change in magnetic flux occurring during time interval Δt

A magnetic flux changing at the rate of one weber per second will induce an emf of one volt for each turn of the conductor. The negative sign in Eq. (37-1) means that the induced emf is in such a direction as to oppose the change that produced it, as will be explained in Sec. 37-3.

Now let us discuss how magnetic flux ϕ linking a conductor may change. In the simple case of a straight wire moving through lines of flux, $\Delta\phi/\Delta t$ represents the rate at which the flux linked by the conductor changes. However, a continuous circuit is necessary for an induced current to exist, and more often we are interested in the emf induced in a loop or coil of wire.

Recall that the magnetic flux ϕ passing through a loop of effective area A is given by

$$\phi = BA \qquad\qquad (37\text{-}2)$$

where B is the magnetic flux density. When B is in *teslas* (*webers per square meter*) and A is in *square meters,* ϕ is expressed in *webers.*

A change in flux ϕ can occur in two principle ways:

1. By changing the flux density B going through a constant loop area A:

$$\Delta\phi = (\Delta B)A \qquad\qquad (37\text{-}3)$$

2. By changing the effective area A in a magnetic field of constant flux density B:

$$\Delta\phi = B(\Delta A) \qquad\qquad (37\text{-}4)$$

Two examples of changing flux density through a constant, stationary coil area are given in Fig. 37-3. In Fig. 37-3*a*, the north pole of a magnet is moved through a circular coil. The changing flux density induces a current in the coil, as indicated by the galvanometer. In Fig. 37-3*b* no current is induced in coil B so long as the current

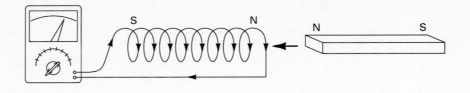

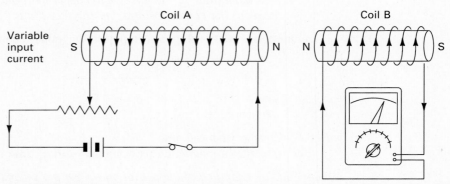

Fig. 37-3 *(a)* Inducing a current by moving a magnet into a coil. *(b)* A changing current in coil *A* induces a current in coil *B*.

in coil *A* is constant. However, by quickly varying the resistance in the left circuit, the magnetic flux density reaching coil *B* can be increased or decreased. While the flux density is changing, a current is induced in the coil on the right.

Note that when the north (N) pole of the magnet is moved into the coil in Fig. 37-3*a*, the current flows in a clockwise direction as viewed toward the magnetic. Therefore, the end of the *coil* near the N pole of the magnet becomes an N pole also (from the right-hand-thumb rule of the last chapter). The magnet and the coil will experience a force of repulsion, making it necessary to exert a force to bring them together. If the magnet is removed from the coil, a force of attraction will exist that makes it necessary to exert a force to separate them. We will see in Sec. 37-3 that such forces are a natural consequence of the conservation of energy.

EXAMPLE 37-1   A coil of wire having an area of 10^{-3} m^2 is placed in a region of constant flux density equal to 1.5 T. In a time interval of 0.001 s, the flux density is reduced to 1.0 T. If the coil consists of 50 turns of wire, what is the induced emf?

Solution   The change in flux density is

$$\Delta B = 1.5 \text{ T} - 1.0 \text{ T} = 0.5 \text{ T}$$

From Eq. (37-3), the change in flux is

$$\Delta\phi = (0.5 \text{ T})(10^{-3} \text{ m}^2) = 5 \times 10^{-4} \text{ Wb}$$

Substituting into Eq. (37-1) yields

$$\mathcal{E} = -N\frac{\Delta\phi}{\Delta t}$$

$$= -50 \text{ turns} \frac{5 \times 10^{-4} \text{ Wb}}{0.001 \text{ s}} = -25 \text{ V}$$

The second general way in which the flux linking a conductor may change is by varying the effective area penetrated by the flux. The following example illustrates this point.

EXAMPLE 37-2 A square coil consisting of 80 turns of wire and having an area of 0.05 m² is placed perpendicular to a field of flux density 0.8 T. The coil is flipped until its plane is parallel to the field in a time of 0.2 s. What is the average induced emf?

Solution The area penetrated by the flux varies from 0.05 m² to zero. Thus the change in flux is

$$\Delta\phi = B(\Delta A)$$
$$= (0.8 \text{ T})(0.05 \text{ m}^2) = 4 \times 10^{-2} \text{ Wb}$$

The induced emf is

$$\mathcal{E} = -N\frac{\Delta\phi}{\Delta t} = (-80 \text{ turns})\frac{4 \times 10^{-2} \text{ Wb}}{0.2 \text{ s}}$$
$$= -16 \text{ V}$$

37-2

EMF INDUCED BY A MOVING WIRE

Another example of a changing area in a constant B field is illustrated in Fig. 37-4. Imagine that a moving conductor of length l slides along a stationary U-shaped conductor with a velocity v. The magnetic flux penetrating the loop increases as the area of the loop increases. Consequently, an emf is induced in the moving wire, and a current passes around the loop.

The origin of the emf can be understood by recalling that a moving charge in a magnetic field experiences a force given by

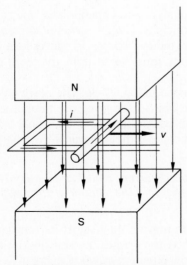

Fig. 37-4 The emf induced by a wire moving perpendicular to a magnetic field.

$$F = qvB$$

For example, in Fig. 37-4, free charges on the conductor are moved to the right through a magnetic field directed into the paper. The magnetic force F acting on the charges moves them through the length l of wire in a direction given by the right-hand-screw rule (away from the reader for conventional current). The work per unit of charge represents the induced emf, which is given by

$$\mathcal{E} = \frac{\text{work}}{q} = \frac{Fl}{q} = \frac{qvBl}{q}$$
$$= Blv \qquad \qquad (37\text{-}5)$$

If the velocity v of the moving wire is directed at an angle θ with the B field, a more general form is needed for Eq. (37-5):

$$\boxed{\mathcal{E} = Blv \sin \theta}$$

EXAMPLE 37-3   A 0.2-m length of wire moves at a constant velocity of 4 m/s in a direction that is 40° with respect to a magnetic flux density of 0.5 T. Calculate the induced emf.

Solution   Direct substitution into Eq. (37-6) yields

$$\mathcal{E} = (0.5 \text{ T})(0.2 \text{ m})(4 \text{ m/s})(\sin 40°)$$
$$= 0.257 \text{ V}$$

The minus sign does not appear in Eq. (37-6) because the direction of the induced emf is the same as the direction of the magnetic force performing work on the moving charge.

37-3
LENZ'S LAW

Throughout the discussions of all physical phenomena, one guiding principle stands out above all the rest: the principle of *conservation of energy*. An emf cannot exist without a cause. Whenever an induced current produces heat or performs mechanical work, the necessary energy must come from the work done in inducing the current.

Recall the example discussed in Fig. 37-3a. The north pole of a magnet pushed into a coil induces a current which itself gives rise to another magnetic field. The second field produces a force which opposes the original force. Withdrawing the magnet creates a force which opposes the removal of the magnet. This is an illustration of *Lenz's law:*

> Lenz's Law: *An induced current will flow in such a direction that it will oppose by its magnetic field the motion of the magnetic field that is producing it.*

The more work that is done in moving the magnet into the coil, the greater will be the induced current and hence, the greater the resisting force. We might have expected this result from the law of conservation of energy. To produce a larger current, we must perform a greater amount of work.

The direction of the current induced in a straight conductor moving through a magnetic field can be determined from Lenz's law. However, there is an easier method, as illustrated in Fig. 37-5. It is known as *Fleming's rule*, or the *right-hand rule:*

> **Fleming's Rule:** *If the thumb, forefinger, and middle finger of the right hand are held at right angles to each other, with the thumb pointing in the direction in which the wire is moving and the forefinger pointing in the field direction (N to S), the middle finger will point in the direction of induced conventional current.*

Fleming's rule is easy to apply and very useful for studying the currents induced by a simple generator. Students sometimes remember the rule by memorizing *motion– flux–current*. These are the directions given by the thumb, forefinger, and middle finger, respectively.

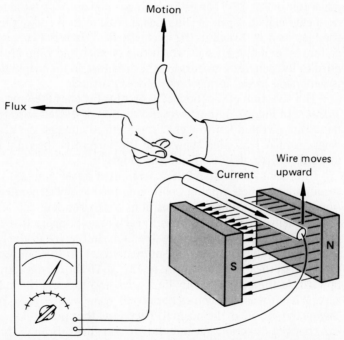

Fig. 37-5 The right-hand rule for determining the direction of induced current.

An electric generator converts mechanical energy into electric energy. We have seen that an emf is induced in a conductor when it experiences a change in flux linkage. When the conductor forms a complete circuit, an induced current can be detected. In a generator, a coil of wire is rotated in a magnetic field, and the induced current is transmitted by wires for long distances from its origin.

The construction of a simple generator is shown in Fig. 37-6. Essentially, there are three components: a *field magnet*, an *armature*, and *slip rings* with *brushes*. The

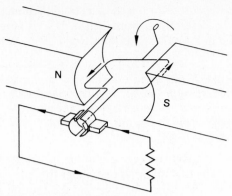

Fig. 37-6 The ac generator.

field magnet may be a permanent magnet or an electromagnet. The armature for the generator in Fig. 37-6 consists of a single loop of wire suspended between the poles of the field magnet. A pair of slip rings is fused to each end of the loop; they rotate with the loop as it is turned in the magnetic field. Induced current is led away from the system by graphite brushes which ride on each slip ring. Mechanical energy is supplied to the generator by turning the armature in the magnetic field. Electric energy is generated in the form of an induced current.

The direction of the induced current must obey Fleming's rule of *motion–flux–current.* In Fig. 37-6, the downward motion of the left wire segment crosses a magnetic flux directed left to right. The induced current is, therefore, toward the slip rings. Similar reasoning shows that the current in the right loop, which is moving upward, will be away from the slip rings.

In order to understand the operation of an ac generator, let us follow the loop through a complete rotation, observing the current generated throughout the rotation. Figure 37-7 shows four positions of the rotating coil and the direction of the current delivered to the brushes in each case. Suppose that the loop is turned mechanically in a counterclockwise direction. In Fig. 37-7a the loop is horizontal, with side M facing the south (S) pole of the magnet. At this point, a maximum current is delivered in the direction shown. In Fig. 37-7b the loop is vertical, with side M facing upward. At this point, no flux lines are being cut, and the induced current drops to zero. When the loop becomes horizontal again, as in Fig. 37-7c, side M is now facing the north (N) pole of the magnet. Therefore, the current delivered to the slip ring M' has changed direction. An induced current flows through the external resistor in a direction opposite to that experienced earlier. In Fig. 37-7d the loop is vertical again, but now side M faces downward. No flux lines are cut, and the induced current again drops to zero. The loop next returns to horizontal as in Fig. 37-7a, and the cycle repeats itself. Thus the current delivered by such a generator alternates periodically, the direction changing twice each rotation.

The emf generated in each segment of a rotating loop must obey the relation

$$\mathcal{E} = Blv \sin \theta \qquad (37\text{-}6)$$

where v is the velocity of a moving wire segment of length l in a magnetic field of flux density B. The direction of the velocity v with respect to the B field at any instant is

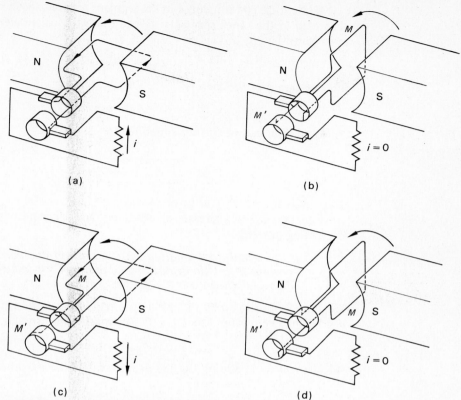

(a)

(b)

(c)

(d)

Fig. 37-7 Production of an alternating current.

denoted by the angle θ. Let us consider the segment M of our rotating current loop when it reaches the position shown in Fig. 37-8. The *instantaneous* emf at that position is given by Eq. (37-6). If the loop rotates in a circle of radius r, the instantaneous velocity v can be found from

where ω is the angular velocity in radians per second. Substituting into Eq. (37-6) gives the instantaneous emf

$$\mathcal{E} = Bl\omega r \sin \theta \qquad (37\text{-}7)$$

Fig. 37-8 Calculating the induced emf.

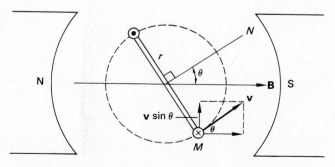

An identical emf is induced in the segment of wire opposite M, and no *net* emf is generated in the other segments. Hence the total instantaneous emf is twice the value given by Eq. (37-7), or

$$\mathcal{E}_{inst} = 2Bl\omega r \sin \theta \qquad (37-8)$$

But the area A of the loop is

$$A = l \times 2r$$

and Eq. (37-8) can be further simplified to

$$\boxed{\mathcal{E}_{inst} = NBA\omega \sin \theta} \qquad (37-9)$$

where N is the number of turns of wire.

Equation (37-9) expresses a very important principle relating to the study of alternating currents:

> *If the armature is rotating with a constant angular velocity in a constant magnetic field, the magnitude of the induced emf varies sinusoidally with respect to time.*

This fact is illustrated by Fig. 37-9. The emf varies from a maximum value when $\theta = 90°$ to a zero value when $\theta = 0°$. The maximum instantaneous emf is therefore

$$\mathcal{E}_{max} = NBA\omega \qquad (37-10)$$

since $\sin 90° = 1$. Stating Eq. (37-9) in terms of the maximum emf, we write

$$\mathcal{E}_{inst} = \mathcal{E}_{max} \sin \theta \qquad (37-11)$$

To see the explicit variation of generated emf with time, we should recall that

$$\theta = \omega t = 2\pi f t$$

where f is the number of rotations per second made by the loop. Thus we can express

Fig. 37-9 Sinusoidal variation of induced emf with time.

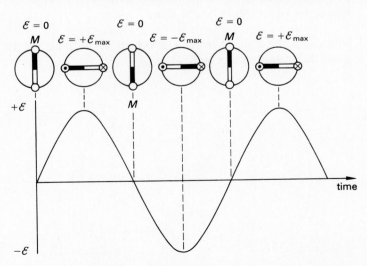

     ELECTRICITY AND MAGNETISM

Eq. (37-11) in the following form:

$$\mathcal{E}_{\text{inst}} = \mathcal{E}_{\text{max}} \sin 2\pi ft \qquad (37\text{-}12)$$

EXAMPLE 37-4   The armature of a simple ac generator consists of 100 turns of wire, each having an area of 0.2 m². The armature is turned with a frequency of 60 rev/s in a constant magnetic field of flux density 10^{-3} T. What is the maximum emf generated?

Solution   We must first calculate the angular velocity of the coil.

$$\omega = (2\pi \text{ rad/rev})(60 \text{ rev/s}) = 377 \text{ rad/s}$$

Substituting this value and other known parameters into Eq. (37-10), we obtain

$$\mathcal{E}_{\text{max}} = NBA\omega$$
$$= (100 \text{ turns})(10^{-3} \text{ T})(0.2 \text{ m}^2)(377 \text{ rad/s})$$
$$= 7.54 \text{ V}$$

Since the induced current is proportional to the induced emf, from Ohm's law, the induced current will also vary sinusoidally according to

$$i_{\text{inst}} = i_{\text{max}} \sin 2\pi ft \qquad (37\text{-}13)$$

The maximum current occurs when the induced emf is a maximum. The sinusoidal variation is similar to that plotted in Fig. 37-9.

The SI unit for frequency is the *hertz* (Hz) which is defined as a cycle per second.

$$1 \text{ Hz} = 1 \text{ cycle/s}$$

Thus a 60-cycle-per-second alternating current has a frequency of 60 Hz.

37-5
THE DC GENERATOR

A simple ac generator can easily be converted to a dc generator by substituting a split-ring commutator for the slip rings, as illustrated in Fig. 37-10. The operation is just the reverse of that discussed earlier for a dc motor (Chap. 36). In the motor, an

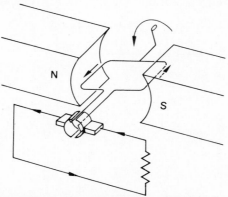

Fig. 37-10 The dc generator.

electric current gives rise to an external torque. In the dc generator, an external torque generates an electric current. The commutator reverses the connections to the brushes twice per revolution. As a result, the current pulsates but never reverses direction. The emf of such a generator varies with time, as shown in Fig. 37-11. Note that the emf is always in the positive direction, but it rises to a maximum and falls to zero twice per complete rotation. Practical dc generators are designed with many coils in several planes so that the generated emf is larger and nearly constant.

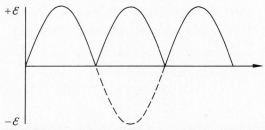

Fig. 37-11 Pulsating emf produced by a dc generator.

37-6

BACK EMF IN A MOTOR

In an electric motor, a magnetic torque turns a current-carrying loop in a constant magnetic field. We have just seen that a coil rotating in a magnetic field will induce an emf that opposes the cause which gave rise to it. This is true even if a current already exists in the loop. Thus *every motor is also a generator.* According to Lenz's law, such an induced emf must oppose the current delivered to the motor. For this reason, the emf induced in a motor is called *back emf* or *counter emf.*

The effect of a back emf is to reduce the net voltage delivered to the armature coils of the motor. Consider the circuit illustrated in Fig. 37-12. The net voltage delivered to the armature coils is equal to the applied voltage V less the induced voltage $\mathcal{E}_b$.

Applied voltage − induced voltage = net voltage

According to Ohm's law, the net voltage across the armature coils is equal to the product of the coil resistance R and the current I. Symbolically, we write

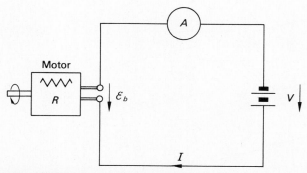

Fig. 37-12 Back emf in a dc motor.

$$V - \mathcal{E}_b = IR \qquad \text{(37-14)}$$

Equation (37-14) tells us that the current through a circuit containing a motor is determined by the magnitude of the back emf. The magnitude of this induced emf, of course, depends on the speed of rotation of the armature. We can show this experimentally by connecting a motor, an ammeter, and a battery in series, as shown in Fig. 37-13. When the armature is rotating freely, a low current is indicated. The back emf reduces the effective voltage. If the motor is stalled by holding the armature stationary, the back emf will drop to zero. The increased net voltage results in a larger circuit current and can cause the motor to overheat and even burn out.

Fig. 37-13 Demonstrating the existence of back emf in a dc motor.

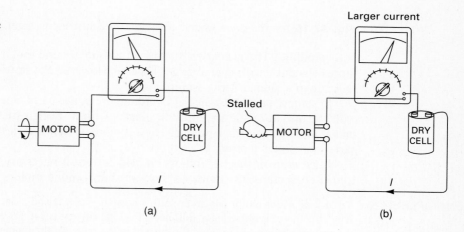

(a)

(b)

37-7

TYPES OF MOTORS

Dc motors are classified according to how the field coils and the armature are connected. When the armature coils and the field coils are connected in series, as shown in Fig. 37-14, the motor is said to be *series-wound*. In this type of motor, the current energizes both the field windings and the armature windings. When the armature turns slowly, the back emf is small and the current is large. Consequently, a large torque is developed at low speeds.

In a *shunt-wound* motor, the field windings and the armature windings are connected in parallel, as illustrated by Fig. 37-15. The entire voltage is applied across

Fig. 37-14 *(a)* The series-wound dc motor. *(b)* Schematic diagram.

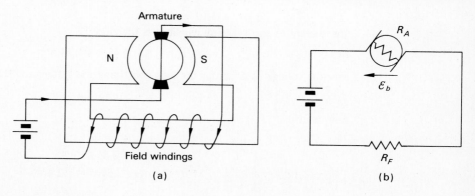

(a)

(b)

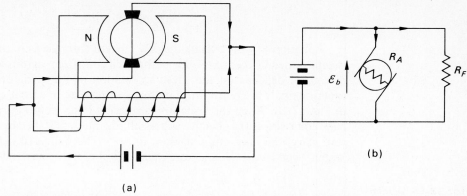

(b)

(a)

Fig. 37-15 *(a)* The shunt-wound dc motor. *(b)* Schematic diagram.

both windings. The primary advantage of a shunt-wound motor is that it produces more constant torque over a range of speeds. However, the starting torque is usually lower than a similar series-wound motor.

In some applications, the field windings are in two parts, one connected in series with the armature and the other in parallel with it. Such a motor is called a *compound motor.* The torque produced by a compound motor varies between that of the series and shunt motors.

In *permanent-magnet* motors no field current is necessary. These motors have torque characteristics similar to those of shunt-wound motors.

EXAMPLE 37-5 A 120-V dc shunt motor has an armature resistance of 3 Ω and a field resistance of 260 Ω. When the motor is operating at full speed, the total current is 3 A. *(a)* What is the back emf of the motor at full speed? *(b)* Find the current in the motor at the moment its switch is turned on.

Solution (a) In order to determine the back emf, we must see how the current is divided between the armature and field windings. The current I_F in the field windings can be found by writing the voltage equation for the outside loop in Fig. 37-15*b*:

$$V = I_F R_F$$

from which

$$I_F = \frac{V}{R_F} = \frac{120 \text{ V}}{260 \text{ }\Omega} = 0.46 \text{ A}$$

The current in the armature windings is therefore

$$I_A = 3 \text{ A} - 0.46 \text{ A} = 2.54 \text{ A}$$

Now the voltage equation for the loop containing the battery and the armature is

$$V - \mathcal{E}_b = I_A R_A$$

from which

$$\mathcal{E}_b = V - I_A R_A$$
$$= 120 \text{ V} - (2.54 \text{ A})(3 \text{ }\Omega)$$
$$= 120 \text{ V} - 7.62 \text{ V} = 112.4 \text{ V}$$

At the instant the switch is turned on, the armature is not yet turning, and consequently $\mathcal{E}_b = 0$. In this case the armature current is

$$I_A = \frac{V}{R_A} = \frac{120\text{ V}}{3\ \Omega} = 40\text{ A}$$

The field current is still 0.46 A, and the total starting current is

$$I = I_A + I_F$$
$$= 40\text{ A} + 0.46\text{ A} = 40.5\text{ A}$$

37-8
THE
TRANSFORMER

It was noted earlier that a changing current in one wire loop will induce a current in a nearby loop. The induced current arises from the changing magnetic field associated with a changing current. Alternating current has a distinct advantage over direct current because of the inductive effect of a current which constantly varies in magnitude and direction. The most common application of this principle is offered by the *transformer*, a device which increases or decreases the voltage in an ac circuit.

A simple transformer is illustrated in Fig. 37-16. There are three essential parts: (1) a primary coil connected to an ac source, (2) a secondary coil, and (3) a soft iron core. As an alternating current is sent through the primary coil, magnetic flux lines move back and forth through the iron core, inducing an alternating current in the secondary coil.

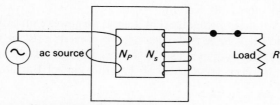

Fig. 37-16 Transformer.

The constantly changing magnetic flux is established throughout the core of the transformer and passes through both primary and secondary coils. The emf $\mathcal{E}_p$ induced in the primary coil is given by

$$\mathcal{E}_p = -N_p \frac{\Delta\phi}{\Delta t} \qquad (37\text{-}15)$$

where N_p = number of primary turns
$\Delta\phi/\Delta t$ = rate at which flux changes

Similarly, the emf $\mathcal{E}_s$ induced in the secondary coil is

$$\mathcal{E}_s = -N_s \frac{\Delta\phi}{\Delta t} \qquad (37\text{-}16)$$

where N_s is the number of secondary turns. Since the same flux changes at the same rate through each coil, we can divide Eq. (37-15) by Eq. (37-16) to obtain

$$\boxed{\frac{\mathcal{E}_p}{\mathcal{E}_s} = \frac{N_p}{N_s}}$$

$$\frac{Primary\ voltage}{Secondary\ voltage} = \frac{primary\ turns}{secondary\ turns}$$

The induced voltage is in direct proportion to the number of turns. If the ratio of secondary turns N_s to primary turns N_p is varied, an input (primary) voltage can provide any desired output (secondary) voltage. For example, if there are 40 times as many turns in the secondary coil, an input voltage of 120 V will be increased to $40 \times 120 = 4800$ V in the secondary coil. A transformer which produces a larger output voltage is called a *step-up transformer*.

A *step-down transformer* can be constructed by making the number of primary turns greater than the number of secondary turns. Using a step-down transformer gives a lower output voltage.

The efficiency of a transformer is defined as the ratio of the power output to the power input. Recalling that electric power is equal to the product of voltage and current, we can write the efficiency E of a transformer as

$$E = \frac{power\ output}{power\ input} = \frac{\mathcal{E}_s i_s}{\mathcal{E}_p i_p}$$

(37-18)

where i_p and i_s are the currents in the primary coil and the secondary coil, respectively. Most electric transformers are carefully designed for extremely high efficiencies, normally above 90 percent.

It is important to recognize that there is no power gain as a result of transformer action. When the voltage is stepped up, the current must be stepped down, so that the product $\mathcal{E}i$ does not increase. To see this more clearly, let us assume that a given transformer is 100 percent efficient. For this perfect transformer, Eq. (37-18) becomes

$$\mathcal{E}_s i_s = \mathcal{E}_p i_p$$

or

$$\frac{i_p}{i_s} = \frac{\mathcal{E}_s}{\mathcal{E}_p}$$

(37-19)

This equation clearly shows the inverse relationship between current and induced voltage.

EXAMPLE 37-6   An ac generator which delivers 20 A at 6000 V is connected to a step-up transformer. What is the output current at 120,000 V if the transformer efficiency is 100 percent?

Solution   From Eq. (37-19),

$$i_s = \frac{\mathcal{E}_p i_p}{\mathcal{E}_s}$$

$$= \frac{(6000\ V)(20\ A)}{120,000\ V} = 1\ A$$

708   ELECTRICITY AND MAGNETISM

Note in the preceding example that the current was reduced to 1 A whereas the voltage was increased twentyfold. Since heat losses in transmission lines vary directly with the square of the current (i^2R), this means that electric power can be transmitted for large distances without significant loss. When the electric power reaches its destination, step-down transformers are used to provide the desired current at lower voltages.

SUMMARY

Electromagnetic induction allows for the production of an electric current in a conducting wire. This is the basic operating principle behind many electrical devices. An understanding of the concepts summarized below is necessary for most applications involving the use of alternating current.

- A magnetic flux changing at the rate of 1 Wb/s will induce an emf of 1 V for each turn of a conductor. Symbolically,

$$\mathcal{E} = -N\frac{\Delta\phi}{\Delta t}$$

Induced emf

- Two principal ways in which the flux changes are

$$\Delta\phi = \Delta BA \qquad \Delta\phi = B\,\Delta A$$

- The induced emf due to a wire of length l moving with a velocity v at an angle θ with a field B is given by

$$\mathcal{E} = Blv \sin\theta$$

Emf Due to Moving Wire

- According to *Lenz's law,* the induced current must be in such a direction that it produces a magnetic force which opposes the force causing the motion.
- *Fleming's rule:* If the thumb, forefinger, and middle finger of the right hand are held at right angles to each other, with the thumb pointing in the direction in which the wire is moving and the forefinger pointing in the field direction (N to S), the middle finger will point in the direction of induced conventional current. (*Motion–flux–current*).
- The instantaneous emf generated by a coil of N turns moving with an angular velocity ω or frequency f is

$$\mathcal{E}_{inst} = NBA\omega \sin \omega t \qquad \mathcal{E}_{inst} = 2\pi fNBA \sin 2\pi ft$$

- The maximum emf occurs when the sin is zero. Thus

$$\mathcal{E}_{max} = NBA\omega \qquad \mathcal{E}_{inst} = \mathcal{E}_{max} \sin 2\pi ft$$

- Since the induced current is proportional to $\mathcal{E}$, we also have

$$\boxed{i_{inst} = i_{max} \sin 2\pi ft}$$ *Instantaneous Current*

- The back emf in a motor is the induced voltage which causes a reduction in the net voltage delivered to a circuit.

Applied voltage − induced back emf = net voltage

$$\boxed{V - \mathcal{E}_b = IR} \qquad \boxed{\mathcal{E}_b = V - IR}$$

- For a transformer having N_p primary and N_s secondary turns

$$\frac{Primary\ voltage}{Secondary\ voltage} = \frac{primary\ turns}{secondary\ turns} \qquad \boxed{\frac{\mathcal{E}_p}{\mathcal{E}_s} = \frac{N_p}{N_s}}$$

- The efficiency of a transformer is

$$\boxed{E = \frac{power\ output}{power\ input} = \frac{\mathcal{E}_p i_p}{\mathcal{E}_s i_s}} \quad \textit{Transformer Efficiency}$$

QUESTIONS

37-1. Define the following terms:
a. Induced emf
b. Lenz's law
c. Ac generator
d. Dc generator
e. Field magnet
f. Armature
g. Slip rings
h. Commutator
i. Electromagnetic induction
j. Back emf
k. Shunt motor
l. Series motor
m. Compound motor
n. Step-up transformer
o. Step-down transformer
p. Transformer efficiency

37-2. Discuss the various factors which influence the magnitude of an induced emf in a length of wire moving in a magnetic field.

37-3. A bar magnet is held in a vertical position with the north pole facing upward. If a closed-loop coil is dropped over the north end of the magnet, what is the direction of the induced current viewed from the top of the magnet?

37-4. A circular loop is suspended with its plane perpendicular to a magnetic field directed from left to right. The loop is removed from the field by moving it upward quickly. What is the direction of the induced current viewed along the field direction? Is a force required to remove the loop from the field?

37-5. An induction coil is essentially a transformer which operates on direct current. As shown in Fig. 37-17, the induction coil consists of a few primary turns wound around an iron core with a large number of secondary turns surrounding the primary. A battery current magnetizes the core so that it attracts the armature of the vibrator and opens the circuit periodically. When the circuit is opened, the field collapses, and a large emf is induced in the secondary coil, producing a spark at the output terminals. What is the function of the capacitor C connected

Fig. 37-17 Induction coil.

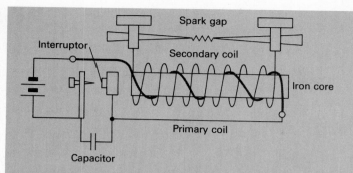

Spark gap

Interruptor

Secondary coil

Iron core

Primary coil

Capacitor

in parallel with the vibrator? Explain how an induction coil is used in the ignition system of an automobile.

37-6. Explain clearly how an ac generator can be converted to a dc generator. How would you proceed to convert an ac generator into an ac motor?

37-7. When the electric motor in a plant is starting, a worker notices that the lights are momentarily dimmed. Explain.

37-8. What type of dc motor should be purchased to operate a winch used to lift heavy objects? Why?

37-9. What type of motor should be used to operate an electric fan where uniform torque is desired at high speeds?

37-10. Explain how the existence of back emf in a motor helps keep its speed constant. *Hint:* What happens to $\mathcal{E}_b$ and I when the armature speed increases or decreases?

37-11. There are three primary ways in which power is lost through the operation of a transformer: (1) wire resistance losses, (2) hysteresis losses, and (3) eddy-current losses. (*Eddy currents* are induced current loops which occur in the mass of a magnetic material resulting from a changing flux.) Explain how energy is wasted by these three processes.

37-12. Explain with the use of diagrams how transformers make it possible to transmit current economically from power installations to homes many miles away.

37-13. An ac generator produces 60 Hz alternating voltage. How many degrees of armature rotation will correspond to one-fourth of a cycle?

37-14. Why is it more economical for power companies to provide alternating current than direct current?

37-15. Prepare a brief report on the following topics and explain the part played by electromagnetic induction.

a. The betatron
b. Induction coil
c. The telephone
d. Eddy currents
e. Magnetohydrodynamic generator

f. Electric power transmission
g. The universal motor
h. Induction motor
i. Synchronous motor

PROBLEMS

Induced Electromotive Force

37-1. A coil of 300 turns moving perpendicular to the flux in a uniform magnetic field experiences a flux linkage of 0.23 mWb in 0.002 s. What is the induced emf?
Ans. −34.5 V

37-2. The magnetic flux linking a loop of wire changes from 5 to 2 mWb in 0.1 s. What is the average induced emf?

37-3. A coil of 120 turns is 90 mm in diameter and has its plane perpendicular to a magnetic field of flux density 60 mT which is produced by a nearby electromagnet. The current in the electromagnet is cut off, and as the field collapses, an emf of 6 V is induced in the coil. How long does it take for the field to disappear?

Ans. 7.63 ms

37-4. A coil of 56 turns has an area of 0.3 m². Its plane is perpendicular to a magnetic field of flux density 7 mT. If this field collapses to zero in 6 ms, what is the induced emf?

37-5. A wire 0.15 m long moves at a constant velocity of 4 m/s in a direction that is 36° with respect to a magnetic field of flux density 0.4 T. The axis of the wire is perpendicular to the magnetic flux lines.

Ans. 0.141 V

37-6. A 0.2-m-long wire moves at an angle of 28° with magnetic flux of density 8 mT. The wire length is perpendicular to the flux. What velocity v is required to induce an emf of 60 mV?

Generators

37-7. The magnetic field in the air gap between the magnetic poles and the armature of an electric generator has a flux density of 0.7 T. The length of the wires on the armature is 0.5 m. How fast must these wires move in order to generate a maximum emf of 1 V in each armature wire?

Ans. 2.86 m/s

* **37-8.** The armature of a simple ac generator has 100 circular turns of wire, each having a radius of 5 cm. The armature turns in a constant 0.06-T magnetic field. What must be the rotational frequency in rpm in order that a maximum emf of 2 V is generated?

37-9. An armature in an ac generator consists of 500 turns, each having an area of 60 cm². The armature is rotated at a frequency of 3600 rpm in a uniform 2-mT magnetic field. (a) What is the frequency of the alternating voltage? (b) What is the maximum emf generated?

Ans. (a) 60 Hz, (b) 2.26 V

37-10. In Prob. 37-9, what is the instantaneous emf at the instant when the plane of the coil makes an angle of 60° with the magnetic flux?

* **37-11.** A circular coil has 70 turns, each 50 mm in diameter. Assume that the coil rotates about an axis that is perpendicular to a magnetic field of 0.8 T. How many revolutions per second must the coil make in order to generate a maximum emf of 110 V?

Ans. 159 rev/s

** **37-12.** The armature of an ac generator has 800 turns, each having an area of 0.25 m². The coil rotates at a constant 600 rpm in a 3-mT field. (a) What is the maximum induced emf? (b) What is the instantaneous emf 0.43 s after the coil passes a position of zero emf?

DC Motors and Back EMF

37-13. The armature coil of the starting motor in an automobile has a resistance of 0.05 Ω. The motor is driven by a 12-V battery, and when the armature is moving at its operating speed, a back emf of 6 V is generated. (a) What is the starting current? (b) What is the current at operating speed? Ignore the internal resistance and other circuit resistances.

Ans. (a) 240 A, (b) 120 A

37-14. A 220-V dc motor draws a current of 10 A in operation and has an armature resistance of 0.4 Ω. (a) What is the back emf when the motor is operating? (b) What is the starting current?

37-15. A 120-V series-wound dc motor has a field resistance of 90 Ω and an armature resistance of 10 Ω. When it is operating at full speed, a back emf of 80 V is generated. (a) What is the total resistance of the motor? (b) What is the initial current? (c) What is the operating current at full speed?

Ans. (a) 100 Ω, (b) 1.2 A, (c) 0.4 A

37-16. The efficiency of the motor in Prob. 37-15 is the ratio of the power output to the power input. Determine the efficiency based on the known data.

Transformers

37-17. A step-up transformer has 400 secondary turns and only 100 primary turns. An alternating voltage of 120 V is connected to the primary coil. What is the output voltage?

Ans. 480 V

37-18. A step-down transformer is used to drop an alternating voltage from 10,000 to 500 V. What must be the ratio of secondary turns to primary turns? If the input current is 1 A and the transformer is 100 percent efficient, what is the output current?

37-19. A step-up transformer has 80 primary turns and 720 secondary turns. The efficiency of the transformer is 95 percent. If the primary draws a current of 20 A at 120 V, what are the current and voltage for the secondary?

Ans. 2.11 A, 1080 V

Additional Problems

37-20. A coil of area 0.2 m² has 80 turns of wire and is suspended with its plane perpendicular to a uniform field. What must be the flux density in order to produce an average emf of −2 V as the coil is flipped parallel to the field in 0.5 s?

37-21. The flux through a coil having 200 turns changes from 0.06 to 0.025 Wb in 0.5 s. The coil is connected to an electric light, and the combined resistance is 2 Ω. What is the average induced emf and what average current is delivered to the light filament?

Ans. 14 V, 7 A

37-22. A 90-mm length of wire moves with an upward velocity of 35 m/s between the poles of a magnet. The flux is directed from left to right and has a density of 80 mT. If the resistance in the wire is 5 mΩ, what are the magnitude and direction of the induced current?

37-23. A generator develops an emf of 120 V and has a terminal potential difference of 115 V when the armature current is 25 A. What is the resistance of the armature?

Ans. 0.2 Ω

37-24. A shunt-wound motor connected across a 117-V line generates a back emf of 112 V when the armature current is 10 A. What is the armature resistance?

* **37-25.** A 110-V shunt-wound motor has a field resistance of 200 Ω connected in parallel with an armature resistance of 10 Ω. When the motor is operating at full speed, the back emf is 90 V. **(a)** What is the starting current? **(b)** What is the operating current?

Ans. (a) 11.6 A, (b) 2.55 A

* **37-26.** A 120-V shunt motor has a field resistance of 160 Ω and an armature resistance of 1 Ω. When the motor is operating at full speed, it draws a current of 8 A. What is the initial starting current? What series resistance must be added to reduce the starting current to 30 A?

* **37-27.** A shunt generator has a field resistance of 400 Ω and an armature resistance of 2 Ω. The generator delivers a power of 4000 W to an external line at 120 V. What is the emf of the generator?

Ans. 187 V

38 Alternating-Current Circuits

After completing this chapter, you should be able to:

1. Determine the instantaneous current for charging and discharging a *capacitor* and for the growth and decay of current in an *inductor*.
2. Write and apply equations for calculating the *inductance* and *capacitance* for inductors and capacitors in an ac circuit.
3. Explain with diagrams the phase relationships for a circuit with (*a*) pure resistance, (*b*) pure capacitance, and (*c*) pure inductance.
4. Write and apply equations for calculating the *impedance*, the *phase angle*, and the *effective current* for a series ac circuit containing resistance, capacitance, and inductance.
5. Write and apply an equation for calculating the resonant frequency for an ac circuit.
6. Define and be able to determine the *power factor* for a series ac circuit.

About 99 percent of the energy generated in the United States is in ac form. There are good reasons for the predominant use of ac circuits. A rotating coil in a magnetic field induces an alternating emf in an extremely efficient manner. Besides, the transformer provides a convenient method of transmitting ac currents over long distances with a minimal power loss.

The only element of importance in the dc circuit (besides a source of emf) was the resistor. Since alternating currents behave very differently from direct currents, additional circuit elements become important. In addition to the normal resistance, electromagnetic induction and capacitance play important roles. In this chapter we present a few elementary aspects of alternating current in electric circuits.

38-1 THE CAPACITOR

In Chap. 32 we discussed the capacitor as an electrostatic device able to store charge. Charging and discharging capacitors in an ac circuit provide an effective means of regulating and controlling the flow of charge. Before discussing the effects of capacitance in an ac circuit, however, it will be useful to describe the growth and decay of charge on a capacitor.

Consider the circuit illustrated in Fig. 38-1, containing only a capacitor and a resistor. When the switch is moved to S_1, the capacitor begins to be charged rapidly

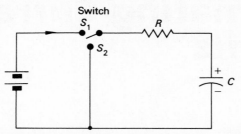

Fig. 38-1 Circuit diagram illustrating a method for charging and discharging a capacitor.

by the current i. However, as the potential difference Q/C between the capacitor plates rises, the rate of flow of charge to the capacitor decreases. At any instant, the iR drop through the resistor must equal the difference between the terminal voltage V_B of the battery and the back emf of the capacitor. Symbolically,

$$V_B - \frac{Q}{C} = iR \tag{38-1}$$

where i = instantaneous current
$\quad\;\; Q$ = instantaneous charge on capacitor

Initially, the charge Q is zero, and the current i is a maximum. Thus, at time $t = 0$,

$$Q = 0 \quad \text{and} \quad i = \frac{V_B}{R} \tag{38-2}$$

As the charge on the capacitor builds up, it produces a back emf Q/C opposing the flow of additional charge; the current i decreases. Both the increase in charge and the decrease in current are exponential functions, as shown by the curves in Fig. 38-2. If it were possible to continue charging indefinitely, the limits at $t = \infty$ would be

$$Q = CV_B \quad \text{and} \quad i = 0 \tag{38-3}$$

The methods of calculus applied to Eq. (38-1) show that the instantaneous charge is given by

$$\boxed{Q = CV_b(1 - e^{-t/RC})} \tag{38-4}$$

and that the instantaneous current is given by

$$\boxed{i = \frac{V_B}{R} e^{-t/RC}} \tag{38-5}$$

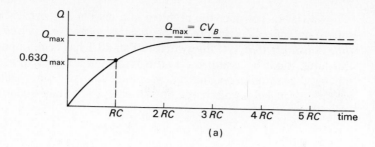

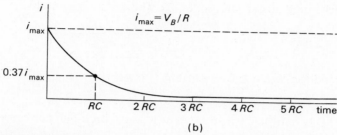

Fig. 38-2 (a) The charge on a capacitor rises, approaching but never reaching its maximum value. (b) The current decreases, approaching zero as the charge builds to its maximum value.

where t is the time. The logarithmic constant $e = 2.71828$ to six significant figures. Substitution of $t = 0$ and $t = \infty$ into the above relations will yield Eqs. (38-2) and (38-3), respectively.

The equations for computing instantaneous charge and current are simplified at the particular instant when $t = RC$. This time, usually denoted by τ, is called the *time constant* of the circuit.

$$\boxed{\tau = RC}$$

$\hspace{2cm}$ *Time Constant* (38-6)

It is seen from Eq. (38-4) that the charge Q rises to $1 - 1/e$ times its final value in one time constant:

$$Q = CV_B\left(1 - \frac{1}{e}\right) = 0.63CV_B \tag{38-7}$$

In a capacitance circuit, the charge on a capacitor will rise to 63 percent of its maximum value after charging for a period of one time constant.

Substituting $\tau = RC$ into Eq. (38-5) shows that the current delivered to the capacitor decreases to $1/e$ times its initial value in one time constant:

$$i = \frac{V_B}{R}\frac{1}{e} = 0.37\frac{V_B}{R} \tag{38-8}$$

In a capacitive circuit, the current delivered to a capacitor will decrease to 37 percent of its initial value after being charged for a period of one time constant.

Now let us consider the problem of a discharging capacitor. For practical reasons, *a capacitor is considered to be fully charged after a period of time equal to five time constants (5RC)*. If the switch in Fig. 38-1 has been in position S_1 for at least this long, it can be assumed that the maximum charge CV_B is on the capacitor. By moving the switch to position S_2 the voltage source is removed from the circuit, and a path is provided for discharge. In this case, the voltage equation (38-1) reduces to

$$-\frac{Q}{C} = iR \tag{38-9}$$

Both the charge and the current decay along curves similar to that shown for the charging current in Fig. 38-2b. The instantaneous charge is found from

$$\boxed{Q = CV_Be^{-1/RC}} \tag{38-10}$$

and the instantaneous current is given by

$$\boxed{i = \frac{-V_B}{R}e^{-t/RC}} \tag{38-11}$$

The negative sign in the current equation indicates that the direction of i in the circuit has been reversed.

After discharging for one time constant, the charge and the current will have decayed to $1/e$ times their initial values. This can be shown by substituting τ into Eqs. (38-10) and (38-11).

> In a capacitance circuit, the charge and the current will decay to 37 percent of their initial values after the capacitor has discharged for a length of time equal to one time constant.

The capacitor is considered to be fully discharged after five time constants ($5RC$).

EXAMPLE 38-1   A 12-V battery having an internal resistance of 1.5 Ω is connected to a 4-μF capacitor through leads having a resistance of 0.5 Ω. (*a*) What is the initial current delivered to the capacitor? (*b*) How long will it take to charge the capacitor fully? (*c*) What is the value of the current after one time constant?

Solution (a)   Initially, there is no back emf from the capacitor. Hence the current delivered to the circuit is the emf of the battery divided by the total resistance in the circuit:

$$i = \frac{\mathscr{E}_B}{R + r} = \frac{12\ \text{V}}{1.5\ \Omega + 0.5\ \Omega} = 6\ \text{A}$$

Solution (b)   The capacitor can be considered fully charged after a time

$$t = 5RC = (5)(2\ \Omega)(4\ \mu\text{F}) = 40\ \mu\text{s}$$

Solution (c)   After one time constant RC, the current will have decayed to 37 percent of its initial value. Thus

$$i_\tau = (0.37)(6\ \text{A}) = 2.22\ \text{A}$$

In the preceding discussions, we simplified the approach by using direct currents. When an alternating voltage is impressed upon a capacitor, there are surges of charge into and out of the capacitor plates. Therefore an alternating current is maintained in a circuit even though there is no path between the capacitor plates. The effect of capacitance in an ac circuit will be discussed later.

38-2
THE INDUCTOR

Another important element in an ac circuit is the *inductor,* which consists of a continuous loop or coil of wire. (See Fig. 38-3.) In Chap. 37 we showed that a change in magnetic flux in the region enclosed by such a coil will induce an emf in the coil.

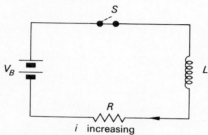

Fig. 38-3 The inductor.

Until now, the flux changes originated from sources outside the coil itself. We now consider the emf induced in a coil as a result of changes in its *own* current. Regardless of how the flux change occurs, the induced emf must be given by

$$\mathscr{E} = -N\frac{\Delta\phi}{\Delta t} \tag{38-12}$$

where N = number of turns
$\Delta\phi/\Delta t$ = rate at which flux changes

When the current through an inductor increases or decreases, a *self-induced* emf arises in the circuit which *opposes* the change. Consider the circuit illustrated in Fig. 38-3. When the switch is closed, the current arises from zero to its maximum value $i = V_B/R$. The inductor responds to this increasing current by setting up an induced back emf. Since the geometry of the inductor is fixed, the rate of change in flux, $\Delta\phi/\Delta t$, or the induced emf $\mathscr{E}$, is proportional to the rate of change in current, $\Delta i/\Delta t$. This proportionality is expressed in the equation

$$\boxed{\mathscr{E} = -L\frac{\Delta i}{\Delta t}} \tag{38-13}$$

The proportionality constant L is called the *inductance* of the circuit. Solving explicitly for the inductance in Eq. (38-13), we write

$$L = -\frac{\mathscr{E}}{\Delta i/\Delta t} \tag{38-14}$$

The unit of inductance is the *henry* (H).

> *A given inductor has an inductance of one henry (H) if an emf of one volt is induced by a current changing at the rate of one ampere per second.*

$$1 \text{ H} = 1 \text{ V} \cdot \text{s/A}$$

The inductance of a coil depends upon its geometry, the number of turns, the spacing of the turns, and the permeability of its core but not upon voltage and current values. In this respect, the inductor is similar to capacitors and resistors.

We shall now consider the growth and decay of current in an inductive circuit. The circuit illustrated in Fig. 38-4 contains an inductor L, a resistor R, and a battery

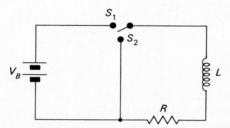

Fig. 38-4 Circuit for studying inductance.

V_B. The switch is positioned so that the battery can be alternately connected and disconnected from the circuit. When the switch is moved to position S_1, a current begins to grow in the circuit. As the current rises, the induced emf $-L(\Delta i/\Delta t)$ is established in opposition to the battery voltage V_B. The net emf must equal the iR drop through the resistor. Thus

$$V_B - L\frac{\Delta i}{\Delta t} = iR \tag{38-15}$$

A mathematical analysis of Eq. (38-15) will show that the rise in current as a function of time is given by

$$i = \frac{V_B}{R}(1 - e^{-(R/L)t}) \tag{38-16}$$

This equation shows that the current i is zero when $t = 0$ and has a maximum of V_B/R when $t = \infty$. The effect of inductance in a circuit is to delay the establishment of this maximum current. The rise and decay of current in an inductive circuit are shown in Fig. 38-5.

The time constant for an inductive circuit is

$$\tau = \frac{L}{R} \tag{38-17}$$

τ is in *seconds* when L is in *henrys* and R is in *ohms*. Insertion of this value into

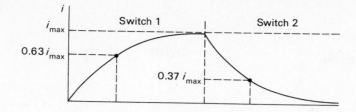

Fig. 38-5 Rise and decay of current in an inductor.

Eq. (38-16) shows that:

In an inductive circuit, the current will rise to 63 percent of its final value in one time constant (L/R).

After the current in Fig. 38-4 has attained a steady value, if the switch is moved to position S_2, the current will decay exponentially, as shown in Fig. 38-5. The equation for the decay is

$$i = \frac{V_B}{R} e^{-(R/L)t}$$

(38-18)

Substitution of L/R into Eq. (38-18) shows that:

In an inductive circuit, the current decays to 37 percent of its initial value in one time constant (L/R).

Once again, for practical reasons the rise or decay time for an inductor is considered to be five time constants (5L/R).

38-3

ALTERNATING CURRENTS

Now that we are familiar with the basic elements in an ac circuit, it is necessary to understand more about alternating currents. The quantitative description of an alternating current is much more complicated than that for direct currents, whose magnitude and direction are constant. An alternating current flows back and forth in a circuit and has no "direction" in the sense direct current has. Additionally, the magnitude varies sinusoidally with time, as we learned in our discussions of the ac generator.

The variation in emf or current for an ac circuit can be represented by a rotating vector or by a sine wave. These representations are compared in Fig. 38-6. The vertical component of the rotating vector at any instant in the instantaneous magnitude of the voltage or the current. One complete revolution of the rotating vector or one complete sine wave on the curve represents one *cycle*. The number of complete cycles per second experienced by an alternating current is called its *frequency* and provides an important description of the current. The relationship between the instantaneous emf $\mathscr{E}$ or the instantaneous current i and the frequency was established in Chap. 37:

$$\mathscr{E} = \mathscr{E}_{max} \sin 2\pi ft$$

(38-19)

$$i = i_{max} \sin 2\pi ft$$

(38-20)

Note that the average value for the current in an ac circuit is zero since the magnitude alternates between i_{max} and $-i_{max}$. However, even though there is no *net*

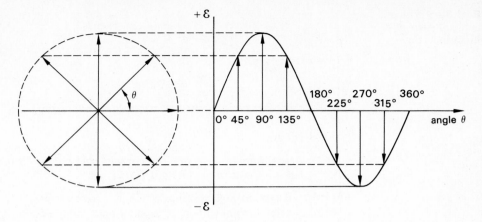

Fig. 38-6 Rotating vector and its corresponding sine wave can be used to represent ac current or voltage.

current, charge is in motion, and electric energy can be released in the form of heat or useful work. The best method of measuring the effective strength of alternating currents is to find the dc value which will produce the same *heating* effect or develop the same *power* as the alternating current. This current value, called the *effective current* i_{eff}, is found to be 0.707 times the maximum current. A similar relation holds for the effective emf or voltage in an ac circuit. Thus

$$i_{eff} = 0.707 i_{max} \qquad (38\text{-}21)$$
$$\mathcal{E}_{eff} = 0.707 \mathcal{E}_{max} \qquad (38\text{-}22)$$

One effective ampere is that alternating current which will develop the same power as one ampere of direct current

One effective volt is that alternating voltage which will produce an effective current of one ampere through a resistance of one ohm.

Ac meters are calibrated to show effective values. For example, if an ac meter measures household voltage to be 120 V at 10 A, Eqs. (38-21) and (38-22) will show that the maximum values of current and voltage are

$$i_{max} = \frac{10 \text{ A}}{0.707} = 14.14 \text{ A}$$

$$V_{max} = \frac{120 \text{ V}}{0.707} = 170 \text{ V}$$

Therefore, the house voltage actually varies between +170 and −170 V while the current ranges from +14.14 to −14.14 A. The usual frequency of voltage variation is 60 Hz.

38-4

PHASE RELATION IN AC CIRCUITS

In all dc circuits, the voltage and the current reach maximum and zero values at the same time and are said to be *in phase*. The effects of inductance and capacitance in ac circuits prevent the voltage and current from reaching maxima and minima at the same time. In other words, the current and voltage in most ac circuits are *out of phase*.

To understand phase relations in an ac circuit, suppose we first consider a circuit containing a *pure* resistor in series with an ac generator, as in Fig. 38-7. This is an idealized circuit in which the inductive and capacitive effects are negligible. Many household devices, such as lights, heaters, and toasters, approximate a condition of pure resistance. In such devices, the instantaneous voltage V and current i are in phase. That is, variations in voltage will result in simultaneous variations of current. When the voltage is a maximum, the current is also a maximum. When the voltage is zero, the current is zero.

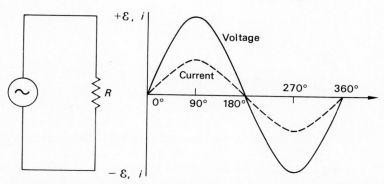

Fig. 38-7 In a circuit containing pure resistance, the voltage and current are in phase.

Next, we consider the phase relation between current and voltage across a *pure inductor*. The circuit illustrated in Fig. 38-8 contains only an inductor in series with the ac generator. We have seen that the presence of inductance in a circuit whose current is changing at the rate $\Delta i/\Delta t$ results in a back emf

$$\mathscr{E} = -L\frac{\Delta i}{\Delta t}$$

which delays the current in reaching its maximum. The voltage reaches a maximum while the current is still at zero. When the voltage reaches a minimum, the current is at a maximum. In a circuit containing only inductance, the voltage is said to lead

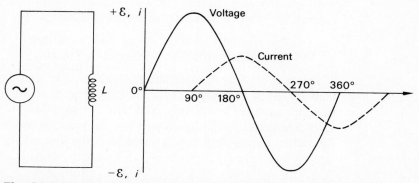

Fig. 38-8 In a pure inductive circuit, the voltage leads the current by 90°.

(occur before) the current by one-fourth of a cycle (or 90°). See the curve in Fig. 38-8.

> *In a circuit containing pure inductance, the voltage leads the current by 90°.*

The effect of capacitance in an ac circuit is opposite to that of inductance. For the circuit shown in Fig. 38-9, the voltage must *lag behind* the current since the flow

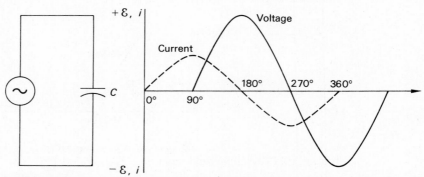

Fig. 38-9 In a circuit containing only capacitance, the voltage lags the current by 90°.

of charge to the capacitor is necessary to build up an opposing emf. When the applied voltage is decreasing, charge flows from the capacitor. The rate of flow of this charge reaches a maximum when the applied voltage is zero.

> *In a circuit containing pure capacitance, the voltage lags the current by 90°.*

This means that the variations in voltage occur one-fourth of a cycle later than the corresponding variations in current.

38-5
REACTANCE

In a dc circuit, the only opposition to current results from the material through which it passes. Resistive heat losses, which obey Ohm's law, also occur in ac circuits. However, in addition to normal resistance, one must contend with inductance and capacitance. Both inductors and capacitors impede the flow of an alternating current, and their effects must be considered along with the opposition of resistance.

> The **reactance** of an ac circuit may be defined as its nonresistive opposition to the flow of alternating current.

We first consider the opposition to the flow of an alternating current through an inductor. Such opposition, called *inductive reactance*, arises from the self-induced back emf produced by a changing current. The magnitude of the inductive reactance X_L is determined by the inductance L of the inductor and by the frequency f of the alternating current and can be found from the formula

$$X_L = 2\pi fL$$

(38-23)

ELECTRICITY AND MAGNETISM

The inductive reactive reactance is measured in *ohms* when the inductance is in *henrys* and the frequency is in *hertz*.

The effective current i in an inductor is determined from its inductive reactance X_L and the effective voltage V by an equation analogous to Ohm's law:

$$V = iX_L \qquad (38\text{-}24)$$

EXAMPLE 38-2 A coil having an inductance of 0.5 H is connected to a 120-V 60-Hz power source. If the resistance of the coil is neglected, what is the effective current through the coil?

Solution The inductive reactance is

$$X_L = 2\pi fL\text{-}(2\pi)(60 \text{ Hz})(0.5 \text{ H})$$
$$= 188.4 \ \Omega$$

The current is found from Eq. (38-24):

$$i = \frac{V}{X_L} = \frac{120 \text{ V}}{188.4 \ \Omega} = 0.637 \text{ A}$$

Opposition to alternating current is also experienced because of the capacitance in a circuit. The *capacitive reactance* X_C is found from

$$\boxed{X_C = \frac{1}{2\pi fC}} \qquad (38\text{-}25)$$

where C = capacitance
f = frequency of alternating current

Capacitive reactance is expressed in *ohms* when C is in *farads* and f is in *hertz*.

Once the capacitance reactance X_C of a capacitor is known, the effective current i can be found from

$$V = iX_C \qquad (38\text{-}26)$$

where V is the applied voltage.

38-6
THE SERIES AC CIRCUIT

In general, an ac circuit contains resistance, capacitance, and inductance in varying amounts. A series combination of these parameters is illustrated in Fig. 38-10. The total voltage drop in a dc circuit is the simple sum of the drop across each element in

Fig. 38-10 Series ac circuit containing resistance, inductance, and capacitance.

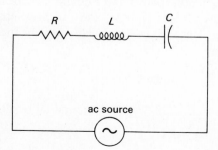

the circuit. However, in the ac circuit, the voltage and current are not in phase with each other. Recall that V_R is always in phase with the current but V_L leads the current by 90° and V_C lags the current by 90°. Clearly, if we are to determine the effective voltage V of the entire circuit, we must develop a means of treating phase differences.

This can best be accomplished by using a vector diagram, called the *phase diagram*. (See Fig. 38-11.) In this method, the effective values of V_R, V_L, and V_C are plotted as rotating vectors. The phase relationship is expressed in terms of the phase angle ϕ, which is a measure of how much the voltage leads the current in a particular circuit element. For example, in a pure resistor, the voltage and the current are in phase and $\phi = 0$. For an inductor, $\phi = +90°$, and in a capacitor $\phi = -90°$. The negative phase angle occurs when the voltage lags behind the current. Following this scheme, V_R appears as a vector along the x axis, V_L is represented by a vector pointing vertically upward, and V_C is directed downward.

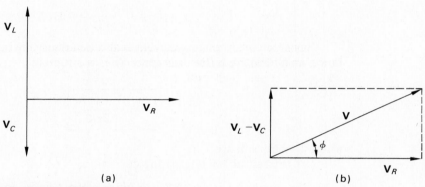

(a) (b)

Fig. 38-11 Phase diagram.

The effective voltage V in an ac circuit can be defined as the vector sum of V_R, V_L, and V_C as they exist on the phase diagram. It can be seen from the figure that the magnitude of V is

$$V = \sqrt{V_R^2 + (V_L - V_C)^2} \tag{38-27}$$

You should verify this equation by applying Pythagoras' theorem to the vector diagram.

Note from the phase diagram that a value of V_L which is greater than V_C results in a positive phase angle. In other words, if the circuit is inductive, the voltage leads the current. In a capacitive circuit, $X_C > X_L$, and a negative phase angle will result, indicating that the voltage lags the current. In any case, the magnitude of the phase angle can be found from

$$\tan \phi = \frac{V_L - V_C}{V_R} \tag{38-28}$$

A more useful form of Eq. (38-27) can be found by recalling that

$$V_R = iR \qquad V_L = iX_L \qquad V_C = iX_C$$

Upon substitution, we find that

$$V = i\sqrt{R^2 + (X_L - X_C)^2} \qquad (38\text{-}29)$$

The quantity multiplied by the current in Eq. (38-29) is a measure of the combined opposition that the circuit offers to alternating current. It is called the *impedance* and is denoted by the symbol Z.

$$Z = \sqrt{R^2 + (X_L - X_C)^2} \qquad (38\text{-}30)$$

The higher the impedance in a circuit, the lower the current for a given voltage. Since R, L_L, and X_C are measured in *ohms*, the impedance is also expressed in *ohms*.

Therefore, the effective current i in an ac circuit is given by

$$i = \frac{V}{Z} \qquad (38\text{-}31)$$

where V = applied voltage
$\quad\quad Z$ = impedance in circuit

It must be remembered that Z depends on the frequency of the alternating current as well as on the resistance, inductance, and capacitance.

Since the voltage across each element depends directly upon resistance or reactance, an alternative phase diagram can be constructed by treating R, X_L, and X_C as vector quantities. Such a diagram can be used to compute the impedance, as indicated in Fig. 38-12. The phase angle ϕ in this representation can be found from

$$\tan \phi = \frac{X_L - X_C}{R} \qquad (38\text{-}32)$$

Of course, this angle is the same as that given by Eq. (38-28).

Fig. 38-12 Impedance diagram.

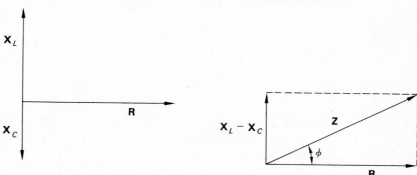

EXAMPLE 38-3   A 40-Ω resistor, a 0.4-H inductor, and a 10-μF capacitor are connected in series with an ac source which generates 120-V 60-Hz alternating current. (*a*) Find the impedance of the circuit. (*b*) What is the phase angle? (*c*) Determine the effective current in the circuit.

Solution (*a*)   We must first compute X_L and X_C as follows:

$$X_L = 2\pi f L = (2\pi)(60 \text{ Hz})(0.4 \text{ H})$$
$$= 151 \ \Omega$$

$$X_C = \frac{1}{2\pi f C} = \frac{1}{(2\pi)(60 \text{ Hz})(10 \times 10^{-6} \text{ F})}$$
$$= 265 \ \Omega$$

The impedance of the circuit is

$$Z = \sqrt{R^2 + (X_L - X_C)^2}$$
$$= \sqrt{(40\ \Omega^2 + (151\ \Omega - 265\ \Omega)^2} = 121\ \Omega$$

Solution (b) From Eq. (38-32), the phase angle is

$$\tan \phi = \frac{151\ \Omega - 265\ \Omega}{40\ \Omega} = -2.85$$

$$\phi = -71°$$

The negative sign is used to indicate that the phase angle is in the fourth quadrant.

Solution (c) Finally, we can determine the effective current from the known impedance.

$$i = \frac{V}{Z} = \frac{120\ V}{121\ \Omega} = 0.992\ A$$

The negative phase angle indicates that the voltage will lag behind this effective current. The circuit is more capacitive than it is inductive.

38-7
RESONANCE

Since inductance causes the current to lag behind the voltage and capacitance causes the current to lead the voltage, their combined effect is to cancel each other. The total reactance is given by $X_L - X_C$, and the impedance in a circuit is a minimum when $X_L = X_C$. When this is true, only the resistance R remains and the current will be a maximum. Setting $X_L = X_C$, we can write

$$2\pi f_r L = \frac{1}{2\pi f_r C}$$

and

$$\boxed{f_r = \frac{1}{2\pi\sqrt{LC}}} \qquad (38\text{-}33)$$

When the applied voltage has this frequency, called the *resonant frequency*, the current in the circuit will be a maximum. Additionally, it should be pointed out that since the current is limited only by resistance, it will be in phase with the voltage.

The antenna circuit in a radio receiver contains a variable capacitor which acts as a tuner. The capacitance is varied until the resonant frequency is equal to a particular signal frequency. The current peaks when this happens, and the receiver responds to the incoming signal.

38-8
THE POWER
FACTOR

In ac circuits, no power is consumed because of capacitance or inductance. Energy is merely stored at one instant and released at another, causing the current and voltage to be out of phase. Whenever the current and voltage are in phase, the power P delivered is a maximum given by

$$P = iV$$

where i = effective current

V = effective voltage

This condition is satisfied when the ac circuit contains only resistance R or when the circuit is in resonance ($X_L = X_C$).

Normally, however, an ac circuit contains sufficient reactance to limit the effective power. In any case, the power delivered to the circuit is a function only of the component of the voltage V which is in phase with the current. From Fig. 38-11, this component is V_R, and we can write

$$V_R = V \cos \phi$$

where ϕ is the phase angle. Thus the effective power consumed in an ac circuit is

$$\boxed{P = iV \cos \phi} \tag{38-34}$$

The quantity $\cos \phi$ is called the *power factor* of the circuit. Note that $\cos \phi$ can vary from zero in a circuit containing pure reactance ($\phi = 90°$) to unity in a circuit containing only resistance ($\phi = 0°$).

Equation (38-21) and Fig. 38-12 show that the power factor can also be found from

$$\boxed{\cos \phi = \frac{R}{Z} = \frac{R}{\sqrt{R^2 + (X_L - X_C)^2}}} \tag{38-35}$$

EXAMPLE 38-4   (a) What is the power factor for the circuit described in Example 38-3? (b) What power is absorbed in the circuit?

Solution (a)   The power factor is found in Eq. (38-35):

$$\cos \phi = \frac{R}{Z} = \frac{40\ \Omega}{121\ \Omega} = 0.33$$

Solution (b)   The power absorbed in the circuit is

$$P = iV \cos \phi = (0.992\ \text{A})(120\ \text{V})(0.33)$$
$$= 39.3\ \text{W}$$

The power factor is sometimes expressed as a percentage instead of as a decimal. For example, the power factor of 0.33 in the above example could be expressed as 33 percent. Most commercial ac circuits have power factors from 80 to 90 percent because they usually contain more inductance than capacitance. Since this requires the electric power companies to furnish more current for a given power, the power companies extend a lower rate to users with power factors above 90 percent. Commercial users can improve their inductive power factors by adding capacitors, for instance.

There are three principal elements in ac circuits: the *resistor,* the *capacitor,* and the *inductor.* A resistor is affected by ac current in the same manner as for dc circuits, and the current is determined by Ohm's law. The capacitor regulates and controls the flow of charge in an ac circuit; its opposition to the flow of electrons is called *capacitive reactance.* The inductor experiences a self-induced emf which adds *inductive reactance* to the circuit. The combined effect of all three elements in opposing electric current is called *impedance.* The major points to remember are summarized below:

- When a capacitor is being charged, the instantaneous values of the charge Q and the current i are found from

$$Q = CV_B(1 - e^{-t/RC})$$

$$i = \frac{V_B}{R}e^{-t/RC}$$

- *The charge on the capacitor will rise to 63 percent of its maximum value as the current delivered to the capacitor decreases to 37 percent of its initial value during a period of one time constant τ.*

$$\tau = RC \qquad \qquad \textit{Time Constant}$$

- When a capacitor is discharging, the instantaneous values of the charge and current are given by

$$Q = CV_B e^{-t/RC}$$

$$i = \frac{-V_B}{R}e^{-t/RC}$$

Both the charge and the current decay to 37 percent of their initial values after discharging for one time constant.

- When alternating current passes through a coil of wire, an inductor, a self-induced emf arises to oppose the change. This emf is given by

$$\mathscr{E} = -L\frac{\Delta i}{\Delta t}$$

$$L = -\frac{\mathscr{E}}{\Delta i/\Delta t}$$

This constant L is called the inductance. An inductance of one henry (H) exists if an emf of 1 V is induced by a current changing at the rate of 1 A/s.

- The rise and decay of current in an inductor are found from

$$i = \frac{V_B}{R}(1 - e^{-(R/L)t}) \qquad \qquad \textit{Current Rise}$$

$$i = \frac{V_B}{R}e^{-(R/L)t} \qquad \qquad \textit{Current Decay}$$

- In an inductive circuit, the current will rise to 63 percent of its maximum value or decay to 37 percent of its maximum in a period of one time constant. For an inductor, the time constant is

$$\tau = \frac{L}{R}$$

- Since alternating currents and voltages vary continuously, we speak of an effective ampere and an effective volt that are defined in terms of their maximum values as follows:

$$i_{\text{eff}} = 0.707 i_{\text{max}}$$

$$\mathscr{E}_{\text{eff}} = 0.707 \mathscr{E}_{\text{max}}$$

- Both capacitors and inductors offer resistance to the flow of alternating current (called reactance), calculated from

$$X_L = 2\pi f L \quad \Omega$$

Inductive Reactance X_L

$$X_C = \frac{1}{2\pi f C} \quad \Omega$$

Capacitive Reactance X_C

The symbol f refers to the frequency of the alternating current in hertz. One hertz is one cycle per second.

- The voltage, current, and resistance in a series ac circuit are studied with the use of phasor diagrams. Figure 38-12 illustrates such a diagram for X_C, X_L, and R. The resultant of these vectors is the effective resistance of the entire circuit called the *impedance Z*.

$$Z = \sqrt{R^2 + (X_L - X_C)^2} \quad \Omega$$

Impedance

- If we apply Ohm's law to each part of the circuit and then to the entire circuit, we obtain the following useful equations. First, the total voltage is given by

$$V = \sqrt{V_R^2 + (V_L - V_C)^2}$$

Voltage

$$V_R = iR$$

$$V_L = iX_L$$

$$V_C = iX_C$$

$$V = iZ$$

$$V = i\sqrt{R^2 + (X_L - X_C)^2}$$

Ohm's Law

- Because the voltage leads the current in an inductive circuit and lags the current in a capacitive circuit, the voltage and current maxima and minima usually do not coincide. The phase angle ϕ is given by

$$\tan \phi = \frac{V_L - V_C}{V_B}$$

or

$$\tan \phi = \frac{X_L - X_C}{R}$$

- The resonant frequency occurs when the net reactance is zero ($X_L = X_C$):

$$f_r = \frac{1}{2\pi\sqrt{LC}}$$
Resonant Frequency

- No power is consumed because of capacitance or inductance. Since power is a function of the component of the impedance along the resistance axis, we can write

$$P = iV\cos\phi$$
Power Factor $= \cos\phi$

$$\cos\phi = \frac{R}{Z}$$

$$\cos\phi = \frac{R}{\sqrt{R^2 + (X_L - X_C)^2}}$$

QUESTIONS

38-1. Define the following terms:

a. Capacitance
b. Inductor
c. Inductance
d. Henry
e. Frequency
f. Impedance
g. Resonance
h. Phase angle

i. Effective current
j. Effective voltage
k. Phase diagram
l. Capacitance reactance
m. Inductive reactance
n. Power factor
o. Resonant frequency

38-2. Inductance in a circuit depends upon its geometry and upon the proximity of magnetic materials. Which of the following coils have higher inductances and why?

a. Closely or widely spaced turns
b. Long coil or short coil
c. Large cross section or small cross section
d. Iron core or air core

38-3. Recalling that the magnetic flux density B within a solenoid is

$$B = \frac{\phi}{A} = \frac{\mu NI}{l}$$

show that the self-inductance of a solenoid is given by

$$L = \frac{N\phi}{I}$$

38-4. Show that the unit for the time constant L/R and RC for inductive and capacitive circuits is the second.

38-5. Sketch a curve of current vs. time for (a) a circuit of high inductance and (b) a circuit of low inductance.

38-6. Plot a curve of voltage vs. time for (a) a charging capacitor and (b) a discharging capacitor. Compare these curves with those in the text for the current.

ELECTRICITY AND MAGNETISM

38-7. By plotting a curve of voltage as a function of time, show how the voltage changes in an inductive circuit. Compare these curves with those in the text for the current.

38-8. Should someone interested in establishing the breakdown voltage of a capacitor in an ac circuit be concerned with the average voltage, maximum voltage, or effective voltage?

38-9. A coil of wire is connected across the terminals of a 110-V dc battery. An ammeter connected in series with the coil indicates a current of 5 A. What happens to the current if an iron core is inserted into the coil? Now, disconnect the dc source, remove the iron core, and reconnect the system to a 110-V ac generator. The ammeter is adjusted to read ac effective amperes. Has the current decreased, increased, or stayed unchanged? What happens to the current if the iron core is inserted? How do you account for your observations?

38-10. An incandescent lamp is connected in series with a 110-V ac generator and a variable capacitor. If the capacitance is increased, will the lamp glow more brightly or will it be dimmed? Explain. What would happen if the generator is replaced by a battery?

38-11. As the capacitance of a circuit increases, what happens to the resonant frequency of the circuit?

38-12. As the frequency is increased in an inductive circuit, what happens to the current in the circuit?

38-13. Inductive reactance depends *directly* on the frequency of the alternating current whereas capacitive reactance varies *inversely* with the frequency. Both oppose the flow of charge in an ac circuit. Explain why their relationship to the frequency differs.

38-14. When a circuit is tuned to its resonant frequency, what is the power factor?

38-15. What is the power factor of a circuit containing (a) pure resistance; (b) pure inductance; (c) pure capacitance?

PROBLEMS

The Capacitor

38-1. A series dc circuit consists of a 4-μF capacitor, a 5000-Ω resistor, and a 12-V battery. (a) What is the time constant for charging this capacitor? (b) What is the initial current? (c) What is the final current? (d) How long is needed to be sure that the capacitor is fully charged?

Ans. (a) 20 ms, (b) 2.40 mA, (c) 0, (d) 100 ms

38-2. A series dc circuit contains a 6-μF capacitor and a 400-Ω resistor connected to a 20-V battery. (a) What is the maximum charge for the capacitor? (b) How much time is required to totally discharge the capacitor?

* 38-3. An 8-μF capacitor is connected in series with a 600-Ω resistor and a 24-V battery. After a time equal to one time constant, what are the charge on the capacitor and the current in the circuit?

Ans. 121 μC, 14.8 mA

* 38-4. The fully charged capacitor of Prob. 38-3 is allowed to discharge by breaking the circuit at the battery. After one time constant, what are the current in the circuit and the charge on the capacitor?

The Inductor

38-5. A series dc circuit contains a 4-mH inductor and an 80-Ω resistor in series with a 12-V battery. (a) What is the initial current? (b) What is the final current? (c) How much time is required for the current to reach 63 percent of its steady-state value?

Ans. (a) 0, (b) 150 mA, (c) 50 μs

38-6. After reaching a steady-state current in Prob. 38-5, the circuit is broken, and the inductor discharges for one time constant. What is the instantaneous current?

38-7. The current in a 25-mH solenoid increases from 0 to 2 A in a time of 0.1 s. What is the magnitude of the self-induced emf?

Ans. 500 mV

38-8. A series dc circuit contains a 0.05-H inductor and a 40-Ω resistor. If the source of emf is 90 V, what are the maximum current in the circuit and the time constant?

Alternating Currents (Effective Current)

38-9. An ac voltmeter when placed across a 12-Ω resistor reads 117 V. What are the *maximum* values for the voltage and current?

Ans. 165 V, 13.8 A

38-10. A capacitor has a maximum voltage rating of 500 V. What is the highest effective ac voltage that can be supplied to it without breakdown?

38-11. A certain appliance is supplied with an effective voltage of 220 V under an effective current of 20 A. What are the maximum values for the voltage and current?

Ans. 311 V, 28.3 A

Reactance

38-12. A 2-H inductor of negligible resistance is connected to a 50-V, 50-Hz ac line. What is the reactance? What is the effective ac current in the coil?

38-13. A 50-mH inductor of negligible resistance is connected to a 120-V, 60-Hz ac line. What is the inductive reactance? What is the effective ac current in the circuit?

Ans. 18.8 Ω, 6.37 A

38-14. A 6-μF capacitor is connected to a 24-V, 50-Hz ac source. What is the current in the circuit?

38-15. A 3-μF capacitor connected to a 120-V, ac line draws an effective current of 0.5 A. What is the frequency of the source?

Ans. 221 Hz

38-16. Find the reactance of a 60-μF capacitor in a 600-Hz ac circuit. What is the reactance if the frequency is reduced to 200 Hz?

38-17. The frequency of an alternating current is 200 Hz and the inductive reactance for a single inductor is 100 Ω. What is the inductance?

Ans. 79.6 mH

38-18. A 20-Ω resistor, a 2-μF capacitor, and a 0.70-H inductor are available. Each of these, in turn, is connected to a 120-V, 60-Hz ac source as the only circuit element. What is the effective ac current in each case?

Series AC Circuits

38-19. A series ac circuit consists of a 100-Ω resistor, a 0.2-H inductor, and a 3-μF capacitor connected to a 110-V, 60-Hz ac source. (a) What is the inductive reactance? (b) What is the capacitive reactance? (c) What is the impedance? (d) What is the phase angle θ? (e) What is the power factor?

Ans. (a) 75.4 Ω, (b) 884 Ω, (c) 815 Ω, (d) $-83°$, (e) 12.3%

* 38-20. Answer the questions of Prob. 38-19 for a circuit containing a 12-mH inductor, an 8-μF capacitor, and a 40-Ω resistor connected to a 110-V, 200-Hz line.

* 38-21. When a 6-Ω resistor and a pure inductor are connected to a 110-V, 60-Hz line, the effective current is 10 A. What is the inductance of the coil? What is the power loss through the resistor? What is the power loss through the inductor?

Ans. 24.5 mH, 600 W, 0

* 38-22. A capacitor is in series with a resistance of 35 Ω and connected to a 220-V ac line. The reactance of the capacitor is 45 Ω. (a) What is the effective ac current? (b) What is the phase angle? (c) What is the power factor?

* 38-23. A coil having an inductance of 0.15 H and a resistance of 12 Ω is connected across a 110-V, 25-Hz line. (a) What is the current in the coil? (b) What is the phase angle? (c) What is the power factor? (d) What power is absorbed by the coil?

Ans. (a) 4.16 A, (b) 63°, (c) 45.4%, (d) 208 W

38-24. What is the resonant frequency for the circuit described by Prob. 38-19?

38-25. What is the resonant frequency for the circuit described by Prob. 38-20?

Ans. 514 Hz

Additional Problems

* 38-26. A 100-V battery is connected in a dc series circuit with a 0.06-H inductor and a 50-Ω resistor. Calculate the maximum current in the circuit and the time constant. If the circuit reaches a steady-state condition and is then broken, what will be the instantaneous current after one time constant?

38-27. An LR dc circuit has a time constant of 2 ms. What is the inductance if the resistance is 2000 Ω? What is the instantaneous current 2 ms after the circuit is connected to a 12-V battery?

Ans. 4 H, 3.78 mA

38-28. A series dc circuit consists of a 12-V battery, a 20-Ω resistor, and an unknown capacitor. The time constant is measured to be 40 ms. What is the capacitance? What is the maximum charge on the capacitor?

* 38-29. A 2-H inductor having a resistance of 120 Ω is connected to a 30-V battery. How much time is required for the current to reach 63 percent of its maximum value? What is the initial rate of current increase in amperes per second? What is the final current?

Ans. 16.7 ms, 15 A/s, 250 mA

* 38-30. Consider the circuit shown in Fig. 38-13. (a) What is the inductive reactance? (b) What is the capacitive reactance? (c) What is the impedance? (d) What is the effective current? (e) What is the power loss in the circuit?

38-31. A tuner circuit contains a 4-mH inductor and a variable capacitor. What must be the capacitance if the circuit is to resonate at a frequency of 800 Hz?

Ans. 9.89 μF

* 38-32. An 8-μF capacitor is in series with a 40-Ω resistor and connected to a 117-V, 60-Hz ac line. (a) What is the reactance? (b) What is the impedance? (c) What is the power factor? (d) What power is lost in the circuit?

* 38-33. An inductor, resistor, and capacitor are connected in series with a 60-Hz ac line. A voltmeter connected to each element in the circuit gives the following read-

Fig. 38-13

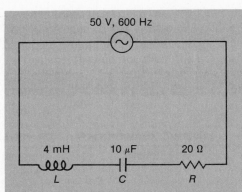

50 V, 600 Hz

4 mH   10 μF   20 Ω

L   C   R

ings: $V_R = 60$ V, $V_L = 100$ V, $V_C = 160$ V. **(a)** What is the total voltage drop in the circuit? **(b)** What is the phase difference between the voltage and the current?

Ans. (a) 84.9 V, (b) −45°

38-34. It is desired to construct a circuit whose resonant frequency is 950 kHz. If a coil in the circuit has an inductance of 3 mH, what capacitance should be added to the circuit?

38-35. A 50-μF capacitor and a 70-Ω resistor are connected in series across a 120-V, 60-Hz ac line. Determine the current in the circuit. What is the phase angle? What is the power loss in the circuit?

Ans. 1.37 A, −37.2°, 131 W

38-36. Refer to Prob. 38-35. What is the voltage across the resistor? What is the voltage across the capacitor? What inductance should be added to the circuit in order to reach resonance?

38-37. The antenna circuit in a radio receiver consists of a variable capacitor and a 9-mH coil. The resistance of the circuit is 40 Ω. A 980-kHz radio wave produces a potential difference of 0.2 mV across the circuit. Determine the capacitance required for resonance. What is the current at resonance?

Ans. 2.93 pF, 5 μA

38-38. A resonant circuit has an inductance of 400 μH and a capacitance of 100 pF. What is the resonant frequency?

PART FOUR
Modern Physics

39 Modern Physics and the Atom

By 1900, physical events had been observed which could not be explained satisfactorily by the laws of classical physics. Light, which was thought to be a wave phenomenon, was found to exhibit properties of particles as well. Newton's concepts of absolute mass, length, and time were found inadequate to describe certain physical events. The light produced by an electric spark in gases did not produce a continuous spectrum on passing through a prism or diffraction grating. These and other unexplained phenomena signaled the beginning of entirely new ways of viewing the world around us.

In 1905, Einstein's first paper on relativity was published, followed in 1916 by a second paper. They put classical physics into a new perspective. Einstein's relativity laid the basis for a universal physics which limited classical newtonian physics to situations involving speeds considerably less than the speed of light.

The work of Einstein stimulated a rash of work by others, which has had immense ramifications; it establishes parameters for maximum energy use, space travel, modern electronics, chemical analysis, x-rays, nuclear weaponry, and many other applications.

Einstein's two papers on relativity dramatically altered physics, but they are not generally understood by many people outside the scientific community. To understand relativity, you must put aside all your preconceived ideas and be willing to view physical events from a new perspective.

The special theory of relativity, published in 1905, is based on two postulates. The first states that every object is in motion compared with something else, that *there is no such thing as absolute rest.*

For example, picture a freight car moving along a railroad track at 40 mi/h. The cargo in relation to the freight car is stationary, but in relation to the earth it is moving at 40 mi/h. According to the first postulate, it is impossible to name anything which is at absolute rest; an object is only at rest (or moving) in relation to some specified reference point.

Einstein's first postulate also tells us that if we see something changing its position with respect to us, we have no way of knowing whether *it* is moving or *we* are moving. If you walk to a neighbor's house, it is just as correct according to this postulate to say that the house came to you. This sounds absurd to us because we are accustomed to using the earth as a frame of reference. Einstein's laws were designed to be completely independent of such a preferred reference frame. From the standpoint of physics, the first postulate is often restated as follows:

The laws of physics are the same for all frames of reference moving at a constant velocity with respect to one another.

Nineteenth-century physicists had suggested that there was a preferred frame of reference, the so-called luminiferous ether. This was the medium through which they thought electromagnetic waves propagated. However, experiments like the famous Michelson–Morley experiment of 1887 (discussed in Chap. 24) and others were unable to detect the existence of the ether. These experiments are the basis for Einstein's second revolutionary postulate:

The free-space velocity of light c is constant for all observers, independent of their state of motion.

To see why this second postulate was so revolutionary, let us consider a bus traveling at 50 km/h, as in Fig. 39-1. A person riding in the bus first throws a baseball with a speed (relative to the person) of 20 km/h toward the front of the bus. Then a second ball is thrown with the same speed toward the rear of the bus. To the person who threw the balls, each ball traveled at 20 km/h, one toward the front of the bus and one toward the rear of the bus. But to an observer on the ground, the velocity of the bus adds to the velocity of the first ball which appears to be traveling at 70 km/h in the same direction as the bus. And when the velocity of the bus is added to the second ball, the observer on the ground sees it traveling at 30 km/h *in the same direction as the bus.* But the speed of light does not change and will appear the same to the person on the bus and the observer on the ground. Light always travels at the same constant speed.

$$c = 3 \times 10^8 \text{ m/s}$$

whether it is traveling with the source or away from the source.

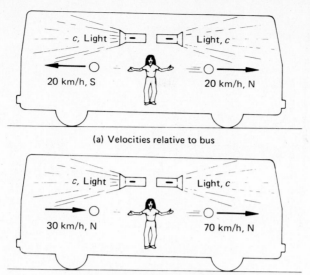

(a) Velocities relative to bus

(b) Velocities relative to ground

Fig. 39-1 The velocity of the bus is 50 km/h to the right (north). Two balls thrown with equal speeds (20 km/h), one to the right and the other to the left, have different speeds relative to the ground. However, the speed of light is independent of a frame of reference.

39-2

SIMULTANEOUS EVENTS: THE RELATIVITY OF TIME

A general understanding of Einstein's postulates does not require extensive mathematics. It only requires that we look at things in a fresh and different way. Our understanding of the world around us depends largely on our experience. The time of flight from Atlanta to New York may be listed as 2 h, and we would assume that an observer on the ground and another on the plane would each record the same time interval. Certainly, our experience seems to support such a conclusion. However, we shall see that time intervals in one reference frame are in general not the same as those measured from a second frame which is in motion relative to the first. Thus, the observer aboard the plane will judge the trip to require less time than that reckoned by the person on the ground. Of course, the apparent discrepancy would not be noticeable at normal speeds.

Our measurements of time are usually based on the occurrence of simultaneous events. For example, the sweeping second hand on a stop watch passes the point marked 0 just as a sprinter leaves the starting block. It later passes the point marked 10 just as the sprinter crosses the finish line. The judgment that the time interval was 10 s is based on the measurement of these simultaneous events. But Einstein's postulates force us to ask whether events judged to be simultaneous in one reference frame would also be judged as simultaneous in a moving frame of reference.

Einstein developed a thought experiment to illustrate the relativity of simultaneous measurements. Imagine that a boxcar is moving along a railroad track with uniform velocity as shown in Fig. 39-2. Two bolts of lightning strike the ends of the boxcar, leaving marks A_0 and B_0 on the boxcar and marks A and B on the ground. An

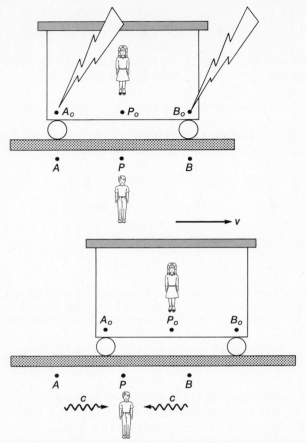

Fig. 39-2 Two lightning bolts strike the ends of a boxcar, making marks on both car and track. The person at point P on the ground observes the events to be simultaneous. The person at point P_0 on the car judges that the event at B_0 occurred *before* the event at A_0.

observer P_0 is at the midpoint of the boxcar, and another observer P is on the ground halfway between points A and B. The events seen by each observer are light signals arriving from the lightning bolts. It must also be noted that the speed c of the light is constant for each observer and is not affected by the motion of the boxcar.

The light signals reaching the ground observer P are judged to arrive at the same time. The signals traveled the same distance, and the events are judged to occur simultaneously. Now consider the same events as observed by the person P_0 aboard the boxcar. By the time the light signal from B_0 has reached the point P on the ground, it has already passed point P_0 on the moving boxcar. The light from A_0 has not yet reached P_0. Thus, the judgment of the person on the boxcar is that events did not occur at the same time. Who is correct? Since according to the principle of relativity, there is no preferred frame of reference, we must conclude that each is correct from his or her own perspective.

RELATIVISTIC LENGTH, MASS, AND TIME

The relativity of simultaneous measurements has far-reaching consequences when material objects approach the speed of light. All physical measurements must account for relative motion. Time, length, and mass measurements will not be the same for all observers. A comparison of measurements made in different frames of reference is easier if we establish what is called a "proper" frame of reference. It might be noted that two events which occur in the *same* frame of reference are more fundamental that those same events as observed from another frame. We will define a *proper time interval* Δt_0 as an interval between two events which occur at the same space point. For example, consider a rocket ship moving in space at a speed v relative to an observer in a laboratory on the earth. Refer to Fig. 39-3. A person aboard this ship measures *proper length* L_0, *proper mass* m_0, and *proper time intervals* Δt_0. The person on the ground who measures these same events which actually occur on the ship will obtain different values L, m, and Δt. Each observer is correct from his or her own viewpoint.

A series of relativistic equations can be developed to predict how measurements are affected by relative motion. In each case, the effect becomes more pronounced as the velocity v of objects approaches the limiting velocity c of light. If we find that the

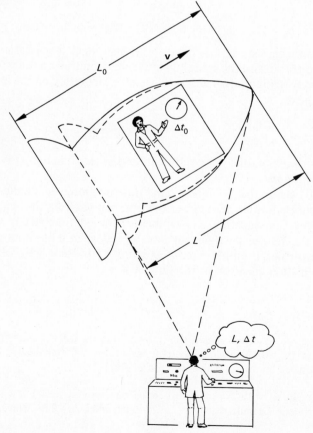

Fig. 39-3 The length of objects and the duration of events are affected by relative motion. The person aboard the rocket measures the length L_0 and the time interval Δt_0; the person in the laboratory observes a shorter length L and records a longer time interval Δt.

proper length of a rocket ship (see Fig. 39-3) is L_0, its length L when it is moving at a relative speed v will be given by

$$L = L_0 \sqrt{1 - \frac{v^2}{c^2}} \qquad \textit{Relativisitic Contraction} \quad (39\text{-}1)$$

Such foreshortening of length in the direction of motion is referred to as *relativistic contraction.*

This means that the length L of a moving object is observed to be shorter by a factor of $\sqrt{1 - v^2/c^2}$ than its length at rest (its proper length). A study of the formula will reveal that the observed length L will be equal to the proper length L_0 when $v = 0$ (the object is at rest). It will begin to shorten as the velocity approaches c.

EXAMPLE 39-1   When a rocket ship is at rest with respect to us, its length is 100 m. How long do we measure its length to be when it moves by us at a speed of 2.4×10^8 m/s, or $0.8c$?

Solution   In this case the proper length L_0 is 100 m. Substitution into Eq. (39-1) gives

$$L = L_0 \sqrt{1 - \frac{v^2}{c^2}} = (100 \text{ m}) \sqrt{1 - \frac{(0.8c)^2}{c^2}}$$

$$= (100 \text{ m}) \sqrt{1 - \frac{0.64c^2}{c^2}} = (100 \text{ m})\sqrt{1 - 0.64}$$

$$= (100 \text{ m})\sqrt{0.36} = (100 \text{ m})(0.6) = 60 \text{ m}$$

As we have seen, time intervals are also affected by relative motion. A clock held by a person aboard the spaceship in Fig. 39-3 indicates a proper time interval Δt_0 that is shorter than that (Δt) observed from the laboratory on the earth. The time interval Δt is given by

$$\Delta t = \frac{\Delta t_0}{1 - v^2/c^2} \qquad \textit{Time Dilation} \quad (39\text{-}2)$$

This slowing of time (longer time intervals) as a function of velocity is referred to as *time dilation.*

To be sure that you understand this equation, you should recognize that Δt and Δt_0 represent time *intervals,* or the time elapsed from the beginning until the end of an event. Consequently, a clock that is running more slowly will record longer time intervals. We can say that time has stopped when it becomes impossible to measure an event; i.e., the time interval is infinite. This is exactly what is predicted by the time-dilation equation in the limit where $v = c$.

$$\Delta t = \frac{\Delta t_0}{\sqrt{1 - c^2/c^2}} = \frac{\Delta t_0}{\sqrt{1 - 1}} = \frac{\Delta t_0}{0} = \infty$$

EXAMPLE 39-2   Suppose we observe a rocket ship moving past us at $0.8c$, as in the previous example. We measure the time between ticks on the rocket-ship clock to be 1.67 s. What time between ticks is measured by the captain of the rocket ship?

Solution   In this case $\Delta t = 1.67$ s, and we are to find Δt_0 from Eq. (39-2). It will be helpful to recall from the previous example that $\sqrt{1 - v^2/c^2} = 0.6$ when $v = 0.8c$. Solving Eq. (39-2) for Δt_0, we have

$$\Delta t_0 = \Delta t \sqrt{1 - v^2/c^2} = (1.67 \text{ s})(0.6) = 1.00 \text{ s}$$

The interval measured by the captain is shorter than that measured by us.

Let us now consider yet another physical quantity which varies with relative velocity. In order for momentum to be conserved independent of a frame of reference, the mass of a body must vary in the same proportion as length and time. If the rest mass of an object is m_0, the mass m of a body moving with a speed v will be measured by

$$\boxed{m = \frac{m_0}{\sqrt{1 - v^2/c^2}}} \qquad \textit{Relativistic Mass} \quad (39\text{-}3)$$

The mass m in this instance is referred to as the *relativistic mass*.

EXAMPLE 39-3 The rest mass of an electron is 9.1×10^{-31} kg. What is its relativistic mass if its velocity is $0.8c$?

Solution In previous examples we found that $\sqrt{1 - v^2/c^2} = 0.6$ when $v = 0.8c$. Using this computation and substituting in Eq. (39-3), we obtain

$$m = \frac{m_0}{\sqrt{1 - v^2/c^2}} = \frac{9.1 \times 10^{-31} \text{ kg}}{0.6} = 1.52 \times 10^{-30} \text{ kg}$$

This represents a 67 percent increase in mass.

It should be clear from Eq. (39-3) that if m_0 is not equal to zero, the value for the relativistic mass m approaches infinity as v approaches c. This would mean that an infinite force would be needed to accelerate a nonzero mass to the velocity of light. Apparently, the free-space velocity of light represents an upper limit for the speed of such masses. However, if the rest mass *is* zero, as for photons of light, the relativistic-mass equation *does* allow for $v = c$.

Startling as the predictions of Einstein's equations are, they are experimentally verified daily in the laboratory. Their conclusions are amazing to us only because we do not have direct experience with such fantastic speeds. The giant atom smashers, the betatrons, and many other devices for accelerating particles are in use all over the country to accelerate atomic and nuclear particles at speeds very close to the speed of light. Protons in the big Brookhaven accelerator have been accelerated to within 99.948 percent of the speed of light. Their mass is shown to be increased precisely as relativity predicts. Electrons can now be accelerated to more than $0.9999999c$, + the speed of light, causing their mass to increase by more than 40,000 times the rest value.

39-4
MASS AND ENERGY

Before Einstein, physicists had always considered mass and energy as separate quantities which must be *conserved* separately. Now mass and energy must be considered as different ways of expressing the same quantity. If we say that mass can be converted to energy and energy to mass, we must recognize that mass and energy are the same thing expressed in different units. Einstein found the conversion factor to be equal to the square of the velocity of light.

$$E_0 = m_0 c^2 \qquad (39\text{-}4)$$

From the way the equation is written, it is easy to see that a tiny bit of mass corresponds to an enormous amount of energy. For example, an object whose rest mass m_0 is 1 kg has a rest energy E_0 of 9×10^{16} J.

A more general discussion of energy must take the effects of relativity into account. The expression for the total energy of a particle of rest mass m_0 and momentum $p = mv$ can be written

$$E = \sqrt{(m_0 c^2)^2 + p^2 c^2} \qquad (39\text{-}5)$$

Now if we substitute for m_0 from the relativistic-mass relationship, Eq. (39-3), the total energy reduces to

$$\boxed{E = mc^2} \qquad (39\text{-}6)$$

where m represents the relativistic mass. This is the most general form for the total energy of a particle.

Note that Eq. (39-5) reduces to $E_0 = m_0 c^2$ when the velocity is zero and hence $p = 0$. Furthermore, if we consider velocities considerably less than c, the equation simplifies to

$$E = \tfrac{1}{2}m_0 v^2 + m_0 c^2 \qquad (39\text{-}7)$$

Here we have the usual expression for kinetic energy with a new term added for the *rest energy*.

The more general expression for the kinetic energy of a particle must consider the effects of relativity. Recall that the kinetic energy E_κ at speed v is defined as the work that must be done to accelerate a particle from rest to speed v. By the methods of calculus it can be shown that the relativistic kinetic energy of a particle is given by

$$\boxed{E_\kappa = (m - m_0)c^2} \qquad (39\text{-}8)$$

This represents the difference between the total energy of a particle and its rest-mass energy.

EXAMPLE 39-4   An electron is accelerated to a speed of $0.8c$. Compare its relativistic kinetic energy with the value based on Newton's mechanics.

Solution   In the previous example we found the relativistic mass of an electron at this speed to be 1.52×10^{-30} kg. Since its rest mass is 9.1×10^{-31} kg, we can determine its kinetic energy by substitution into Eq. (39-8).

$$
\begin{aligned}
E_\kappa &= (m - m_0)c^2 \\
&= (15.2 \times 10^{-31} \text{ kg} - 9.1 \times 10^{-31} \text{ kg})(3 \times 10^8)^2 \\
&= (6.1 \times 10^{-31} \text{ kg})(9 \times 10^{16}) \\
&= 5.49 \times 10^{-14} \text{ J}
\end{aligned}
$$

The newtonian value is based on $\frac{1}{2}m_0 v^2$, where $v = 0.8c$.

$$\frac{1}{2}m_0 v^2 = \frac{1}{2}(9.1 \times 10^{-31} \text{ kg})(0.8c)^2$$
$$= (4.55 \times 10^{-31} \text{ kg})(0.64c^2)$$
$$= (4.55 \times 10^{-31} \text{ kg})(0.64)(3 \times 10^8 \text{ m/s})^2$$
$$= 2.62 \times 10^{-14} \text{ J}$$

The relativistic kinetic energy is more than twice its newtonian value.

39-5
QUANTUM THEORY AND THE PHOTO-ELECTRIC EFFECT

Recall from Chap. 24 that the photoelectric effect led to the establishment of a dualistic theory of light. (See Fig. 39-4.) The ejection of electrons as a result of incident light could not be accounted for in terms of the existing electromagnetic theory.

In an attempt to bring experiment into agreement with theory, Maxwell Planck postulated that electromagnetic energy is absorbed or emitted in discrete packets, or

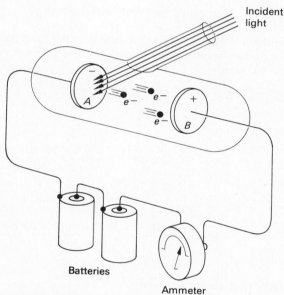

Fig. 39-4 The photoelectric effect.

quanta. The energy of such quanta, or *photons,* is proportional to the frequency of the radiation. Planck's equation can be written

$$\boxed{E = hf}$$

(39-9)

where h is the proportionality constant known as *Planck's constant.* Its value is

$$h = 6.63 \times 10^{-34} \text{ J/Hz}$$

Einstein used Planck's equation to explain the photoelectric effect. He reasoned that if light is emitted in photons of energy hf, it must also travel as photons. When a quantum of light strikes a metallic surface, it has an energy equal to hf. If all this energy is transferred to a single electron, the electron might be expected to leave the metal with energy hf. However, at least an amount of energy W is needed to remove the electron from the metal. The term W is called the *work function* of the surface. Thus, the ejected electron leaves with a maximum kinetic energy given by

$$\boxed{E_\kappa = \tfrac{1}{2}mv^2_{\max} = hf - W} \qquad (39\text{-}10)$$

This is *Einstein's photoelectric equation.*

As the frequency of incident light is varied, the maximum energy of the ejected electron varies. The lowest frequency f_0 at which an electron is emitted occurs when $E_\kappa = 0$. In this case,

$$f_0 = \frac{W}{h} \qquad (39\text{-}11)$$

The quantity f_0 is often called the *threshold frequency.*

EXAMPLE 39-5   Light of wavelength 650 nm is required to cause electrons to be ejected from the surface of a metal. What is the kinetic energy of the ejected electrons if the surface is bombarded with light of wavelength 450 nm?

Solution   The work function W of the surface is equal to the energy of the 650-nm light. Hence,

$$W = hf_0 = h\frac{c}{\lambda_0}$$

$$= 6.63 \times 10^{-34}\,\text{J} \cdot \text{s}\left(\frac{3 \times 10^8\,\text{m/s}}{650 \times 10^{-9}\,\text{m}}\right)$$

$$= 3.06 \times 10^{-19}\,\text{J}$$

The energy of the 450-nm light is

$$hf = \frac{hc}{\lambda} = \frac{(6.63 \times 10^{-34}\,\text{J} \cdot \text{s})(3 \times 10^8\,\text{m/s})}{450 \times 10^{-9}\,\text{m}}$$

$$= 4.42 \times 10^{-19}\,\text{J}$$

From Einstein's photoelectric equation,

$$E_\kappa = hf - w$$

$$= 4.42 \times 10^{-19}\,\text{J} - 3.06 \times 10^{-19}\,\text{J} = 1.36 \times 10^{-19}\,\text{J}$$

39-6
WAVES AND PARTICLES

Electromagnetic radiation has a dual character in its interaction with matter. Sometimes it exhibits wave properties, as demonstrated by interference and diffraction. At other times, as in the photoelectric effect, it behaves like particles, which we have called photons. In 1924, Louis de Broglie was able to demonstrate this duality of matter by deriving a relationship for the wavelength of a particle.

This relationship can be seen by looking at two expressions for the energy of a photon. We have already seen from Planck's work that the energy of a photon can be expressed as a function of its wavelength λ.

$$E = hf = \frac{hc}{\lambda}$$

Another expression for the energy of a photon was given earlier in the section on relativity:

$$E = \sqrt{(m_0c^2)^2 + p^2c^2}$$ (39-5)

This equation shows that photons have momentum $p = mv$ due to their relativistic mass. However, the rest mass m_0 of a photon is zero, so that Eq. (39-5) becomes

$$E = \sqrt{p^2c^2} = pc$$ (39-12)

Since we also know that $E = hc/\lambda$, we can write

$$\frac{hc}{\lambda} = pc$$

from which the wavelength of a photon is given by

$$\lambda = \frac{h}{p}$$ (39-13)

De Broglie proposed that all objects have wavelengths related to their momentum—whether the objects are wavelike or particlelike. For example, the wavelength of an electron or any other particle is given by de Broglie's equation, which can be rewritten

$$\boxed{\lambda = \frac{h}{mv}}$$ *de Broglie Wavelength* (39-14)

EXAMPLE 39-6 What is the de Broglie wavelength of an electron which has a kinetic energy of 100 eV?

Solution Recalling that $1 \text{ eV} = 1.6 \times 10^{-19}$ J, we see that 100 eV is equivalent to 1.6×10^{-17} J, which must be equal to $\frac{1}{2}mv^2$. Thus,

$$\tfrac{1}{2}mv^2 = 1.6 \times 10^{-17} \text{ J}$$

from which

$$v^2 = \frac{2(1.6 \times 10^{-17} \text{ J})}{m} = \frac{3.2 \times 10^{-17} \text{ J}}{9.1 \times 10^{-31} \text{ kg}} = 3.52 \times 10^{13} \text{ m}^2/\text{s}^2$$

The velocity of the electron is therefore

$$v = \sqrt{3.52 \times 10^{13} \text{ m}^2/\text{s}^2} = 5.93 \times 10^6 \text{ m/s}$$

Now the wavelength can be found from Eq. (39-14).

$$\lambda = \frac{h}{mv} = \frac{6.63 \times 10^{-34} \text{ J} \cdot \text{s}}{(9.1 \times 10^{-31} \text{ kg})(5.93 \times 10^6 \text{ m/s})}$$
$$= 1.23 \times 10^{-10} \text{ m} = 0.123 \text{ nm}$$

Note from de Broglie's equation that the higher the velocity of the particle, the shorter its wavelength. Remember that the relativistic mass must be used if the velocity of a given particle is large enough to warrant it.

Early attempts to explain the structure of an atom were based on a model attributed to J. J. Thomson, the discoveror of the electron. In this model the electrons are implanted in a spherical space of positive charge, as illustrated in Fig. 39-5. Such a model explained the observed neutrality of atoms by postulating equal amounts of positive and negative charge.

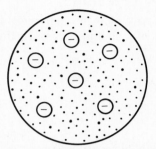

Fig. 39-5 The Thomson model of the atom consisted of electrons stuck in a sphere of positive charge.

Thomson's model was short-lived, however, because of the work of Ernest Rutherford in 1911. He bombarded a thin metal foil with a stream of alpha particles, as illustrated in Fig. 39-6. An alpha particle is a tiny positively charged particle emitted by a radioactive substance such as radium. Most of the positively charged particles penetrated the foil easily, as indicated by a flash of light when they struck the zinc sulfide screen, and a few were deflected slightly. Much to Rutherford's surprise, however, others were deflected at extreme angles. Some were even deflected backward. (Rutherford himself said it was like firing a cannon-ball through tissue paper and having it come back and hit you in the face.) These extreme deflections could not be explained in terms of the Thomson model for the atom. If charge really was scattered through the atom, the electrical forces would be far too weak to repulse the alpha particles at the large angles actually observed.

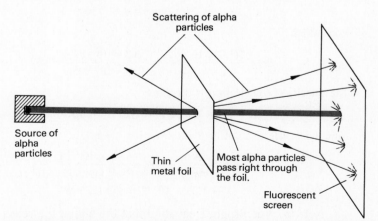

Fig. 39-6 Rutherford scattering of alpha particles provided the first evidence for the atomic nucleus.

Rutherford explained these results by assuming that all the positive charge of an atom is concentrated into a very small region, called the *nucleus* of the atom. The electrons were assumed to be distributed in the space around this positive charge. With the atom consisting largely of empty space, the fact that most of the alpha particles passed right through the foil is easily explained. Furthermore, Rutherford found the large-angle scattering to be a consequence of electrostatic repulsion. Once the alpha particle has penetrated the electrons surrounding the positive charge, it is close to a large positive charge of great mass. Large electrostatic forces are therefore expected.

Rutherford has been credited with the discovery of the *nucleus* because he was able to develop formulas to predict the observed scattering of alpha particles. On the basis of his calculations, the diameter of the nucleus was estimated to be approximately one ten-thousandth of the diameter of the atom itself. The positive part of the atom was seen to be concentrated in the nucleus, approximately 10^{-5} nm in diameter. Rutherford believed the electrons to be grouped around the nucleus so that the diameter of the whole atom is about 0.1 nm.

39-8 ELECTRON ORBITS

An immediate difficulty accompanying the Rutherford atom was related to the stability of the atomic electrons. We know from Coulomb's law that the electrons should be attracted to the nucleus. One possible explanation is that the electrons are moving in circles around the nucleus much as planets revolve around the sun. The necessary centripetal force would be provided by coulomb attraction.

Consider, for example, the hydrogen atom, which consists of a single proton and a single electron. We might expect the electron to remain in a constant orbit about the nucleus, as illustrated in Fig. 39-7a. The charge of the electron is labeled $-e$, and the equal but opposite charge of the proton is labeled $+e$. At a distance r from the nucleus, the electrostatic force of attraction on the electron is given by Coulomb's law,

$$F_e = \frac{e^2}{4\pi\epsilon_0\, r^2} \tag{39-15}$$

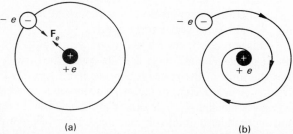

(a)   (b)

Fig. 39-7 *(a)* A stable electron orbit in which the centripetal force is provided by the electrostatic force F_e. *(b)* Instability due to electromagnetic radiation which should cause the electron to lose energy and spiral into the nucleus.

where each charge has the magnitude e. For a stable orbit, this force must exactly equal the centripetal force, given by

$$F_c = \frac{mv^2}{r} \qquad (39\text{-}16)$$

where m is the mass of the electron traveling with a velocity v. Setting $F_e = F_c$, we have

$$\frac{e^2}{4\pi\epsilon_0\,r^2} = \frac{mv^2}{r} \qquad (39\text{-}17)$$

Solving for the radius r, we obtain

$$\boxed{r = \frac{e^2}{4\pi\epsilon_0\,mv^2}} \qquad (39\text{-}18)$$

According to classical theory, Eq. (39-18) should predict the orbital radius r of the electron as a function of its speed v.

The problem with this approach is that the electron must be continually accelerated under the influence of the electrostatic force. According to classical theory, an accelerated electron must radiate energy. The total energy of the electron would therefore gradually decrease, reducing the speed of the electron. As can be seen from Eq. (39-18), gradual reduction in electron speed v results in smaller and smaller orbits. Thus, the electron should spiral into the nucleus, as shown in Fig. 39-6b. The fact that it does not compels us to acknowledge a fundamental inconsistency in Rutherford's atom.

39-9
ATOMIC SPECTRA

All substances radiate electromagnetic waves when they are heated. Since each element is different, such emitted radiation can be expected to provide clues to atomic structure. These electromagnetic waves are analyzed by a *spectrometer,* which uses a prism or a diffraction grating to organize the radiation into a pattern called a *spectrum.* For an incandescent source of light, the spectrum is *continuous;* i.e., it contains all wavelengths and is similar to a rainbow. If, however, the source of light is a very hot gas under low pressure, the spectrum of the emitted light consists of a series of bright lines separated by dark regions. Such spectra are called *line emission spectra.* The chemical composition of a vaporized material can be determined by comparing its spectrum with known spectra.

A line emission spectrum for hydrogen is shown in Fig. 39-8. The sequence of lines, called a *spectral series,* has a definite order, the lines becoming more and more crowded as the limit of the series is approached. Each line corresponds to a characteristic frequency or wavelength (color). The line of longest wavelength, 656.3 nm, is in the red and is labeled H_α. The others are labeled, in order, as H_β, H_γ, and so on.

It is also possible to obtain similar information from a gas or vapor in an unexcited state. When light is passed through a gas, certain discrete wavelengths are *absorbed.* These *absorption spectra* are like those produced by emission except that the characteristic wavelengths appear as *dark* lines on a light background. The

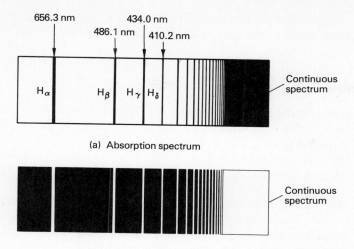

656.3 nm   434.0 nm

486.1 nm   410.2 nm

H_α H_β H_γ H_δ

Continuous spectrum

(a) Absorption spectrum

Continuous spectrum

(b) Emission spectrum

Fig. 39-8 Line spectra for the Balmer series of the hydrogen atom.

absorption spectrum for hydrogen is compared with the emission spectrum in Fig. 39-8.

As early as 1884, Johann Jakob Balmer found a simple mathematical relationship for predicting the characteristic wavelengths of some of the lines in the hydrogen spectrum. His formula is

$$\frac{1}{\lambda} = R\left(\frac{1}{2^2} - \frac{1}{n^2}\right) \tag{39-19}$$

where λ = wavelength
R = Rydberg constant
$n = 3, 4, 5, \ldots$

If λ is measured in meters, the value for R is

$$R = 1.097 \times 10^7 \text{ m}^{-1}$$

The series of wavelengths predicted by Eq. (39-19) is called the *Balmer series*.

EXAMPLE 39-7   Using Balmer's equation, determine the wavelength of the H_α line in the hydrogen spectrum. (This first line occurs when $n = 3$.)

Solution   Direct substitution yields

$$\frac{1}{\lambda} = 1.097 \times 10^7 \text{ m}^{-1}\left(\frac{1}{2^2} - \frac{1}{3^2}\right) = 1.524 \times 10^6 \text{ m}^{-1}$$

from which

$$\lambda = 656.3 \text{ nm}$$

Other characteristic wavelengths are found by setting $n = 4, 5, 6$, and so on. The limit of the series is found by setting $n = \infty$ in Eq. (39-19).

Since the discovery of Balmer's equation, several other series spectra have been discovered for hydrogen. In general, all these discoveries can be summarized by the single equation

$$\frac{1}{\lambda} = R\left(\frac{1}{l^2} - \frac{1}{n^2}\right) \qquad (39\text{-}20)$$

where l and n are integers with $n > l$. The series predicted by Balmer corresponds to $l = 2$ and $n = 3, 4, 5, \ldots$. The *Lyman series* is in the ultraviolet region and corresponds to $l = 1$, $n = 2, 3, 4, \ldots$; the *Paschen series* is in the infrared and corresponds to $l = 3$, $n = 4, 5, 6, \ldots$; and the *Brackett series,* also in the infrared, corresponds to $l = 4$, $n = 5, 6, 7, \ldots$.

39-10 THE BOHR ATOM

Observation of atomic spectra have indicated that atoms emit only a few rather definite frequencies. This fact does not agree with the Rutherford model, which predicts an unstable atom emitting radiant energy of all frequencies. Any theory of atomic structure must account for the regularities observed in atomic spectra.

The first theory to explain the line spectrum of the hydrogen atom satisfactorily was offered by Niels Bohr in 1913. He assumed, like Rutherford, that the electrons were in circular orbits about a dense positively charged nucleus, but he decided that electromagnetic theory cannot be strictly applied on the atomic level. Thus, he avoided the problem of orbital instability due to emitted radiation. Bohr's first postulate is as follows:

An electron may exist only in those orbits where its angular momentum is an integral multiple of h/2π.

Thus, contrary to classical prediction, electrons may be in certain specified orbits without the emission of radiant energy.

The basis for Bohr's first postulate can be seen in terms of de Broglie wavelengths. The stable orbits are those in which an integral number of electron wavelengths can be fitted into the circumference of the Bohr orbit. Such orbits would allow for standing waves, as illustrated in Fig. 39-9 for four wavelengths. The condi-

Fig. 39-9 A stable electron orbit, showing a circumference equal to four de Broglie wavelengths.

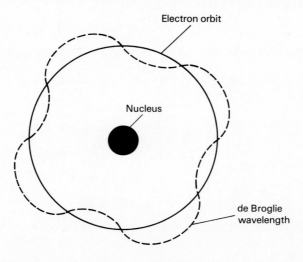

Electron orbit

Nucleus

de Broglie wavelength

tions for such standing waves would be given by

$$n\lambda = 2\pi r \qquad n = 1, 2, 3, \ldots \qquad (39\text{-}21)$$

where r is the radius of an electron orbit which contains n wavelengths.
Since $\lambda = h/mv$, we can rewrite Eq. (39-21) as

$$n\frac{h}{mv} = 2\pi r$$

from which it can be shown that the angular momentum mvr is given by

$$mvr = \frac{nh}{2\pi} \qquad (39\text{-}22)$$

The number n, called the *principal quantum number*, may take on the values $n = 1$, 2, 3, ...

A *second postulate* given by Bohr places even further restrictions on atomic theory by incorporating the quantum theory.

> *If an electron changes from one stable orbit to any other, it loses or gains energy in discrete quanta equal to the difference in energy between the initial and final states.*

In equation form, Bohr's second postulate can be written

$$hf = E_i - E_f$$

where hf = energy of emitted or absorbed photon
E_i = initial energy
E_f = final energy.

Let us return to our discussion of the hydrogen atom to see whether Bohr's postulates will help to bring theory into harmony with observed spectra. Recall that application of Coulomb's law and Newton's law resulted in Eq. (39-18) for the radius r of the orbiting electron.

$$r = \frac{1}{4\pi\epsilon_0}\frac{e^2}{mv^2}$$

According to Bohr's theory,

$$mvr = \frac{nh}{2\pi}$$

Solving these two equations simultaneously for the radius r and for the velocity v, we obtain

$$r = n^2\frac{\epsilon_0 h^2}{\pi me^2} \qquad (39\text{-}23)$$

$$v = \frac{e^2}{2\epsilon_0 nh} \qquad (39\text{-}24)$$

These equations predict the possible radii and velocities for the electron, where $n = 1, 2, 3, \ldots$.

We now derive an expression for the total energy of a hydrogen atom for any electron orbit.

$$E_T = E_k + E_p$$

The kinetic energy is found by substituting from Eq. (39-24),

$$E_k = \tfrac{1}{2}mv^2 = \frac{me^4}{8\epsilon_0 \, n^2 h^2} \tag{39-25}$$

The potential energy of the atom for any orbit is

$$E_p = \frac{1}{4\pi\epsilon_0} \frac{e^2}{r} = -\frac{me^4}{4\epsilon_0^2 \, n^2 h^2} \tag{39-26}$$

after substitution of r from Eq. (39-23). The potential energy is negative because outside work is necessary to remove the electron from the atom. Adding Eq. (39-25) to Eq. (39-26), we find the total energy to be

$$E_T = -\frac{me^4}{8\epsilon_0^2 \, n^2 h^2} \tag{39-27}$$

Returning to Bohr's second postulate, we are now in a position to predict the energy of an emitted or absorbed photon. Normally, the electron is in its ground state corresponding to $n = 1$. If the atom *absorbs* a photon, the electron may jump to one of the outer orbits. From an *exited* state, it will soon fall back to a lower orbit, *emitting* a photon in the process.

Suppose an electron is in an outer orbit of quantum number n_i and then returns to a lower orbit of quantum number n_f. The decrease in energy must be equal to the energy of the emitted photon.

$$E_i - E_f = hf$$

Substituting the total energy for each state from Eq. (39-27), we obtain

$$hf = E_i - E_f$$
$$= -\frac{me^4}{8\epsilon_0^2 \, n_i^2 \, h^2} + \frac{me^4}{8\epsilon_0^2 \, n_f^2 \, h^2}$$
$$= \frac{me^4}{8\epsilon^2 \, h^2} \left(\frac{1}{n_f^2} - \frac{1}{n_i^2} \right)$$

Dividing both sides of this relation by h and remembering that $f = c/\lambda$, we can write

$$\frac{1}{\lambda} = \frac{me^4}{8\epsilon^2 \, h^3 \, c} \left(\frac{1}{n_f^2} - \frac{1}{n_i^2} \right) \tag{39-28}$$

Substituting in the values, we obtain

$$\frac{me^4}{8\epsilon_0^2 \, h^3 \, c} = 1.097 \times 10^7 \; m^{-1}$$

which is equal to Rydberg's constant R. Therefore Eq. (39-28) is simplified to

$$\frac{1}{\lambda} = R\left(\frac{1}{n_f^2} - \frac{1}{n_i^2}\right)$$

(39-29)

The above relationship is exactly the same in form as Eq. (39-20), which was established by Balmer and others from experimental data. Thus, the Bohr atom brings theory into harmony with observation.

EXAMPLE 39-8 Determine the wavelength of the photon emitted from a hydrogen atom when the electron jumps from the first excited state to the ground state.

Solution In this case, $n_f = 1$ and $n_i = 2$. Thus

$$\frac{1}{\lambda} = R\left(\frac{1}{n_f^2} - \frac{1}{n_i^2}\right)$$

$$= (1.097 \times 10^7\, m^{-1})\left(\frac{1}{1^2} - \frac{1}{2^2}\right) = 8.23 \times 10^6\, m^{-1}$$

from which

$$\lambda = 1.22 \times 10^{-7}\ \text{m} = 122\ \text{nm}$$

39-11
ENERGY
LEVELS

From Bohr's work we now have a picture of an atom in which orbital electrons can occupy a number of energy levels. The total energy at the nth level is given by Eq. (39-27).

$$E_n = -\frac{me^4}{8\epsilon_0^2\, n^2 h^2}$$

(39-30)

When the hydrogen atom is in its stable *ground* state, the quantum number n is equal to 1. The possible *excited* states are given for $n = 2, 3, 4, \ldots$.

EXAMPLE 39-9 Determine the energy of an electron in the ground state for a hydrogen atom.

Solution The following constants are needed:

$$\epsilon_0 = 8.85 \times 10^{-12}\, c^2/\text{N} \cdot \text{m}^2 \qquad h = 6.63 \times 10^{-34}\ \text{J} \cdot \text{s}$$
$$m = 9.1 \times 10^{-31}\ \text{kg} \qquad e = 1.6 \times 10^{-19}\, c$$

Direct substitution into Eq. (39-30), with $n = 1$, yields

$$E_1 = -\frac{(9.1 \times 10^{-31}\ \text{kg})(1.6 \times 10^{-19})^4}{8[8.85 \times 10^{-12}(\text{C}^2/\text{N} \cdot \text{m})^2]^2(1)^2(6.63 \times 10^{-34}\ \text{J} \cdot \text{s})^2}$$

$$= -2.17 \times 10^{-18}\ \text{J}$$

A more convenient unit for measuring energy at the atomic level is the electron-volt (eV). Recall from Chap. 29 that an electronvolt is the energy acquired by an electron accelerated through a potential difference of 1 V. Consequently

$$1\ \text{eV} = 1.6 \times 10^{-19}\ \text{J}$$

In applications where a larger unit of energy is required the megaelectron volt (MeV) is more appropriate.

$$1 \text{ MeV} = 10^6 \text{ eV}$$

From the previous example, the energy of the ground-state electron can be expressed in electronvolts as follows:

$$E_1 = -2.17 \times 10^{-18} \text{ J} \frac{1 \text{ eV}}{1.6 \times 10^{-19} \text{ J}}$$

$$= -13.6 \text{ eV}$$

This fact can be used to write Eq. (39-30) in a simpler form:

$$E_n = \frac{-13.6 \text{ eV}}{n^2} \qquad (39\text{-}31)$$

Similar calculations will yield smaller negative values for the outer orbits. If the electron were to be entirely removed from the atom, a case where $n = \infty$, 13.6 eV of energy would be required. ($E_\infty = 0$.) Similarly, a photon of energy 13.6 eV could be emitted if an electron were to be captured by an ionized hydrogen atom and wind up in the ground state.

The atomic spectra observed for hydrogen is now understood in terms of energy levels. The Lyman series results from electrons returning from some excited state to the ground state, as seen in Fig. 39-10. The Balmer, Paschen, and Brackett series occur when the final states are orbits for which $n = 2$, $n = 3$, and $n = 4$, respectively.

An energy-level diagram for hydrogen is shown in Fig. 39-11. Such diagrams are often used to describe the various energy states of atoms.

39-10 The Bohr atom and a description of the origin of the Lyman, Balmer, Paschen, and Brackett spectral series.

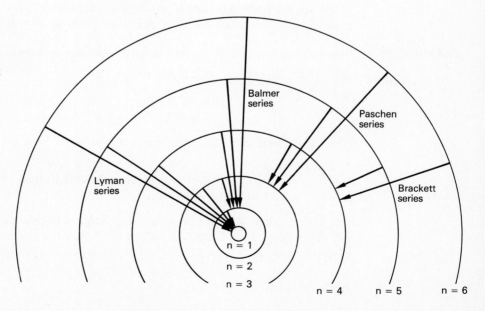

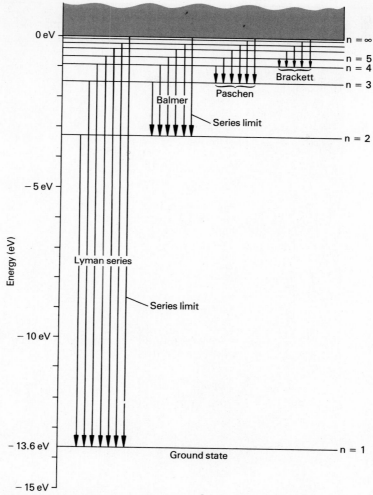

Fig. 39-11 Energy-level diagram for the hydrogen atom.

Although the Bohr atom remains a convenient way to describe the atom, a much more refined theory has been found necessary. The model of the electron as a point particle moving in a perfectly circular orbit does not explain many atomic phenomena. Additional quantum numbers have been established to describe the shape and orientation of the electron cloud about the nucleus, as well as the spin motion of the electrons. It has also been established that no two electrons of the same atom can exist in exactly the same state, even though they may exist at the same Bohr energy level. A more complete description of modern atomic theory can be found in textbooks in the field of atomic physics.

The works of Einstein, Bohr, de Broglie, Balmer, and many others have led to a much clearer understanding of nature. We no longer view the world as though all phenomena can be seen, touched, and observed in traditional ways. Greater understanding of the atom has led to many industrial applications based on the principles discussed in this chapter. A summary of the major topics is given below:

- According to Einstein's equations of relativity, length, mass, and time are affected by relativistic speeds. The changes become more significant as the ratio of an object's velocity v to the free space velocity of light c becomes larger.

$$L = L_0\sqrt{1 - \frac{v^2}{c^2}}$$ *Relativistic Contraction*

$$m = \frac{m_0}{\sqrt{1 - v^2/c^2}}$$ *Relativistic Mass*

$$t = \frac{t_0}{\sqrt{1 - v^2/c^2}}$$ *Time Dilation*

In the above equations, $c = 3 \times 10^8$ m/s.
- The total energy of a particle of rest mass m_0 and speed v can be written in either of the following forms:

$$E = mc^2 \qquad E = \sqrt{m_0^2c^4 + p^2c^2}$$ *Total Energy*

In these equations m is the relativistic mass as determined by the speed v, and p is the momentum mv.
- The *relativistic kinetic energy* is found from

$$E_\kappa = (m - m_0)c^2$$ *Relativistic Kinetic Energy*

- The quantum theory of electromagnetic radiation relates the energy of such radiation to its frequency f or wavelength λ.

$$E = hf \qquad E = \frac{hc}{\lambda} \qquad h = 6.63 \times 10^{-34} \text{ J/Hz}$$

- In the photoelectric effect, the kinetic energy of the ejected electrons is the energy of the incident radiation hf less the work function of the surface W.

$$E_\kappa = \tfrac{1}{2}mv^2 = hf - W$$ *Photoelectric Equation*

- The lowest frequency f_0 at which a photoelectron is ejected is the threshold frequency. It corresponds to the work-function energy W.

$$f_0 = \frac{W}{h}$$

$$\boxed{W = hf_0} \qquad \textit{Threshold Frequency}$$

- By combining wave theory with particle theory, de Broglie was able to give the following equation for the wavelength of any particle whose mass and velocity are known:

$$\boxed{\lambda = \frac{h}{mv}} \qquad \boxed{h = 6.63 \times 10^{-34} \text{ J/Hz}} \qquad \begin{array}{l}\textit{de Broglie}\\ \textit{Wavelength}\end{array}$$

- Bohr's first postulate states that the angular momentum of an electron in any orbit must be a multiple of $h/2\pi$. His second postulate states that the energy absorbed or emitted by an atom is in discrete amounts equal to the difference in energy levels of an electron. These concepts are given as equations below:

$$\boxed{mvr = \frac{nh}{2\pi}} \qquad \boxed{hf = E_f - E_i} \qquad \textit{Bohr's postulates}$$

- Absorption and emission spectra for gases verify the discrete nature of radiation. The wavelength λ or frequency f which corresponds to a change in electron energy levels is given by

$$\boxed{\frac{1}{\lambda} = R\left(\frac{1}{n_f^2} - \frac{1}{n_i^2}\right)} \qquad \boxed{f = Rc\left(\frac{1}{n_f^2} - \frac{1}{n_0^2}\right)}$$

$$R = \frac{me^4}{8\epsilon_0^2 hc} = 1.097 \times 10^7 \, m^{-1} \qquad \textit{Rydberg's Constant}$$

- The total energy of a particular quantum state n for the hydrogen atom is given by:

$$\boxed{E_n = -\frac{me^4}{8\epsilon_0^2 n^2 h^2}} \qquad \text{or} \qquad \boxed{E_n = -\frac{13.6 \text{ eV}}{n^2}}$$

where $\epsilon_0 = 8.85 \times 10^{-12} \text{ C}^2/\text{N} \cdot \text{m}^2$
$e = 1.6 \times 10^{-19} \text{ C}$
$m_e = 9.1 \times 10^{-31} \text{ kg}$
$h = 6.63 \times 10^{-34} \text{ J/Hz}$

QUESTIONS

39-1. Define the following terms:

a. Relativisitic mass
b. Relativistic contraction
c. Time dilation
d. Einstein's postulates
e. Planck's postulate
f. Work function
g. Photoelectric effect
h. Spectrometer
i. Line emission spectra
j. Absorption spectra
k. Spectral series
l. Bohr atom
m. Principal quantum number
n. Energy level
o. Excited atom

39-2. If it were possible for a material object to travel past you at the speed of light, describe the mass, length, and time interval you would observe.

39-3. An astronaut holds a clock of mass m and length L. Assume that the astronaut passes you at relativistic speed. Compare your measurements of m, L, and Δt with those made by the astronaut for the same clock.

39-4. Recalling Newton's second law of motion, what happens to the thrust requirements for propelling rockets to higher and higher relativistic speeds? Theoretically, what force would be required to achieve the velocity of light?

39-5. You are enclosed in a box with six opaque walls. Are there any experiments you can perform inside the box to prove that you are (a) moving with constant linear velocity, (b) accelerating, or (c) rotating with constant angular velocity?

39-6. Combine Eqs. (39-5) and (39-3) to obtain Einstein's equation for the total relativistic energy, $E = mc^2$.

39-7. Suppose you want photoelectrons to have a kinetic energy E_κ and you know the work function W of the surface; how would you determine the required wavelength λ of the incident light?

39-8. Describe an experiment you might perform to determine the work function of a surface assuming you have a light source of varying wavelength.

39-9. Sketch on a single diagram the Balmer series and the Lyman series for the hydrogen emission spectrum. What is meant by the *series limit?*

39-10. Explain clearly what is meant when we say that the energy of the ground state is -13.6 eV for the hydrogen atom. What is the significance of the minus sign?

39-11. Describe an experiment which will demonstrate (a) a line emission spectrum, (b) a line absorption spectrum, and (c) a continuous spectrum.

39-12. Hydrogen atoms in their ground state are bombarded by electrons which have been accelerated through a potential difference of 12.8 V. Which lines of the Lyman series will be emitted by the hydrogen atoms?

PROBLEMS

Relativity

39-1. Three metersticks travel past an observer at speeds of $0.1c$, $0.6c$, and $0.9c$. What lengths would be recorded by the observer?

Ans. 99.5 cm, 80.0 cm, and 43.6 cm

* 39-2. Two identical space ships A and B pass each other in space with a relative velocity of 6×10^7 m/s. The length of each ship is 23.6 m. A person aboard ship A will measure what length for ship B? What will a person aboard ship B observe as the length of ship A? If each measures his own ship, what lengths result?

* 39-3. A space ship A travels past an observer B with a relative velocity of two-tenths of the velocity of light. The observer B determines that it takes a person on spaceship A exactly 3.96 s to perform a given task. What time will be measured for the same event by observer A?

Ans. 3.88 s

39-4. Elementary particles called mu-mesons rain down through the atmosphere at 2.97×10^8 m/s. At rest the mu-meson would decay, on average, 2 μs after it came into existence. What is the lifetime of the atmospheric mu-meson from the viewpoint of an observer on the earth?

39-5. What mass is required to run about 1 million 100-W electric light bulbs for 1 year?

Ans. 35 g

39-6. At a cost of 8 cents per kWh, what is the cost of the maximum energy to be released from a 1-kg mass?

39-7. What is the mass of an electron traveling at a speed of 2×10^8 m/s? The rest mass of an electron is 9.1×10^{-31} kg. What is the total energy of the electron? What is its relativistic kinetic energy?

Ans. 12.2×10^{-31} kg, 1.10×10^{-13} J, 0.280×10^{-13} J

39-8. Compute the mass and the speed of electrons having a relativistic kinetic energy of 1.2 MeV.

The Photoelectric Effect

39-9. The first photoelectrons are emitted from a copper surface when the wavelength of incident radiation is 282 nm. What is the threshold frequency for copper? What is the work function for a copper surface?

Ans. 1.06×10^{15} Hz, 4.40 eV

39-10. If the photoelectric work function of a material is 4.0 eV, what is the minimum frequency of light required to eject photoelectrons? What is the threshold wavelength?

39-11. When light of frequency 1.6×10^{15} Hz strikes a material surface, electrons just begin to leave the surface. This is the threshold frequency. What is the maximum kinetic energy of photoelectrons emitted from this surface when light of frequency 2.0×10^{15} Hz falls on the material?

Ans. 1.66 eV

39-12. The work function of a nickel surface is 5.01 eV. If a nickel surface is illuminated by light of wavelength 200 nm, what is the kinetic energy of an emitted photoelectron?

Waves and Particles

39-13. What is the de Broglie wavelength of a proton ($m = 1.67 \times 10^{-27}$ kg) when it is moving with a speed of 2×10^7 m/s?

Ans. 1.99×10^{-14} m

39-14. Determine the kinetic energy of an electron if its de Broglie wavelength is 2×10^{-11} m.

39-15. The charge on a proton is $+1.6 \times 10^{-19}$ C and its rest mass is 1.67×10^{-27} kg. What is the de Broglie wavelength of a proton if it is accelerated from rest through a potential difference of 500 V?

Ans. 1.28 pm

39-16. What is the de Broglie wavelength of the waves associated with an electron which has been accelerated from rest through a potential difference of 160 V?

Atomic Spectra and Energy Levels

39-17. Determine the wavelength of the first three spectral lines of atomic hydrogen in the Balmer series.

Ans. 656, 486, and 434 nm

39-18. Determine the wavelengths of the first three spectral lines of atomic hydrogen in the Paschen series.

39-19. Determine the radius of the $n = 4$ Bohr level of the classical Bohr hydrogen atom.

Ans. 847 pm

39-20. What is the classical radius of the first Bohr orbit in the hydrogen atom?

39-21. Determine the wavelength of the photon emitted from a hydrogen atom when the electron jumps from the $n = 3$ Bohr level to the ground level.

Ans. 103 nm

39-22. What is the maximum wavelength of an incident photon if it can ionize a hydrogen atom originally in its second excited state ($n = 3$)?

Additional Problems

39-23. An event which occurs on a spaceship traveling at $0.8c$ relative to the earth is observed by a person on the ship to last for 3 s. What time would be observed by a person on the earth? How far will the person on earth judge that the spaceship has traveled during this event?

Ans. 5 s, 1.2×10^6 km

39-24. The rest mass of a proton is 1.67×10^{-27} kg. What is the total energy of a proton which has been accelerated to a velocity of 2.5×10^8 m/s? What is its relativistic kinetic energy?

39-25. Compute the mass and the speed of protons having a relativistic kinetic energy of 235 MeV. The rest mass of a proton is 1.67×10^{-27} kg.

Ans. 2.09×10^{-27} kg, 1.8×10^8 m/s

39-26. How much work is required to accelerate a 1-kg mass from rest to a speed of $0.1c$? How much work is required to accelerate this mass from an initial speed of $0.3c$ to a final speed of $0.9c$? *Hint:* Recall that work is equal to the change in kinetic energy.

39-27. A particle of mass m is traveling at nine-tenths of the speed of light. By what factor is its relativistic kinetic energy greater than its Newtonian kinetic energy?

Ans. 3.19

39-28. What is the momentum of a 40-eV photon? What is the wavelength of an electron which has the same momentum as this photon?

39-29. In a photoelectric experiment shown as Fig. 39-12, a source of emf is connected

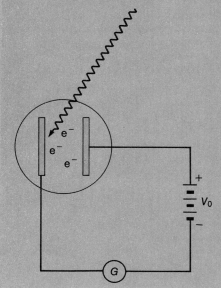

Fig. 39-12 The stopping potential V_0 is that potential difference which stops the transition of the most energetic photoelectrons ejected by the incident radiation.

in series with a galvanometer G. Light falling on the metal cathode produces photoelectrons. The source of emf is biased against the flow of electrons retarding their motion. The potential difference V_0 just sufficient to stop the most energetic photoelectrons is called the stopping potential. Assume that a surface is irradiated with light of wavelength 450 nm, causing electrons to be ejected from the surface at a maximum speed of 6×10^5 m/s. **(a)** What is the work function of the surface? **(b)** What is the stopping potential?

> **Ans.** (a) 2.78×10^{-19} J, (b) 1.02 V

39-30. When monochromatic light of wavelength 450 nm strikes a cathode, photoelectrons are emitted with a velocity of 4.8×10^5 m/s. What is the work function for the surface in electronvolts? What is the threshold frequency?

* 39-31. In a photoelectric experiment, 400-nm light falls on a certain metal ejecting photoelectrons. The potential required to stop the flow of electrons (stopping potential) is found to be 0.20 V. What is the energy of the incident photons in electronvolts? What is the work function of the surface? What is the threshold frequency?

> **Ans.** 3.11 eV, 2.91 eV, 7.02×10^{14} Hz

39-32. What is the velocity of a neutron ($m = 1.675 \times 10^{-27}$ kg) which has a de Broglie wavelength of 0.1 nm. What is its kinetic energy in electronvolts?

39-33. In the hydrogen atom an electron falls from the $n = 3$ level to the $n = 2$ level and emits a photon in the Balmer series. What is the wavelength of the emitted light?

> **Ans.** 656 nm

* 39-34. What is the maximum frequency of the Balmer series for the hydrogen atom?

39-35. Calculate the frequency and the wavelength of the H_B line of the Balmer series. The transition is from the fourth Bohr level to the second Bohr level.

> **Ans.** 6.16×10^{14} Hz, 486 nm

40 Nuclear Physics and the Nucleus

After completing this chapter, you should be able to:

1. Define the *mass number* and the *atomic number,* and demonstrate your understanding of the nature of fundamental nuclear particles.
2. Define *isotopes* and discuss the use of a mass spectrometer to separate isotopes.
3. Calculate the mass defect and the binding energy per nucleon for a particular isotope.
4. Demonstrate your understanding of radioactive decay and nuclear reactions; describe alpha particles, beta particles, and gamma rays, listing their properties.
5. Calculate the activity and the quantity of radioactive isotope remaining after a period of time if the half-lives and the initial values are given.
6. State the various conservation laws, and discuss their application to nuclear reactions.
7. Draw a rough sketch of a nuclear reactor describing the various components and their function in the production of nuclear power.

The work of Rutherford and Bohr left us with a picture of the atom as a dense, positively charged nucleus surrounded by a cloud of electrons at distinct energy levels. From this point of view, the nucleus is the center of an atom, containing most of the atom's mass. The behavior of the atom is also affected by the nucleus because in the neutral atom the total number of positive charges in the nucleus must equal the number of electrons.

In this chapter, we look at the basic internal structure of the nucleus. We shall find that classical physics is not adequate to describe interactions at this level. Topics to be discussed include nuclear binding energy, radioactivity, and nuclear energy. The emphasis will be on providing a broad understanding of the atomic nucleus and its behavior.

Nuclear technology has grown enormously since its beginning in the early 1940s. The study of the atomic nucleus once was a subject reserved mainly for physicists, but today there are few people whose lives are not touched by some aspect of nuclear science. As patients we see the doctor use radioactive materials to diagnose a condition or treat it. As citizens, we are concerned with the promises and dangers of large-scale nuclear power production. More than ever before, technicians and engineers need a better understanding of the atomic nucleus and its potential.

<table>
<tr><td>

40-1

THE ATOMIC NUCLEUS

</td><td>

All matter is composed of different combinations of *at least* three fundamental particles: protons, neutrons, and electrons. For example, a beryllium atom (Fig. 40-1) consists of a nucleus which contains four protons and five neutrons. The fact that beryllium is electrically neutral requires that four electrons surround the nucleus. The two inner electrons are at a different energy level ($n = 1$) than the outer two electrons ($n = 2$).

</td></tr>
</table>

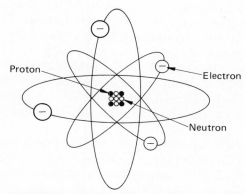

Fig. 40-1 A model of a beryllium atom. The nucleus consists of four protons and five neutrons surrounded by four electrons. The positive charge of the protons is exactly balanced by the negatively charged electrons in the neutral atom.

Rutherford's scattering experiments demonstrated that the nucleus contains most of the mass of an atom and that the nucleus is only about one ten-thousandth of the diameter of the atom. Thus, a typical atom with a diameter of 10^{-10} m (100 pm) would have a nucleus about 10^{-14} m (10 fm) in diameter. The prefixes *pico* (10^{-12}) and *femto* (10^{-15}) are useful for nuclear dimensions. Since the diameter of the atom is 10,000 times that of its nucleus, the atom, and therefore matter, consists largely of space that is almost empty.

Let us review what is known about the fundamental particles. The electron has a mass of 9.1×10^{-31} kg and a charge of $e = -1.6 \times 10^{-19}$ C. The proton is the nucleus of a hydrogen atom. It has a mass of 1.673×10^{-27} kg and a positive charge equal in magnitude to the charge of an electron ($+e$). Since the mass of an electron is extremely small, the mass of a proton is approximately the same as the mass of a hydrogen atom, which consists of one proton and one electron. The proton has a diameter of approximately 3 fm.

The other nuclear particle, the neutron, is present in the nuclei of all elements except hydrogen. It has a mass of 1.675×10^{-27} kg, which is slightly greater than that of the proton, but it has no charge. Thus, while neutrons contribute to the mass of a nucleus, they do not affect the net positive charge of the nucleus, which is due only to protons. The neutron also has a diameter of approximately 3 fm. Table 40-1 summarizes the data we have discussed for three fundamental particles.

Table 40-1 Fundamental Particles

Particle	Symbol	Mass, kg	Charge, C
Electron	e	9.1×10^{-31}	-1.6×10^{-19}
Proton	p	1.673×10^{-27}	$+1.6 \times 10^{-19}$
Neutron	n	1.675×10^{-27}	0

From what we now know about the fundamental particles, it is clear that diagrams like Fig. 40-1 cannot be taken too seriously. Distances are not normally presented to scale in such schematic representations. Moreover, classical laws of physics often do not apply for the microworld of the nucleus.

A true understanding of atomic and nuclear events will require a new way of thinking. For example, one might ask what holds the nucleus together. Clearly if Coulomb's electrostatic repulsion applies in the nucleus, it must be overcome by a much larger force. Both this much larger force and the electrostatic force are immense compared with the gravitational force. This third force is called the *nuclear force*.

The nuclear force is a very strong, short-range force. If two nucleons (which are protons or neutrons) are separated by approximately 1 fm, a strong attractive force occurs which quickly drops to zero as their separation becomes larger. The force appears to be the same, or nearly the same, between two protons or two neutrons or between a neutron and a proton. It is an attractive force until the nucleons cannot occupy the same space at the same time. If one nucleon is completely surrounded by other nucleons, its nuclear force field will be saturated, and it cannot exert any force on nucleons outside those surrounding it.

40-2
THE ELEMENTS

For many centuries, scientists have been studying the various elements found on the earth. A number of attempts have been made to organize the different elements according to their chemical and/or physical properties. The modern grouping of elements is the periodic table. One form of the periodic table is printed in Table 40-2.

Each element is assigned a number that distinguishes it from any other element. For example, the number for hydrogen is 1, the number for helium is 4, and the number for oxygen is 8. These numbers equal the number of protons in the nucleus of that element. The number is given the symbol Z and is called the *atomic number*.

The **atomic number** Z of an element is equal to the number of protons in the nucleus of an atom of that element.

The atomic number indirectly determines the chemical properties of an element because Z determines the number of electrons needed to balance the positive charge

Table 40-2 The Periodic Table (Adapted from *General Chemistry* by Frederick Longo. Copyright 1974 by McGraw-Hill, Inc. Used with permission of McGraw-Hill Book Company.)

Group

Period	IA	IIA	IIIB	IVB	VB	VIB	VIIB	VIIIB			IB	IIB	IIIA	IVA	VA	VIA	VIIA	0
1	1 H 1.00797																	2 He 4.0026
2	3 Li 6.939	4 Be 9.0122											5 B 10.811	6 C 12.01115	7 N 14.0067	8 O 15.9994	9 F 18.9984	10 Ne 20.183
3	11 Na 22.9898	12 Mg 24.312											13 Al 26.9815	14 Si 28.086	15 P 30.9738	16 S 32.064	17 Cl 35.453	18 Ar 39.948
4	19 K 39.102	20 Ca 40.08	21 Sc 44.956	22 Ti 47.90	23 V 50.942	24 Cr 51.996	25 Mn 54.9380	26 Fe 55.847	27 Co 58.9332	28 Ni 58.71	29 Cu 63.54	30 Zn 65.37	31 Ga 69.72	32 Ge 72.59	33 As 74.9216	34 Se 78.96	35 Br 79.909	36 Kr 83.80
5	37 Rb 85.47	38 Sr 87.62	39 Y 88.905	40 Zr 91.22	41 Nb 92.906	42 Mo 95.94	43 Tc (99)	44 Ru 101.07	45 Rh 102.905	46 Pd 106.4	47 Ag 107.870	48 Cd 112.40	49 In 114.82	50 Sn 118.69	51 Sb 121.75	52 Te 127.60	53 I 126.9044	54 Xe 131.30
6	55 Cs 132.905	56 Ba 137.34	57 La 138.91	72 Hf 178.49	73 Ta 180.948	74 W 183.85	75 Re 186.2	76 Os 190.2	77 Ir 192.2	78 Pt 195.09	79 Au 196.967	80 Hg 200.59	81 Tl 204.37	82 Pb 207.19	83 Bi 208.980	84 Po (210)	85 At (210)	86 Rn (222)
7	87 Fr (223)	88 Ra (226)	89 Ac (227)	104 Ku (260)														

Transition elements
d
p
s

Lanthanide series

58 Ce 140.12	59 Pr 140.907	60 Nd 144.24	61 Pm (147)	62 Sm 150.35	63 Eu 151.96	64 Gd 157.25	65 Tb 158.924	66 Dy 162.50	67 Ho 164.930	68 Er 167.26	69 Tm 168.934	70 Yb 173.04	71 Lu 174.97

Actinide series

90 Th 238.03	91 Pa (231)	92 U 238.03	93 Np (237)	94 Pu (242)	95 Am (243)	96 Cm (247)	97 Bk (249)	98 Cf (251)	99 Es (254)	100 Fm (253)	101 Md (256)	102 No (254)	103 Lw (257)

Key
Atomic number → 6
Symbol → C
Atomic mass (weight) → 12.01115

Atomic weight values listed in parentheses are approximate.

of the nucleus. The chemical nature of an atom depends on the number of electrons, in particular the outermost, or valence, electrons.

As the number of protons in a nucleus increases, so does the number of neutrons. In lighter elements, the increase is approximately one to one, but heavier elements may have more than 1½ times more neutrons than protons. For example, oxygen has 8 protons and 8 neutrons, whereas uranium has 92 protons and 146 neutrons. The total number of nucleons in a nucleus is called the *mass number A*.

> The **mass number** A of an element is equal to the total number of protons and neutrons in its nucleus.

If we represent the number of neutrons by N, we can write the mass number A in terms of the atomic number Z and the number of neutrons.

$$A = Z + N \qquad (40\text{-}1)$$

Thus, the mass number of uranium is 92 + 146 or 238.

A general way of describing the nucleus of a particular atom is to write the symbol for the element with its mass number and atomic number shown as follows:

$$\text{mass number} \atop \text{atomic number}[\text{symbol}] = {}^{A}_{Z}X \qquad (40\text{-}2)$$

For example, the uranium atom has the symbol ${}^{238}_{92}U$.

The structures and symbols for the first four elements are shown in Fig. 40-2. An alphabetical listing of all the elements is given in Table 40-3.

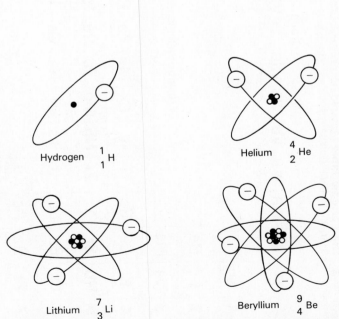

Hydrogen ${}^{1}_{1}H$

Helium ${}^{4}_{2}He$

Lithium ${}^{7}_{3}Li$

Beryllium ${}^{9}_{4}Be$

Fig. 40-2 The structure and symbols of the first four elements.

TABLE 40-3 INTERNATIONAL ATOMIC WEIGHTS (BASED ON CARBON 12)*

Element	Symbol	Atomic number	Atomic weight	Element	Symbol	Atomic number	Atomic weight	Element	Symbol	Atomic number	Atomic weight
Actinium	Ac	89	(227)	Hafnium	Hf	72	178.49	Promethium	Pm	61	(145)
Aluminum	Al	13	26.9815	Hahnium	Ha	105	(260)	Protactinium	Pa	91	(231)
Americium	Am	95	(243)	Helium	He	2	4.0026	Radium	Ra	88	(226)
Antimony	Sb	51	121.75	Holmium	Ho	67	164.930	Radon	Rn	86	(222)
Argon	Ar	18	39.948	Hydrogen	H	1	1.00797	Rhenium	Re	75	186.22
Arsenic	As	33	74.9216	Indium	In	49	114.82	Rhodium	Rh	45	102.91
Astatine	At	85	(210)	Iodine	I	53	126.9044	Rubidium	Rb	37	85.47
Barium	Ba	56	137.34	Iridium	Ir	77	192.2	Ruthenium	Ru	44	101.07
Berkelium	Bk	97	(247)	Iron	Fe	26	55.847	Rutherfordium	Rf	104	(260)
Beryllium	Be	4	9.0122	Krypton	Kr	36	83.80	Samarium	Sm	62	150.35
Bismuth	Bi	83	208.980	Lanthanum	La	57	138.91	Scandium	Sc	21	44.956
Boron	B	5	10.811	Lawrencium	Lw	103	(257)	Selenium	Se	34	78.96
Bromine	Br	35	79.904	Lead	Pb	82	207.19	Silicon	Si	14	28.086
Cadmium	Cd	48	112.40	Lithium	Li	3	6.939	Silver	Ag	47	107.868
Calcium	Ca	20	40.08	Lutetium	Lu	71	174.97	Sodium	Na	11	22.9898
Californium	Cf	98	(251)	Magnesium	Mg	12	24.312	Strontium	Sr	38	87.62
Carbon	C	6	12.01115	Manganese	Mn	25	54.9380	Sulfur	S	16	32.064
Cerium	Ce	58	140.12	Mendelevium	Md	101	(256)	Tantalum	Ta	73	180.948
Cesium	Cs	55	132.905	Mercury	Hg	80	200.59	Technetium	Tc	43	(97)
Chlorine	Cl	17	35.453	Molybdenum	Mo	42	95.94	Tellurium	Te	52	127.60
Chromium	Cr	24	51.996	Neodymium	Nd	60	144.24	Terbium	Tb	65	158.924
Cobalt	Co	27	58.9332	Neon	Ne	10	20.183	Thallium	Tl	81	204.37
Columbium	(see Niobium)			Neptunium	Np	93	(237)	Thorium	Th	90	232.038
Copper	Cu	29	63.546	Nickel	Ni	28	58.71	Thulium	Tm	69	168.934
Curium	Cm	96	(247)	Niobium	Nb	41	92.906	Tin	Sn	50	118.69
Dysprosium	Dy	66	162.50	(Columbium)				Titanium	Ti	22	47.90
Einsteinium	Es	99	(254)	Nitrogen	N	7	14.0067	Tungsten			
Erbium	Er	68	167.26	Nobelium	No	102	(254)	(Wolfram)	W	74	183.85
Europium	Eu	63	151.96	Osmium	Os	76	190.2	Uranium	U	92	238.03
Fermium	Fm	100	(257)	Oxygen	O	8	15.9994	Vanadium	V	23	50.942
Fluorine	F	9	18.9984	Palladium	Pd	46	106.4	Wolfram			
Francium	Fr	87	(223)	Phosphorus	P	15	30.9738	(Tungsten)	W	74	183.85
Gadolinium	Gd	64	157.25	Platinum	Pt	78	195.09	Xenon	Xe	54	131.30
Gallium	Ga	31	69.72	Plutonium	Pu	94	(244)	Ytterbium	Yb	70	173.04
Germanium	Ge	32	72.59	Polonium	Po	84	(209)	Yttrium	Y	39	88.905
Gold	Au	79	196.967	Potassium	K	19	39.102	Zinc	Zn	30	65.37
				Praseodymium	Pr	59	140.907	Zirconium	Zr	40	91.22

*Values in parentheses are mass numbers of longest-lived or best-known isotopes.

EXAMPLE 40-1 How many neutrons are in the nucleus of an atom of mercury $^{201}_{80}$Hg?

The symbol shows that the atomic number is 80 and the mass number is 201. Thus, from Eq. (40-1), the number of neutrons is

$$N = A - Z = 201 - 80 = 121$$

40-3
THE ATOMIC MASS UNIT

The very small masses of nuclear particles call for an extremely small unit of mass. Scientists normally express atomic and nuclear masses in atomic mass units/(u).

> One **atomic mass unit** (1 u) is exactly equal to one-twelfth of the mass of the most abundant form of the carbon atom.

In terms of the kilogram, the atomic mass unit is

$$1 \text{ u} = 1.6606 \times 10^{-27} \text{ kg} \qquad (40\text{-}3)$$

The mass of a proton is 1.007276 u, and that of a neutron is 1.008665 u.

EXAMPLE 40-2 The periodic table shows the average atomic mass of barium to be 137.34 u. What is the average mass of the barium nucleus?

Solution The mass of the nucleus is the atomic mass less the mass of the surrounding cloud of electrons. Thus, we must first determine the mass of an electron in atomic mass units.

$$m_e = 9.1 \times 10^{-31} \text{ kg} \frac{1 \text{ u}}{1.6606 \times 10^{-27} \text{ kg}}$$

$$= 5.5 \times 10^{-4} \text{ u}$$

Since the atomic number Z of barium is 56, there must be the same number of electrons. This total mass is

$$m_T = 56(5.5 \times 10^{-4} \text{ u}) = 3.08 \times 10^{-2} \text{ u}$$

The average atomic mass was given as 137.34 u. Thus, the nuclear mass is

$$137.34 \text{ u} - 0.0308 \text{ u}$$

or

$$137.31 \text{ u}$$

The masses given for atoms in the periodic table include the electron masses.

It should be remembered that mass can be equated to units of energy from Einstein's relation

$$E = mc^2$$

Consequently, a mass of 1 u corresponds to an energy given by

$$(1 \text{ u})c^2 = (1.66 \times 10^{-27} \text{ kg})(3 \times 10^8 \text{ m/s})^2 = 1.49 \times 10^{-10} \text{ J}$$

In the more convenient units of electronvolts, we can write

$$(1 \text{ u})c^2 = 1.49 \times 10^{-10} \text{ J} \frac{1 \text{ eV}}{1.6 \times 10^{-19} \text{ J}}$$

$$= 9.31 \times 10^8 \text{ eV} = 931 \text{ MeV}$$

The conversion factor from mass units (u) to energy units (MeV) is therefore

$$c^2 = 931 \text{ MeV/u} \qquad (40\text{-}3)$$

As an exercise you might verify that the electron and the proton have rest-mass energies of 0.511 and 938 MeV, respectively.

40-4

ISOTOPES

It is possible for two atoms of the same element to have nuclei containing different numbers of neutrons. Such atoms are called *isotopes*.

Isotopes are atoms which have the same atomic number Z but different mass numbers A.

For example, naturally occurring carbon is a mixture of two isotopes. The most abundant form, $^{12}_{6}C$, has six protons and six neutrons in its nucleus. Another form, $^{13}_{6}C$, has an extra neutron. Some elements have as many as 10 different isotopic forms.

Experimental verification of the existence of isotopes is accomplished with a mass spectrometer. This device, illustrated in Fig. 40-3, is used to separate the isotopes of an element. A source of singly ionized atoms of a particular element is positioned above the velocity selector. These ions are missing one electron and therefore have a charge of $+e$. They are propelled at varying speeds into the crossed E and B fields of the velocity selector. Ions with velocities sufficient to make the magnetic force $\mathbf{F}_m$ equal and opposite to the electric force $\mathbf{F}_e$ will pass through the bottom slit

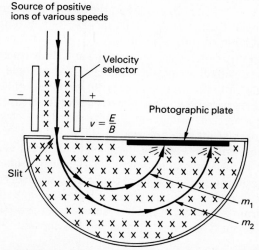

Fig. 40-3 The mass spectrometer is used to separate isotopes of different mass; the crosses indicate that the direction of the magnetic field is into the paper.

undeflected. Recalling that $F_e = eE$ and $F_m = evB$, we write

$$evB = eE$$

from which

$$\boxed{v = \frac{E}{B}}$$ (40-4)

Only ions with this velocity will pass through the slit at the bottom of the selector.

The fast-moving positive ions next pass into the lower region, where another B field causes them to experience a perpendicular magnetic force. The magnitude of this force will be constant and equal to evB, but its direction will always be at right angles to the velocity of the ion. The result is a circular path of radius R. The magnetic force provides the necessary centripetal force. In this case, we have

$$F_{\text{magnetic}} = F_{\text{centripetal}}$$

or

$$evB = \frac{mv^2}{R}$$

where m is the mass of the ion of charge e. Solving for R, we find that the radius of the semicircular path is given by

$$\boxed{R = \frac{mv}{eB}}$$ (40-5)

Since v, e, and B are constant, Eq. (40-5) gives the radius as a function of the mass of given ions. Ions of different mass will strike the photographic plate at different positions because their semicircular paths are different. Wherever a beam of ions strikes the plate, a darkened line will be produced. The distance of a particular line from the slit is twice the radius in which that beam of ions moves. In this manner the mass can be determined from Eq. (40-5).

The mass spectrometer is used to separate and study isotopes. Most elements are mixtures of atoms with different mass numbers. For example, if an ion beam of pure lithium is injected into the mass spectrometer, two types of atoms are observed. The darker bank occurs because around 92 percent of the atoms have a mass of 7.016 u. The remaining 8 percent, producing a lighter band, are atoms with mass 6.015 u. These two isotopes of lithium are written $_3^7$Li and $_3^6$Li.

Since some elements, e.g., tin, have many different isotopic forms, it is not surprising that the average atomic mass for elements is often not very close to an integer. Average atomic mass is affected by the mass numbers and relative abundance of each isotopic form. For example, chlorine has an average atomic mass of 35.453 u, which results from a mixture of the two isotopes $_{17}^{35}$Cl and $_{17}^{37}$C. The lighter chlorine isotope occurs about 3 times as often as the heavier one.

EXAMPLE 40-3   While studying chlorine with the mass spectrometer, it is noted that a very intense line occurs 24 cm from the entrance slit. Another lighter line appears at a distance of 25.37 cm. If the mass of the ions which form the first line is 34.980 u, what is the mass of the other isotope?

Solution   Since the given distances represent diameters, the radii of the two paths are

$$R_1 = 12 \text{ cm} \quad \text{and} \quad R_2 = 12.685 \text{ cm}$$

Now from Eq. (40-5),

$$R_1 = \frac{m_1 v}{eB} \quad \text{and} \quad R_2 = \frac{m_2 v}{eB}$$

Since e, v, and B are constant, we have

$$\frac{m_1}{m_2} = \frac{R_1}{R_2}$$

from which the mass m_2 is found to be

$$m_2 = \frac{m_1 R_2}{R_1} = \frac{(34.980 \text{ u})(12.685 \text{ cm})}{(12 \text{ cm})}$$

$$= 36.977 \text{ u}$$

40-5 THE MASS DEFECT AND BINDING ENERGY

One of the startling results which can be demonstrated with the mass spectrometer is that the mass of a nucleus is not exactly equal to the sum of the masses of its nucleons. Let us consider, for example, the helium atom, ^4_2He, which has two electrons about a nucleus containing two protons and two neutrons. The atomic mass is found from the periodic table to be 4.0026 u.

Now let us compare this value with the mass of all the individual particles which make up the atom:

$$2\,p = 2(1.0007276 \text{ u}) = 2.014552 \text{ u}$$
$$2\,n = 2(1.008665 \text{ u}) = 2.017330 \text{ u}$$
$$2\,e = 2(0.00055 \text{ u}) = \underline{0.001100 \text{ u}}$$
$$\text{Total mass} = 4.032982 \text{ u}$$

The mass of the parts (4.0331 u) is apparently greater than the mass of the atom (4.0026 u).

$$m_{\text{parts}} - m_{\text{atom}} = 4.0330 \text{ u} - 4.0026 \text{ u}$$
$$0.0304 \text{ u}$$

When protons and neutrons join to form a helium nucleus, the mass is decreased in the process. This difference is called the *mass defect*. A mass defect can be shown to exist for atoms of all elements.

> The mass defect is defined as the difference between the rest mass of a nucleus and the sum of the rest masses of its constituent nucleons.

We have seen from Einstein's work that mass and energy are equivalent. We might suppose, then, that the mass decrease in joining nucleons together will result in an energy decrease. Since energy is conserved, a decrease in the energy of the system means that energy must be released in joining the system together. In the case of helium, this energy would come from a mass of 0.0304 u and would be equal to

$$E = mc^2 = (0.0304 \text{ u})\frac{931 \text{ MeV}}{1 \text{ u}} = 28.3 \text{ MeV}$$

The total energy which would be released if we could build a nucleus from protons and neutrons is called the *binding energy* of the nucleus. As we have just seen, the binding energy of $^4_2$He is 28.3 MeV, as illustrated in Fig. 40-4a.

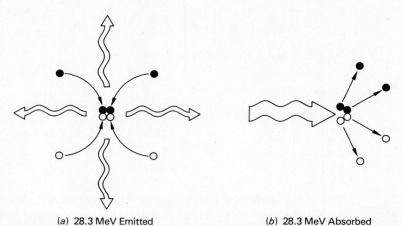

(a) 28.3 MeV Emitted   (b) 28.3 MeV Absorbed

Fig. 40-4 *(a)* When two protons and two neutrons are fused together to form a helium nucleus, energy is released. *(b)* The same amount of energy is required to break the nucleus apart into its constituent nucleons.

We can also reverse the above process and state that the binding energy is the energy required to break a nucleus apart into its constituent particles.

> The **binding energy** of a nucleus is defined as the energy required to separate a nucleus into its constituent nucleons.

In our example, an energy of 28.3 MeV must be supplied to $^4_2$He in order to separate the nucleus into two protons and two neutrons (Fig. 40-4b).

An isotope of atomic number Z and mass number A consists of Z protons, Z electrons, and $N = (A - Z)$ neutrons. If we neglect the binding energy of the electrons, a neutral isotope would have the same mass as Z neutral hydrogen atoms plus the mass of the neutrons. The masses of $^1_1$H and m_n are

$$m_H = 1.007825 \text{ u} \qquad m_n = 1.008665 \text{ u} \qquad (40\text{-}6)$$

If we represent the atomic mass by M, the binding energy E_B can be approximated by

$$E_B = [(Zm_H + Nm_n) - M]c^2 \qquad \begin{matrix} Binding \\ Energy \end{matrix} \quad (40\text{-}7)$$

In applying this equation, we should remember that $N = A - Z$ and that $c^2 = 931$ MeV/u.

EXAMPLE 40-4   Determine the total binding energy and the binding energy per nucleon for the $^{14}_7$N nucleus.

MODERN PHYSICS

Solution   For nitrogen, $Z = 7$, $N = 7$, and $M = 14.003074$ u.

$$E_B = [(Zm_H + Nm_n) - M]c^2$$
$$= \{[7(1.007825 \text{ u}) + 7(1.008665 \text{ u})] - 14.003074 \text{ u}\}(931 \text{ MeV/u})$$
$$= (0.112356 \text{ u})(931 \text{ MeV/u}) = 104.6 \text{ MeV}$$

Since $^{14}_{7}\text{N}$ contains 14 nucleons, the binding energy per nucleon is

$$\frac{E_B}{A} = \frac{104.6 \text{ MeV}}{14 \text{ nucleons}} = 7.47 \text{ MeV/nucleon}$$

The atomic mass M used in Eq. (40-7) must be taken for the particular isotope of the element, not from the periodic table (Table 40-2) or from Table 40-3. These tables give the atomic masses of the naturally occurring mixture of isotopes for each element. The atomic mass of $^{12}_{6}\text{C}$, for example, is exactly 12.0000 u by definition. The periodic table gives a value of 12.01115 u because naturally occurring carbon contains very small amounts of $^{13}_{6}\text{C}$ in addition to the more abundant $^{12}_{6}\text{C}$ atoms. The term *nuclide* is used to refer to a particular isotope which has a particular number of nuclear particles and, hence, a specific mass. The masses of several common nuclides are given in Table 40-4. These are the masses which should be used in determining mass defects and binding energies. Electron masses are included so that the given masses are basically the mass of the nucleus plus Z atomic electrons.

The binding energy per nucleon, as computed in Example 40-4, is an important way of comparing the nuclei of various elements. A plot of the binding energy per nucleon as a function of increasing mass number is shown in Fig. 40-5 for many stable nuclei. Note that the mass numbers toward the center (50 to 80) yield the highest binding energy per nucleon. Elements ranging from $A = 50$ to $A = 80$ are the most stable.

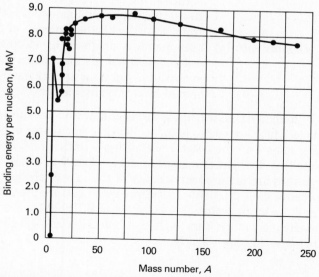

Fig. 40-5 The average binding energy per nucleon for the most stable nucleus at each mass number.

The strong nuclear force holds the nucleons tight in the nucleus, overcoming the coulomb repulsion of protons. However, the balance of forces is not always maintained, and sometimes particles or photons are emitted from the nuclei of atoms. Such unstable nuclei are said to be *radioactive* and have the property of *radioactivity*.

All naturally occurring elements with atomic numbers greater than 83 are radioactive. They are slowly decaying and disappearing from the earth. Uranium and radium are two of the better-known examples of naturally radio-active elements. A few other naturally occurring, lighter, and less active elements have also been discovered.

Unstable nuclei are also produced artificially as by-products of nuclear reactors, for study in laboratories, or for other purposes. Additionally, some elements are made radioactive naturally by bombardment with high-energy photons.

There are three major forms of radioactive emission from atomic nuclei:

1. *Alpha particles* (α) An alpha particle is the nucleus of a helium atom and consists of two protons and two neutrons. It has a charge of $+2e$ and a mass of 4.001506 u. Because of their positive charges and relatively low speeds ($\approx 0.1c$), alpha particles do not have great penetrating power.

2. *Beta particles* (β) There are two kinds of beta particles, a beta minus particle (β^-) and a beta plus particle (β^+). The beta minus particle is simply an electron of charge $-e$ and mass equal to 0.00055 u. A beta plus particle, also called a *positron,* has the same mass as an electron but the opposite charge ($+e$). These particles are generally emitted at speeds near the velocity of light. The beta minus particles are much more penetrating than alpha particles, but beta plus particles easily combine with electrons; then rapid annihilation of both the positrons and electrons occurs, with the emission of gamma rays.

3. *Gamma rays* (γ) A gamma ray is a high-energy electromagnetic wave similar to heat and light but of much higher frequency. These rays have no charge or rest mass and are the most penetrating radiation emitted by radioactive elements.

To understand why these forms of radiation are emitted, it is helpful to look at nuclei which are relatively stable. Plotting the number of neutrons N vs. the number of protons Z for these stable nuclei gives a rough graph like that shown in Fig. 40-6. Note that the light elements are stable when the ratio of Z to N is close to 1. More neutrons are required for stability in the heavier elements. The additional nuclear forces of the extra neutrons are needed to balance the higher electric forces which result as more protons are collected together. Whenever a nucleus occurs which deviates very much from the line, it is unstable and will emit some form of radiation, thereby achieving stability.

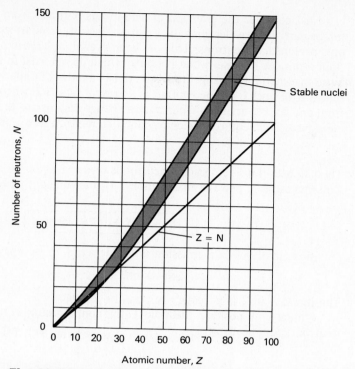

Fig. 40-6 A comparison of the number of neutrons as a function of the atomic number. Notice that nuclei of higher Z have the greater proportion of neutrons.

40-7

RADIO-ACTIVE DECAY

Let us look at radioactive decay by alpha, beta, and gamma radiation and see what occurs during each process. The emission of an alpha particle $^4_2\alpha$ reduces the number of protons in the parent nucleus by 2 and the number of nucleons by 4. Symbolically, we write

$$^A_Z X \rightarrow ^{A-4}_{Z-2} Y + ^4_2\alpha + \text{energy} \qquad (40\text{-}8)$$

The energy term results from the fact that the rest energy of the products is less than that of the parent atom. The difference in energy is carried away primarily by the kinetic energy imparted to the alpha particle. The recoil kinetic energy of the much more massive daughter atom is small by comparison.

EXAMPLE 40-5   Write the reaction which occurs when $^{226}_{88}\text{Ra}$ decays by alpha emission.

Solution   Applying Eq. (40-8), we write

$$^{226}_{88}\text{Ra} \rightarrow ^{222}_{86}\text{Rn} + ^4_2\alpha + \text{energy}$$

Notice that the unstable element radium has been transformed into a new element, radon, which is closer to the stability line.

Next consider the emission of beta-minus particles from the nucleus. If beta-minus particles are electrons, how can an electron come from a nucleus containing only protons and neutrons? This can be answered, at least in part, by analogy to the Bohr atom. We have seen that photons, which do not exist in the atom, are emitted by atoms when they change from one state to another. Similarly, electrons, which do not exist in nuclei, can be emitted as a form of radiation when the nucleus changes from one state to another. When such a change does occur, the total charge must be conserved. This requires the conversion of a neutron into a proton and an electron.

$$_0^1n \rightarrow {}_1^1p + {}_{-1}^0e$$

Thus, in beta-minus emission a neutron is replaced by a proton. The atomic number Z increases by 1, and the mass number is unchanged. Symbolically.

$$_Z^A X \rightarrow {}_{Z+1}^A Y + {}_{-1}^0\beta + \text{energy} \tag{40-9}$$

An example of beta emission is the decay of an isotope of neon into sodium:

$$_{10}^{23}Ne \rightarrow {}_{11}^{23}Na + {}_{-1}^0\beta + \text{energy}$$

The increase in Z is necessary to conserve charge.

Similarly, in positron (beta-plus) emission, a proton in the nucleus decays to a neutron and a positron.

$$_1^1p \rightarrow {}_0^1n + {}_{+1}^0e$$

The atomic number Z decreases by 1, and the mass number A is unchanged. Symbolically.

$$_Z^A X \rightarrow {}_{Z-1}^A Y + {}_{+1}^0\beta + \text{energy} \tag{40-10}$$

An example of positron emission is the decay of an isotope of nitrogen into an isotope of carbon:

$$_7^{13}N \rightarrow {}_6^{13}C + {}_{+1}^0\beta + \text{energy}$$

In both types of beta emission, the kinetic energy is shared mostly by the beta particle and another particle called a *neutrino*. The neutrino has no rest mass and no electric charge, but it can have both energy and momentum.

In gamma emission, the parent nucleus maintains the same atomic number Z and the same mass number A. The gamma photon simply carries away energy from an unstable nucleus. Frequently, a succession of alpha and beta decays is accompanied by gamma decays, which carry off excess energy.

The radioactive disintegration of $_{92}^{238}U$ is shown in Fig. 40-7 as a series of decays through a number of elements until it becomes a stable $_{82}^{206}Pb$ nucleus.

40-8
HALF-LIFE

A radioactive material continues to emit radiation until all the unstable atoms have decayed. The number of unstable nuclei decaying or disintegrating every second for a given isotope can be predicted on the basis of probability. This number is referred to

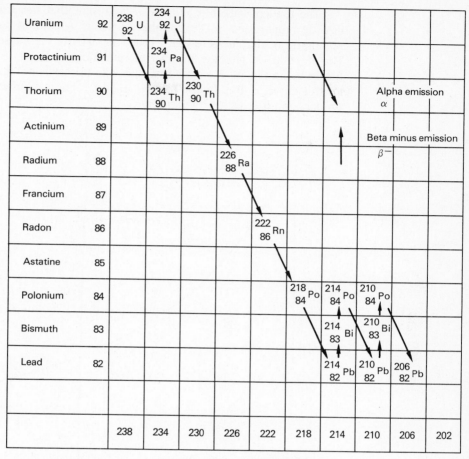

Fig. 40-7 The uranium series of disintegration. Uranium decays, through a series of alpha- and beta-minus emissions, from $^{238}_{92}U$ to $^{206}_{82}Pb$.

as the *activity R,* given by

$$R = \frac{-\Delta N}{\Delta t} \qquad (40\text{-}11)$$

where N is the number of undecayed nuclei. The negative sign is included because N is decreasing with time. The units for R are inverse seconds (s^{-1}).

In practice, the activity in disintegrations per second is so large that a more convenient unit, the *curie* (Ci), is defined as follows.

One **curie** (Ci) is the activity of a radioactive material which decays at the rate of 3.7×10^{10} disintegrations per second.

$$1 \text{ Ci} = 3.7 \times 10^{10} \text{ s}^{-1} \qquad (40\text{-}12)$$

The activity of 1 g of radium is slightly less than 1 Ci.

The random nature of nuclear decay means that the activity R at any time is directly proportional to the number of nuclei remaining; i.e., as the number of remaining nuclei decreases with time, the activity must also decrease with time. Therefore, if we plot the number of remaining nuclei as a function of time, as illustrated in Fig. 40-8, we see that radioactive decay is not linear. The time it takes for this curve to drop to one-half its original value is different for each radioactive isotope, and it is called the *half-life*.

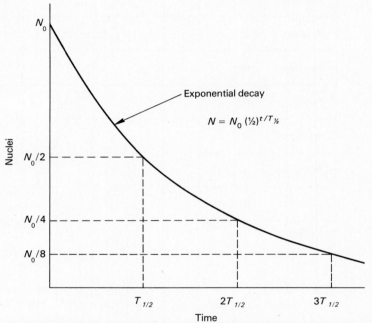

Fig. 40-8 The radioactive decay curve illustrating the half-life as the time $T_{1/2}$ required for one-half of the unstable nuclei, present at time $t = 0$, to decay.

> The **half-life** $T_{1/2}$ of a radioactive isotope is the length of time in which one-half of its unstable nuclei will decay.

For example, the half-life of radium 226 is 1620 yr; 1 g of this isotope will decay to 0.5 g in 1620 yr, to 0.25 g in 2(1620 yr), to 0.125 g in 3(1620 yr), and so on.

We can use this definition of the half-life to determine how many nuclei are present at a time t. If we start out at time $t = 0$ with a number N_0 of unstable nuclei, then after n half-lives have passed, there will be left a number of nuclei N given by

$$N = N_0(\tfrac{1}{2})^n \qquad (40\text{-}13)$$

The number n of half-lives in the period of time t is, of course, $t/T_{1/2}$. Thus, a more applicable form of the above relation is

$$\boxed{N = N_0(\tfrac{1}{2})^{t/T_{1/2}}}$$

Since the amount of radioactive material is determined by the number of nuclei present, an equation similar to Eq. (40-13) can be used to compute the mass of remaining radioactive material after a number of half-lives.

The same idea applied to the activity R of a radioactive sample yields the relation

$$\boxed{R = R_0(\tfrac{1}{2})^{t/T_{1/2}}} \qquad (40\text{-}14)$$

EXAMPLE 40-6   The worst by-product of ordinary nuclear reactors is the radioactive isotope plutonium 239, which has a half-life of 24,400 yr. Suppose the initial activity of a sample containing 1.64×10^{20} $^{239}_{94}\text{Pu}$ nuclei is 4 mCi. (*a*) How many of these nuclei remain after 73,200 yr? (*b*) What will be the activity at that time?

Solution (a)   Substitution into Eq. (40-12) yields

$$N = N_0\left(\frac{1}{2}\right)^{t/T_{1/2}} = 1.64 \times 10^{20}\left(\frac{1}{2}\right)^{73,200 \text{ yr}/24,400 \text{ yr}}$$

$$= 1.64 \times 10^{20}(\tfrac{1}{2})^3 = 1.64 \times 10^{20}(\tfrac{1}{8})$$

$$= 2.05 \times 10^{19} \text{ nuclei}$$

Solution (b)   We obtain the remaining activity from Eq. (40-14).

$$R = R_0(\tfrac{1}{2})^{t/T_{1/2}} = 4 \text{ mCi}(\tfrac{1}{8}) = 0.5 \text{ mCi}$$

Both these calculations assume that no new $^{239}_{94}\text{Pu}$ nuclei are being created by other processes. It is easy to see from this example why disposal of some radioactive materials is such a difficult problem.

40-9
NUCLEAR REACTIONS

In a chemical reaction the atoms of two molecules react to form different molecules. In a *nuclear reaction,* nuclei, radiation, and/or nucleons collide to form different nuclei, radiation, and nucleons. If the colliding objects are charged, at least one of the colliding masses must be accelerated to a relatively high velocity. Normally the bombarding particle is very light, e.g., a proton $^1_1 p$ or an alpha particle $^4_2 \alpha$. These nuclear projectiles are accelerated with many different devices, e.g., Van de Graaff generators, cyclotrons, and linear accelerators.

In the nuclear reactions we shall study, several conservation laws must be observed, primarily *conservation of charge, conservation of nucleons,* and *conservation of mass-energy.*

Conservation of charge requires that the total charge of a system neither increase nor decrease in a nuclear reaction.

Conservation of nucleons requires that the total number of nucleons in the interaction remain unchanged.

Conservation of mass-energy requires that the total mass-energy of a system remain unchanged in a nuclear reaction.

Now let us observe what happens when an alpha particle $^4_2 \alpha$ strikes a nucleus in a sample of nitrogen gas $^{14}_7\text{N}$. (Refer to Fig. 40-9.) The first step is the entry of the alpha particle, which adds 2 protons and 2 neutrons to the nucleus. The atomic

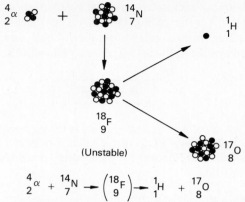

$$\frac{4}{2}\alpha + \frac{14}{7}N \longrightarrow \left(\frac{18}{9}F\right) \longrightarrow \frac{1}{1}H + \frac{17}{8}O$$

Fig. 40-9 Striking a nitrogen-14 nucleus with an alpha particle.

number Z is increased by 2, and the mass number A is increased by 4. The resulting nucleus is an *unstable* compound nucleus of fluorine $^{18}_{9}F$. This unstable nucleus quickly disintegrates into the final products, oxygen $^{17}_{8}O$ and hydrogen $^{1}_{1}H$. The overall reaction can be written

$$^{4}_{2}\alpha + ^{14}_{7}N \rightarrow ^{1}_{1}H + ^{17}_{8}O \tag{40-15}$$

Note how charge and nucleons are conserved in these reactions. There was a net charge of $+9e$ before the reaction and a net charge of $+9e$ after the reaction, and there are 18 nucleons before and after the reaction.

40-10
NUCLEAR FISSION

Before the discovery of the neutron in 1932, alpha particles and protons were the primary particles used to bombard atomic nuclei, but as charged particles they have the disadvantage of being repelled electrostatically by the nucleus. Consequently, very large energies are required before nuclear reactions occur.

Since neutrons have zero electric charge, they can easily penetrate the nucleus of an atom with no coulomb repulsion. Very fast neutrons may pass completely through a nucleus or may cause it to disintegrate. Slow neutrons may be captured by a nucleus, creating an unstable isotope, which may disintegrate.

Whenever the absorption of an incoming neutrons causes a nucleus to split into two smaller nuclei, the reaction is called *fission* and the product nuclei are called *fission fragments.*

> **Nuclear fission** is the process by which heavy nuclei are split into two or more nuclei of intermediate mass numbers.

Whenever a slow neutron is captured by a uranium nucleus $^{235}_{92}U$, an unstable nucleus ($^{236}_{92}U$) is produced which may decay in several ways into smaller product nuclei (Fig. 40-10). Such fission reactions may produce fast neutrons, beta particles, and gamma rays in addition to the product nuclei. For this reason the products of a fission process, including fallout from a nuclear explosion, are highly radioactive.

The fission fragments have a smaller mass number and therefore about 1 MeV more binding energy per nucleon (see Sec. 40-5). As a result, fission releases a large

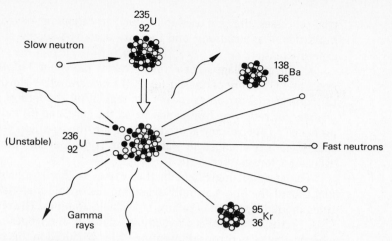

Fig. 40-10 Nuclear fission of ^{235}U by capture of a slow neutron.

amount of energy. In the above example approximately 200 MeV per fission is produced.

Because each nuclear fission releases more neutrons, which may lead to additional fission, a *chain reaction* is possible. As seen in Fig. 40-11, the three neutrons released from the fission of $^{235}_{92}$U produce three additional fissions. Thus, starting with one neutron, we have liberated nine after only two steps. If such a chain reaction is not controlled, it can lead to an explosion of enormous magnitude.

40-11
NUCLEAR REACTORS

A *nuclear reactor* is a device which controls the nuclear fission of radioactive material, producing new radioactive substances and large amounts of energy. These devices are used to furnish heat for electric power generation, propulsion, and indus-

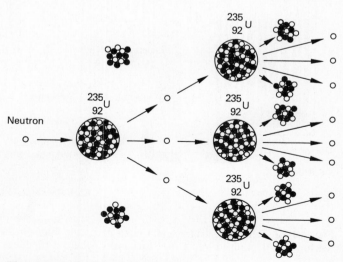

Fig. 40-11 Nuclear chain reaction.

trial processes; they are used to produce new elements or radioactive materials for a multitude of applications; and they are used as a supply of neutrons for scientific experimentation.

A schematic diagram of a typical reactor is given in Fig. 40-12. The basic components are; (1) a *core* of nuclear fuel, (2) a *moderator* for slowing down fast neutrons, (3) *control rods* or other means for regulating the fission process, (4) a heat exchanger for removing heat generated in the core, and (5) radiation shielding. Steam produced by the reactor is used to drive a turbine, which generates electricity. The spent steam is changed to water in the condenser and pumped back to the heat exchanger for another cycle.

The essential ingredient in the reactor is the fissionable material, or nuclear fuel. About the only naturally occurring fissionable material is $^{235}_{92}U$, which constitutes about 0.7 percent of naturally available uranium. The remaining 99.3 percent is $^{238}_{92}U$. Fortunately, $^{238}_{92}U$ is a *fertile* material, i.e., it changes to a fissionable material when struck by neutrons. Plutonium $^{239}_{94}Pu$ produced in this manner can provide new fuel for the reactor.

The production of additional fuel as a part of the reactor's operation has led to the design of *breeder* reactors, in which there is a net increase in fissionable material. In other words, the reactor produces more fuel than it consumes. This does not

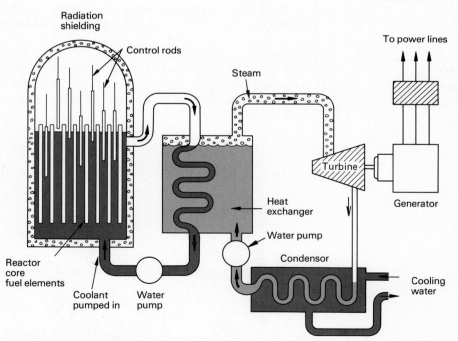

Fig. 40-12 Schematic diagram of a nuclear reactor. Water heated under pressure in the reactor core is pumped into a heat exchanger, where it produces steam to operate a turbine.

violate the law of conservation of energy. It only provides for the production of fissionable material from fertile materials.

The fissionable fuel in most reactors depends on the availability of slow neutrons, which are more likely to produce fission. Fast neutrons liberated by the fission of nuclei must therefore be slowed down. For this reason reactor fuel is embedded in a suitable substance called a *moderator*. The function of this substance is to slow neutrons without capturing them.

Neutrons have a mass about the same as that of a hydrogen atom. It might be expected, then, that substances containing hydrogen atoms would be effective as moderators of neutrons. The neutron is analogous to a marble, which can be stopped by a collision with another marble but will merely bounce off a cannonball because of the great mass difference. Water (H_2O) and heavy water, containing ^2_1H instead of ^1_1H, are often used as moderators. Other suitable materials are graphite and beryllium.

In order to control the nuclear furnace, it is necessary to regulate the number of neutrons which initiate the fission process. Substances like boron and cadmium capture neutrons efficiently and are excellent control materials. A typical reactor has control rods which can be inserted into the reactor at variable distances. By adjusting the position of these rods, the activity of the nuclear furnace is controlled. A supplementary set of rods is available to allow the reactor to be shut down completely in an emergency.

40-12 NUCLEAR FUSION

In our earlier discussion on mass defect we calculated that 28.3 MeV of energy is released in the formation of ^4_2He from its component nucleons. This joining together of light nuclei into a single heavier nucleus is called *nuclear fusion*. This process provides the fuel for stars like our own sun, and it is also the principle behind the hydrogen bomb. Many consider the fusion of hydrogen into helium as the ultimate fuel.

The use of nuclear fusion as a controlled source of energy is not without problems. It is still believed by most physicists that extremely high temperatures will be necessary to sustain nuclear fusion. The fusing nuclei would require millions of electronvolts of kinetic energy to overcome their coulombic repulsion. In the hydrogen bomb, this enormous energy is supplied by an atomic explosion, which then triggers the fusion process. The peaceful production of fusion by this method presents the problem of containment. The nuclear fuel would need to be so hot that it would instantly disintegrate any known substance. Present research methods involve containment by magnetic fields or rapid heating by powerful lasers. It is easy to see why the idea of "cold fusion" through an electrolytic process has generated so much excitement.

If the problems of fusion are ever solved, this energy source could provide a solution to our formidable problem of dwindling resources. The deuterium commonly found in seawater could provide us with a virtually inexhaustible supply of fuel. It would represent more than a billion times the energy available in all our coal and oil reserves. In addition, it appears that fusion reactors would have much less of a problem with radioactive residue than that currently experienced with fission reactors.

In this chapter we have studied the fundamental particles which make up the nuclei of atoms. Protons and neutrons are held together in the nucleus by very strong nuclear forces which are active only within the nucleus. It was seen that when such particles are joined together, the resulting mass is less that of the constituents parts. For very massive nuclei it was also pointed out that energy results from tearing these nuclei apart. In either case, there is an enormous potential for useful energy. The major concepts which must be remembered in this chapter are as follows:

- The fundamental nuclear particles discussed in this chapter are summarized in the following table. The masses are given in atomic mass units (u) and the charge is in terms of the electronic charge $+e$ or $-e$, which is 1.6×10^{-19} C.

Fundamental Particles

Particle	Symbol	Mass, u	Charge
Electron	$-_{1}^{0}e$, $-_{1}^{0}\beta$	0.00055	$-e$
Proton	$_{1}^{1}p$, $_{1}^{1}H$	1.007276	$+e$
Neutron	$_{0}^{1}n$	1.008665	0
Positron	$+_{1}^{0}e$, $+_{1}^{0}\beta$	0.00055	$+e$
Alpha particle	$_{2}^{4}\alpha$, $_{2}^{4}He$	4.001506	$+2e$

The atomic masses of the various elements are given in the text.
- The atomic number Z of an element is the number of protons in its nucleus. The mass number A is the sum of the atomic number and the number of neutrons N. These numbers are used to write the nuclear symbol:

$$\boxed{A = Z + N}$$

Symbol: $_{Z}^{A}X$

- One *atomic mass unit* (1 u) is equal to one-twelfth the mass of the most abundant carbon atom. Its value in kilograms is given below. Also, since $E = mc^2$, we can write the conversion factor from mass to energy as c^2.

$$\boxed{1 \text{ u} = 1.6606 \times 10^{-27} \text{ kg}} \qquad \boxed{c^2 = 931 \text{ MeV/u}}$$

$$1 \text{ MeV} = 10^6 \text{ eV} = 1.6 \times 10^{-13} \text{ J}$$

In the mass spectrometer, the velocity v and the radius R of the singly ionized particles are

$$\boxed{v = \frac{E}{B}} \qquad \boxed{R = \frac{mv}{eB}}$$

Mass Spectrometer

- The *mass defect* is the difference between the rest mass of a nucleus and the sum of the rest masses of its nucleons. The *binding energy* is obtained by multiplying the mass defect by c^2.

$$E_B = [(Zm_H + Nm_n) - M]c^2 \qquad \textit{Binding Energy}$$

where $m_H = 1.007825$ u
$m_n = 1.008665$ u
$c^2 = 931$ MeV/u
$M =$ atomic mass
$N = A - Z$
$Z =$ atomic number

- Several general equations for radioactive decay are

$$^A_Z X \rightarrow {}^{A-4}_{Z-2} Y + {}^4_2 \alpha + \text{energy} \qquad \textit{Alpha Decay}$$

$$^A_Z X \rightarrow {}^{A}_{Z-1} Y + {}^{0}_{-1} \beta + \text{energy} \qquad \textit{Beta-Minus Decay}$$

$$^A_Z X \rightarrow {}^{A}_{Z-1} Y + {}^{0}_{+1} \beta + \text{energy} \qquad \textit{Beta-Plus Decay}$$

- The *activity R* of a sample is the rate at which the radioactive nuclei decay. It is generally expressed in curies (Ci).

One *curie* (1 Ci) $= 3.7 \times 10^{10}$ disintegrations per second (s^{-1})

- The *half-life* of a sample is the time $T_{1/2}$ in which one-half the unstable nuclei will decay.
- The number of unstable nuclei remaining after a time t depends on the number n of half-lives that have passed. If N_0 nuclei exist at time $t = 0$, then a number N exists at time t. We have

$$N = N_0 (\tfrac{1}{2})^n \qquad \text{where } n = \frac{t}{T_{1/2}}$$

- The activity R and mass m of the radioactive portion of a sample are found from similar relations:

$$R = R_0 (\tfrac{1}{2})^n \qquad m = m_i (\tfrac{1}{2})^n$$

- In any nuclear equation, the number of nucleons on the left side must equal the number of nucleons on the right side. Similarly, the net charge must be the same on each side.

40-1. Define the following terms:
 a. Nuclear force
 b. Nucleon
 c. Atomic number
 d. Mass number
 e. Atomic mass unit

 l. Beta particles
 m. Gamma particles
 n. Half-life
 o. Activity
 p. Curie

f. Isotopes
g. Mass spectrometer
h. Mass defect
i. Binding energy
j. Radioactivity
k. Alpha particles

q. Nuclear fission
r. Chain reaction
s. Nuclear reactor
t. Nuclear fusion
u. Moderator

40-2. Write the symbol $_Z^A X$ for the most abundant isotopes of (a) cadium, (b) silver, (c) gold, (d) polonium, (e) magnesium, and (f) radon.

40-3. From the curve describing the binding energy per nucleon (Fig. 40-5) would you expect the mass defect to be greater for chromium $_{24}^{52}Cr$ or uranium $_{92}^{238}U$? Why?

40-4. The binding energy is greater for the mass numbers in the central part of the periodic table. Discuss the significance of this in relation to nuclear fission and nuclear fusion. How do you explain the release of energy in both fusion and fission in view of the fact that one process brings nuclei together and the other tears them apart?

40-5. How is the stability of an isotope affected by the ratio of the mass number A to the atomic number Z? Does the element whose ratio is closest to 1 always appear to be the more stable?

40-6. Define and compare alpha particles, beta particles, and gamma rays. Which are likely to do the most damage to human tissue?

40-7. Given a source which emits alpha, beta, and gamma radiation, draw a diagram showing how you could demonstrate the charge and penetrating power of each type of radiation. Assume you have at your disposal a source of a magnetic field and several thin sheets of aluminum.

40-8. Describe and explain, step by step, the decay of $_{92}^{238}U$ to the stable isotope of lead, $_{82}^{206}Pb$. (Refer to Fig. 40-7.)

40-9. Write in the missing symbol, in the form $_Z^A X$, for the following nuclear disintegrations:

a. $_{90}^{234}TH \rightarrow _{91}^{234}Pa + \underline{\quad}$

b. $_{15}^{32}P \rightarrow \underline{\quad} + _{+0}^{1}e$

c. $_{94}^{239}Pu \rightarrow _{90}^{234}Th + \underline{\quad}$

d. $_{92}^{238}U \rightarrow \underline{\quad} + _{2}^{4}\alpha$

40-10. Write the missing symbol for the following nuclear reactions:

a. $_1^2H + _1^3H \rightarrow _2^4H + \underline{\quad}$

b. $_{12}^{25}Mg + \underline{\quad} \rightarrow _{13}^{28}Al + _1^1H$

c. $_4^9Be + _2^4\alpha \rightarrow _6^{12}C + \underline{\quad}$

d. $_1^1H + \underline{\quad} \rightarrow _6^{12}C + _2^4He$

40-11. Explain the function of the following components of a nuclear reactor: (a) uranium, (b) radiation shielding, (c) moderator, (d) control rods, (e) heat exchanger, and (f) condenser.

40-12. Give examples to show how beta decay and alpha decay tend to bring unstable nuclei closer to the stability curve of Fig. 40-6.

40-13. Radon has a half-life of 3.8 days. Consider a sample of radon having a mass m and an activity R. What mass of radioactive radon remains after 3.8 days? Does this mean that the activity is reduced to one-half in the time of one half-life?

40-14. Radioactive carbon ^{14}C has a half-life of 5570 hr. In a living organism, the relative concentration of this isotope is the same as it is in the atmosphere because of the interchange of materials between the organism and the air. When an organism dies, this interchange stops and radioactive decay begins without replacement from the living organism. Explain how this principle can be used to determine the age of fossil remains.

NOTE: *Unless otherwise given, the masses of particular nuclides should be taken from Table 40-4.*

Table 40-4 Atomic Masses for Several Nuclides

Nuclide	Atomic number	Mass number	Atomic mass
Hydrogen	1	1	1.007825 u
Deuterium	1	2	2.014102 u
Tritium	1	3	3.016050 u
Helium 3	2	3	3.016030 u
Helium 4	2	4	4.002603 u
Lithium 6	3	6	6.015125 u
Lithium 7	3	7	7.016930 u
Beryllium 9	4	9	9.012186 u
Boron 11	5	11	11.009305 u
Carbon 12	6	12	12.000000 u
Carbon 13	6	13	13.003354 u
Nitrogen 14	7	14	14.003074 u
Oxygen 16	8	16	15.994915 u
Neon 20	10	20	19.992440 u
Copper 64	29	64	63.929759 u
Tin 120	50	120	119.902108 u
Gold 197	79	197	196.966541 u
Mercury 204	80	204	203.973865 u
Thallium 206	81	206	205.976104 u
Polonium 216	84	216	216.001922 u
Radon 222	86	222	222.017531 u
Radium 224	88	224	224.020218 u
Radium 226	88	226	226.025360 u
Thorium 233	90	233	233.041469 u
Thorium 234	90	234	234.043630 u
Protactinium 233	91	233	233.040130 u
Uranium 238	92	238	238.050790 u

The Elements

40-1. How many neutrons are in the nucleus of $^{208}_{82}$Pb? How many protons? What is the ratio N/Z?

Ans. 126, 82, 1.54

40-2. The nucleus of a certain isotope contains 143 neutrons and 92 protons. Write the symbol for this nucleus.

40-3. From a stability curve it is determined that the ratio of neutrons to protons for a cesium nucleus is 1.49. What is the mass number for this isotope of cesium?

Ans. 137

40-4. Most nuclei are approximately spherical in shape and have a radius that may be approximated by

$$r = r_0 A^{1/3}$$

where $r_0 = 1.2 \times 10^{-15}$ m. What is the approximate radius of the nucleus of a gold atom ($^{197}_{79}$Au)?

The Atomic Mass Unit

40-5. Consider a 2-kg cylinder of copper. What is the mass in atomic mass units? In MeV? In joules?

Ans. 1.20×10^{27} u, 1.12×10^{30} MeV, 1.80×10^{17} J

40-6. A certain nuclear reaction releases an energy of 5.5 MeV. How much mass (in atomic mass units) is required to produce this energy?

40-7. The periodic table gives the average mass of a silver atom as 107.842 u. What is the average mass of the silver nucleus?

Ans. 107.816 u

40-8. Consider the mass spectrometer as illustrated by Fig. 40-3. A uniform magnetic field of 0.6 T is placed across both upper and lower sections of the spectrometer and the electric field in the velocity selector is 120,000 V/m. A singly charged neon atom ($+1.6 \times 10^{-19}$ C) of mass 19.992 u passes through the velocity selector and into the spectrometer. (a) What is the velocity of the neon atom as it emerges from the velocity selector? (b) What is the radius R of its circular path?

40-9. When passing a stream of ionized lithium atoms through a mass spectrometer, it is noticed that the radius of the path followed by $^{7}_{3}$Li (7.0169 u) is 14.00 cm. A lighter line is formed by the $^{6}_{3}$Li (6.0151 u). What is the radius of the path followed by the $^{6}_{3}$Li isotopes?

Ans. 12.00 cm

Mass Defects and Binding Energy

40-10. Calculate the mass defect and the binding energy for the neon 20 atom ($^{20}_{10}$Ne).

40-11. Calculate the binding energy of tritium $^{3}_{1}$H. What is the binding energy per nucleon? How much energy in joules is required to tear the nucleus apart into its constituent nucleons?

Ans. 8.48 MeV, 2.83 MeV/nucleon, 1.36×10^{-12} J

40-12. Calculate the mass defect of $^{7}_{3}$Li. What is the binding energy per nucleon?

40-13. Determine the binding energy per nucleon for carbon 12 ($^{12}_{6}$C).

Ans. 7.68 MeV/nucleon

40-14. What are the mass defect and the binding energy per nucleon for tin 120 ($^{120}_{50}$Sn)?

Radioactivity and Nuclear Decay

40-15. The cobalt 60 ($^{60}_{27}$Co) nucleus emits gammas rays of approximately 1.2 MeV. How much mass is lost by the nucleus when it emits a gamma ray of this energy?

Ans. 0.00129 u

40-16. The half-life of the radioactive isotope indium 109 is 4.30 h. If the activity of a sample is 1 mCi at the start, how much activity remains after 4.30, 8.60, and 12.9 h?

40-17. The initial activity of a sample containing 7.7×10^{11} bismuth 212 nuclei is 4.0 mCi. The half-life of this isotope is 60 min. (a) How many bismuth 212 nuclei remain after 30 min? (b) What is the activity then?

Ans. 5.44×10^{11} nuclei, (b) 2.83 mCi

40-18. Strontium 90 is produced in appreciable quantities in the atmosphere during the test of an atom bomb. If this isotope has a half-life of 28 years, how long will it take for the initial radioactivity to drop to one-fourth its original activity?

Nuclear Reactions

* **40-19.** Determine the minimum energy released in the nuclear reaction

$$^{19}_{9}F + ^{1}_{1}H \rightarrow ^{4}_{2}He + ^{16}_{8}O + \text{energy}$$

(The atomic mass of $^{19}_{9}F$ is 18.998403 u.)

Ans. 8.11 MeV

* **40-20.** Determine the approximate kinetic energy imparted to the alpha particle when $^{226}_{88}Ra$ decays to form $^{222}_{86}Rn$. Neglect the energy imparted to the radon nucleus. The nuclidic masses are in Table 40-4.

* **40-21.** Find the energy involved in the production of two alpha particles in the reaction

$$^{7}_{3}Li + ^{1}_{1}H \rightarrow ^{4}_{2}He + ^{4}_{2}He + \text{energy}$$

Ans. 18.2 MeV

* **40-22.** Compute the kinetic energy released in the beta minus decay of thorium 233 ($^{233}_{90}Th$). The nuclidic masses are in Table 40-4.

Additional Problems

* **40-23.** Calculate the energy required to separate the nucleons in mercury 204 ($^{204}_{80}Hg$).

Ans. 1607 MeV

* **40-24.** What is the de Broglie wavelength of an electron having a momentum of 4.1×10^{-25} kg · m/s?

* **40-25.** The velocity selector in a mass spectrometer has a magnetic field of 0.2 T perpendicular to an electric field of 50 kV/m. The same magnetic field is across the lower region. (a) What is the velocity of singly charged $^{7}_{3}Li$ ions as they leave the selector? (b) If the radius of the circular path in the spectrometer is 9.10 cm, what is the atomic mass of the $^{7}_{3}Li$ ion?

Ans. (a) 2.5×10^{5} m/s, (b) 7.024 u

* **40-26.** What are the mass defect and the binding energy per nucleon for boron 11 ($^{11}_{5}B$)?

* **40-27.** How much energy in MeV is required to tear apart a deuterium atom ($^{2}_{1}H$)?

Ans. 2.22 MeV

* **40-28.** Plutonium 232 decays by alpha emission with a half-life of 30 min. How much of this substance remains after 4 h if the original sample had a mass of 4 g? Write the equation for this radioactive decay.

* **40-29.** If 32×10^{9} atoms of a radioactive isotope are reduced to only 2×10^{9} atoms in a time of 48 h, what is the half-life for this isotope?

Ans. 12 h

* **40-30.** A certain radioactive isotope retains only 10 percent of its original activity after a time of 4 h. What is the half-life of this material?

* **40-31.** When a $^{6}_{3}Li$ nucleus is struck by a proton, an alpha particle and a product nucleus are released. Write the equation for this reaction. What is the net energy remaining after the reaction?

Ans. 4.02 MeV

* **40-32.** Uranium 238 undergoes alpha decay. Write the equation for the reaction and calculate the disintegration energy.

* **40-33.** A nuclear reactor operates at a power level of 2 MW. Assuming that approximately 200 MeV of energy is released for a single fission of $^{235}_{92}U$, how many fission processes are occurring each second in the reactor?

Ans. 6.25×10^{16}

Electronics

41

OBJECTIVES

After completing this chapter, you should be able to:

1. Explain with the use of diagrams the operation of a vacuum-tube diode, a cathode-ray tube, and an x-ray tube.
2. Distinguish, by diagrams and discussion, between N- and P-type semiconductors, describing how they can be produced commercially.
3. Explain with the use of diagrams how a *transistor* operates when it is connected with a common base, a common emitter, and a common collector.
4. Solve transistor application problems similar to those presented in this chapter.
5. Define and discuss applications of the *zener diode* the *LED, photodiodes, silicon-controlled rectifiers,* and *integrated circuits.*

We are surrounded by electronics, ranging from stereos to wristwatches to automobile ignitions. The number of devices which are in some way electronic is constantly increasing. In fact, many of them would have been considered miracles 20 years ago. We now discuss some of the basic principles underlying this growing field of technology.

**41-1
THERMIONIC
EMISSION**

The roots of the electronic revolution go back to the late nineteenth century, when, in 1883, Thomas Edison noticed that black deposits were forming inside the glass light bulbs he was experimenting with. Further experiment showed that if a metal plate was also sealed inside the evacuated bulb and connected as shown in Fig. 41-1, the gavanometer would indicate a current between the filament and the plate. This current only flowed when the positive terminal of the battery was connected to the plate. Several years later this current was found to be the result of emission of electrons from the hot filament. The negatively charged electrons were attracted to the positive charge of the plate, generating an electric current between the plate and the filament.

Fig. 41-1 The Edison effect.

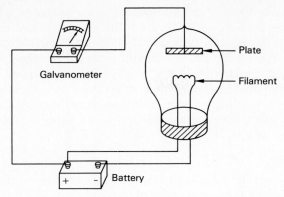

Galvanometer

Plate

Filament

Battery

The phenomenon which causes the emission of electrons from a heated filament is known as *thermionic emission.* It can only be observed in the absence of oxygen. A filament hot enough to cause thermionic emission will burn up in air.

41-2 VACUUM TUBES

The current flow observed by Edison occurred only when the plate (anode) was positive with respect to the filament (cathode). If the battery was reversed in the circuit, the current ceased. The negatively charged plate repelled the negatively charged electrons instead of attracting them, and the electrons which did escape the filament were attracted back to it by its positive charge.

Thermionic emission led to the development of the vacuum tube, which was the heart of all electronic devices until the 1950s. What made vacuum tubes so useful was the singular property that current could flow through them only in one direction.

A typical vacuum tube had a cathode heated by a separate heater filament and a plate anode. Tubes of this type, known as *diodes,* were the simplest form of the vacuum tube. A more sophisticated version, the *triode,* is shown in Fig. 41-2. The third element in the triode, the *grid,* was an electrode made up of a mesh of very thin wires; electrons could easily pass through the many holes in it. When the tube was in operation, if a small negative voltage was applied to the grid, the electrons emitted by the cathode were repelled by the grid and did not pass through it. The result was that

Fig. 41-2 Vacuum-tube triode.

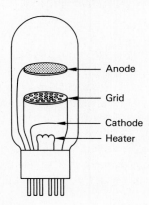

Anode

Grid

Cathode

Heater

no current flowed to the anode. However, if no voltage or a positive voltage was applied to the grid, the electrons from the cathode passed through the grid and current flowed between the cathode and the anode. Small voltages applied to the grid were able to control the flow of current in the tube completely.

These and similar tubes made possible the development of radio, television, computers, and numerous other modern devices. Now, the development and spread of *solid-state* technology utilizing *semiconductors* has made them virtually extinct. Two vacuum tubes, however, are still very widely used, the cathode-ray tube (CRT) and the x-ray tube.

41-3
THE CATHODE-RAY TUBE

The cathode-ray tube is the key to electronically produced visual displays in oscilloscopes (Fig. 41-3), the computers, and television sets. The tube consists of an *electron gun, deflection plates,* and the *screen.* These three parts are encased in a heavy-duty, evacuated, glass case. The electron gun is in the long, thin part of the tube, as shown in Fig. 41-4. It is made up of a cathode and two grids and anodes. The anodes are shaped to allow a stream of electrons to pass through them into the larger section of the tube.

Fig. 41-3 A general oscilloscope. *(Thornton Associates, Inc.)*

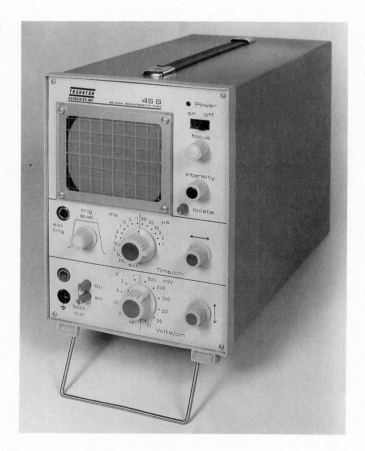

MODERN PHYSICS

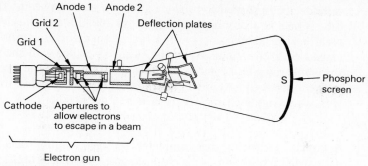

Fig. 41-4 Cathode-ray tube.

This beam of electrons passes a set of flat metal deflection plates, which can be charged like a capacitor, creating an electric field between them. Depending on the direction of the electric field, the electrons are deflected toward one plate or the other. In this way the beam can be deflected up or down or from side to side and aimed at any point on the screen at the front of the tube.

In a television set, the horizontal and vertical deflections are produced by magnetic fields rather than by electric fields. By sending high-frequency alternating current through coils, it is possible to make the electron beam sweep back and forth across the tube. The screen itself is coated with *phosphor,* a substance that emits light when it is struck by high-energy electrons. Wherever the electron beam hits the screen, a bright spot appears. Wherever the picture should be light, the grid voltage is positive and the beam is strong, resulting in a bright spot. Wherever the picture should be dark, the beam is retarded by the grid and fewer electrons (or none) strike the screen, resulting in a dark spot. By covering the screen with very closely spaced lines varying in brightness the picture is produced on the screen. The lines are drawn at a rate of about 16,000 per second, and the entire screen is redrawn 60 times per second. Since both these speeds are much too great for the human eye to perceive, we see what appears to be a continuous, moving image.

41-4
THE X-RAY TUBE

Virtually all modern x-ray tubes (Fig. 41-5) are refinements of a tube designed in 1913 by William Coolidge, of the General Electric Company. As in other vacuum tubes, electrons are thermionically emitted by a hot cathode. In these tubes the

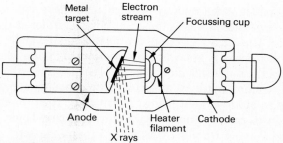

Fig. 41-5 Commercial x-ray tube.

cathode is cup-shaped to produce a narrow electron beam. The anode in the x-ray tube is held at a very high voltage (20 to 100 kV), and the electrons reach it with a great deal of kinetic energy. The highly energetic electrons strike a metal target embedded in the anode. When the electrons collide with the metal atoms, many of the bombarding electrons knock inner electrons out of the metal atoms leaving holes in the inner shells. As outer electrons fall into these holes, they emit x-ray photons with wavelengths characteristic of the target metal.

41-5
SEMI-CONDUCTORS

Solid-state technology has completely replaced vacuum tubes in all but a few highly specialized applications. This has been possible because of the unique characteristics of *semiconductors.* These materials consist of a few elements and compounds which are not electrical insulators but do not exhibit the extremely high conductivity of true conductors either.

What makes something a conductor? The outer electrons of an atom, known as the *valence electrons,* are bound to the atom less tightly than the electrons which are closer to the nucleus. In a conductor they are really held very loosely, and in the solid state many of them break away from the atoms and roam freely throughout the solid. Thus they are called *free electrons.* When a battery or other means is used to place a potential difference across the solid, these free electrons are attracted to the positive potential and a current flows, as shown in Fig. 41-6.

Insulators, like rubber and glass, behave just oppositely. The outer valence electrons are held very tightly to the atoms, and they do not move freely about the solid. Unless an extremely high voltage is applied to an insulator, no current flows through it.

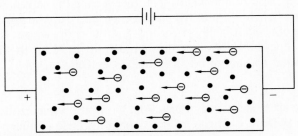

Fig. 41-6 Flow of free electrons in a conductor.

Semiconductors fall midway between conductors and insulators. Their valence electrons are neither as free as those in a conductor nor as tightly bound as those in an insulator. The valence electrons are actually shared by the atoms in the semiconductor. This process is known as *covalent bonding* and accounts for many of the properties of the semiconductors. Figure 41-7 shows how the electrons are shared by the hydrogen atoms in the H_2 molecule. Covalent bonding permits electrons to move from atom to atom without being entirely free. The process is much like moving from person to person around your set in a square dance. You are always with a partner, yet you are traveling completely around the set.

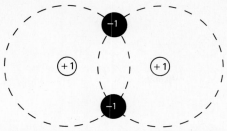

Fig. 41-7 Two covalently bonded hydrogen atoms.

Another way to look at the difference between conductors, insulators, and semiconductors is the *energy-band theory*. Since electrons can occupy a discrete number of energy levels, they can have only those energies which fall into *allowed bands*. The energy band in which the valence electrons normally move is called the *valence band*. Electrons which are free to move and conductor current are in the *conduction band*. As can be seen in Fig. 41-8 the conduction and valence bands overlap in conductors, are very far apart in insulators, and are separated by a narrow energy gap in semiconductors. The overlap makes it very easy for electrons to move into the conduction band in conductors. The large gap makes it almost impossible for them to do so in insulators, and the small gap makes it fairly easy in semiconductors.

Let us look at a real semiconductor. Because of its availability, silicon is the most commonly used semiconductor, although germanium is also used. In its natural state silicon is a *crystal*. This means that in a piece of silicon the atoms are located at specific points in an ordered *lattice*. Figure 41-9 shows a diagram of the silicon lattice. There are four valence electrons surrounding each atom. In a perfect crystal each of these electrons would be shared with a neighboring atom. The crystal would be an insulator because no electrons would be available for current flow. However, any imperfection will cause the silicon to conduct some current.

The most common impurity occurs when an atom of another substance is placed at a point in the crystal lattice that would normally be occupied by a silicon atom. Two important types of impurities are *N-type donors* and *P-type acceptors*.

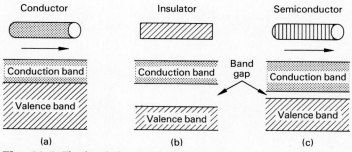

Fig. 41-8 The band theory of conduction.

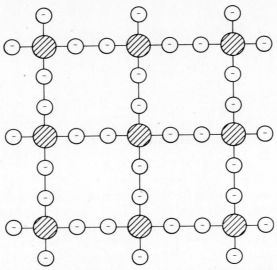

Fig. 41-9 The atomic structure and shared electrons in pure silicon.

When an impurity atom has more than the four valence electrons needed to pair with the neighboring silicon atom, the extra unpaired electrons are not covalently bonded and can move freely. Thus the impurity donates extra negatively charged electrons to the crystal. This explains the names donor and N-type (for negative). A typical N-type impurity can be seen in Fig. 41-10.

Commercial N-type semiconductor material is made by adding very small controlled quantities of a selected impurity to the silicon crystal. The intentional impuri-

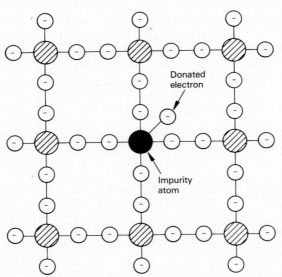

Donated electron

Impurity atom

Fig. 41-10 In an N-type semiconductor, an impurity atom joins the silicon structure, donating an extra electron.

ties are known as *dopants*. Phosphorus, arsenic, and antimony are common N-type dopants. Each has five valence electrons and therefore adds one free electron to the crystal. The P-type silicon is also commercially produced by doping. In this case the dopant has one less valence electron than silicon. Thus P-type dopants have three valence electrons, and common P-type dopants are aluminum, boron, gallium, and indium. Any of these dopants yields a semiconductor like the one in Fig. 41-11.

The *hole* marked in the figure is where the fourth valence electron would be if the spot were occupied by a silicon atom and not an impurity atom. When attracted by a positive potential, electrons near neighboring holes can jump over and fill the holes. Each time an electron moves in one direction, the hole moves in the opposite direction, to the atom from which the electron came. In this way a movement of electrons one way causes a movement of holes the other way. Since negative and positive charges always behave oppositely, it is just as if the holes were positive charges and the current can be treated as if it were a movement of positive charges (the holes).

41-7
PN JUNCTION

Some interesting effects result from the combination of P-type and N-type materials. The most basic of these is formation of a *PN junction* by placing a slice of P-type material in contact with a slice of N-type material. When the two semiconductors are first joined, some of the free electrons from the N-type material jump past the junction and fill some of the holes in the P-type material. This leaves a thin layer of positively charged ions along the junction of the N side. These positive ions are the atoms which have lost an electron. Similarly, the atoms on the P side which have acquired electrons form a layer of negative ions. Once these layers are formed, the negative ions repel other electrons, preventing any more electrons from jumping over, and the positive ions prevent any holes from jumping over.

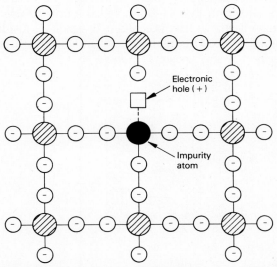

Fig. 41-11 In a P-type semiconductor, the absence of a valence electron produces an electronic hole.

In this state the PN junction is rather stable. But things change when a battery is connected across the junction. We first look at what happens when the battery is connected with its negative terminal on the P side of the junction and its positive terminal on the N side, as in Fig. 41-12. The positive holes in the P-type material move toward the negative potential of the battery, and the negative electrons move toward the positive potential of the battery. The result is that both holes and electrons move *away* from the junction and there is no current flow across the junction. When connected in this manner, the junction is said to be *reverse-biased*.

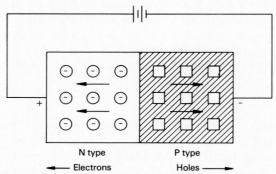

Fig. 41-12 Reverse-biased PN junction.

If the battery is connected across the junction in the opposite direction, positive to the P side and negative to the N side, the behavior is quite different. The electrons from the donor atoms in the N-type material are attracted to the positive potential and flow toward the junction. The holes flow toward the junction in the opposite direction toward the negative potential. In short, there is a current flow. This type of connection, shown in Fig. 41-13, is called *forward biasing* of the junction. The semiconductor device we have been discussing behaves just like the simplest vacuum tube discussed in Sec. 41-2; it allows current flow in only one direction. Because of this similarity it is also known as a *diode*.

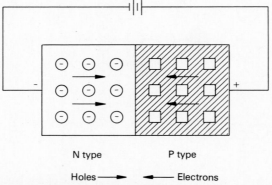

Fig. 41-13 Forward-biased PN junction.

The significant advantages of the semiconductor diode over its vacuum-tube predecessor are shared by all other semiconductor devices. It is very small; it does not require high voltage to operate; and it is more durable.

The semiconductor diode also exhibits another very interesting characteristic. It does not obey Ohm's law. Figure 41-14 shows a graph of current vs. voltage for an ordinary conductor and the same information for a forward-biased semiconductor diode. The conductor shows the straight line that results from the relation $V = IR$. The diode does not. Instead, the current rises very slowly at the start and then rises very rapidly. The slow rise at the beginning occurs because the forces resulting from the ion layers at the junction must be overcome. Once they are overcome, however, there is very little resistance to current flow. Diodes do not follow a simple rule like Ohm's law. Each diode has its own characteristic curve, which depends on the material used to make it and the dopants used in it.

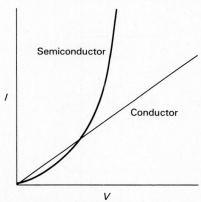

Fig. 41-14 The relationship between voltage and current for a conductor and for a semiconductor.

41-8
DIODE APPLICATIONS

We have already seen that a PN junction diode passes current only when forward-biased. This immediately makes it an excellent choice to convert alternating current to direct current (rectifying). Figure 41-15 shows a diode inserted in a circuit as a *half-wave rectifier*. (Notice the symbol representing the diode. The triangle can be seen as an arrow pointing in the direction of allowed current flow.) The output of this circuit is not a true, constant direct current but a current which flows in pulses, only

Fig. 41-15 The semiconductor diode used as a half-wave rectifier.

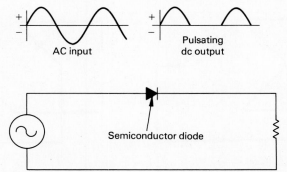

half of the time. The remainder of the current, which would normally flow the other way, is blocked.

Often the pulsed output of the half-wave rectifier is not sufficient to meet the needs of a particular application. When this is the case, it is possible to *filter* the output of the rectifier to smooth out the pulses and make it look more like the continuous output of a battery. A possible circuit for such filtering is shown in Fig. 41-16. The filter consists of two capacitors and an inductor. The inductor opposes changes in current; thus it opposes the pulsed current passing through the diode. Excess current is diverted by the inductor during periods of high current flow from the diode and charges capacitor C_1. When the current flow between pulses is zero, C_1 discharges through the inductor, maintaining current flow through the inductor at a fairly constant level. Capacitor C_2, at the output end of the inductor, similarly smooths out any variations in the current output from the inductor.

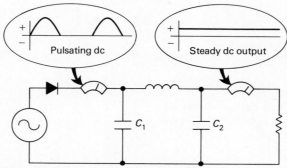

Fig. 41-16 An electronic filtering circuit.

The half-wave rectifier is not the only means of accomplishing that purpose. More common is a *bridge rectifier,* which uses four diodes arranged as in Fig. 41-17. The ac input is connected to the bridge at points 1 and 3, and the output load is connected to points 2 and 4. When point 1 is positive and point 2 negative, the current flows as in Fig. 41-17*a*. Current entering the bridge at point 1 cannot pass through diode D_4, which is reverse-biased, but it can, and does, pass through diode D_2. It then leaves the bridge at point 2 and goes through the load R_L. It reenters the bridge at point 4, again facing two possible paths. D_4 remains reverse-biased and

Fig. 41-17 Bridge rectifier.

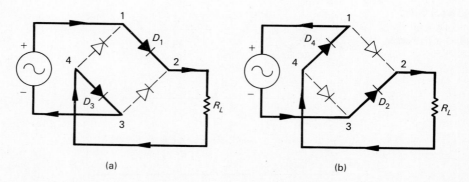

(a)   (b)

nonconducting, and so the current goes through D_3 and leaves at point 3. During the other half of the cycle, when the input polarity is reversed, the current flows the path shown in Fig. 41-17b using the other pair of diodes, D_2 and D_4, in a similar manner. In this way the diode pairs which are conducting alternate according to the alternations in the input polarity. Such a rectifier is called a *full-wave rectifier* since all of the input signal is transmitted.

A major advantage of the full-wave rectifier is its output form, shown in Fig. 41-18 in comparison with the ac input and the output of a half-wave rectifier. The figure shows that the full-wave rectifier leaves no gaps where the current output is zero. Instead, the current continuously varies between zero and its peak value in the forward direction. Thus the output requires much less smoothing to make it look like the constant output of a battery, and the device does not in effect ignore half the source current.

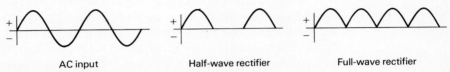

AC input   Half-wave rectifier   Full-wave rectifier

Fig. 41-18 A comparison of waveforms.

41-9
THE ZENER DIODE

So far we have been discussing diodes as purely one-direction devices; when forward-biased, they conduct, and when reverse-biased, they do not. A look at a complete characteristic curve for a typical semiconductor diode (Fig. 41-19) shows that this is not quite the case. When the diode is reverse-biased, the current remains virtually zero as the reverse voltage is increased until a critical voltage known as the *zener voltage* is reached. At that point the resistance of the diode abruptly drops, and the diode becomes an excellent conductor in the reverse direction.

This apparently strange effect is easily explained by examining the behavior of the electrons and holes at the junction in the diode. When the diode is reverse-

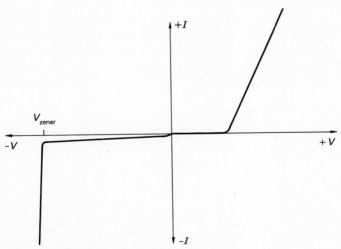

Fig. 41-19 Diode characteristic curve.

biased, the electrons and holes all move away from the junction, leaving it virtually free of charges of either type. There are, however, always a few electrons and holes created by random thermal motions in the crystal. These few created free holes and electrons flow away from the junction and constitute a nonzero *reverse current*. This reverse current, of a few microamperes, is always present when the diode is reverse-biased but can usually be ignored.

If the reverse-bias voltage is increased sufficiently, a point is reached where the electrons accelerated by this voltage have enough energy to knock more electrons loose from atoms with which they collide. These freed electrons knock more electrons loose, creating a large flow of electrons, known as the *avalanche,* or *zener effect*. Once the zener voltage is reached, the voltage across the diode does not increase to any extent, even though the current increases dramatically. This is another clear violation of Ohm's law.

We might expect that a diode operated at such comparatively high current in its reverse direction would burn itself up, as vacuum-tube rectifiers do. (If enough voltage is connected across them for current to flow in the reverse direction, they destroy themselves.) Special zener diodes are made for just this purpose, and as long as their specifications are not exceeded, they can be run perfectly well for long times. Figure 41-20 shows the circuit symbol for a zener diode.

Fig. 41-20 Zener diode symbol.

The tendency for a diode operating in the zener region of its characteristic curve to maintain a constant voltage regardless of current flow through it makes the zener diode especially useful in applications where a constant voltage at a particular point is critical, particularly in regulated power supplies. A regulated power supply is a voltage-current source whose output voltage remains constant at a specified value regardless of the output current.

Figure 41-21 shows a way of using two identical zener diodes connected back to back to limit the output of an ac source. When the two diodes are connected in this way, one of them is reverse-biased and voltage-limiting, while the other is forward-biased and conducting. When the ac source reverses its polarity, the diodes exchange roles and the conducting diode becomes the limiter and the limiter becomes the

Fig. 41-21 Zener diode regulator.

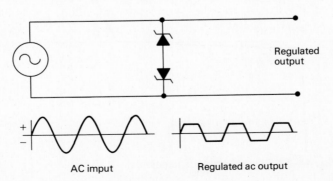

MODERN PHYSICS

conductor. The peak output of this *voltage regulator* remains less than or equal to the zener voltage of the two diodes.

A single PN junction device is a diode. The next step is the *transistor,* which consists of a slice of N-type semiconductor sandwiched between two slices of P-type semiconductor (called a PNP transistor) or a slice of P-type semiconductor sandwiched between two slices of N-type semiconductor (called an NPN transistor). This transistor was invented in 1948 by Shockley, Bardeen, and Brattain at Bell Laboratories, and in the years since then has completely revolutionized the field of electronics and many associated fields.

Let us consider a typical NPN junction transistor biased as shown in Fig. 41-22. The left section is made of heavily doped N-type semiconducting material and is called the *emitter.* The center section is made of a very thin and lightly doped P-type semiconducting slice. It is called the *base.* The right section is made of a medium doped N-type semiconducting material and is called the *collector.* Note that the emitter and base form an NP junction, and the base and collector form a PN junction.

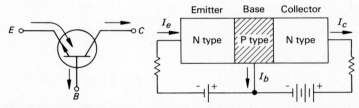

Fig. 41-22 An NPN transistor and its symbol. Arrows indicate the direction of electron currents. (Note that electron flow is opposite to conventional current.)

The emitter-base NP junction in Fig. 41-22 is forward biased allowing electrons to flow easily from the N side to the P side. The base region is very thin and has relatively few holes, allowing most of the electrons to diffuse right across it to the collector. But some of the electrons are trapped in the holes of the base region. If they are not removed from the base, the negative charge of the trapped electrons will repel the electrons from the emitter and rapidly shut off the electron current from the emitter to the collector. By removing or not removing this buildup of electrons from the base region, the transistor can be used as a switch.

When the electrons are removed from the base, there is an electron current from the base. The emitter current I_e is divided into base current I_b and a collector current I_c. That is,

$$I_e = I_b + I_c \tag{41-1}$$

The ratio of the collector current to the emitter current is called α (alpha).

$$\alpha = \frac{I_c}{I_e} \tag{41-2}$$

This number is a specification given by the manufacturer with each transistor. Since 95 to 99 percent of the electrons from the emitter travel through the base into the collector without being trapped, α typically ranges from 0.95 to 0.99.

If we know α, we can calculate the relation of the very small base current to the emitter current. Using Eq. (39-1), we can solve for I_b in terms of I_e and I_c.

$$I_b = I_e - I_c$$

But $I_c = \alpha I_e$, as computed from Eq. (39-2). Substitution gives

$$I_b = I_e - \alpha I_e$$

Finally, an expression for the base current of a transistor is

$$\boxed{I_b = I_e(1 - \alpha)} \qquad \text{(41-3)}$$

EXAMPLE 41-1 Given a transistor with $\alpha = 0.97$, calculate I_b and I_c when $I_e = 50$ mA and the transistor is connected as in Fig. 41-22.

Solution Substitution into Eq. (39-3) gives

$$I_b = I_e(1 - \alpha) = (50 \text{ mA})(1 - 0.97)$$
$$= (50 \text{ mA})(0.03)$$
$$= 1.5 \text{ mA}$$

Now the collector current is found from Eq. (39-1).

$$I_c = I_e - I_b = 50 \text{ mA} - 1.5 \text{ mA}$$
$$= 48.5 \text{ mA}$$

The PNP transistor is constructed just like the NPN transistor, except that the N-type material is sandwiched between two slices of P-type material. Because of this reversal of the positions of the two material types, the primary current carriers in the transistor are holes instead of electrons. Since holes are effectively positive charges, the biases are necessarily reversed. For the emitter junction to be forward-biased the emitter voltage must be negative with respect to the base, and for the collector junction to be reverse-biased, the collector must be positive with respect to the base.

The biasing for a PNP transistor is shown in Fig. 41-23. The change in biasing does not affect the way the junctions operate. It merely accommodates the rearrangement of materials and their properties. In operation, most holes flow from the emitter junction to the base and then into the collector.

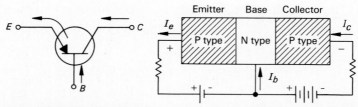

Fig. 41-23 A PNP transistor and its symbol. Arrows indicate electron flow.

In an amplification device we want to be able to input a varying signal with low-magnitude voltage or current characteristics and output a similarly varying signal with much higher voltage or current. A vacuum-tube triode is controlled by voltage. A very small voltage on the grid of the triode controls a similar but much higher voltage in the plate circuit. The transistor, on the other hand, depends on the currents in it to control the amplification of the voltage or power in the output circuit.

Figure 41-24 shows a simple amplifier circuit utilizing the transistor. Normally ac circuits have two terminals for input and two for output. Because the transistor is a three-terminal device, one of its terminals must be common to both the input and output circuits. Any of the three terminals may be selected as the common terminal. In Fig. 41-24 the base is the common terminal.

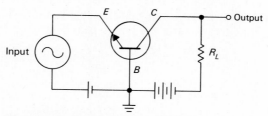

Fig. 41-24 Common-base amplifier circuit.

As we discussed in Sec. 41-10, the emitter current and the collector current are almost equal. For this reason the ratio of output current to input current, called the *current gain,* is approximately 0.95 to 0.99 when the transistor is used in this configuration and is equal to α. In this case, collector current is considered as output and emitter current is considered as input.

Normally the voltage across the forward-biased region is small. This region represents a very low resistance, and little voltage is needed to send a relatively high current through it. The second diode region, which is reverse-biased, represents a much higher resistance to the passage of current. Since voltage is the product of current and resistance, the resulting voltage must be high.

The ratio of output resistance to input resistance in a typical transistor is usually quite high. A typical input resistance might be 500 Ω, and a typical output resistance might be as much as 500 kΩ. This results in a ratio R_{out}/R_{in} equal to 1000.

EXAMPLE 41-2 Assume a transistor to have an input resistance of 500 Ω and an output resistance of 400 kΩ. Determine the input and output voltages if the emitter current is 3 mA and $\alpha = 0.98$.

Solution The input voltage is found from Ohm's law.

$$V_{in} = I_e R_{in} = (10 \times 10^{-6} \, A)(500 \, \Omega)$$
$$= 5 \times 10^{-3} \, V = 5 \, mV$$

To determine the output voltage, we must first determine the collector current from α. Recalling that $\alpha = I_c/I_e$, we can solve for I_c as follows:

$$I_c = \alpha I_e = 0.98(3 \times 10^{-3} \, A) = 2.94 \times 10^{-3} \, A$$

Thus, the output voltage is

$$V_{out} = I_c R_{out} = (2.94 \times 10^{-3} \text{ A})(400 \times 10^3 \text{ } \Omega)$$
$$= 1176 \text{ V}$$

The *voltage amplification factor* A_v for a transistor is defined as the ratio of output voltage to input voltage.

$$\boxed{A_v = \frac{V_{out}}{V_{in}}} \qquad \begin{array}{c} \textit{Voltage} \\ \textit{Amplification Factor} \end{array} \quad (41\text{-}4)$$

This ratio is also referred to as the *voltage gain*.

Along with an increase in voltage comes an increase in power. Recall that the power dissipated by a circuit component is equal to the voltage across it times the current through it. In any case, the power gain G is given by

$$G = \frac{\text{power out}}{\text{power in}} = \frac{V_{out} I_c}{V_{in} I_e} \qquad (41\text{-}5)$$

But since $\alpha = I_c/I_e$, we can write

$$\boxed{G = \alpha \frac{V_{out}}{V_{in}} = \alpha A_v} \qquad \textit{Power Gain} \quad (41\text{-}6)$$

EXAMPLE 41-3 Determine the voltage amplification factor and the power gain for the transistor of Example 41-2.

Solution In the previous example the input voltage was 1.5 mV, and the output voltage was 1176 V. This gives an amplification factor of

$$A_v = \frac{V_{out}}{V_{in}} = \frac{1176 \text{ V}}{1.5 \times 10^{-3} \text{ V}} = 784{,}000$$

Now the power gain can be found from Eq. (41-6),

$$G = \alpha A_v = (0.98)(784{,}000) = 768{,}320$$

where we used $\alpha = 0.98$ from the previous example

Transistors can also be connected with the emitter as the common terminal, i.e., connected to input and output circuits. A simple *common emitter* amplifier circuit is shown in Fig. 41-25. As before, the input is across the emitter and base contacts, but the output in this case is from the emitter and collector terminals. The emitter junction is forward-biased, and the collector junction is reverse-biased. If the bias on the emitter junction is increased, the current flowing through the collector junction will increase as a result of the increase in current through the emitter junction. However, the base current will retain its usual low value. This low base current is also the current flowing in the input circuit. Thus in this configuration we get a considerable current gain in the output circuit (collector circuit) over the current in the input circuit (base). We already know that the base current is given by

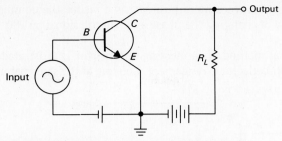

Fig. 41-25 Common-emitter amplifier circuit.

$$I_b = I_e(1 - \alpha)$$

and that the output, or collector, current is

$$I_c = \alpha I_e$$

These relations allow us to compute an expression for the current gain, or current amplification factor, A_i

$$A_i = \frac{I_{\text{out}}}{I_{\text{in}}} = \frac{\alpha I_e}{(1 - \alpha)I_e} = \frac{\alpha}{1 - \alpha} \qquad (41\text{-}7)$$

A special symbol β is used to represent the current gain in the above form. Thus, for a transistor having a known α the current gain can be found from

$$\beta = \frac{\alpha}{1 - \alpha} \qquad (41\text{-}8)$$

This is also known as the *base-collector current gain*.

EXAMPLE 41-4   In a common-emitter amplifier the value of α is 0.96. What is the current gain?

Solution   Substitution into Eq. (41-8) yields

$$\beta = \frac{\alpha}{1 - \alpha} = \frac{0.96}{1 - 0.96} = 24$$

Typically, the values for β range from 20 to several hundred.

To determine the voltage gain we must again look at the effective input and output resistances. As in the common-base circuit, the input resistance is across the forward-biased emitter junction and is very low. The output resistance is high but not quite as high as in the common-base circuit. The result is a voltage gain nearly as high as in the common-base circuit.

The power gain depends on the products of the current gain and voltage gain. While the voltage gain is slightly lower, the current gain β is much, much higher, and the power gain is also much higher.

The common-emitter circuit has one other very interesting feature. When the incoming signal causes the base potential to become more negative, increasing the forward bias of the emitter junction, the collector potential moves in the opposite direction and becomes more positive. As a result, the common-emitter circuit has the property of changing the phase angle of the signal by 180° and acting as a *signal inverter*.

The final method of connecting a transistor is illustrated in Fig. 41-26. This is the common-collector arrangement. Looking at the same parameters as for the other

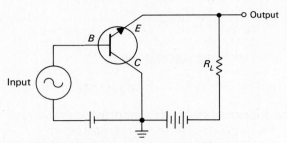

Fig. 41-26 Common-collector amplifier circuit.

two circuits, we get the following results. The input current is the base current, and the output current is the emitter current. The current gain is consequently

$$A_i = \frac{I_{\text{out}}}{I_{\text{in}}} = \frac{I_e}{I_e(1 - \alpha)}$$

from which

$$A_i = \frac{1}{1 - \alpha} \tag{41-9}$$

In this case the base voltage is greater than the output voltage, and the net gain is less than 1, the lowest of the three arrangements. Related to this, the input resistance is high and the output resistance is low. Finally, because of the very low voltage gain, the power gain is also the lowest of the three configurations. Table 41-1 compares the various gain characteristics of the three circuits.

Table 41-1 A Comparison of the Three Possible Transistor Amplifier Arrangements

	Common base	Common emitter	Common collector
Current gain	α (low)	β (high)	$\dfrac{1}{1-\alpha}$ (highest)
Voltage gain	Highest	High	Low
Power gain	Moderate	Highest	Lowest

There are two points to remember when using transistors in circuits: (1) Transistors require a dc voltage source for biasing before an ac signal can be applied to them for amplification or other modification. This dc voltage is often supplied by a regulated power supply using a diode bridge and zener diodes. (2) Since transistors are easily damaged by high temperature, the current must be rather low.

MODERN PHYSICS

Transistor manufacturers publish data sheets and graphs of transistor characteristics. These documents are useful for circuit design and for verifying that none of the relevant parameters will be exceeded, destroying the transistor or other sensitive semiconductor circuit elements.

41-12
OTHER
SEMICON-
DUCTOR
DEVICES

Since the invention of the transistor, many other semiconductor devices have been developed. Some of them have interesting and unusual attributes which make possible many of the small electronic devices we enjoy today. One of the newest devices is the light-emitting diode (LED), used to provide visual displays in electronic devices. When arranged as shown in Fig. 41-27, a set of seven LEDs can be used to form

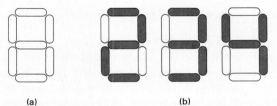

(a)   (b)

Fig. 41-27 LED display. *(a)* Arrangement of LEDs. *(b)* Pattern of lighting to form the number 234.

numbers or letters. They are common in pocket-calculator and wristwatch displays. They are simply junction diodes which emit light due to the current flowing through them. The diodes are operated with a forward bias, allowing current to flow across the PN junction. At the junction, however, some of the electrons and holes combine to form electron-hole pairs. The electrons fill the holes, and the two are canceled out as current carriers. The energy of the recombined electrons and holes is released in the form of photons. Materials which behave this way are used to make LEDs, e.g., gallium arsenide, gallium arsenide phosphide, and gallium phosphide. The first two emit visible light, and the last emits infrared light.

Most visible-light LEDs emit red light, although some emit green and yellow light. They are not powerful light sources like incandescent light bulbs and cannot be used for general illumination, but they are bright enough to be visible under most conditions and ideal for indicator lights and digital displays. Their advantages are that they are very small and rugged; they use little current and require low voltages; and their useful lifetimes are extremely long. Some LEDs may last as long as 100 years.

Similar to the LED but functioning in an inverse manner are photosensitive semiconductor devices. One of these is the photodiode. These diodes are run reverse-biased. When struck by light, the photons are absorbed by the atoms. This raises the energy level of the valence electrons sufficiently to cause them to jump into the conduction band. Thus, an electron-hole pair is created for each absorbed photon, and the electrons and holes are both swept away by the bias voltage, producing a current flow. Just as the light output of the LED is related to the current flow in the diode, the current flow in the photodiode is related to the intensity of the light striking the diode. The circuit symbols for the LED and the photodiode and a set of characteristic curves for a photodiode are shown in Fig. 41-28.

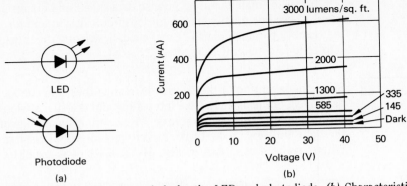

Fig. 41-28 *(a)* Circuit symbols for the LED and photodiode. *(b)* Characteristic curves for a photodiode.

An interesting application of LEDs and light-sensitive diodes is in the construction of a type of semiconductor *relay*. A relay is a device for indirectly switching an electric current by using another current. Before the advent of semiconductor technology relays were all electromechanical devices of various designs. One of the great advantages of a relay is that very high, and consequently dangerous, currents can be switched without handling them directly. A relay for this purpose is found in the starting circuit of a car. A small current in the starting switch activates the solenoid switch; it actually turns on the very high current which goes to the starting motor.

The optical semiconductor relay, which cannot be used for very high-current applications, is shown in Fig. 41-29. In operation, when a current is passed through the LED, it emits light. The light from the LED strikes the photodiode, which then conducts as long as the LED remains activated. In this way the photodiode acts as a switch turned on and off by the LED.

Another semiconductor device which may serve as a type of switch is the *silicon-controlled rectifier* (SCR) (Fig. 41-30). It is similar to a transistor with an extra layer of semiconductor material added to the stack. Just as a transistor can be viewed as two diodes in series, the SCR can be viewed as three diodes in series. For the SCR in the figure they are, from left to right, an NP junction, a PN junction, and another NP junction. In addition, the electrodes are renamed. The end electrodes are the *anode* and *cathode,* and the center electrode is called the *gate*.

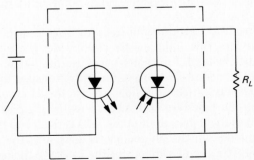

Fig. 41-29 Simple LED-photodiode relay.

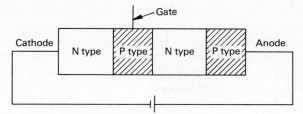

Fig. 41-30 Silicon-controlled rectifier.

When a low bias voltage is applied between the anode and cathode, the center diode is reverse-biased and there is no current flow. In this state the device is said to be switched *off.* As the bias voltage is increased, a point is reached where the reverse-biased diode breaks down and the device begins to conduct. At this point the device has switched *on.* The point at which the SCR switches on, called the *breakover voltage,* is controlled by the positive bias applied to the gate. Once this point is reached, the gate loses control and the SCR must be disconnected in some way to restore control to the gate. The circuit symbol for the SCR and a set of characteristic curves for a typical SCR are shown in Fig. 41-31.

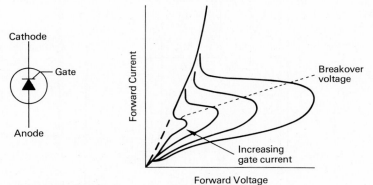

Fig. 41-31 SCR symbol and characteristic curves.

Two other simple semiconductor devices exhibit properties different from any others discussed up to this point, the *photoresistor* and the *thermistor.* The photoresistor is the key component of all modern light meters. In fact almost all cameras which use batteries use a photoresistor to measure light intensity. Most photoresistors in use today are made of cadmium sulfide (CdS). Unlike the other semiconductor devices we have discussed the photoresistor has no junction. It is a uniform crystal of CdS. When a voltage is applied across it, a current flows in proportion to the intensity of the light striking it. Conduction is not by free electrons but by electron-hole pairs, which are created when photons knock electrons loose from atoms in the crystal and into the conduction band. The more intense the light, the greater the number of photons hitting the crystal and the greater the conductance.

The last device to be discussed in this section is the thermistor. It takes advantage of the fact that the resistance of a semiconductor decreases as the temperature increases. The effect is opposite to, and much greater than, the change in the resistance of a conductor with temperature. This can be seen clearly in Fig. 41-32. The y axis, showing the resistance, is logarithmic to accommodate the large changes in resistance. When a thermistor is maintained with a constant voltage across it, there will be large changes in the current which flows through the device as a result of changes in the temperature of the device. This means that a thermistor can be used to measure the temperature of its environment.

It also may be used as a safety device in complex solid-state components. The thermistor is placed in the parts of such components where high temperatures may be experienced. The circuit is then designed so that when the current through the thermistor rises to a critical value, the component shuts off, protecting the semiconductor devices within it from being destroyed by overheating. In this way the thermistor acts as a temperature-controlled fuse.

Fig. 41-32 Structure symbol and characteristic of a typical thermistor.

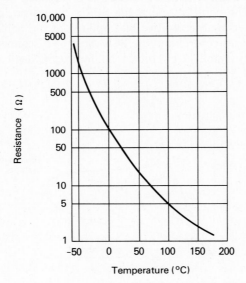

41-13
INTEGRATED CIRCUITS

As great as the electronic revolution the transistor caused in the 1950s and 1960s is a new revolution due to the *integrated circuit* (IC). These devices combine numerous circuit elements (resistors, capacitors, transistors, diodes, and others) on tiny slices of ultrapure silicon. Each IC represents a complete circuit of a particular type, and ICs can be combined with each other and with other semiconductor devices to make large-scale components. They have also made possible certain devices like electronic wristwatches, which depend on the IC to pack all the necessary electronics in a case small enough to wear on the wrist.

The most obvious attribute of the IC is its incredibly small size. A typical IC is 0.05 in. square and about 0.01 in. thick. This wafer may contain thousands of separate devices, enough to serve numerous complex purposes. It is protected by a container which usually measures only a small fraction of an inch in any direction.

Although the size of the IC is remarkable, it is not its only advantage over conventional devices. Because the entire circuit is fabricated at the same time, it is a highly reliable device, more so than any of its predecessors. This characteristic is invaluable in such things as spaceship guidance systems and pacemakers.

The individual devices which make up the IC function on the same principles as the semiconductor devices already discussed. They depend on the PN junction and the action of both electrons and holes for their various properties.

The silicon wafer which makes up the IC is known as the *substrate,* and the two processes of fabrication are *diffusion* and *etching.* Diffusion, as in the manufacture of junction-dependent devices, is the controlled injection of appropriate dopants into the silicon. Etching is the use of hydrofluoric acid to remove materials from the surface of the substrate to expose a region for further processing. These steps are combined with several steps which deposit materials on the substrate to make the necessary P-type and N-type regions forming diodes, transistors, and related devices.

Let us outline the steps needed to create an area of P-type silicon in the substrate. The first step is the oxidation of the top surface of the silicon to form a thin layer of silicon oxide. This oxide layer is impervious to most contaminants and serves to isolate and protect the substrate.

After the oxide layer comes the *photoresist* layer. The photoresist is a material which becomes acid-resistant when it is exposed to ultraviolet light. Exposure to ultraviolet light changes its chemical properties so that the areas which are exposed cannot be dissolved by acids.

The exposure to ultraviolet light is made after the coated substrate has been covered with a *mask.* This is a pattern which is opaque in the areas into which the dopants are eventually to be diffused. Masks are produced photographically by taking a picture of the appropriate pattern and then reducing it to the microminiature size needed for the IC.

After the placement of the mask and the exposure, the exposed photoresist is washed away with a solvent. This leaves a layer of exposed silicon oxide covering the area which will become the P-type region. Etching with hydrofluoric acid removes the oxide and leaves bare silicon substrate. The various steps in this process, in order, are shown in Fig. 41-33.

The bare silicon can now be doped with the desired dopant to produce a P-type region in the substrate. If the same process is repeated on an immediately adjacent area with an N-type dopant, a P-type region will border on an N-type region and create a PN junction. The addition of metal contacts on top of these regions gives a usable semiconductor diode like the one in Fig. 41-34.

Similar processes can be used to form all other necessary components in the IC. A P-type resistor on a P-type substrate is illustrated in Fig. 41-35. The upper N-type region is formed by diffusing enough N-type dopant into the already formed P-type region to convert it to an N-type region in that area. The plus sign indicates a high concentration of the N-type dopant. The purpose of the two N-type regions is to isolate the resistor from the substrate and the rest of the IC. The resistance is controlled by adjusting the length and cross-sectional area of the P-type conductor region and the amount of dopant diffused into the region.

A capacitor is formed similarly to a resistor. A diffused conducting layer acts as one of the plates of the capacitor, a metal contact acts as the other plate, and the oxide layer between them acts as the dielectric. A possible arrangement is shown in Fig. 41-36.

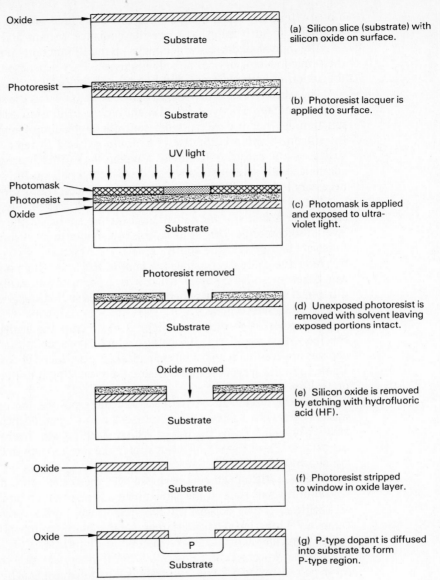

Oxide → | (a) Silicon slice (substrate) with silicon oxide on surface.

Substrate

Photoresist → | (b) Photoresist lacquer is applied to surface.

Substrate

UV light

Photomask →
Photoresist →
Oxide → | (c) Photomask is applied and exposed to ultra-violet light.

Substrate

Photoresist removed | (d) Unexposed photoresist is removed with solvent leaving exposed portions intact.

Substrate

Oxide removed | (e) Silicon oxide is removed by etching with hydrofluoric acid (HF).

Substrate

Oxide → | (f) Photoresist stripped to window in oxide layer.

Substrate

Oxide → | (g) P-type dopant is diffused into substrate to form P-type region.

P

Substrate

Fig. 41-33 The steps in producing an integrated circuit.

Of course ICs could not be the highly sophisticated circuits they are without the ability to contain transistors. Several types of transistors can be formed in ICs, and they can be formed in several ways. The basic and the preferred structures for IC transistors are shown in Fig. 41-37.

ICs are manufactured from circular silicon wafers about 1.5 in. in diameter. Each wafer is large enough to produce about 500 ICs. After fabrication they are individually tested, and defective circuits are marked. The wafer is then cut apart and

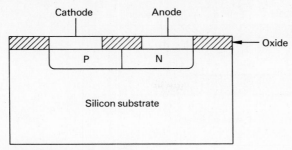

Fig. 41-34 The IC diode.

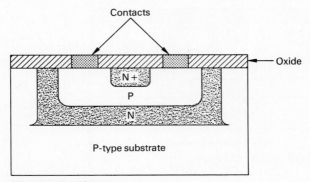

Fig. 41-35 The IC resistor.

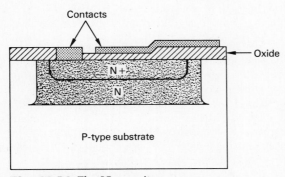

Fig. 41-36 The IC capacitor.

the defective circuits are discarded. The yield of good circuits is only about 10 to 40 percent, but since so many circuits are made at one time, the IC is one of the cheapest, smallest, and most reliable electronic devices produced. This fact has made pocket calculators, small computers, and similar high-quality electronic products as common and inexpensive as they are today, and it is hard even to imagine where the next step in the advance of electronics technology may lead is.

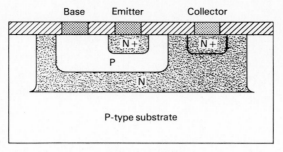

(a) Basic structure

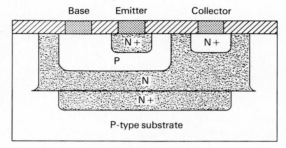

(b) Preferred structure

Fig. 41-37 The IC transistor.

SUMMARY

Electronics is much too broad a subject to give extensive preparation in one chapter. The intent of this unit of study was merely to cover in a descriptive way many of the concepts important for today's industrial workers. The major ideas are summarized below:

- A *diode* is an electronic device consisting of two elements. In a vacuum tube these elements are the plate and the filament; in the semiconductor diode, the two elements are formed by a PN junction.
- A *semiconductor* is a solid or liquid material with a resistivity between that of a conductor and an insulator. An N-type semiconductor contains *donor* impurities and free electrons. A P-type semiconductor consists of *acceptor* atoms and electron holes. A *PN junction* is the union of a P-type crystal with an N-type crystal in such a way that current is conducted in only one direction.
- A *transistor* is a semiconductor device consisting of three or more electrodes; usually the emitter, the collector, and the base.
- For a common-base transistor amplifier

Emitter current = base current + collector current

$$I_e = I_b + I_c \qquad a = \frac{I_c}{I_e}$$

Current Gain

- The *current gain* for other connections is

$$\beta = \frac{\alpha}{1 - \alpha}$$
Common-Emitter Amplifier

$$A_i = \frac{1}{1 - \alpha}$$
Common-Collector Amplifier

- The voltage gain A_v and the power gain G are given by

$$A_v = \frac{V_{out}}{V_{in}}$$
Voltage Gain

$$G = \frac{power\ in}{power\ out} = \frac{V_{out}\ I_c}{V_{in}\ I_e}$$

$$G = \alpha\frac{V_{out}}{V_{in}} = \alpha A_v$$
Power Gain

QUESTIONS

41-1. In which case is there a larger current between anode and cathode in a vacuum-tube triode: when the grid is positive or negative?

41-2. Why is it necessary for vacuum tubes to be evacuated?

41-3. In terms of band theory, under what circumstances might an insulator be made to conduct? How can this be done?

41-4. In a semiconductor, why is the movement of a hole like that of a positive charge?

41-5. Why is doping necessary to make semiconductors conduct?

41-6. Explain the difference between P-type and N-type doping.

41-7. Discuss how a missing atom at a lattice site contributes to the conductivity of a semiconductor.

41-8. Which semiconductor device operates most like a check valve in a fluid system?

41-9. How could a change in the base bias be used to turn off a transistor so that no current would flow through it?

41-10. Transistors are sometimes mounted on finned metal heat sinks. Why is this necessary?

41-11. Which light source would be hotter, an LED or an incandescent light bulb? Why?

41-12. Which semiconductor device is best suited to ensuring that a specific point in a circuit is maintained at a specified voltage?

41-13. How does a photoresistor differ from a photodiode? Explain.

ELECTRONICS

821

Transistors and Applications

41-1. Given an NPN transistor with a common-base connection and $\alpha = 0.98$, determine the base current and the collector current when the emitter current is 40 mA.

Ans. $I_b = 0.8$ mA, $I_c = 39.2$ mA

41-2. If the base current of a transistor with a common-base connection is 1.6 mA and the emitter current is 60 mA, what is α?

41-3. For a common-base amplifier, the input resistance is 800 Ω and an output resistance is 600 kΩ. (a) Determine the voltage gain if the emitter current is 12 mA and $\alpha = 0.97$. (b) What is the power gain?

Ans. (a) 727 (b) 706

41-4. The power gain is 800 for a common-base amplifier, and the voltage amplification factor is 840. Determine the collector current when the base current is 1.2 mA.

41-5. Determine the current gain in a common-emitter amplifier circuit when α is known to be 0.98.

Ans. 49

Additional Problems

41-6. In the previous problem, what is the collector current if the emitter current is 20 μA? What is the base current?

41-7. A transistor with an effective input resistance of 400 Ω, an effective output resistance of 900 kΩ, and $\alpha = 0.96$ is connected in a common-base circuit. (a) What is the voltage gain when the input current I_e is 8 μA? (b) What is the power gain?

Ans. (a) 2160 (b) 2074

41-8. Calculate I_c and I_b for the conditions described in Prob. 41-7.

41-9. For a transistor with $I_e = 8$ μA and $\alpha = 0.97$ connected with a common emitter calculate β, I_{in} and I_{out}.

Ans. 32.2, 0.24 μA, 7.76 μA

41-10. For a transistor with $\alpha = 0.99$ connected with a common collector calculate the current gain A_i.

41-11. A transistor has $\beta = 99$; what is the value of α?

Ans. 0.99

41-12. For the transistor of Prob. 41-11, if $I_b = 0.1$ mA, what are the values of I_e and I_c?

APPENDIX A

Mathematics Review

The brief review of mathematics presented in this appendix is not intended to be rigorous. It is assumed that you have already gained experience in high school algebra and have a passing acquaintance with right-triangle trigonometry. The topics discussed below will provide a review or references for the mathematics necessary for the solution of physical problems.

A-1
EXPONENTS AND RADICALS

It is often necessary to multiply a quantity by itself a number of times. A shorthand method of indicating the number of times a quantity is taken as a product is through the use of a superscript number called the *exponent*. This notation works according to the following scheme:

$$\begin{array}{ll} \text{For any number } a: & \text{For the number 2:} \\ a = a^1 & 2 = 2^1 = 2 \\ a \times a = a^2 & 2 \times 2 = 2^2 = 4 \\ a \times a \times a = a^3 & 2 \times 2 \times 2 = 2^3 = 8 \\ a \times a \times a \times a = a^4 & 2 \times 2 \times 2 \times 2 = 2^4 = 16 \end{array}$$

The powers of the number a are read as follows: a^2 is read "a squared;" a^3 is read "a cubed;" a^4 is read "the fourth power of a," or "a to the fourth power." More generally, we speak of a^n as "a to the nth power." In these examples, the letter a is referred to as the *base,* and the superscripts 1, 2, 3, 4, and n are called the *exponents.*

If $a^n = b$, then not only is b equal to the nth power of a, but, by definition, a is said to be the nth *root of b*. For example, $2^2 = 4$ means that 2 is the square root of 4. Similarly, $2^3 = 8$ means that 2 is the cube root of 8, and $2^5 = 32$ means that 2 is the fifth root of 32.

In general, the nth root of the quantity b is written

$$\sqrt[n]{b} \qquad \text{the } n\text{th root of } b$$

In the case of the square root, we simply write the radical, omitting the number 2:

$$\sqrt{b} \qquad \text{the square root of } b$$

It can be shown that a radical may be expressed differently by using a fractional exponent such that

$$\sqrt[n]{b} = b^{1/n}$$

For example, $\sqrt[3]{4} = 4^{1/3}$, and $\sqrt{10} = 10^{1/2}$.

Six important rules regarding exponents and radicals are given below, with examples following each rule:

$$\boxed{\text{Rule 1} \quad a^m \times a^n = a^{m+n}}$$

Example: $\quad 4^3 \times 4^2 = 4^{3+2} = 4^5$

$$\boxed{\text{Rule 2} \quad a^{-n} = \frac{1}{a^n}}$$

Examples: $\quad 2^{-3} = \dfrac{1}{2^{+3}}; \qquad 10^2 = \dfrac{1}{10^{-2}}$

$$\boxed{\text{Rule 3} \quad (a^m)^n = a^{mn}}$$

Examples: $\quad (10^2)^3 = 10^6; \qquad (10^{-3})^4 = 10^{-12}$

$$\boxed{\text{Rule 4} \quad (ab)^n = a^n b^n}$$

Examples: $\quad (2 \times 3)^2 = 2^2 \times 3^2 = 4 \times 9 = 36$

$\qquad\qquad\quad (2 \times 10^2)^3 = 2^3 \times 10^6 = 8 \times 10^6$

$$\boxed{\text{Rule 5} \quad \sqrt[n]{a^m} = a^{m/n}}$$

Examples: $\quad \sqrt[3]{2^9} = 2^{9/3} = 2^3 = 8$

$\qquad\qquad\quad \sqrt{10^{-4}} = 10^{-4/2} = 10^{-2} = \dfrac{1}{100}$

$$\boxed{\text{Rule 6} \quad \sqrt[n]{ab} = \sqrt[n]{a} \times \sqrt[n]{b}}$$

Examples: $\quad \sqrt{4 \times 10^4} = \sqrt{4} \times \sqrt{10^4} = 2 \times 10^2$

$\qquad\qquad\quad \sqrt[3]{8 \times 10^{-6}} = \sqrt[3]{8} \times \sqrt[3]{10^{-6}} = 2 \times 10^{-2}$

A-2

SCIENTIFIC NOTATION

Frequently in scientific work you may encounter very large or very small numbers. A convenient shorthand notation allows you to express any number as a number between 1 and 10 times an integral power of 10. Some multiples of 10 are:

$$
\begin{array}{ll}
0.0001 = 10^{-4} & 2.34 \times 10^{-4} = 0.000234 \\
0.001 = 10^{-3} & 2.34 \times 10^{-3} = 0.00234 \\
0.01 = 10^{-2} & 2.34 \times 10^{-2} = 0.0234 \\
0.1 = 10^{-1} & 2.34 \times 10^{-1} = 0.234 \\
1 = 10^{0} & 2.34 \times 10^{0} = 2.34 \\
10 = 10^{1} & 2.34 \times 10^{1} = 23.4 \\
100 = 10^{2} & 2.34 \times 10^{2} = 234.0 \\
1{,}000 = 10^{3} & 2.34 \times 10^{3} = 2{,}340.0 \\
10{,}000 = 10^{4} & 2.34 \times 10^{4} = 23{,}400.0
\end{array}
$$

Thus the number 456,000 may be written in scientific notation by determining the number of times the decimal point must be moved to the left in order to arrive at the shorthand notation. Examples are:

$$467 = (4\ 6\ 7.) = 4.67 \times 10^2$$

$$30 = (3\ 0.) = 3.0 \times 10^1$$

$$35{,}700 = (3\ 5\ 7\ 0\ 0.) = 3.57 \times 10^4$$

Similarly, any small decimal number can be written as a number between 1 and 10 times a *negative* power of 10. The negative exponent in this case will be the number of times the decimal point is moved to the right. This will always be one more than the number of zeros that separate the first digit from the decimal. Examples are:

$$0.24 = (0.2\ 4) = 2.4 \times 10^{-1}$$

$$0.0032 = (0.0\ 0\ 3\ 2) = 3.2 \times 10^{-3}$$

$$0.0000469 = (0.0\ 0\ 0\ 0\ 4\ 6\ 9) = 4.69 \times 10^{-5}$$

To transfer from scientific notation back to decimal notation, the procedure is simply reversed.

Recalling the laws of exponents, scientific notation can be used for multiplication, division, and addition of very large or very small numbers. When two numbers are multiplied, the exponents of 10 are added. For example, 200×4000 may be written $200 \times 4000 = (2.0 \times 10^2) \times (4.0 \times 10^3) = 8.0 \times 10^5$. Other examples are:

$$2200 \times 40 = (2.2 \times 10^3) \times (4.0 \times 10^1) = 8.8 \times 10^4$$

$$0.0002 \times 900 = (2 \times 10^{-4}) \times (9 \times 10^2) = 18 \times 10^{-2}$$

$$1002 \times 3 = (1.002 \times 10^3) \times 3 = 3.006 \times 10^3$$

Similarly, when one number is divided by another, the exponent of the latter is subtracted from the exponent of the former. Examples are:

$$\frac{7000}{350} = \frac{7 \times 10^3}{3.5 \times 10^2} = 2 \times 10^{3-2} = 2 \times 10^1$$

$$\frac{1200}{0.003} = \frac{1.2 \times 10^3}{3 \times 10^{-3}} = 0.4 \times 10^{3-(-3)} = 0.4 \times 10^6$$

$$\frac{0.008}{400} = \frac{8 \times 10^{-3}}{4 \times 10^2} = 2 \times 10^{-3-2} = 2 \times 10^{-5}$$

When adding two numbers in powers of 10, care must be taken to convert all numbers to be added so that they have identical exponents. Then addition is performed as usual. Examples are:

$$100 + 300 = (1 \times 10^2) + (3 \times 10^2) = 4 \times 10^2$$
$$2000 + 400 = (2 \times 10^3) + (0.4 \times 10^3) = 2.4 \times 10^3$$

Of course, in the above examples, the scientific notation is not very useful. However, in the following examples, the longhand method would be very cumbersome:

$$(4.75 \times 10^{18}) + (6 \times 10^{19}) = (4.75 \times 10^{18}) + (60 \times 10^{18})$$
$$= 64.75 \times 10^{18}$$
$$(1.4 \times 10^{-19}) + (4 \times 10^{-21}) = (140 \times 10^{-21}) + (4 \times 10^{-21})$$
$$= 144 \times 10^{-21}$$

It should be noticed that, when two numbers differ by more than three powers of 10, the smaller can usually be ignored in the process of addition. For example,

$$(1.6 \times 10^{24}) + (2 \times 10^{21}) = (1600 \times 10^{21}) + (2 \times 10^{21})$$
$$= 1602 \times 10^{21}$$

The number 1602 does not differ from the number 1600 by sufficient margin to be significant.

<table>
<tr><td>

A-3

LITERAL EQUATIONS AND FORMULAS

</td><td>

A literal equation is an equation in which some or all of the known quantities, usually constants, are represented by letters instead of numbers. Therefore, the roots of a literal equation are also expressed in terms of letters. For instance, the equation

</td></tr>
</table>

$$ax - 5b = c$$

must be solved for x in terms of a, b, and c. Transposing, we have

$$ax = 5b + c$$

Dividing by a,

$$x = \frac{5b + c}{a}$$

Perhaps the most common occurrence of the literal equation is the formula. A *formula* is a mathematical statement of equality in which letter symbols are combined with numbers to express a physical relationship. For example, the volume of a cone is expressed by the formula

$$V = \frac{\pi r^2 h}{3} \tag{A-1}$$

in which it is assumed that the radius r and the height h are known and the volume V is to be found.

In formulas, the various letters and symbols can be thought of as variables, the value of any one depending on the values assigned to the others. For example, in Eq.

MODERN PHYSICS

(A-1), the volume V varies in accordance with values assigned for h and r. Should r and V be given and the value of h desired, the formula can be solved explicitly for h in accordance with the axioms of the preceding section. For instance, clearing Eq. (A-1) of fractions yields

$$3V = \pi r^2 h$$

Dividing by the literal coefficient of h, πr^2,

$$\frac{3V}{\pi r^2} = \frac{h\pi r^2}{\pi r^2} \quad \text{or} \quad h = \frac{3V}{\pi r^2}$$

A-4 TRIGO-NOMETRY

Often it is necessary to determine the components of forces or a resultant of two or more concurrent forces more accurately than graphs will allow. The graphical approaches are also time-consuming. By learning a few principles that apply to all right triangles, you can significantly improve your ability to work with vectors. Moreover, hand-held calculators make many of the calculations relatively simple.

First, let's review some of the things we already know about right triangles. We will follow the convention that uses Greek letters for angles and Roman letters for sides. Commonly used Greek symbols are

α alpha β beta γ gamma
θ theta ϕ phi δ delta

In the right triangle drawn as Fig. A-1, the symbols R, x, and y refer to the side

Fig. A-1

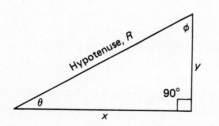

The Pythagorean theorem
$R^2 = x^2 + y^2$

dimensions, and θ, ϕ, and $90°$ are the angles. You should recall that the sum of the smaller angles in a right triangle is $90°$:

$$\phi + \theta = 90° \qquad \qquad \textit{Right Triangle}$$

There is also a relationship between the sides, which is known as the pythagorean theorem:

The pythagorean theorem: The square of the hypotenuse is equal to the sum of the squares of the other two sides.

$$\boxed{R^2 = x^2 + y^2}$$
The Pythagorean Theorem   (A-2)

The *hypotenuse* is defined as the longest side. It is conveniently located as that side directly opposite the right angle—the line joining the two perpendicular sides.

EXAMPLE A-1   What length of guy wire is needed to stretch from the top of a 40-ft telephone pole to a ground stake located 60 ft from the foot of the pole?

Solution   Draw a rough sketch, such as Fig. A-2. Identify the length R as the hypotenuse of a right triangle, then from the pythagorean theorem

$$R^2 = (60 \text{ ft})^2 + (40 \text{ ft})^2$$
$$= 3600 \text{ ft}^2 + 1600 \text{ ft}^2 = 5200 \text{ ft}^2$$

Taking the square root of both sides gives

$$R = \sqrt{5200 \text{ ft}^2} = 72.1 \text{ ft}$$

In general, to find the hypotenuse we could express the pythagorean theorem as

$$R = \sqrt{x^2 + y^2} \qquad \qquad \textit{Hypotenuse} \quad \text{(A-3)}$$

On some electronic calculators, the sequence of entries might be as follows:

$$ x \quad \boxed{x^2} \quad \boxed{+} \quad y \quad \boxed{x^2} \quad \boxed{=} \quad \boxed{\sqrt{x}} $$

Fig. A-2

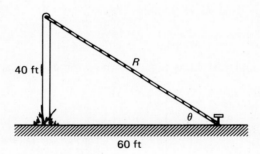

40 ft

R

θ

60 ft

In this instance, x and y are the values of the shorter sides, and the boxed symbols are operation keys on the calculator. You should verify the solution to the previous problem using $x = 40$ and $y = 60$. (The input procedure varies depending on the make of calculator.)

Of course, the pythagorean theorem can also be used to find either of the shorter sides if the remaining sides are known. Solution for x or for y yields

$$x = \sqrt{R^2 - y^2} \qquad y = \sqrt{R^2 - x^2} \qquad \qquad \text{(A-4)}$$

Trigonometry is the branch of mathematics which takes advantage of the fact that similar triangles are proportional in size. For example, in two similar right triangles, the ratio of any two sides will be the same numerical value regardless of the dimensions of either triangle.

Once an angle is identified in a right triangle, the sides *opposite* and *adjacent* to that angle may be labeled. The meanings of opposite, adjacent, and hypotenuse are given in Fig. A-3. You should study this figure until you understand fully the meaning of these terms. Verify that the side opposite to θ is y and that the side adjacent to θ is x. Also notice that the sides described by "opposite" and "adjacent" change if we refer to angle ϕ.

Fig. A-3

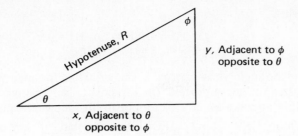

In a right triangle, there are three side ratios that are very important. They are the *sine,* the *cosine,* and the *tangent,* defined as follows for angle θ:

$$\sin \theta = \frac{\text{opp } \theta}{\text{hyp}}$$

$$\cos \theta = \frac{\text{adj } \theta}{\text{hyp}} \qquad \text{(A-5)}$$

$$\tan \theta = \frac{\text{opp } \theta}{\text{adj}}$$

To make sure that you understand these definitions, you should verify the following for the triangles in Fig. A-4:

$$\sin \theta = \frac{9}{15} \qquad \cos \gamma = \frac{m}{H} \qquad \tan \alpha = \frac{y}{x}$$

$$\sin \alpha = \frac{y}{R} \qquad \cos \beta = \frac{n}{H} \qquad \tan \phi = \frac{12}{9}$$

Fig. A-4

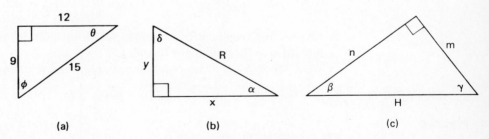

(a)　　　　　(b)　　　　　(c)

You should first identify the right angle, then label the longest side (opposite to the 90° angle) as the hypotenuse. Then, for a particular angle, the sides opposite and adjacent may be identified.

The constant values for the trigonometric functions of all angles between 0 and 90° have been calculated. They may be found from tables or, more conveniently, with a small calculator. Suppose we wish to know the cosine of 47°, for example. In a table of cosines we would see the number 0.682 adjacent to the angle 47°, and we would write

$$\cos 47° = 0.682$$

With some calculators, we would enter the angle 47, then strike the $\boxed{\cos}$ key to obtain the same result. From either a table or from your calculator, you should verify the following:

$$\tan 38° = 0.781 \qquad \cos 31° = 0.857$$
$$\sin 22° = 0.375 \qquad \tan 65° = 2.147$$

To find the angle whose tangent is 1.34 or to find the angle whose sine is 0.45, we would reverse the above process. On a calculator, for example, we would enter the number 1.34, then we would look for one of the following sequences, depending on the calculator, $\boxed{\text{INV}}$ $\boxed{\text{tan}}$, $\boxed{\text{ARC}}$ $\boxed{\text{tan}}$, or $\boxed{\text{tan}^{-1}}$. Any of these will give the angle whose tangent is the entered value. In the above examples, we find that

$$\tan \theta = 1.34 \qquad \theta = 53.3°$$
$$\sin \theta = 0.45 \qquad \theta = 26.7°$$

You should now be able to apply trigonometry to find unknown angles or sides in a right triangle. The following procedure will be helpful:

APPLICATION OF TRIGONOMETRY

1. Draw the right triangle from the stated conditions of the problem. (Label all sides and angles with either their known values or a symbol for an unknown value.)
2. Isolate an angle for study; if an angle is known, choose that one.
3. Label each side according to whether its relation to the chosen angle is opp, adj, or hyp.
4. Decide which side or angle is to be found.
5. Recall the definitions of the trigonometric functions:

$$\sin = \frac{\text{opp}}{\text{hyp}} \qquad \cos = \frac{\text{adj}}{\text{hyp}} \qquad \tan = \frac{\text{opp}}{\text{adj}}$$

6. Choose that trigonometric function which involves (*a*) the unknown quantity, and (*b*) no other unknown quantity.
7. Write the trigonometric equation and solve for the unknown.

EXAMPLE A-2 What is the length of the rope segment *x* in Fig. A-5?

Fig. A-5

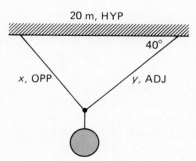

20 m, HYP

40°

x, OPP   *y*, ADJ

Solution The sketch is already drawn, so we proceed with step 2. The 40° angle is selected for reference, and opp, adj, and hyp are labeled on the figure. The decision is made to solve for *x* (opp). Since the sine function involves opp and hyp, we choose that function and write the equation.

$$\sin 40° = \frac{x}{20 \text{ m}}$$

Solving for x by multiplying both sides by 20 m, we obtain

$$x = (20 \text{ m}) \sin 40°$$

If we use tables, we determine that $\sin 40° = 0.643$, and then

$$x = (20 \text{ m})(0.643) = 12.9 \text{ m}$$

On some calculators, we would calculate x as follows:

20   ☐ ×   40   ☐ sin   ☐ =

It is easy to see what an advantage the calculator is in the application of trigonometry. As an additional exercise, you might show that $y = 15.3$ m.

EXAMPLE A-3   Refer back to Fig. A-2. What is the angle θ that the guy wire makes with the ground? Note the location of the angle θ, then label opp, adj, and hyp. For the angle θ, we may write

$$\tan \theta = \frac{\text{opp}}{\text{adj}} = \frac{40 \text{ ft}}{60 \text{ ft}}$$

The tangent function was the only one that involved the two *known* sides. If tables are to be used, then θ is the angle whose tangent is $^{40}/_{60}$ or 0.667. In this case, we find that

$$\tan \theta = 0.667 \quad \text{and} \quad \theta = 33.7°$$

The sequence on some calculators might be as follows:

40   ☐ ÷   60   ☐ =   ☐ INV   ☐ tan

Of course, the procedure varies. Study your manual and work with your calculator until you understand the process thoroughly.

Conversion Equivalents

LENGTH
1 meter (m) = 39.37 in. = 3.281 ft = 6.214×10^{-4} mi = 10^{10} Å = 10^{15} fermis
1 in. = 0.02540000 m; 1 ft = 0.3048 m; 1 mi = 1609 m
1 nautical mi = 1852 m = 1.1508 mi = 6076.10 ft
1 angstrom (Å) = 10^{-10} m; 1 mil = 10^{-3} in.; 1 rod = 16.5 ft; 1 fathom = 6 ft

AREA
1 m^2 = 10.76 ft^2 = 1550 in.^{2}; 1 hectare = 10^4 m^2 = 2.471 acres
1 ft^2 = 929 cm^2; 1 in.^{2} = 6.452 cm^2 = 1.273×10^6 circular mils; 1 acre = 43,560 ft^2

VOLUME
1 m^3 = 35.31 ft^3 = 6.102×10^4 in.^{3}
1 ft^3 = 0.02832 m^3; 1 U.S. gallon = 231 in.^{3}; 1 liter = 1.000028×10^{-3} m^3 = 61.02 in.^{3}

TIME AND FREQUENCY
1 year = 365.2422 days = 8.766×10^3 h = 5.259×10^5 min = 3.156×10^7 s
1 sidereal day (period of earth's revolution) = 86,164 s
1 hertz (Hz) = 1 cycle/s

SPEED
1 m/s = 3.281 ft/s = 3.6 km/h = 2.237 mi/h = 1.944 knots
1 km/h = 0.2778 m/s = 0.9113 ft/s = 0.6214 mi/h
1 mi/h = 1.467 ft/s = 1.609 km/h = 0.8689 knot

MASS
1 kg = 2.205 lb_m = 0.06852 slug
1 lb_m = 0.4536 kg = 0.03108 slug; 1 slug = 32.17 lb_m = 14.59 kg

DENSITY
1 g/cm^3 = 1000 kg/m^3 = 62.43 lb_m/ft^3 = 1.940 $slug/ft^3$
1 lb_m/ft^3 = 0.03108 $slug/ft^3$ = 16.02 kg/m^3 = 0.01602 g/cm^3

FORCE
1 newton (N) = 10^5 dynes = 0.1020 kg wt = 0.2248 lb
1 lb (force) = 4.448 N = 0.4536 kg wt = 32.17 poundals

PRESSURE
1 N/m^2 = 9.869×10^{-6} atm = 1.450×10^{-4} lb/in.^{2} = 0.02089 lb/ft^2
= 7.501×10^{-4} cmHg = 4.015×10^{-3} in. of water = 10^{-5} bar
1 lb/in.^{2} = 144 lb/ft^2 = 6895 N/m^2 = 5.171 cmHg = 27.68 in. of water
1 atm = 406.8 in. of water = 76 cmHg = 1.013×10^5 N/m^2
= 10,330 kg wt/m^2 = 2116 lb/ft^2 = 14.70 lb/in.^{2} = 760 torr

WORK, ENERGY, HEAT	1 joule (J) = 0.2389 cal = 9.481×10^{-4} Btu = 0.7376 ft · lb = 10^7 ergs = 6.242×10^{18} eV
	1 kcal = 4186. joules = 3.968 Btu = 3087 ft · lb
	1 eV = 1.602×10^{-19} joule; 1 unified amu = 931.48 MeV
	1 kW · h = 3.6×10^6 joules = 3413 Btu = 860.1 kcal = 1.341 hp · h
POWER	1 hp = 2545 Btu/h = 550 ft · lb/s = 745.7 watts = 0.1782 kcal/s
	1 watt (W) = 2.389×10^{-4} kcal/s = 1.341×10^{-3} hp = 0.7376 ft · lb/s
ELECTRIC CHARGE	1 faraday = 96,487 coulombs; 1 electron charge = 1.602×10^{-19} coulomb
MAGNETIC FLUX	1 weber (Wb) = 10^8 maxwells = 10^8 lines
MAGNETIC INTENSITY	1 tesla (T) = 1 newton/amp · m = 1 weber/m^2 = 10,000 gauss

INDEX

Beta particle, 778, 780–781
Bevel gears, 219–220
Bimetallic strip, 310
Binding energy, 775–777
Biot-Savart law, 669
Blackbody, 348
Blackbody radiation, 348–349
Block-and-tackle, 215–216
Bohr, N., 447
Bohr model of atom, 447–448, 754–757
Boiling, 331, 367, 369
Boiling point, 330–331, 369
Bonding, covalent, 798–799
Boundary conditions, 411–412
Boyle, R., 358
Boyle's law, 357–362, 366
Brackett series, 754, 758
Breakover voltage, 815
Breeder reactor, 786–787
Bridge rectifier, 804–805
Brightness, 452–453
British thermal unit, 321–322
Brush, 699–703
Bulk modulus, 234–238
Buoyant force, 273–276

Calipers, 14–15
Caloric theory, 320–321
Calorie (unit), 321–322
Calorimeter, 327–329
Candela (unit), 8, 10, 455
Capacitance, 9, 595, 604
    computing of, 599–601
Capacitive reactance, 725
Capacitor, 594, 597–599, 605–606, 715–719, 724–726
    charged, energy of, 609
    charging and discharging of, 616–617, 715–719
    IC, 817, 819
    in parallel, 605–609
    parallel-plate, 598–601
    in series, 605–609
    variable, 601
Carnot, S., 387
Carnot cycle, 387–388
Carnot engine, 387
Cathode, 814
Cathode-ray tube (CRT), 796–797
Celsius, A., 300
Celsius scale, 300–303, 306–308
Center
    of curvature, 470
    of gravity, 76–77

Centimeters of mercury (unit), 270
Centrifugal force, 172
Centripetal acceleration, 168–170, 191–192
Centripetal force, 170–172
Chain hoist, 208
Chain reaction, nuclear, 785
Characteristic frequency, 411–413
Charge (see Electric charge)
Charging, 541, 594–597
    of battery, 639–641
    of capacitor, 715–719
    by contact, 547
    by induction, 549–550
Charles, J., 359
Charles' law, 359–362
Chromatic aberration, 514–515
Circuit (see Electric circuit)
Circuit diagram, 605, 621
Circular mil (unit), 626
Circular motion, 167–169
Circular waves, 441
Coefficient
    of area expansion, 312
    of convection, 347
    of kinetic friction, 52–53
    of linear expansion, 308–310
    of performance, 392
    of restitution, 158–159
    of static friction, 52–53
    of volume expansion, 312–313
Coherent source, 522, 524
Cohesion, 238
Collector, of transistor, 807
Collision, 153, 155–161
Combustion, heat of, 334
Common-base amplifier circuit, 809–810, 812
Common-collector amplifier circuit, 812
Common-emitter amplifier circuit, 810–812
Commutator, split-ring, 688, 703–704
Compass, 656
Component method
    of calculating torque, 71
    of vector addition, 29–31
Compound microscope, 512–513
Compound motor, 706
Compressibility, 238
Compression, 230, 232–233
Compression ratio, 391–392
Compression stroke, 390–391
Compressions, 418–419
Compressor, 393–394
Concave mirror, 470–477, 509

Concurrent forces, 25, 29–31, 43
Condensation, 332, 368–369, 385
    heat of, 332
Condensation pulse, 402–403
Condenser, 393–394
Conduction, thermal, 339–344
Conduction band, 799
Conductivity
    electric, 239
    thermal, 10, 342–344
Conductor, 545–546, 624–627, 798–799
    charging of, 594–597
Conical pendulum, 174–176
Conservation
    of angular momentum, 200–201
    of charge, 783
    of energy, 141–143, 209, 286, 380, 698
    of mass-energy, 783
    of mechanical energy, 141–142
    of momentum, 155–157
    of nucleons, 783
    of thermal energy, 327
Constant acceleration, 88–93
Constant-pressure thermometer, 304–305
Constant speed, 85
Constant-volume thermometer, 304
Constructive interference, 409–411, 428–429, 522–523, 526
Contact force, 45
Continuous spectrum, 752–753
Contraction, relativistic, 744
Control rod, 786–787
Convection, thermal, 339–340, 345–348
Convection coefficient, 347
Convection current, 345–347
Converging lens, 503–511
Converging mirror (see Concave mirror)
Conversion factors, 16–17, 833–834
Convex lens, 509
Convex mirror, 470–474, 476
Coolidge, W., 797
Core, of nuclear reactor, 786
Corona discharge, 597
Corpuscular theory of light, 440–443, 447
Cosine, 25, 829
Cosmic photon, 445–446
Coulomb, C.A. de, 550
Coulomb (unit), 9, 551
Coulomb's law, 550–553, 558, 561, 751, 755

Newton's third law, 41–42
Node, 411–412
Nonconcurrent forces, 67
Normal boiling point, 369
NPN transistor, 807–808
Nuclear chain reaction, 785
Nuclear fission, 784–785
Nuclear force, 768, 778
Nuclear fuel, 786
Nuclear fusion, 787
Nuclear reaction, 783–784
Nuclear reactor, 778, 785–787
Nucleon, 768, 783
Nucleus (*see* Atomic nucleus)
Nuclide, 777

Object distance, 468
Objective lens, 512–513
Ocular, 513
Odd harmonics, 422
Oersted, H., 662–663
Ohm, G.S., 621, 625
Ohm (unit), 7, 9, 622, 725, 727
Ohm's law, 621–624, 632
Oil-drop experiment, 586–588
Opaque body, 450–451
Optical instruments, 493, 502–515, 528–531
Optical semiconductor relay, 814
Orbit, electron, 751–752
Oscillating field, 532–533
Oscillation
    fundamental mode of, 412
    higher modes of, 412
Oscilloscope, 796
Otto cycle, 390–391
Output power, 209–210
Output work, 208–211, 272, 378
Overtones, 422–423, 428
Oxide layer, in IC fabrication, 817–818

P-type semiconductor, 799–801
P-V diagram, 380–382, 388
Pain threshold, 426–427
Parabolic reflector, 478–479
Parallel connection, 607–608, 635–637
Parallel-plate capacitor, 598–601
Parallelogram method of vector addition, 21–23
Paramagnetism, 662
Particle theory of light, 440–443, 447
Particles, waves and, 748–749

Pascal, B., 269, 281
Pascal (unit), 9, 233, 265
Pascal's law, 269, 271
Paschen series, 754, 758
Pendulum, 174–176, 254–257
Penumbra, 451–452
Performance, coefficient of, 392
Period, 170, 246, 252–254, 405
    of pendulum, 255–256
Periodic motion, 244–246
Periodic table, 768–769
Periodic wave motion, 405–408
Permanent-magnet motor, 706
Permeability, 660–662
Permittivity, 567, 601–605
Phase, 329–334, 357
Phase angle, 726–727
Phase diagram, 370, 726–727
Phase relations in ac circuits, 722–724
Phosphor, 797
Photodiode, 813–814
Photoelectric effect, 446–447, 747–748
Photometer, grease-spot, 453
Photon, 445–447, 747–749, 756
Photoresist layer, 817–818
Photoresistor, 815
Physical quantity, 7–8
Physics, definition of, 3–4
Picofarad (unit), 599
Picometer (unit), 767
Piston, 381, 383–384, 390–392
Pitch
    of screw, 222
    of sound, 424–425, 427–429
Pith-ball electroscope, 541–543
Pith balls, 541–542
Planck, M., 447, 747
Planck's constant, 747
Plane angle, 8
Plane mirror, 467–469
Plane-polarized wave, 531–533
Planetary gears, 219–220
Planetary motion, 178
Plutonium, 786
PN junction, 801–803
    forward-biased, 802
    reverse-biased, 802
PNP transistor, 807–808
Point source, 450
Polarization, 531–533
Polarizer, 532–533
Polaroid filter, 533
Pole, magnetic, 656–657
Polygon method of vector addition, 21–23

Positive charge, 543–544, 579
Positive work, 136–139
Positron, 778, 780
Potential (*see* Electric potential)
Potential energy, 136–142, 285–286, 578
    electric, 577–581, 609
    gravitational, 577–579
Pound (unit), 10, 23, 118, 833
Pound-feet (unit), 69
Pound-second (unit), 154
Power, 9, 132, 144–146, 425, 834
    average, 198
    electric, 623–624
    input, 209–210
    output, 209–210
    radiant, 348
    rotational, 197–199
    of wave, 408
Power factor, 728–729
Power gain, 810–811
Power stroke, 390–392
Pressure, 8, 262, 833
    absolute, 270, 286, 358
    atmospheric, 269–271
    fluid, 265–269
    fluid velocity and, 283–285
    gauge, 270
    measurement of, 269–271
    vapor, 368–369, 371
Prevost's law of heat exchange, 350
Principle quantum number, 755
Prism, 445, 493, 502–503
Projectile, 104
Projectile motion, 104–110
    angular, 107–110
    horizontal, 105–107
Proton, 544–545, 767–768
Pulley, 208, 214–216
Pulse, 402
Pythagorean theorem, 25, 827–828

Quality of sound, 424–428
Quantum, 447, 747
Quantum number, 759
    principal, 755
Quantum theory, 446–447, 747–748, 755

R-value, 342–345
Radial magnetic field, 684
Radian (unit), 8, 186–187
Radiant flux, 451
Radiant intensity, 10
Radiant power, 348